WITHDRAWN
FROM
CPS LIBRARY

B2A/CPB/TC

HANDBOOKS IN OPERATIONS RESEARCH
AND MANAGEMENT SCIENCE
VOLUME 11

Handbooks in Operations Research and Management Science

Advisory Editors

M. Florian
Université de Montréal

A.M. Geoffrion
University of California at Los Angeles

R.M. Karp
University of California at Berkeley

T.L. Magnanti
Massachusetts Institute of Technology

D.G. Morrison
University of California at Los Angeles

S.M. Pollock
University of Michigan at Ann Arbor

A.F. Veinott, Jr.
Stanford University

P. Whittle
University of Cambridge

Editors

J.K. Lenstra
Georgia Institute of Technology

G.L. Nemhauser
Georgia Institute of Technology

Volume 11

ELSEVIER

Amsterdam – Boston – Heidelberg – London – New York – Oxford – Paris
San Diego – San Francisco – Singapore – Sydney – Tokyo

Supply Chain Management: Design, Coordination and Operation

Edited by

A. G. de Kok
Technische Universiteit Eindhoven

Stephen C. Graves
Massachusetts Institute of Technology

2003
ELSEVIER
Amsterdam – Boston – Heidelberg – London – New York – Oxford – Paris
San Diego – San Francisco – Singapore – Sydney – Tokyo

ELSEVIER B.V.
Sara Burgerhartstraat 25
P.O. Box 211, 1000 AE Amsterdam, The Netherlands

© 2003 Elsevier B.V. All rights reserved.

This work is protected under copyright by Elsevier, and the following terms and conditions apply to its use:

Photocopying
Single photocopies of single chapters may be made for personal use as allowed by national copyright laws. Permission of the Publisher and payment of a fee is required for all other photocopying, including multiple or systematic copying, copying for advertising or promotional purposes, resale, and all forms of document delivery. Special rates are available for educational institutions that wish to make photocopies for non-profit educational classroom use.

Permissions may be sought directly from Elsevier's Science & Technology Rights Department in Oxford, UK: phone: (+44) 1865 843830, fax: (+44) 1865 853333, e-mail: permissions@elsevier.com. You may also complete your request on-line via the Elsevier homepage (http://www.elsevier.com), by selecting 'Customer Support' and then 'Obtaining Permissions'.

In the USA, users may clear permissions and make payments through the Copyright Clearance Center, Inc., 222 Rosewood Drive, Danvers, MA 01923, USA; phone: (+1) (978) 7508400, fax: (+1) (978) 7504744, and in the UK through the Copyright Licensing Agency Rapid Clearance Service (CLARCS), 90 Tottenham Court Road, London W1P 0LP, UK; phone: (+44) 207 631 5555; fax: (+44) 207 631 5500. Other countries may have a local reprographic rights agency for payments.

Derivative Works
Tables of contents may be reproduced for internal circulation, but the permission of Elsevier is required for external resale or distribution of such material.
Permission of the Publisher is required for all other derivative works, including compilations and translations.

Electronic Storage or Usage
Permission of the Publisher is required to store or use electronically any material contained in this work, including any chapter or part of a chapter.

Except as outlined above, no part of this work may be reproduced, stored in a retrieval system or transmitted in any form or by any means, electronic, mechanical, photocopying, recording or otherwise, without prior written permission of the Publisher.
Address permissions requests to: Elsevier's Science & Technology Rights Department, at the phone, fax and e-mail addresses noted above.

Notice
No responsibility is assumed by the Publisher for any injury and/or damage to persons or property as a matter of products liability, negligence or otherwise, or from any use or operation of any methods, products, instructions or ideas contained in the material herein. Because of rapid advances in the medical sciences, in particular, independent verification of diagnoses and drug dosages should be made.

First edition 2003

Library of Congress Cataloging in Publication Data
A catalog record from the Library of Congress has been applied for.

British Library Cataloguing in Publication Data
A catalogue record from the British Library has been applied for.

ISBN: 0 444 51328 0
ISSN: 0927-0507

⊖ The paper used in this publication meets the requirements of ANSI/NISO Z39.48-1992 (Permanence of Paper). Printed in The Netherlands.

Contents

CHAPTER 1
Introduction
A.G. de Kok and Stephen C. Graves — 1
1. Introduction — 1
2. Main business trends that created SCM — 2
3. Outline of the volume — 11

PART I Supply Chain Design

CHAPTER 2
Supply Chain Design and Planning – Applications of Optimization Techniques for Strategic and Tactical Models
Ana Muriel and David Simchi-Levi — 17
1. Introduction — 17
Part I: Production/Distribution Systems
2. Introduction — 19
3. Piece-wise linear concave costs — 22
4. All-unit discount transportation costs — 39
Part II: Pricing to improve supply chain performance
5. Introduction — 65
6. Coordinating pricing and inventory decisions — 66
7. Pricing models with production capacity limits — 74
8. Computational results and insights — 76
Part III: Supply chain design models
9. Introduction — 77
10. The single-source capacitated facility location problem — 78
11. A distribution system design problem — 82
12. Conclusions — 88
References — 89

CHAPTER 3
Supply Chain Design: Safety Stock Placement and Supply Chain Configuration
Stephen C. Graves and Sean P. Willems — 95

1. Introduction	95
2. Approaches to safety stock placement	97
3. Model formulation	102
4. Heavy industry and consumer packaged goods example	107
5. Supply-chain configuration	120
6. Conclusion	129
References	131

CHAPTER 4
Supply Chain Design: Flexibility Considerations
J.W.M. Bertrand — 133

1. Introduction	133
2. Conceptual research on manufacturing flexibility	135
3. The flexible machine investment problem; volume and mix flexibility	142
4. Resource flexibility, range, mobility, uniformity and throughput time	152
5. Empirical research on flexibility	161
6. Supply chain flexibility	164
7. Design for supply chain flexibility	172
8. Conclusion	191
References	193

CHAPTER 5
Design for Postponement
Jayashankar M. Swaminathan and Hau L. Lee — 199

1. Introduction	199
2. Postponement enablers	201
3. Process standardization	202
4. Process resequencing	213
5. Component standardization	218
6. Related strategies and other benefits	221
7. Conclusions	223
References	224

PART II Supply Chain Coordination

CHAPTER 6
Supply Chain Coordination with Contracts
Gérard P. Cachon — 229

1. Introduction	229
2. Coordinating the newsvendor	233

3. Coordinating the newsvendor with price-dependent demand ... 257
4. Coordinating the newsvendor with effort-dependent demand ... 264
5. Coordination with multiple newsvendors ... 271
6. Coordinating the newsvendor with demand updating ... 285
7. Coordination in the single-location base-stock model ... 292
8. Coordination in the two-location base-stock model ... 297
9. Coordination with internal markets ... 312
10. Asymmetric information ... 316
11. Conclusion ... 329
References ... 332

CHAPTER 7
Information Sharing and Supply Chain Coordination
Fangruo Chen — 341

1. Introduction ... 341
2. Value of information ... 343
3. Incentives for sharing information ... 377
4. Future research ... 410
References ... 413

CHAPTER 8
Tactical Planning Models for Supply Chain Management
Jayashankar M. Swaminathan and Sridhar R. Tayur — 423

1. Introduction ... 423
2. Notations used ... 427
3. Stationary and independent demand ... 428
4. Alternative demand assumptions ... 436
5. Generalizations ... 440
6. Applications ... 446
7. Conclusions and future directions ... 448
References ... 449

PART III Supply Chain Operations

CHAPTER 9
Planning Hierarchy, Modeling and Advanced Planning Systems
Bernhard Fleischmann and Herbert Meyr — 457

1. Types of supply chains ... 458
2. Supply chain planning ... 468

3. Advanced planning systems: General structure 480
4. Advanced planning systems: Particular systems 508
References 519

CHAPTER 10
Supply Chain Operations: Serial and Distribution Inventory Systems
Sven Axsäter 525

1. Introduction 525
2. Different ordering policies 528
3. Serial systems 536
4. Order-up-to-S policies in distribution systems 541
5. Batch-ordering in distribution systems 551
6. Conclusions 554
References 556

CHAPTER 11
Supply Chain Operations: Assemble-to-Order Systems
Jing-Sheng Song and Paul Zipkin 561

1. Introduction 561
2. One-period models 563
3. Multi-period, discrete-time models 567
4. Continuous-time models 575
5. Research on system design 590
6. Summary and future directions 592
References 593

CHAPTER 12
Planning Supply Chain Operations: Definition and Comparison of Planning Concepts
Ton G. de Kok and Jan C. Fransoo 597

1. Introduction 597
2. The hierarchical nature of SCP 608
3. Constraints for SCOP 618
4. Mathematical programming models for supply chain planning 626
5. Stochastic demand models for supply chain planning 633
6. Comparison of supply chain planning concepts for general supply chains 655
7. Summary and issues for further research 667
References 671

CHAPTER 13
Dynamic Models of Transportation Operations
Warren B. Powell — 677

1. Operational challenges in transportation	682
2. A general modeling framework	698
3. Algorithmic strategies	707
4. Modeling operational problems	724
5. Implementation issues for operational models	752
6. Summary remarks	753
References	754

Subject Index — 757

A.G. de Kok and S.C. Graves, Eds., *Handbooks in OR & MS, Vol. 11*
© 2003 Elsevier B.V. All rights reserved.

Chapter 1

Introduction

A.G. de Kok
Technische Universiteit, Eindhoven

Stephen C. Graves
Massachusetts Institute of Technology

1 Introduction

Supply Chain Management (SCM) has been a very visible and influential research topic in the field of operations research (OR) over the course of the last decade of the twentieth century. The problems and experiences that have emerged from business practices have stimulated many researchers to contribute to a deeper understanding about underlying phenomena and causal relationships. Supply Chain Management has also served as an application area, where existing OR methods and techniques have been applied to new models for new problems, to new models for old problems that regained attention and to existing models for old problems. In the last case we find that progress has been made to extend existing results, stimulated by the apparent need for such extensions.

One might naturally start a handbook on SCM with a definition of the term Supply Chain Management. We have decided to resist this temptation as there are already too many competing definitions, and we do not see value in attempting to create a new definition or synthesize one from the current contenders. SCM has developed into a notion that covers strategic, tactical and operational management issues. We have made an attempt to structure the area by means of the chapters in this handbook. By no means do we claim to deal with all management issues commonly understood as being part of Supply Chain Management. Nevertheless, we do believe that this handbook covers a broad range of SCM issues that lend themselves to being formulated and analysed with mathematical models.

As appropriate for an OR handbook, this volume focuses primarily on supply chains as a context to apply OR methods and models. As a consequence, we are concerned with the decision-making processes that arise in SCM and are derived from managerial and economic considerations. In particular, we investigate and explore how OR can support decisions in the design, planning and operation of a supply chain. By doing so, we identify the richness of SCM as an OR application field, which promises another 'Golden Decade' of research.

In this introduction we provide an overview of SCM as an OR application area. Since many of the chapters in this handbook carefully position a particular aspect of SCM in a business and economic context, we deliberately restrict the introduction to a high-level of abstraction. This allows us to discuss a number of relevant trends in the business environment that proved to be the main impetus for the prospering of SCM during the last decade of the twentieth century. The added value of such an overview should be to position SCM in its business context and to provide a framework to understand and position the subsequent chapters of this handbook in relation to each other.

2 Main business trends that created SCM

In this section we discuss the main business trends during the late eighties and nineties of the twentieth century that provided the fertile soil from which SCM developed.

2.1 Core competencies

Prahalad and Hamel (1990) argue that a number of companies have achieved significantly better results than their competitors by focussing on only a few competencies, so-called core competencies, and by outsourcing other non-core activities to companies that have a core competence on those activities. This reasoning has gained a lot of attention from large, highly vertically-integrated companies, such as Philips Electronics, Unilever, P&G, General Motors, etc. and has been adopted at a surprisingly fast pace. Whereas implementation of a company-wide information systems, such as an Enterprise Resource Planning (ERP) system, typically has taken three to seven years within these large companies, the implementation of the core-competency strategy has often been accomplished within one or two years.

In our effort to understand the success of the core-competency strategy in terms of its adoption by global companies, we identify a number of circumstances that seem to characterize the late eighties business environment.

2.1.1 Short-term focus

In the Western economic world the eighties were a decade of relatively low economic growth and high unemployment rates. In that climate a short-term focus prevailed. The core-competency strategy allowed firms to increase their return on investments (ROI) and related business performance indicators almost instantaneously: outsourcing non-core competencies eliminated the associated fixed cost in the denominator of ROI, which typically resulted in increasing the ROI. The economic climate permitted big multinational companies to outsource high-cost operations to companies with lower costs;

for instance, companies with union operations and expensive labour contracts would outsource these operations to non-union companies with more cost flexibility. Thus, firms could substantially reduce not only fixed costs, but also the variable product costs as well.

The first companies adopting the core-competency strategy showed immediate improvements in their balance sheets, resulting in rapid increases in their stock market value. Many companies decided to reap similar benefits and started outsourcing, as well.

2.1.2 Technological improvements require high capital investments

By the end of the eighties, multinational companies with a tradition of capital-intensive manufacturing, such as electronics, white goods, automotive and consumer packaged goods, had invested for three decades in manufacturing mechanization and automation. This process replaced labour with capital, to the point that their capital–labour ratio approached that of primary industries, such as chemicals and metals. In fact, most of these vertically integrated companies found that more and more of the added value from their manufacturing had shifted upstream in their supply chains, from assembly to fabrication. Consequently, the investments required for further improvements in labour productivity and process capabilities kept increasing. A sector that was archetypical for such capital investment requirements was the semiconductor industry that emerged in the early seventies and matured in the eighties. Many multinational electronics manufacturers had their own semiconductor division.

These capital investment requirements demanded a strategic assessment. Most companies decided to concentrate on their brands, implying that they concentrated on Marketing and Sales, and Research and Development of their product portfolio as well as on Purchasing in order to leverage their buying power. Upstream manufacturing activities were outsourced to subcontractors. Interestingly, but logically, a number of these subcontractors decided to consider manufacturing their core competence and started a process of acquisitions that continues to date. In the electronics industry these companies are currently called Electronics Manufacturing Services (EMS) companies; in the semiconductor industry these companies are known as foundries. Apparently it is possible for them to carry the burden of large capital investments that could not be carried by the global multi-billion brand-owners. A possible explanation can be found in the stock market, again. The stock market analysts seem to have lower ROI expectations of these new manufacturing conglomerates than of the brand-owners.

Whether the current situation with multinational brand-owners focussing on Marketing and Sales, Research and Development, and Purchasing and multinational 'service companies' focussing on manufacturing and logistics is a stable economic equilibrium remains to be seen. In his thought-provoking book Clockspeed, Fine (1998), provides empirical evidence of his theory that the business environment shows a constant process of vertical integration and disintegration, stimulated by competition based on technological breakthroughs and fostered by internal inertia of large vertically-integrated companies.

2.1.3 SCM as core competence

In the early nineties a number of companies, such as Hewlett-Packard (HP), recognized that SCM was one of their core competencies. Although HP was in the test and measurement industry and the computer industry since the 1950's, by the early nineties the company had evolved from a business-to-business company into a business-to-consumer company delivering PCs and printers via a dealer network to the consumers. In parallel to concentration on Research and Development (in particular software and printing technology), Marketing and Sales, and outsourcing manufacturing, HP developed the skills for 'worldclass' SCM. HP recognized that one of its key differentiators could be to offer both speed of delivery and product diversity to the market, and that this could be done without owning traditional manufacturing assets. Lee and Billington (1993, 1995) discuss the main ideas behind the HP approach. They introduce the term postponement in the SCM field, implying that product diversity is created as close as possible to the consumer, thereby allowing for efficiencies upstream in the supply chain. The postponement concept is developed and explored in Chapter 5.

Another company that has made SCM its core competence is Dell. Prior to Dell, PC manufacturers sold PCs through their dealer network, implying substantial capital investments in inventory by the dealers and exposure to obsolescence risk for the manufacturer. In contrast, Dell decided to sell direct to the customer using the Internet as its marketing and sales channel. Dell is then able to assemble to order each client's PC, thereby eliminating the need for final product inventory. The Dell business model requires that Dell's suppliers hold stocks of components in consignment at or near Dell's assembly factories. Thus Dell operates its supply chain with minimal inventories on its books.

Whereas the above is a somewhat idealized description, Dell does operate its supply chain with considerably less inventory than its competitors, while providing customized products with short delivery lead times. The Dell example should be considered a showcase of 'worldclass' SCM. It shows the potential for operating low-inventory, high-flexibility and customized-product supply chains. In many industrial sectors the potential must be huge, given the fact that many sectors have much lower market diversity than the PC sector.

2.1.4 Relevance for Operation Research applied to supply chains

The disintegration of the brand-owning companies has led to an enormous increase in the number of contractual relationships between brand-owners and their subcontractors and suppliers, as well as between brand-owners and their downstream channel partners. Contracts are the mechanisms by which the brand-owner can leverage its buying power, yielding lower purchase prices, higher product quality and greater delivery reliability and speed. As such, the careful design of contracts is paramount to the profitability of the brand-owner. These contracts are also critical to assuring the sustainability of the supplier or subcontractor, recognizing their need to obtain sufficient economies of scale and scope. Furthermore, these contracts are essential mechanisms for finding effective ways to spread the risk across a supply chain. In Chapters 6 and 7 of this handbook supply chain contracts are extensively discussed. Chapter 6 focuses on the design of contracts in general with an emphasis on risk sharing, while Chapter 7 examines the design of contracts with respect to sharing of demand and supply information.

The myriad of relationships between legally independent companies operating a supply chain from commodities to consumers poses structurally complex network design problems to each of these companies. Whereas contractual relationships are one-to-one by definition, the network design problem is a many-to-many problem. Apart from questions concerning locations of factories and warehouses, tactical issues of safety stock positioning, capacity slack positioning and transportation mode selection have to be addressed. Operations research has wide applicability to these issues, and has provided very useful decision support. In this volume, Chapter 2 covers the application of optimization models and methods to supply chain design. Chapter 3 discusses the strategic positioning of safety stocks, while Chapter 4 focuses on investments in resources across the supply chain so that a strategic trade-off between customer service, market diversity and supply chain flexibility investments can be made.

2.2 The Bullwhip effect

One particular phenomenon that has attracted great attention in industry and academia is the Bullwhip effect. In the late fifties Forrester (1958) conducted experimental research that revealed that demand variations amplify from link to link going upstream in the supply chain, i.e., from consumers to raw materials. By means of simulation, he identified the root causes of this variation amplification: information distortion and information delay.

Lee, Padmanabhan and Whang (1997) built upon and extended the ideas of Forrester to identify common business practices that led to information distortion and information delays. This paper stimulated a large amount of work on understanding the phenomenon and developing counter measures. This work drew upon and applied concepts from the OR literature, including echelon stock concepts, inventory pooling and forecasting processes that induced the best estimates of future demand. The latter seems obvious, but in many situations incentives are not aligned between business functions, yielding *wishful thinking* forecasts or *target sales* forecasts. Echelon stock concepts and inventory pooling stimulated the implementation of Vendor Managed Inventory (VMI) concepts.

The implementation and dissemination of these concepts improved the overall knowledge base on Supply Chain Management. In general, one may conclude that communication of the Bullwhip effect and its root causes across all business function has increased mutual understanding between different business functions and between different companies. For many companies it became clear that they were only one of the many players involved in the game of satisfying the customer with a service or a product.

2.2.1 Relevance for Operations Research applied to supply chains

It is interesting to remark here that the Bullwhip effect is by now well understood, yet it poses a challenging mathematical problem when incorporating the underlying dynamics into commonly studied multi-echelon inventory management problems and multi-site production planning problems. The main reason for this is the fact that the dynamics related to the Bullwhip effect entail non-stationary random demands and dynamic capacity availability amongst others, as well as analysis of transient processes. Such non-stationary stochastic processes typically do not allow for a straightforward and rigorous analysis. Still, the issue of non-stationarity must be addressed. In this handbook first efforts are reported in several chapters. In Chapter 2 the impact of non-stationarity on supply chain design and pricing is discussed. Chapter 4 deals with the impact of non-stationarity on investments in flexibility, i.e., slack resources. In Chapters 9 and 12 the rolling schedule concept, commonly used in practice to deal with non-stationarities, is discussed extensively. And in Chapter 13 dynamic models of transportation operations are formulated and solved by a new generic method. Substantial research efforts are required to provide models and methods that can be applied to real-world problems.

2.3 Manufacturing as a global commodity

2.3.1 Final assembly is simple

Although capital has replaced quite a number of labour-intensive activities, e.g. welding in automotive and printed circuit board mounting in electronics, still a number of manual activities remain before a product can be delivered to the customer. Most of these activities relate to the final assembly, test and packaging of the product. For a while time manufacturers believed that even these activities could be substituted by automation, giving birth to the concept of Flexible Assembly Systems [cf. Suri, Sanders and Kamath (1993)], but soon they discovered that such flexible systems are economically viable only in complex assembly activities with very high requirements on consistent product quality, or assembly activities that are no longer acceptable to be performed by human beings. What remained was a collection of relatively simple labour-intensive assembly activities, whose output quality could be controlled and supported by the common-sense Japanese manufacturing concepts and technology [cf. Chase, Aquilano and Jacobs (1999)], that have now been embedded in best practice manufacturing.

From this observation many companies concluded that their final assembly activities could be outsourced as well, or that they could be treated as nomadic activities, i.e., final assembly activities are started-up in a particular region of the globe if labour is cheap and abandoned as soon as another region has lower labour rates. This is an economically viable manufacturing concept because the fixed investments for such facilities are perceived to be low and the transportation costs for inbound and outbound shipments are thought to be low relative to the other costs. Apart from labour rates, the reason for abandoning final assembly activities in a region can be governmental support and incentives from other regions that outweigh possible labour rate disadvantages.

The major impact of portable manufacturing is the geographical spread of manufacturing activities. This has increased the complexity of physical distribution activities, and hence the complexity of supply chain planning and control activities. Where normally face-to-face contact enables fast and informal communication, nowadays planners, schedulers, expediters, group leaders and many others involved in supply chain activities have to rely on information systems and formal communication. Furthermore, the location of production is typically quite distant from the point of consumption or demand; thus the logistics function is more complex.

2.3.2 Physical distribution is cheap

The outsourcing of the physical distribution function and its increased impact on customer service have stimulated the emergence of third party logistics (3PL)service providers, that take over the actual planning and control functions involved in physical distribution from the Original Equipment Manufacturers (OEM). By doing so, these 3PL service providers should be able to improve the performance of the physical distribution function, while leveraging scale to reduce physical distribution costs.

The emergence of 3PL service providers creates another interface between two legally independent entities, i.e., the manufacturer (or supplier) and the customer. This requires contractual relationships to assure performance. In this context the difficulty lies in the fact that the 3PL provider indeed leverages scale by engaging in several contractual relationships with OEMs, so that the actual cost of a service towards each OEM cannot be separated from the costs of services towards other OEMs. Typically 3PLs operate according to some tariff structure combined with customer-specific rebates based on the power of the customer. Issues related to the tariff structure 3PLs are discussed, amongst other 3PL issues, in Chapter 2.

2.3.3 Relevance for Operations Research applied to supply chains

The complexity of planning and control of a geographically dispersed supply chain, crossing multiple organizational boundaries, is huge and today largely unsolved in practical terms. Though OR has contributed to the design and planning of supply chains, there has been less success implementing the control principles due to the lack of information systems that seamlessly connect the various organizational entities, so that full transparency of information is achieved. Most supply chains still consist of informational silo's that exchange information periodically. The exchange of information is at best imperfectly orchestrated, requiring quite some management attention. Although companies like Cisco and Dell claim to have IT architectures that provide such seamless integration, one should be aware that this relates only to the integration with 1st tier suppliers. This OEM-1st tier supplier interface is responsible for only a small portion of the added value created in the supply chain, albeit that the cumulated value at this interface is almost 100% of the final product cost.

The control principles underlying the planning and control of supply chains are discussed in Chapters 9–13. The strategic and tactical issues involved in asset management in geographically dispersed supply chains are discussed in Chapters 3–5 and 8.

2.4 Information technology

2.4.1 Enterprise resource planning systems

From the mid-eighties onwards, company-wide implementation of so-called Enterprise Resource Planning (ERP) systems was used as a means to introduce new business processes. A lot of attention was paid to the identification of best practices across the company and at other (competing) companies. External consultants supported the implementation process. The typical throughput time of such implementation projects ranged from two to six years, depending on the size and the change management culture of the company. During the nineties many horror stories were published in both scientific journals and the media about the problems occurring during the ERP implementation process. In many cases it was stated that the benefits obtained from the implementation did not have much to do with the IT system itself, but rather from improvements in business processes. Yet, it should be emphasized here, that without the information and transaction processing capabilities of ERP systems, global companies would not be able to operate effectively and efficiently. Without ERP systems implemented across a globally operating company, information would not be available for taking the appropriate measures. On top of that, ERP software vendors have shown that software standardization and maintenance is possible, even for such functionally and architecturally complex systems. The core competence of ERP software vendors, i.e., developing and maintaining standard software to support business processes across a wide range of industrial and public sectors, requires an investment in human resources, that individual companies cannot afford.

Enterprise Resource Planning systems are systems that enable the execution of all business processes, such as order processing, invoicing, transportation, warehouse picking, work order release and purchase order release. Enterprise Resource Planning systems are transactional systems that also support various decision-making processes, such as inventory management, production planning, forecasting, etc. This mixture of transactional system and decision-support system makes it hard to define an ERP system in a rigorous manner. The emergence of so-called Advanced Planning and Scheduling (APS) systems that focus entirely on decision-support permits one to view the ERP systems as being primarily the transactional IT backbone of a company.

Enterprise Resource Planning systems are a 'conditio sine qua non,' a prerequisite for implementation of intra- and inter-company Supply Chain Management. Enterprise Resource Planning systems in their role of transactional backbone provide the required data about future sales plans, customer orders, actual inventories and work-in-process, available resources and cost and pricing information. However, ERP systems are not sufficient for true inter-company SCM. SCM requires information exchange between ERP systems of different companies. From an IT perspective this implies standardization of interfaces and the associated data models. In the late eighties initiatives such as EDIFACT focussed on exchanging transactional data, such as invoices and purchase orders. Only recently the concept of Collaborative Planning, Forecasting and Replenishment (CPFR) requires the exchange of planning data, such as sales plans and production plans. Technically speaking this is similar to the exchange of transactional data. However, planning data contain information about a company's strategy. Most companies are quite reluctant to share this information with suppliers or customers, since this data might, accidentally or not, be shared with competitors. The problem of information privacy has not been resolved and it is quite likely that it cannot be resolved.

2.4.2 Advanced planning systems

During the seventies and eighties OR applications led to the implementation of tailor-made Decision Support Systems (DSS) for supply chains. Initially such DSSs were run on mainframes, but soon after the emergence of the PC such applications were run on this platform. DSSs supported production planning, inventory management and transportation planning. The required inputs were downloaded from IT backbone systems and the outputs were uploaded again, either manually or using an IT interface. Companies such as Manugistics and Numetrix originate from the early eighties. However, these DSSs never raised the same interest with top management as ERP systems. Despite this, we should remark here that DSSs are widely spread across all business function, yet not recognized as such. Virtually any planner, product manager, R&D manager or controller, has developed some sort of DSS with spreadsheet programs, such as Excel. In particular, planning functions are often supported by homemade spreadsheets. In many cases such spreadsheets support the planner in 'solving' extremely complex planning problems.

The lack of attention of top management with respect to DSSs changed fundamentally in the early nineties when the notion of a DSS was replaced by the notion of an APS. One of the keys to the initial success of APS software was the claim of the APS software vendors that they sold, similar to ERP software vendors, standard software. Furthermore, APS software vendors were *top management-geared*. Statements were made about the huge profits that could be gained with the company-wide implementation of APS systems. In late 2001, AMR Research concluded that the promises made were not realized and that APS implementations were restricted to implementation of stand-alone modules (e.g. production planning module and supply chain planning module) instead of integrated APS suites supporting multiple business functions.

For OR researchers this conclusion did not come as a surprise. APS systems *are* DSSs. Decision support requires a careful study of the business processes to be supported, including all peculiarities. In most cases such peculiarities translate into constraints on decision variables that make the problem to be solved NP-hard, when assuming all relevant inputs are known, and even impossible to formulate properly, when assuming some relevant inputs are stochastic. As a consequence, to develop a DSS for such problems entails careful engineering of tailor-made algorithms and requires very scarce human resources. In Chapter 9 APS systems are discussed extensively. In all chapters we will be confronted with the complexity of relevant SCM problems and learn that many questions are left for further research. We should be aware that the promises of APS software vendors led many top managers to believe that all relevant SCM decision support problems can be routinely solved: everything can be optimised, and there is no need for investments in problem-driven research, as might be done by operations researchers.

Despite this scepticism, APS software vendors have drawn the attention of top management to OR. Furthermore, APS software vendors are employers of OR researchers, either directly or indirectly. APS software implementation has given a boost to the development of *solver engines* based on LP and MIP, requiring state-of-the-art scientific OR knowledge to solve large-scale problems. Many researchers in stochastic OR filled in the gap left by the leading APS software vendors related to addressing business issues under uncertainty.

2.4.3 Internet and World Wide Web

A discussion on Information Technology related to SCM is not complete without addressing the impact of the Internet and World Wide Web. The World Wide Web enabled companies to reach out directly to consumers. In fact, consumers have taken over in-company activities, such as order configuration and order entry. Despite the meltdown of the New Economy, sales over the Web contribute considerably to the revenue of many companies and will increase in the future. The direct contact with consumers has allowed firms to acquire individual consumer profiles. In turn, such profiles enable improved forecasting of sales in parallel to mass customization (cf. Chapter 5). Furthermore, with the consumer profiles, a firm can do dynamic pricing, so as to set the right price for the right product, aimed at an increase in turnover and a reduction in product obsolescence. In the business-to-consumer markets, the World Wide Web has created the means to create many-to-many markets, such as auctions.

In the business-to-business environment, the World Wide Web has provided similar opportunities to reduce costs of customer service and purchase order processing, and to reach out to new customers. The Web has also made it possible to share information across companies during joint R&D projects. But most importantly, the Internet has created a low-cost standard public IT infrastructure that enables communication around the globe. Problems of information security have been addressed by applying methods from cryptography. The remaining problem is the problem of standardized messages and interfaces. In that sense the problems mentioned above in relation to EDI still stand. Much effort is put into making progress here by developing voluntary standards, such as XML, and companies join in consortia developing the required standards, such as RosettaNet.

The above clearly shows that much more effort is needed to create a seamless, secure and low-cost IT infrastructure, yet principally IT need not hamper SCM improvements.

2.4.4 Relevance for Operations Research applied to supply chains

Most interesting problems in OR require a substantial amount of data, either due to structural complexity or due to uncertainty for which the probability distribution of random variables and processes must be determined or validated. One might say that only during 1990s has the required data been available at a reasonable cost in time and effort. The implementation of ERP systems, implying centralized databases and data warehouses, made access to detailed transaction data possible.

The Internet has been important in particular for the implementation of SCM. Supply Chain Management implies in many cases that information must be exchanged between different organizations and companies. Nowadays this can be done at low cost and with high security. Exchange of data through the Internet also occurs when an OR application is offered as a service. Typically the application is hosted at a server. Customers using the service have to send their input data to this server and receive output data after processing. *Application Service Providers* (ASP) often have their roots in OR. The OR research discussed in this handbook is likely to be incorporated in such services in the near future.

3 Outline of the volume

This volume consists of three parts. Part I deals with Supply Chain Design. In Chapter 2, Muriel and Simchi-Levi discuss the optimal location of warehouses and factories as well as some tactical problems related to pricing and integrated production, inventory and transportation policies. These models yield the infrastructure from which Chapter 13 departs to develop operational transportation policies. In Chapter 3, Graves and Willems discuss various strategic and tactical issues that must be addressed when deciding on investments in inventory capital to hedge against uncertainty. Also the issue of supplier selection in the context of the trade-off between supplier flexibility and variable material costs is discussed in detail. Chapter 4 by Bertrand provides an overview of the literature on flexibility in the context of Supply Chain Design. The literature review reveals that most of the flexibility concepts from the literature do not provide insight into the issue of allocation of assets across the supply chain, so that flexibility is created at the right links. Bertrand proposes a modelling framework that addresses this issue. This modelling framework shows a close resemblance with the modelling concepts from the design of contracts, which are reviewed in Chapters 6 and 7. Part I closes with Chapter 5, where Swaminathan and Lee examine the relationship between product and process design and Supply Chain Design. One key notion is postponement, which we briefly addressed above.

Part II deals with Supply Chain Coordination. In this context coordination refers to the design of contracts between suppliers and buyers, as well as the information that is exchanged between them. The different incentives of supplier and buyer are formalized in a game–theoretic context, showing that without proper incentive schemes the supply chain becomes inefficient in comparison to a supply chain with centralized control. Relatively simple models reveal fundamental insights on Supply Chain Coordination and already have had a great impact in the business practice of today. In Chapter 6, Cachon focuses on contracts that allow for various kinds of transfer payments and identifies conditions under which such transfer payments yield a properly coordinated supply chain. In Chapter 7, Chen studies the value of information exchange and sharing. By comparing alternatives for sharing information between the links in the supply chain, we obtain insights about which information is most valuable and under what circumstances. The results from Chapters 6 and 7 provide inputs in terms of costs and prices, as well as available information, for the coordination of the supply chain. Still, many other parameters are required to execute the supply chain. In Chapter 8, Swaminathan and Tayur provide a framework for understanding the role of tactical planning parameters, such as forecast accuracy, mean and variance of lead times and capacity utilization. They also emphasize the issue of the structural complexity of a supply chain. Real-world problems have such an enormous structural complexity that there is hardly any hope for solving them cleanly with a closed-form formula. Thus, Swaminathan and Tayur propose alternative routes to cope with this complexity.

The complexity of SCM becomes even more apparent in Part III, which is dedicated to Supply Chain Operations. In Chapter 9, Fleischmann and Meyr provide an overall Supply Chain Planning (SCP) framework. This framework shows the hierarchical nature of real-world SCM and further reveals the structural complexity already discussed by Swaminathan and Lee. The SCP framework provides the means to assess the state-of-the-art of Advanced

Planning and Scheduling systems. In Chapter 10, Axsater discusses the progress made during the nineties with respect to the analysis of multi-echelon serial and divergent inventory systems. The fact that the structure of the optimal policy for divergent systems remains unknown, even for the most benign random demand processes, motivates the development and analysis of various control policies. As discussed above, Dell has introduced a new business model in the consumer market that was normally only used in business-to-business environments, i.e., Assemble-To-Order. This revitalized the interest in the models that describe the control of inventories in such an environment. Song and Zipkin report in Chapter 11 on the substantial progress made in this area. Following up on Chapter 9, De Kok and Fransoo discuss Supply Chain Operations Planning (SCOP) applied to arbitrary multi-echelon inventory systems, i.e., many-to-many relationships between items (links) to be controlled. They propose a framework that enables the assessment of the feasibility of supply chain control concepts proposed in the literature and provide some quantitative results that reveal the counter-intuitive behaviours of such systems. Finally, Chapter 13 discusses the role of the logistics service providers for effective Supply Chain Management. Powell presents a general framework (vocabulary) for modelling a wide range of problems that arise when dealing with transportation optimization under uncertainty in demand, pricing, etc. The models emerging from this framework are tackled with a generic method, called adaptive dynamic programming. The underlying idea is the concept of incomplete states and approximate value functions that allow for the development of approximation methods. Some test problems show promising results. Powell also addresses issues of data quality that are relevant for all problems discussed in this handbook.

Acknowledgements

Many researchers have contributed to this volume. First of all we would like to thank the authors of the chapters who have invested their great skills to deliver a synthesis of a topic area, as well as their perspectives on what has been accomplished and what remains to be done. A decentralized process initiated by the various authors involved numerous colleagues that commented on draft versions of the chapters. As editors of this volume, we have benefited a great deal from this spontaneous process. Finally we would like to thank our publisher North-Holland for their patience. Quality comes first!

References

Chase, R.B., N.J. Aquilano, F.R. Jacobs (1999). *Production and Operations Management: Manufacturing and Services*, McGraw-Hill, Irwin.

Fine, C.H. (1998). *Clockspeed: Winning Industry Control in the Age of Contemporary Advantage*, Perseus Books, Reading, Massachusetts.

Forrester, J.W. (1958). System dynamics: a major breakthrough for decision makers. *Harvard Business Review* 36(4), 37–66.

Lee, H.L., C. Billington (1993). Hewlett-Packard gains control of inventory and service through design for localization. *Interfaces* 23, 1–11.

Lee, H.L., C. Billington (1995). The evolution of supply-chain-management models and practice at Hewlett-Packard. *Interfaces* 25, 42–63.

Lee, H.L., V. Padmanabhan, S. Whang (1997). Information distortion in a supply chain. *Management Science* 43, 546–558.

Prahalad, C.K. and G. Hamel (1990). The core competence of the corporation. *Harvard Business Review* 68, May–June, 79–91.

Suri, R., J.L. Sanders, M. Kamath (1993). Performance evaluation of Production Networks, in: S.C. Graves, A.H.G. Rinnooy Kan, P.H. Zipkin (eds.), *Logistics of Production and Inventory*, North-Holland, Amsterdam, pp. 199–286.

PART I

Supply Chain Design

A.G. de Kok and S.C. Graves, Eds., *Handbooks in OR & MS, Vol. 11*
© 2003 Elsevier B.V. All rights reserved.

Chapter 2

Supply Chain Design and Planning – Applications of Optimization Techniques for Strategic and Tactical Models

Ana Muriel
University of Massachusetts, Amherst, MA 01003, USA

David Simchi-Levi
Massachusetts Institute of Technology, Cambridge, MA 02139, USA

1 Introduction

In recent years, there has been renewed interest in the area of logistics among both industry and academia. A number of major forces have contributed to this trend. First, industry has realized the magnitude of savings that can be achieved by better planning and management of complex logistics systems. Second, many companies have started breaking traditional organizational barriers leading to the cooperation among different functional departments, and thus expanding the scope and size of these systems. At the same time, information and communication systems have been widely implemented and provide access to data from all components of the supply chain. Finally, deregulation of the transportation industry has led to the development of a variety of transportation modes and reduced transportation costs, while significantly increasing the complexity of logistics systems.

These developments call for the implementation of *optimization based* Decision Support Systems that take into account the interaction between the various levels of the logistics network, and utilize the wealth of available information.

Unfortunately, like many other complex business systems, logistics and supply chain management problems are *not* so rigid and well defined that they can be entirely delegated to computers. Instead, in almost every case, the flexibility, intuition, and wisdom that are unique characteristics of humans are essential to effectively manage the systems. However, there are many aspects of these systems which can only be effectively analyzed and understood with the aid of a computer. It is exactly this type of assistance which Decision Support Systems are designed to provide. As the name implies, these systems do not make decisions. Instead, they assist and support the human decision-maker in his or her decision making process.

Within the various disciplines that make up supply chain management, optimization based Decision Support Systems are used to address a wide range of problems, from strategic problems like logistics network design, to tactical problems like the coordination of inventory and transportation decisions, all the way through day-to-day operational problems like production scheduling, delivery mode selection, and vehicle routing. The inherent size and complexity of many of these problems make optimization based Decision Support Systems essential for effective decision making. Indeed, optimization based Decision Support Systems have been used extensively in the last few years to radically improve logistics and supply chain efficiencies.

This chapter describes optimization models that effectively address the coordination of various decisions concerning the planning and design of the supply chain, and are promising foundations for the development of Decision Support Systems in this field. The chapter is divided into three parts, each of which focuses on a different problem area:

Production/Distribution Systems: Part I introduces models which are designed to help determine the appropriate production, inventory, and transportation policies for a set of manufacturing plants, warehouses and retailers. Given plant, warehouse and retailer locations, production, inventory and transportation costs, as well as demand forecasts for each retail outlet, the objective is to determine policies which minimize system-wide costs. As we demonstrate, realistic production and transportation cost functions that exhibit economies of scale make solving these problems challenging.

Of course, forecast demand is not enough to determine an effective inventory policy; uncertainty in demand also needs to be incorporated in the analysis. In practice, this is typically done by decomposing the problem into two parts: The first is identifying an inventory policy that balances holding and fixed costs assuming forecast demand over a given planning horizon, see Stenger (1994). The second is determining safety stock levels and incorporating these in the inventory level that should be maintained at the beginning of each period. Thus, the models analyzed in this part of the chapter help optimize inventory decisions associated with the first part of the decomposition approach used in practice.

Pricing to improve Supply Chain Performance: Dynamic pricing techniques such as yield management have been successfully applied to a variety of industries, e.g., airlines or rental car agencies, with a focus on those that have perishable inventory. In Part II of this chapter, we extend dynamic pricing techniques to a more general supply chain setting with non-perishable inventory. Specifically, we consider pricing, production, and inventory decisions simultaneously in a finite and an infinite horizon single product environment. The objective is to maximize profit under conditions of periodically varying inventory holding and production costs, and price-sensitive, stochastic demand.

Unfortunately, concepts such as convexity or k-convexity, that have been proven effective for classical inventory models, are not applicable for supply chain models with general, price-dependent, stochastic demand processes. Thus, to analyze models that incorporate pricing decisions, we introduce the notion of *symmetric k-convex* functions. This notion allows us to characterize the structure of the optimal policy for finite and infinite horizon, single product, periodic review models with general price-dependent stochastic demand functions.

Interestingly, we demonstrate that dynamic pricing strategies have the potential to radically improve supply chain performance. Indeed, the computational results reported in Part II suggest that companies that experience variability in demand curves or have limited production capacity may significantly benefit from dynamic pricing.

Logistics Network Design: Network configuration may involve issues relating to plant, warehouse and retailer location. These are strategic decisions since they have a long-lasting effect on the firm. In Part III of this chapter, we concentrate on the following key strategic decisions:

1. determining the appropriate number of warehouses,
2. determining the location of each warehouse,
3. determining the size of each warehouse, and
4. determining which products customers will receive from each warehouse.

We therefore assume that plant and retailer locations will not be changed. The objective is to design or reconfigure the logistics network so as to minimize annual system-wide costs including production and purchasing costs, inventory holding costs, facility costs (storage, handling, and fixed costs), and transportation costs, subject to a variety of *service level* requirements.

PART I: PRODUCTION/DISTRIBUTION SYSTEMS

2 Introduction

In the last decade many companies have recognized that important cost savings and improved service levels can be achieved by effectively integrating production plans, inventory control and transportation policies throughout their supply chains. The focus in this and the following two sections is on planning models that integrate decisions across the supply chain for companies that rely on third party carriers.

The models described in these sections are motivated in part by the great development and growth of many competing transportation modes, mainly as a consequence of deregulation of the transportation industry. This has led to a significant decrease in transportation costs charged by third party distributors and, therefore, to an ever-growing number of companies that rely on third party carriers for the transportation of their goods.

One important mode of transportation used in the retail, grocery and electronic industries is the LTL (Less-than-TruckLoad) mode, which is attractive when shipment sizes are considerably less than truck capacity. Typically, LTL carriers offer volume, or quantity, discounts to their clients to encourage demand for larger, more profitable shipments (Fig. 1).

Volume discounts can be of two types: (1) incremental discounts, which can be modeled as a piece-wise linear concave function of the quantity shipped, and (2) all-unit discounts, which, as we demonstrate later, result in the piece-wise linear continuous function depicted below. These cost functions are supported by the industry standard transportation rating engine, called CZAR (Southern Motor Carrier's Complete Zip Auditing and Rating engine), which most LTL carriers use.

Similarly, production costs can often be approximated by piece-wise linear and concave functions in the quantity produced, e.g., set-up plus linear manufacturing costs. These economies of scale motivate the *shipper* to coordinate the production, routing and timing of shipments over the transportation network to minimize system-wide costs. In what follows, we refer to this general problem as the *Shipper Problem*.

This planning model, while quite general, is based on several assumptions which are consistent with the view of modern logistics networks. Indeed, the model deals with situations in which all facilities are part of the same logistics network, and information is available to a central decision-maker whose objective is to optimize the entire system. Thus, distribution problems in the retail and grocery industries are special cases of our model where the logistics network does not include manufacturing facilities.

The model also applies to situations in which suppliers and retailers are engaged in *strategic partnering*. For instance, in a Vendor Managed Inventory (VMI) partnership, point-of-sales data is transmitted to the supplier, which is

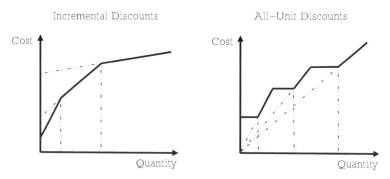

Fig. 1. Common LTL quantity discount cost structures.

responsible for the coordination of production and distribution including managing retail inventory and shipment schedules. Hence, in this case, the model includes manufacturing facilities, warehouses and retail outlets.

Related models analyzing the distribution problem from the carriers point of view are discussed in Farvolden, Powell, and Lustig (1993) and Farvolden and Powell (1994). The first paper develops a fast algorithm for solving large-scale linear programming multi-commodity network flow problems with capacity constraints. The second suggests a heuristic strategy for the problem of determining the number of vehicles the carrier should use in different links of the service network. For a survey of the practical challenges faced by LTL carriers in the design and management of their networks and various solution approaches, the reader is referred to Crainic and Laporte (1997), Crainic and Roy (1992), Braklow, Graham, Hassler, Peck, and Powell (1992), Powell and Sheffi (1989), Powell (1986), Crainic and Rosseau (1986) and Chapter 13 of this handbook.

For completion, we briefly review other commonly used transportation and/or distribution models. Models integrating inventory control policies and vehicle-routing strategies have been analyzed extensively in the literature. See Bramel and Simchi-Levi (1997), Anily and Bramel (1999) and Toth and Vigo (2001) for recent reviews on vehicle routing and inventory/routing problems. These models are quite different from the models analyzed here due to the structure of the transportation cost and the fact that most of them assume that the shipper operates its own fleet of vehicles. This is also the case for the model recently studied by Lee, Çentikaya, and Jaruphongsa (2000), which focuses on the coordination of inventory replenishments and dispatch schedules at a warehouse that serves a single retailer. The warehouse orders incur a fixed cost and the outbound transportation cost function consists of a fixed cost per delivery plus a cost per vehicle dispatched. More general piece-wise linear transportation costs, which include both the ones studied below and those just mentioned, have been considered in Croxton, Gendron and Magnanti (2000a) to model the selection of different transportation modes and shipment routes in merge-in-transit operations. In this case, a set of warehouses coordinates the flow of goods from a number of suppliers to multiple retailers with the objective of reducing costs through consolidation.

Finally, a new trend in distribution management is the acquisition of TL (TruckLoad) transportation services through auction; see Caplice (1996). Specifically, various transportation exchange sites link together shippers, third party logistics intermediaries and carriers, and allow for economic efficiencies through an auction or bidding process. Depending on the exchange, either the carriers bid and the shipper assigns carriers to individual shipments, or the shippers bid and the carrier selects the shipments to serve. In the former case, the carrier must select the set of loads on which to bid, determine the appropriate bidding cost, and be prepared to adjust in real time its current operations to accommodate the new loads. Given the bidding costs, the shipper must determine the cost minimizing assignment of carriers to loads. In the latter case, the carrier must determine which loads and prices to accept and how to adjust its operations to service these loads while maximizing profitability. Examples of shippers that allow carriers to bid on transportation loads include companies such as Sears Roebuck, Ford Motor Company, Wal-Mart and K-Mart, see de Vries and Vohra (2000). The literature on combined value auctions is rapidly growing, see e.g., DeMartini, Kwasnica, Ledyard, and Porter (1999), Rothkopf, Pekeč, and Harstad (1998), Ledyard (2000), Ledyard, Olson, Porter, Swanson, and Torma (2000), Sandholm (1999, 2000), Fujishima, Leyton-Brown, and Shoham (1999), Leyton-Brown, Shoham, and Tennenholtz (2000), Kelly and Steinberg (2000).

The following sections describe our modeling approach and results for the *Shipper Problem* under each of the two common transportation cost functions described above.

3 Piece-wise linear concave costs

In this section, we focus on the *Shipper Problem* under piece-wise linear and concave production and transportation costs, and use properties resulting from the concavity of the cost function to devise an efficient algorithm.

The objective of the shipper is to find a production plan, an inventory policy and a routing strategy so as to minimize total cost and satisfy all the demands. Backlogging of demands may be allowed, incurring a known penalty cost which is a function of the length of the shortage period and the level of shortage. In this case, four different costs must be balanced to obtain an overall optimal policy: production costs, LTL shipping charges, holding costs incurred when carrying inventory at some facility and penalty costs for delayed deliveries.

Chan, Muriel, and Simchi-Levi (1999) formulate this tactical problem as a concave cost multi-commodity network flow problem. Unfortunately, most of the literature on network flows is devoted to the analysis of minimum-cost network flow problems for which the cost is a linear function of the amount shipped on an arc, see Ahuja, Magnanti, and Orlin (1993). In practice, however, situations in which there is a set-up charge, or a discount due to economies of scale give rise to concave cost functions. In this case, an exhaustive search of all extreme points would provide an optimal flow, since a concave function achieves its minimum at an extreme point of the convex feasible region. However, such an approach is impractical for all but the simplest of problems. This, of course, is not surprising since the fixed-charge network design model, in which the cost of using an edge is simply a fixed charge independent of the quantity shipped, is a special case of the concave-cost network flow problem and is NP-Complete, see Johnson, Lenstra, and Rinnooy Kan (1978). Consequently, the exact algorithms that have been developed are either valid only for networks with special structures or run in exponential time in the general case.

For instance, Zangwill (1968) is one of the first authors to analyze the minimum-concave-cost problem. He presents an algorithm with complexity $O(an^d)$, for acyclic networks with a single source (or a single destination), a arcs, n nodes, and $d+1$ destinations (or sources in the single destination case). This algorithm can also be applied to the multi-commodity case, again with either a single source or a single destination, since the problem can be reduced to a single-commodity network flow problem. For the general single-commodity minimum-concave-cost problem, Erickson, Monma, and Veinott (1987) develop a dynamic-programming procedure, called the send-and-split method. The algorithm runs in polynomial time for planar networks in which all demand nodes lie in a bounded number of faces. When the underlying network enjoys the strong-series-parallel property, Ward (1999) develops a polynomial time algorithm to solve the multi-commodity network flow problem with aggregate concave cost. This appears to be the first algorithm to solve the problem in polynomial time.

While all algorithms mentioned above are exact and share a dynamic programming approach, Falk and Soland (1969) and Soland (1971) present branch and bound heuristics based on approximations of the concave functions by linear ones. Gallo and Sodini (1979) find local optimality conditions for the concave-cost multi-commodity network flow problem on uncapacitated networks, and propose a vertex following algorithm to determine the local minima. Yaged (1971) proposes a different method to find local optima; in this case, the point satisfying the Kuhn-Tucker conditions is found by a successive-approximation, fixed-point algorithm. The quality of the local optimum can be improved by using stronger optimality conditions and a greedy-type algorithm; see Minoux (1989) and Guisewite and Pardalos (1990) for a survey of results and solution techniques.

Balakrishnan and Graves (1989) consider a multi-commodity network flow problem, very similar to the one analyzed in this section, in which the arc costs are piece-wise linear concave functions. They develop a composite algorithm that combines good lower bounds and effective heuristic solutions based on solving the Lagrangian relaxation of a specific formulation of the problem. Similarly, Amiry and Pirkul (1997) use a Lagrangian decomposition of the same problem to obtain slightly tighter bounds. However, as for fixed-charge network problems [see Gendron and Crainic (1994)], Muriel and Munshi (2002) show that the lower bounds generated by these Lagrangian relaxation and decomposition methods are no better than that provided by the linear programming relaxation of the problem, in both capacitated and uncapacitated networks.

Finally, we must point out that the multi-commodity network flow problem with piece-wise linear concave costs generalizes the fixed-charge network design problems that arise in various applications in telecommunications, transportation, logistics and production planning, see, e.g., Magnanti and Wong (1984), Balakrishnan, Magnanti, and Mirchandani (1997), Balakrishnan, Magnanti, Shulman, and Wong (1991), Gavish (1991) and Minoux (1989). These models have been extensively studied, especially in the telecommunications literature in the context of the network loading problem. In this case, capacitated facilities are to be installed on edges of a telecommunication network to support prescribed point-to-point demand flow, see for instance Stoer and Dahl (1994) or Bienstock, Chopra, and Günlük (1998). For a review, we refer the reader to Gendron, Crainic, and Frangioni (1999). A common approach used to solve these network design problems is Lagrangian relaxation, together with dual ascent, subgradient optimization and/or bundle methods to optimize the Lagrangian dual. Crainic, Frangioni, and Gendron (1999) report on the performance of different relaxations and dual optimization methods.

In what follows, we first incorporate the time dimension into the model by constructing the so-called expanded network. This expanded network is used to formulate the Shipper Problem as a set-partitioning problem. The formulation is found to have surprising properties, which are used to develop an efficient algorithm and to show that the linear programming relaxation of the set-partitioning formulation is tight in certain special cases (Section 3.4). Computational results, demonstrating the performance of the algorithm on a set of test problems, are reported in Section 3.5.

3.1 The LTL shipper model

Consider a generic transportation network, $G = (N, A)$, with a set of nodes N representing the suppliers, warehouses and customers. Customer demands for the next T periods are assumed to be deterministic and each of them is considered as a separate *commodity*, characterized by its origin, destination, size and the time period when it is demanded. Our problem is to plan production and route shipments over time so as to satisfy these demands while minimizing the total production, shipping, inventory and penalty costs.

A standard technique to efficiently incorporate the time dimension into the model, see for instance, Farvolden et al. (1993), is to construct the following *expanded network*. Let $\tau_1, \tau_2, \ldots, \tau_T$ be an enumeration of the relevant time periods of the model. In the original network, G, each node i is replaced by a set of nodes $i_1, i_2, \ldots, i_T$. We connect node i_u with node j_v if and only if $\tau_v - \tau_u$ is exactly the time it takes to travel from i to j. Thus, arc $i_u \to j_v$ represents freight being carried from i to j starting at time τ_u and ending at time τ_v. We call such arcs *shipping links*. In order to account for penalties associated with delayed shipments, a new node is created for each commodity and serves as its ultimate sink. For a given commodity, a link between a node representing its associated retailer at a specific time period, and its corresponding sink node, represents the penalty cost of delivering a specific shipment in that time

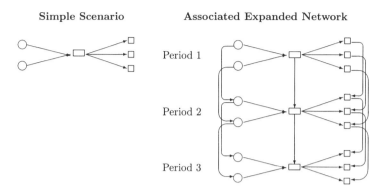

Fig. 2. Example of expanded network.

period, and is called *penalty link*. Finally, we add links (i_l, i_{l+1}) for $l = 1, 2, \ldots, T-1$, referred to as *inventory links*. Let $G_T = (V, E)$ be the *expanded network*. Figure 2 illustrates the expanded network for a simple scenario where the shipping and inventory costs have to be balanced over a time horizon of just three periods and shortages are not allowed. For simplicity, we assume that travel times are zero.

Observe that, using the expanded network, the shipper problem can be formulated as a concave-cost multi-commodity network flow problem. Production decisions can be easily incorporated into this model. For this purpose, in the expanded network, each production facility at a specific time is represented by two nodes connected by a single link whose cost represents the concave (e.g., set-up plus linear) manufacturing costs. This link is not different from the shipping links in our original model and, consequently, we can restrict the discussion, without loss of generality, to the pure distribution problem.

3.2 A set-partitioning approach

To describe our modeling approach, we introduce the following notation. Let $\mathcal{K} = \{1, 2, \ldots, K\}$ be the index set of all commodities, or different demands with fixed origin and destination, and let w_k, $k = 1, 2, \ldots, K$, be their corresponding size. For instance, commodity $k = 1$ may correspond to a demand of $w_1 = 100$ units that needs to be shipped from a certain supplier to a certain retailer and must arrive by a particular period of time or incur delay penalties. Let the set of all possible paths for commodity k be P_k and let c_{pk} be the sum of inventory and penalty costs incurred when commodity k is shipped along path $p \in P_k$. Observe that the shipping cost associated with a path will depend on the total quantity of all commodities being sent along each of its shipping links and, consequently, it can't be added to the path cost a priori. Thus, each shipping edge, whose cost must be globally computed, needs to be considered separately. Let the set of all shipping edges be SE and for each edge $e \in SE$, let z_e be the total sum of weight of the commodities traveling on that edge.

We assume that the cost of a shipping edge e, $e \in SE$, of the expanded network $G_T(V, E)$, is $F_e(z_e)$, a *piece-wise linear and concave cost function* which is non-decreasing in the total quantity, z_e, of the commodities sharing edge e. As presented in Balakrishnan and Graves (1989), this special cost structure allows for a formulation of the problem as a mixed integer linear program. For this purpose, the piece-wise linear concave functions are modeled as follows. Let R be the number of different slopes in the cost function, which we assume, without loss of generality, is the same for all edges to avoid cumbersome notation. Let M_e^{r-1}, M_e^r, $r = 1, \ldots, R$, denote the lower and upper limits, respectively, on the interval of quantities corresponding to the rth slope of the cost function associated with edge e. Note that $M_e^0 = 0$ and M_e^R can be set to the total quantity of all commodities that may use arc e. We associate with each of these intervals, say r, a variable cost per unit, denoted by α_e^r, equal to the slope of the corresponding line segment, and a fixed cost, f_e^r, defined as the y-intercept of the linear prolongation of that segment. See Fig. 3 for a graphical representation. Observe that the cost incurred by any quantity on a certain range is the sum of its associated fixed cost plus the cost of sending all units at its corresponding linear cost. That is, we can express the arc flow cost function, $F_e(z_e)$, as

$$F_e(z_e) = f_e^r + \alpha_e^r z_e,$$

if $z_e \in (M_e^{r-1}, M_e^r]$. Clearly,

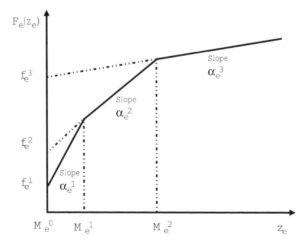

Fig. 3. Piece-wise linear and concave cost structure.

Property 1. *The concavity and monotonicity of the function F_e implies that,*

1. $\alpha_e^1 > \alpha_e^2 > \cdots > \alpha_e^R \geq 0$,
2. $0 \leq f_e^1 < f_e^2 < \cdots < f_e^R$,
3. $F_e(z_e) = \min_{r=1,\ldots,R}\{f_e^r + \alpha_e^r z_e\}$. *The minimum is achieved at a unique index s, unless $z_e = M_e^s$, in which case the two consecutive indexes s and $s+1$ lead to the same minimum cost.*

We are now ready to introduce an integer linear programming formulation of the Shipper Problem for this special cost structure. Recall that z_e denotes the total flow on edge e and let z_{ek} be the quantity of commodity k that is shipped along that edge. For all $e \in SE$ and $r = 1, \ldots, R$ define the *interval* variables,

$$x_e^r = \begin{cases} 1, & \text{if } z_e \in (M_e^{r-1}, M_e^r], \\ 0, & \text{otherwise,} \end{cases}$$

and, in addition, for every k, $k \in \mathcal{K}$, let the *quantity* variables be

$$z_{ek}^r = \begin{cases} z_{ek}, & \text{if } z_e \in (M_e^{r-1}, M_e^r], \\ 0, & \text{otherwise.} \end{cases}$$

In order to relate these edge flows to path flows we define, for each $e \in SE$ and $p \in \bigcup_{k=1}^K P_k$,

$$\delta_p^e = \begin{cases} 1, & \text{if shipping link } e \text{ is in path } p, \\ 0, & \text{otherwise.} \end{cases}$$

Finally, let variables

$$y_{pk} = \begin{cases} 1, & \text{if commodity } k \text{ follows path } p \text{ in the optimal solution} \\ 0, & \text{otherwise,} \end{cases}$$

for each $k \in \mathcal{K}$ and $p \in P_k$. These variables are referred to as *path flow* variables. Observe that defining these variables as binary variables implies that for every commodity k only one of the variables y_{pk} takes a positive value. This reflects a common business practice in which each commodity, that is, items originated at the same source and destined to the same sink in the expanded network, is shipped along a single path. These integrality constraints are, however, not restrictive, as pointed out in Property 2 below, since the problem is uncapacitated and the cost functions concave.

In the *Set-Partitioning* formulation of the LTL Shipper Problem, the objective is to select a minimum cost set of feasible paths. Thus, we formulate the LTL shipper problem for piece-wise linear concave edge costs as the following mixed integer linear program, which we denote by Problem P.

$$\text{Problem } P: \quad \text{Min} \sum_{k=1}^{K} \sum_{p \in P_k} y_{pk} c_{pk} + \sum_{e \in SE} \sum_{r=1}^{R} \left[f_e^r x_e^r + \alpha_e^r \left(\sum_{k=1}^{K} z_{ek}^r \right) \right]$$

s.t.

$$\sum_{p \in P_k} y_{pk} = 1, \quad \forall k = 1, 2, \ldots, K, \tag{3.1}$$

$$\sum_{p \in P_k} \delta_p^e y_{pk} w_k = \sum_{r=1}^{R} z_{ek}^r, \quad \forall e \in SE, k = 1, \ldots, K, \tag{3.2}$$

$$z_{ek}^r \leq w_k x_e^r \quad \forall e, r, k, \tag{3.3}$$

$$\sum_{k=1}^{K} z_{ek}^r \leq M_e^r x_e^r, \quad \forall e \in SE, r = 1, \ldots, R, \tag{3.4}$$

$$\sum_{k=1}^{K} z_{ek}^r \geq M_e^{r-1} x_e^r, \quad \forall e \in SE, r = 1, \ldots, R, \tag{3.5}$$

$$\sum_{r=1}^{R} x_e^r \leq 1 \quad \forall e \in SE, \tag{3.6}$$

$$y_{pk} \in \{0, 1\}, \quad \forall k = 1, 2, \ldots, K, \text{ and } p \in P_k, \tag{3.7}$$

$$x_e^r \in \{0, 1\}, \quad \forall e \in SE, \text{ and } r = 1, 2, \ldots, R,$$
$$z_{ek}^r \geq 0, \quad \forall e \in SE, \forall k = 1, 2, \ldots, K,$$
$$\text{and } r = 1, 2, \ldots, R. \tag{3.8}$$

In this formulation, constraints (3.1) ensure that exactly one path is selected for each commodity and constraints (3.2) set the total flow on an edge e to be equal to the total flow of all the paths that use that edge. Constraints (3.3)–(3.6) are used to model the piece-wise linear concave function. Constraints (3.3) specify that if some commodity k is shipped on edge e using cost index r, the associated interval variable, x_e^r, must be 1. Constraints (3.4) and (3.5) make sure that if cost index r is used on edge e, then the total flow on that edge must fall in its associated interval, $[M_e^{r-1}, M_e^r]$.

Finally, constraints (3.6) indicate that at most one cost range can be selected for each edge.

Let Z^* be the optimal solution to Problem P. Let Z_{R_x} and Z_{R_y} be the optimal solutions to relaxations of Problem P where the integrality constraints of interval (x) and path flow (y) variables, respectively, are dropped. A consequence of Property 1 is the following result.

Property 2. *We have*,

$$Z^* = Z_{R_x} = Z_{R_y}.$$

To find a robust and efficient heuristic algorithm for Problem P, we study the performance of a relaxation of Problem P that drops integrality and redundant constraints. Although constraints (3.3) are not required for a correct mixed-integer programming formulation of the problem, we keep them because they improve significantly the performance of the linear programming relaxation of Problem P. In fact, Croxton, Gendron and Magnanti (2000b) show that, without them, the linear programming relaxation of this model approximates the piece-wise linear cost functions by their lower convex envelope. Furthermore, keeping these constraints makes constraints (3.4)–(3.6) redundant in the correct mixed-integer programming formulation, as a direct consequence of Property 1 part 3, and in the linear programming relaxation of problem P as well, as Lemma 3 below shows. This will be useful to considerably reduce the size of the formulation of the problem, while preserving the tightness of its linear programming relaxation.

Let Problem P_{LP}^R be the linear program obtained from Problem P by relaxing the integrality constraints and constraints (3.4)–(3.6). That is,

$$\text{Problem } P_{LP}^R : \text{Min} \sum_{k=1}^{K} \sum_{p \in P_k} y_{pk} c_{pk} + \sum_{e \in SE} \sum_{r=1}^{R} \left[f_e^r x_e^r + \alpha_e^r \left(\sum_{k=1}^{K} z_{ek}^r \right) \right]$$

s.t. (3.1)–(3.3)

$y_{pk} \geq 0, \quad \forall k = 1, 2, \ldots, K, \text{ and } p \in P_k,$

$x_e^r \geq 0, \quad \forall e \in SE, \text{ and } r = 1, 2, \ldots, R,$

$z_{ek}^r \geq 0, \quad \forall e \in SE, \forall k = 1, 2, \ldots, K,$

and $r = 1, 2, \ldots, R.$

Chan et al. (1999) prove the following.

Lemma 3. *The optimal solution value to Problem P_{LP}^R is equal to the optimal solution value to the linear programming relaxation of Problem P.*

3.3 Structural properties

To analyze the relaxed problem, we start by fixing the fractional path flows and study the behavior of the resulting linear program. Let $y = (y_{pk})$ be the vector of path flows in a feasible solution to the relaxed linear program, Problem P_{LP}^R.

Observe that, given the vector of path flows y, the amount of each commodity sent on each edge is known and, thus, Problem P_{LP}^R can be decomposed into multiple subproblems, one for every edge. Each subproblem determines the cost that the linear program associates with the corresponding edge flow. We refer to the subproblem associated with edge e as the Fixed-Flow Subproblem on edge e, or Problem FF_y^e.

Let the proportion of commodity k shipped along edge e be

$$\gamma_{ek} = \sum_{p \in P_k} \delta_p^e y_{pk}.$$

Using Eq. (3.2), the equality $\sum_{r=1}^{R} z_{ek}^r = w_k \gamma_{ek}$ must clearly hold; that is, the sum of all the flows of commodity k on the different cost intervals on edge e must be equal to the total quantity, $w_k \gamma_{ek}$, of commodity k that is shipped on that edge.

For each edge e, the total shipping cost on e, as well as the value of the corresponding variables z_{ek}^r and x_e^r, that Problem P_{LP}^R associates with the vector of path flows y, can be obtained by solving the Fixed-Flow Subproblem on edge e:

$$\text{Problem } FF_y^e : \text{Min} \sum_{r=1}^{R} \left[f_e^r x_e^r + \alpha_e^r \sum_{k=1}^{K} z_{ek}^r \right]$$

s.t.

$$z_{ek}^r \leq w_k x_e^r \quad \forall k = 1, \ldots, K, \text{ and } r = 1, \ldots, R, \quad (3.9)$$

$$\sum_{r=1}^{R} z_{ek}^r = w_k \gamma_{ek}, \quad \forall k = 1, \ldots, K, \quad (3.10)$$

$$z_{ek}^r \geq 0, \quad \forall k = 1, \ldots, K, \text{ and } r = 1, \ldots, R,$$
$$x_e^r \geq 0, \quad \forall r = 1, \ldots, R.$$

Let $C_e^*(y) \equiv C_e^*(\gamma_{e1},\ldots,\gamma_{eK})$ be the optimal solution to the Fixed-Flow Subproblem on edge e for a given vector of path flows y, or, equivalently, for given corresponding proportions $\gamma_{e1},\ldots,\gamma_{eK}$, of the commodities shipped on that edge.

The following Theorem determines the solution to the subproblem.

Theorem 4. *For any given edge $e \in SE$, let the proportion γ_{ek} of commodity k to be shipped on edge e be known and fixed, for $k = 1,2,\ldots,K$, and let the commodities be indexed in non-decreasing order of their corresponding proportions, that is,*

$$\gamma_{e1} \leq \gamma_{e2} \leq \cdots \leq \gamma_{eK}.$$

Then, the optimal solution to the Fixed-Flow Subproblem on edge e is

$$C_e^*(\gamma_{e1},\ldots,\gamma_{eK}) = \sum_{k=1}^{K} F_e\left(\sum_{i=k}^{K} w_i\right) [\gamma_{ek} - \gamma_{ek-1}], \tag{3.11}$$

where $\gamma_{e0} := 0$.

Intuitively, the above Theorem just says that in an optimal solution to the Fixed-Flow Subproblem associated with any edge e, fractions of commodities are consolidated to be shipped at the cheapest possible cost per unit. At first, a fraction γ_{e1} of all commodities $1,2,\ldots,K$ is available. Thus, these commodities get consolidated to achieve a cost per unit of $F_e(\Sigma_{k=1}^{K} w_k)/\Sigma_{k=1}^{K} w_k$, i.e., the cost per unit associated with sending the full K commodities on that edge, and the available fraction γ_{e1} is sent incurring a cost of $\gamma_{e1} F_e(\Sigma_{k=1}^{K} w_k)$. At that point, none of commodity 1 is left and a fraction $(\gamma_{e2} - \gamma_{e1})$ is the maximum available simultaneously from all commodities $2,3,\ldots,K$. Again these commodities get consolidated and that fraction, $(\gamma_{e2} - \gamma_{e1})$, from each commodity is sent at a cost $(\gamma_{e2} - \gamma_{e1}) F_e(\Sigma_{k=2}^{K} w_k)$. This process continues until the desired proportion of each commodity has been sent.

A generalization of this result to capacitated networks has recently been derived, see Muriel and Munshi (2002).

3.4 Solution procedure

Theorem 4 provides a simple expression of the cost that the relaxed problem, Problem P_{LP}^R, assigns to any given fractional path flows and thus it allows for the efficient computation of the impact of modifying the flow in a particular path. This is the key to the algorithm developed in this section. Indeed, the algorithm transforms an optimal fractional solution to the linear program P_{LP}^R into an integer solution by modifying path flows, choosing for each commodity the path that leads to the lowest increase in the objective of the linear program.

3.4.1 The linear programming based heuristic
Step 1: Solve the linear program, Problem P_{LP}^R. Initialize $k=1$.
Step 2: For each arc compute a marginal cost which is the increase in cost incurred in the Fixed-Flow Subproblem by augmenting the fractional flow of commodity k to 1. Note that this is easy to compute using Theorem 4.
Step 3: Determine a path for commodity k by finding the minimum cost path on the expanded network with edge costs equal to the marginal costs.
Step 4: Update the flows and the costs on each link (again employing Theorem 3.4) to account for commodity k being sent along that path.
Step 5: Let $k=k+1$ and repeat steps (2)–(5) until $k=K+1$.

Evidently, the effectiveness of this heuristic depends on the tightness of the linear programming relaxation of Problem P. For this reason, we study the difference between integer and fractional solutions to Problem P. Chan et al. (1999) show that in some special cases an integer solution can be constructed from the optimal fractional solution of Problem P_{LP}^R without increasing its cost. In particular, using Theorem 4, they prove the following result.

Theorem 5. *In the following cases:*

1. *Single period, multiple suppliers, multiple retailers, two warehouses,*
2. *Two periods, single supplier, multiple retailers, single warehouse,*
3. *Two periods, multiple supplier, multiple retailers, single warehouse using a cross-docking strategy,*
4. *Multiple periods, single supplier, single retailer, single warehouse that uses a cross-docking strategy.*

The solution to the linear programming relaxation of problem P is the optimal solution to the shipper problem. That is,

$$Z^* = Z^{\text{LP}}.$$

Furthermore, in the first three cases, all extreme point solutions to the linear program are integer.

The *cross-docking* strategy referred to in the last two cases, is a strategy in which the stores are supplied by central warehouses which do not keep any stock themselves. That is, in this strategy, the warehouses act as coordinators of the supply process, and as transshipment points for incoming orders from outside vendors.

The Theorem thus demonstrates the exceptional performance of the linear programming relaxation, and consequently of the heuristic, in some special cases. A natural question at this point is whether these results can be generalized. The answer is no in general. To show this, Chan et al. (1999) construct examples with a single supplier, a single warehouse and multiple retailers and time periods, for which

$$\frac{Z^*}{Z^{LP}} \to \infty,$$

as the number of retailers and time periods increases.

Lemma 6. *The linear programming relaxation of Problem P can be arbitrarily weak, even for a single-supplier, single-warehouse, multi-retailer case in which demand for the retailers is constant over time.*

It is important to point out that the instances in which the heuristic solution is found to be arbitrarily bad are characterized by the unrealistic structure of the shipping cost. In these instances, the shipping cost between two facilities is a pure fixed charge (regardless of quantity shipped) in some periods, linear (with no fixed charges) in others, and yet prohibitively expensive so that nothing can be shipped in the remaining periods. The following examples illustrate this structure.

Example of weak linear programming solution: Consider a three-period single-warehouse model in which a single supplier delivers goods to a warehouse which, in turn, replenishes inventory of three retailers over time. The warehouse uses a cross-docking strategy and, thus, it does not keep any inventory. Let transportation cost be a fixed charge of 100 for any shipment from the supplier to the warehouse at any period. Transportation from the warehouse to retailer i, $i = 1, 2, 3$ is very large for shipments made in period i (in other words, retailer i cannot be reached in period i) and negligible for periods $j \neq i$. Let inventory cost be negligible for all retailers at all periods, and let demand for each retailer be 0 units in periods 1 and 2 and 100 units in period 3.

Observe that, in order to reach the three retailers, shipments need to be made in at least two different periods. Thus, the optimal integer solution is 200. However, in the solution to the linear program 50 units are sent to each of the 'reachable' retailers in each period, and a transportation cost of 50 is charged at each period (as stated in Theorem 4, since only a fraction of 1/2 of the commodities is sent on any edge, exactly that fraction of the fixed cost is charged). Thus, the optimal fractional solution is 150 and the ratio of integer to fractional solutions is 3/2.

In this instance, even if fractional and integer solutions are different, the linear programming based heuristic generates the optimal integer solution. However, we can easily extend the above scenario to instances for which the difference between the solution generated by the heuristic and the optimal integer solution is arbitrarily large.

Example of weak heuristic solution: For that purpose, we add n new periods to the above setting. In period 4, the first of the new periods, the cost for shipping from supplier to warehouse is linear at a rate of $1/3$ and the cost for shipping from the warehouse to each of the 3 retailers is 0. On all the other $n-1$ periods the cost of shipping is very high and thus no shipments will be made after period 4. Inventory costs at all retailers and all periods are negligible. Demand for each of the three retailers at each of the new n periods is 100, while demand during the first 3 periods is 0. It is easy to see that the optimal integer and fractional solutions are identical to those in the 3-period case, with costs of 200 and 150, respectively. However, the heuristic algorithm will always choose to ship each commodity in period 4, since the increase in cost in the corresponding path would be $1/3 \times 100$ while it is at least 50 in any of the first 3 periods. Thus, the total cost of the heuristic solution is $1/3 \times 100 \times n$ and the gap with the optimal integer solution arbitrarily large.

The following section reports the practical performance of the algorithm on a set of randomly generated instances.

3.5 *Computational results*

The computational tests carried out are divided into three categories:

1. Single-period layered networks.
2. General networks.
3. Multi-period single-warehouse distribution problems:
   - Pure distribution instances.
   - Production/distribution instances.

The first two categories are of special interest because they allow us to compare our results with those reported by Balakrishnan and Graves (1989), henceforth B&G (1989). The third set of problems models practical situations in which each of the retailers is assigned to a single warehouse and production and transportation costs have to be balanced with inventory costs over time.

In the three categories the tests were run on a Sun SPARC20 and CPLEX was used to solve the linear program, Problem P_{LP}^R, using an equivalent formulation where path flow variables are replaced by flow-balance constraints. During our computational work, we observed that the dual simplex method is more efficient than the primal simplex method in solving these highly degenerate problems, an observation also made by Melkote (1996). This is usually the case for programs with variable upper bound constraints, such as our constraints $z_{ek}^r \leq w_k x_e^r$. We should also point out that most of the CPU time reported in our tests is used in solving the linear program. Thus, to enhance the computational performance of our algorithm and increase the size of the problems that it is capable of handling, future research focused on efficiently solving the linear program is needed. For instance, the original set-partitioning formulation, Problem P_{LP}^R, could be solved faster using column generation techniques. In these tests, however, we focused on evaluating the quality of the integer solutions provided by the heuristic and the tightness of the linear programming relaxation.

We now discuss each class of problems and the effectiveness of our algorithm.

3.5.1 Single-period layered networks

B&G (1989) present exceptional computational results for single-period layered networks. In these instances, commodities flow from the manufacturing facilities to distribution centers, where they are consolidated with other shipments. These shipments are then sent to a number of warehouses, where they are split and shipped to their final destinations. Thus, every commodity must go through two layers of intermediate points: *consolidation points*, also referred to as distribution centers, and *breakbulk points*, or warehouses.

To test the performance of our algorithm and to compare it with that of B&G (1989), we generated instances of the layered networks following the details given in their paper. In this computational work, five different problem classes, referred to as LTL1–LTL5, are considered.

Table 1 shows the sizes of the different classes of problems. For each of these classes, the first column (B&G) of Table 2 presents the average ratio between the upper bounds generated by the heuristic proposed by B&G (1989) and a lower bound on the optimal solution, over 5 randomly generated instances. The numbers are taken from their paper. We do not include, though, their average CPU times because the machines they use are completely different than ours and, in addition, they do not report total computational time for the entire algorithm. The second and third columns report the average deviation from optimality and computational performance of the Linear Programming Based Heuristic (LPBH) over 10 random

Table 1
Test problems generated as in Balakrishnan and Graves (1989)

Number of nodes	Problem class				
	LTL1	LTL2	LTL3	LTL4	LTL5
SOURCE	4	5	6	8	10
CONSOLIDN	5	10	12	15	20
BREAKBULK	5	10	12	15	20
DESTN	4	5	6	8	10
Arcs	42–47	131–141	190–207	309–312	358–372
Commodities	10	20	30	50	60

Table 2
Computational results for layered networks

Problem class	B&G	LPBH	
	LB/UB (%)	LP/Heurisitic (%)	Avg. CPU time (sec)
LTL1	99.8	100	1.04
LTL2	100	100	7.94
LTL3	99.6	100	20.74
LTL4	99.1	100	55.72
LTL5	99.5	100	100.48

Balakrishnan and Graves results (B&G) versus those of our Linear Programming Based Heurisitic (LPBH).

Table 3
Computational results for general networks

Problem class	Size			B&G	LPBH	
	No. of nodes	No. of arcs	No. of comm.	LB/UB (%)	LP/Heuristic (%)	Avg. CPU time (sec)
GEN1	10	47–54	10	99.9	100	2.18
GEN2	15	109–136	20	98.7	99.53	24.04
GEN3	20	196–235	30	98.4	99.88	139.83
GEN4	30	364–428	50	96.2	98.59	1313.06
GEN5	40	340–370	60	98.5	99.98	159.57

Balakrishnan and Graves results (B&G) versus those of our Linear Programming Based Heuristic (LPBH).

instances, for each of the problem classes. In all of them, our algorithm *finds the optimal integer solution*; furthermore, the solution to the linear program in the first step of our algorithm is integer, providing the optimal solution to the problem.

Of course, since in all the previous instances the linear program provided the optimal integer solution, the performance of our procedure has not really been tested. In the following subsections we present computational results for problem classes in which the solution to the linear program is not always integer.

3.5.2 *General networks*

In this subsection, we report on the performance of our algorithm on general networks, in which every node can be an origin and/or a destination, generated exactly as they are generated by B&G (1989). These results together with those of B&G (1989) are reported in Table 3. In this category, B&G (1989) consider five different problem classes, referred to as GEN1,...,GEN5, and generate five random instances for each of them.

Ch. 2. Supply Chain Design and Planning   37

We, in turn, solve 10 different randomly generated instances for each of the problem classes. Again, we do not include their average CPU times due to the reasons mentioned above.

3.5.3 Multi-period single-warehouse distribution problems

Here we consider a single-warehouse model where a set of suppliers replenishes inventory of a number of retailers over time. We test two different types of instances: pure distribution instances in which the routing and timing of shipments are to be determined, and production/distribution instances in which the production schedule is also integrated with the transportation and inventory decisions.

Pure distribution instances. We assume that shortages are not allowed and analyze three different strategies:

1. *Classical inventory/Distribution strategy:* Material flows always from the suppliers through a single warehouse where it can be held as inventory.
2. *Cross-docking strategy:* All material flows through the warehouse where shipments are reallocated and immediately sent to the retailers.
3. *A distribution strategy that allows for direct shipments:* Items may be sent either through the warehouse or directly to the retailer. The warehouse may keep inventory.

For each strategy, we analyze different situations where the number of suppliers is either 1, 2, or 5, the number of retailers is 10, 12, or 20 and the number of periods is 8 or 12. For each combination of the number of suppliers, retailers and periods presented in Table 6, 10 instances are generated. The retailers and suppliers are randomly located on a 1000×1000 grid, while the warehouse is randomly assigned to the 400×400 subgrid at the center. Demand is generated for each retailer–supplier pair at each time period, except for the cases with five suppliers in which each of these pairs has an associated demand with probability 1/3. These demands are generated from a uniform distribution on the integers in the interval [0, 100).

All suppliers and retailers are linked to the warehouse and the distance associated is the corresponding Euclidean distance between the nodes of the grid. In the case of a *Distribution Strategy that Allows for Direct Shipments*, shipping edges from each of the suppliers to each of the retailers are added. The holding costs per unit of inventory are different at the warehouses and retailer facilities and are presented in Table 5. All holding costs at the suppliers are set to zero. Two shipping-cost functions, representing cost per item per unit distance, are considered: The first is assigned to shipments from the suppliers to the warehouse. The second is incurred by the material flowing from the warehouse to the retailers. The cost function (dollars per mile per unit) associated with direct shipments is equal to that of shipments from the warehouse to a retailer. Both functions have an initial set-up cost for using the link and three different linear rates depending on the quantity shipped, see Table 4. However, the ranges to which those linear costs correspond are different for the different Problem classes. This is done so that, in an optimal solution, shipments are consolidated and thus the concave cost function plays an important role in the analysis. These ranges and the corresponding problem classes are presented in Table 5.

Observe, see Table 6, that in most of the instances tested, the linear program is tight and it provides the optimal integer solution. Only in three out of the 150 instances generated, the solution to the linear program is not integer and, in such cases, our algorithm finds a solution which is within 0.8% from the optimal fractional solution.

Production/distribution instances. This section demonstrates the effectiveness of the algorithm when applied to production/distribution systems, i.e., systems

Table 4
Linear and set-up costs used for all the test problems

Type of arc	α_e^1	α_e^2	α_e^3	Set-up
Supplier–warehouse	0.15	0.105	0.084	25
Warehouse–retailer	0.25	0.20	0.16	10

Table 5
Inventory costs and different ranges for the different test problems

Problem class	Inventory cost		Supplier–warehouse cost		Warehouse–retailer cost	
	Warehouse	Retailer	Range 1	Range 2	Range 1	Range 2
I1	5	10	800	1500	200	400
I2					300	600
I3					300	600
I4	10	20	1000	2000	150	300
I5					200	400
I6					200	400
C1	10	20	800	1500	200	400
C2					300	600
C3					300	600
C4	10	20	1000	2000	150	300
C5					200	400
C6					200	400
D1	10	20	500	1000	150	300
D2					200	400
D3					200	400

Table 6
Computational results for a single warehouse

Strategy	Problem class	Number of suppliers	Number of stores	Number of periods	LP/Heuristic (%)	CPU time (sec)
Classical inventory/ distribution strategy	I1	1	10	12	100	65.21
	I2	2			100	187.37
	I3	5			100	163.23
	I4	1	20	8	99.946	83.5
	I5	2			100	210.51
	I6	5			99.953	200.68
Cross-docking strategy	C1	1	10	12	100	60.0
	C2	2			100	174.13
	C3	5			100	159.06
	C4	1	20	8	100	79.73
	C5	2			100	202.83
	C6	5			100	186.0
Direct shipments allowed	D1	1	12	8	100	51.23
	D2	2			100	165.83
	D3	5			99.921	117.27

in which one needs to coordinate production planning, inventory control and transportation strategies over time. For that purpose, we consider the same set of problems, I1–I3, as in the *Classical Inventory/Distribution Strategy* described in the previous section and add production decisions at each of the supplier sites. This is incorporated into the model as explained in Section 3.1.

We consider a fixed set-up cost for producing at any period plus a certain cost per unit. The set-up cost is varied in the set $\{50, 100, 500, 1000\}$ and the linear production cost is set to 1. Inventory holding rate at the supplier site (after production) is set to half of that at the warehouse. For the 60 different instances generated, the linear programming relaxation gave an *integer* solution every time.

4 All-unit discount transportation costs

In this section, we study coordination of production, inventory and transportation activities under the all-unit discount transportation cost structure. Specifically, this cost function, described in Fig. 4, implies that if Q units are shipped, the transportation cost function is

$$G(Q) = \begin{cases} 0, & \text{if } Q = 0, \\ c, & \text{if } 0 < Q < M_1, \\ \alpha_1 Q, & \text{if } M_1 \leq Q < M_2, \\ \alpha_2 Q, & \text{if } M_2 \leq Q < M_3, \\ \vdots \\ \vdots \end{cases}$$

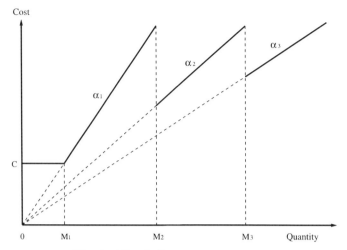

Fig. 4. All-unit discount cost structure.

where $\alpha_1 > \alpha_2 > \cdots \geq 0$ and $\alpha_1 M_1 = c$. Thus, c is a *minimum charge* for shipping a small volume, i.e., c is the total cost when the number of units shipped is no more than M_1. Interestingly, in practice, when the shipper is planning to ship Q units, $M_i \leq Q < M_{i+1}$, the cost is calculated as

$$F(Q) = \min\{G(Q), G(M_{i+1})\} = \min\{\alpha_i Q, \alpha_{i+1} M_{i+1}\}.$$

That is, if the order quantity is greater than a certain value, the shippers pay as if they were shipping M_{i+1} units. This is called in the industry *shipping Q but declaring M_{i+1}*.

This commonly used practice implies that the true transportation cost function, $F(\cdot)$, has the structure described by the solid line in Fig. 5. As the dashed lines indicate, the associated solid lines originate at point $(0,0)$.

We refer to such cost functions as *modified all-unit discount cost functions*. Notice that such a cost function satisfies the following properties:

(p1) it is a non-decreasing function of the amount shipped,
(p2) the cost per unit is non-increasing in the amount shipped.

As indicated in the next section, these two properties are sufficient to derive the results presented below.

To justify considering this cost function, we should point out again that most LTL carriers use an industry standard transportation rating engine called CZAR (Southern Motor Carrier's Complete Zip Auditing and Rating engine). This engine allows the shipper to find the transportation cost of every shipment, which is a function of the source, destination, product class and discount. The carrier and the shipper contractually agree on the product class

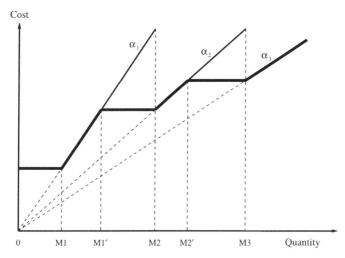

Fig. 5. Modified all-unit discount cost structure.

(typically *class 100*) and on the level of discount which implies that the shipper will pay only a given fraction, say 90%, of the cost generated by the rating engine. Now, given this input, the transportation cost as a function of the amount shipped enjoys the structure of the cost function described in our model.

The lack of concavity in this case significantly increases the complexity of the problem. For this reason, our approach is to pursue a better understanding of the problem by considering two simple scenarios: (1) the Single-item Lot Sizing Problem, in which a single retailer places orders from a warehouse to satisfy its demand over time, and (2) the Single-warehouse Multi-retailer Problem, in which the warehouse orders from an outside supplier and replenishes inventory of a number of retailers.

4.1 The single-item economic lot sizing problem

The *Single-item Economic Lot Sizing Problem* can be stated as follows: A facility, possibly a warehouse or a retail outlet, faces known demands over a finite planning horizon. At each period, the order cost function (or transportation cost function in our application) and the holding cost function are given and they can be different from period to period. Backlogging is not allowed. The objective is to decide when and how many units to order so as to minimize total ordering and holding costs over the finite horizon without any shortages.

The Single-item Economic Lot Sizing Problem, initially analyzed by Wagner and Whitin (1958), has recently been the subject of intensive research. Most of the work has focused on ordering (or transportation) costs, which are assumed to be concave in the amount ordered. For instance, Aggarwal and Park (1993), Federgruen and Tzur (1991) and Wagelmans, Van Hoesel, and Kolen (1992) have shown that it is possible to take advantage of the special cost structure of the model and use it to develop fast, exact, algorithms. A few authors have considered more general cost structures. Federgruen and Lee (1990) consider both 'all-unit' (different from the one considered here) and incremental discount cost structures. Specifically, in the case of 'all-unit' discount cost structures the authors characterize structural properties of optimal solutions that lead to the development of a polynomial time algorithm. However, it is easy to show that these properties do not hold in our case. Shaw and Wagelmans (1998) and Van Hoesel and Wagelmans (2001) develop pseudopolynomial and approximation algorithms for more general capacitated versions of the Economic Lot Sizing problem. Specifically, Shaw and Wagelmans (1998) develop an algorithm that solves the problem with general piece-wise linear cost functions to optimality in time proportional to the average demand and the square of the number of time periods, but is independent of the capacity limit. Thus, this algorithm can be used to solve the Economic Lot Sizing problem with modified all-unit discount cost function. Van Hoesel and Wagelmans (2001) develop a fully polynomial approximation scheme for the capacitated Economic Lot Sizing problem which only requires monotonicity of the ordering cost function. In this case, the complexity of the algorithm is proportional to the number of time periods, the logarithm of total demand and the logarithm of the sum of the capacity limits for each period. This dependency on the capacity limit make the algorithm inappropriate to solve the uncapacitated problem considered here.

In approaching the Single-Item Economic Lot Sizing Problem with modified all-unit discount costs, the first challenge is to determine its complexity. Chan, Muriel, Shen, and Simchi-Levi (2002) show that the 2-partition problem can be reduced to a single-item lot sizing problem with non-stationary modified all-unit discount ordering cost functions that have a bounded number of breakpoints. A similar reduction procedure can be performed in the case of holding and ordering cost functions that do not change over time, but the number of breakpoints in the cost function grows with the number of items in the 2-partition problem. In either case, since the 2-partition problem is NP-hard, see Karp (1972) and Garey and Johnson (1979), we have the following result.

Theorem 7. *The Single-item Economic Lot Sizing Problem with modified all-unit discount ordering cost functions is NP-hard.*

The theorem thus suggests that research on this problem should be focused on identifying easily implementable policies with an attractive worst-case performance. One such class of policies is the class of *Zero-Inventory-Ordering*

(ZIO) policies, in which a retailer orders only when its inventory is down to zero. It is well known that the best ZIO policy can be found in $O(T^2)$ time using a shortest path algorithm [see Bramel and Simchi-Levi (1997), pp. 166–167]. The question, of course, is how far can the cost of the optimal ZIO policy be from the cost of the optimal policy? This question is answered in the next subsection.

4.1.1 Worst-case analysis of ZIO policies

Consider a single facility facing time varying demand for the next T periods. Demand in period t, $t = 1, 2, \ldots, T$, is known and denoted by d_t. A holding cost, at a rate $h_t \geq 0$, is charged on all items carried over from period t to period $t + 1$, $t = 1, 2, \ldots, T-1$. Let $F_t(Q)$ be the ordering cost associated with an order of size Q placed at the beginning of period t. We assume that the ordering cost function $F_t(\cdot)$ belongs to the class of modified all-unit discount cost functions.

The objective is to determine when to order and how many units to order so as to minimize ordering and holding costs over the planning horizon, without any shortages. In what follows we refer to this problem as the lot sizing problem, with Z^* being the cost of the optimal strategy. That is, Z^* is the minimum system-wide cost, associated with the best inventory ordering strategy for the lot sizing problem. In addition, we let Z^{ZIO} be the cost of the best ZIO policy and, given a particular inventory ordering policy S, we denote its associated cost by $Z(S)$.

We start by identifying structural properties of the solutions to the lot sizing problem.

Given any feasible policy S, let R be the number of orders placed over the horizon and let t_j be the jth period in which an order is placed, $j = 1, 2, \ldots, R$. In what follows, we assume without loss of generality that orders are used to satisfy demand in a first-in-first-out basis. Thus, the order placed at period t_j covers a portion of demand at a certain period s_j, full demand $d_{s_j+1} + d_{s_j+2} + \cdots + d_{s_j+r_j-1}$ of the consecutive periods $s_j + 1, s_j + 2, \ldots, s_j + r_j - 1$, and a portion of the amount demanded at the following period, $s_j + r_j$, for $j = 1, 2, \ldots, T$ and some $r_j \in \{0, 1, \ldots, T-s_j\}$. Obviously, s_j is the first period whose demand is not fully covered, i.e., satisfied, by orders placed in periods previous to t_j.

Given policy S, let Q_j be the quantity ordered at period t_j, for $j = 1, 2, \ldots, R$, and let TC_j denote the cost per unit associated with that order.

In the remainder of this subsection, we consider the ordering cost at each period to be distributed evenly among the quantity ordered. That is, each unit of demand satisfied from the order of Q_j units in period t_j is assumed to have an ordering cost of $TC_j \equiv [(F_{t_j}(Q_j))/Q_j]$. Finally, let H_j be the cost of holding one item in inventory from period t_j to t_{j+1}.

That is, $H_j = h_{t_j} + h_{t_j+1} + \cdots + h_{t_{j+1}-1}$.

Some of the parameters just introduced are associated with a particular policy S, and their values vary from policy to policy. For simplicity, we drop the correspondence to the specific policy, except when we refer to the optimal policy. In this case, parameters associated with the optimal policy, S^*, are indicated by adding an * to the notation.

Lemma 8. *Given any feasible policy S, there exists a feasible policy with lower or equal cost satisfying the following properties,*

1. *For all $j, j = 1, 2, \ldots, R-1$, $s_j < t_{j+1}$; that is, the first period the jth order is consumed is earlier than the time the $(j+1)$th order is placed.*
2. *If $s_j + r_j \geq t_{j+1}$ then $TC_j + H_j > TC_{j+1}$; that is, if the order at t_j covers demands occurring after the next order has been placed, at period t_{j+1}, then the cost per unit associated with ordering and holding those units in inventory in the earlier period must be higher.*

Proof. Suppose that the current policy S does not satisfy the first property; that is, there exists an ordering period j such that $s_j \geq t_{j+1}$. Since both the orders at periods t_j and t_{j+1} cover demand occurring on or after period t_{j+1}, the two orders can be combined and placed in either t_j or t_{j+1}, whichever leads to the overall lower cost. The total cost associated with ordering the combined quantity and holding the units in inventory until period t_{j+1} is no more than

$$(Q_j + Q_{j+1}) \operatorname{Min} \{TC_j + H_j, TC_{j+1}\} \leq Q_j(TC_j + H_j) + Q_{j+1}TC_{j+1},$$

which is the cost associated with ordering those quantities and holding the units in inventory until period t_{j+1} in the current policy S. Since all other costs remain the same when combining the orders, the above argument shows that we can always obtain a policy with lower or equal cost satisfying the property.

Similarly, we can show that if $s_j + r_j \geq t_{j+1}$ and $TC_j + H_j \leq TC_{j+1}$ the quantity ordered at period t_{j+1} could be added to the order at period t_j without increasing cost, which proves the second property. ∎

To prove the worst-case result, we break up the quantity Q_j ordered at period $t_j, j = 1, 2, \ldots, R$, when following a policy S that satisfies the properties in Lemma 3, in two:

$$Q_j = \alpha_j Q_j + (1 - \alpha_j) Q_j,$$

where $\alpha_j Q_j$, $0 < \alpha_j \leq 1$, denotes the portion of the jth shipment that is used to satisfy demands from some time $s_j < t_{j+1}$ until the $(j+1)$th order is placed. Similarly, $(1-\alpha_j)Q_j$ is the quantity destined to satisfy demands on or after period t_{j+1}.

Following this notation, the total cost associated with an optimal policy S^* can be written as

$$Z^* = Z(S^*) = TC_1^* \alpha_1^* Q_1^* + \sum_{j=2}^{R} ((TC_{j-1}^* + H_{j-1}^*)(1 - \alpha_{j-1}^*)Q_{j-1}^*$$
$$+ TC_j^* \alpha_j^* Q_j^*) + H^*, \tag{4.1}$$

where H^* denotes the total inventory cost incurred by policy S^* minus the cost of carrying, for each $j = 1, 2, \ldots, R-1$, the portion of demand ordered in period t_j^* but used no earlier than t_{j+1}^*, from period t_j^* to period t_{j+1}^*. That is, $H^* = $ total holding cost $- \sum_{j=1}^{R-1} H_j^*(1 - \alpha_j^*)Q_j^*$.

Let S^* be an optimal solution which satisfies the conditions in Lemma 8. We construct a ZIO policy, $\overline{S}$, by modifying the optimal policy S^* as follows.

4.1.2 Transformation procedure
Step 0: Let $S = S^*$.
Step 1: Find the smallest index k, such that $\alpha_{k-1} < 1$; that is, t_k is the earliest period in which an order is placed before inventory has been fully depleted.
Step 2: Either,

- Combine 1: move $(1 - \alpha_{k-1})Q_{k-1}$ from the order at period t_{k-1} to that at period t_k, or,
- Combine 2: move $\alpha_k Q_k$ from the order at period t_k to that at period t_{k-1} and $(1-\alpha_k)Q_k$ from the order at period t_k to that at period t_{k+1},

whichever results in a lower cost.
Step 3: If combining orders in Step 2 causes the second property in Lemma 8 to be violated, combine orders without increasing total cost, as shown in the proof of the Lemma, until the current policy satisfies the condition.
Step 4: Let S be the new policy. If all periods in policy S satisfy the ZIO property, then $\overline{S} = S$. Otherwise, repeat Steps 1 to 3.

See Figs. 6–8 for illustration of the *Combine* procedures. Observe that $Z(\overline{S}) \geq Z^{ZIO}$, since Z^{ZIO} denotes the best solution value among all ZIO policies. Thus, to bound the worst-case performance of Z^{ZIO} it suffices to study the ratio $[Z(\overline{S})/Z(S^*)]$.

Note that, at each iteration of the transformation procedure, if the ZIO property is not satisfied at a certain period t in the current policy S, then it was not satisfied in S^* either. Hence, for each period t_k considered at *combining step k* (Step 2 with index k) there exists an index $j \geq k$, such that in the optimal solution, S^*, the order at period $t_k = t_j^*$ is placed before all earlier inventory has been used.

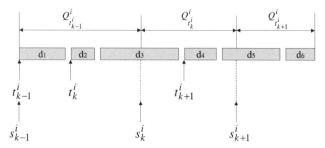

Fig. 6. Initial policy.

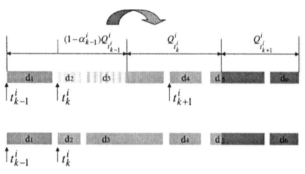

Fig. 7. Policy obtained when *Combine 1* is performed.

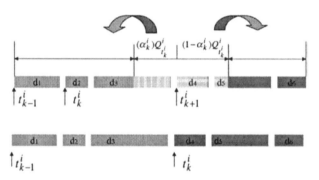

Fig. 8. Policy obtained when *Combine 2* is performed.

The following three lemmas demonstrate that, at each iteration of Step 2, the increase in cost accrued can be bounded.

Lemma 9. *The increase in cost at combining step k is not more than*

1. $TC_k(1 - \alpha_{k-1})Q_{k-1}$ *if Combine 1 is executed.*
2. $(TC_{k-1} + H_{k-1} - TC_k)\alpha_k Q_k$ *if Combine 2 is executed.*

This result is a consequence of properties (p1) and (p2) of the cost function. Thus, at each iteration of Step 2 in the *Transformation Procedure*, the increase in cost from the original policy S^* to the current policy S is no more than

$$\text{Min}\{TC_k(1 - \alpha_{k-1})Q_{k-1}, (TC_{k-1} + H_{k-1} - TC_k)\alpha_k Q_k\}.$$

This increase in cost can be bounded using the following result.

Lemma 10.

$$\min(A\alpha, B\beta) \leq \frac{1}{3}(A(\alpha + \beta) + B\alpha), \quad \text{for all } B, A, \alpha, \beta \geq 0.$$

Proof.

$$\min(A\alpha, B\beta) \leq \frac{\alpha - \beta + \max(\alpha, \beta)}{3\max(\alpha, \beta)} B\beta + \frac{2\max(\alpha, \beta) - \alpha + \beta}{3\max(\alpha, \beta)} A\alpha.$$

Note that

$$\{\alpha - \beta + \max(\alpha, \beta)\}\beta \leq \max(\alpha, \beta)\alpha,$$

and

$$\{2\max(\alpha, \beta) - \alpha + \beta\}\alpha \leq (\alpha + \beta)\max(\alpha, \beta).$$

Hence it follows that

$$\min(A\alpha, B\beta) \leq \frac{1}{3}(B\alpha + A(\alpha + \beta)). \qquad \blacksquare$$

Lemma 11. *At each iteration of combining step k, the increase in cost is no more than one third of the corresponding jth term, i.e., j such that $t_k = t_j^*$, in the expression of the optimal cost, Equation (4.1). That is,*

$$\text{Min}\{TC_k(1 - \alpha_{k-1})Q_{k-1}, (TC_{k-1} + H_{k-1} - TC_k)\alpha_k Q_k\}$$
$$\leq \frac{1}{3}\left[(TC_{j-1}^* + H_{j-1}^*)(1 - \alpha_{j-1}^*)Q_{j-1}^* + TC_j^* \alpha_j^* Q_j^*\right].$$

Proof. The increase in cost in the combine step associated with the kth order is no more than,

$$\min\{TC_k(1 - \alpha_{k-1})Q_{k-1}, (TC_{k-1} + H_{k-1} - TC_k)\alpha_k Q_k\}.$$

Let $A = Q_{k-1}(1-\alpha_{k-1})$, $B = Q_k \alpha_k$, $\alpha = TC_k$ and $\beta = TC_{k-1} + H_{k-1} - TC_k$ and apply Lemma 10 to get,

$$\min\{TC_k(1 - \alpha_{k-1})Q_{k-1}, (TC_{k-1} + H_{k-1} - TC_k)\alpha_k Q_k\}$$
$$\leq \frac{1}{3}[(TC_{k-1} + H_{k-1})(1 - \alpha_{k-1})Q_{k-1} + TC_k \alpha_k Q_k]$$
$$\leq \frac{1}{3}\left[(TC^*_{j-1} + H^*_{j-1})(1 - \alpha^*_{j-1})Q^*_{j-1} + TC^*_j \alpha^*_j Q^*_j\right].$$

The last inequality is easy to show by comparing the current solution S with the optimal solution, S^*, and realizing that:

1. The two solutions are identical from period $t_k = t^*_j$ onwards, since the orders placed on those periods have not been modified in previous iterations. Thus, $TC_k \alpha_k Q_k = TC^*_j \alpha^*_j Q^*_j$. In addition, observe that the demand on or after period t_k not covered by those orders must be satisfied with units from the previous order in both solutions; that is, $(1 - \alpha_{k-1})Q_{k-1} = (1 - \alpha^*_{j-1})Q^*_{j-1}$.
2. $t_{k-1} = t^*_{j-1}$ and the quantity ordered at that period may have been increased in the previous iteration leading to $TC_{k-1} + H_{k-1} \leq TC^*_{j-1} + H^*_{j-1}$. ∎

This proves that $Z(\bar{S}) \leq (4/3)Z^*$, since the index j is strictly increasing in the number of iterations performed and the sum of those terms for all j is no larger than the optimal value, Z^*. Thus,

Theorem 12. *For every instance of the lot sizing problem, i.e., the Single-item Economic Lot Sizing Problem with modified all-unit discount cost function, we have*

$$Z^{ZIO} \leq \frac{4}{3} Z^*,$$

and this bound is tight.

It remains to show that the bound is tight. This is proved by constructing instances for which the worst-case ratio converges to $4/3$.

Lemma 13. *There exist instances of the lot sizing problem for which the ratio Z^{ZIO}/Z^* is arbitrarily close to $4/3$.*

Proof. Consider an Economic Lot Sizing problem where demands of $d_1 = \delta$ for period 1 and $d_2 = 2$ for period 2 must be satisfied. There is no charge for carrying inventory from one period to the next. The ordering cost is described by a different modified all-unit discount cost function for each period. In period 1, there is a fixed charge of 2 for any shipment of size no greater than 1 and a rate of 2 per unit for all other shipments. The cost function in period 2 is linear with a rate of 1 per unit.

In the optimal solution to this instance of the lot sizing problem, 1 unit is ordered at period 1 and $1 + \delta$ units at period 2, with a total cost of $Z^* = 3 + \delta$. However, the best ZIO policy consists of ordering δ units at period 1 and 2 units at period 2, with a total cost of $Z^{ZIO} = 4$. Hence,

$$\frac{Z^{ZIO}}{Z^*} = \frac{4}{3+\delta} \to \frac{4}{3} \text{ as } \delta \to 0. \qquad \blacksquare$$

We note that the only properties of the modified all-unit discount cost function used in Chan et al. (2002) to prove Theorem 12 are that it is non-decreasing in the quantity shipped and that the cost per unit is non-increasing in that quantity, i.e., properties (p1) and (p2). Hence, the theorem holds true for any transportation function satisfying those properties. In a similar way, holding costs can be generalized to be any function of the quantity held that satisfies those two properties.

Observe also that the example developed in Lemma 13, which shows that the worst-case bound is tight, makes use of different ordering cost functions for the two periods. This suggests that it may be possible to improve the worst-case bound when the ordering cost function does not vary over time. Indeed, in that case, Chan et al. (2002) again transform an optimal policy which satisfies the properties developed in Lemma 8 into a ZIO policy and show that the increase in cost due to the transformation is no more than 1/4.6 times the cost of the optimal policy.

Theorem 14. *For every instance of the lot sizing problem in which the ordering cost function is the same for all periods in the planning horizon, we have*

$$Z^{ZIO} \leq \frac{5.6}{4.6} Z^*.$$

A natural question at this point is whether the worst-case bound $(Z^{ZIO}/Z^*) \leq (5.6/4.6) \approx 1.217$ is tight. Although this question is still open, Chan et al. (2002) describe instances of the problem for which the ratio of the solution generated by the heuristic to the optimal solution converges to $1/[2(\sqrt{2} - 1)] \approx 1.207$.

Lemma 15. *There exist instances of the lot sizing problem with a stationary ordering cost function for which the ratio Z^{ZIO}/Z^* is arbitrarily close to $1/[2(\sqrt{2}-1)]$.*

4.1.3 Computational results

To complete our study, we need to analyze the empirical performance of the best ZIO policy. The objective is to answer two major questions: (i) How far is the cost of the best ZIO policy from that of an optimal one for various test problems? (ii) What are the advantages of using the ZIO policy compared to known pseudo-polynomial time algorithms that can generate optimal solutions to the problem? How does the performance of the best ZIO policy compare with that of polynomial approximation algorithms based on these exact pseudo-polynomial algorithms?

To answer these questions, we evaluate the performance of the *best ZIO policy*, from total cost and speed points of view, relative to that of the best, to our knowledge, exact algorithm. The exact algorithm is based on a dynamic programming approach developed by Shaw and Wagelmans (1998), whose running time is only linearly dependent on the magnitude of the data. More specifically, the procedure has complexity $O(T^2 \bar{q} \bar{d})$, where T is the number of periods, $\bar{q}$ is the average number of linear pieces required to represent the cost function and $\bar{d}$ is the average demand. This algorithm can be applied to the Economic Lot Sizing problem with any type of piece-wise linear ordering cost functions.

We consider four different classes of instances representing a variety of problem sizes, as described in Table 7. For each of these classes, we generated 10 random instances of the Economic Lot Sizing problem with modified all-unit discount transportation costs. Demand at each period was generated from a normal distribution with the mean and standard deviation given in Table 7. The transportation cost functions considered in each case are described in Table 8. They do not change over time. Holding costs are generated according to a uniform distribution in the interval $[0.2, 0.7]$.

Table 9 exhibits, for each of the problem classes, the average and maximum ratios of the cost of the best ZIO policy to that of the optimal cost, and the average computation time in seconds for both the heuristic and

Table 7
Sizes of the different problem classes

Problem classes	Number of periods	Number of transportation cost function breakpoints	Average demand	Standard deviation
Class 1	10	4	40	20
Class 2	12	4	300	100
Class 3	12	8	1500	500
Class 4	12	8	6000	800

Table 8
Transportation cost functions considered for the different problem classes

Cost interval		1	2	3	4	5	6	7	8
Class 1	breakpoint	0	40	80	120				
	fixed cost	40	0	80	0				
	slope	0	1	0	0.6667				
Class 2	breakpoint	0	300	600	900				
	fixed cost	600	0	1200	0				
	slope	0	2	0	1.3333				
Class 3	breakpoint	0	1500	3000	4500	6000	7500	9000	10,500
	fixed cost	2250	0	4500	0	6000	0	7200	0
	slope	0	1.5	0	1	0	0.8	0	0.6857
Class 4	breakpoint	0	1000	5000	10,000	15,000	20,000	25,000	30,000
	fixed cost	2500	0	12,500	0	18,750	0	23,437.5	0
	slope	0	2.5	0	1.25	0	0.9375	0	0.78125

Table 9
Computational results showing the performance of ZIO policies

Problem classes	Average ratio Z^{ZIO}/Z^*	Maximum ratio	Average exact algorithm's CPU time	Average heuristic's CPU time
Class 1	1.0104	1.0344	0.12	< 0.01
Class 2	1.0044	1.0179	1.13	< 0.01
Class 3	1.0029	1.0080	10.14	< 0.01
Class 4	1.0001	1.0004	42.71	< 0.01

exact algorithms (using a Sun Workstation Ultra 1). We should point out that in order to solve Class 4 problems to optimality, we had to incorporate a dynamic memory allocation subroutine in Shaw and Wagelmans' algorithm.

These computational results indicate that the restriction to ZIO policies is especially effective when average demand size is high: As demand size grows, the relative error decreases while the time to find the best ZIO policy remains negligible, less than 0.01 sec in all cases. The running time of Shaw and Wagelmans' algorithm, on the other hand, increases drastically as a function of the demand size.

We thus conclude that restricting the search to ZIO policies allows us to deal effectively with cases of large demands and numerous breakpoints in the transportation cost function. These are exactly the cases for which the exact algorithm becomes computationally expensive. Indeed, this is an important issue because many hierarchical planning models for multi-item, multi-stage logistics systems are solved via Lagrangian Relaxation. Using this technique, the problem is typically decomposed into many lot-sizing problems, each of which needs to be solved for numerous combinations of the Lagrangian multipliers. See Federgruen and Tzur (1991) for a discussion of the relationship between the classical Economic Lot Sizing problem and multi-item, multi-stage production problems.

An alternative approach to efficiently solve the Economic Lot Sizing problem when demands are large is to scale the demands by a certain factor, use the exact algorithm by Shaw and Wagelman's on the scaled demands and transform the solution to accommodate the original demands. We refer to this procedure as the scaling algorithm. In what follows we compare the performance of such a heuristic procedure with the performance of the best ZIO policy. To implement the scaling algorithm, we divide the period demands by a selected scaling factor and round up to ensure feasibility in the original problem. In addition, once the solution to the Economic Lot Sizing problem with scaled demands has been found, a solution to the problem with the original demands is found by multiplying the order quantities by the scaling factor and using the following routine to remove the excess quantity ordered as a consequence of rounding up.

Let R be the number of ordering periods in which an order is placed in the scaled solution and let t_k denote the kth such period and Q_{t_k} the quantity ordered. In addition, let $D_{t_k,T}$ denote the total demand, and $Q_{t_k,T}$ be the total quantity ordered, from period t_k through the end of the horizon. Starting with $k = R$ and moving backwards in time in the set of ordering periods do: if $Q_{t_k,T} > D_{t_k,T}$ then set $Q_{t_k} = Q_{t_k} - (Q_{t_k,T} - D_{t_k,T})$.

We tested 10 instances randomly generated according to a normal distribution with a mean demand of 60,000 units and standard deviation 8000 units. Table 10 shows the average performance results for both the ZIO policy and the solution generated by the scaling algorithm for different scaling factors.

Clearly, the ability of the scaling algorithm to provide solutions close to the optimal solution depends on the magnitude of the scaling factor and the solution deteriorates rapidly as the factor grows. Even for large scaling factors, the computational time to find the best ZIO policy is significantly lower than that to run the scaling algorithm.

Table 10
Computational results comparing the performance of ZIO policies with that of the scaling algorithm

Policy	ZIO	Scaling algorithm		
Scaling factor		100	1000	10,000
Average relative error	0.02%	0%	0.01%	0.89%
CPU time	0.003	3.4	0.35	0.037

4.2 The single-warehouse multi-retailer problem

In this section, we study a class of multi-period distribution problems with transportation cost structures that model both the incremental and all-unit discount cost functions. Specifically, we consider a classical inventory–distribution model in which a single warehouse receives inventory from a single supplier and replenishes the inventory of n retailers. In these situations, shipments from the supplier to the warehouse are often delivered by TruckLoad (TL) carriers whose costs can be approximated by piece-wise linear concave functions. Henceforth, we assume that the transportation cost function from the supplier to the warehouse is of the incremental discount type. Of course, this function may also include piece-wise linear concave production costs. By contrast, since shipment sizes from the warehouse to a retailer are relatively small, these shipments are typically delivered by LTL carriers whose costs follow the modified all-unit discount cost structure. The objective is to find an optimal shipment plan that exploits the quantity discount effect and, at the same time, controls the inventory holding cost at the retailers end.

We assume that shortages and backlogging are not allowed either at the warehouse or at the retailers. Furthermore, we assume that the warehouse uses a common logistic strategy, referred to as cross-docking, in which the warehouse acts merely as a coordinator of the supply process, and as a transshipment point for incoming orders from the supplier, but does not hold any stocks. Extensions to systems with central stock are discussed at the end of this subsection.

Observe that the Single-Warehouse Multi-Retailer Problem described here can also be used to model the *joint replenishment problem*, see Joneja (1990). In this problem, a single facility replenishes a set of items over a finite horizon. Whenever the facility places an order for a subset of the items, two types of costs are incurred: A joint set-up cost and an item-dependent set-up cost. The objective in the joint replenishment problem is to decide when and how many units to order for each item so as to minimize inventory holding and ordering costs over the planning horizon. Evidently, the concave fixed-charge ordering cost functions in this problem are a special case of the modified all-unit discount cost functions. Since the joint replenishment problem is NP-hard, see Arkin, Joneja, and Roundy (1989), the Single-Warehouse Multi-Retailer Problem is also NP-hard, even if all transportation cost functions are concave.

An interesting question is whether it is NP-hard for a single, or fixed number of retailers. This question was answered in the previous section where we show that a special case of our problem, in which a single retailer is replenished by a single warehouse with zero transportation cost for shipments to the warehouse and modified all-unit discount transportation costs for shipments to the retailer, is NP-hard. Thus, the Single-Warehouse Multi-Retailer Problem described above is NP-hard even for a fixed number of retailers.

Let n be the number of retailers served by the warehouse and T be the length of the planning horizon under consideration. For each $t = 1, 2, \ldots, T$, we let $K_0^t(\cdot)$ be the piece-wise linear concave transportation cost function associated with shipping items from the supplier to the warehouse at time t. Similarly, for each $i = 1, 2, \ldots, n$ and $t = 1, 2, \ldots, T$, we denote by $K_i^t(\cdot)$ the modified all-unit discount transportation cost function associated with shipping items from the warehouse to retailer i at time t. Finally, for each $i = 1, 2, \ldots, n$ and $t = 1, 2, \ldots, T$, let h_t^i denote the cost of holding an item at retailer i at the end of period t, and d_t^i the demand of retailer i at time t.

Again, our objective is to find the size and timing of shipments so as to minimize total transportation and inventory costs while satisfying all demands without shortages. In what follows, we will refer to this problem as the Single-Warehouse Multi-Retailer Problem. Let Z^* be the optimal solution to the Single-Warehouse Multi-Retailer Problem and for any heuristic H, let Z^H be the cost of the solution generated by heuristic H.

We first show that unless $P = NP$, it is not possible to develop an algorithm that runs in polynomial time and generates, for any instance of the problem, a solution which is within a factor of $O(\log n)$ from optimality.

Theorem 16. *Suppose there exists a $\gamma > 0$ and a polynomial time heuristic, H, for the Single-Warehouse Multi-Retailer Problem such that for all instances*

$$\frac{Z^H}{Z^*} \leq \gamma \log n,$$

then $P = NP$.

Proof. The proof is based on showing that the set covering problem can be reduced to the Single-Warehouse Multi-Retailer Problem. It is well known, see Feige (1998) or Arora and Sudan (1997), that there is no polynomial time algorithm for the set-covering problem with worst-case bound better than $\gamma \log n$, for $\gamma > 0$, unless $P = NP$.

Consider an instance of the set covering problem: $\min(\Sigma_{t=1}^m x_t : Ax \geq 1)$, where $A = (a_{i,t})$ is a $n \times m$ 0–1 matrix. It can be reduced to the Single-Warehouse Multi-Retailer problem with n retailers and $m+1$ periods as follows. Let

$$K_i^t(x) = \begin{cases} M\delta(x) & \text{if } a_{i,t} = 0 \\ 0 & \text{if } a_{i,t} = 1 \end{cases} \quad \text{for all } i \text{ and } t = 1, 2, \ldots, m;$$

$$K_i^{m+1}(x) = M\delta(x) \quad \text{for all } i;$$

$$K_0^t(x) = \delta(x) \quad \text{for } t = 1, 2, \ldots, m, m+1;$$

$$d_t^i = \begin{cases} 0 & \text{if } t = 1, 2, \ldots, m, \\ 1 & \text{if } t = m+1, \end{cases} \quad \text{for all } i;$$

$$h_t^i = 0 \quad \text{for all } i, t,$$

where M is some large number no less than m, and $\delta(x) = 1$ when $x > 0$, and 0 otherwise.

The high set-up cost at time $m + 1$ forces retailers to order in earlier periods. In addition, an order for retailer i is placed at time t only if $a_{i,t} = 1$, since there is a large fixed cost associated with shipments in periods in which $a_{i,t} = 0$. Thus, finding the best inventory ordering policy in this situation is equivalent to finding the minimum number of ordering periods, which is determined by clustering retailers that will be served together at a certain time. ∎

This shows that the Single-Warehouse Multi-Retailer Problem contains the set covering problem as a special case and, consequently, that there is no polynomial time algorithm with a fixed worst-case bound. Thus, Chan, Muriel, Shen, Simchi-Levi, and Teo (2002) focus on simple policies that can be found in pseudo-polynomial time and provide a solution that is within a certain fixed percentage from optimality. In particular, they consider ZIO policies in which orders are placed only at times when on-hand inventory has been fully depleted. Let Z^{ZIO} be the cost associated with the optimal ZIO policy. Using arguments similar to those presented in Section 4.1, Chan et al. (2002) show the following results.

Theorem 17. *For every instance of the Single-Warehouse Multi-Retailer Problem, we have*

$$Z^{ZIO} \leq \frac{4}{3} Z^*,$$

and this bound is tight.

In practice, the ordering cost function does not vary from period to period, i.e., for all t, $K_0^t(\cdot) = K_0(\cdot)$ and $K_i^t(\cdot) = K_i(\cdot)$, $i = 1, 2, \ldots, n$. In this case, Chan et al. (2002) show that the worst-case ratio of the cost of the solution generated by the algorithm to the optimal cost is no more than $(5.6/4.6) \approx 1.22$. That is,

Theorem 18. *For every instance of the Single-Warehouse Multi-Retailer Problem in which the transportation cost functions are stationary, we have*

$$Z^{ZIO} \leq \frac{5.6}{4.6} Z^*.$$

The optimal ZIO policy can be found in polynomial time for any fixed number of retailers using the algorithm presented below. Not surprisingly, however, the computational complexity of this method grows exponentially as the number of retailers increases. To overcome this problem, we subsequently propose a linear programming based heuristic that runs in polynomial time. This algorithm is shown to be very efficient in our computational study.

4.2.1 Optimal ZIO policy

When the number of retailers is fixed, we can find the best ZIO policy in time which is polynomial in T and exponential in the number of retailers n, by formulating an associated shortest path problem.

Let $T = \{1, 2, \ldots, T+1\}$ be the set of different time periods, where $T+1$ is used for notational convenience. Let $N = \{1, 2, \ldots, n\}$ be the set of retailers. Construct an acyclic graph $G = (V, A)$, where

$$V = \{\bar{u} = \langle u_1, \ldots, u_n\rangle | u_i \in T, i = 1, \ldots, n\} = \underbrace{T \times T \times \cdots \times T}_{n \text{ times}},$$

$$A = \{\langle u_1, \ldots, u_n\rangle \to \langle v_1, \ldots, v_n\rangle | v_i \geq u_i \quad \text{for all } i;$$

there is at least one component i such that $u_i < v_i$; for every i with $u_i < v_i$ we have $u_i = \min_{\{j=1,2,\ldots,n\}} u_j \equiv u$ (i.e., all the components that changed had the same value, u)$\}$.

Given an arc $\langle u_1, \ldots, u_n\rangle \to \langle v_1, \ldots, v_n\rangle$, or $\bar{u} \to \bar{v}$, let k be the number of components that are different in $\bar{u}$ and $\bar{v}$, and $I = \{i_1, i_2, \ldots, i_k\}$ be the set of indices of those components; that is, for $l = 1, 2, \ldots, k$, i_l is such that $u_{i_l} < v_{i_l}$. Observe that $k \geq 1$ and by construction $u_{i_1} = u_{i_2} = \cdots = u_{i_k} = u$. The arc $\bar{u} \to \bar{v}$ represents ordering at period u to satisfy demands of each retailer i_l, $l = 1, 2, \ldots, k$, from period u through $v_{i_l} - 1$. Thus, the cost associated with this arc is the cost of ordering those units at period u and holding them in inventory until their consumption. Specifically, the cost of this arc is

$$K_0^u(d_{u,v_{i_1}}^{i_1} + d_{u,v_{i_2}}^{i_2} + \cdots + d_{u,v_{i_k}}^{i_k}) + C_{u,v_{i_1}}^{i_1} + C_{u,v_{i_2}}^{i_2} + \cdots + C_{u,v_{i_k}}^{i_k},$$

where,

- $d_{u,v}^i$ is the total demand faced by retailer i from period u to $v-1$;
- $K_0^u(d_{u,v_{i_1}}^{i_1} + d_{u,v_{i_2}}^{i_2} + \cdots + d_{u,v_{i_k}}^{i_k})$ is the cost of shipping $d_{u,v_{i_1}}^{i_1} + d_{u,v_{i_2}}^{i_2} + \cdots + d_{u,v_{i_k}}^{i_k}$ units from the supplier to the warehouse at time u; and,
- $C_{u,v}^i$ is total shipping and holding costs for retailer i if we order at period u to cover the demands in periods $u, u+1, \ldots, v-1$, i.e., $C_{u,v}^i = K_i^u(d_{uv}^i) + \sum_{t=u}^{v-2} h_t^i d_{t+1,v}^i$.

It is easy to see that the shortest path from $\langle 1,1,\ldots,1\rangle$ to $\langle T+1, T+1, \ldots, T+1\rangle$ in $G = (V, A)$ corresponds to finding the best ZIO policy.

To illustrate our method, consider the following example with a single warehouse and two retailers. In this case, G has nodes $\langle i,j\rangle$ for $i,j = 1, 2, \ldots, T+1$ and three different types of arcs:

Type a: $(i,j) \to (i,k)$ $\forall i > j$ and $k > j$,
Type b: $(i,j) \to (k,j)$ $\forall i < j$ and $k > i$,
Type c: $(i,i) \to (k,j)$ $\forall k > i$ and $l > i$,

The cost of a type a arc is $K_0^j(d_{j,k}^2) + C_{j,k}^2$, while the cost of a type b arc is $K_0^i(d_{i,k}^1) + C_{i,k}^1$. Finally, the cost of a type c arc equal to $K_0^i(d_{i,k}^1 + d_{i,l}^2) + C_{i,k}^1 + C_{i,l}^2$.

Observe that a path in this network, G, can be interpreted as a feasible solution to the One-Warehouse Multi-Retailer Problem. Indeed, type a (resp. b) arcs correspond to the situation when only retailer 2 (resp. 1) places an order at a specific time period, whereas type c arcs correspond to situations in which both retailers place an order. Figure 9 provides an example of the network when $T = 4$. The path depicted in this figure corresponds to the

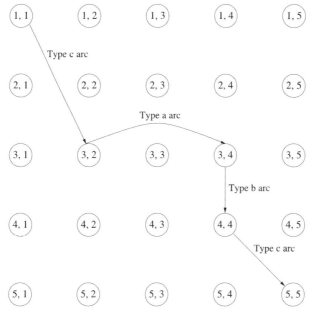

Fig. 9. An example of the shortest path algorithm.

following ordering strategy. Retailer 1 orders in periods 1, 3 and 4 while retailer number 2 orders in periods 1, 2 and 4.

The shortest path algorithm runs in time $O(|V|\log|V|+|A|)$, where $|V| = O((T+1)^n)$, $|A| = O((T)^{2n})$, where n is the number of retailers. The first term, $O(|V|\log|V|)$, bounds the complexity of constructing the network and the second, $O(|A|)$, is the time to find the shortest path on a topologically sorted network. Unfortunately, this exact algorithm grows to be computationally expensive as the number of retailers increases. Thus, the next step is to develop a heuristic that finds a good ZIO policy in polynomial time.

4.2.2 Linear programming based algorithm

We introduce a linear programming based heuristic that generates close-to-optimal ZIO policies and, thus, effective solutions to the Single-Warehouse Multi-Retailer Problem.

We start by formulating the problem of finding an optimal ZIO policy as an integer program. The algorithm is based on solving the linear programming relaxation of the resulting model and transforming the fractional solution obtained into an integer solution in a similar spirit to that of the algorithm presented in Section 3 for the Shipper Problem with piece-wise linear concave costs.

The piece-wise linear concave costs associated with shipments from the supplier to the warehouse are modeled as in Section 3.1, see Fig. 3. In this case, we have only T concave shipping arcs, representing shipments from supplier to warehouse at each period of time. Thus, the more general index e in the model in Section 3.1 will be substituted by t. let Q_t^0 denote the warehouse order at time t. We can express associated transportation cost, $K_0^t(Q_t^0)$, as $K_0^t(Q_t^0) = f_t^r + \alpha_t^r Q_t^0$, where r is such that $Q_t^0 \in (M_t^{r-1}, M_t^r]$.

We define the following variables (analogous to the interval and quantity variables introduced in Section 3.1). For each $t = 1, 2, \ldots, T$ and $r = 1, 2, \ldots, R$, let

$$X_t^r = \begin{cases} 1, & \text{if } Q_t^0 \in (M_t^{r-1}, M_t^r] \\ 0, & \text{otherwise.} \end{cases}$$

For each retailer $i = 1, 2, \ldots, n$, and periods $1 \leq t \leq k \leq T$, let Z_{tk}^i =quantity ordered by retailer i at time t to satisfy demand at period $k \geq t$ and

$$Z_{tk}^{ir} = \begin{cases} Z_{tk}^i, & \text{if } Q_t^0 \in (M_t^{r-1}, M_t^r] \\ 0, & \text{otherwise,} \end{cases}$$

for each $r = 1, 2, \ldots, R$. In what follows we refer to the X variables as interval variables and to the Z variables as *quantity variables*.

In order to model ordering and inventory costs at the retailer level, we consider a dummy period $T+1$ and define for each retailer $i=1,2,\ldots,n$ and periods $1 \leq t < k \leq T+1$, c^i_{tk} =total cost of ordering at period t to satisfy demand for periods t through $k-1$ and holding the units in inventory until their consumption. That is,

$$c^i_{tk} = K^i_t \left(\sum_{j=t}^{k-1} d^i_j \right) + \sum_{j=t}^{k-2} h^i_j \left(\sum_{l=j+1}^{k-1} d^i_l \right).$$

Observe that a ZIO policy for retailer i can be interpreted as a path from 1 to $T+1$ on a network with nodes $\{1, 2, \ldots, T+1\}$ and arcs (t, k), for $1 \leq t < k \leq T+1$, with associated cost c^i_{tk}. In what follows we refer to this network as the ith retailer's network or G_i.

Thus, to calculate ordering and inventory costs at retailer i we formulate a shortest path model on G_i using variables

$$Y^i_{tk} = \begin{cases} 1, & \text{if an order is placed by retailer } i \text{ at time } t \\ & \text{to satisfy demands for period } t \text{ through } k-1, \\ 0, & \text{otherwise}, \end{cases}$$

and flow conservation constraints. We refer to $Y = (Y^i_{tk})$ as the vector of path flows.

The best ZIO policy can be found by solving the following integer program.

$$\text{Problem } SW: \text{Min} \sum_{t=1}^{T} \sum_{r=1}^{R} \left[f^r_t X^r_t + \alpha^r_t \sum_{i=1}^{n} \sum_{k=t}^{T} Z^{ri}_{tk} \right] + \sum_{i=1}^{n} \sum_{t=1}^{T} \sum_{k=t+1}^{T+1} c^i_{tk} Y^i_{tk}$$

s.t.

$$Z^{ri}_{tk} \leq d^i_k X^r_t, \quad \forall r = 1, 2, \ldots, R, \quad i = 1, 2, \ldots, n \text{ and } 1 \leq t \leq k \leq T \tag{4.2}$$

$$\sum_{r=1}^{R} Z^{ri}_{tk} = d^i_k \sum_{l=k+1}^{T+1} Y^i_{tl}, \quad \forall 1 \leq t \leq k \leq T, \quad i = 1, 2, \ldots, n \tag{4.3}$$

$$\sum_{j: j > l} Y^i_{lj} - \sum_{j: j < l} Y^i_{jl} = \begin{cases} 1 & \text{if } l = 1 \\ -1 & \text{if } l = T+1 \\ 0 & \text{if } 1 < l \leq T \end{cases} \quad \forall i = 1, 2, \ldots, n, \tag{4.4}$$

$$Z_{tk}^{ri} \geq 0 \quad \forall r = 1, 2, \ldots, R, \quad i = 1, 2, \ldots, n \quad \text{and} \quad 1 \leq t \leq k \leq T$$

$$X_t^r \in \{0, 1\} \quad \forall r = 1, 2, \ldots, R \quad \text{and} \quad t = 1, 2, \ldots, T$$

$$Y_{tk}^i \in \{0, 1\} \quad \forall i = 1, 2, \ldots, n \quad \text{and} \quad 1 \leq t \leq k \leq T+1.$$

The first set of constraints, (4.2), specifies that if some quantity is ordered at time t by any retailer and shipped on interval r of the transportation cost function then the associated interval variable, X_t^r, must be 1. These, together with the integrality of the X variables, are the only constraints needed to model the piece-wise linear concave costs, since Lemma 3 in Section 3 can easily be extended to this formulation. Obviously, constraints (4.2) could be aggregated for all k. However, this would considerably weaken the linear programming relaxation of problem SW. Eq. (4.3) guarantees that if a positive amount is shipped to retailer i at time t to satisfy demand at the retailer at period k then the retailer must order at period t to cover demands for period t through some $l-1 \geq k$. Observe that these constraints (4.4) correspond to finding, for each retailer i, a path from 0 to $T+1$ on the retailer's network, G_i.

Unfortunately, solving this integer program is computationally intractable for all but small size problems. To overcome this difficulty, we observe the great similarity between this formulation and that of the Shipper Problem under piece-wise linear concave costs in Section 3 and make use of the structural properties derived there, namely Theorem 4. This allows us to develop a polynomial time heuristic that finds an effective ZIO policy based on the solution to the linear programming relaxation of Problem SW. Theorem 4 will be extensively used by the algorithm in order to compute the increase in costs in the solution to the linear program when the vector Y is modified in the search for an integer solution.

Of course, the effectiveness of such a heuristic depends on the strength of the linear programming relaxation of Problem SW. Let Z^{LP} be the optimal solution value of the linear programming relaxation of Problem SW. Recently, Shen, Simchi-Levi, and Teo (2000) applied randomized rounding techniques and a novel way to approximate the piece-wise linear concave cost functions to prove the following result.

Theorem 19. *For every instance of the Single-Warehouse Multi-Retailer Problem we have,*

$$Z^{LP} \leq 8(\log 2 + \log T + \log n) Z^* / 3.$$

Observe that Theorem 16 implies that for any polynomial time heuristic H for the Single-Warehouse Multi-Retailer Problem and any $\gamma > 0$, there exists an instance of the problem such that $Z^H > [\gamma \log n] Z^*$, unless $P \equiv NP$. This, together with Theorem 19, implies that for that instance,

$$\frac{Z^H}{\gamma \log n} > Z^* \geq \frac{3Z^{LP}}{8(\log 2 + \log T + \log n)}.$$

Thus, the lower bound developed in Theorem 19 is the best possible bound on the optimal solution of the Single-Warehouse Multi-Retailer Problem unless $P \equiv NP$.

We are now ready to present the algorithm for the Single-Warehouse Multi-Retailer Problem.

Linear-programming based algorithm:
Step 1: Solve the linear programming relaxation of problem *SW*. Let $Y^* = (Y_{tk}^{i*})$ be the optimal solution. Initialize $i=1$.
Step 2: For each arc $t \to k$, $1 \leq t < k \leq T+1$, in network G_i compute a marginal cost, $c_{t,k}^i$, as follows. The marginal cost is the total increase in cost in the solution to the linear program incurred when augmenting the flow on that arc from the fractional Y_{tk}^{i*} to 1. That is,

$$c_{t,k}^i = W_{tk}^i + (1 - X_{t,k}^{i*})c_{t,k}^i,$$

where W_{tk}^i is the increase in transportation cost to the Warehouse resulting from modifying flow in the linear program from Y_{tk}^{i*} to 1. This cost increase can be easily calculated using Theorem 4.
Step 3: Determine the ordering epochs of retailer i by finding the minimum cost path from 1 to $T+1$ on network G_i with edge costs equal to the marginal costs.
Step 4: Update the amount and costs of warehouse orders at each period to account for retailer i's ordering strategy. Costs are updated using Theorem 4.
Step 5: Let $i = i+1$ and repeat steps (2)–(5) until $i = n+1$.

4.2.3 Computational results

We test the performance of the linear programming based algorithm in terms of both computational time and relative deviation from the optimal ZIO policy. For this purpose we apply the algorithm to two types of problems. The first is the Single-Warehouse Multi-Retailer Problem with retailer ordering cost represented by the modified all-unit discount cost function. The second is the Single-Warehouse Multi-Retailer Problem with concave ordering cost functions for the retailers. Of course, in both type of problems, warehouse ordering cost is a piece-wise linear concave function. Observe that there exists an optimal ZIO policy in the concave case (Type 2) and thus an optimal integer solution to Problem SW is optimal for the associated Single-Warehouse Multi-Retailer Problem.

Type 1 Instances. We consider four problem classes corresponding to 5, 25, 50 and 100 retailers. The planning horizon is 12 periods and demands for each retailer are generated from a normal distribution with mean 100 and standard deviation 20. Holding costs are randomly generated in the interval [0.1, 0.6]. Supplier–Warehouse transportation costs are described by piecewise linear concave functions with either 3 or 5 breakpoints between 0 and the maximum amount that could possibly be ordered from the supplier to satisfy retailer demands. These breakpoints and some initial fixed costs and slopes are given in Table 11. We consider the breakpoints fixed and randomly vary fixed costs and slopes over time by multiplying the initial values by a parameter generated from a uniform distribution on the interval (0.5, 3). Similarly, the warehouse–retailer transportation cost function for a particular retailer is a modified all-unit discount function with either 5, 6 or 7 breakpoints and again fixed costs and slopes are randomly varied (in the same manner) over time. Observe that it is enough to specify the breakpoints and the fixed cost of the first interval to construct the entire all-unit discount cost function. The breakpoints are given in Table 12 and the initial fixed cost on the first interval is randomly generated according to a uniform distribution on the interval (50, 500).

Table 13 shows, for each problem class, the average computation time of the linear programming based algorithm over 200 instances generated. For this moderate-size instances tested, the optimal ZIO policy can be calculated by solving the integer program, Problem SW, and used to evaluate the performance of the heuristic solution. The associated average computation times are given in the fourth column of Table 13. The cost of the optimal ZIO policy obtained is compared to the heuristic solution in the last two columns: The first column reports the average ratio for cases in which the solution to the linear programming relaxation of Problem SW was not integer. The second reports the average over all problems tested.

Table 11
Concave transportation cost functions considered in Type 1 Instances

Cost interval		1	2	3	4	5	6
5–25 retailers	breakpoint	0	1000	3000	12,000		
	fixed cost	500	1000	1750	2350		
	slope	1	0.5	0.25	0.2		
50 retailers	breakpoint	0	6000	10,000	20,000	35,000	50,000
	fixed cost	550	9550	10,590	13,190	17,090	20,990
	slope	1.5	1.24	0.98	0.72	0.46	0.2
100 retailers	breakpoint	0	12,000	20,000	40,000	70,000	100,000
	fixed cost	700	2780	5380	10,580	18,380	26,180
	slope	1.5	1.24	0.98	0.72	0.46	0.2

Table 12
Breakpoints associated with the all-unit discount cost functions in Type 1 Instances

Cost interval	1	2	3	4	5	6	7
7 breakpoints	200	300	450	600	750	900	1050
6 breakpoints	200	300	450	750	900	1050	
5 breakpoints	200	400	650	850	1000		

Table 13
Computational results for all-unit discount warehouse–retailer costs (Type 1)

Problem class	Number of retailers	CPU time (sec)	CPU time with IP (sec)	Frequency of fractional solution	Z^H/Z^{ZIO} fractional cases	Z^H/Z^{ZIO} all cases
Class 1	5	≈0	3	4/200	1.010	1.0002
Class 2	25	≈0	27	5/200	1.013	1.0004
Class 3	50	2	124	3/200	1.037	1.0006
Class 4	100	23	507	2/200	1.025	1.0003

Type 2 Instances: Here we study the performance of the linear programming based algorithm for instances in which all the transportation costs are piece-wise linear and concave.

We again consider different problem classes, with normally and independently, identically distributed retailer demands with mean 100 and standard deviation 20, and generated 10 instances for each class. Holding costs are set to 0.2 per unit per period. The piece-wise linear concave transportation costs considered have three price breaks (i.e., four segments with different slope) in the range from 0 to the maximum possible demand that could be satisfied using that link. Associated fixed cost and variable costs are randomly generated over time, as in the Type 1 instances, by multiplying the initial values given in Table 14 by a parameter uniformly distributed in (0.5, 5).

Table 15 describes the six problem classes tested and reports the average computation time and the average ratio of heuristic to optimal solutions over the five instances tested for each class. We observe that, in all the instances tested, the solution to the linear programming relaxation coincides with the optimal integer solution.

4.2.4 Extension to system with central stock

The bounds on the performance of ZIO policies developed can be easily extended to a more general distribution problem with central stocks, in which the warehouse is allowed to carry inventory. To show this, we observe that, since the transportation charges from supplier to warehouse are concave, it is optimal for the warehouse to follow a ZIO policy. Thus an order, $Q_x^0 > 0$,

Table 14
Initial concave transportation cost functions considered in Type 2 Instances

Cost interval		1	2	3	4
Supplier–warehouse	breakpoint	0	1000	3000	12,000
	fixed cost	500	1000	1750	2350
	slope	1	0.5	0.25	0.2
Warehouse–retailer	breakpoint	0	200	400	800
	fixed cost	100	200	400	480
	slope	1.5	1	0.5	0.4

Table 15
Computational results for concave transportation costs on all links (Type 2)

Problem class	Number of periods	Number of retailers	CPU time (sec)	Z^H/Z^{ZIO} $= Z^H/Z^*$
Class 1	6	5	≈ 0	1
Class 2	12	5	1	1
Class 3	6	10	≈ 0	1
Class 4	12	10	4	1
Class 5	6	25	2	1
Class 6	12	25	5	1

placed by the warehouse in period x, will cover all of the retailers' orders from a certain period $a(x) \geq x$ to a period $b(x) \geq a(x)$ and can be expressed as

$$Q_x^0 = \sum_{i=1}^{n} \sum_{t=a(x)}^{b(x)} Q_t^i.$$

In this way, each order placed by the retailers in periods $a(x)$ through $b(x)$ is associated with the warehouse order at time x. Using this observation, the proofs of Theorem 12 and Theorem 14 follow in much a similar way as for the original case.

To put this extension in perspective, it is appropriate to point out that the model with central stock is directly related to the seminal work of Roundy (1985). In his work, Roundy analyzed the Single-Warehouse Multi-Retailer model with concave ordering cost functions, infinite time horizon and constant demand rates. For this problem, Roundy shows that Power-of-Two policies, which belong to the class of ZIO policies, are highly effective. Our results indicate that in the case of a finite horizon, time varying demand and modified all-unit discount costs, ZIO policies are very effective as well. Indeed, by restricting the solution set to ZIO policies we can obtain a solution whose cost is no higher than 4/3 times the optimal cost and this bound is tight. If the transportation cost functions do not change from period to period, then there exists a ZIO policy whose cost is no higher than 5.6/4.6 times the optimal cost. Unfortunately, finding the optimal ZIO policy in our case is an NP-hard problem. This is in contrast to the model analyzed by Roundy where finding the best Power-of-Two policy can be done very efficiently.

PART II: PRICING TO IMPROVE SUPPLY CHAIN PERFORMANCE

5 Introduction

In recent years, scores of retail and manufacturing companies have started exploring innovative pricing strategies in an effort to improve their operations and ultimately the bottom line. Firms are employing methods such as dynamically adjusting price over time based on inventory levels or production schedules as well as segmenting customers based on their sensitivity to price and lead time.

For instance, no company underscores the impact of the Internet on product pricing strategies more than Dell Computers. The exact same product is sold at different prices on Dell's Web site, depending on whether the purchase is made by a private consumer, a small, medium or large business, the federal government, an education or health care provider. A more careful review of Dell's strategy, see Agrawal and Kambil (2000), suggests that even the price of the same product for the same industry is not fixed; it may change significantly over time.

Dell is not alone in its use of a sophisticated pricing strategy. Consider:

- Boise Cascade Office Products sells many products on-line. Boise Cascade states that prices for the 12,000 items ordered most frequently on-line might change as often as daily [Kay, 1998].
- Ford Motor Co. uses pricing strategies to match supply and demand and target particular customer segments. Ford executives credit the effort with $3 billion in growth between 1995 and 1999 [Leibs, 2000].

These developments call for models that integrate production decisions, inventory control and pricing strategies. Such models and strategies have the potential to radically improve supply chain efficiencies in much the same way as revenue management has changed the airline industry, see Belobaba (1987) or McGill and Van Ryzin (1999). Indeed, in the airline industry, revenue management provided growth and increased revenue by 5%, see Belobaba, 1987. In fact, if it were not for the combined contributions of revenue management and airline schedule planning systems, American Airlines [Cook, 2000] would have been profitable only one year in the decade beginning in 1990. In the retail industry, to name another example, dynamically pricing commodities can provide significant improvements in profitability, as shown by Gallego and van Ryzin (1994).

The coordination of pricing, production and distribution decisions is consistent with recent efforts in industry to cut across traditional organizational barriers. Indeed, in most companies, pricing and promotional decisions are typically made by marketing and sales units within the company, usually with very little regard to the impact of these decisions on supply chain performance. However, as observed earlier, more and more companies are exploring innovative pricing strategies in an effort to boost their profit and improve supply chain efficiencies. Thus, models similar to those described in the following sections are clearly important in supporting this new trend.

6 Coordinating pricing and inventory decisions

Many papers address the coordination of replenishment strategies and pricing policies, starting with the work of Whitin (1955) who analyzed the celebrated newsvendor problem with price-dependent demand. For a review, the reader is referred to Eliashberg and Steinberg (1991), Petruzzi and Dada (1999), Federgruen and Heching (1999) or Chan, Simchi-Levi, and Swann (2001).

To date, the literature has confined itself mainly to either: (i) models with variable ordering costs but no fixed costs; (ii) models in which inventory cannot be carried over from one period to the next; or (iii) models in which replenishment decisions are made only at the beginning of the planning horizon, see Federgruen and Heching (1999). Recently, however, Chen and Simchi-Levi (2002a,b) analyzed a fairly general inventory/pricing model. Specifically, Chen and Simchi-Levi (2002a) consider a finite horizon, periodic review, single product model with stochastic demand. Demands in different periods are independent of each other and their distributions depend on the product price. Pricing and ordering decisions are made at the beginning of each period, and all shortages are backlogged. The ordering cost includes both a fixed cost and a variable cost proportional to the amount ordered. Inventory holding and shortage costs are convex functions of the inventory level carried over from one period to the next. The objective is to find an inventory policy and pricing strategy maximizing expected profit over the finite horizon.

The model is similar to the model analyzed by Federgruen and Heching (1999), except that the latter assumes that the ordering cost is *proportional* to the amount ordered and thus does not include a fixed cost component. In addition, the demand function is assumed to be a linear function of the price, see Lemma 1 in Federgruen and Heching (1999).

The paper by Thomas (1974) also considers a model similar to the one by Chen and Simchi-Levi (2002a), namely, a periodic review, finite horizon model with a fixed ordering cost and stochastic, price-dependent demand. The paper postulates a simple policy, referred to by Thomas as (s, S, p), which can be described as follows. The inventory strategy is an (s, S) policy: If the inventory level at the beginning of period t is below the reorder point, s_t, an order is placed to raise the inventory level to the order-up-to level, S_t. Otherwise, no order is placed. The price, p, depends on the initial inventory level at the beginning of the period. Thomas provides a counter example which shows that with a 'few prices' (i.e., when price is restricted to a discrete set) this policy may fail to be optimal. Thomas goes on to say:

> If all prices in an interval are under consideration, it is conjectured that an (s, S, p) policy is optimal under fairly general conditions.

In Section 6.1, we review the main assumptions of the model analyzed by Chen and Simchi-Levi (2002a). In Section 6.1.1 we characterize the optimal inventory and pricing policies for *additive demand functions*. We show that in this case the policy proposed by Thomas is indeed optimal. In Section 6.1.2 we analyze general demand functions which may be *non-additive*. We demonstrate that in this case the profit-to-go function is not necessarily k-concave and an (s, S, p) policy is not necessarily optimal. We introduce the concept of symmetric k-convex functions and apply it to provide a characterization of the optimal policy. In Section 6.2 we extend the results obtained by Chen and Simchi-Levi (2002a) for the finite horizon model to the infinite horizon case under both discounted and average cost criteria. Finally, in Section 6.3 we apply the results to the model with zero fixed-cost and illustrate that the techniques developed in Chen and Simchi-Levi (2002a,b) allow to extend the results of Federgruen and Heching to more general demand processes.

6.1 The finite horizon model

Consider a firm that has to make inventory and pricing decisions over a finite time horizon with T periods. Demands in different periods are independent of each other. For each period t, $t = 1, 2, \ldots, T$, let d_t be the demand in period t, p_t the selling price in period t, and $\underline{p}_t$, $\bar{p}_t$ are lower and upper bounds on p_t, respectively.

Chen and Simchi-Levi (2002a) concentrate on demand functions of the following form:

Assumption 20. For $t = 1, 2, \ldots, T$, the demand function satisfies

$$d_t = D_t(p_t, \varepsilon_t) := \alpha_t D_t(p_t) + \beta_t, \qquad (6.1)$$

where $\varepsilon_t = (\alpha_t, \beta_t)$, and α_t, β_t are two random variables with $E\{\alpha_t\} = 1$ and $E\{\beta_t\} = 0$. The random perturbations, ε_t, are independent across time.

Observe that, by scaling and shifting, the assumptions $E\{\alpha_t\} = 1$ and $E\{\beta_t\} = 0$ can be made without loss of generality. A special case of this demand function, the *additive* demand function, is analyzed in Section 6.1.1. In this case, the demand function is of the form $d_t = D_t(p) + \beta_t$. This implies that only β_t is a random variable while $\alpha_t = 1$. In Section 6.1.2 we analyze the general demand functions (6.1). Observe that a special case of the model analyzed in Section 6.1.2 is a model with the *multiplicative* demand function. In this case, the demand function is of the form $d_t = \alpha_t D_t(p)$, where α_t is a random variable. Finally observe that special cases of the function $D_t(p)$ include $D_t(p) = b_t - a_t p$ ($a_t > 0$, $b_t > 0$) in the additive case and $D_t(p) = a_t p^{-b_t}$ ($a_t > 0$, $b_t > 1$) in the multiplicative case; both are common in the economics literature [see Petruzzi & Dada, 1999].

Chen and Simchi-Levi (2002a) assume the following.

Assumption 21. For all t, $t = 1, 2, \ldots, T$, the inverse function of D_t, denoted by D_t^{-1}, is continuous and strictly decreasing. Furthermore, the expected revenue

$$R_t(d) := d D_t^{-1}(d)$$

is a concave function of expected demand d.

The assumption thus implies that expected demand is a monotone decreasing function of price, an assumption satisfied by many products, except perhaps for some luxury products, see Federgruen and Heching (1999). Both the monotonicity and concavity assumptions are satisfied by many demand functions that are common in the marketing or economics literature.

Let x_t be the inventory level at the beginning of period t, just before placing an order. Similarly, y_t is the inventory level at the beginning of period t after placing an order. The ordering cost function includes both a fixed cost and a variable cost and is calculated for every t, $t = 1, 2, \ldots, T$, as

$$k_t \delta(y_t - x_t) + c_t(y_t - x_t),$$

where

$$\delta(u) := \begin{cases} 1, & \text{if } u > 0, \\ 0, & \text{otherwise}. \end{cases}$$

As is common in standard inventory management models, we assume that the fixed cost, k_t, is a *non-increasing function* of time.

Unsatisfied demand is backlogged. Let x be the inventory level carried over from period t to period $t+1$. Since we allow backlogging, x may be positive or negative. A cost $h_t(x)$ is incurred at the end of period t which represents inventory holding cost when $x > 0$ and penalty cost if $x < 0$. The following assumption is common to most inventory models.

Assumption 22. For each t, $t = 1, 2, \ldots, T$, $h_t(x)$, is a convex function of the inventory level x at the end of period t.

The objective is to decide on ordering and pricing policies so as to maximize total expected profit over the entire planning horizon. That is, the objective is to choose y_t and p_t so as to maximize

$$E\left\{\sum_{t=1}^{T} -k_t\delta(y_t - x_t) - c_t(y_t - x_t) - h_t(x_{t+1}) + p_t D_t(p_t, \varepsilon_t)\right\}, \quad (6.2)$$

where $x_{t+1} = y_t - D_t(p_t, \varepsilon_t)$.

Denote by $v_t(x)$ the profit-to-go function at the beginning of time period t with inventory level x. A natural dynamic program for the above maximization problem is as follows. For $t = T, T-1, \ldots, 1$,

$$v_t(x) = c_t x + \max_{y \geq x, \bar{p}_t \geq p \geq \underline{p}_t} -k_t\delta(y - x) + f_t(y, p), \quad (6.3)$$

where

$$f_t(y, p) = -c_t y + E\{p D_t(p, \varepsilon_t) - h_t(y - D_t(p, \varepsilon_t)) + v_{t+1}(y - D_t(p, \varepsilon_t))\}, \quad (6.4)$$

and $v_{T+1} = 0$. Let

$$p_t(y) \in \operatorname{argmax}_{\bar{p}_t \geq p \geq \underline{p}_t} f_t(y, p). \quad (6.5)$$

Then

$$v_t(x) = c_t x + \max_{y \geq x} -k_t\delta(y - x) + f_t(y, p_t(y)).$$

We now relate our problem to the celebrated stochastic inventory control problem discussed by Scarf (1960). In that problem, demand is assumed to be exogenously determined, while in our problem demand depends on price. Other assumptions regarding the framework of the model are similar to those made by Scarf (1960).

For the classical stochastic inventory problem, Scarf (1960) showed that an (s, S) policy is optimal. In this policy, the optimal decision in period t is characterized by two parameters, the reorder point, s_t, and the order-up-to level, S_t. An order of size $S_t - x_t$ is made at the beginning of period t if the initial inventory level at the beginning of the period, x_t, is smaller than s_t. Otherwise, no order is placed.

To prove that an (s, S) policy is optimal, Scarf (1960) uses the concept of k-convexity.

Definition 23. A real-valued function f is called k-convex for $k \geq 0$, if for any $z \geq 0$, $b > 0$ and any y we have

$$k + f(z+y) \geq f(y) + \frac{z}{b}(f(y) - f(y-b)). \tag{6.6}$$

A function f is called k-concave if $-f$ is k-convex.

For the purpose of the analysis of problem (6.3), Chen and Simchi-Levi (2002a) find it useful to introduce another, yet equivalent, definition of k-convexity.[1]

Definition 24. A real-valued function f is called k-convex for $k \geq 0$, if for any $x_0 \leq x_1$ and $\lambda \in [0, 1]$,

$$f((1-\lambda)x_0 + \lambda x_1) \leq (1-\lambda)f(x_0) + \lambda f(x_1) + \lambda k. \tag{6.7}$$

Proposition 25. *Definitions 23 and 24 are equivalent.*

Definition 24 emphasizes the difference between k-convexity and traditional convexity (which is also 0-convexity). It is clear from this definition that one significant difference between k-convexity and traditional convexity is that (6.7) is not symmetric with respect to x_0 and x_1.

It turns out that this asymmetry is the main barrier when trying to identify the optimal policy to problem (6.3) for non-additive demand functions. Indeed, in Section 6.1.2 we indicate that the profit-to-go function is not necessarily k-concave and an (s, S, p) policy is not necessarily optimal for general demand processes. This motivates the development of a new concept, the symmetric k-concave function, which allows Chen and Simchi-Levi (2002a) to characterize the optimal policy in the general demand case.

However, under the additive demand model analyzed in Section 6.1.1, this concept is not needed. Specifically, Chen and Simchi-Levi (2002a) prove that for additive demand processes, the profit-to-go function is k-concave and hence the optimal policy for problem (6.3) is an (s, S, p) policy, precisely the policy conjectured by Thomas (1974).

---

[1] Professor Paul Zipkin pointed out to us that this equivalent characterization of k-convexity has appeared in Porteus (1971).

6.1.1 Additive demand function

In this section, we focus on additive demand functions, i.e., demand functions of the form

$$d_t = D_t(p_t) + \beta_t,$$

where β_t is a random variable.

To characterize the optimal policy in this case, Chen and Simchi-Levi (2002a) prove the following property.

Lemma 26. *Suppose there is a finite value $p_t(y)$ that maximizes (6.5) for any value of y. Then, $y - D_t(p_t(y))$ is a non-decreasing function of y.*

The lemma thus implies that the higher the inventory level at the beginning of time period t, y_t, the higher the expected inventory level at the end of period t, $y_t - D_t(p(y_t))$. Using this property, together with the new definition of k-convex functions, see Definition 24, Chen and Simchi-Levi (2002a) prove,

Theorem 27. *For any t, $t = T, T-1, \ldots, 1$, we have*

  a. *$f_t(y, p_t(y))$ and $v_t(x)$ are k-concave.*
  b. *There exist s_t and S_t with $s_t \leq S_t$ such that it is optimal to order $S_t - x_t$ and set the selling price $p_t = p_t(S_t)$ when $x_t < s_t$, and not to order anything and set $p_t = p_t(x_t)$ when $x_t \geq s_t$.*

The theorem thus implies that the (s, S, p) policy introduced by Thomas (1974) is indeed optimal for additive demand processes. An interesting question is whether $p_t(y)$ is a non-increasing function of y. Unfortunately, this property, which holds for the model with no fixed cost, see Section 6.3, does not hold for our model.

Proposition 28. *The optimal price, $p_t(y)$ is not necessarily a non-increasing function of y.*

6.1.2 General demand functions

In this section, we focus on general demand functions (6.1). Our objective in this section is two-fold. First, we demonstrate that under the general demand functions, $v_t(x)$ may not be k-concave and an (s, S, p) policy may fail to be optimal for problem (6.3). Second, we characterize the structure of the optimal policy for the general demand functions (6.1).

Specifically, the Lemma 29, proved in Chen and Simchi-Levi (2002a), illustrates that the profit-to-go function is not k-concave in general.

Lemma 29. *There exists an instance of problem (6.3) with a multiplicative demand function and time independent parameters such that the functions $f_{T-1}(y, p_{T-1}(y))$ and $v_{T-1}(x)$ are not k-concave.*

Of course, it is entirely possible that even if the functions $f_t(y, p_t(y))$ and $v_t(x)$ are not k-concave for some period t, the optimal policy is still an (s, S, p) policy. The Lemma 30, proved in Chen and Simchi-Levi (2002a), shows that this is not true in general.

Lemma 30. *There exists an instance of problem* (6.3) *with multiplicative demand functions where an* (s, S, p) *policy is not optimal.*

To overcome these difficulties, Chen and Simchi-Levi (2002a) propose a weaker definition of k-convexity, referred to as symmetric k-convexity:

Definition 31. A real-valued function f is called sym-k-convex for $k \geq 0$, if for any x_0, x_1 and $\lambda \in [0, 1]$,

$$f((1 - \lambda)x_0 + \lambda x_1) \leq (1 - \lambda)f(x_0) + \lambda f(x_1) + \max\{\lambda, 1 - \lambda\}k. \quad (6.8)$$

A function f is called sym-k-concave if $-f$ is sym-k-convex.

Observe that k-convexity, and hence convexity, is a special case of sym-k-convexity. Interestingly, our analysis of sym-k-convex functions reveals that these functions have properties that are parallel to those of k-convex functions, see Bertsekas (1995). Specifically, based on properties of these functions, Chen and Simchi-Levi (2002a) prove the following results.

Theorem 32. *For any* t, $t = T, T-1, \ldots, 1$, *we have*

 a. $f_t(y, p_t(y))$ *and* $v_t(x)$ *are sym-k-concave.*
 b. *There exists* s_t *and* S_t *with* $s_t \leq S_t$ *and a set* $A_t \subset [s_t, (s_t + S_t)/2]$, *such that it is optimal to order* $S_t - x_t$ *and set* $p_t = p_t(S_t)$ *when* $x_t < s_t$ *or when* $x_t \in A_t$ *and not to order anything and set* $p_t = p_t(x_t)$ *otherwise.*

Theorem 32 thus implies that an (s, S, A, p) policy is the optimal policy for problem (6.3) under general demand processes. In such a policy, the optimal inventory strategy is characterized by two parameters s_t and S_t and a set $A_t \subset [s_t, (s_t + S_t)/2]$, possibly empty. When the inventory level, x_t, at the beginning of period t is less than s_t or if $x_t \in A_t$ an order of size $S_t - x_t$ is made. Otherwise, no order is placed. Thus, it is possible that an order will be placed when the inventory level $x_t \in [s_t, (s_t + S_t)/2]$, depending on the problem instance. On the other hand, if $x_t \geq (s_t + S_t)/2$ no order is placed. Price depends on the initial inventory level at the beginning of the period.

6.2 The infinite horizon case

The finite horizon models analyzed in the previous section are clearly appropriate for products with short life cycles, e.g., personal computers (PC), printers or fashion items. However, these models are less appropriate for products with long life cycles, e.g., non-fashion items. In this case, it is important to characterize the optimal policy in the infinite horizon case.

We thus consider a model similar to the one analyzed in the previous sections except that in the infinite horizon case all parameters are assumed to be time independent. Of course, it is tempting to try and extend the results of Theorem 32, which establishes the optimality of an (s, S, A, p) policy for the finite horizon general demand model, to the infinite horizon case. Surprisingly, Theorem 33, proved in Chen and Simchi-Levi (2002b), shows that this intuition can be misleading.

Theorem 33. *A stationary (s, S, p) policy is optimal for both the additive demand model and the general demand model under average and discounted cost criteria.*

Thus, the theorem suggests that in the infinite horizon case, the optimal policy is an (s, S, p) policy, independent of whether demand is additive or not. Interestingly, our proof of the optimal policy for the general demand model is based on two key results: The first is that the long-run average (or discounted) profit function is symmetric k-concave, suggesting that a stationary (s, S, A, p) policy is optimal. Surprisingly, our second result shows that in the infinite horizon case the set A is an empty set.

6.3 Special case: zero fixed-cost

The results described in the previous sections also apply to the special case in which the ordering cost function includes only variable but no fixed cost, i.e., $k_t = 0$ for all t, $t = 1, \ldots, T$. Indeed, by Theorem 32, the functions v_t and $f_t(y, p_t(y))$, $t = 1, 2, \ldots, T$, are symmetric 0-concave, and hence, from Definition 31, they are concave. Furthermore, and unlike the model with fixed cost, in this case, Chen and Simchi-Levi show that $p_t(y)$ is a non-increasing function of y. Thus,

Corollary 34. *Consider problem (6.3) with zero fixed-cost and general demand functions (6.1). In this case, a base-stock list price policy is optimal.*

The base-stock list price policy is a policy described by Federgruen and Heching (1999). Here, in each period the optimal policy is characterized by an order-up-to level, referred to as the base-stock level, and a price which depends on the initial inventory level at the beginning of the period. If the initial inventory level is below the base-stock level, an order is placed to raise the inventory level to the base-stock level. Otherwise, no order is placed, and a discount price is offered. This discount price is a non-increasing function of the initial inventory level.

Thus, Corollary 34 extends the results of Federgruen and Heching (1999) to more general demand processes. Indeed, Federgruen and Heching analyzed the zero fixed-cost model both in the finite horizon and infinite horizon cases. A key assumption in their paper implied by their Lemma 26 is that the demand function, d_t, is a linear function of the price. Corollary 34 suggests that this policy holds under much more general assumptions on the demand process.

7 Pricing models with production capacity limits

Very few pricing models have explicitly considered production capacity limits. One exception is the work by Chan et al. (2001) who analyzed *partial update* strategies, namely *Delayed Production* and *Delayed Pricing* strategies. In the first, decisions about the pricing policy are determined at the beginning of the planning horizon while production and inventory decisions are made period by period. Thus, in this case, the planner uses periodic production and inventory decisions as tools to better match supply and demand. In a Delayed Pricing strategy, on the other hand, decisions about production levels are made at the beginning of the planning horizon while pricing and inventory decisions are made period by period.

The following examples, see Chan et al. (2001), illustrate situations under which the two planning strategies are appropriate.

1. A retailer whose primary distribution channel is through catalogs, determines prices in advance in order to advertise and print catalogs. Production decisions are determined period by period, based on demand distribution in present and future periods as well as inventory from previous periods.
2. A supplier faces non-stationary demand and initially determines period prices to better match expected demand and supply in each period. The supplier contracts with a manufacturer over a time horizon, offering the manufacturer these fixed prices in advance for planning purposes, but allowing orders to be placed in each period due to the manufacturer's high inventory holding cost and unpredictable demand. The supplier adjusts production in each time period based on previous inventory and expected orders.
3. A manufacturer needs to determine a procurement strategy at the beginning of the year for the next 12 months. For this purpose, the manufacturer decides a priori on her monthly production levels and commits the supplier to deliver components just-in-time. The manufacturer sells products over the phone and web and determines price in each period so as to set the demand level to approximately match production and to clear previous inventory.

These examples demonstrate the two planning models described earlier. In the first two examples, the firm determines prices for a planning horizon a priori and makes production decisions based on the state of the system and future demand. The firm varies production levels based on inventory left over from previous periods as well as current and future demand.

The third example illustrates the second planning model. The manufacturer plans production at the beginning of the horizon but makes the price decision on a period by period basis. In this case, price can be used as a market clearing mechanism to deal with inventory from previous periods.

In the examples described above, it may be profitable to set aside inventory to satisfy future demand, even if the decision means losing sales in the current period. Although choosing to lose sales may seem counter to making profit, the inventory set aside is likely to generate a larger income in the future. This would typically occur if the price in the future is higher or if the future production costs were high. This is the intuition behind the concept of 'Save-Up-To Level' which we will introduce below.

Consider now a Delayed Production strategy in which pricing decisions have already been made at the beginning of the horizon. The challenge is to identify properties of the optimal production and inventory policy such that the firm maximizes expected profit. Specifically, consider a production-inventory model in which demand in period t depends on price according to a general stochastic demand function $D_t(p_t, \varepsilon_t(p_t))$ where $\varepsilon_t(p_t)$ is a random variable with a known distribution. Let $\mathbf{P}$ be the price vector chosen at the beginning of the horizon, that is, $\mathbf{P} = \{p_1, p_2, \ldots, p_T\}$ where T is the length of the planning horizon.

The production facility has a limited capacity, q_t, $t = 1, 2, \ldots, T$, production cost includes only a variable component but no fixed cost, and inventory holding cost is charged on inventory carried from one period to the next. All parameters are time dependent. Finally, shortages are lost and demand does not have to be satisfied even if inventory exists; that is, the decision-maker may decide to forgo immediate revenue for potentially higher revenue in the future.

Chan et al. (2001) prove the following result.

Lemma 35. *Given a vector of prices* $\mathbf{P}$, *there exists an optimal policy for the Delayed Production strategy with an optimal order-up-to level,* y_t^*, *and an optimal save-up-to level,* S_t^*.

Thus, at the beginning of time period t, the amount produced, X_t, should raise the available inventory to the optimal order-up-to level, Y_t^*, or as close as possible to it if the production capacity constraint is reached (i.e., $X_t = q_t$). The save-up-to level, S_t^*, is the amount that should be saved in period t to satisfy demand in future periods even if sales are lost in the current period. Observe that the Lemma implies that both the order-up-to and the save-up-to policies are independent of the inventory level at the beginning of the period, I_{t-1}.

Of course, the Lemma also applies to the special case in which the decision-maker adopts a fixed price policy, i.e., a policy in which the product is sold for the same price in all time periods. In this case, the Lemma allows the decision-maker to determine the best production-inventory policy maximizing expected profit for a given fixed price policy. Thus, a search on all possible prices determines the optimal fixed price.

Consider now the Delayed Pricing policy in which production quantities, X_t, are determined at the beginning of the horizon. The objective is to determine a pricing policy and an inventory strategy so as to maximize expected profit. The inventory policy will specify, period by period, the amount of available product to be sold as well as the minimum amount of inventory to be transferred to the next period.

Unfortunately, this case is more complex than the previous one and indeed the following observations can be made:

- The save-up-to level in a specific period depends on the initial inventory level in that period.
- Price does not necessarily increase as a function of decreasing inventory. That is, unlike the model with backlogging, it is possible that as initial inventory increases, price also increases.

A related work is the paper by Van Mieghem and Dada (1999), in which the authors explicitly consider price postponement versus production postponement strategies. They focus on a single-period, two-stage process with an initial decision, e.g., production decision, followed by a realization of demand, followed by another decision, e.g., pricing decision. Thus, price (production) postponement as outlined by Van Mieghem and Dada is different from Delayed Pricing (Production) in Chan et al. (2001). Specifically, in the model analyzed by Van Mieghem and Dada, the postponed decisions are made after demand is realized.

8 Computational results and insights

The key challenge when considering dynamic pricing strategies is to identify conditions under which this strategy provides significant profit benefit over (the best) fixed price strategy. For this purpose, Federgruen and Heching (1999) and Chan et al. (2001) performed extensive computational studies. In both papers, the focus is on periodic review models with variable ordering costs but no fixed costs. These computational studies provide the following insights.

- *Available Capacity:* Assuming everything else being equal, the smaller the production capacity relative to average demand, the larger the benefit from dynamic pricing [Chan et al., 2001].
- *Demand Variability:* The benefit of dynamic pricing increases as the degree of demand uncertainty, measured by the coefficient of variation, increases [Federgruen and Heching, 1999].
- *Seasonality in Demand Pattern:* The benefit of dynamic pricing increases as the level of demand seasonality increases [Federgruen and Heching, 1999, Chan et al., 2001].

- *Length of the planning horizon:* The longer the planning horizon the smaller the benefit from dynamic pricing [Federgruen and Heching, 1999].

All in all, research [Federgruen and Heching, 1999, Chan et al., 2001] indicates that, depending on the data and the model assumptions, dynamic pricing may increase profit by 2–6%. This increase in profit due to dynamic pricing is very significant for industries with low profit margins, e.g., retail and computer industries.

To determine the effectiveness of the planning models developed in Section 7, Chan et al. (2001) conducted an extensive computation study. The objective of the study was two-fold:

1. Identify situations where partially delayed planning, i.e., either Delayed Production or Delayed Pricing, provides significant increase in expected profit relative to a fixed price strategy, and
2. Determine conditions under which one strategy outperforms the other.

Below we provide a summary of the insights obtained from the computational study.

- Delayed Pricing and Delayed Production provide significant increase in expected profit in most of the cases analyzed.
- The performance of partial update strategies, either Delayed Pricing or Delayed Production, tends to increase as seasonality increases and as capacity becomes more constrained.
- Delayed Pricing usually outperforms Delayed Production. Exceptions occur when production cost is high or under certain types of seasonality.

The last insight, concerning the performance of Delayed Pricing versus Delayed Production, is in agreement with the one obtained by Van Mieghem and Dada (1999) for a somewhat related two stage problem. As observed earlier, they consider a single period model where the postponed decisions are made after demand is realized. They found that in many instances Pricing Postponement outperformed Production Postponement; one exception was in a case with high production cost.

PART III: SUPPLY CHAIN DESIGN MODELS

9 Introduction

One of the most important aspects of logistics is deciding where to locate new facilities, such as retailers, warehouses or factories. These strategic decisions are a crucial determinant of whether or not materials will flow efficiently through the distribution system.

In this section we consider two important warehouse location problems: the Single-Source Capacitated Facility Location Problem and a distribution system design problem. In each case, the problem is to locate a set of warehouses in a distribution network. We assume that the cost of locating a warehouse at a particular site includes a *fixed* cost (e.g., building costs, rental costs, etc.) and a *variable* cost for transportation. This variable cost includes the cost of transporting the product to the retailers as well as possibly the cost of moving product from the plants to the warehouse. In general, the objective is to locate a set of facilities so that total cost is minimized subject to a variety of constraints which might include:

- each warehouse has a capacity which limits the area it can supply,
- each retailer receives shipments from *one and only one* warehouse,
- each retailer must be within a fixed distance of the warehouse that supplies it, so that a reasonable delivery lead time is ensured.

Location analysis has played a central role in the development of the operations research field. In this area lie some of the discipline's most elegant results and theories. We note here the paper of Cornuéjols, Fisher, and Nemhauser (1977) and the two excellent books devoted to the subject by Mirchandani and Francis (1990) and Daskin (1995).

This section closely follows the material in Bramel and Simchi-Levi (1997) and is organized as follows. We first present an efficient algorithm for the Single-Source Capacitated Facility Location Problem. In this problem a set of retailers needs to be served by a number of warehouses with limited capacity. In Section 11, we present a more general model where all levels of the distribution system, i.e., plants and retailers, are taken into account when deciding warehouse locations.

All of the algorithms developed in this section are based on Lagrangian relaxation techniques which have been applied successfully to a wide range of location problems.

10 The single-source capacitated facility location problem

Consider a set of retailers geographically dispersed in a given region. The problem is to choose where in the region to locate a set of warehouses. We assume there are m sites that have been preselected as possible locations for these warehouses. Once the warehouses have been located, each of n retailers will get its shipments from a single warehouse. We assume:

- If a warehouse is located at site j:
  ○ a fixed cost f_j is incurred, and
  ○ there is a capacity q_j on the amount of demand it can serve.

Let the set of retailers be N where $N = \{1, 2, \ldots, n\}$, and let the set of potential sites for warehouses be M where $M = \{1, 2, \ldots, m\}$. Let w_i be the demand or flow between retailer i and its warehouse for each $i \in N$. We assume that the cost of transporting the w_i units of product from warehouse j to retailer i is c_{ij}, for each $i \in N$ and $j \in M$.

The problem is to decide where to locate the warehouses and then how the retailers should be assigned to the open warehouses in such a way that total cost is minimized. It is easy to verify that the capacity constraint implies that a retailer will not always be assigned to its nearest warehouse.

This problem is called the single-source Capacitated Facility Location Problem (CFLP), or sometimes the Capacitated Concentrator Location Problem (CCLP).

To formulate the problem as an integer linear program, define the following decision variables:

$$Y_j = \begin{cases} 1, & \text{if a warehouse is located at site } j, \\ 0, & \text{otherwise,} \end{cases}$$

for $j \in M$, and

$$X_{ij} = \begin{cases} 1, & \text{if retailer } i \text{ is served by a warehouse at site } j, \\ 0, & \text{otherwise,} \end{cases}$$

for $i \in M$, and $j \in M$.

The Single-Source Capacitated Facility Location Problem can be formulated as follows:

$$\text{Problem } P: \text{ Min } \sum_{i=1}^{n} \sum_{j=1}^{m} c_{ij} X_{ij} + \sum_{j=1}^{m} f_j Y_j$$

s.t.
$$\sum_{j=1}^{m} X_{ij} = 1 \quad \forall i \in N \tag{10.1}$$

$$\sum_{i=1}^{n} w_i X_{ij} \leq q_j Y_j \quad \forall j \in M \tag{10.2}$$

$$X_{ij}, Y_j \in \{0, 1\} \quad \forall i \in N, j \in M \tag{10.3}$$

Constraints (10.1) (along with the integrality conditions (10.3)) ensure that each retailer is assigned to exactly one warehouse. Constraints (10.2) ensure that the warehouse's capacity is not exceeded, and also that if a warehouse is not located at site j, no retailer can be assigned to that site.

Let Z^* be the optimal solution value of the Single-Source Capacitated Facility Location Problem. Note we have restricted the assignment variables

(X) to be integer. A related problem, where this assumption is relaxed, is simply called the (multiple-source) Capacitated Facility Location Problem. In that version, a retailer's demand can be *split* between any number of warehouses. In the Single-Source Capacitated Facility Location Problem, it is required that each retailer have only *one* warehouse supplying it. In many logistics applications, this is a realistic assumption since without this restriction optimal solutions might have a retailer receive many deliveries of the same product (each for, conceivably, a very small amount of the product). Clearly, from a managerial, marketing and accounting point of view, restricting deliveries to come from only one warehouse is a more appropriate delivery strategy.

Several algorithms have been proposed to solve the CFLP in the literature; all are based on the Lagrangian relaxation technique. This includes Neebe and Rao (1983), Barcelo and Casanovas (1984), Klincewicz and Luss (1986), and Pirkul (1987). The one we derive here is similar to the algorithm of Pirkul which seems to be the most effective.

We apply the Lagrangian relaxation technique by including constraints (10.1) in the objective function. For any vector $\lambda \in R^n$, consider the following problem, P_λ:

$$\text{Min} \sum_{i=1}^{n} \sum_{j=1}^{m} c_{ij} X_{ij} + \sum_{j=1}^{m} f_j Y_j + \sum_{i=1}^{n} \lambda_i \left(\sum_{j=1}^{m} X_{ij} - 1 \right)$$

subject to (10.2)–(10.3).

Let Z_λ be its optimal solution and note that

$$Z_\lambda \leq Z^*, \quad \forall \lambda \in R^n.$$

To solve P_λ we separate the problem by site. For a given $j \in M$, define the following problem R_λ^j, with optimal objective function value Z_λ^j:

$$\text{Min} \sum_{i=1}^{n} (c_{ij} + \lambda_i) X_{ij} + f_j Y_j$$

$$\text{s.t.} \sum_{i=1}^{n} w_i X_{ij} \leq q_j Y_j$$

$$X_{ij} \in \{0, 1\} \quad \forall i \in N$$

$$Y_j \in \{0, 1\}.$$

10.1 Solving P_λ^j

Problem P_λ^j, can be solved efficiently. In the optimal solution to P_λ^j, Y_j is either 0 or 1. If $Y_j = 0$, then $X_{ij} = 0$ for all $i \in N$. If $Y_j = 1$, then the problem is no more difficult than a constraint 0–1 Knapsack Problem, for which efficient algorithm exist; see, e.g., Nauss (1976). If the optimal knapsack solution is less than $-f_j$, then the corresponding optimal solution to P_λ^j is found by setting $Y_j = 1$ and X_{ij} according to the knapsack solution, indicating whether or not retailer i is assigned to site j. If the optimal knapsack solution is more than $-f_j$, then the optimal solution to P_λ^j is found by setting $Y_j = 0$ and $X_{ij} = 0$ for all $i \in N$.

The solution to $\mathcal{R}_\lambda$ is then given by

$$Z_\lambda \equiv \sum_{j=1}^{m} Z_\lambda^j - \sum_{i=1}^{n} \lambda_i.$$

For any vector $\lambda \in \mathcal{R}^n$, this is a lower bound on the optimal solution Z^*. To find the best such lower bound we use a subgradient procedure described in Bramel and Simchi-Levi (1997).

10.2 Upper bounds

For a given set of multipliers, if the values $\{X\}$ satisfy (10.1), then we have an optimal solution to Problem P, and we stop. Otherwise, we perform a simple subroutine to find a feasible solution to P. The procedure is based on the observation that the knapsack solutions found when solving P_λ give us some information concerning the benefit of setting up a warehouse at a site (relative to the current vector λ). If, for example, the knapsack solution corresponding to a given site is 0, i.e., the optimal knapsack is empty, then this is most likely not a 'good' site to select at this time. In contrast, if the knapsack solution has a very negative cost, then this is a 'good' site. Given the values Z_λ^j for each $j \in M$, let π be a permutation of $1, 2, \ldots, m$ such that

$$Z_\lambda^{\pi(1)} \leq Z_\lambda^{\pi(2)} \leq \cdots \leq Z_\lambda^{\pi(m)}.$$

The procedure we perform allocates retailers to sites in a myopic fashion. Let W be the minimum possible number of warehouses used in the optimal solution to Problem P. This number can be found by solving the bin-packing problem defined on the values w_i with bin capacities q_j. Starting with the 'best' site, in this case site $\pi(1)$, assign the retailers in its optimal knapsack to this site. Then, following the indexing of the knapsack solutions, take the next 'best' site (say site $j \equiv \pi(2)$) and solve a new knapsack problem: one defined with costs $\bar{c}_{ij} \equiv c_{ij} + \lambda_i$ for each retailer i still unassigned. Assign all retailers in this knapsack solution to site j. If this optimal knapsack is empty, then a warehouse is not located at that site, and we go on to the next site. Continue in this manner until W warehouses are located.

The solution may still not be a feasible solution to Problem P since some retailers may not be assigned to a site. In this case, unassigned retailers are assigned to sites that are already chosen where they fit with minimum additional cost. If needed, additional warehouses may be opened following the ordering of π. A local improvement heuristic can be implemented to improve on this solution, using simple interchanges between retailers.

10.3 Computational results

We now report on various computational experiments using this algorithm. The retailer locations were chosen uniformly over the unit square. For simplicity, we made each retailer location a potential site for a warehouse, thus $m = n$. The fixed cost of a site was chosen uniformly between 0 and 10. The cost of assigning a retailer to a site was the Euclidean distance between the two locations. The values of w_i were chosen uniformly over the unit interval. We applied the algorithm mentioned above to many problems and recorded the relative error of the best solution (upper bound) to the best lower bound (maximum Z_λ) found, and the computation time required. The algorithm is terminated when the relative error is below 1% or when a prespecified number of iterations is reached. In Table 16 the numbers below 'Error' are the relative errors averaged over five randomly generated problem instances. The numbers below 'CPU Time' are the CPU times averaged over the five problem instances. All computational times are on an IBM Risc 6000 Model 950.

11 A distribution system design problem

So far the location model we have considered has been concerned with minimizing the costs of transporting products between warehouses and

Table 16
Computational results for the single-source capacitated facility location problem

n	m	Error (%)	CPU time (sec)
10	10	1.1	10.2
20	20	1.9	21.3
50	50	3.4	192.8
100	100	4.8	426.7

retailers. We now present a more realistic model that considers the cost of transporting the product from manufacturing facilities to the warehouses as well.

Consider the following warehouse location problem. A set of plants and retailers are geographically dispersed in a region. Each retailer experiences demands for a variety of products which are manufactured at the plants up to their capacity limits. A set of warehouses with limited capacities must be located in the distribution network from a list of potential sites.

The cost of locating a warehouse includes the transportation cost per unit from warehouses to retailers and also the transportation cost from plants to warehouses. In addition, as in the Single-Source Capacitated Facility Location Problem, there is a site-dependent fixed cost for locating each warehouse.

The data for the problem are the following:

- L = number of plants; we will also let $L = \{1, 2, \ldots, L\}$
- J = number of potential warehouse sites; also let $J = \{1, 2, \ldots, J\}$
- I = number of retailers; also let $I = \{1, 2, \ldots, I\}$
- K = number of products; also let $K = \{1, 2, \ldots, K\}$
- W = number of warehouses to locate
- $c_{\ell j k}$ = cost of shipping one unit of product k from plant l to warehouse site j
- d_{jik} = cost of shipping one unit of product k from warehouse site j to retailer i
- f_j = fixed cost of locating a warehouse at site j
- v_{lk} = supply of product k at plant l
- w_{ik} = demand for product k at retailer i
- s_k = volume of one unit of product k
- q_j = capacity (in volume) of a warehouse at site j

We make the additional assumption that a retailer gets delivery for a product from one warehouse only. This does not preclude solutions where a retailer gets shipments from different warehouses, but these shipments must be for different products. On the other hand, we assume that the warehouse can receive shipments from any plant and for any amount of product (within its capacity limit).

The problem is to determine where to locate the warehouses, how to ship product from the plants to the warehouses, and also how to ship the product from the warehouses to the retailers. This problem is similar to the one analyzed by Pirkul and Jayaraman (1996).

We again use a mathematical programming approach. Define the following decision variables:

$$Y_j = \begin{cases} 1, & \text{if a warehouse is located at site } j \\ 0, & \text{otherwise,} \end{cases}$$

and

U_{ljk} = amount of product k shipped from plant, ℓ to warehouse j,

for each $\ell \in L, j \in J$ and $k \in K$. Also define:

$$X_{jik} = \begin{cases} 1, & \text{if retailer } i \text{ receives product } k \text{ from warehouse } j \\ 0, & \text{othrewise,} \end{cases}$$

for each $j \in J, i \in I$ and $k \in K$.

Then, the Distribution System Design Problem can be formulated as the following integer program:

$$\min \sum_{\ell=1}^{L}\sum_{j=1}^{J}\sum_{k=1}^{K} c_{\ell jk} U_{\ell jk} + \sum_{i=1}^{I}\sum_{j=1}^{J}\sum_{k=1}^{K} d_{jik} w_{ik} X_{jik} + \sum_{j=1}^{J} f_j Y_j$$

s.t. $\quad \displaystyle\sum_{j=1}^{J} X_{jik} = 1 \quad \forall i \in I, k \in K$ \hfill (11.1)

$$\sum_{i=1}^{I}\sum_{k=1}^{K} s_k w_{ik} X_{jik} \leq q_j Y_j \quad \forall j \in J \tag{11.2}$$

$$\sum_{i=1}^{I} w_{ik} X_{jik} = \sum_{\ell=1}^{L} U_{\ell jk} \quad \forall j \in J, k \in K \tag{11.3}$$

$$\sum_{j=1}^{J} U_{\ell jk} \leq v_{\ell k} \quad \forall \ell \in L, k \in K \tag{11.4}$$

$$\sum_{j=1}^{J} Y_j = W \tag{11.5}$$

$Y_j, X_{jik} \in \{0,1\} \quad \forall i \in I, j \in J, k \in K$ \hfill (11.6)

$U_{\ell jk} \geq 0 \quad \forall \ell \in L, j \in J, k \in K.$ \hfill (11.7)

The objective function measures the transportation costs between plants and warehouses, between warehouses and retailers and also the fixed cost of locating the warehouses. Constraints (11.1) ensure that each retailer/product pair is assigned to one warehouse. Constraints (11.2) guarantee that the capacity of the warehouses is not exceeded. Constraints (11.3) ensure that there is a conservation of the flow of products at each warehouse, that is, the amount of each product arriving at a warehouse from the plants is equal to the amount being shipped from the warehouse to the retailers. Constraints (11.4) are the supply constraints. Constraint (11.5) ensures that we locate exactly W warehouses.

Observe that in the model described here, transportation cost is a linear function of the amount shipped. Indeed, at this strategic level, and unlike the tactical level described earlier, annual transportation cost is based on average shipment size. That is, at the strategic level, the model only approximates the transportation cost functions and thus transportation costs are linear.

The model can handle several extensions like a warehouse handling fee or a limit on the distance of any link used. Another interesting extension is when there are a fixed number of possible warehouse types to choose from. Each type has a specific cost along with a specific capacity. The model can be easily extended to handle this situation.

As in the previous problem, we will use Lagrangian relaxation. We relax constraints (11.1) (with multipliers λ_{ik}) and constraints (11.3) (with multipliers θ_{jk}). The resulting problem is:

$$\min \sum_{\ell=1}^{L}\sum_{j=1}^{J}\sum_{k=1}^{K} c_{\ell jk} U_{\ell jk} + \sum_{j=1}^{J}\sum_{i=1}^{I}\sum_{k=1}^{K} d_{jik} w_{ik} X_{jik} + \sum_{j=1}^{J} f_j Y_j$$

$$+ \sum_{j=1}^{J}\sum_{k=1}^{K} \theta_{jk}\left[\sum_{i=1}^{I} w_{ik} X_{jik} - \sum_{\ell=1}^{L} U_{\ell jk}\right] + \sum_{i=1}^{I}\sum_{k=1}^{K} \lambda_{ik}\left[1 - \sum_{j=1}^{J} X_{jik}\right],$$

subject to (11.2), (11.4)–(11.7).

Let $Z_{\lambda,\theta}$ be the optimal solution to this problem. This problem can be decomposed into two separate problems, P_1 and P_2. They are the following:

Problem P_1 : $Z_1 \equiv \operatorname{Min} \sum_{\ell=1}^{L}\sum_{j=1}^{J}\sum_{k=1}^{K} [c_{\ell jk} - \theta_{jk}] U_{\ell jk}$

s.t. $\sum_{j=1}^{J} U_{\ell jk} \leq v_{\ell k}, \quad \forall \ell \in L, k \in K$ \hfill (11.8)

$U_{\ell jk} \geq 0, \quad \forall \ell \in L, j \in J, k \in K.$

Problem P_2 : $Z_2 \equiv \operatorname{Min} \sum_{j=1}^{J}\sum_{i=1}^{I}\sum_{k=1}^{K} [d_{jik} w_{ik} - \lambda_{ik} + \theta_{jk} w_{ik}] X_{jik} + \sum_{j=1}^{J} f_j Y_j$

s.t. $\sum_{i=1}^{I}\sum_{k=1}^{K} s_k w_{ik} X_{jik} \leq q_j Y_j, \quad \forall j \in J$ \hfill (11.9)

$\sum_{j=1}^{J} Y_j = W, \quad Y_j, X_{jik} \in \{0,1\}, \quad \forall i \in I, j \in J, k \in K.$ \hfill (11.10)

11.1 Solving P_1

Problem P_1 can be solved separately for each plant/product pair. In fact, the objective functions of each of these subproblems can be improved (without loss in computation time) by adding the constraints:

$$s_k U_{\ell jk} \leq q_j, \quad \forall \ell \in L, j \in J, k \in K. \quad (11.11)$$

For each plant/product combination, say plant 1 and product k, sort the J values $\bar{c}_j \equiv c_{\ell jk} - \theta_{jk}$. Starting with the smallest value of $\bar{c}_j$, say $\bar{c}_{j'}$, if $\bar{c}_{j'} \geq 0$, then the solution is to ship none of this product from this plant. If $\bar{c}_{j'} < 0$, then ship as much of this product as possible along arc (ℓ, j') subject to satisfying constraints (11.8) and (11.11). Then if the supply v_{1k} has not been completely shipped, do the same for the next cheapest reduced cost ($\bar{c}$), as long as it is negative. Continue in this manner until the entire product has been shipped or the reduced costs are no longer negative. Then proceed to the next plant/product combination repeating this procedure. Continue until all the plant/product combinations have been scanned in this fashion.

11.2 Solving P_2

Solving Problem P_2 is similar to solving the subproblem in the Single-Source Capacitated Location Problem. For now we can ignore constraint (11.10). Then, we separate the problem by warehouse. In the problem corresponding to warehouse j, either $Y_j = 0$ or $Y_j = 1$. If $Y_j = 0$, then $X_{jik} = 0$ for all $i \in N$ and $k \in K$. If $Y_j = 1$, then we get a Knapsack Problem with NK items, one for each retailer/product pair. Let Z_2^j be the objective function value when Y_j is set to 1 and the resulting knapsack problem is solved. After having solved each of these, let π be a permutation of the numbers $1, 2, \ldots, J$ such that

$$Z_2^{\pi(1)} \leq Z_2^{\pi(2)} \leq \cdots \leq Z_2^{\pi(J)}.$$

The optimal solution to P_2 is to choose the W smallest values:

$$Z_2 \equiv \sum_{j=1}^{W} Z_2^{\pi(j)}.$$

For fixed vectors λ and θ, the Lagrangian lower bound is

$$Z_{\lambda,\theta} = Z_1 + Z_2 + \sum_{i=1}^{I}\sum_{k=1}^{K} \lambda_{ik}.$$

To maximize this bound, i.e., $\max_{\lambda,\theta}\{Z_{\lambda,\theta}\}$, we again use the subgradient optimization procedure.

11.3 Upper bounds

At each iteration of the subgradient procedure, we attempt to construct a feasible solution to the problem. Consider Problem P_2. Its solution may have a retailer/product combination assigned to several warehouses. We determine the set of retailer/product combinations that are assigned to one and only one warehouse and fix these. Other retailer/product combinations are assigned to warehouses using the following mechanism. For each remaining retailer/product combination, we determine the cost of assigning it to a particular warehouse. After determining that this assignment is feasible (from a warehouse capacity point of view), the assignment cost is calculated as the cost of shipping all of the demand for this retailer/product combination through the warehouse plus the cost of shipping the demand from the plants to the warehouse (in the cheapest possible manner while satisfying plant capacity constraints, possibly along one or more arcs from the plants to the warehouse). For each retailer/product combination we determine the penalty associated with assigning the shipment to its second best warehouse instead of its best warehouse. We then assign the retailer/product combination with the highest such penalty and update all arc flows and remaining capacities. We continue in this manner until all retailer/product combinations have been assigned to warehouses.

11.4 Computational results

In Table 17, we report running times, in seconds, on an IBM PC 166 MHz machine for a variety of problem sizes. The results are given as a function of various parameters. In all cases, the number of potential locations for warehouses is 32, the number of suppliers is 9, the numbers of products is also 9, and we require that the distance between a customer and a

Table 17
Running times

Number of customers	Number of warehouses	Running time 5% (sec)	Running time 1% (sec)
144	6	64	106
144	5	95	209
144	4	99	227
73	6	31	60
73	5	19	54
73	4	20	37

warehouse serving it will be no more than 100 miles. The optimization was terminated when the relative difference between the cost of the solution generated and the lower bound was within a specified gap. Thus, the column 'Running Time 5%' provides the running times when the gap is 5% while 'Running Time 1%' provides the running times when the gap is 1%. Finally, these six test problems represent real-world data that we have received from a producer and distributor of soft drinks in the Northeastern part of the U.S.

12 Conclusions

The last few years have been marked by considerable progress in the development and implementation of information and communication systems. These systems allow companies to track customer demand, inventory, and the availability of production facilities. Of course, as pointed out in Shapiro (1998), ready access to transactional data does not automatically lead to better decision making. Optimization models and new solution techniques that use the wealth of information to better design and manage the supply chain are key to improving supply chain performance.

This chapter describes a variety of optimization models and solution methods for the integration of various tactical and strategic decisions within the supply chain. The problems addressed range from the coordination of production, inventory and transportation, through the determination of pricing and production strategies, to supply chain design models. Most of the models are deterministic models while some incorporate uncertainty in customer demand. In all cases we utilize the inherent structure of the optimal strategies to develop computationally efficient algorithms and solve realistic instances. Moreover, for some of the problems, we can theoretically prove optimality or develop attractive worst-case bounds on the performance of various algorithms. The results demonstrate the power of optimization techniques and the great potential of these methods when implemented in Decision Support Systems.

Of course, many challenges still remain! For instance, in the Production/Distribution models analyzed in Part I, these challenges include incorporating production, warehousing and transportation capacities in some of the models, extending the tactical models to assembly systems, analyzing different modes of transportation, and most importantly, extending the analysis to practical situations in which the decision-maker is faced with uncertain demand or supply. Similarly, the pricing models and results should be extended to models with multiple classes of customers differentiated by their sensitivity to price and lead-time, models with discretionary sales, as well as multi-stage supply chains. Finally, the supply chain design problems analyzed in Part III should be generalized to incorporate demand uncertainty.

In this chapter, we have focused on a single decision-maker that has full control and access to information over the entire logistics network. In many practical situations, however, an important issue in supply chain design and planning is the management of information flows. How information is shared among different locations and organizations, and how decision power is distributed among multiple agents has a significant impact on supply chain performance. This effect has been observed in industry and rigorously analyzed in the academic literature under various settings, see Tayur, Ganeshan, and Magazine (2000), Lee, So, and Tang (2000) or Chen, Drezner, Ryan, and Simchi-Levi (2000). The flow of information and the coordination of distributed decision making within the supply chain are the subject of much current research, but are beyond the scope of this chapter.

Acknowledgements

Research supported in part by ONR Contracts N00014-90-J-1649 and N00014-95-1-0232, and by NSF Contracts DDM-9322828, DMI-9732795 and DMI-0134175.

References

Aggarwal, A., J. K. Park (1993). Improved algorithms for economic lot-size problems. *Operations Research* 41, 549–571.

Agrawal, V., A. Kambil (2000). *Dynamic Pricing Strategies in Electronic Commerce*, Working paper. Stern Business School, New York University.

Ahuja, R. K., T. L., Magnanti, J. B. Orlin (1993). *Network Flows: Theory, Algorithms and Applications*, Prentice Hall, Englewood Cliffs, New Jersey.

Amiry, A., H. Pirkul (1997). New formulation and relaxation to solve a concave-cost network flow problem. *Journal of the Operational Research Society* 48, 278–287.

Anily, S., J. Bramel (1998). S. Tayur, R. Ganeshan, M. Magazine (eds.), *Vehicle Routing and the Supply Chain. Quantitative Models for Supply Chain Management*, Kluwer Academic Publishers, Boston. Chapter 6, pp. 147–196.

Arkin, E., D. Joneja, R. Roundy (1989). Computational complexity of uncapacitated multi-echelon production planning problems. *Operations Research Letters* 8, 61–66. North-Holland.

Arora, S., M. Sudan (1997). Improved low-degree testing and its applications. *Proceedings of the 29th Annual ACM Symposium on the Theory of Computing*, 485–496.

Balakrishnan, A., S. Graves (1989). A composite algorithm for a concave-cost network flow problem. *Networks* 19, 175–202.

Balakrishnan, A., T. L. Magnanti, P. Mirchandani (1997). Network design, in: M. Del'Amico, F. Maffioli, S. Martello (eds.), *Annotated Bibliographies in Combinatorial Optimization*, John Wiley & Sons, New York, pp. 311–334. Chapter 18.

Balakrishnan, A., T. L. Magnanti, A. Shulman, R. T. Wong (1991). Models for planning capacity expansion in local access telecommunication networks. *Annals of Operations Research* 33, 239–284.

Barcelo, J., J. Casanovas (1984). A heuristic lagrangian algorithm for the capacitated plant location problem. *European Journal of Operations Research* 15, 212–226.

Belobaba, P.P. (1987). Airline yield management: An overview of seat inventory control. *Transportation Science* 21, 63–73.

Bertsekas, D. (1995). *Dynamic Programming and Optimal Control*, Volume One, Athena Scientific.

Bienstock, D., S. Chopra, O. Günlük (1998). Minimum cost capacity installation for multi-commodity network flows. *Mathematical Programming Series B* 81(3-1), 177–199.

Braklow, J. B., W. Graham, S. Hassler, K. Peck, W. B. Powell (1992). Interactive optimization improves service and performance for yellow freight system. *Interfaces* 22(1), 147–172.

Bramel, J., D. Simchi-Levi (1997). *The Logic of Logistics: Theory, Algorithms, and Applications for Logistics Management*. Springer Series in Operations Research, Springer-Verlag, New York.

Caplice, C.G. (1996). *An Optimization Based Bidding Process: A New Framework for Shipper–Carrier Relationships*. Ph.D. Thesis, Massachusetts Institute of Technology.

Chan, L.M.A., A. Muriel, D. Simchi-Levi (1999). *Production/Distribution Planning Problems with Piece-Wise Linear and Concave Cost Structures*. Northwestern University.

Chan, L. M. A., A. Muriel, Z. J. Shen, D. Simchi-Levi (2002). An approximation algorithm for the economic lot sizing model with piece-wise linear cost structures. *Operations Research* 50, 1058–1067.

Chan, L. M. A., A. Muriel, Z. J. Shen, D. Simchi-Levi, C. P. Teo (2002). Effective zero inventory ordering policies for the single-warehouse multi-retailer problem with piecewise linear cost. *Management Science* 48, 1446–1460.

Chan, L.M.A., D. Simchi-Levi, J. Swann (2001). *Effective Dynamic Pricing Strategies with Stochastic Demand*. Massachusetts Institute of Technology.

Chen, Y. F., Z. Drezner, J. K. Ryan, D. Simchi-Levi (2000). Quantifying the Bullwhip effect in a simple supply chain: the impact of forecasting, lead times and information. *Management Science* 46, 436–443.

Chen, X., D. Simchi-Levi (2002a). Coordinating inventory control and pricing strategies with random demand and fixed ordering cost: The finite horizon case. Massachusetts Institute of Technology.

Chen, X., D. Simchi-Levi (2002b). Coordinating Inventory Control and Pricing Strategies with Random Demand and Fixed Ordering Cost: The Infinite Horizon Case. Massachusetts Institute of Technology.

Cook, T. (2000) Creating Competitive Advantage in the Airline Industry. Seminar sponsored by the MIT Global Airline Industry Program and the MIT Operations Research Center.

Cornuéjols, G., M. L. Fisher, G. L. Nemhauser (1977). Location of bank accounts to optimize float: An analytical study of exact and approximate algorithms. *Management Science* 23, 789–810.

Crainic, T.G., A. Frangioni, B. Gendron (2001). Bundle-based relaxation methods for multi-commodity capacitated fixed charge network design problems. *Discrete Applied Mathematics*, 112, 73–99.

Crainic, T. G., G. Laporte (1997). Planning models for freight transportation. *European Journal of Operational Research* 97(3), 409–438.

Crainic, T. G., J. M. Rosseau (1986). Multicommodity, multimode freight transportation: A general modeling and algorithmic framework for the service network design problem. *Transportation Research B: Methodology* 20B, 225–242.

Crainic, T. G., J. Roy (1992). Design of regular intercity driver routes for the LTL motor carrier industry. *Transportation Science* 26(4), 280–295.

Croxton, K. L., B. Gendron, T. L. Magnanti (2000). A comparison of mixed-integer programming models for non-convex piecewise linear cost minimization problems. Publication CRT-2000-31, Centre de recherche sur les transports, Université de Montréal.

Croxton K. L., B. Gendron, T. L. Magnanti (2003). Models and methods for merge-in-transit operations. *Transportation Science* 37, 1–22.

Daskin, M. (1995). *Network and Discrete Location: Models, Algorithms and Applications*, New York, John Wiley & Sons.

DeMartini, C., A. M. Kwasnica, J. O. Ledyard, D. Porter (1999). A New and Improved Design for Multi-Object Iterative Auctions. Working paper 1054, Division of the Humanities and Social Sciences, California Institute of Technology.

Eliashberg, J., R. Steinberg (1991). Marketing-production joint decision making, in: J. Eliashberg, J.D. Lilien (eds.), *Management Science in Marketing*, Volume 5 of *Handbooks in Operations Research and Management Science*, North Holland, Amsterdam.

Erickson, R., C. Monma, Veinott, A., Jr. (1987). Send-and-split method for minimum-concave-cost network flow. *Mathematics of Operations Research* 12, 634–664.

Falk, J., R. Soland (1969). An algorithm for separable nonconvex programming problems. *Management Science* 18, B378–B387.

Farvolden, J. M., W. B. Powell, I. L. Lustig (1993). A primal partitioning solution for multicommodity network flow problem. *Operations Research* 41, 669–693.

Farvolden, J. M., W. B. Powell (1994). Subgradient methods for the service network design problem. *Transportation Science* 28, 256–272.

Federgruen, A., A. Heching (1999). Combined pricing and inventory control under uncertainty. *Operations Research* 47(3), 454–475.

Federgruen, A., C. Y. Lee (1990). The dynamic lot size model with quantity discount. *Naval Research Logistics* 37, 707–713.

Federgruen, A., M. Tzur (1991). A simple forward algorithm to solve general dynamic lot sizing models with n periods in $O(n \log n)$ or $O(n)$ time. *Management Science* 37, 909–925.

Feige, U. (1998). A threshold of $\ln n$ for approximating set cover. *Journal of the ACM* 45(4), 634–652.

Fujishima, Y., K. Leyton-Brown, Y. Shoham (1999). Taming the Computational Complexity of Combinatorial Auctions: Optimal and Approximate Approaches. Working paper, Stanford University.

Gallego, G., G. van Ryzin (1994). Optimal dynamic pricing of inventories with stochastic demand over finite horizons. *Management Science* 40, 999–1020.

Gavish, B. (1991). Topological design of telecommunications networks-local access design methods. *Annals of Operations Research* 33, 17–71.

Gallo, G., C. Sodini (1979). Concave cost minimization on networks. *European Journal of Operations Research* 3, 239–249.

Garey, M. R., D. S. Johnson (1979). *Computers and Intractability*, New York, W.H. Freeman and Company, p. 223.

Gendron, B., T. G. Crainic (1994). Relaxations for multicommodity capacitated network design problems, publication CRT-965, Centre de recherche sur les transports. Université de Montréal, 1994.

Gendron, B., T. G. Crainic, A. Frangioni (1999). Multicommodity capacitated network design, in: B. Sansó, P. Soriano (eds.), *Telecommunications Network Planning*, Kluwer Academic Publishers, Boston, pp. 1–19.

Guisewite, G. M., P. M. Pardalos (1990). Minimum concave-cost network flow problems: applications, complexity, and algorithms. *Annals of Operations Research* 25, 75–100.

Johnson, D. S., J. K. Lenstra, A. H. G. Rinnooy Kan (1978). The complexity of the network design problem. *Networks* 8, 279–285.

Joneja, D. (1990). The joint replenishment problem: new heuristics and worst case performance bounds. *Operations Research* 38(4), 711–723.

Karp, R. M (1972). Reducibility among combinatorial problems, in: R. E. Miller, J. W. Thatcher (eds.), *Complexity of Computer Computations*, New York, Plenum Press, pp. 85–103.

Kay, E. (1998). Flexed pricing. *Datamation* 44(2), 58–62.

Kelly, F., R. Steinberg (2000). A combinatorial auction with multiple winners for universal service. *Management Science* 46(4), 586–596.

Klincewicz, J. G., H. Luss (1986). A lagrangian relaxation heuristic for capacitated facility location with single-source constraints. *Journal of the Operational Research Society* 37, 495–500.

Ledyard, J.O. (2000). A Brief History of Combined Value Auction Mechanisms: Theory and Practice. California Institute of Technology.

Ledyard, J.O., M. Olson, D. Porter, J.A. Swanson, D.P. Torma (2000). The First Use of a Combined Value Auction for Transportation Services. Working paper 1093, Division of the Humanities and Social Sciences, California Institute of Technology.

Lee, C.-Y., S. Çentikaya, W. Jaruphongsa (2000). A Dynamic Model for Inventory Lot-Sizing and Outbound Shipment Consolidation at a Third Party Warehouse. Texas A&M.

Lee, H. L., K. C. So, C. Tang (2000). The value of information sharing in a two-level supply chain. *Management Science* 46(5), 626–643.

Leibs, S. (2000). Ford heads the profits. *CFO The Magazine* 16(9), 33–35.

Leyton-Brown, K., Y. Shoham, M. Tennenholtz (2000). An Algorithm for Multi-Unit Combinatorial Auctions. *Conference Proceedings of the American Association for Artificial Intelligence*. Austin, Texas.

Magnanti, T. L., R. T. Wong (1984). Network design and transportation planning: models and algorithms. *Transportation Science* 18, 1–55.

Melkote, S. (1996). Integrated Models of Facility Location and Network Design. Ph.D. Thesis, Northwestern University.

McGill, J. I., G. J. Van Ryzin (1999). Revenue management: research overview and prospects. *Transportation Science* 33, 233–256.

Minoux, M. (1989). Network synthesis and optimum network design problems: models, solution methods and applications. *Networks* 19, 313–360.

Mirchandani, P. B., R. L. Francis (1990). *Discrete Location Theory*, New York, John Wiley & Sons.

Muriel, A., F. Munshi (2002). Capacitated Multicommodity Network Flows with Piecewise Linear Concave Costs. Working paper, University of Massachusetts, Amherst.

Nauss, R. M. (1976). An efficient algorithm for the 0-1 knapsack problem. *Management Science* 23, 27–31.

Neebe, A. W., M. R. Rao (1983). An algorithm for the fixed-charged assigning users to sources problem. *Journal of the Operational Research Society* 34, 1107–1113.

Petruzzi, N. C., M. Dada (1999). Pricing and the newsvendor model: a review with extensions. *Operations Research* 47, 183–194.

Pirkul, H. (1987). Efficient algorithms for the capacitated concentrator location problem. *Operations Research* 14, 197–208.

Pirkul, H., V. Jayaraman (1996). Production, transportation and distribution planning in a multi-commodity tri-echelon system. *Transportation Science* 30, 291–302.

Porteus, E. (1971). On the optimality of the generalized (s, S) policies. *Management Science* 17, 411–426.

Powell, W. B. (1986). A local improvement heuristic for the design of less-than-truckload motor carrier networks. *Transportation Science* 20(4), 246–257.

Powell, W. B., Y. Sheffi (1989). Design and implementation of an interactive optimization system for network design in the motor carrier industry. *Operations Research* 37(1), 12–29.

Rothkopf, M. H., A. Pekeč, R. M. Harstad (1998). Computationally manageable combinational auctions. *Management Science* 44(8), 1131–1147.

Roundy, R. (1985). 98%-Effective integer-ratio lot-sizing for one-warehouse multi-retailer systems. *Management Science* 31, 1416–1430.

Sandholm, T. (1999). An algorithm for winner determination in combinatorial auctions. *Proceedings of the Sixteenth International Joint Conference on Artificial Intelligence (JCAI)*, Stockholm, Sweden.

Sandholm, T. (2000). Approaches to winner determination in combinatorial auctions. *Decision Support Systems* 28(1–2), 165–176.

Scarf, H. (1960). The optimality of (s, S) policies for the dynamic inventory problem. *Proceedings of the 1st Stanford Symposium on Mathematical Methods in the Social Sciences*, Stanford University Press, Stanford, CA.

Shapiro, J.F. (1998). Bottom-up vs. top-down approaches to supply chain modeling, in: Tayur, Ganeshan, Magazine (eds.), *Quantitative Models for Supply Chain Management*, Kluwer Academic Publishers, Boston, pp. 737–760.

Shaw, D. X., A. P. M. Wagelmans (1998). An algorithm for single-item capacitated economic lot sizing with piecewise linear production costs and general holding costs. *Management Science* 44(6), 831–838.

Shen, Z.J., D. Simchi-Levi, C.P. Teo (2000). Approximation Algorithms for the Single-Warehouse Multi-Retailer Problem with Piecewise Linear Cost Structures. National University of Singapore.

Soland, R. (1971). Optimal facility location with concave costs. *Operations Research* 22, 373–382.

Stenger, A.J., 1994. Distribution resource planning, in: Robeson and Copacino (eds.), *The Logistics Handbook*, Free Press, New York, pp. 391–410.

Stoer, M., G. Dahl (1994). A polyhedral approach to multicommodity survivable network design. *Numerische Mathematik* 68, 149–167.

Tayur, S., R. Ganeshan, M. Magazine (2000). *Quantitative Models in Supply Chain Management*, Kluwer Academic Publishers, Boston.

Thomas, L. J. (1974). Price and production decisions with random demand. *Operations Research* 26, 513–518.

Toth, P., D. Vigo (2001). The Vehicle Routing Problem. *SIAM Monographs on Discrete Mathematics and Applications*.

Van Hoesel, C. P. M., A. P. M. Wagelmans (2001). Fully polynomial approximation schemes for single-item capacitated economic lot-sizing problems. *Mathematics of Operations Research* 26(2), 339–357.

Van Mieghem, J. A., M. Dada (1999). Price versus production postponement: capacity and competition. *Management Science* 45(12), 1631–1649.

de Vries, S., R. Vohra (2000). Combinatorial auctions: A survey. *INFORMS Journal of Computing* 15.

Wagelmans, A., S. Van Hoesel, A. Kolen, (1992). Economic lot sizing – An O(n log n) algorithm that runs in linear time in the Wagner–Whitin case. *Operations Research* 40, S145–S156 Suppl. 1, Jan–Feb 1992.

Wagner, H. M., T. M. Whitin (1958). Dynamic version of the economic lot sizing model. *Management Science* 5, 89–96.

Ward, J. A. (1999). Minimum-aggregate-concave-cost multicommodity flows in strong-series-parallel networks. *Mathematics of Operations Research* 24(1), 106–129.

Whitin, T. M. (1955). Inventory control and price theory. *Management Science* 2, 61–80.

Yaged, B. (1971). Minimum cost routing for dynamic network models. *Networks* 1, 139–172.

Zangwill, W. I. (1968). Minimum concave cost flows in certain networks. *Management Science* 14, 429–450.

A.G. de Kok and S.C. Graves, Eds., *Handbooks in OR & MS, Vol. 11*
© 2003 Elsevier B.V. All rights reserved.

Chapter 3

Supply Chain Design: Safety Stock Placement and Supply Chain Configuration

Stephen C. Graves
Leaders for Manufacturing Program and A. P. Sloan School of Management, Massachusetts Institute of Technology, Cambridge MA 02139-4307, USA
E-mail: sgraves@mit.edu

Sean P. Willems
Boston University School of Management, 595 Commonwealth Avenue, Boston MA 02215, USA; E-mail: willems@bu.edu

1 Introduction

The focus of this chapter is on safety stock placement in the design of a supply chain, as well as on the optimal configuration of the supply chain to minimize total supply chain cost. As our intent is not to cover all of supply chain design, we first need to position this chapter relative to the other work in this handbook on supply chain design. We also need to position our treatment of the safety stock placement problem relative to other chapters that address multi-echelon inventory systems.

There is a great range of decisions associated with the design of a supply chain. One might group the design decisions into three broad categories.

First, there are the traditional decisions of network design as applied to the design of a supply chain. The choice of nodes corresponds to questions about the number, location and sizing of facilities. The choice of arcs corresponds to setting the general logistics strategy in terms of who serves whom and by what transportation or production mode. Muriel and Simchi-Levi cover these models in Chapter 2 of this book.

Second, we mention the decisions that are made in product design that determine the topology, as well as the key economics, of the supply chain. Ideally, one would like to concurrently design the product and its supply chain so as to meet the market objectives for the product with the best performance of the supply chain. Lee and Swaminathan look at the impact of product design decisions on the supply chain, with a particular focus on understanding the tactic of postponement as a way to achieve product proliferation with a well performing supply chain.

Third, we note the design decisions that allow the supply chain to be responsive to uncertainty and variability. In Chapter 4, Bertrand addresses the general question of how to accomplish flexibility in a supply chain, for instance by means of having flexible facilities and/or capacity buffers, as well as through contracting mechanisms. In the current chapter, we examine a completely different tactic, the deployment of inventory as safety stock for addressing demand uncertainty. In particular we look at the strategic placement of safety stocks across a supply chain.

In this chapter, we also introduce a new design consideration for how to configure the supply chain. The configuration decision entails choosing how to source each step or stage in the supply chain, where there might be several options that vary in terms of lead-time and cost. For instance, the configuration decision includes decisions about choice of suppliers for raw materials, choice of transportation modes, and choice of processing options, which might vary in terms of technology and capacity.

As the majority of the chapter is on safety stock placement in a supply chain, there is a close connection between this chapter and the body of literature on multi-echelon inventory systems. In this handbook, there are three chapters that focus, to some degree, on multi-echelon inventory models, namely Axsäter (Chapter 10), Song and Zipkin (Chapter 11) and de Kok and Fransoo (Chapter 12). We see three distinctions between the focus of this chapter and the general literature on multi-echelon inventory systems, as treated in these other chapters.

First, the primary emphasis on the approaches studied in this chapter is in terms of providing decision support for supply chain design, rather than supply chain operation. By this, we mean that the intent is to determine the best overall strategy for deploying safety stock across the supply chain so as to buffer it against demand uncertainty. In particular, we are concerned with questions about where are the best places in the supply chain to position a safety stock, and how much is needed to protect the chain. In contrast with much of the multi-echelon inventory literature, the intent is not to find the inventory control policy for operating the supply chain.

Second, much of the multi-echelon literature focuses on specific network topologies such as serial, assembly or distribution systems. We know, however, that de Kok and Fransoo in Chapter 12 do explicitly describe a multi-echelon algorithm that applies to general network structures. The purpose of this chapter is to consider multi-echelon models that have been specifically designed for optimizing the placement of safety stocks in real-world supply chains. As such, we find that the network topologies of most supply chains are neither an assembly nor distribution system, and thus require different approaches. Admittedly, in order to make progress on these more complex systems, these approaches for safety stock placement require simplifications and at times, strong assumptions. As a consequence, these safety stock models lack some of the rigor found in the literature for multi-echelon systems. But, on the plus side, they have had substantial success in being applied in practice.

Third, we assume that the inventory policies throughout the supply chain just rely on local information and make local decisions in terms of their inventory management and replenishment. In contrast, the models in these other three chapters allow for a central decision maker to coordinate and control the actions at all stages in the supply chain.

The structure of the chapter is to consider first two approaches to safety stock placement, which we term the stochastic-service model and the guaranteed-service model. These two approaches provide an interesting contrast to how one models and analyzes a supply chain for the purposes of setting safety stocks. We then address the issue of how to optimally configure the supply chain. We introduce the notion of options for each stage in the supply chain, where the options differ in terms of lead-time and cost. We show how this work builds on the safety-stock placement models and we formulate an optimization model that finds the best choice of options and safety stock placement to minimize the total supply chain cost. We conclude the chapter with some reflections on this material and its applicability and value to practice, as well as reflections on opportunities for research.

2 Approaches to safety stock placement

In this chapter, we consider two approaches to optimizing safety stock levels in multi-echelon supply chains. Our intent is to compare and contrast these approaches in terms of their underlying assumptions, computational and modeling implications, and the nature of the results produced. Both approaches adopt a network representation of the supply chain, where nodes in the network correspond to stages in the supply chain and arcs denote the precedence relationship between stages. A stage represents a processing or transformation activity in the supply chain. Depending on the scope and granularity of the analysis being performed, the stage could represent anything from a single step in a manufacturing or distribution process to a collection of such steps to an entire assembly and test operation. Regardless of the level of detail chosen by the modeler, a stage corresponds to the material flow of a single item or a single family of items, and each stage is a candidate location for the placement of a safety stock of inventory. When it is necessary to distinguish the safety stocks for different items at the same location, then we need to replicate the stages. For example, if two products flow through a distribution center, we might model each product in the supply chain map by a stage that corresponds to that SKU at that distribution center.

The approaches also assume decentralized control throughout the supply chain. There is no central decision maker that coordinates and controls the actions at all of the stages in the supply chain. Instead, for the purposes of determining safety stocks, we assume that each stage in the supply chain manages its inventory with a simple control policy that takes inputs from adjacent upstream and downstream stages. Thus, in order to be implemented in practice, the final recommendations of the model with regard to safety stocks must be translated into the control policies that are in use throughout the supply chain. As a final note on this issue, saying the supply chain is subject to decentralized control is not equivalent to saying the supply chain is locally optimized. In an optimization context, the models attempt to find the safety stock levels, under the assumption of decentralized control, that minimize the total safety stock cost for the supply chain. For the approaches presented here, this requires global access to information to run the optimization and calculate the system's performance measures. In particular, demand information is passed from finished goods stages through the chain to raw materials and cost and lead-time data is passed in the opposite direction.

The two approaches differ in how they model the replenishment mechanism between stages in the supply chain. We refer to the two approaches as the *stochastic-service model* and *guaranteed-service model*. The stochastic-service model assumes the delivery or service time between stages can vary based on the material availability at the supplier stage. The guaranteed-service model assumes that each stage can quote a delivery or service time that it can always satisfy.

In the stochastic-service model, each stage in the supply chain maintains a safety stock sufficient to meet its service level target. In this setting, a stage that has one or more upstream-adjacent supply stages has to characterize its replenishment time taking into account the likelihood that these suppliers will meet a replenishment request from stock. Because the upstream suppliers will not always meet demand requests immediately from stock, each stage will occasionally experience a delay in obtaining its supplies from its upstream suppliers. Due to this stochastic delay, the replenishment time for the stage is also stochastic, even when the processing time at the stage is deterministic. The inventory level required at each stage to meet its service level target depends on its replenishment time. And the challenge in this work is in how to characterize these replenishment times given that a stage might have multiple upstream suppliers, and given that each of the upstream stages might also be dependent upon unreliable suppliers.

In the guaranteed-service model, each stage provides guaranteed service to its customer stages. In this setting, a supply stage sets a service time to its downstream customer and then must hold sufficient inventory so that it can always satisfy the service-time commitment. A key assumption in this model is to assume that demand is bounded for the purposes of making the service-time guarantee. As a consequence, the service-time guarantee can be accomplished with a finite stock of inventory. The guaranteed nature of these service times assures that the replenishment time for downstream stages is predictable and deterministic. This then allows the downstream stage to plan its inventory so that it can also make a service-time guarantee to its customers. In this work, the challenge is determining the best choice of service times within the supply chain that minimize the total supply-chain inventory and meet the service requirements for the supply-chain's customer.

The stochastic-service and guaranteed-service approaches both require strong assumptions in order to produce tractable models. The stochastic-service model assumes that the system behaves the same under all demand conditions. That is, each stage reacts in a predictable way whether there is ample inventory or there is a stock-out, and inventory is the only countermeasure available to deal with demand and supply uncertainty in the supply chain. If one were to make an analogy to a checkout clerk in a grocery store, the stochastic-service model assumes that the clerk behaves the same way if the line is one person or fifty people.

The guaranteed-service model makes an equally strong assumption. In order to provide guaranteed service, the guaranteed-service model assumes that the safety stock policy is only being designed to meet some portion of the demand, as specified by the demand bound. When demand exceeds the bound, the model does not attempt to address how the system will react. In effect, the guaranteed-service model assumes that inventory is held to handle some nominal level of uncertainty and that other responses or tactics are available to address demand or supply uncertainty beyond this nominal level. Continuing the grocery-checkout analogy, if the system were designed to process a maximum of 20 customers in a one-hour interval, then when 25 customers show up in an hour, the model does not say how exactly the additional customers would be served. In effect, the model just assumes outside measures are adopted to serve these customers in the specified time frame. (We note here that the control framework proposed in Chapter 12 assumes that other countermeasures are applied when planned lead times are threatened.)

The next two sections discuss in more detail the papers that have appeared in both streams of work.

2.1 Stochastic-service model approach

Lee and Billington (1993) develop a multi-echelon inventory model to reflect the decentralized supply chain structure they witnessed in Hewlett-Packard's DeskJet printer supply chain. Their goal was to produce a model that manufacturing and materials managers could use to evaluate different strategic decisions involved with the creation of a new-product supply chain. They model a supply chain as a collection of SKU-locations where each stage in the supply chain accepts as an exogenous input a service level target or a base stock policy. In the case where service level targets were inputs, the authors develop a single-stage base-stock calculation that, while approximate, is tractable. The single-stage base-stock level is a function of the replenishment lead-time at the stage, which includes the production lead-time, plus the effects from production downtime and random delays due to component shortages. Lee and Billington show how to propagate the single-stage model to multiple stages by developing expressions for the random delays induced on downstream stages from shortages from the base-stock policy of upstream stages.

Ettl, Feigin, Lin and Yao (2000) also consider a supply chain context that is quite similar in spirit to the work of Lee and Billington (1993). The single-stage base-stock model in Ettl et al. (2000) makes a distinction between the *nominal* lead-time a stage quotes and the *actual* lead-time the stage experiences. The actual lead-time will exceed the nominal lead-time when there is a stock-out at a supplier. The authors develop an approximate characterization of the random variable for the actual lead-time. This approximation is based on assuming that at most one supplier is out of stock at any time instant, and then determining the stock-out probability for each supplier, given their service targets and this assumption. The authors use an $M/M/\infty$ model of the supplier's replenishment process to develop a bound on the expected delay induced by a stock-out. Weighting these delays by the stock-out probabilities for each supplier, and combining with the nominal lead-time provides the characterization of the actual lead-time for a single stage. Given this lead-time, a base-stock level is determined to assure a given service target for the stage. As in the case of Lee and Billington (1993), the single-stage model extends immediately to a multiple-stage supply chain. Indeed, given service level targets for every stage in the supply chain, it is possible to decompose the performance analysis of a multiple-stage system into the analysis of a series of single-stage base-stock systems.

In addition to performance analysis, Ettl et al. (2000) go on to place their supply chain model into an optimization context. The authors' objective function is to minimize the total inventory investment in the supply chain, defined as work-in-process inventory plus safety stock inventory. The decision variables are the safety factor (or service level) at each stage. The authors then develop expressions for the partial derivative of the objective function with respect to the safety factors. This formulation allows the authors to solve the resulting nonlinear programming problem using conjugate gradient methods.

Glasserman and Tayur (1995) consider a context very similar to that of Lee and Billington (1993) and Ettl et al. (2000) but go on to introduce capacity limits into their multi-echelon model. The introduction of production capacity requires each stage to operate a modified base-stock policy where at each period the stage will order the minimum of its capacity and the amount to bring its inventory position back to the base-stock level. The problem formulation of Glasserman and Tayur (1995) follows the framework of Clark and Scarf (1960) with the addition of capacity. The authors first develop recursions for stage inventories, production levels, and pipeline inventories. The authors develop estimates of the derivatives of the inventory requirements with respect to the base-stock levels, based on an infinitesimal perturbation analysis. They use these estimates to generate the gradient of the cost function, with which they can conduct a gradient-based search to find the optimal base-stock policy.

2.2 Guaranteed-service model approach

The guaranteed service-time approach traces its lineage back to the 1955 manuscript, which was later reprinted in 1988 (Kimball, 1988). In that paper, Kimball describes the mechanics of a single stage that operates a base-stock policy in the face of random but bounded demand. In particular, beyond the deterministic production time assumed at the stage, there is an incoming service time that represents the delivery time quoted from the stage's supplier and an outgoing service time representing the delivery time the stage quotes to its customer. Kimball further assumes that demand over any interval of time is bounded. Given this characterization, the base-stock level at the stage is set equal to the maximum demand over the net replenishment time, which is defined as the incoming service time plus the production time minus the outgoing service time.

Simpson (1958) develops a model to determine the optimal safety stocks in a serial supply chain. Simpson uses Kimball's work as the building block, coupling adjacent stages together through the use of service time. In particular, the incoming (or inbound) service time of a downstream stage is equal to the outgoing (or outbound) service time of the upstream stage, namely its supplier. The optimal stocking locations in the supply chain can then be found by determining the optimal service times in the supply chain. Simpson proves that for a serial supply chain the optimal service times satisfy an extreme point property where the outgoing service time at a stage will equal either zero or its incoming service time plus its production time. In terms of inventory, the optimal policy is an 'all or nothing' policy, in which a stage either has no safety stock or carries a decoupling safety stock, namely enough stock to decouple the downstream stages from the upstream stages. Simpson suggests an enumeration procedure to find the optimal service times.

Simpson also provides a rich interpretation for the bounded demand process. Rather than saying that bounded demand reflects the maximum demand the stage will see, the bound can instead reflect the maximum amount of demand the company wants to satisfy from safety stock. Under this interpretation, when demand exceeds the bound, the stage will have to resort to extraordinary measures, like expediting and overtime, to meet the demand requirement by means other than using safety stock.

Graves (1988) observes that the serial-line problem, as formulated by Simpson (1958), can be solved as a dynamic program. Inderfurth (1991), Inderfurth and Minner (1998), Graves and Willems (1996, 2000) extend Simpson's work to supply chains modeled as assembly networks, distribution networks, and spanning trees. In each case, the optimization problem is still to determine the service times that minimize the total cost for safety stock in the supply chain. The challenge is to determine an efficient approach to traverse the state space of the dynamic program. The definition of service time must also be expanded to include the cases where a stage can see more than one upstream or downstream stage. In the case of multiple upstream stages feeding a downstream stage, the papers assume that the downstream stage has to wait until the item with the longest service time arrives. In the case of an upstream stage supplying multiple downstream stages, the papers assume the upstream stage quotes the same service time to all of its adjacent downstream stages.

3 Model formulation

The single-stage base-stock policy is the common building block for all of the papers in this chapter. The differences in the approaches deal with how adjacent stages interact with one another and the assumptions about the operating behavior of the individual stages. This section develops both the underlying base-stock equations and the resulting multi-echelon problem formulations for both the guaranteed and stochastic-service models.

The goal of this section is to distill the two models into their simplest elements so that the reader can clearly see the similarities and differences between the two approaches. The goal is not to replicate the contents of the papers surveyed in the literature review nor is it to provide a careful development of the features of each model. Rather, the intent is to give a self-contained development that highlights the essence of each approach. We refer the reader to the specific papers for the critical details and refinements that are necessary for the successful implementation of each model.

To get started, we note the key similarities of the two approaches. In each model, each stage in the supply chain operates according to a base-stock policy. In each period, the stage observes demand and places a replenishment order on its suppliers equal to the observed demand. There are no capacity constraints. There is a common underlying review period for all stages in the supply chain. Demand is stationary and independent across nonoverlapping intervals, with mean demand per period of μ and a standard deviation of σ. Associated with each stage is a deterministic processing time (or lead-time) that includes all of the time required to transform the item at the stage. Once all of the stage's required inputs are available, the processing time includes any waiting time, manufacturing time and transportation time at the stage.

3.1 Stochastic-service model

In the stochastic-service model, each stage sets its base stock to meet a service level target, i.e., an upper bound on the probability that a stage is out of stock in any period and thus cannot meet customer demand directly from stock on hand. The service level target for external customers is an exogenous input to the model, usually dictated by market conditions. The service level target for internal customers is an input when the model is used for performance evaluation; alternatively, the service level target for internal customers is a decision variable when the model is placed within an optimization. We say that a stage provides stochastic service because a demand order will receive immediate service when stock is on hand, but will be subject to a random delay when the stage is out of stock.

The replenishment time at a stage equals the processing time at the stage plus any delay from upstream stages. If we denote the replenishment time at stage j as a random variable τ_j, the processing time as a constant L_j, and the delay for supplier i as a random variable Δ_i, then the replenishment time at stage j equals:

$$\tau_j = L_j + \max_{i:(i,j)\in A} \{\Delta_i\} \tag{3.1}$$

where A is the set of directed arcs in the network representation of the supply chain.

In the worst case, this delay might equal the entire replenishment time from its slowest supplier, e.g.,

$$\tau_j = L_j + \max_{i:(i,j)\in A} \{\tau_i\} \tag{3.2}$$

The development of an exact characterization of τ_j is extremely challenging. To illustrate, with N suppliers, there are 2^N-1 combinations of suppliers that might be out of stock in any period. For each stage that is out of stock, determining its delay requires considering where its first unallocated unit is in its replenishment process. Finally, there are the multi-echelon ramifications when a supplier's supplier is out of stock. As a consequence, one must make some simplifications to make the analysis of this model more tractable. We describe an approach here, which is loosely based on the development in Ettl et al. (2000).

For purposes of illustration, we assume that at most one supplier will stock out per period and the delay will equal the supplier's processing time. (This assembly assumption is also discussed in Chapter 12.) This allows us to express the expected replenishment time at stage j as:

$$E[\tau_j] = L_j + \sum_{i:(i,j)\in A} \pi_{ij} L_i \tag{3.3}$$

where π_{ij} is the probability that in a period stage i is causing a stock-out at stage j. Ettl et al. (2000) use this form of equation for the expected replenishment time, but use a bound on the expected delay, rather than the supplier's processing time as we have done in (3.3). They derive the bound on the expected delay by means of applying an M/M/∞ model to the supplier's replenishment process.

We assume the demand over the replenishment time is normally distributed with mean $\mu_j E[\tau_j]$ and with standard deviation $\sigma_j \sqrt{E[\tau_j]}$. We assume that the base stock is given by $B_j = \mu_j E[\tau_j] + k_j \sigma_j \sqrt{E[\tau_j]}$, where k_j is the safety factor necessary to achieve the service level target for the stage.

To determine the expected replenishment time, given by (3.3), we need to find an expression for π_{ij}. Ettl et al. (2000) propose the following calculation for π_{ij}:

$$\pi_{ij} = \frac{1 - \Phi(k_i)}{\Phi(k_i)} \left(1 + \sum_{h:(h,j) \in A} \frac{1 - \Phi(k_h)}{\Phi(k_h)} \right)^{-1} \tag{3.4}$$

where k_i denotes the safety factor at stage i and $\Phi(k_i)$ represents the cumulative distribution function for a standard normal random variable. In (3.4), the term in parentheses acts to normalize the probability that a stock-out does or does not occur, and the first part of the expression calculates the fraction of occurrences that are attributable to stage i.

Given the assumptions of normally distributed demand over the replenishment lead-time, we follow the development in Ettl et al. (2000) to get the following expression for the expected on-hand inventory for stage j:

$$E[I_j] = k_j \sigma_j \sqrt{E[\tau_j]} + \sigma_j \sqrt{E[\tau_j]} \int_{z=k_j}^{\infty} (z - k_j) \phi(z) \, dz \tag{3.5}$$

where τ_j is defined by (3.3) and (3.4), and $\phi()$ is the probability density function for a standard normal. The first term is the expected inventory level at stage j, equal to the base stock level net the expected demand. Since the on-hand inventory level cannot be negative, we need to augment the first term with the second term, which corresponds to the expected number of shortages or backorders.

We can now develop an expression for the total safety stock cost across the supply chain. We let C_j^S denote the per unit holding cost of safety stock at stage j. C_j^S is typically determined by multiplying the cumulative cost of the product at stage j by a holding cost rate. Given this cost characterization, we let $\mathbf{C}^{\text{ssm}}$ denote the total safety stock cost of the stochastic-service model. Then,

$$\mathbf{C}^{\text{ssm}} = \sum_{j=1}^{N} C_j^S \sigma_j \sqrt{E[\tau_j]} \left(k_j + \int_{z=k_j}^{\infty} (z - k_j) \phi(z) \, dz \right) \tag{3.6}$$

We can use (3.6) for performance evaluation in a supply chain, namely to find the inventory requirements and costs for a given set of service level targets or safety factors. We can also place (3.6) in an optimization context, where the objective is to minimize safety stock cost and the decision variables are the service level targets, or equivalently the safety factors, at each stage in the supply chain.

3.2 Guaranteed-service model

In the guaranteed-service model, each stage sets its base stock so as to guarantee that it can meet its service-time commitment to its customers. That is, each stage will quote a guaranteed service or delivery time to its downstream customers, who know that this commitment will be met with certainty. The service time for external customers is an exogenous input, just as with the service level target for the stochastic-service model. The service time for internal customers can be either an input or a decision variable, depending upon whether the model is being used for performance evaluation or for optimization.

As we have noted earlier, the guarantee applies to a bounded demand process. We specify for each stage j a function $D_j(t)$ that represents the maximum demand over t consecutive periods for which we will guarantee the service commitment. For each stage j, the model finds the base stock that satisfies the stage's service time commitment, provided that the demand time series is always within the demand bound given by $D_j(t)$. In a typical application, similar to the stochastic-service model, one might assume that actual demand at stage j is normally distributed with mean demand per period of μ_j and a standard deviation of σ_j. Then a common way to set the demand bound is as follows:

$$D_j(t) = t\mu_j + k_j\sigma_j\sqrt{t},$$

where k_j is a given safety factor. When demand exceeds the demand bound, then the safety stock in the system will not be adequate to assure the service times. We assume that in this case of extraordinary demand, some correspondingly extraordinary measures are taken to augment the safety stock so that the demand can be served. Alternatively, one might view demand in excess of this bound as being lost or somehow being served from another source.

For the guaranteed-service model, the replenishment time at a stage does not drive base stock requirements, but, rather, it is the net replenishment time that is of importance. In order to understand net replenishment time, we first define the concept of service time. Service time is the amount of time that elapses between when a downstream stage places an order on an upstream stage and when the order is delivered by the upstream stage to the downstream stage and is available to begin processing at that stage. Each stage in the supply chain quotes a service time to its downstream (customer) stages and it is quoted service times from its upstream (supplier) stages. We describe the service time that stage j quotes its customers as the outbound service time, denoted by s_j^{out}. The inbound service time at stage j is denoted by s_j^{in}. Since stage j cannot start its processing activities until it receives all of the required inputs, we can state the inbound service time at stage j in terms of the outbound service times for its suppliers:

$$s_j^{in} = \max_{i:(i,j) \in A} \{s_i^{out}\}. \qquad (3.7)$$

Now, for the guaranteed-service model the replenishment time at stage j is:

$$\tau_j = s_j^{in} + L_j. \qquad (3.8)$$

Since both the service time and processing time (by assumption) are deterministic constants, we have that the replenishment time for this model is also deterministic. The net replenishment time for stage j is the replenishment time minus the stage's outbound service time, i.e., $s_j^{in} + L_j - s_j^{out}$. In this model, we set the base stock for each stage to cover the maximum demand over its net replenishment time, as will be shown next.

We assume each stage j starts at time 0 with initial inventory $I_j(0) = B_j$. Given the assumptions of guaranteed service and the definition of the service times, the inventory at time t, $I_j(t)$, equals

$$I_j(t) = B_j - \sum_{v=0}^{t-s_j^{out}} d_j(v) + \sum_{w=0}^{t-L_j-s_j^{in}} d_j(w), \qquad (3.9)$$

where $d_j(t)$ denotes the demand in period t. In period t, stage j completes into its inventory the replenishment order that was placed in period $t - L_j - s_j^{in}$. Correspondingly, in period t, stage j must serve the replenishment orders placed by its customers in period $t - s_j^{out}$. We can simplify (3.9) as,

$$I_j(t) = B_j - \sum_{v=t-L_j-s_{j+1}^{in}}^{t-s_j^{out}} d_j(v). \qquad (3.10)$$

In order to satisfy the service-time guarantee, we need to set the base stock B_j so that the inventory on hand $I_j(t)$ is always non-negative. That is, we will want to set

$$B_j \geq \sum_{v=t-L_j-s_{j+1}^{in}}^{t-s_j^{out}} d_j(v). \qquad (3.11)$$

In words, we need for the base stock to equal (or exceed) the maximum possible demand over the net replenishment time. But given the assumption of a bounded demand process, then we can set $B_j = D_j(s_j^{in} + L_j - s_j^{out})$ and be assured that Eq. (3.11) holds.

For illustration, assume we set the demand bound as given earlier. Then we choose $B_j = (s_j^{in} + L_j - s_j^{out})\mu_j + k_j\sigma_j\sqrt{s_j^{in} + L_j - s_j^{out}}$. We can immediately find that the expected inventory on-hand at stage j equals:

$$E[I_j] = k_j\sigma_j\sqrt{s_j^{in} + L_j - s_j^{out}}. \tag{3.12}$$

We can use (3.12) to determine the inventory requirements for a given setting of the service times in a supply chain. We can also incorporate (3.12) into an optimization to find the best choice of service times. The objective function for the optimization could be the total holding cost for safety stocks, which we denote by $\mathbf{C}^{gsm}$ and state as:

$$\mathbf{C}^{gsm} = \sum_{j=1}^{N} C_j^S k_j \sigma_j \sqrt{s_j^{in} + L_j - s_j^{out}}. \tag{3.13}$$

In the optimization model, one minimizes the objective (3.13) with the decision variables being the service times at the stages in the supply chain and subject to constraints (3.7) to relate the inbound to the outbound service times, and non-negativity constraints on the net replenishment times.

4 Heavy industry and consumer packaged goods example

In this section we apply the two approaches for safety stock placement to two examples. Our intent is two-fold. First, we wish to show the applicability of these approaches to different industries. The examples presented in Lee and Billington (1993); Ettl et al. (2000); Graves and Willems (2000) are all drawn from the high-technology industry. Here we will present examples from two other industries, heavy industry and consumer packaged goods, to demonstrate the characteristics of their supply chains and the differences in the structure of their optimal solutions. Our second purpose is to illustrate how the results of the two approaches can differ, and then to discuss the implications for implementing these models.

4.1 Bulldozer assembly and manufacturing

In this section we present the assembly and manufacturing process for a bulldozer. Figure 1 presents the bulldozer's supply chain map.

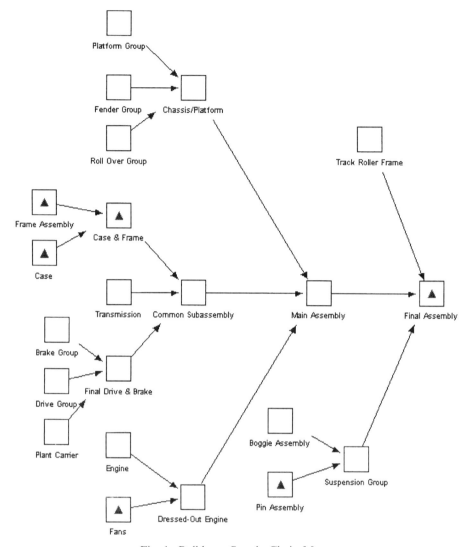

Fig. 1. Bulldozer Supply Chain Map.

At a high level, a bulldozer is put together in three operations. In the common subassembly step, the transmission, drivetrain, and brake system are attached to the case and frame. The main assembly process then installs the chassis and the engine to the common subassembly. In final assembly the track and suspension are installed. A real-world map of this supply chain exceeds 1000 stages with many of the stages shown in Fig. 1 spawning their own large supply chains. Whereas we have combined many stages in order to present the supply chain, the structure of the supply chain accurately represents the manufacturing process and flow for bulldozers. In addition, we have

Table 1
Parameters for Bulldozer Supply Chain

Stage name	Nominal time	Stage cost ($)
Boggie assembly	11	575
Brake group	8	3850
Case	15	2200
Case & frame	16	1500
Chassis/platform	7	4320
Common subassembly	5	8000
Dressed-out engine	10	4100
Drive group	9	1550
Engine	7	4500
Fans	12	650
Fender group	9	900
Final assembly	4	8000
Final drive & brake	6	3680
Frame assembly	19	605
Main assembly	8	12,000
Pin assembly	35	90
Plant carrier	9	155
Platform group	6	725
Roll over group	8	1150
Suspension group	7	3600
Track roller frame	10	3000
Transmission	15	7450

made two major simplifications to the supply chain. First, we ignore the customization process in this analysis; in effect, we are modeling the supply chain for a stock bulldozer that will be modified at the dealer. Second, we have not modeled all of the different variations the bulldozer can come in. Besides adding complexity to the example, after adding the different variations, the network is no longer a spanning tree, and thus requires a solution technique presented in Humair and Willems (2003). Table 1 provides the cost and nominal lead-time information for the supply chain.

The cumulative cost of the product is already $28,990 when the common subassembly stage is complete. The chassis/platform and engine subcomponents each contribute $7095 and $9250, respectively, to the cost of the product. The total cost for the bulldozer, upon completion of final assembly, is $72,600. Because of the lean manufacturing initiative at the company, lead-times are quite low given the complexity of the product. The average daily demand is 5 and the daily standard deviation is 3. Assuming 260 days per year, the cost of goods sold is $94,380,000. The company applies an annual holding cost rate of 30% when calculating inventory costs.

For the guaranteed-service model, we set the demand bound to correspond to the 95th percentile of demand, and thus, set the safety factor as $k = 1.645$. Figure 1 graphically presents the optimal solution to the guaranteed-service model; a triangle within a stage designates that the stage

Table 2
Optimal Service Times and Safety Stock Costs under Guaranteed-Service Model

Stage name	Service time	Stage safety stock cost ($)
Boggie assembly	11	0
Brake group	8	0
Case	0	12,614
Case & frame	15	6373
Chassis/platform	16	0
Common subassembly	20	0
Dressed-out engine	20	0
Drive group	9	0
Engine	7	0
Fans	10	1361
Fender group	9	0
Final assembly	0	607,969
Final drive & brake	15	0
Frame assembly	0	3904
Main assembly	28	0
Pin assembly	21	499
Plant carrier	9	0
Platform group	6	0
Roll over group	8	0
Suspension group	28	0
Track roller frame	10	0
Transmission	15	0

holds a safety stock. Since the underlying network is a spanning tree with 22 nodes and 21 arcs, we can optimize the network with the algorithm from Graves and Willems (2000). The resulting optimal service times and holding costs for the safety stock are displayed in Table 2.

The optimal inventory policy does not demonstrate the clear decoupling policy that one often sees in the guaranteed-service model. There is a large safety stock at final assembly, which is necessary to provide immediate service to the distribution department of the company. The safety stock at final assembly is sized to cover the demand variability over the net replenishment time for the stage of 28 days. The remaining stages, for the most part, carry no safety stock; the exceptions are a few long lead-time stages where safety stock is held so as to keep the net replenishment time for final assembly to 28 days. In total, the annual holding cost for the safety stock in the supply chain is $633,000.

To understand the solution better, we repeated the optimization but with a constraint that forced the common subassembly to have a service time of zero and thus to hold a safety stock. A priori one might suspect that having a safety stock at the common subassembly stage would lead to a good if not optimal solution, as this would seem to be a logical point to decouple the chain. From the resulting optimization, we found that the chassis/platform and dressed-out engine stages also quote service times of zero and are thus decoupling points. However, the annual holding cost for safety stock increases by nearly 10%, from $633,000 to $693,000. In contrast to our experience with supply chains for high-tech products, we observe that the incredibly expensive nature of the components and the relatively short lead-times make it uneconomical to develop local decoupling points.

We begin the analysis of the stochastic-service model by determining the range of allowable service targets per stage. To maintain consistency with the presentation of the guaranteed-service model, we again assumed a 95% service level at the final assembly stage. For the other stages, the service level is a parameter that is to be set or serve as a decision variable in an optimization. We need for the resulting service levels to be consistent with the assumptions that were made in the development of (3.3), our calculation of the expected replenishment time. In particular, we assumed that for each stage, at most one of its supplier stages stocks out in any period. In order to make this assumption operational, we set an upper bound on the probability that two or more of a stage's suppliers stock out in a period, namely 0.10. Thus, we restrict the choice of service levels so that the probability that a stage has two or more suppliers out of stock is no more than 0.10.

In the bulldozer supply chain, stages have between one and three suppliers. For stages with three suppliers, we impose a lower bound of 0.80 on the service level for each supplier. For stages with two suppliers, we impose a lower bound of 0.68 on the service level for each supplier. As justification for these lower bounds, we observe that with an assumption that the stock-out events of the suppliers are independent, then setting the service levels to these lower bounds results in the stage having a probability of 0.90 that at most one supplier is out of stock. There is no claim that this is the best way to implement the stochastic-service model; rather, we argue that this seems a reasonable way to proceed with the model based on the assumptions that underlie its development, and given the purposes of this chapter. Finally, we will also use 68% as the lower bound on the service level for the case of a sole supplier.

In Table 3 we report the results for the stochastic-service model when the service level for each stage is set to its lower bound. The table displays the expected lead-time and annual holding cost for the safety stock for each stage in the bulldozer supply chain. We cannot guarantee that this is the best solution for the stochastic-service model for this example. However, we did conduct an extensive grid search over the service levels and found the lower bounds on the service levels always to be binding.

As expected, every stage carries a safety stock sufficient to cover the expected lead-time. On a percentage basis, two types of stages have expected lead-times that differ significantly from their nominal times. First, there are those stages that have short nominal lead-times; final assembly is an

Table 3
Nominal and Expected Lead-times for the Stochastic-Service Model

Stage name	Nominal lead-time	Service level (%)	Expected lead-time	Stage safety stock cost ($)
Boggie assembly	11	68	11.00	1160
Brake group	8	80	8.00	9342
Case	15	68	15.00	5181
Case & frame	16	80	24.24	18,184
Chassis/platform	7	80	10.29	19,521
Common subassembly	5	80	10.29	79,764
Dressed-out engine	10	80	14.61	30,328
Drive group	9	80	9.00	3989
Engine	7	68	7.00	7240
Fans	12	68	12.00	1369
Fender group	9	80	9.00	2316
Final assembly	4	95	7.57	299,472
Final drive & brake	6	80	9.71	24,693
Frame assembly	19	68	19.00	1604
Main assembly	8	80	11.14	164,194
Pin assembly	35	68	35.00	324
Plant carrier	9	80	9.00	399
Platform group	6	80	6.00	1524
Roll over group	8	80	8.00	2791
Suspension group	7	80	18.15	15,589
Track roller frame	10	80	10.00	8139
Transmission	15	80	15.00	24,754

example of this kind of stage. Second, there are stages with suppliers that have long nominal lead-times; the suspension group is an example of this kind of stage.

Whereas the safety stock in final assembly is now much less than in the case of the guaranteed-service model, overall we find that there is about 12% more inventory cost with the stochastic-service model. In the stochastic-service model, safety stock is the single countermeasure to address the demand variability in the supply chain. In the guaranteed-service model, safety stock is used to protect against demand variability up to the demand bound. The model assumes that other countermeasures, including expediting and overtime, are utilized when demand exceeds the demand bound. While the guaranteed-service model does not quantify the cost of these other countermeasures, understanding the gap between the two models presented gives some indication of the relative benefit of using only safety stock as a countermeasure versus other operational tactics. Figure 2 displays the total annual holding cost as a function of the service level for the external customer for two different types of policies for each model.

The top two lines represent what one could reasonably consider being the base case for each of the models. For the stochastic-service model, each stage maintains a service level equal to final assembly's service level. For the

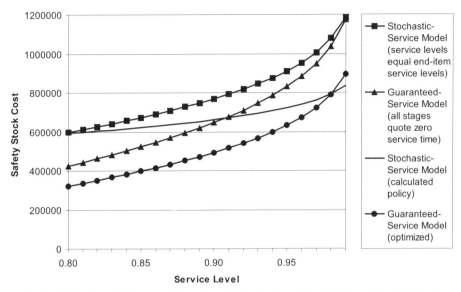

Fig. 2. Safety Stock Cost as a Function of Service Level in Bulldozer Supply Chain.

guaranteed-service model, each stage quotes a zero service time. For each model, the base case has each stage holding significant safety stock so as to decouple it from the other stages in the supply chain. For extremely high service levels, the two models are virtually identical, which is not a surprise given the assumptions in the models. As service levels decrease, there is a difference in cost between the models, due to the fact that the replenishment times for the stochastic-service model increase due to delays from the suppliers. In this example, an increase in the replenishment time at many stages has a large impact on the stage's inventory requirement, due to the expensive nature of the product and the relatively short nominal lead-times.

The lower two lines represent the best inventory policy identified for the stochastic-service model and the optimal inventory policy for the guaranteed-service model. The difference between the two policies allows the manager to quantify the cost of using countermeasures other than inventory. For example, at the 95% service level the cost difference is $80,000 and at 85% service the difference is $224,000; Table 4 provides the numerical values for each of the four policies.

Table 4 also helps determine the appropriate demand bound. These calculations allow us to trade off the cost of safety stock against the cost of other tactics, like expediting and subcontracting, which can also be employed to satisfy demand. If these other tactics are cheaper than holding the higher levels of safety stock, then it makes financial sense to adopt a safety stock policy that only meets 95%, or less, of the possible demand realizations.

Table 4
Safety Stock Cost as a Function of Service Level in Bulldozer Supply Chain

Service level	Stochastic-service model (service levels equal end-item service levels)	Guaranteed-service model (all stages quote zero service time)	Stochastic-service model (calculated policy)	Guaranteed-service model (optimized)
0.80	596,618	425,062	593,788	323,743
0.81	611,063	443,382	599,044	337,697
0.82	625,931	462,306	604,532	352,110
0.83	641,281	481,902	610,279	367,035
0.84	657,215	502,252	616,330	382,534
0.85	673,780	523,452	622,711	398,680
0.86	691,037	545,616	629,459	415,562
0.87	709,091	568,885	636,623	433,284
0.88	728,073	593,428	644,268	451,977
0.89	748,150	619,459	652,469	471,803
0.90	769,527	647,249	661,322	492,969
0.91	792,481	677,150	670,952	515,742
0.92	817,409	709,633	681,531	540,483
0.93	844,880	745,350	693,297	567,686
0.94	875,651	785,240	706,570	598,068
0.95	911,043	830,735	721,877	632,719
0.96	953,156	884,186	740,048	673,429
0.97	1,006,022	949,896	762,629	723,477
0.98	1,078,523	1,037,248	792,955	790,007
0.99	1,185,142	1,174,924	836,583	894,866

4.2 Battery manufacturing and distribution

Figure 3 presents a supply chain map for a single battery product line. This supply chain depicts the manufacturing and packaging process for one size of battery that is sold in three regions in three types of packaging. In this setting, the battery manufacturing stage represents the manufacturing of a single size like AAA, AA, C, or D. The battery size is produced in a single bulk manufacturing facility. Finished batteries are then sent to three pack locations that produce specialty battery packages. For example, the package that comprises an end-item SKU is distinguished by the number of batteries included, the artwork on the package, and the inclusion or exclusion of items like RFID tags, hangers, bar codes, and price labels. Each SKU is sent to the company's three distribution centers (DCs) in the United States for distribution to regional markets. The nominal lead-time and direct cost added for each stage are displayed in Table 5. Table 5 demonstrates the commodity nature of the business. Materials have relatively short lead-times and the cost per item is extremely low. Whereas the cumulative cost of a bulldozer is $72,600, the total unit cost of a battery is less than one dollar. Indeed, the material and process cost to package the battery is on the same order of magnitude as the cost of the battery.

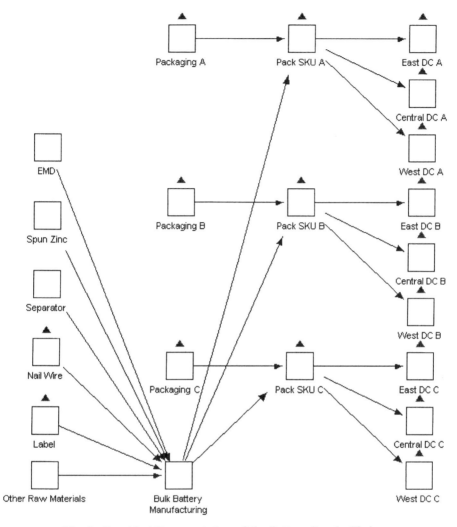

Fig. 3. Graphical Representation of the Battery Supply Chain.

The company's holding cost rate is 25%. The daily demand parameters are given in Table 6. Assuming 260 days per year, the expected cost of goods sold per year is $53,569,000. On a daily basis the demand is highly variable with the coefficient of variation being more than two in several instances. Clearly, the daily demand of a SKU at a DC is not well modeled as being from a normal distribution. Nevertheless, we note that for each model we are effectively assuming that the demand over a stage's replenishment time is normally distributed, which seems more plausible given the range of nominal lead-times. Furthermore, for all of the upstream stages, the

Table 5
Parameters for Battery Supply Chain

Stage name	Nominal time	Stage cost ($)
Bulk battery manufacturing	5	0.07
Central DC A	6	0.02
Central DC B	6	0.01
Central DC C	4	0.01
East DC A	4	0.00
East DC B	4	0.01
East DC C	4	0.01
EMD	2	0.13
Label	28	0.06
Nail wire	24	0.02
Other raw materials	1	0.24
Pack SKU A	11	0.07
Pack SKU B	11	0.12
Pack SKU C	9	0.24
Packaging A	28	0.16
Packaging B	28	0.24
Packaging C	28	0.36
Separator	2	0.02
Spun zinc	2	0.05
West DC A	5	0.01
West DC B	8	0.03
West DC C	6	0.06

demand is pooled over several regions and, except for packaging, over all SKUs. Hence, the assumption of normality seems defensible for the purposes at hand.

For the guaranteed-service model, we set the demand bound again to correspond to the 95th percentile of the demand process. Figure 3 graphically displays the optimal safety stock locations when the guaranteed-service model is optimized; a triangle within a stage denotes safety stock being held at the stage. The resulting optimal service times are displayed in Table 7. The interesting result here is that the bulk manufacturing facility does not hold any safety stock. Instead the three packing locations are the decoupling points in the supply chain. The intuition is that the packing locations are able to pool the demand variability for the three DCs and also pool the variability over the lead-time from the bulk manufacturing plant. This is more cost effective than holding inventory at the bulk plant but then having the regional DCs holding a safety stock that covers not only their lead-time but the pack lead-times as well. The optimal annual holding cost for safety stock is $853,000.

For the stochastic service method, we have a 95% service level target for each SKU at each regional DC. As we did with the bulldozer example, we assess a lower bound on the service level provided by each supplier to a stage, where the lower bound depends on the number of suppliers. In the battery

Table 6
Demand Information for Battery Supply Chain

Stage name	Mean demand	Standard deviation of demand
Central DC A	43,422	67,236
Central DC B	16,350	39,552
Central DC C	5536	11,213
East DC A	67,226	109,308
East DC B	15,765	34,079
East DC C	6416	14,125
West DC A	65,638	119,901
West DC B	10,597	23,277
West DC C	3519	6576

Table 7
Optimal Service Times using Guaranteed-Service Model

Stage name	Service time	Stage safety stock cost ($)
Bulk battery manufacturing	7	0
Central DC A	0	56,889
Central DC B	0	38,245
Central DC C	0	11,066
East DC A	0	73,716
East DC B	0	26,907
East DC C	0	13,940
EMD	2	0
Label	2	23,361
Nail wire	2	7163
Other raw materials	1	0
Pack SKU A	0	251,253
Pack SKU B	0	94,741
Pack SKU C	0	37,573
Packaging A	7	52,953
Packaging B	7	25,852
Packaging C	7	13,022
Separator	2	0
Spun zinc	2	0
West DC A	0	91,507
West DC B	0	26,531
West DC C	0	8279

supply chain, six suppliers supply the bulk battery manufacturing plant. To maintain a probability of 0.90 that at most one stage will be out of stock in a period, each of the six supplier stages must have a service level of at least 0.91. The bulk battery is combined with packaging at each pack location. We set the lower bound on the service level for these two inputs for the pack location to be 0.68. Since the three pack locations are themselves

Table 8
Expected Lead-times and Safety Stock Costs for the Stochastic-Service Model

Stage name	Nominal lead-time	Service level (%)	Expected lead-time	Stage safety stock cost ($)
Bulk battery manufacturing	5	68	8.66	54,467
Central DC A	6	95	6.55	60,191
Central DC B	6	95	6.55	40,465
Central DC C	4	95	4.32	11,649
East DC A	4	95	4.55	79,616
East DC B	4	95	4.55	29,060
East DC C	4	95	4.32	14,674
EMD	2	91	2.00	11,799
Label	28	91	28.00	20,375
Nail wire	24	91	24.00	6288
Other raw materials	1	91	1.00	15,402
Pack SKU A	11	95	19.00	261,404
Pack SKU B	11	95	19.00	98,568
Pack SKU C	9	95	17.00	39,220
Packaging A	28	68	28.00	25,117
Packaging B	28	68	28.00	12,263
Packaging C	28	68	28.00	6177
Separator	2	91	2.00	1815
Spun zinc	2	91	2.00	4538
West DC A	5	95	5.55	97,628
West DC B	8	95	8.55	27,775
West DC C	6	95	6.32	8606

finished goods locations, we assume that both the pack locations and the nine DC-SKU pairs all set a service level equal to the customer service level target, namely 0.95. Again, we find that the best solution for the stochastic-service model seems to set the service level to the lower bound at each stage. Table 8 displays the nominal and expected lead-times for the battery supply chain.

On a percentage basis, the bulk manufacturing stage and the three packaging stages have the greatest increase between nominal and expected lead-time. The difference is attributed to the fact that these stages each have one or more suppliers with significant nominal lead-times. The total safety stock cost under the stochastic-service model is $927,000.

To gain more insight into the role that different countermeasures may play in this supply chain, Figure 4 displays the cost for each policy under different end-item service levels. Varying the service levels at all end-item stages does not change the structure of the optimal policy under the guaranteed-service model. For the stochastic-service model, the nine DCs and the three pack locations had their service levels changed to equal the current end-item service level while all the other stages maintained their existing levels from Table 8. The safety stock costs are shown graphically in Figure 4 and reported in Table 9. One can view the cost difference as being the additional

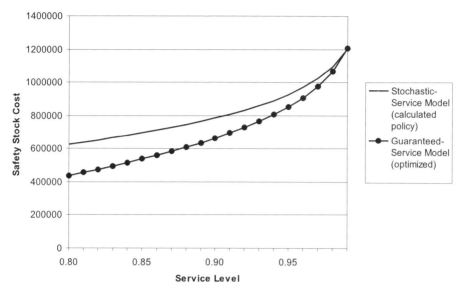

Fig. 4. Safety Stock Cost as a Function of Service Level in Battery Supply Chain.

Table 9
Safety Stock Cost as a Function of Service Level in Battery Supply Chain

Service level	Stochastic-service model (calculated policy)	Guaranteed-service model (optimized)
0.80	627,155	436,454
0.81	639,637	455,266
0.82	652,624	474,696
0.83	666,184	494,818
0.84	680,421	515,713
0.85	695,400	537,481
0.86	711,203	560,239
0.87	727,950	584,132
0.88	745,791	609,333
0.89	764,911	636,061
0.90	785,539	664,596
0.91	807,974	695,298
0.92	832,635	728,652
0.93	860,105	765,326
0.94	891,159	806,285
0.95	927,098	853,000
0.96	969,960	907,883
0.97	1,023,570	975,355
0.98	1,096,198	1,065,047
0.99	1,201,556	1,206,414

cost for the stochastic-service model to handle the full range of demand realizations by means of safety stock.

5 Supply-chain configuration

Lee and Billington (1993), Ettl et al. (2000), Graves and Willems (2000) all report on industrial applications in which their work was used to optimize safety stocks in a supply chain. These applications provide compelling evidence of the financial impact that optimizing inventory levels can have in practice, as reductions of 25–50% in the safety-stock holding cost are common. However, this work starts after most of the design decisions for the supply chain have been set, namely the topology of the network and the key cost and lead-time parameters. In effect, these models are being applied to existing supply chains where the only design options available are in terms of whether to locate safety stocks at a stage, and if so, how much to maintain.

In Chapter 2 of this handbook, Muriel and Simchi-Levi present and consider one category of supply-chain design problems, referred to as network design problems. The intent of the network design problem is typically to determine the optimal manufacturing and distribution network for a company's entire product line. The most common approach is to formulate a large-scale mixed-integer linear program that captures the relevant fixed and variable operating costs for each facility and each major product family. The fixed costs are usually associated with the investment and/or overhead costs for opening and operating a facility, or with placing a product family in a facility. The variable costs include not only the manufacturing, procurement and distributions costs, but also the tariffs and taxes that depend on the network design. Network design focuses on the design of two or three major echelons in the supply chain for multiple products. Due to the nature of the problem being solved, network design is typically solved every two to five years.

In this section, we consider another type of supply-chain design problem that arises after the network design for the supply chain has been set. These decisions determine the *total supply chain cost*, which we define to be the cost of goods sold (COGS), plus the inventory holding costs for the pipeline inventory and for the safety stock. For example, in the bulldozer supply chain presented earlier, the total supply chain cost consists of the annual COGS of $94,380,000, plus the annual holding cost for pipeline stock of $2,007,000, plus the annual holding cost for safety stock of $633,000. For the battery supply chain, there is a similar cost breakdown with annual COGS of $55,364,000, and annual holding costs for pipeline and safety stock of $942,000 and $853,000. In both examples COGS is an order of magnitude larger than the total inventory cost, while the pipeline stock exceeds the safety stock.

For every supply chain a company launches, there is a set of decisions that are made after the network design and that act to configure each stage in the network. In particular, these decisions result in determining the key operating parameters for each stage, including the lead-time and cost added at the stage. These decisions determine the total supply chain cost. For instance, the company must decide whether to source a part locally or globally. The company must decide whether to dedicate machinery to a manufacturing process or to conduct the manufacturing process on shared equipment. The company must decide the transportation mode to move product into a distribution channel. These decisions affect more than just safety stock cost. They also affect the cost of goods sold, pipeline stock cost, quality costs, and time-to-market costs.

To inform these decisions, we introduce a problem that we refer to as the supply-chain configuration problem. For this problem, we address how to configure the supply chain for a new product for which the product's design has already been decided and the topology for the supply-chain network has been set. The central question is to determine what suppliers, parts, processes, and transportation modes to select at each stage in the supply chain. For each stage, we have a set of options that are differentiated, at a minimum, by their lead-times and their direct costs added. Our supply-chain design framework considers the total supply chain cost, equal to the cost of goods sold, plus the inventory holding costs for both safety stock and pipeline stock. The supply chain configuration problem chooses a sourcing option for each stage of the supply chain so as to minimize the sum of these costs.

In the next section, we expand our discussion of options as the fundamental construct in the supply chain configuration problem. We then present the model formulation for the supply-chain configuration problem. The final section presents an example of the approach applied to the bulldozer example presented earlier.

5.1 Option definition

The supply-chain configuration problem is based on the same assumptions made for the safety stock optimization problem, with one significant difference. In the safety stock optimization problem, a stage represents a processing or transformation activity. That is, it is a defined task that will take a certain amount of time at a certain cost. In the supply-chain configuration problem, we still model the supply chain as a network of stages but now a stage represents functionality that must be provided. The essence of the configuration problem is to decide how best to satisfy this functionality in the context of the overall supply chain.

For each stage, we assume that we can specify one or more options that can satisfy the stage's required activity. For example, if a stage represents the procurement of a metal housing, then one option might be a locally based high-cost provider and another option could be a low-cost international supplier.

For each stage, we assume that we will select a single option. Thus, we do not permit the possibility of having dual or multiple sources for a single activity or stage; this might be a topic for further research.

We characterize an option at a stage by its direct cost added and its processing time or lead-time. When a stage reorders, the processing or lead-time is the time to process an item at the stage, provided all of the inputs are available. An option's direct cost represents the direct material and direct labor costs associated with the option. If the option were the procurement of a raw material from a vendor, then the direct costs would be the purchase price including transportation and the labor cost to unpack and inspect the product.

In practice, there might be other dimensions or attributes upon which different options are evaluated. For instance, different suppliers might differ in terms of the quality of the product they supply. Similarly, different options for a manufacturing activity might differ in terms of the amount of capacity that could be made available to the supply chain. We do not consider these other attributes in this presentation. In effect, we assume that the different options at a stage are the same on all attributes except for lead-time and cost. Admittedly, this is a simplification of reality. We leave it to future research to extend the work presented here to address this additional complexity.

We will present the configuration problem for the case of guaranteed service. One could also develop the supply-chain configuration model with the assumption that stages provide stochastic service, but we do not do this here. Rather, we will follow the development of the supply-chain configuration problem for the guaranteed-service model, as given by Graves and Willems (2002).

The model assumptions for the supply-chain configuration problem are the same as for the guaranteed-service safety-stock problem presented earlier in this chapter. We assume that each stage j promises a guaranteed service time s_j^{out} by which the stage will satisfy its demand, either from internal or external customers. Similarly, we define s_j^{in} to be the inbound service time for stage j, which equals the maximum of the service times quoted to stage j by its suppliers. We assume each stage operates according to a periodic review policy with a common review period. We assume the demand process for any finished good is stationary with mean demand per period of μ and a standard deviation of σ. For the purpose of determining the safety stock, we assume that we are given a bound on the demand process for each stage.

If we let n denote the nth option at stage j, then L_{jn} and C_{jn} represent the processing time and cost added associated with the nth option at stage j. The choice of option at a stage will have an impact on the cost of goods sold, on the amount of safety stock and pipeline stock at the stage, and the holding costs. For a given option n, the stage's contribution to the COGS is $C_{jn}\mu_j$

per period. As before, we can calculate the stage's safety stock to be $k_j \sigma_j \sqrt{s_j^{in} + L_{jn} - s_j^{out}}$. Since the choice of an option at a stage decides the lead-time at the stage, the work-in-process or pipeline stock now depends on the option chosen. In particular, the pipeline stock at a stage will equal $\mu_j L_{jn}$ when option n is selected. Finally, the option choice will also affect the holding cost rate because it depends on the cumulative cost at the stage; the option choices at the stage and at any upstream suppliers determine the stage's cumulative cost, and thus the holding cost rate.

5.2 Model formulation

We can formulate the supply chain configuration problem as a non-linear mixed-integer optimization problem where the decision variables are the binary variable for option selection and the services times.

P

$$\min \sum_{i=1}^{N} \left[\alpha c_i \left[D_i \left(s_i^{in} + t_i - s_i^{out} \right) - \left(s_i^{in} + t_i - s_i^{out} \right) \mu_i \right] + \alpha \left(c_i - \frac{x_i}{2} \right) t_i \mu_i + \beta x_i \mu_i \right]$$

s.t.

$$\sum_{n=1}^{O_i} L_{in} y_{in} - t_i = 0 \qquad \text{for } i = 1, 2, \ldots, N \qquad (5.1)$$

$$\sum_{n=1}^{O_i} C_{in} y_{in} - x_i = 0 \qquad \text{for } i = 1, 2, \ldots, N \qquad (5.2)$$

$$c_i - \sum_{j:(j,i)\in A} c_j - x_i = 0 \qquad \text{for } i = 1, 2, \ldots, N \qquad (5.3)$$

$$s_i^{in} \geq s_j^{out} \qquad \text{for } i = 1, 2, \ldots, N, j : (j, i) \in A \qquad (5.4)$$

$$s_i^{in} + t_i - s_i^{out} \geq 0 \qquad \text{for } i = 1, 2, \ldots, N \qquad (5.5)$$

$$s_j^{out} \leq S_j \qquad \text{for all demand nodes } j \qquad (5.6)$$

$$s_i^{in}, s_i^{out} \geq 0 \text{ and integer} \qquad \text{for } i = 1, 2, \ldots, N \qquad (5.7)$$

$$\sum_{n=1}^{O_i} y_{in} = 1 \qquad \text{for } i = 1, 2, \ldots, N \qquad (5.8)$$

$$y_{in} \in \{0, 1\} \qquad \text{for } i = 1, 2, \ldots, N, 1 \leq n \leq O_i \qquad (5.9)$$

where O_i, number of options to choose from at stage i; C_{in}, direct cost added of the nth option at stage i; L_{in}, lead-time of the nth option at stage i; $D_i()$, maximum demand function for stage i; α, scalar representing the holding cost rate; β, scalar converting the model's underlying time unit into the company's time interval of interest; μ_i, mean demand rate at stage i; c_i, cumulative cost at stage i; t_i, selected option's lead-time at stage i; x_i, selected option's cost at stage i; y_{in}, indicator variable which equals 1 if stage i's nth option is selected and 0 otherwise; S_j, maximum service time permitted for demand node j.

The objective function has three terms, each corresponding to a component of the total supply chain cost. The first term represents stage i's safety stock cost, which is a function of the stage's net replenishment time and demand characterization. The holding cost at stage i equals the cumulative cost of the product at stage i times the holding cost rate. The second term expresses the pipeline stock cost as the product of the holding cost rate, the average cost of the product at the stage, and the expected amount of pipeline stock. The third term, cost of goods sold (COGS), represents the total cost of all the units that are delivered to customers during a company-defined interval of time. The incremental contribution to COGS is calculated at each stage by a product of the average demand at the stage, the option's cost, and a scalar β, which expresses COGS in the same units as pipeline stock cost and safety stock cost. (For instance, one might set α and β so that all terms are expressed as annual costs or as the total costs over the lifetime of the supply chain.)

The cost and time associated with the option chosen at each stage is given in (5.1) and (5.2). Constraint (5.3) calculates the cumulative cost at each stage. Constraints (5.4–5.7) assure that the service times are feasible. In particular, the incoming service time at every stage is at least as large as the largest service time quoted to the stage, the net replenishment time of each stage is non-negative, the maximum service times to the customer must be no greater than the user-defined maximums, and service times must be non-negative and integer. The last two constraints, (5.8) and (5.9), enforce the sole sourcing of options.

Graves and Willems (2002) describe how to solve **P** by dynamic programming when the underlying network is a spanning tree.

While **P** clearly uses safety stock optimization as a building block, it also exhibits behavior that is far more complex than just optimizing safety stock. For safety stock optimization, inventory stocking decisions at one stage in the supply chain affect adjacent downstream stages in the supply chain through the downstream stage's net replenishment time. In the supply chain configuration problem, inventory decisions again affect downstream adjacent stages, but the cost at the current stage has an impact on all stages that are downstream of the current stage, not just those that are adjacent. On an intuitive level, **P** is balancing the increase in COGS against the decrease in inventory-related costs. One can reduce inventory related costs by choosing more responsive options, but at the cost of an increase to the COGS. A key realization is that this tradeoff cannot be properly considered by solving **P** one stage at a time, in isolation; rather, one needs to consider the impact of configuration decisions on the entire supply chain in order to produce the globally optimal solution. The benefit to **P** comes from globally balancing the potential increase in COGS with the benefits one gets from being able to reduce inventory costs.

5.3 Example

To gain more insight into the supply chain configuration problem, we will revisit the bulldozer supply chain discussed earlier in the chapter. There are two types of stages in the bulldozer supply chain: procurement and assembly stages. Procurement stages are stages that do not have any incoming arcs; they represent the purchase of materials outside the supply chain. All of the other stages in the supply chain are assembly stages, at which one or more components are combined in the process.

For the example, there are two options per stage. The stage lead-times and costs from the original presentation correspond to the standard option at each stage. If the stage is a procurement stage, this is the existing procurement arrangement. If the stage is an internal assembly stage, this is the traditional manufacturing method at the stage. All procurement stages also have a consignment option where the supplier is responsible for providing immediate delivery to the bulldozer line. Each assembly stage has an expedited option that corresponds to the company investing in process improvement opportunities to decrease the stage's lead-time. These second options are not based on actual data at the company, but they are indicative of the kinds of option costs we have seen in similar supply chains.

We calculate the cost of the consignment option by the following formula: for each one-week reduction in the supplier lead-time, the supplier will increase the purchase price of the part by 0.75%. This is a similar structure to the kind of arrangements that we have encountered before in practice; see Graves and Willems (2002) and Willems (1999). Typically, the cost increase for a week's reduction ranges from 0.5% to 1% of the original purchase price. The increase in price represents the cost to the supplier for bearing the additional inventory holding cost.

For the expedited assembly option, we classify the required improvement activity at a stage as easy, medium or hard. An easy improvement activity might include the assignment of additional labor resources to the task or the dedication of some minor equipment. For the purposes of this analysis, the cost of an easy improvement is $97,500. By dividing this by the average annual demand, we convert this into a per unit approximation of the cost increase, namely $75 per unit. Most of the upstream assembly stages fall into the camp of assembly operations that could be easily improved. Medium improvement activities cost $150 per unit and hard improvements cost $300 per unit. As the cost of the improvement increases, significant redesign and additional human labor are often required. Final assembly is the only stage that is classified as hard to redesign.

The costs and associated lead-times for each option are presented in Table 10.

Figure 5 depicts the service times that correspond to optimizing the supply chain configuration problem.

Among the 22 stages, the optimal supply chain configuration selects the higher cost, shorter lead-time option for only six of the stages. The procurement stages with the higher cost, shorter lead-time option are the brake group, fender group, and plant carrier. The assembly stages with the higher cost, shorter lead-time option are the common subassembly, dressed-out engine, and main assembly. All of the other stages continue to use their original option.

The optimal inventory policy has also changed in reaction to the different options selected. In the original safety stock optimization, the common subassembly and dressed-out engine held no safety stock and both quoted a service time of twenty days to the main assembly. The chassis/platform also held no safety stock, but quoted its maximum possible service time of sixteen days. With the reduction in the processing time at the common subassembly, the brake group and the plant carrier, the common subassembly is now able to quote a service time of eight days to the main assembly. To achieve this eight-day service time requires that two of its suppliers, transmission and case & frame, must now hold a safety stock. Furthermore, stages in the other two sub-networks that supply main assembly are also holding safety stock so that the inbound service time to main assembly remains at eight days. This is a good example of how subtle changes in the configuration of some stages have a dramatic impact on the resulting safety stock policy

Table 11 summarizes the results from optimizing the supply chain configuration and compares the results to those for a solution that keeps the original option at each stage and optimizes the resulting guaranteed-service model.

We observe that when we optimize the safety stock, there are savings in annual holding cost of $198,000 relative to a base case in which each stage holds a safety stock and quotes a service time of zero. When we optimize the configuration, we find an additional savings in total supply chain costs of $371,000. We see, as expected, that when we optimize the supply-chain configuration we actually increase the COGS but get an overall savings due to lower inventory holding costs.

Based on Graves and Willems (2002) and the work presented here, we are able to formulate some initial hypotheses about the behavior of optimal supply chain configurations. First, the further upstream the supply chain, the less likely is it that we choose a higher cost option. Higher-cost options increase the cost at a stage, which not only increases the COGS but also the holding cost for all of the pipeline and safety stock at downstream stages. Furthermore, since the cumulative cost at an upstream stage is typically

Table 10
Option values for Bulldozer Supply Chain Configuration

Stage name	Option description	Option time	Option cost ($)
Boggie assembly	Standard procurement	11	575
	Consignment	0	584
Brake group	Standard procurement	8	3850
	Consignment	0	3896
Case	Standard procurement	15	2200
	Consignment	0	2250
Case & frame	Standard assembly	16	1500
	Expedited assembly	4	1575
Chassis/platform	Standard assembly	7	4320
	Expedited assembly	2	4395
Common subassembly	Standard assembly	5	8000
	Expedited assembly	2	8075
Dressed-out engine	Standard assembly	10	4100
	Expedited assembly	3	4175
Drive group	Standard procurement	9	1550
	Consignment	0	1571
Engine	Standard procurement	7	4500
	Consignment	0	4557
Fans	Standard procurement	12	650
	Consignment	0	662
Fender group	Standard procurement	9	900
	Consignment	0	912
Final assembly	Standard assembly	4	8000
	Expedited assembly	1	8300
Final drive & brake	Standard assembly	6	3680
	Expedited assembly	2	3755
Frame assembly	Standard procurement	19	605
	Consignment	0	622
Main assembly	Standard assembly	8	12,000
	Expedited assembly	2	12,150
Pin assembly	Standard procurement	35	90
	Consignment	0	95
Plant carrier	Standard procurement	9	155
	Cosnsignment	0	157
Platform group	Standard procurement	6	725
	Consignment	0	732
Roll over group	Standard procurement	8	1150
	Consignment	0	1164
Suspension group	Standard assembly	7	3600
	Expedited assembly	2	3675
Track roller frame	Standard procurement	10	3000
	Consignment	0	3045
Transmission	Standard procurement	15	7450
	Consignment	0	7618

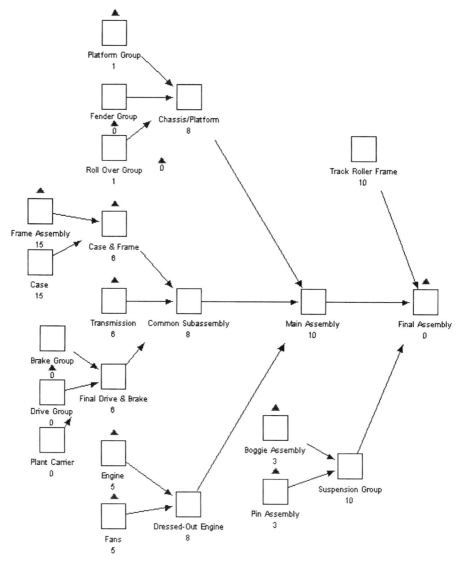

Fig. 5. Optimal Service Times for Bulldozer Supply Chain.

relatively small, it is not that costly to just hold a decoupling safety stock at the upstream stage, thereby making its effective lead-time to the rest of the supply chain zero. Therefore, when choosing a higher cost option at an upstream stage, the inventory savings will have to be truly dramatic to justify the higher cost to the supply chain.

Conversely, we note that a higher-cost, shorter lead-time option is more likely to be attractive at a downstream stage. For instance, we find for stages that represent the transportation of a finished good to an end customer, a

Table 11
Comparison of Optimal Safety Stock and Supply Chain Cofiguration

Cost category	Results from safety stock optimization ($)	Optimal supply chain configuration ($)	Numerical difference ($)	Percentage difference (%)
Cost of goods sold	94,380,000	94,848,000	468,000	0.50%
Total safety stock cost	632,719	499,786	(132,933)	−21.01%
Total pipeline stock cost	2,006,843	1,300,328	(706,514)	−35.21%
Total supply chain cost	97,019,561	96,648,114	(371,447)	−0.38%

faster but more expensive transportation mode is able to pay for itself in terms of inventory savings.

Second, the greatest potential for supply chain configuration occurs when the supply chain has a structure where the selection of options across the chain makes different sub-networks of the supply chain similarly responsive, that is, have the same service times. As an example, making one subassembly very responsive will not likely be cost-effective unless the other components can either be equally responsive, or it is cost-effective to decouple them with inventory so that effectively the subassemblies in the echelon are all similarly responsive. If one of the stages cannot be made more responsive, then the high-cost subassembly could be produced with cheaper options at no penalty to the supply chain's performance.

6 Conclusion

In this chapter we have presented two general approaches to safety stock placement, and have introduced the supply chain configuration problem. The safety stock placement work, as evidenced by the applications cited in the papers, has proven itself to be of value to practice. Both approaches quantify the impact that demand uncertainty has in supply chains. By taking a system-wide view of the problem, these models are able to mitigate the impact of this uncertainty in a cost-effective manner.

We find that there is a significant opportunity to improve the total supply chain cost by jointly optimizing sourcing and inventory decisions during the configuration of the supply chain. The earlier that these supply chain considerations can be incorporated into product and sourcing decisions, the more leverage we have – we see this in this chapter in terms of value of getting the configuration right vis-à-vis solely optimizing safety stocks.

Nevertheless, we wish to conclude this chapter with some thoughts about research opportunities.

Product life cycles are increasingly short, with products within a product family continually being introduced and terminated. Supply chains need to be designed to accommodate this. In particular, demand is never stationary and there is huge uncertainty and risk over a product life cycle. In particular, the risk applies not just to the inventory holding cost in the chain, but also to the inventory investment required to fill the chain since enough demand may not materialize to empty out the chain. There is a need for good models and approaches to determine how to evolve the supply chain to handle generations of products.

Many products see seasonal or cyclic demand. Furthermore, there are often different service targets, holding cost rates, and/or costs for stock outs in different periods. Characterizing both optimal and reasonable approaches for planning safety stocks across a supply chain and across a seasonal cycle is worthy of research.

Supply chains are most certainly not limited to just demand uncertainty. Other types of uncertainty, such as lead-time, capacity, and yield uncertainty, can be equally important. While this chapter has not covered these kinds of uncertainty, the models presented can adapt to these issues, albeit with additional assumptions and often in an ad hoc fashion. Nevertheless, there are opportunities for the development of more general and comprehensive methods for handling the full range of supply chain uncertainties.

Properly designing contracts is another opportunity. This can be thought of as a different form of the configuration decision. In this case, we are looking to establish contracts throughout the supply chain so as to get the best overall performance. In particular, one would expect to design contracts with, say, suppliers so that there is some consistency in how the supply chain is able to respond to upswings (or downswings) in demand. Furthermore, one would hope to understand how to design and coordinate contracts across a number of suppliers or channel partners so as to distribute the risks and rewards in the most economic way.

We have found empirically an interesting analog between supply chains and project management networks. As with a project management network, we find when applying the guaranteed-service model, there is a critical path that underlies the optimal safety stock policy. The difference is that instead of a lead-time-weighted critical path, there is a critical path that is driven by cumulative cost, maximum replenishment time and safety stock policy. For example, appropriately buffering a long lead-time part makes its effective lead-time to the system zero. Identifying and characterizing the components of the critical path in the supply chain is a potentially fertile area to begin the development of new solution approaches.

As a final opportunity, we would hope to see the continuing deployment of models in this chapter to practice. This should provide an opportunity to examine, test, and validate the underlying assumptions of these models. To wit, the stochastic service and guaranteed-service models offer two different perspectives on how the world works. Can we determine which is right? Can we say anything about which is more common? Is either of them right? Or is there a better perspective? We hope that some future research will be able to conduct a careful empirical study of how well these models match reality, as well as how good is the decision support that they provide.

Acknowledgement

This research has been supported in part by the MIT Leaders for Manufacturing Program, a partnership between MIT and major U.S. manufacturing firms; and by the Singapore-MIT Alliance, an engineering education and research collaboration among the National University of Singapore, Nanyang Technological University, and MIT. The second author was also supported by the Boston University School of Management Junior Faculty Research Program.

References

Axsäter, S. (2003). Supply Chain Operations: Serial and Distribution Inventory Systems, in: A. G. de Kok, S. C. Graves (eds.), *Handbooks in Oper. Res. And Management Sci, Supply Chain Management: Design, Coordination and Operation*, Vol. 11. Amsterdam, The Netherlands, North-Holland Publishing Company. Chapter 10.

Bertrand, W. (2003). Supply Chain Design: Flexibility Considerations, in: A. G. de Kok, S. C. Graves (eds.), *Handbooks in Oper. Res. And Management Sci. Supply Chain Management: Design, Coordination and Operation*, Vol. 11. Amsterdam, The Netherlands, North-Holland Publishing Company. Chapter 4.

Clark, A. J., H. Scarf (1960). Optimal Policies for a Multi-Echelon Inventory Problem. *Management Science* 6, 475–490.

de Kok, A. G., J. Fransoo (2003). Planning Supply Chain Operations: Definition and Comparison of Planning Concepts, in: A. G. de Kok, S. C. Graves (eds.), *Handbooks in Oper. Res. And Management Sci. Supply Chain Management: Design, Coordination and Operation*, Vol. 11. Amsterdam, The Netherlands, North-Holland Publishing Company. Chapter 12.

Ettl, M., G. E. Feigin, G. Y. Lin, D. D. Yao (2000). A Supply Network Model with Base-Stock Control and Service Requirements. *Operations Research*. 48, 216–232.

Glasserman, P., S. Tayur (1995). Sensitivity Analysis for Base-Stock Levels in Multiechelon Production-Inventory Systems. *Management Science* 41, 263–281.

Graves, S. C. (1988). Safety Stocks in Manufacturing Systems. *Journal of Manufacturing and Operations Management* 1, 67–101.

Graves, S. C., S. P. Willems (1996). Strategic Safety Stock Placement in Supply Chains, *Proceedings of the 1996 MSOM Conference*, Hanover, NH.

Graves, S. C., S. P. Willems (2000). Optimizing Strategic Safety Stock Placement in Supply Chains. *Manufacturing & Service Operations Management* 2, 68–83.

Graves, S. C., S. P. Willems (2002). Optimizing the Supply-Chain Configuration for New Products, MIT Working Paper, January, 32 pages.

Humair, S., S. P. Willems (2003). Optimal Inventory Placement in Networks with Clusters of Commonality, Working Paper, January, 19 pages.

Inderfurth, K. (1991). Safety Stock Optimization in Multi-Stage Inventory Systems. *International Journal of Production Economics* 24, 103–113.

Inderfurth, K., S. Minner (1998). Safety Stocks in Multi-Stage Inventory Systems under Different Service Levels. *European Journal of Operational Research* 106, 57–73.

Kimball, G. E. (1988). General Principles of Inventory Control. *Journal of Manufacturing and Operations Management* 1, 119–130.

Lee, H. L., C. Billington (1993). Material Management in Decentralized Supply Chains. *Operations Research* 41, 835–847.

Muriel, A., D. Simchi-Levi (2003). Supply Chain Design and Planning – Applications of Optimisation Techniques for Strategic and Tactical Models, in: A. G. de Kok, S. C. Graves (eds.), *Handbooks in Oper. Res. And Management Sci. Supply Chain Management: Design, Coordination and Operation*, Vol. 11. The Netherlands, North-Holland Publishing Company, Amsterdam. Chapter 2.

Simpson, K. F. (1958). In-Process Inventories. *Operations Research* 6, 863–873.

Song, J.-S., P. Zipkin (2003). Supply Chain Operations: Assemble-to-Order Systems, in: A. G. de Kok, S. C. Graves (eds.), *Handbooks in Oper. Res. And Management Sci. Supply Chain Management: Design, Coordination and Operation*, Vol. 11. Amsterdam, The Netherlands, North-Holland Publishing Company. Chapter 11.

Willems, S. P. (1999), Two papers in supply-chain design: supply-chain configuration and part selection in multigeneration products. MIT Ph.D. Dissertation, Cambridge MA.

Chapter 4

Supply Chain Design: Flexibility Considerations

J.W.M. Bertrand

Department of Technology Management, Technische Universiteit Eindhoven 5600 MB, Eindhoven, The Netherlands

1 Introduction

After costs, quality and reliability, flexibility has emerged in the last two decades as the fourth important performance indicator for operational systems. Researchers unanimously agree that the importance of operational flexibility stems from the dramatic change that has taken place since 1980s in the market places in virtually all sectors of industry. As markets saturated, firms started to compete on product differentiation and product innovation. As a result, both the number of product variants simultaneously offered to the market, and the new product introduction rate, substantially increased. Product innovation and new product development became important research fields and concurrent engineering was developed as a new approach in structuring and managing the product innovation process, aimed at delivering better quality products at lower costs in shorter time (Clark and Wheelwright, 1993; Chapters 7, 8, and 9). At the beginning of the 1990s 'time-to-market' was an established concept in the industry, indicating the industry-wide awareness of innovation speed as a competitive weapon (Stalk and Hout, 1990; Chapter 1). However, during the same period, firms that decided to compete on product diversity and product innovation were faced with the downside of this strategy, being a large increase in demand uncertainty at the product variant level.

For each of the product variants to be offered, production capacity has to be reserved in order to be able to deliver the product up to a certain level, and work-in-process and inventory has to be built up. Thus upfront investments have to be made for setting up a supply chain for a product family. These investments have to be recovered from the revenues from sales.

The initial lack of flexibility to adapt the supply chain to emerging demand for product variants frequently led to lost sales for some product variants, and product markdowns and writing-off of excess inventory for other product variants. Thus, supply flexibility became an important feature of industrial production systems. In supply chains that experienced an imbalance between demand uncertainty and supply chain flexibility, manufacturers and suppliers were put under high pressure to increase their flexibility.

Flexibility in manufacturing mainly stems from three sources. First, there is the variety of the manufacturing technologies employed. Technological variety allows for a large variety of different products to be made. Second, there is the flexibility in the amount of capacity available for production. Inflexible capacity limits the volume of products that can be delivered to the market. Third, there is the inflexibility in the timing and frequency of production. Timing and frequency are often restricted on economic grounds (due to change-over and set-up costs). Inflexibility in the timing and frequency of production of a product leads to high levels of work-in-process and inventory, and to long lead times for introducing new product variants.

In the context of this chapter we deal with three dimensions of flexibility: volume flexibility, mix flexibility and new product flexibility. As a rule, flexibility has to be designed into a system: it requires investments and should be justified on the basis of the potential benefits to be obtained from being able to exercise the flexibility. We therefore deal with the flexibility subject from the investment point of view: What is the value of volume flexibility, mix flexibility and new product flexibility in a specific supply chain, and how much should be invested in (flexible) capacity?

Supply chain flexibility is not a subject that has been researched extensively. In fact most research on flexibility has dealt with the more general issue of manufacturing flexibility at the plant level. Research relevant from the supply chain perspective has developed along rather independent research lines. This state of affairs is reflected in this chapter; it discusses the research results obtained in a number of different research areas that are relevant for understanding how to design a supply chain for flexibility, specifically volume flexibility, mix flexibility and new product flexibility for a product family. We start with definitions found in literature of the flexibility concept, and single out the flexibility dimensions and flexibility aspects that are relevant from the supply chain perspective. Then we summarize the research findings related to these flexibility dimensions and aspects, and finally we develop a model of the supply chain flexibility decision problem, using these research findings as main inputs. Throughout the chapter suggestions for promising research are given.

1.1 Overview of the chapter

In the Section 2 through Section 5 we give a short overview of research on flexibility based on a selection of published papers in the field. Section 2 deals with conceptual research on manufacturing flexibility and serves to position the contingent and multi-dimensional character of the flexibility concept. Section 3 deals with results of model-based research on flexibility. A selection of papers are discussed that deal with the problem of investing in flexible production capacity to cope with uncertainty about future demand levels for products, i.e., to create product volume flexibility and product mix flexibility.

In Section 4 a selection of papers are discussed that deal with the effects of machine, routing, and worker flexibility, and the use of these flexibilities, on the time-flexibility of production, i.e., the responsiveness of production to short-term demand. Section 5 discusses a selection of papers that report on empirical research on manufacturing flexibility. These papers provide information about the use of flexibility as a strategic option, and about relationships between flexibility and observed performance.

The Sections 6 and 7 deal with supply chain flexibility. In Section 6 we discuss a small selection of papers that provide essential results for the discussion of the supply chain design problem. In Section 7 we combine elements discussed in the Sections 2–6 to formulate the supply chain design problem from the perspective of creating volume flexibility, mix flexibility and new product flexibility. Section 8 gives the conclusions.

2 Conceptual research on manufacturing flexibility

Important contributions to our conceptual understanding of manufacturing flexibility can be found in Slack (1983, 1987), Browne, Dubois, Rathmill, Sethi, and Stecke (1984), Gerwin (1987, 1993), Carlsson (1989), De Meyer, Nakana, Miller, and Ferdows (1989), Gupta and Buzacott (1989), Gupta and Goyal (1989), Mandelbaum and Brill (1989), Sethi and Sethi (1990), Ramasesh and Jayakumar (1991), Hyun and Ahn (1992), Chandra and Tombak (1992), Chen, Calantone, and Chung (1992), Dixon (1992), Gaimon and Singhal (1992), Nagurar (1992), Gupta and Somers (1992, 1996), Das and Nagendra (1993), De Groote (1994a), Upton (1994), De Toni and Tonchia (1998), Parker and Wirth (1999), Koste and Malhotra (1999), Beach, Muhlemann, Price, Paterson, and Sharp (2000), D'Souza and Williams (2000) and Pereira and Paulré (2001). Each of these papers builds on results obtained in earlier papers and provides a contribution, either in terms of the theoretical analysis of the flexibility concept, the identification of the various dimensions and properties of flexibility in a production environment, or in identifying the relationships between the flexibility dimensions. In this section we summarize the main findings in this line of research using a selection of the above papers as main reference.

Conceptual research on manufacturing flexibility had started off in the 1980s. Slack (1987) interviewed managers in 10 manufacturing organizations, ranging from mass assembly industry to batch/jobbing industry, on their perception of manufacturing flexibility. He finds that most managers had only limited view on their companies flexibility needs. Managers who supply resources stress flexibility as a means for coping with unplanned disturbances. Manufacturing managers see flexibility as an aid to greater productivity. Managers on the demand side see enhanced availability of supply, either by widening the range of what can be made or by shortening supply lead time. The managers also indicate the flexibility dimensions they distinguish. From this information Slack (1987) derives a hierarchical framework of flexibility dimensions and flexibility aspects.

A general and abstract definition of flexibility is given in Upton (1994), who characterizes flexibility as "the ability to change or react with little penalty in time, effort, cost or performance." Change may take place in different areas and may pertain to different aspects of a firm's environment or its internal processes. Moreover, change may come at different levels of magnitude and with different levels of surprise. De Groote (1994a) states that "A particular technology is said to be more flexible than another, if an increase in the diversity of the environment yields a more desirable change in performance, than the change that would be obtained with the other technology under the same conditions." Thus flexibility is contingent upon the environmental diversity that one wants to cope with, and on how one prefers the one output of the system over the other. The implication is that the flexibility of a specific technology can only be evaluated in the context of a particular environment and a particular output preference.

Upton (1994) states that the flexibility concept must be specified dependent on the context in which the flexibility issue has been raised. He states that the boundaries of the system must be clearly defined (machine, shop, firm, supply chain), and he proposes that flexibility of the system be characterized on three aspects:

- The dimensions of flexibility – What exactly is it that flexibility is required over – what needs to change or be adapted?
- Time horizon – What is the general period over which changes must occur? Minute-by-minute, days, weeks, or years?
- Elements – Which of the following are we trying to manage or improve: range, uniformity across the range, or mobility? As examples of the dimension of flexibility Upton (1994) mentions the thickness of the slab in a rolling mill, the changes in input material that a chemical process should tolerate, and the mix of products being manufactured.

Range, uniformity and mobility are different ways in which a system can be flexible. The range increases with the size of the set of options or alternatives, which may be accommodated or effected. Examples are: the range of sizes of components that can be processed, and the range of volume of output for which a plant is profitable.

Mobility refers to transition penalties for moving within the range. Low values of transition penalties imply high mobility. Mobility may be measured by time or costs of change. For instance the mobility required for a product line can be measured by the set-up times and set-up costs required for changing between product types, and the mobility of output volume of plant can be measured by the cost and time it takes to change the output volume from one level to the other within the range.

Uniformity refers to the extent to which general performance measures such as product quality and production costs are indifferent to at what particular point within the range the system operates. For instance, a production line that can produce each of the products within the range at the same costs per unit, is viewed as being more flexible than a line that can produce the same product range, at the same average costs per product, but some products are produced at lower than average cost and others are produced at higher than average costs.

Flexibility is needed to cope with uncertainty and change. Gerwin (1993) distinguishes four types of market uncertainty:

- market acceptance of kinds of products,
- length of product life cycle,
- demand for specific product options,
- aggregate product demand,

and two types of process uncertainty:

- resource availability,
- material availability.

He relates each of these six types of uncertainty to six different dimension of flexibility. These flexibility dimensions are: mix flexibility, product innovation flexibility, product modification flexibility, volume flexibility, process routing flexibility and material flexibility. As noticed by Slack (1987) and Upton (1994) each of these dimensions of flexibility has three aspects: range, uniformity and mobility.

Manufacturing flexibility requires investments and therefore it should be carefully investigated as which are the uncertainties that a firm faces. At the strategic level a firm has to decide how to cope with these uncertainties; one of the possibilities being not to invest in flexibility at all. In this context Gerwin (1993) states that "an unintentional bias exists in favor of recommending more flexibility than is economically appropriate." Gerwin (1993) distinguishes four generic strategies for coping with uncertainty:

(1) Reduction of uncertainty, i.e., by investing in variance reduction such as long term contracts with customers and suppliers, preventive maintenance and total quality control, and design for manufacturing.
(2) Banking, i.e., the use of flexibility to accommodate types of uncertainty that are known in a probabilistic sense such as variations in demand (the so-called 'known unknowns'). The use of safety stocks in production–inventory systems is an example of banking.
(3) Adaptations, i.e., the defensive use of flexibility to accommodate uncertainties that are not known in a probabilistic sense, (the so-called 'unknown unknowns') such as the break-out of strike.
(4) Redefinition, i.e., the pro-active use of flexibility to raise customer expectations, to increase uncertainty for rivals and to gain competitive advantages. The decrease of the new product introduction lead time is an example of redefinition.

Each of these strategies may lead to a required flexibility at the systems level, which in turn results in a required manufacturing flexibility in combination with methods for delivering flexibility (such as organization of work, information systems, and operations control methods).

Gerwin (1993) positions manufacturing flexibility in a strategic decision-making framework as an element of manufacturing strategy. Manufacturing strategy decision-making is viewed as an evolving process that takes as input the environmental uncertainties and the systems performance resulting from the use of existing manufacturing flexibility, and that formulates manufacturing flexibility requirements and measures to reduce or redefine the environmental uncertainties. At the operational level, methods for delivering flexibility are used to convert existing flexibility into systems' performance. According to Gerwin (1993) flexibility has also a pro-active function in creating uncertainties that competitors cannot deal with. Thus the decision about the relevant flexibility dimensions, the appropriate flexibility levels (range, mobility, uniformity) per dimension, and the decision about methods for delivering flexibility have to be made at the strategic level, since investments may be required, and effects on performance can be long term. Using the available flexibility to realize the required performance is the realm of tactical and operational decision-making.

The relationships in detail between strategic choices and flexibility options are still unclear and seem to depend on type of industry (mass production, batch production, one-of-a-kind production), and on the innovativeness of the markets operated in. However, efforts have been made to identify the qualitative relationships between technological choices and the effects for the flexibility at the various flexibility dimensions. Such taxonomies of the flexibility concept have been developed by Browne et al. (1984), Gupta and Goyal (1989), Sethi and Sethi (1990), and Beach et al. (2000).

Koste and Malhotra (1999) provide a theoretic framework for analyzing the dimensions of manufacturing flexibility. In their paper they present a set of definitions of flexibility dimensions and discuss the causal relationships between these flexibility dimensions. Each of the dimensions is further characterized by a range of values that can be attained without incurring high transition penalties or large changes in performance outcomes. Thus each of the flexibility dimensions can be characterized by the transition costs or speed of changing from one state to another state within the range, also known as mobility, and by the uniformity of the performance of production across the range.

Koste and Malhotra (1999) conclude to the following relationships between the various flexibility dimensions:

- The basis of the flexibility of a manufacturing system is formed by the *machine flexibility*, the *labor flexibility* and the *materials handling flexibility*. Machine, labor and materials handling flexibility all pertain to the range of processing routings that can be used to make a product;

a process routing being a sequence of operations on machines that lead to a product.
- *Operation flexibility* and *routing flexibility* are conditional on the machine flexibility, the labor flexibility and the materials handling flexibility, but are also influenced by product design and process design. Thus, product and process design can be important sources of manufacturing flexibility. Also, given the routing flexibility and the operations flexibility desired, machines, labor, and materials handling systems can be chosen in order to achieve this routing and operation flexibility.
- Given the types and numbers of machines, labor and materials handling systems available, and given the process routings for the products, a shop has a certain *volume flexibility* and a certain *mix flexibility*. Automation levels, labor contracts, work organization and management techniques are important factors for achieving volume flexibility and mix flexibility. These form what Gerwin (1993) calls the 'system for delivering flexibility.'
- Volume flexibility and mix flexibility are dimensions of output flexibility. Two other dimensions of output flexibility are the *product modification flexibility* and the *new product flexibility*. Both are dimensions of product flexibility and refer to the range of different products that can be made. A product modification refers to a product change which leaves the functional characteristics unchanged, but leads to an improved design; a new product will have changed functional characteristics. Both flexibilities are conditional on the machine flexibility, the labor flexibility and the materials handling flexibility, but also depend on the capabilities and work organization of the product design and the process design departments.

  Apart from the range, also the speed of new product flexibility is important. This refers to the time and costs needed to deliver a new product, or a modified product, to the market. Speed and costs of a product introduction are determined by the time and resources needed for the product and process design phase, and also by the time and costs required to fill the production system with work-in-process and inventory of the new product, up to the level needed to support the planned supply level in the market. Thus, a production system that can economically work at lower levels of work-in-process and inventory can have a higher ability at the product flexibility dimension.
- The last dimension of output flexibility mentioned by Koste and Malhotra (1999) is the *expansion flexibility*. Expansion flexibility implies adding more or new resources (machines, labor, materials handling systems) to the production system in order to be able to expand the overall output of the system. Here also, range, mobility and uniformity are relevant aspects. Expansion flexibility is created at the strategic level in a company and requires the planning and control of options for acquiring capital goods, material supply, expanding human resources,

expanding subcontracting, and requires the planning and control of the financing resources needed for the expansion.

2.1 Discussion

Overlooking the published research over the last decades, it seems that researchers agree on the set of relevant dimensions of flexibility and the relevant aspects per dimension. We therefore do not foresee many new contributions in this area on the short-term. Quantitative research is needed to identify the exact relationships between the flexibility dimensions and between flexibility and performance.

The relevant dimensions of flexibility at the plant level are: volume flexibility, mix flexibility, new product or product modification flexibility and expansion flexibility. Machine, labor, materials handling, operation and routing flexibility are all internal or shop floor flexibility dimensions of a plant. The internal flexibility dimensions do have an impact on the flexibility dimensions at the plant level. However, the exact relationships between the internal and external flexibility dimensions will depend on the detailed flexibility characteristics of the manufacturing systems (machines, labor, materials handling system), the detailed flexibility characteristics of the activities that are needed to produce a product (operations, routings), and the detailed characteristics of the operations planning and control system of a plant. These detailed characteristics are plant specific, and therefore different plants will be difficult to compare unless we can quantify the flexibility dimensions, the flexibility aspects and their relationships.

Quantitative research on manufacturing flexibility aims at identifying the exact relationships between the amount of flexibility on one or more flexibility dimensions of a system, and the performance of that system. Since formal methods are used to obtain results, the models studied are small scale, entail only one or two flexibility dimensions, and study the impact of flexibility on one or two performance dimensions.

Model-based research on flexibility generally deals with models of production systems consisting of a set of different machines, and a number of workers. Materials handling systems are sometimes mentioned but seldom explicitly modeled. An exception to this is the vast literature on Flexible Manufacturing Systems (FMSs), where the production capabilities of integrated FMSs are often constrained by the characteristics of the connecting materials handling system. In this chapter we will not pay special attention to (integrated) FMSs.

In model-based research it is generally assumed that each machine can perform a range of operations, and each worker can work at a certain range of machines. Further, an operation is defined as a specific transformation of a specific piece of material. The definition of operation flexibility in Koste and Malhotra (1999) implies that a product has operation flexibility if there exist alternate sequencing plans for producing the product, i.e., alternate sequences of operations, each operation to be executed on one of the machines in the system. In model-based research [e.g., Jordan and Graves (1995)] such a sequence of operations that leads to a product, is called a process. In industrial engineering literature, a process plan of a product is defined as a planned series of operations which advances a material from one stage of completion to another [Tanner (1985), p. 322]. Since operations are carried out on machines, and each type of machine has a limited range of operations that it can perform, the set of machines in a production system completely determines the range of operations that can be performed, and therefore the range of products that can be made. However, there is often a one-to-many relationship between products and operations. A product is uniquely defined by its technical product specifications, but various sequences of operations may exist in a production system that, for a given raw material, lead to the same product.

Depending on the level of automation, workers are needed for carrying out the operations. This can range from operations that are carried out by workers using a machine, to operations that are carried out by machines that operate on palletized parts that are positioned and transported by a fully automated materials handling system, only requiring a worker for loading and unloading the pallets. Worker flexibility serves as a hedge against the disturbing effects of worker absenteeism and worker turnover, just as routing flexibility can serve as a hedge against the disturbing effects of machine failures. However, worker flexibility can also be used to response to variations in demand for capacity across the different machine types in a production system. In this chapter, we will focus on this latter aspect of worker flexibility.

The different technologies (different machines, different materials handling systems, different workers) available in a production system determine the range of different products that can be made; the broader the range of different technologies, the broader the range of products. A general machine shop consisting of a wide range of different machine tools can clearly produce a much wider range of different products than specialized parts manufacturing shop of a gearbox plant. It follows that the broader the technology range, the higher the product modification and product innovation flexibility that can be supported by a production system.

A large technological range however comes at a price, since it increases the costs per unit produced by the production system relative to the cost per unit of a more specialized or more focused plant. When deciding about the range of different technologies of a production system, a trade-off has therefore to be made. Focusing the production system on a specific range of products, requiring only a specific range of operations and routings, allows for the selection of specialized machines, workers, tools, materials handling systems, and working methods. This leads to increased learning effects, to increased output for a given amount of capital investment, and therefore to a decrease in cost per unit produced. These are known as the benefits to be obtained from focused factories [Skinner (1974)].

The decision about the technology range of a production system has to be based on a trade-off between the benefits to be obtained from low costs per unit produced, and the risks of not finding sufficient demand for the products that can be made on the system, over the economic lifetime of the system. Technology range therefore primarily caters for uncertainty regarding process requirements that follow from future product modification and future product innovations. To our knowledge no research has been reported in operations management literature on the strategic question about the range of technologies that should be available in a production system. This clearly belongs to the realm of operation strategy and not to operations management.

A supply chain consists of a network of transformation processes, each process performed in a plant. Thus, in a supply chain context we are primarily interested in the manufacturing flexibility dimensions at the plant level, for each of the plants involved in the supply chain. When designing a supply chain for flexibility, we have to consider the relevant flexibility dimensions of each plant. Flexibility at the plant level is expressed as constraints on the volume, mix and timing of the items that are produced by the plant. In the next section we will discuss model-based research on the use of machine flexibility to cope with uncertainties in the level of demand (volume and mix). New product and product modification flexibility depends also on the throughput time of production in the plants involved in the supply chain. In Section 4 we will discuss the impact of machine, worker, and routing flexibility on production throughput time, and therefore on new product flexibility.

3 The flexible machine investment problem; volume and mix flexibility

We discuss the research into the optimal use of machine flexibility in order to cope with demand level uncertainty. We restrict our discussion to the research that has been performed by Andreou (1990), Fine and Freund (1990), Gupta (1993), Jordan and Graves (1995), Boyer and Keong Leong (1996), and Van Mieghem (1998). All this research studies models with multiple products and multiple machines where each product requires only one operation (i.e., a process for the product consists of one operation only) and where some of the products can be made on different machines. Thus, some of the machines can perform more than one operation or process. In all this literature, machines are referred to as 'resources,' a convention that we will adopt in this section. The problem is studied as an investment problem, asking the question to what extent to invest in resources that can perform more than one operation, given the uncertainty about the *level* of demand for the products to be produced.

Andreou (1990) presents an investment model to calculate the dollar value of flexible resource for a two-product production system with (correlated) stochastic future demand levels per product. He considers a production system consisting of a mix of dedicated and flexible resources and calculates the option value of flexible capacity as a function of the uncertainty in demand level, the costs of dedicated and flexible capacity and the revenues from sales. The analysis shows that value of flexibility can be substantial, especially under high uncertainty in demand. The analysis also shows that most of the benefits to be obtained from product mix flexibility are captured by having only a certain percentage of total capacity to be flexible, the optimal percentage being dependent on the variability of the demand level and the correlation in demand levels between the two products.

Fine and Freund (1990) present a model of the cost-flexibility trade-offs involved in investing in product-flexible manufacturing resources. They formulate the problem as a two-stage stochastic program. In the first stage they make a capacity decision, before the resolution of uncertainty in demand level. In the second stage, after demand levels for products are known, the firm takes its production decisions, constrained by the first stage resource decisions. They consider a situation with n different product families with each product family having a dedicated resource, and one flexible resource that can be used for any of the n product families. The problem consists of deciding about the amount of dedicated resource, K_j, $j=1,\ldots,n$, and the amount of flexible resource, K_F,

- assuming per resources linear acquisition costs as a function of amount of resource installed, K_j,
- assuming linear production costs as a function of amount of product produced on dedicated or flexible resource,
- assuming that revenues are strictly concave as a function of the amount of product sold,
- assuming linear, technology-independent variable production costs, and
- assuming probabilistic information about the demand level per product family.

Demand level uncertainty is modeled as k possible states, $i=1,\ldots,k$, each state corresponding to a vector of demand levels per product, and each state occurring with probability $p_i > 0$ with $\sum_i p_i = 1$. Demand uncertainty is thus modeled as a discrete set of separate demand level and product profitability scenarios.

Fine and Freund (1990) present an optimization model with linear constraints that can solve the above problem and that simultaneously provides the optimal amount of products to be produced under each demand scenario. They show that for the two-product problem, the allocation of production to dedicated resource and to flexible resource is unique; uniqueness cannot be guaranteed for more than two products. The model also is used to show that flexible capacity should be acquired when the expected value of its best use for each realization of demand, summed over all possible realizations of demands, exceeds its costs. Given the model used this should not come as a surprise but it is an important managerial insight, implying that the value of an option must be evaluated on the basis of all future scenarios including their probabilities of realization, instead of on the basis of its value under the most likely scenario only.

Fine and Freund (1990) also provide results for the optimal capacity levels. For the two-product family situation, family A and family B, they prove that an increase in capacity costs for dedicated resource A leads to a decrease of K_A, an increase in amount of flexible resource, K_{AB}, and a decrease in K_B. Hence the flexible resource substitutes for dedicated resource B as well as for dedicated resource A. Furthermore the magnitude of decrease in K_A exceeds the magnitude of increase in K_{AB}, which in turn, exceeds the magnitude of decrease in K_B. These results illustrate the complex interactions between the problem parameters, on the one hand, and the optimal levels of investments in dedicated and flexible resource, on the other hand.

For two numerical cases, Fine and Freund (1990) also investigate the sensitivity of the solutions of their model to correlation and variability in demand. They consider a symmetric two-product case with for each product, A and B, either high, medium or low demand. They model perfectly negative correlated, perfectly positive correlated and non-correlated demand, and for each of these three cases they investigate different levels of riskiness (uncertainty) in demand. Their results illustrate that the need for flexible capacity increases relative to level of riskiness in the presence of perfectly negative correlated demand, and is zero, regardless of the level of riskiness, in the presence of perfectly positive correlated demand. For uncorrelated demand the analysis of the case shows that the need for flexible capacity is zero for moderate levels of riskiness, then increases for increasing riskiness up to a maximum value and then decreases again for a further increase in riskiness. This rather counterintuitive result points out that the direction of correlation in demand and the level of risk do not by themselves constitute enough information to predict the need for flexible capacity. A final remarkable result of the analysis of the two-product model is that for each of the three cases, the optimal value of the objective function is increasing in the level of riskiness.

Fine and Freund (1990) studied a model where resources are either dedicated to one of the products, or can process all products (the flexible resource). Gupta (1993) studies a model where all resource can be flexible. Building on the results obtained in Gupta and Buzacott (1991), he considers the situation where N products with uncertain demand levels are to be made on M resources, where each of the resources has the same capacity, Q, and can process at most $1 \leq K \leq N$ different products. He develops a two-stage stochastic programming formulation of the problem to determine the optimal levels of M, Q and K, as a function of uncertainty in demand, sales revenues and resource costs.

The model consists of two formulations: P_1 and P_2. The first model serves to find values for P_{ij}, the amount of product i produced on resource j, in order to maximize the total profit, given a realization of demand, $d_1, \ldots, d_N$, and given values for M, Q and K. It is assumed that r is the revenue earned, net of all costs, per unit of any product regardless of type.

$$P_1: \max \phi(d_1, d_2, \ldots, d_N) = \sum_i \left\{ r \sum_j P_{ij} \right\}$$

subject to:

$$g_1: K - \sum_i \delta(P_{ij}) \geq 0 \qquad \forall j = 1, \ldots, M$$

$$g_2: d_i - \sum_j P_{ij} \geq 0 \qquad \forall i = 1, \ldots, N$$

$$g_3: Q - \sum_i P_{ij} \geq 0 \qquad \forall j = 1, \ldots, M$$

$$g_4: \sum_i \sum_i P_{ij} = \min\left(MQ, \sum_i d_i\right)$$

$$g_5: P_{ij} \geq 0 \qquad \forall i = 1, \ldots, N, \quad j = 1, \ldots, M$$

The second model serves to find values of M, Q and K to maximize the expected total profit resulting from optimally solving the problem P_1 after the realization of demand, given that demand is only known in probabilistic terms, denoted as $D_1, \ldots, D_N$.

$$P_2: \max[\Psi(M, Q, K)] = E[\Psi(D_1, D_2, \ldots, D_N] - C(M, Q, K)$$

where $E(\cdot)$ is the total expected profit, and $C(\cdot)$ is the investment costs. In view of the large amount of literature dealing with economies of scale, it is reasonable to assume that $C(\cdot)$ is concave in the arguments M and Q; however, the same cannot be said about the argument K. We may expect that the investment costs increase with an increasing rate as a function of K, the number of products that can be produced on a resource.

Analysis of P_1 shows that the problem is combinatorial hard. Gupta (1993) develops a heuristic to solve problem P_1 and uses simulation to calculate the 95% confidence interval for average revenue as a function of M, Q and K. Next, adaptive random search is used to find the best combinations of M, Q and K, for a given investment function $C(M, Q, K)$.

The procedure is applied to a number of 10-product problem instances with the following investment cost function

$$C(M, Q, K) = M^a(\beta_0 + \beta_1 Q + \beta_2 \ln K),$$

which is concave in all arguments. For the example cases studied, it appears that the relative benefit of flexibility depends on scale economies. In particular if scale economies is high (low value of a) the benefit of flexibility is limited. The data furthermore show that with certain cost structures, duplication of resources with limited flexibility may provide just as good an ability to cope with variations in demand volume and demand mix, as investments in highly flexible resources.

In the research of Fine and Freund (1990) all resource flexibility is concentrated in one resource only, whereas in the research of Gupta (1993) all of the resources are equally flexible. Gupta (1993) notices that the problem of deciding about the optimal level of resource flexibility is combinatorial hard and therefore difficult to solve to optimality for large problems. Jordan and Graves (1995) specifically investigate how to configure resource flexibility, based on intuitively developed principles about optimal configurations. To understand how to create good flexibility configurations they consider a 10-product 10-resource example. Expected demand for each product is 100 and capacity of each resource is 100. Demand follows a truncated normal distribution with a standard deviation of 40 and minimum and maximum possible demands of 20 and 180 units, respectively. Product demands are independent. A simulation model is used to randomly sample demand for each product, allocate demand to resources to maximize the demand filled (unfilled demand is lost) subject to capacity constraints, and collect statistics on sales, lost sales, and capacity utilization. They first investigate the no-flexibility case, where each resource is exclusively dedicated to one product, and the total flexibility case, where each resource can produce all products. For the no-flexibility case, expected sale is 853 units and expected utilization is 85.3%. Total flexibility results are 954 units and 95.4%, respectively. Starting from the no-flexibility case, Jordan and Graves (1995) add flexibility incrementally to the system. They add one 'link' (e.g., product 1 can also be produced on resource 2) at a time and measure the impact on sales and utilization. They first add product 1 to resource 2, then add product 2 to resource 3, then add product 3 to resource 4, and so on. The tenth link is to add product 10 to resource 1. Their results show that, if well configured, limited flexibility can achieve almost all of the benefits with respect to sales and capacity utilization that can be obtained from full flexibility.

Jordan and Graves (1995) introduce the concept of 'chaining' to formulate the principles of how to configure resource flexibility. A chain is a group of resources and products that all are connected directly or indirectly by assignments of products to resources. No product in a chain can be produced by a resource outside that chain; no resource in a chain can produce a product from outside that chain. Jordan and Graves (1995) show that for a given number of 'links,' the performance of a system is maximized by creating one chain of maximum length. Thus, for the example case, with 10 additional links available creating one chain of length 10 gives a better performance than creating 5 chains of length 2 each.

Jordan and Graves (1995) formulate the following guidelines for creating resource flexibility in a multi-product multi-resource system:

- try to equalize the number of resources (measured in total units of capacity) to which each resource in the chain is directly connected,
- try to equalize the number of products (measured in total units of expected demand) to which each resource in the chain is directly connected,
- try to create a circuit that encompasses as many resources and products as possible.

The key behind chaining is that all products in the chain effectively share all the resource capacity in the chain. In the 10-product 10-resource case situation studied, the expected sale was equal to capacity. To see how the value of flexibility changes when resource capacities are varied, Jordan and Graves (1995) varied the total capacity of the 10 resources between 500 and 1500 units where capacity is equally split among the resources. Their results show that the benefits of adding flexibility are substantial for a wide range of total capacity. Even when capacity is 25% above or below expected demand, expected sales increases by more than 5%.

Jordan and Graves (1995) also develop a measure for the inflexibility in a given product-resource configuration. The measure, $\pi(M^*)$, is defined as the maximal probability over all groupings of products (M) that there will be unfilled demand for a group of products, while simultaneously there is excess capacity at resources making other products. The measure indicates whether adding more flexibility to the configuration is likely to lead to higher expected output.

Their results suggest that for high levels of demand uncertainty, and many products and resources, limited resource flexibility with only two products per resource may not provide the same ability to cope with demand uncertainty as total flexibility. However, even with relatively high demand uncertainty and many products and resources, limited flexibility with not more than four products per resource can provide almost all of the benefits of total flexibility.

The model studied by Jordan and Graves (1995) does not consider costs of capacity nor costs of flexibility. Boyer and Keong Leong (1996) expand the model used by Jordan and Graves (1995) to also include the costs of changing over a resource from one product to another. Such a system has therefore restricted mobility [Upton (1994)]. Change-over costs are modeled as a loss of capacity; a percentage of available capacity is lost if the resource is used for two products. They develop a binary integer programming formulation of the problem that is embedded in a simulation model to maximize the expected output under stochastic demand levels for the different products, and for a given configuration of resources. For two case problems taken from the automobile industry, the model is used to investigate the effect of different levels of resource flexibility and change-over costs on expected output. For the 10-product 10-resources configuration studied by Jordan and Graves (1995) with the demand per product equal to 100 and capacity per resource equal to 100, Boyer and Keong Leong (1996) show that the percentage improvement in expected output for resource flexibility over no resource flexibility decreases for increasing change-over costs, and show that, for change-over costs less than 50%, the decrease in improvement is roughly linear with increases in change-over costs. Even when change-over costs are quite high, the benefits of limited process flexibility are quite large; for change-over costs up to 100%, configurations with resource flexibility have a higher expected output than configurations without resource flexibility. This is because with resource flexibility, it is still possible to decide which of the products is going to be made on a flexible resource, which is not possible with inflexible resource.

The results of Boyer and Keong Leong (1996) confirm that it is rarely beneficial to pursue total flexibility, since limited flexibility, if configured in one product-resource chain, offers approximately 95% of the output benefits of total flexibility.

Van Mieghem (1998) studies the optimal investment decisions in flexible manufacturing resources as a function of product margins, investment costs and multi-variate demand level uncertainty. His contribution to existing knowledge is that he studies the case where the products have different margins. He considers a two-product firm that has the option to invest in product-dedicated resources and/or in a flexible resource. He models the problem as a two-stage multi-dimensional newsvendor problem. First the firm must decide on a non-negative vector of resource capacity levels, $K \in \mathfrak{R}_+^3$, $K = (k_1, k_2, k_3)$ before the product demand vector, $D \in \mathfrak{R}_+^2$, $D = (d_1, d_2)$, is observed. After demand is observed the firm decides on production quantities per resource $x = (y_1, y_2, z_1, z_2) \in \mathfrak{R}_+^4$, where $y_i + z_i$ is the total amount produced of product i, y_i is the amount of product i produced on the dedicated resource, and z_i is the amount of product i produced on the flexible resource. The firm chooses its production vector x so as to maximize operating profit

$$\max_{y,z \in \mathfrak{R}_+^2} p_1(y_1 + z_1) + p_2(y_2 + z_2)$$

subject to :

$$y_1 \leq k_1 \tag{3.1}$$

$$y_2 \leq k_2 \tag{3.2}$$

$$z_1 + z_2 \leq k_3 \tag{3.3}$$

$$y_1 + z_1 \leq d_1 \tag{3.4}$$

$$y_2 + z_2 \leq d_2 \tag{3.5}$$

where $p \in \mathfrak{R}_+^2$ is a prize vector.

It is assumed that D is a continuous random vector that has a joint probability density function, g, which is positive over its support. The investment costs are linear in the capacity, $C(K) = c'K$, where $c \in \Re_+^3$ is a vector of marginal investment costs.

The investment decision is modeled as:

$$\max_{K \in \Re_+^3} V(K) = E\pi(K, D) - C(K)$$

where $E\pi(K, D)$ is the expected value of the operation profits.

Mathematically analyzing the properties of this problem, Van Mieghem (1998) shows how optimal investment depends on costs and prices. He first solves for the optimal contingent production decisions $x(K, D)$ and the associated three vector $\lambda(K, D)$ of optimal dual variables of the capacity constraints (3.1) through (3.3) in the product mix problem. Given a capacity vector $K \in \Re_+^3$ the demand space $\Re_+^2$ is partitioned into five domains $\Omega_0, \Omega_1, \Omega_2, \Omega_3$ and Ω_4 where Ω_0 denotes the firm's capacity region, and Ω_1 through Ω_4 denote disjunct regions that completely cover the set of demand realizations that the firm cannot serve (for details we refer to the paper).

Van Mieghem (1998) formulates the optimality equation in terms of the dual variables:

An investment vector $K^* \in \Re_+^3$ is optimal if and only if there exists a $V \in \Re_+^3$ such that:

$$\begin{pmatrix} 0 \\ p_2 \\ p_2 \end{pmatrix} P(\Omega_1(K^*)) + \begin{pmatrix} p_2 \\ p_2 \\ p_2 \end{pmatrix} P(\Omega_2(K^*)) + \begin{pmatrix} p_1 \\ p_2 \\ p_1 \end{pmatrix} P(\Omega_3(K^*))$$

$$+ \begin{pmatrix} p_1 \\ 0 \\ p_1 \end{pmatrix} P(\Omega_4(K^*)) = C - V, \tag{3.6}$$

$$V'K^* = 0 \tag{3.7}$$

where $P(\Omega_j(K^*))$ denotes the total demand probability mass in the domain $\Omega_j(K^*)$.

The optimal investment is found by superimposing the multi-variate demand distribution on the capacity model and adjusting the lines that constrain the feasibility region such that the probabilities of the four domains $\Omega_1, \ldots, \Omega_4$ offset the marginal investment cost C as in the optimality Eq. (3.6). Van Mieghem (1998) derives the conditions under which it is optimal to invest only in dedicated resources, under which conditions it is optimal to invest in one dedicated resource and in the flexible resource, and under which conditions it is optimal to invest in all three resources, under uncorrelated demand. Furthermore he shows that the optimal value $V^*(K)$ is a non-increasing convex function of the price vector, p. He also studies the effects of the parameters of demand uncertainty in the optimal investment. For both perfectly positively and perfectly negatively correlated demand he derives the conditions regarding marginal prices and costs under which it is optimal to only invest in dedicated resources, in one dedicated resource and the flexible resource, and in all three resources. He shows that price conditions exist under which it is optimal to invest in flexible resource, also under positively correlated demand, which is not the case for this problem when the products have equal prices. For details about these results we refer to the paper.

3.1 Discussion

The literature discussed above shows the value of resource flexibility for coping with demand *level* uncertainty for a group of products. Resource flexibility can be a substitute for dedicated capacity. The optimal split between dedicated and flexible capacity depends on the cost difference between flexible and dedicated capacity, price differences between the products, the riskiness of demand, and the correlation between product demands. Change-over costs between products produced over the same resource decrease the benefits obtained from resource flexibility but do not eliminate them. For a given flexibility configuration the marginal value of adding more flexibility to the system is a decreasing function. In other words, the major part of the benefits to be obtained from resource flexibility is already obtained with a limited amount of flexibility. The flexibility configuration has been shown to be an important factor; for a given number of product-resource links, the best performance is obtained with creating the longest chain possible; the performance of a fully chained system comes very close to the performance of a totally flexible system.

The papers discussed provide important managerial insight and some of them also demonstrate the use of the results in real life industrial applications. An important direction for further research in this area can be found in extending the optimal investment problem with the option to sell unused capacity under various conditions to other parties. Another interesting extension would be to study the case where the investor has to invest in a configuration of resources on which he produces products that are supplied to different customers under different contracts. This situation occurs when different supply chains use a common supplier and supply contracts have to be negotiated in parallel to the investment decision. This research would combine knowledge from contract theory, game theory and investment theory. A third promising research avenue would be to include the trade-off between investments in change-over cost reduction and inventory carrying costs in the flexible resource investment decision problem. In the presence of change-over costs on a flexible resource, production will take place in batches, and as a result the total costs of operating the system must include change-over costs and inventory carrying costs. To our knowledge such flexible resource models have not been studied yet.

What is the relevance of this research for supply chain design? The research deals with the investment decision in flexible resources in anticipation of future demand for a group of products, where future demands are uncertain. The relevant flexibility measures at the supply chain design level are volume flexibility, mix flexibility and modification or new product flexibility. Volume and mix flexibility for a product family can be created by installing excess capacity, or by contracting excess capacity and excess material supply in each of the plants involved in the supply chain. This would be costly, and therefore would not be economically justified, unless the demand uncertainties can be pooled at the resource level in each of the plants. Pooling can be achieved by investing in flexible resources. Thus when setting up a supply chain the resource configuration of each of the plants involved in the chain should be carefully evaluated for the volume and mix flexibility that it can provide to the supply chain. It is therefore not only important to know what production output can be realized with a given resource configuration, but also what is the optimal resource configuration to invest in, given the resource costs, the product margins and uncertainty in demand. For the investment decision for final assembly plant for a product family, the models discussed can be directly applied. Most of the models were inspired by industrial cases of this type. However, a supply chain not only consists of a final assembly plant but also may encompass many first and second tier suppliers. In each of these supplying plants capacity has to be installed or reserved for the supply chain. Different options in terms of resource configurations may be available for realizing specific potential for supply in the chain. Designing a supply chain requires the coordinated design of the resource configuration across the plants in the chain. Information about the resource configuration options per plant and its consequences for supply is input to this design.

The third flexibility dimension at the supply chain level is the product modification or new product flexibility. This refers to the time needed for making a modification or a new product available in the market (the new product lead time). In the context of this chapter we restrict new product lead time to the time needed for introducing new variants of a product family (the new product lead time), assuming that the resource configuration used in the supply chain can perform the process needed for the modified or new products. An important factor in the new product lead time is the production throughput time in the supply chain; if production throughput times are short, the new product lead time can be short. Resource flexibility has been shown to have a high impact on production throughput times. In the next section we discuss literature on the relationships between resource flexibility, including range, mobility and uniformity aspects, and production throughput times.

4 Resource flexibility, range, mobility, uniformity and throughput time

In this section we discuss the research on the effects of machine and worker flexibility, on production order throughput time in relation to the costs and capacity losses incurred by changing over a resource from processing one product to another product.

In the literature in this field the general assumption is made that demand per product can be modeled as a stationary stochastic variable *with a known mean*. Demand manifests itself over time as a random variable, either as a random interarrival time between arrivals of demand for a product, or as a random amount of products demanded per time period. Also it is assumed that the capacity of the production system is given, and that capacity is larger than the capacity needed to serve the average demand. Thus, in the long run, all demand can be served, and the flexibility of the production system mainly serves to make the system responsive to the short-term variations in demand. It should be noted that for some production systems, demand responsiveness can also be created by keeping stocks. In such production systems, both resource flexibility and stocks are means to achieve short-term demand responsiveness.

The effects of resource flexibility on short-term demand responsiveness has been amply studied in operations management literature. In this section we discuss a selection of papers from this literature. Specifically we discuss research performed by Wayson (1965), Nelson (1967, 1970), Fryer (1973, 1974, 1975), Treleven and Elvers (1985), Porteus (1985), Park and Bobrowski (1989), Vander Veen and Jordan (1989), Malhotra and Ritzman (1990), Malhotra, Fry, Kher, and Donohue (1993), Malhotra and Kher (1994), Kher (2000) and Garavelli (2001).

Other research results can be found in: Weeks and Fryer (1976), Hogg, Philips, and Maggard (1979), Gunther (1981), Treleven (1989), Corbey (1991), Park (1991), Bernardo and Mohamed (1992), Bobrowski and Park (1993), Felan, Fry, and Philipoom (1993), Nandkeolyar and Christy (1992), Wisner and Pearson (1993), Morris and Terinze (1994), Benjaafar (1994), Hutchinson and Pflughoeft (1994), Fry, Kehr, and Malhotra (1995), Benjaafar and Ramakrishnan (1996), Daniels, Hooper, and Mazzola (1996), Ho and Moody (1996), Jensen, Malhotra, and Philipoom (1996), Das and Nagendra (1997), Shafer and Charnes (1997), Kher, Malhotra, Philipoom, and Fry (1999), Jensen (2000), Smunt and Meredith (2000), Garg, Vrat, and Kanda (2001) and Nam (2001).

Most research on this subject uses systematic computer simulation as a research tool. A short description of the simulation model used will be given for each study.

One of the first to research the effects of resource flexibility on throughput time is Wayson (1965) who performed a simulation study of a simple job shop. The shop consists of nine work centers, each containing one machine. Orders arrive according to a Poisson process. The number of operations per order, g, follows a geometric distribution function on the domain $g \in N_+/0$ with probabilities $(1/9)(8/9)^{g-1}$. Planned order routings are generated as follows: Each work center has equal probability of being the first work center to be visited. The work center for each next operation is selected with equal probability on the condition that immediate succeeding operations visit different work centers. This procedure models the situation where there is a high diversity in order routings.

The model further assumes zero transportation times, no labor constraints, and 100% resource availability. At each work center, operation processing times are negative exponentially distributed with the same parameter. The order arrival rate is such that the shop faces a 90% utilization rate. We recognize this as a typical job shop model. Assuming first come first served sequencing at the work centers, work order throughput time characteristics can be calculated for this system using basic queuing theory.

Resource flexibility is modeled as a real variable, z, $0 \leq z \leq 8$ where for instance $z = 2.4$ means that each operation can be performed at two work centers other than the planned work center, and that there is a probability of 0.4 that an operation can be performed at three work centers other than the planned work center. Thus $z = 0$ means no resource flexibility; each operation must be performed at the planned work center; $z = 8$ means total resource flexibility.

The resource flexibility is used as follows: if upon completion of an operation of an order at a work center there is a next operation that has to be performed the next work center is selected among the set of work centers where the next operation can be performed. Among the possible work centers, the work center is selected that, at that time, has the least number of orders in queue. At each work center, orders are processed on a first come first served basis. It is assumed that execution of an operation at alternative work centers takes the same processing time as executing it at the planned work center (perfect uniformity).

Performance is measured with the average queue length (which is proportional to average throughput time). The results of Wayson (1965) demonstrate the strong impact that resource flexibility can have on average order throughput time. If each operation can be performed at two work centers (the planned and one other work center) the average queue length is only 3.4, as compared to 9.8 without resource flexibility. Even if there is only a 40% probability that an operation can be performed at an alternative resource ($z = 0.4$), the average queue length goes down from 9.8 to 5.4. His results also show the strong decrease in marginal benefits to be obtained from an incremental increase in resource flexibility; there is hardly any decrease in average queue length if the flexibility is increased from 3 to 8 alternative work centers. As noticed before, if well-configured, most of the benefits of flexibility can already be obtained with a little flexibility. Wayson (1965) also measured the fraction of times that an alternative resource is used for processing of an operation (which indicates control effort). For a flexibility of 0.4, an alternative work center is used for about 20% of the operations. Thus even with limited flexibility, using this flexibility in an unconstrained way can lead to substantial control costs.

The results of Wayson (1965) are obtained for a model that only captures a few elements of real life production systems. For instance, it is assumed that alternative resources are equally efficient as the planned resource; the costs of exercising flexibility, such as forgetting and relearning are neglected, and it is assumed that machine capacity is the only limiting resource.

Nelson (1967) was one of the first to study the use of worker flexibility in worker and machine constrained production systems. Using computer simulation he investigates a 2 work center job shop with two identical machines per work center. The shop has characteristics in terms of arrival times, processing times and routings similar to the model used by Wayson (1965). Nelson (1967) varies the design of the system by studying the system with 1, 2, 3, and 4 workers, where each worker can work with equal efficiency in each work center. He studies centralized control, where each worker after completing his job returns to a central pool to be allocated to his next job, and decentralized control where each worker remains to work at his current work center until he runs out of work and then goes to the other work center, if work is available there. Three queue disciplines are used: First Come First Served, First in System First Served, and Shortest Processing Time, in combination with five labor allocation rules, among which 'random' and 'most work in queue.' The results of the simulation study indicate that worker flexibility can strongly decrease the mean and variance of the order throughput time, and that the magnitude of the effect depends on the labor allocation procedure and queue discipline used. Centralized labor allocation performs consistently better than decentralized labor allocation.

Centralized worker allocation is especially important in case workers have different efficiencies in different work centers. Workers then should preferably work in the work center were they are most efficient, unless no work is available there, and should return to this work center as soon as sufficient work is available there. Labor transfer costs (change-over time, forgetting effects) however, would in turn limit the frequency of worker transfer [Nelson (1970)].

Fryer (1973, 1974, 1975) investigates the effect of various labor allocation rules on order throughput time, for a three-department production system, with each department consisting of four work centers with two identical machines. The production system has 12 workers who all can work with equal efficiency on all machines. Order arrival times, order routings and order processing times are all random variables with parameters such that the average worker utilization rate is 90%. Fryer (1973) distinguishes inter-departmental labor (re)allocation from intra-departmental labor (re)allocation, and also studies the effect of a (re)allocation delay on the effectiveness of labor flexibility. Workers are fully flexible over all work centers. He finds that interdepartmental flexibility has a pronounced impact on performance, as opposed to intra-departmental flexibility, and that the average order throughput time could decrease with about 40%, as compared with the reference situation where to each of the 12 work centers one worker is allocated who is never reallocated. This result is obtained with a zero reallocation delay. He also finds that the decrease in average order throughput time strongly depends on the reallocation delay. A delay equal to the average operation processing time still results in an average throughput time decrease of 23%. However, if it takes an amount of time equal to two times the average operation processing time to reallocate a worker after his current work center has become idle, only a decrease in throughput time of 3% remains. Thus, labor (re)allocations should be fast (or pre-planned) in order to be effective.

Treleven and Elvers (1985) investigate the effect of 11 different labor reallocation rules (where to allocate) on mean and variance of queue time, mean and variance of lateness, percentage of late jobs and total number of labor transfers, for a 9 work center job shop with two machines per work center, 9 and 12 workers, random order arrivals, random routings and random operation processing times, with parameters set such that labor utilization is 90%. Workers are equally efficient in work centers were they can work. Statistical analysis of a comprehensive simulation study of this model reveals no significant differences in performance between the shop performance under any of the eleven labor allocation rules, expect for the performance criteria 'total number of labor transfers.' Thus the decision about *where* to allocate a worker seems to have hardly any impact on the order-related shop performance measures. Therefore, the allocation rule should be chosen that minimize the number of reallocations, since reallocations come at a cost. The best rule in this respect is to allocate a worker to the work center with the longest queue.

Park and Bobrowski (1989) investigate the effect of order release mechanisms on the effectiveness of labor flexibility in a simulation study of a 5 work center, 2 machines per work center, 5 workers job shop. They consider centralized and decentralized labor (re)allocation for four different levels of worker flexibility. Their results indicate that the order release mechanisms has no significant impact on the effectiveness of worker flexibility on performance, and confirm the earlier findings regarding the strongly decreasing marginal value of increasing flexibility.

Malhotra and Ritzman (1990) investigate the environmental factors of a production system that determine the effectiveness of using resource flexibility to improve the throughput time performance of the system. They use three different shop configurations, each containing a number of fabrication shops and one or two assembly shops, to represent different levels of machine flexibility and labor flexibility. They investigate the impact that yield uncertainty, capacity tightness and lot sizes have on the effectiveness of using machine flexibility to achieve a good shop performance in terms of customer satisfaction, work in process and inventory, and labor costs, in an MRP driven production system. Computer simulation is used as a research tool. The results indicate that machine flexibility is especially helpful in environments characterized by high uncertainties, tight capacities and large batch sizes. In particular, as lot size and shop utilization increase, gains achieved by having machine or worker flexibility are high for the customer service measure; with small batch sizes and low shop utilizations improvements are moderate. Finally, simultaneous introduction of both machine and worker flexibility improves performance only marginally above that of either machine or worker flexibility alone.

Malhotra et al. (1993) investigate the impact of learning and labor attrition on the effect of worker flexibility on performance in machine and worker constrained job shops. They study a 6 work center shop with each work center containing 4 identical machines, and with 12 workers that have different levels of flexibility. Orders arrive randomly and visit each work center just once, whereas operation processing times are exponentially distributed. Parameters are set such that a worker utilization of 85% is achieved. Orders are assigned a due date proportional to the total work content of the order. A worker is reallocated as soon as there is no more work in the queue at his current work center, and is allocated to the work center that contains the job that has been in the system for the longest time. Finally, orders are dispatched according to the earliest due date. Malhotra et al. (1993) investigate the cost impact of acquiring worker flexibility under two levels of the learning rate, 75 and 85%, and two levels of time required to process the first order at a work center: two times the standard operation processing time and four times the standard operation processing time. This was combined with three levels of worker turnover and attrition, 0, 8 and 16%, and six degrees of worker flexibility, representing the number of work centers that a worker can work in. As performance criteria were used: the mean order throughput time, the average mean tardiness, the percentage of jobs tardy and the percentage of time workers spent in learning new tasks. Their results indicate that turn-over and attrition has a significant impact on the performance measures, where this impact was primarily present in the high learning loss environment (low learning rate and high initial processing time). They also found that for high learning rate, and for low learning rate in combination with low initial processing time, mean order throughput time decreased with increasing levels of flexibility (although at a decreasing rate, as found earlier). However, for low learning rate in combination with high initial processing time, the mean flow time increased again for flexibility levels larger than three (each worker can work at three resources). This is explained by the large fraction of time spent on learning in this learning environment. With full flexibility, time spent on learning may go up from 5% for zero attrition to 30%, for 16% attrition.

In a next study, Malhotra and Kher (1994) used the same model, the same experimental conditions and the same performance measures as in Malhortra et al. (1993) to investigate the effects of worker allocation policies in the presence of differences in efficiency when performing operations in different work centers, and in the presence of finite transfer delays. Efficiency ranges between 0.75 and 1.00 and transfer delays of 15% and 30% of the average operation processing times are considered. They investigated centralized and decentralized decisions about *when* to reallocate and five rules about *to which* work center to (re)allocate. They found that, with zero transfer delay, centralized decision-making performs best with hardly any difference between the allocation rules (which confirms the results found by Treleven and Elvers (1985)). However, in the presence of transfer delays the best performance is obtained with decentralized control (which limits the number of transfers) in combination with either allocation to the most efficiency work center, or the work center with the longest queue. Allocation to the most efficient work center resulted for all conditions in an acceptable low mean order throughput time of about 182 in a range 178 to 225.

In a sequel to the research of Malhotra and Kher (1994), Kher (2000) investigates the impact of flexibility on shop performance in machine and worker constrained job shops with simultaneous learning and forgetting effects. He uses the same shop model and performance criteria as in Malhotra et al. (1993). Learning is modeled with a log-linear model; forgetting is modeled by adjusting downwards the number of jobs processed, in function of the time passed since the last time this type of job had been performed. Furthermore it is assumed that for each work center, an upfront amount of training is required before the worker can start his first job at that resource. In the experimental design three flexibility policies, three forgetting rates and five worker turnover and attrition rates are combined. The results reveal that at a high forgetting rate (85%) acquiring and using worker flexibility has a negative impact on average work order throughput time and average tardiness. For lower forgetting rates (90 and 95%) work order throughput times and tardiness did improve when employing flexibility, even for high attrition rates. These results show that the benefits of worker flexibility strongl depend on the magnitude of learning and forgetting effects.

In all the research discussed above, the effects of set-up time and batch sizes are not considered. Orders are taken as given from outside, and for each of the operations of an order the processing time is given. Furthermore, the utilization of the shop is considered as an input parameter. However, it is known that batch sizes and shop utilization have a large impact on the throughput time. Therefore, when designing a supply chain for new product flexibility, the batch sizes and the utilization are important design parameters, that should be considered in parallel to machine flexibility and worker flexibility.

There is an abundance of research on the impact of batch sizes and capacity utilization on work order throughput time. In most of this research queuing models are used. For an overview we refer to Suri et al., Chapter 5, and Karmarkar, Chapter 6, in Graves, Rinnooy Kan, and Zipkin (1993) [see also Lambrecht, Ivens, and Vandaele (1998)]. In this chapter we discuss three papers that deal with the optimal setting of capacity, capacity use and batch sizes in an economic setting in the systems design phase.

Porteus (1985) studies the problem of whether or not to invest in set-up time reduction, taking into account reduced inventory related operating costs, but neglecting other advantages such as increased flexibility and increased effective capacity. He considers a single product situation with a deterministic sales rate, m, set-up costs, k, unit production costs, c, fractional per unit time opportunity costs of capital, b, non-financial per unit time inventory holding costs, h and fractional per unit opportunity cost of capital, i. He assumes that set-up costs can be influenced at a cost:

$$a(k) = a - b \ln(k) \qquad 0 < k \leq k_0,$$

where k_0 is the initial set-up costs.

Porteus shows that for this costs model the total relevant costs is a convex–concave function over the interval $[0, k_0]$ with a unique local minimum

$$k^* = \min\left(k_0, \frac{2b^2 i^2}{m(ic+h)}\right)$$

This result implies that high volume firms should invest more in set-up cost reduction than low volume firms. Furthermore, for high volume products, the optimal quantity is independent of the sales rate, and total cost is a strictly concave function of the sales rate, implying the usual economies of scale. Porteus (1985) also derives results for the simultaneous setting of optimal sales rate and optimal set-up costs, for a linear relationship between sales rate and price.

The result of Porteus (1985) could explain why over the last decades reduction of set-up time and set-up costs has predominantly taken place in mass production, specifically in mass assembly industry. In low volume capital goods industry, set-up times and set-up costs have hardly changed. In low volume supply chains, large batch sizes will remain the rule and short throughput time must be achieved by either excess capacity or resource flexibility [Malhotra and Ritzman (1990)].

Vander Veen and Jordan (1989) analyze machine investment decisions, considering machine flexibility, machine capacity, production forecasts and costs related to investment, inventory, set-up, material and labor. The paper focuses on the trade-offs between machine investment and utilization decisions.

Assuming machine speed as a decision variable, and assuming demand per product as a given, they develop a method to identify the optimal number of machines, M, to produce any number, N, of different products, also accounting for product allocation decisions, as well as production batch size decisions. The method considers investment costs in machines, where investment costs depend on machine speed and set-up costs. Inventory costs and labor costs in turn depend on machine speed.

The method is demonstrated with data from a sheet metal press shop, including a sensitivity analysis for the set-up time. It appears that, in this example, set-up times significantly affect investment decisions and total costs. The data provided by the method indicate how much a company should be willing to invest in reducing the set-up times.

Building on the results obtained by Jordan and Graves (1995), Garavelli (2001) studies the operating costs incurred by three different flexibility configurations of a production system with N resources that have to produce N product families. He considers the no-flexibility (NF) configuration, where each resource is dedicated to one of the product families, the total-flexibility (TF) configuration, where each resource can product each product family, and the limited flexibility (LF) configuration, where each product family can be produced on two resources in a full chain between products and resources. In Jordan and Graves (1995) the LF conzfiguration is shown to provide close to complete product demand mix flexibility, at lower costs then the TF configuration. Garavelli (2001) studies the differences in order throughput time between these three configurations. He studies the situation where orders for product families arrive at a given rate with exponential interarrival time. All product families have the same arrival rates. All orders require exponentially distributed processing time, with the same mean, independent of the resource that processes the orders. Changing over from producing one product family to another on a resource requires a deterministic set-up time. Allocation of orders to resources for the configurations TF and LF works as follows. There exists a one-to-one allocation of product families to resources, indicating per family the standard resource. Upon arrival an order is allocated to the standard resource of the product family it belongs to, unless the order queue of that resource exceeds a given threshold, TV. In the latter case, the order is allocated to the resource which can process that order and has the shortest queue of orders waiting. Orders are processed at the resource in order of arrival.

This process is studied for 5 levels of order arrival rates, implying 60, 70, 80 and 90% resource use (excluding set-up times), set-up times equal to $ST = 0$% (which is used as a benchmark) and 30% of average order processing time, two values of the threshold, $TV = 3$ and 10, and two values for the number of product families and number of resources, $N = 5$ and 10. Garavelli (2001) uses computer simulation to determine the performance in terms of throughput and average throughput time for each of the 80 cases. The research focuses on the effects of set-up times. For the $ST = 0$, $TV = 3$, $N = 5$ cases, the LF and TF configurations show decreases in average throughput time ranging from about 10% for the 60% net resource use case, to about 60% for the 90% net resource use case, as compared to the NF configuration. For the $ST = 0$, $TV = 10$, $N = 5$ cases, these numbers range from about 1% to about 38%. As may be expected, resource flexibility leads to shorter order throughput times and thus improves mobility. This drastically changes for the $ST = 0.3$ cases. Then for $TV = 3$ the throughput time for the TF configuration gets extremely large for 80% resource use and is infinite (not sufficient resource available to process all orders) for 90% resource use. The LF configuration does not show this poor performance, but performance differences between the LF and NF configuration are smaller than with $ST = 0$; they range from 0 to 28%. Similar performance effects are obtained for the cases with $N = 10$. These results demonstrate that in the presence of set-up times, resource flexibility should be used with caution, and provide another argument for considering investments in set-up time reduction when configuring a production system.

4.1 Discussion

The research discussed in this section provides insights into the relationships between resource flexibility, specifically range, mobility and uniformity, and the order throughput times of a production system. These insights are important when designing a supply chain. First, the order throughput times of the plants that are involved in a supply chain, determine to a large extent the product modification and the new product flexibility of the supply chain. Second the order throughput times, together with the order batch sizes, determine to a large extent the work-in-process and inventory that will be present in the supply chain to support a specific level of supply to the market. The costs of the capital tied up in work-in-process and inventory in the supply chain must be included in the investment costs of setting up of the supply chain and must be weighed against the costs of investing in flexible resources.

For production systems that face high capacity utilization, high demand variations and large batch sizes, the research on machine and worker flexibility reveals that large throughput time reductions can be achieved with a little resource flexibility, if well configured. The benefits of resource flexibility, although at a lower level, remain in the presence of moderate learning and forgetting effects. However in case of worker flexibility, allocation decisions should not be delayed, and the reallocation frequency should be under control. Thus the operations planning and control system (the system for delivering flexibility) should be designed to realize the benefits of the available resource flexibility, while avoiding possible negative effects such as high (re)learning, high reallocating rates, and high set-up costs. Research has revealed that relatively simple operations planning and control systems exist that can do the job.

Resource slack and resource flexibility also play a role in the design of the supply chain system for providing volume flexibility and mix flexibility, as discussed in the previous section. Supply chain design decisions are taken long before the actual demand levels become known. The need for short-term resource slack and short-term resource flexibility follows from the actual demand levels in relation to available sources on the long run, and from short-term measures that can be taken in response to these demand levels, such as adjusting prices, promotional actions, and accepting of only a part of the demand, leading to the actual sales levels that the system has to deal with. Using short-term measures to optimally adapt the use of available resources to sales and vice versa, is another important area for future research.

5 Empirical research on flexibility

In this section we discuss a selection of papers that report on empirical research on flexibility. These papers provide information about the use of flexibility as a strategic option and about the relationships between the use of flexibility of various types and firm performance.

Swamidass and Newell (1987) performed an empirical study in which they collected data from 35 manufacturers in the U.S. machinery and machine tool industry, in order to investigate the relationships between manufacturing strategy, environmental uncertainty and performance. One of their findings was that the greater the manufacturing flexibility, the better the performance, regardless of the type of manufacturing process used.

Ettlie and Penner-Hahn (1994) conducted a survey study in U.S. durable goods industry in order to investigate the relationships between manufacturing strategy and the various types of manufacturing flexibility found in the plants. They selected firms that had recently introduced flexible manufacturing systems or flexible assembly systems. They found that the more flexibility is emphasized in strategic focus, the more likely a plant is to have a shorter average change-over time per part family. They also found that firms focus on flexibility by concentrating on more part families per change-over time in the production planning. Realizing the benefits of FMSs seems to require partly redesign of the products.

Suarez, Cusumano, and Fine (1996) studied 31 PCB plants belonging to 14 electronic firms in the United States, Japan and Europe. They postulated five factors: product technology, production management techniques, relationships with suppliers and subcontractors, human resource management, and product development process, to have relationships with three dimensions of flexibility: mix flexibility, volume flexibility and new product flexibility. They found that newer, more automated processes were associated with less mix flexibility and with less new product flexibility, and with more volume flexibility. Lean production management techniques did positively correlate with mix and new production flexibility. Close relationships with suppliers and subcontractors had a positive impact on all three flexibility dimensions. Furthermore, plants with wage structures linked to plant performance had better volume flexibility, and plants that followed design for manufacturability principles (in particular, reuse of components) had greater mix flexibility and new product flexibility.

Upton (1995, 1997) investigated the product change-over flexibility in 52 plants of 11 companies in the uncoated fine paper industry, based on detailed structured interviews with managers and operators, followed up with a one-day wrap-up conference per plant. He found that most of the variance in change-over flexibility (or process mobility) across plants could be explained by the work experience of the people in the plant, and the emphasis that their managers lay on change-over flexibility. The size and the computer technology of a plant were not important determinants of its mobility; it even looked like computer integration could be detrimental to the flexibility of a plant.

Gupta and Somers (1996) developed three hypotheses regarding the relationships between strategy, flexibility and performance, and tested them on the basis of survey data collected from 269 firms from precision machinery, electrical and electronics, industrial machinery, metal products and automobile and auto part firms. Survey data were collected about the opinion of the respondents regarding strategy, flexibility and their relationships in their firm. They found that the aggressiveness dimension in business strategy is significantly related to all flexibility dimensions. Aggressive organizations report that they tend to sacrifice short-term profitability for going for market share, and therefore develop various forms of flexibility to be able to respond to changing market conditions. They also find that organizations pursuing a defensive strategy tend to seek very little flexibility. They furthermore found that the application of flexible manufacturing technology impacted negatively on both growth and financial performance, and that product flexibility had a negative relationship with growth performance. Finally, volume flexibility was found to have a positive relationship with growth performance. This indicates that firms that seek profitability with growth seem to invest in volume flexibility, whereas firms that seek profitability with product differentiation in the market, seem to invest in product and process flexibility.

Cagliano and Spina (2000) investigated the role of advanced manufacturing technologies in achieving strategically flexible production. Strategic flexibility is defined as the ability to shift competitive and manufacturing priorities rapidly from one set of goals to another, within the same manufacturing system.

Strategic flexible production is based on multi-focusedness, process integration and process ownership. They developed six hypotheses regarding the relationships between strategically flexible production strategies, and the application of advanced manufacturing technology. The hypotheses were tested in a survey of 392 firms in 20 countries in Europe, America and Japan. They found that the adoption of strategically flexible production does not correlate with the more intense use of computerized equipment or software applications. However, the higher level of organizational integration required for a complete orientation to strategically flexible production often went along with greater computer based cross-functional integration. Furthermore they found a positive relationship between the use of assembly robots and manufacturing quality, and between the use of MRP-II software and improvement in manufacturing lead time. However, no independent effect was found of the use of automated manufacturing technology on performance improvement; technology alone seems unable to improve manufacturing performance.

The adoption of strategically flexible production was found to drive most of the improvements in manufacturing lead time, which was further reinforced by the use of cross-functional computer integration. It was also found that computer integration is the main influence on significantly higher improvements in product variety.

For a comprehensive review of empirical research on manufacturing flexibility, we refer to Vorkurka and O'Leary-Kelly (2000). This paper synthesizes the body of empirical research regarding content-related issues and identifies possible avenues for future research. Furthermore, the paper examines several important methodological issues regarding manufacturing flexibility research, and indicates repeated methodological problems with regard to measurement validity, measurement reliability and general design, and suggest solutions.

5.1 Conclusions

Wrapping up the main finding in the reviewed literature in the Sections 3–5 we can state that:

- A small amount of resource flexibility, if well configured, can achieve almost all of the benefits to be obtained from total resource flexibility. This goes for volume, mix as well as new product flexibility.
- Transfer delays, change-over times and costs, and learning and forgetting may seriously decrease the benefits to be obtained from resource flexibility or inhibit its application. This especially pertains to new product flexibility. Thus investments in change-over time reduction should be considered, simultaneously with investments in type and amount of resources. Moreover when considering the use of worker

flexibility, the effect of its use on worker efficiency and on labor costs, should be modeled in the investment decision problem, simultaneous with the decisions about investments in type and amount of machine resources.
- Operations planning and control policies can play an important role in the impact of resource and routing flexibility, change-over costs and time, and worker efficiency on the performance. Control policies especially affect product flexibility and efficiency given the strongly decreasing marginal contribution of flexibility to performance. Control policies can effectively be used to constrain the negative impact of change-over times and costs and differences in labor efficiency on performance, while still realizing a large part of the benefits to be obtained from resource and routing flexibility.
- Strategic focus on flexibility seems to be an important condition for creating the resource and routing flexibility required, and for delivering the systems flexibility aimed at. Strategic focus seems to impact both managerial attitude, leading to investments in resource and routing flexibility (Section 3), and worker attitude. It also seems to mobilize the knowledge acquisition and knowledge deployment processes needed to convert potential resource and routing flexibility into output flexibility via an effective 'system for delivering flexibility,' as discussed in Section 4.

6 Supply chain flexibility

With one exception, all research reviewed in the previous section pertains to flexibility that can be delivered by a production system consisting of a set of resources that has been set up to produce a group of products, where each product requires one or more operations on the resources. No published papers have been found that study the flexibility of a supply chain as such. This would require the modeling of the supply relationships between the plants (or production units) involved in the production of a product or a product family, and the study of the relationship between, on the one hand, the flexibility of each of the production units with respect to the volume, mix and timing of their part of the production and, on the other hand, the flexibility of the supply chain as a whole. Although no models are available, the literature on flexibility does reveal the main causal relationship between the various dimensions of manufacturing flexibility at the plant level and at the supply chain flexibility with respect to volume, mix and new product flexibility.

In Section 7 we will present a supply chain model that can be used to analyze the decision options in designing a supply chain for volume, mix and new product flexibility. We will use views and approaches similar to those used in the literature reviewed in the Sections 3 and 4. For volume and mix flexibility, we will build on the types of models used in the literature to study investment problems. For new product flexibility we will build on the literature on production order throughput time in relation to resource flexibility.

Relationships between plants (or production units) involved in a supply chain can be of different kinds. Some plants may be part of the same company, other plants may belong to different companies. Some plants may deliver standard items that are produced to stock and supply to different customers. Other plants may supply make-to-order or engineer-to-order items and may or may not supply to different customers. Relationships and supply contracts between plants and the supply chains they are involved in will depend on relative power, market position etc., and these relationships and contracts will influence the possibilities to coordinate the supply chain. Therefore, in the next section we will review a number of papers published in operations management literature on supply contracts that provide information about supply chain design and flexibility. A comprehensive overview of this literature can be found in Cachon, 'Supply Chain Coordination with Contracts' (Chapter 6 of this volume) and in Chen, 'Information Sharing and Supply Chain Coordination' (Chapter 7 of this volume).

To date, virtually all research on output flexibility has been concerned with flexibility at the plant level. Only a few papers have been published that explicitly deal with flexibility at the supply chain level [Tsay and Lovejoy (1999), Cachon and Lariviere (2001)]. Nevertheless, the results of most research that pertains to flexibility at the plant level are also relevant at the supply chain level. However, the units of analysis when researching supply chain flexibility are different from the units of analysis when researching flexibility at the plant level. To be able to put the various research results in perspective, we position our subject as follows:

- At the supply chain level, the units of analysis are:
  - a product family consisting of a set of product variants,
  - a production system consisting of the set of plants that performs the processes needed in the supply chain (end-product manufacturing, semi-finished product manufacturing, component manufacturing),
  - a set of suppliers that deliver raw materials to the plants.
- At the plant level, the units of analysis are:
  - the set of items produced by the plant, which may be end-products, semi-finished products or components, and which may belong to more than one product family,
  - the resource configuration that of plant,
  - the processes that are used to produce the items on the resource configuration in the plant.

Recall that the relevant flexibility dimensions at the supply chain level are: volume flexibility, mix flexibility and new product flexibility. The relevant flexibility dimensions at the plant level are machine flexibility, worker flexibility and material handling flexibility.

Flexibility that is available at the plant level (machine, worker, materials handling flexibility) can be made available at the supply chain level in the form of volume flexibility, mix flexibility and new product flexibility for the items produced in that plant for the supply chain. However, it should be realized that each plant that is involved in a specific supply chain, is not necessarily involved in only this supply chain. Plants may be involved in various supply chains, and as such they must build up and employ flexibility that is geared to the requirements of each of the supply chains they are involved in. Thus, a plant must carefully select the supply chains that it is involved in, in order to match their requirements with the manufacturing focus of the plant regarding cost, quality, speed and flexibility. Similarly, when selecting a plant to be involved in a supply chain, one must carefully consider the cost, quality, speed and flexibility that can be delivered by the plant, in the light of the requirements at the supply chain level.

Although supply chain flexibility is dependent on plant flexibility, it should be carefully distinguished from plant flexibility. Since a supply chain cannot be stronger than its weakest link, the plant in the supply chain with the least flexibility determines the (volume, mix and new product) flexibility of the supply chain as a whole. A supply chain is therefore quite vulnerable to the performance of the individual plants in the chain; one plant consistently failing to live up to its promised level of supply, may simultaneously cause lost sales for the supply chain in the market place, and large stocks of unfinished products, leaving the 'supply chain owner' with a double loss. Thus a supply chain should be designed such that the flexibilities of its plants are balanced. However, what does it mean that the flexibility of the plants in a supply chain must be balanced? This is not clear at the outset, since the concept flexibility has different dimensions and each plant in the supply chain can have a different impact on realizing end-product flexibility in the market.

In Section 7 we will present a conceptual framework that can be used for the design of a supply chain for a product family during maturity phase of its product life cycle. We will assume that the design decisions are made in anticipation of the product family introduction in the market. We will assume probabilistic knowledge about the levels of demand for the product variants in the product family during the maturity phase, and we will assume that for each of the plants in the supply chain, investments in resources can be made to accommodate different levels of supply. We will consider three dimensions of flexibility: volume flexibility and mix flexibility with respect to the demand level for the product variants and (new) product flexibility for introducing new product variants during the product family life cycle. We will not consider short delivery flexibility, i.e., the responsiveness to short-term variations in demand around the demand levels. An important source of short-term delivery flexibility is inventory. The proper use of inventory of components, semi-finished products and end products for dealing with short-term demand variations is treated in de Kok and Fransoo (Chapter 12 of this volume).

We complete this section with a short discussion of supply chain research in relation to supply chain design and flexibility. A comprehensive overview of supply chain research, spanning the period 1956–1998, can be found in Ganeshan et al., Chapter 27 in Tayur, Ganeshan, and Magazine (1999).

6.1 Research on supply chain flexibility

Supply chain management has emerged in the 1990s as a new research field in production and operations management. Although the term was new, the subject that was studied was not. Chains of manufacturing activities had been studied before, in particular in the 1950s. Clark and Scarf (1960) studied multi-echelon inventory systems and developed base-stock control as a way to control a chain that faces stationary stochastic demand and constant stock replenishment lead times. The approach of Clark and Scarf (1960) can be typified as normative rational decision theory. Forrester (1961) studied the dynamic behavior of chains of production–inventory systems facing various kinds of dynamics in demand, including non-stationary demand. Whereas Clark and Scarf (1960) studied a chain under the assumption of centralized control, where the central controller has complete knowledge of the inventory positions in the chain, Forrester (1961) also studied chains under the assumption that each element of the chain is locally controlled, with decisions aiming for local objectives and based on locally available information about the inventory positions in the system. Using basic knowledge from control theory, Forrester (1961) was able to explain the upstream amplification of small variations in end-product demand, from the decentralized control structure of the system. This was named 'the Forrester effect,' after him. In the 1990s the Forrester effect has been renamed as the Bullwhip effect, see Lee, Padmarabhan, and Whang (1997), who also provide an analysis of the root causes of the Bullwhip effect. The approach of Forrester (1961) can be typified as rational explanatory research. In fact, for almost 30 years, Forrester remained one of few researchers in operations management who had studied supply chains under the assumption that each element in the chain would pursue its own local objectives.

In the 1980s increased competition due to market saturation forced consumer goods manufacturers to increase the number of product variants in a product family and to increase the product innovation rate, leading to shorter product life cycles. At the start of the 1990s the economic globalization led to an increase of outsourcing of manufacturing activities to third parties. Product families containing hundreds of product variants with short life cycles, many of them being region- or country-specific, were being made on assembly lines, with first and second tier supply of product specific components and subassemblies being outsourced to external companies. As a result, the complexity of the supply chains increased tremendously, and researchers started to shift the emphasis to new aspects of the emerging supply chain coordination problem. These were:

- the buyer–supplier relationships in the supply chain,
- information asymmetry in the supply chain.

Within the scope of this chapter, which is the design of flexibility in supply chains, the research on buyer–supplier relationships is most important. Research of buyer–supplier relationships concentrates on supply contracts. Using concepts from game theory, different types of supply contracts have been studied in order to determine how effective a specific contract is in coordinating the supply chain. Perfect (channel) coordination is obtained if, when both the supplier and the buyer behave rationally under the contract, the total supply chain profit is equal to the total profit obtained under optimal centralized control.

When a supplier wants to produce and deliver customer-specific parts under a contract, he usually has (at least partly) to (re)design the production processes and he may have to invest in new manufacturing equipment, in new tools and in operator training and education. The terms of the contract determine the risk that the supplier runs in view of the uncertainty regarding the orders that will be placed in the future. Investment risks, demand uncertainty and price must be in balance in order for the contract to be sufficiently attractive for the supplier. In research, the buyer–supplier situation has therefore been characterized as a two-period two-player decision system, where either the buyer, or the supplier can take the initiative. When initiative is with the buyer he communicates in the first period information about future (uncertain) demand to the supplier and negotiates a supply contract. The supplier has to decide whether or not to accept the contract, given the price, the terms of delivery, and his costs. The contract may include fixed or variable prices, contract down payments or quantity discounts, purchase commitments or delivery commitments. Alternatively, the supplier may offer a contract to the buyer. If a contract is agreed the supplier prepares for the second period by installing production capacity, buying materials and preparing for production (often forced compliance to the contract is assumed).

In the second period the buyer is faced with real demand (or acquires better information about future demand) and places orders under the agreed contract. Next the supplier delivers as far as he is able to, and given the amounts delivered, the buyer serves the real demand.

Supply chain literature mentions the following elements that should be arranged in a supply contract – horizon length, pricing, periodicity of ordering, quantity commitments, flexibility, delivery commitments, quality, and information sharing. A comprehensive review of literature on contracts in supply chain research is given in Agrawal et al., Chapter 10, in Tayur et al. (1999), and in Cachon (Chapter 6 of this volume).

In the context of designing flexible supply chains, we are especially interested in quantity commitments, delivery commitments and flexibility aspects of contracts. Research on quantity commitments and/or quantity flexible contracts has been published by Bassok and Anupindi (1997), Eppen and Iyer (1997), Parlar and Weng (1997), Graves, Kletter, and Hetzel (1998), Tsay and Lovejoy (1999), Li and Kouvelis (1999), and Cachon and Lariviere (2001). Most of this research investigates the channel optimality of the contract under various conditions regarding delivery commitment, pricing and demand information sharing. The results show that decentralized decision-making in a supply chain can be detrimental to the overall performance of the supply chain. The main reasons for the poor performance being the double marginalization and the different parties making decisions under different states of information.

Tsay (1999) shows that in a buyer–supplier situation where end-customer demand is sensitive to price, these problems can be partially remedied by a quantity-flexible contract in which the buyer commits to a minimum purchase and the supplier guarantees a maximum coverage, both stated as a percentage deviation from the buyer's initial forecasts, and a fixed price is agreed per unit delivered. This type of contract corresponds with business practice in technology intensive industry [Farlow, Schmidt, and Tsay (1995)]. Other researchers have investigated contracts which contain differential prices for exercising revisions of an initial order placed in the first period, or allow for product returns in the second period at a certain price. These contracts seem to be more in line with the business practice in retail chains, in particular in the relationship between manufacturing and retail in consumer products with short life cycles [Parlar and Weng (1997), Iyer and Bergen (1997)]. Other contract forms that may coordinate the supply chain are buy-back contracts and revenue sharing contracts (see Cachon, Chapter 6 of this volume).

6.2 Forced versus voluntary compliance

The assumption of forced compliance to the contract that is made in most contracting research is a very strong one from a model analysis point of view, but can be quite unrealistic in many industrial situations. It takes away the freedom of the supplier after the contract has been arranged. This may be justified in situations where the buyer is very powerful, but in most situations suppliers deliver to more than one customer and will tend to install capacity such that they can pool the uncertainty in demand from their customers. In fact, the decision to subcontract part of the production is based either on technological capabilities of a supplier that the buyer does not possess, in which case the buyer is captive, or on the expectation that the external supplier will be able to work cheaper, and to be more responsive because of his capability to realize economies of scale and to pool demand uncertainty.

Thus voluntary compliance is a more realistic assumption regarding supplier behavior under a supply contract.

Van Mieghem (1999) studies a two-stage dynamic stochastic program of a manufacturer and a subcontractor that decide separately on their capacity investment levels. After demand uncertainty is resolved both parties have the option to subcontract under a given contract or supply when deciding about their production (implying voluntary compliance). He presents outsourcing conditions for three contract types: (1) price-only contracts, where an exact transfer price is set for each unit supplied by the subcontractor; (2) incomplete contracts, where both parties negotiate about the division of surplus capacity based on ex-post bargaining power; (3) state-dependent price-only contracts. Van Mieghem (1999) shows that only state-dependent price-only contracts can eliminate all decentralization costs and coordinate the supply chain with respect to capital investment decisions. He furthermore finds that sometimes firms are better off by leaving some contract parameters unspecified ex-ante and agreeing to negotiate ex-post. He shows that a price focused strategy for managing subcontractors may backfire because a lower transfer price may lead to lower investments by the subcontractor, and decreased manufacturers profit. Finally it is shown that the option value of subcontracting increases as markets are more uncertain or more negatively correlated.

Cachon and Lariviere (2001) study the sharing of demand forecast information in a two party supply chain under both forced compliance and voluntary compliance, were the buyer offers the supplier a contract consisting of firm commitments and options. They show that under full information sharing and voluntary compliance, the buyer will neither offer firm commitments nor options, but only use the price in the contract. They also show that under voluntary compliance and under mild conditions regarding capacity, cost and demand uncertainty, the price that maximizes the buyer's profit creates a supplier capacity that is smaller than the capacity under forced compliance. Furthermore they show that under asymmetric information and sharing demand forecasts, firm commitments can play a role as signaling instrument, but at a price; although the need to purchase forecast credibility in this way can lower the manufacturers profit, total supply chain profitability increases because more capacity is built.

6.3 Discussion

Almost all research on supply contracts takes the coordinated channel, that is the overall supply chain optimal supply decision, as reference point. From an overall economic point of view this may make sense. However, if total supply chain profit is optimized and most or even all of this profit is gained by one of the parties (which happens in some of the models studied in literature), then the question is whether this leads to a viable situation. Each party in the chain needs to make sufficient profit in order to be able to invest, improve and expand. Thus, optimal supply chain coordination as implied by the concept of the coordinated supply channel might to be too restricted to cover all relevant aspects of a buyer–supplier relationship in, for instance, technology intensive industries. There we can see manufacturers who have an interest in improving the capabilities of their suppliers so that they will become better suppliers in the future. Incorporating these aspects in the models, would substantially complicate the analysis, since the buyer would have to evaluate the effects on future costs of leaving more profit to the supplier on the short-term.

Another aspect not considered in current research on contracts in supply chains is the multiple supplier situation and the multi-echelon nature of most manufacturing supply chains. Usually a manufacturer who is organizing his supply chain arranges supply contracts with one or more suppliers per item in the chain [see Kamien and Li (1990)]. Some of the items may be end-products; others may be standard parts; some items may be expensive, others may be cheap; some of the suppliers may be powerful, others may lack power. Thus, within one supply chain different types of dependencies may exist between manufacturer and his suppliers. Voluntary compliance may be the dominating mode in buyer–supplier relationships, and end-customer demand information may have different criticality to different suppliers. Given the fact that a manufacturing firm setting up a supply chain for a product family will generally have many suppliers and each of these suppliers will generally have many customers, each also involved in other supply chains, the concept of 'coordinated channel' will have little direct meaning to any of the participants in the supply chain for a particular product family. However, the coordinated channel is a very elegant theoretical concept for analyzing the efficiency of contracts for single buyer–supplier relationships.

In this chapter we focus on the design of the supply chain for a product family. In line with existing literature, we model the supply chain from the perspective of the manufacturer and assume the supply chain to consist of a network of plants which supply components, semi-finished products and end-products, and which require deliveries of raw materials, components and semi-finished products for their manufacturing processes. We only consider critical items such as the product-specific items and the expensive standard items; items in the category 'bolts and nuts' are left out because their supply needs not to be contracted yet in the design phase of the product family.

We will model the supply chain under the assumptions of centralized control, full information sharing and forced compliance. This seems unrealistic but this chapter is about design, and the assumptions of centralized control, full information sharing and forced compliance, provide a good reference point for the design of the processes and supply relationships in the supply chain. Moreover, the research of Van Mieghem (1999) shows that supply contracts exist that coordinate the supply chain with respect to capital investment decisions, and in view of the many contracts types that can coordinate the supply chain we may expect that also in supply chains involving multiple buyer–supplier relationships, configurations of contracts between parties in the chain will be found that can coordinate the channel. In the remainder of this chapter we anticipate on this development and consider the design of the supply chain as a centralized decision problem.

In the next section we will formulate the supply chain design problem, building on the research results discussed in the previous section. But first we start with a few remarks on design, and on the positioning of the supply chain design problem in the new product design process.

7 Design for supply chain flexibility

Design can be viewed as an information processing activity resulting in the creation of an object. Design can also viewed as a multi-criteria decision-making process (see Kusiak, 1999, Chapter 9). Design methods and design methodologies have been developed for the design of products and processes, ranging from methods for structuring the design process, such as concurrent engineering, to design tools such as quality function deployment, the analytical hierarchical process and constraint-based design. In the product design process we generally distinguish the following phases:

- customer requirements specification
- functional product specification
- technical product specification.

The difficulty of design is due to the complexity of the decision space, which in essence consists of all possible functional product specifications and all possible technical product specifications. The problem consists of, first, knowing what is the set of all viable technical product specifications that satisfy a given set of functional specifications, and second, choosing among this set one that optimizes the functional requirements, given the constraints on the resources and the time to be spent on the design process.

At the start of the design process, initial functional product specifications are set, which are managed during the design process, taking into account the expected market pay-off of offering a product with specific functionally within a given time frame, and the expected amount of resource and time needed for developing a product with this specific functionality [Huchzermeier and Loch (2001)]. Given the complex and multi-criteria nature of the design process both the conversion of customer requirements into functional product specifications, and the conversion of functional product specifications into technical product specifications are iterative overlapping processes which have to be managed from all relevant perspectives such as product costs, product quality, product manufacturability, product maintainability and product recovery [Kusiak, 1999; Chapter 9].

Supply chain design is an element of the product design process for a product family. Fine (1998; Chapter 8) divides supply chain development into the supply chain architecture decisions and the logistics/coordination system decisions. Supply chain architecture decisions include decisions on whether to make or buy a component, sourcing decisions and contracting decisions. In the choice for a specific supplier or manufacturer in the product supply chain, a number of different aspects will play a role, such as region of settlement, currency risks, company continuity, the current company customer base, product quality risks, codevelopment capabilities etc. All these aspects will have to be taken into account in the multi-criteria decision-making that constitutes the design process. Thus, the supply chain design problem for a new product family is embedded in the overall new product design problem, and can be addressed from all the different angles mentioned above. In his book Fine (1998; Chapter 8) presents the FAT 3-DCE decision model that describes the product, process and supply chain decision functions and illustrates the interdependence between product design and process design via the choice of technology, the interdependence between product and supply chain design via the choice of product architecture, and the interdependence between process design and supply chain design via the decision regarding the focusedness of the manufacturing system.

In this chapter we study the supply chain design problem for a new product family, assuming that the decisions regarding the main technical specification of the materials, components, semi-finished products and end-products have been made, and that the product design process is in the phase of selecting the manufacturing technologies, selecting the suppliers, and installing the production capacity

Our supply chain design problem therefore consists of answering two major questions:

1. What should be the supply levels, including flexibility options, for each of the items in the product structure?
2. What technology and what capacity should be installed to produce each of the items in the product structure?

As in most design problems, these two questions cannot be answered independently. For setting the supply level per item, information is needed about the capabilities and costs of technologies that can perform the processes to produce the items. Also, the choice of technology and capacity per process requires information about the supply level required, which again depends on end-product demand and the supply level of all other items in the chain. Thus, the design of a supply chain is an iterative process and must be structured as such (see Fig. 1).

Therefore, we first discuss the coordination of the supply levels of the items; this we call supply chain modeling (Section 7.1). Then we discuss the selection of technologies for producing an item (Section 7.2). This is

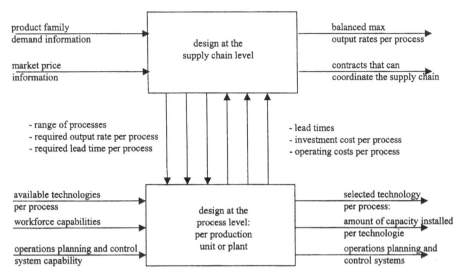

Fig. 1. The iterative supply chain design process.

concluded by a more detailed discussion of the design at the supply chain level (Section 7.3).

7.1 Supply chain modeling

Supply chains can be studied from different perspectives, and for each perspective, different aspects of the supply chain must be incorporated into the supply chain model. For instance, when studying the dynamic behavior of a supply chain (the bullwhip effect) aspects such as (de)centralized decision making, replenishment lead times, information sharing and demand uncertainty must be modeled. When studying optimal material allocation in a supply chain, the material requirements structure, the replenishment lead times, inventory carrying costs, ordering costs and demand uncertainty must be modeled. In this chapter we model the problem of creating delivery flexibility in the supply chain. We therefore largely abstract from the operational control of the supply chain, and assume that the operational supply chain control system will be able to fully employ the flexibility that has been created in the supply chain design phase. Thus we focus on decisions about manufacturing technology and production capacity in the supply chain for a product family. The modeling approach used in this section has been inspired by the models used for analyzing investments in flexibility in Fine and Freund (1990), Jordan and Graves (1995) and Van Mieghem (1998). We first model the product family, then we model the demand process, and finally we model the supply chain.

7.1.1 The product family

Our unit of analysis is the supply chain for a product family. A product family is defined as a group of product variants, each variant being a derivative of a platform product. Two characteristics of product families are important from a supply chain design point of view.

(1) Demand for product variants within a product family can be correlated. Specifically negative correlation may exist between relative demand per variant in the product family.
(2) Semi-finished product, component and material commonality may exist between product variants within a product family.

Correlated demand and commonality can have a large impact on product family costs and on the operational control costs of a product family [Baker (1985), Baker, Magazine, and Nuttle (1986), Lee (1996), see also Swaminathan and Lee, Chapter 5 of this volume], but also on the flexibility that is available in the supply chain. Therefore, when considering component commonality during product design, also the impact of commonality on supply chain design and supply chain flexibility should be taken into account.

7.1.2 Product demand model

We consider a product family consisting of $i = 1, \ldots, |I|$ product variants, where I denotes the set of product variants. We consider the problem as a two-phase decision problem. At the start of phase one the product design process has developed to the stage were:

- the product variants are known,
- the product structure (bill of materials) for each product variant is known,
- for each product variant knowledge about future demand during phase two is available in the form of a joint distribution function $F(D_1, D_2, D_3, \ldots, D_{n_I})$ where D_i denotes the level of demand for product variant i during phase two. We further assume F to be continuous and differentiable, that $F(D_1, \ldots, D_i, \ldots, D_{n_I}) = 0$ for any $D_i \leq 0$, and that $F(D_1, \ldots, D_i, \ldots, D_{n_I}) > 0$ for $0 < D_i < D_{\max}$. (Negative) correlation between demand for product variant is contained in the joint demand distribution function.

We assume that the second phase starts after introduction of the product family in the market, and covers the maturity phase of the product life cycle. We assume that the maturity phase for each product variant is characterized by a constant demand level, d_i. Actual demand per period during the maturity phase can be modeled as:

$$d'_i(t) = d_i + \varepsilon_i(t) \quad t = 1, \ldots, H, \tag{7.1}$$

where H denotes the number of periods in the maturity phase. The period length would typically be a week or a month, and be the basis for operational control of the supply chain. The variable $\varepsilon_i(t)$ represents the short-term demand uncertainty for product variant i. The operational control of the supply chain mainly pertains to setting safety stocks and to placing replenishment orders in response to the realizations of the short-term demand $d'_i(t)$. We assume that an operational control system will be in place that is able to satisfy any demand pattern during the maturity phase under the condition that $d'_i(t) \leq d_i^m$ where d_i^m denotes the maximum demand level of product variant i that can be satisfied during the maturity phase, given the processing capacity installed at the start of maturity phase. Operational Planning and Control concepts for supply chains are discussed (Chapter 12 of this volume) by de Kok and Fransoo.

7.1.3 Supply chain model

The supply chain for the product family can be modeled as an ordered set of processes. Each process converts materials, components and/or semi-finished products into higher order items. Material supply relationships exist between the processes, induced by the bill of material of the product family. Let J denote the set of processes in the supply chain, and $j \in J$ denote an arbitrary process in the supply chain. In our supply chain model definition, a one-to-one correspondence exists between items and processes. Each item (raw material, component, semi-finished product or end-product) is produced by one process, and each process produces one and only one item. In this definition a process is an abstraction of the way in which an item in the supply chain is made out of its component(s) (conform Jordan and Graves (1995)). The term 'process' refers to the conversion of items into higher level items in the bill of material structure. A process can be realized in various ways, using a variety of technologies, each with its own characteristics regarding investments needed, operating costs, and operating flexibility. We will develop a supply chain model that enables us to relate decisions about technologies to be used for manufacturing of the various items in the supply chain, to decisions about the delivery flexibility for end-products to be made available in the market place. The abstract concept 'process' enables us to do so.

We assume that each process j can be realized by any of a set of technologies where each technology $\tau \in K_j$ is characterized by its investment costs, processing costs, and the flexibility aspects, range, mobility and uniformity.

Since the processes correspond to items in the supply chain, the number of different items to be distinguished in the supply chain is equal to the number of processes and the index j denotes both the item and the process that leads to the item. We can model the end-item-process requirement structure in the supply chain as an $|I| \times |J|$ matrix R, where $r_{i,j}$ denotes the amount of units of item j that need to be produced by process j in order to get one unit of end product variant i.

Now consider an arbitrary end-product variant $i \in I$. The row vector $\underline{r}_i$ gives the amounts of units that need to be processed in all processes in the supply chain (including the buy processes) in order to be able to deliver one unit of the product variant, i. Thus $r_{i,j} = 0$ implies that the item produced with process j is not part of product variant i. By definition $r_{k,k} = 1$. The commonality between two product variants, i_1, i_2 is completely determined by the correspondence between the row vectors r_{i_1}, r_{i_2}. Corresponding non-zero elements of these row vectors indicate material and process commonality.

Now suppose that each of the processes j in the supply chain is constrained by the maximum output rate, that is the maximum amount of units of that item j that can be processed per time period. Let y_j denote this maximum. Let s_i denote the amount of units of end-product variant i that can be sold per time unit. Then s_i, the amounts of end-product variants that can be sold per time unit, is constrained by:

$$\sum_{i \in I} r_{ij} s_i \leq y_j \quad \forall j \in J \tag{7.2}$$

The supply chain design problem also entails the decision about what technology, and how much, to install for each of the processes j (or what supply contract to arrange if a buy processes is concerned). This problem has to be considered at two levels. The first level is the supply chain level. At this level the operational results that can be obtained during the maturity phase, given the supply chain constraints $\{y_j\}$, has to be evaluated. The main concern here is the alignment of the constraints $\{y_j\}$ such that the expected operational result from end-products sold are weighed against the investment costs incurred by the constraints $\{y_j\}$. The second level is the process level, where for each process, j, the available technologies are evaluated in order to select the optimal technology as a function of the maximum output rate y_j of that process. Like in any design process, overall design and detailed design (in this case, design at the supply chain level and design at the plant level) are interdependent and the final design is realized in an iterative way (see Fig. 1). In the next subsection we elaborate on the nature of the design problem at the plant level.

For ease of reasoning we assume that the maximum output rates, y_j, can be set only once, and are fixed during the entire maturity phase of the product family. Resetting of the maximum output rates may be possible during the maturity phase (e.g., in response to new demand information) but has to be done in a coordinated way, taking into account the effectuation lead times of implementing the new maximum output rate, as will be discussed in the rest of this chapter.

7.2 Design at the process level

Manufacturing processes are realized in manufacturing systems. A manufacturing system is defined as 'a unified assemblage of hardware, which includes trained workers, production facilities (including tools, jigs and fixtures), materials handling equipment, and other supplementary devices, which is able to converse process factors of production, particularly the raw materials, into finished products, aiming at maximum productivity' [Dorf and Kusiak, (1994), Chapter 21]. The main decision problems in process design are the selection of technologies, in relation to the set-up of the workflow (sequence of operations), the selection and training of the work force, and the set-up of the production planning and control system. Technologies are selected on economic grounds by choosing the alternative, which gives the best performance terms of quality, costs, lead time and flexibility. At the process design level we can only determine the optimal type of technology as a function of the output rate y_j (amounts of units processed per period) that should be made available at the supply chain level. The theory on production economics provides a basic relationship between the costs per unit produced as a function of the production rate for different technologies. Specialized technology requires higher investments than general purpose technology, but the production speed is higher. So the output rate determines which technology is optimal. Figure 2 illustrates this effect on cost per unit produced, for the choice between three different technology levels for manufacturing components in mechanical machine shop. The three technologies differ in the investment costs needed for being able to realize the required output rate, and in the variable cost per unit produced.

Technologies that have smaller variable production costs require higher investments. As a result technologies can be ordered according to the output rate. If the required production rate is in region A, then general purpose machines should be chosen. In the region B the numerically controlled machine should be chosen. In region C, the special purpose machines. The figure suggests that total cost is a concave function of production rate, if we only consider the set of Pareto-optimal technologies with respect to investment costs and variable production costs.

If similar products are produced on the same technology in order to achieve economies of scale, an important aspect of this technology is its flexibility in terms of range, mobility and uniformity. These flexibility aspects have to be considered in the supply chain design phase because they determine the new product and product modification flexibility of the supply chain. A technology that has a high range, high mobility and high uniformity can carry out a wide diversity of processes, without change-over time, at a uniform costs across the processes. Mobility determines production batch sizes, work-in-process and production throughput times in the supply chain.

7.2.1 Range

It will be clear that general purpose technology can perform a wide range of processes, whereas specialized technology can only perform one or a

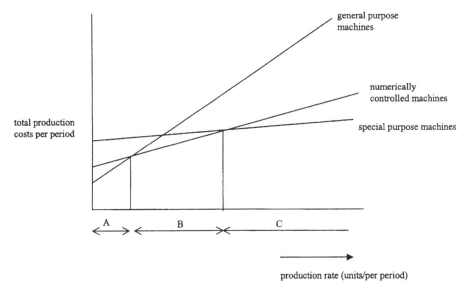

Fig. 2. Total production cost as a function of production rate for three different technologies.

small range of processes. In general, the higher the specialization, the smaller the range of processes that economically can be performed on a technology. On the other hand products and processes can be codesigned in such a way that they can share the same manufacturing technology, resulting in economies of scale. Thus if two products and their production processes can be designed such that the same technology can be used for both processes, then a higher level of technological specialization may be economically justified on the basis of the combined output rate for the two processes. This is exactly what has happened in 1980s when flexible manufacturing technology was introduced to produce different but technologically similar low-volume parts on one FMS. Empirical research into the effects of the use of FMSs revealed that the real economical benefits were only obtained some time after installation, when products and processes were redesigned in order to better fit with the capabilities of the FMS [Ettlie and Penner-Hahn (1994)].

Also for assembly processes, technology provides such opportunities. In the 1980s many car manufacturers started to codesign body parts and end-assembly lines such that a variety of end-products could be assembled on the same line, without any change-over time between the product variants. Thus the assembly line not only can perform multiple processes (this was already available before) but can do this without any relevant change-over time or change-over costs. This brings us to the second aspect of resource flexibility; change-over flexibility or mobility.

7.2.2 Mobility

In the presence of change-over times or costs, a technology can still show range flexibility but more capacity is needed, or additional investments in work-in-process and inventories must be made, in order to achieve the same output rate as with a technology that does not require change-over time or costs [Gupta (1993), Boyer and Keong Leong (1996)].

As discussed in Section 4 investments in work-in-process or inventories that are needed to realize a certain output rate, should be added to the investments needed for the technology itself, in order to get the total investments needed for being able to perform a process at output rate, y_j. The reason for this is the following. In the presence of change-over costs and change-over time, production will take place in certain batch sizes. These batch sizes result from an overall minimization of total variable costs, and lead, for a given investment in technology and capacity, to process operating costs, inventory carrying costs, work-in-process costs, and a maximum output rate. Methods and techniques for the setting of optimal production batch sizes have been developed over the last decades [see e.g., Karmarkar (1987), Suri et al., Chapter 5 in Graves et al. (1993), Buzacott and Shanthikumar (1993 Chapters 5, 6, 7 and 8) and Lambrecht et al. (1998)].

Being able to perform a process j at output rate y_j requires a combination of investments in technology, in worker training, in production planning and control, and in inventory and work-in-process. The work-in-process and inventory level that support an output rate y_j, must be built up before the start of the maturity phase, and must be maintained during the maturity phase as long as the output rate y_j stands as a rate that must be achievable. The costs involved in building up the work-in-process and inventory must be considered as an investment related to the choice of a specific technology. Moreover, work-in-process and inventory determine to a large extent the new product flexibility, since it will be necessary to first build up the work-in-process and inventory of the new product variant in the supply chain to the level that supports the output rate y_j, before the new product can be made available in the market.

In the design phase, early involvement of suppliers and manufacturers in the product and process design is of utmost importance. Early involvement is needed in order to take maximum advantage of available knowledge at the supplier or manufacturer about technologies. Much advantage can be gained from designing the items such that their processes can be combined with the processes of items that the supplier or comaker already produces, or will produce for other supply chains. For a supplier or manufacturer that can combine different but technologically similar processes for different customers, manufacturing system designs are possible which combine low costs and high flexibility, which is to the advantage of each of the supply chains. The studies of Jordan and Graves (1995) and Boyer and Keong Leong (1996) show that technologies with limited process flexibility may perform virtually equal to a set of technologies that show full process flexibility (that is, each process can be performed on each technology). Suppliers should be selected then can add the process to a 'fully chained' configuration of production resources, with a flexibility level that is in balance with the demand uncertainty faced by all of the items made on the resource configuration [Jordan and Graves (1995)].

During the design of the product family, a design guideline should be to design components, semi-finished products and end-products such that their manufacturing processes can be easily performed on a joint technology. Opportunities for this will be amply available when designing a product family, especially at the higher levels of the production structure where semi-finished products and end-products are made. For instance, postponement of product specificity [see Lee (1996)] leads to high levels of item commonality, and thus to a high level of process commonality in a product structure, which in turn may lead to low total costs and increased flexibility. Avoidance of technological specificity for producing different but technologically similar products may further lead to decreased production costs and increased flexibility. First, it decreases the number of constraints on sales to the market. Second, due to the decrease in production costs, it makes it possible to have higher process output rates, y_j, available for the same amount of capital investment. This follows directly from the concavity of production costs as a function of output rate for processes performed on a joint technology. Third, a high output rate of a manufacturing system allows for lager investments in set-up time and set-up costs reduction [see Porteus (1985), De Groote (1994b)], and will result in smaller batch sizes and shorter production throughput times, which improves new product flexibility.

De Groote (1994b) investigates the role of resource flexibility (range and mobility) in a capacitated multi-product lot-sizing model. His analysis provides valuable insights for decision-making about the range and mobility of a manufacturing system. As an illustration of how range and mobility aspects of resources can be taken into account in the process design phase we will give a short summary of his model and his results.

De Groote (1994b) considers a product line that is characterized by n different products $i=1,\ldots,n$, where d_i denotes the average demand. The technology is characterized by a nominal set-up time, S, the direct set-up cost per unit of set-up time, c_s, and the fixed cost (per unit of time), f. Each product has a finite production rate, r_i, and a relative duration of set-up q_i. Thus set-up time and set-up costs per product are q_iS or c_sq_iS. Finally, the unit cost of production (labor and material) is c_i, and the opportunity cost of capital per unit of time is r.

This problem consists of finding the batch sizes Q_i that minimize the average costs per unit time, while meeting the constraints on the capacity. De Groote (1994b) defines: $\alpha = \sum_{i=1}^{n} m_i/r_i$ as the fraction of time needed for production. As a result, the maximal number of nominal set-ups per unit time is $(1-\alpha)/S$.

The above problem can be formulated as:

$$\min_{Q_i} \sum_{i=1}^{n} \left(\frac{c_s S d_i q_i}{Q_i} + \frac{rc_i Q_i}{2} + c_i d_i \right)$$

$$\text{subject to}: \sum_{i=1}^{n} \frac{d_i q_i}{Q_i} \leq \frac{1-\alpha}{S}$$

The solution to this problem is given by Parsons (1966). De Groote (1994b) studies the optimal costs per unit of time as given by Spence and Porteus (1987). The optimal cost is:

$$\sum_{i=1}^{n} \sqrt{2 c_s S r c_i d_i q_i} + \sum_{j=1}^{n} c_i d_i + f \quad \text{when} \quad \sum_{i=1}^{n} \sqrt{\frac{rc_i d_i q_i}{2 c_s S}} \leq \frac{1-\alpha}{S}$$

$$\frac{\left(\sum_{i=1}^{n} \sqrt{2 S r c_i m_i q_i} \right)^2}{4(1-\alpha)} + c_s(1-\alpha) + \sum_{i=1}^{n} c_j d_j + f \quad \text{otherwise.}$$

De Groote (1994b) performs a sensitivity analysis of the optimal costs subject to changes in aggregate characteristics of this production system. These aggregate characteristics are:

$$d = \sum_{i=1}^{n} d_i$$

$$v = \left[\sum_{i=1}^{n} \sqrt{c_i d_i q_i} \right]^2 / \left[\sum_{i=1}^{n} c_i d_i \right]$$

The index v can be interpreted as a measure of product variety (mixing uniformity aspects and change-over aspects).

De Groote (1994b) shows that in this model the minimal cost is a concave function of the variety measure v. Thus variety comes at a price. He furthermore compares three technologies: a labor intensive job shop (*JS*), an automatic transfer line (*TL*), and a flexible manufacturing system (*FMS*). The job shop is characterized by low capital investment, relatively fast change-overs and high unit costs of production. The transfer line exhibits higher capital investment, much longer change-over times and much lower unit cost of production. The *FMS* exhibits the largest capital investments, but has instantaneous change-overs and low unit costs of production. Note that the main difference between the transfer line and the FMS is the change-over time and change-over costs, which causes the investment costs for the *FMS* to be higher than for the transfer line.

De Groote (1994b) shows for which combinations of d and v each of the three technologies has the least minimal costs (assuming of course optimal batch sizes $Q_i, i, \ldots, n$). He demonstrates that job shop technology is to be preferred for low demand, the transfer line is to be preferred for high demand and low variety, and that the FMS is to be preferred for high demand and high variety.

Aggregate models of plant design, like the ones used by Porteus (1985), Spence and Porteus (1987) and De Groote (1994b) can be used in the supply chain design phase to estimate the effects on total resource costs of different technologies, as a function of product (or process) variety and output rate, anticipating that in a later phase optimal batch sizes will be used in the operational control of the supply chain. Models like the ones discussed in Section 4 can be used to estimate the production throughput times per plant.

The process design phase results in the investment costs $\psi_j^k(y_j)$ the operating costs δ_j^k, and the throughput time per item j, $j \in J$, that result from performing process j with technology k at output rate y_j. The cost functions are used at the supply chain level to determine the best technology and the optimal output rates $\hat{y}_j$, $j \in J$. This is the subject of the next subsection.

7.3 Design at the supply chain level[1]

In this subsection we wrap up the supply chain design problem and for ease of reading we repeat a number of terms and formulations that have been introduced in previous (sub)sections.

We consider a supply chain that delivers a product family to a market. The supply chain design problem concerns the decisions relating to the installation of resources throughout the network of production units. It is our assumption that the installation of resources aims at enabling a maximum supply rate for each variant within the product family during some period of time during the maturity phase of the product family. We abstract from the possibility that during this maturity phase several versions of a variant supersede each other. This is no restriction if we may assume that each of these variants require the same amount of resource per unit during the maturity phase.

The maximum supply rate per variant to be enabled during the maturity phase is derived from information about the demand rate per variant during the maturity phase. Let us define the set I of product variants. Let $(D_1, D_2, \ldots, D_{|I|})$ denote the demand levels of variants 1 to $|I|$. We assume $(D_1, D_2, \ldots, D_{|I|})$ to be multi-variate normally distributed, thereby taking into account possible substitution and complementarity effects. Each product

---
[1] The author is indebted to A.G. de Kok for the formulation of the supply chain design problem given in this subsection.

variant consists of a number of items. We define J as the set of items. Let (r_{ij}) denoted the 'flat-BOM,' i.e.,

r_{ij} number of items j required to produce one variant i, $i \in I, j \in J$

Let us assume that we decide to maintain a supply rate s_i for product variant i. In order to enable this supply rate we must maintain a supply rate $r_{ij}s_i$ for item j dedicated to variant i. Since we want to maintain the supply rates s_i simultaneously during the maturity phase we must maintain a supply rate y_j for item j that satisfies

$$\sum_{i \in I} r_{ij}s_i \leq y_j \quad j \in J. \tag{7.3}$$

Assuming that we know the demand levels $\{d_i\}_{i \in I}$ at the start of the maturity phase, the product variant supply rates $\{s_i\}_{i \in I}$ should satisfy

$$s_i \leq d_i, \quad i \in I \tag{7.4}$$

We assume that item j is produced on some technology. In the supply chain design phase we must decide about the process technology that produces item j. In most situations a limited number of candidate technologies is available for producing item j. In the sequel we assume that the technology choice has been made. For given technology choices we want to determine the maximal supply rates $\{y_j\}_{j \in J}$ that maximize the profits during the maturity phase. The optimal technology choices can be found by searching the space of technology options for each item. Such a search should yield both the optimal technology options and the optimal supply rates $\{y_j\}_{j \in J}$. It is important to notice here that the only decisions taken during the design phase relate to technology options and maximal supply rates. The decisions about the actual supply rates of the variants are taken only when the maturity phase is about to start and more reliable information is available about the actual demand for each variant during the maturity phase. Implicitly we have assumed that we only consider items j for which we have to decide about their supply rates due to the effectuation lead time of the installation of the resources that produce these items.

Another important assumption made in the sequel is that items and processes producing these items are equivalent, i.e., each item is 1-1 related to a process. Clearly, a technology installed may be capable of accommodating multiple processes, thereby introducing mutual relationships between items through costs, in particular related to economies-of-scale, i.e., concavity of cost functions. To clarify this further, define the following cost functions

$\hat{\psi}_k(y)$ fixed cost during the maturity phase of technology k when maintaining a supply rate y

$\psi_j(y)$ fixed cost during the maturity phase of process j when maintaining a supply rate y

Letting J_k denote the items processed by technology k and assuming a supply rate y_j for all $j \in J_k$, the total cost associated with technology k equals $\hat{\psi}_k(\sum_{j \in I_k} y_j)$. In order to simplify the analysis we assume that

$$\hat{\psi}_k\left(\sum_{j \in J_k} y_j\right) = \sum_{j \in J_k} \psi_j(y_j).$$

The main idea behind this assumption is that experienced engineers responsible for the supply chain design decisions can incorporate the appropriate economy-of-scale effects in the functions $\psi_j(y_j)$, even though the true economies-of-scale relate to $\hat{\psi}_k(y)$. Thus we assume that the fixed costs associated with a supply rate y_j for item j equals $\psi_j(y_j)$. We assume that ψ_j is concave, thereby representing the economies-of-scale.

The variable production costs associated with item j are denoted by δ_j. The variable costs depend on y_i through Eq. (7.3). Given supply rates $\{y_j\}_{j \in J}$ and $\{s_i\}_{i \in I}$ the total costs during the maturity phase equal

$$\tilde{c}(\{y_j\}_{j \in J}, \{s_i\}_{i \in I}) = \sum_{j \in J} \left(\psi_j(y_j) + \delta_j \sum_{i \in I} r_{ij} s_i\right). \tag{7.5}$$

We assume that the selling price of variant i equals β_i. Then we obtain the following expression for the overall product family profit during the maturity phase, Π, given supply rates $\{y_j\}_{j \in J}$ and $\{s_i\}_{i \in I}$,

$$\Pi = \sum_{i \in I} \beta_i s_i - \sum_{j \in J} \left(\psi_j(y_i) + \delta_j \sum_{i \in I} r_{ij} s_i\right)$$

$$= \sum_{i \in I} \left(\beta_i - \sum_{j \in J} \delta_j r_{ij}\right) s_i - \sum_{j \in J} \psi_j(y_j). \tag{7.6}$$

Let us assume that the product variants are numbered in descending order of $\beta_i - \sum_{j \in J} \delta_j r_{ij}$. Then it is clear that we want to satisfy as much as possible demand from product variant 1, next demand from product variant 2, etc. Thus it follows from equation (7.3)–(7.5) that given the item supply rates $\{y_j\}_{j \in J}$ and the actual demand levels $\{d_i\}_{i \in I}$, the optimal product variant supply rates $\{s_i^*(\{y_j\}_{j\in J})\}_{i\in I}$ should satisfy

$$s_i^*(\{y_j\}_{j\in J}) = \min\left(d_i, \min_{j\in J}\left(\frac{y_j - \sum_{m=1}^{i-1} r_{jm}s_m^*}{r_{ij}}\right)\right), \quad i=1,\ldots,|I|. \quad (7.7)$$

Thus, the optimal variant supply rates can be expressed in terms of $\{d_i\}_{i\in I}$ and the supply chain design decision variables $\{y_j\}_{j\in J}$. This enables us to formulate the supply chain design problem as follows.

7.3.1 Supply chain design problem

$$\max_{\{y_j\}_{j\in J}} E_{(D,\ldots,D_{|I|})}\left[\sum_{i\in I}\left(\beta_i - \sum_{j\in J}\delta_j r_{ij}\right) s_i^*(\{y_j\}_{j\in J}) - \sum_{j\in J}\psi_j(y_j)\right]$$

This supply chain investment decision problem can be formulated as a two phase decision problem, like the one studied in Van Mieghem (1998).

If different processes are performed on a joint technology e.g., all product variants are assembled at a joint assembly line, then constraints of the following type must be used.

Let J_s denote a subset of processes that are performed on a joint technology with maximum output rate y_s. Then for this subset we use a constraint:

$$\sum_{j\in J_s} y_j \leq y_s,$$

and the joint investment cost functions would be $\psi_s^k(y_s)$.

If the joint technology for a subset of processes, J_s, is not an assembly line, but a configuration of dedicated resources with maximal output rate y_j and a flexible resources with output rate y_F, such as discussed in Andreou (1990) and in Fine and Freund (1990), we should use constraints of the following type:

$$\sum_{j\in J} r_{ij}s_i \leq y_j'$$

$$y_j' = y_j + y_F \quad j \in J_s$$

$$y_j' = y_j \quad j \notin J_s$$

$$\sum_{j\in J_s} y_j' = \sum_{j\in J_s} y_j + y_F$$

Also a more complex resource configurations that can perform a subset of the processes in J, such as the one considered in Jordan and Graves (1995), can be included in the model. However, the complexity of the constraint set would rapidly increase. A more practical approach therefore is to make use of the observation that a little flexibility, if well-positioned (e.g., fully chained) can achieve almost all of the performance of a totally flexible resource configuration. This would allow for the use of only one aggregate constraint on the output of all the processes in J_s, in the optimization.

Graves and Tomlin (2001) extend in the single plant model studied in Jordan and Graves (1995) to a multi-stage supply chain, consisting of F stages and I different products, with each product requiring processing at each stage. Stage f, $f = 1, \ldots, F$, has K_f different plants, where the term plant refers to any capacitated processing resource. Product-plant links (i, k) at stage f are represented by an arc set A_f. At stage f plant k can process the items in this stage of product i iff $(i, k) \in A_f$. $P^f(i)$ defines the set of plants of stage f that can process the items of product i, i.e., $k \in P^f(i)$ iff $(i, k) \in A_f$. Similarly they define the set of plants of stage f that can process items of one or more of the products in set M as $P^f(M) = \{k : (i, k) \in A_f, i \in M\}$. They assume that all products i such that $(i, k) \in A_f$ require the same amount of plant f's capacity per unit processed, and define c_k^f to be the number of product units expressed in end-product equivalents, that can be processed in plant k of phase f over the planning horizon.

Graves and Tomlin (2001) assume a two-phase sequential decision process. In the first phase it is decided which product items can be processed in each of the plants in each stage. In the second phase demand is realized and production is allocated to plants to meet demand.

To evaluate a flexibility configuration Graves and Tomlin (2001) define a single-period production-planning problem that minimizes the amount of demand that cannot be met by the supply chain. For a given demand realization $\boldsymbol{d} = \{d_1, \ldots, d_I\}$ and flexibility configuration $\boldsymbol{A} = \{A_1, \ldots, A_F\}$, the production-planning problem is the following linear program

$$x(\boldsymbol{d}, \boldsymbol{A}) = \text{Min} \left\{ \sum_{i=1}^{I} x_i \right\}$$

subject to

$$\sum_{(i,k) \in A_f} s_{i,k}^f + x_i \geq d_i \quad i = 1, \ldots, I, \ f = 1, \ldots, F$$

$$\sum_{(i,k) \in A_f} s_{i,k}^f \leq c_j \quad k = 1, \ldots, K_f, \ f = 1, \ldots, F$$

$$s_{i,k}^f, x_i \geq 0.$$

where $x(\boldsymbol{d}, \boldsymbol{A})$ is the total demand shortfall, x_i is the shortfall for product i, and $s_{i,k}^f$ is the amount items for product i processed in plant k at stage f.

A given resource configuration A is evaluated by the expected shortfall $E[X(\boldsymbol{D},A)]$, where the expectation is over the random demand vector $\boldsymbol{D} = \{D_1,\ldots,D_I\}$ which has a known distribution.

Graves and Tomlin (2001) introduce the concept of configuration inefficiency, CI, as

$$\text{CI} = 100 \times \left(\frac{E[X(\boldsymbol{D},A)] - E(X_1(\boldsymbol{D},A_1))]}{E[X_1(\boldsymbol{D},A_1)]}\right)$$

where $E[X_1(\boldsymbol{D},A_1)]$ denotes the expected stand-alone shortfall that would result from the stage with the larger shortfall, if it were the only stage in the chain.

According to Graves and Tomlin (2001) configuration inefficiency is caused by two phenomena, floating bottlenecks and stage-spanning bottlenecks. Floating bottlenecks are the direct result of demand uncertainty; stage-spanning bottlenecks can manifest itself even if demand is certain: they result from non-matching flexibility configuration for the different stages in the supply chain. Graves and Tomlin (2001) develop a flexibility measure for supply chains, g, based on the excess capacity available to any subset of products, relative to an equal share allocation of capacity. Using this measure, they develop analytic measures for the spanning inefficiency and for the floating inefficiency to gain insights in the effects of different resource characteristics per stage on configuration inefficiency. They use computer simulation to test these insights. From their analytic and simulation results, they conclude that supply chains with g-values greater than 1 perform very well, in particular if in each stage the plant-products links are chained. Chaining has an additional advantage; as long as each stage uses a chaining policy, there is no need to coordinate the exact capacity between stages to have a supply chain that performs well. Thus chaining per design stage can be used as an alternative to coordinating the amount of capacity installed per product (or product variant) between the stages in a supply chain.

7.4 Discussion

The supply chain design problem consists of choosing maximal output rates y_j, such that the total expected profit over the maturity phase, minus the total investment costs, is maximized. Under linear investment costs as a function of y_j the supply chain investment problem in principle can be solved with the methods and techniques used in Fine and Freund (1990), or Van Mieghem (1998). Further research is needed for the solving of the problem under concave investment cost functions, since economic literature reveals that investment costs tend to be concave in the output rate.

Also empirical research is needed into the structure of the investment costs in flexible resource configurations for different types of processes. In particular, working out in more detail the cost structure as a function of the production rate, the range of processes and the mobility. We may expect that these cost structures will be different for different types of supply chains, e.g., a supply chain for consumer electronics will use different processes and have different costs structures than a supply chain for fast moving consumer goods in food industry, or a supply chain for capital goods. These differences in cost structure may explain differences in dynamics between these industries.

Another important research subject is the inclusion in the investment costs of the work-in-process and inventory that results from lack of mobility (change-over times and change-over costs). This would require the extension of models studied by Porteus (1985) and by De Groote (1994b) with relationships between batch sizes, production rate, and capacity utilizations, on the one hand, and production lead times and work-in-process or workload, on the other hand [see Karmarkar (1987), Lambrecht et al. (1998)].

Research is also needed into the impact of product development decisions on the structure of the constraint set:

$$\sum_{i \in I} r_{ij} s_i \leq y_j \qquad j \in J$$

7.4.1 Managing the constraint set of a supply chain

For a product family containing $|I|$ product variants, the maximum number of constraints that can be active in the supply chain consists of the sum of all possible combination of elements out of the set $\{1, 2, \ldots, |J|\}$. This can be a very large number. Fortunately however, product families are generally developed in a systematical way, applying design principles like modular design, design for manufacturability, postponement of specificity, etc. If applied with an open eye for the supply chain consequences, the use of these principles can lead to rather transparent supply chain process architectures. The supply chain process architecture is to a large extent determined by the product architecture [see Fine (1998), Chapter 8].

In the ideal supply chain, all product variants of a family would require the same raw materials, and all product variants would require the same set of (generic) processes. However, different variants of a product family will often differ with respect to the buy items. Then, to each separate product variant, at least one specific constraint would apply.

Balancedness of constraints requires that all processes that are specific for a product variant, must have the same maximum output rate. For that reason it would make sense to bring control, both for the implementation and use of the manufacturing technology used for these processes, in one hand. The supply chain then ideally consist of production systems that takes care of all common processes, and a number of production systems which take care of the variant-specific processes. This would result in a constraint structure that only contains one constraint at the product family level and a set of constraints at the product variant level. Such a simple constraint structure facilitates easy economic evaluation of the supply chain design. However, it may occur that processes exist that are common for a subset of product variants, the subset not being the complete product family. Supply chain transparency is improved if such subset-commonality applies to disjoint groups of product variants, such that the structure is kept is simple as possible.

The supply chain model studied in Graves and Tomlin (2001) is a special case. Following their approach the supply chain design problem would be solved in two phases. In the first phase it is decided for each stage in the supply chain, how much resource flexibility must be created between the resources (plants) in that stage, in order to eliminate configuration inefficiency. The assumption then would be that the investment needed to create this level of resource flexibility is justified by the demand shortfall that later on can be avoided by using the resource flexibility. In the second phase for each stage in the supply chain it is decided how much capacity in total must be installed. In terms of end-product equivalents, total capacity in each stage must be the same, leading to balanced constraints on output rate per stage. Thus the capacity investment problem can be modeled as a single stage capacity investment problem with aggregate demand and total sales price at the product family, and total investment costs and total variable production costs summed over all stages as parameters.

7.4.2 Supply chain design and new product lead time

In the discussion in Section 4 we have assumed that a supply chain is designed for a product family where all variants are simultaneously available in the market and remain available during the product lifecycle maturity phase. This is not a very realistic assumption regarding real-life product families. Most product families start with an initial set of variants, gradually introduce new variants and take variants out of the market. If new product variants require new processes that cannot be performed (efficiently) on the existing technologies in the supply chain, the speed of new product introduction may be seriously hampered. Therefore, a supply chain should also be flexible enough to be able to cater for the different processes that will result from the future product innovations that can be foreseen for the product family. This requires that in the conceptual design phase of a product family, the boundaries are set for all processes that during the entire lifecycle of the product family must be performed. These processes do not pertain to specific items, but to generic items with parameters that indicate the range over which item features and the range of associated processes can vary. The manufacturing technology chosen for this generic process must be able to perform the associated processes in an efficient way. Supply chain design for a product family must therefore be based on a generic description of the processes needed for the manufacturing of the product family. The supply chain can then be set-up as a generic manufacturing channel, and new product variants can be developed within the range of parameter values of the processes available in the supply chain. This greatly enhances speed of new product introduction, and fosters production continuity for the production systems in the supply chain. If new product development takes place within the boundaries of available technology (both in terms of capabilities and maximum output rates), new product design lead time and new product risks are reduced, and capacity use can be much more smooth, specifically the phasing out of the old product variant and the phasing in of the new product variant. Research in this area is starting up. For instance, Peters and McGinnis (2000) provide results for the dynamic product (re)assignment problem for a number of resources that each can be used for a family of products with overlapping life cycles. They assume non-negative reassignment costs and also assume that all resources are able to perform the processes for all products equally efficient. Thus, their study assumes that the technology is able to efficiently produce all product variants that are developed within the product family.

In an empirical study of a number of firms, Suarez et al. (1996) found a positive correlation between finding mix flexibility and finding new product flexibility. This is in line with the analysis of flexibility in this chapter. Mix flexibility and new product flexibility both require investments in technology that can perform a well-defined range of processes. Volume flexibility, on the other hand, can only be achieved by installing slack capacity, as a hedge against demand uncertainty at the product family level.

8 Conclusion

In this chapter we have discussed the design of supply chains for flexibility. In this context we have focused on volume flexibility, mix flexibility and new product flexibility for a supply chain of a product family. We have considered the problem as an investment problem, where in the product design phase for a product family, decisions must be taken regarding investments in resources that will be used to produce the items in the Bill of Material of the product family. Since flexibility at the supply chain level results from the flexibility of the manufacturing systems that are involved in the supply chain, we have first reviewed the literature on the manufacturing flexibility concept. Manufacturing flexibility has many dimensions, between which hierarchical relationships can be identified. In particular, volume, mix and new product flexibility has been shown to depend on machine, labor and materials handling flexibility. Each flexibility dimension is further characterized by its range (the different options available), the mobility within the range (the ease of changing from one option to another), and the uniformity within the range (the extent to which each of the options within the range is equally cost effective).

Starting with these concepts we have reviewed the literature on the resource investment problem for the single-plant product-family resource investment problem under demand level uncertainty. This research revealed the intricate relationships between investments in dedicated and flexible resource, its dependence on the level of demand uncertainty, on the correlation between demands, and on the sales margin differences between products, and provided basic insights for the solving of the single-plant resource investment problem under specific conditions such as linear investment costs. Important insights from this research are that a little flexibility, if configured well, delivers almost all the benefits from full flexibility, and that for flexibility to be maximally effective, products and resources should be as highly interconnected as possible.

Next we have reviewed the literature on the effects of resource flexibility (machine and labor), and its aspects range, mobility and uniformity, on the production throughput time under uncertain demand around a given known level. The literature revealed that a small amount of resource flexibility can lead to a large decrease in production throughput time. However, the magnitude of this effect is limited by the presence of learning and forgetting effects, efficiency differences between using different resources for the same product, and delays in reallocating workers to alternative resources. Moreover, costs may be incurred when transferring work or workers to other resources. In the presence of these effects, the production planning and control system must be designed such that the resource flexibility is used with caution in order to realize a maximal effect with minimal costs.

Empirical research on manufacturing flexibility revealed that firms that have high product mix flexibility also tend to have high new product flexibility, which is consistent with the findings of model-based research, and that the use of FMSs seems to support high volume flexibility, but not high mix and new product flexibility. Moreover, apart from the positive effect of cross-functional computer integration on new product flexibility, the use of information technology seems not related to high flexibility. Finally, empirical research reveals that strategic focuses on flexibility seems to impact both managerial attitude, leading to investments in resource flexibility and worker attitude, and seems to mobilize the knowledge acquisition and deployment processes needed to convert potential resource flexibility into actual output flexibility.

Supply chains may subcontract part of their processes to external parties, and may need to buy parts and materials from external sources. The output of the supply chain can be constrained by each of the plants and sourcing units involved in the chain, including the external sourcing units. Therefore, we have discussed supply contracts at the hand of a selection of papers from literature. The literature revealed that for the two-player situation, also under decentralized decision-making and voluntary compliance, supply contracts exist that can coordinate the supply chain, that is, achieve the supply chain optimal performance. This observation encouraged us to restrict our discussion of the supply chain design problem to centralized decision-making under forced compliance, which is a simpler problem to solve.

Building on the results reported in literature we have formulated the supply chain design problem as a iterative decision process, consisting of choosing technologies for performing processes for producing the items in the Bill of Materials, and coordinating the maximum supply rates for all the items in the Bill of Material in view of the investment costs, the variable production costs, the demand level uncertainty and the product sales prices. This latter problem has been formulated as a two phase stochastic decision problem. We have discussed properties of this design problem, such as balancedness of constraints on process output rates, and the relationship between technology choice and the volume and diversity of products made on a technology. Throughout this chapter, we have indicated promising subjects for future research.

References

Andreou, S.A. (1990). A capital budgeting model for product-mix flexibility. *Journal of Manufacturing and Operations Management* 3, 5–23.

Baker, K.R. (1985). Safety stocks and component commonality. *Journal of Operations Management* 6(1), 13–32.

Baker, K.R., M.J. Magazine, H.L.W. Nuttle (1986). The effects of commonality on safety stock in a simple inventory model. *Management Science* 32, 982–988.

Bassok, Y.B., R. Anupindi (1997). Analysis of supply contracts with total minimal commitment. *IIE Transactions* 29, 373–381.

Beach, R., A.P. Muhlemann, D.H.R. Price, A. Paterson, J.A. Sharp (2000). A review of manufacturing flexibility. *European Journal of Operational Research* 122, 41–57.

Benjaafar, S. (1994). Models for performance evaluation of flexibility in manufacturing systems. *International Journal of Production Research* 32, 1383–1402.

Benjaafar, S., R. Ramakrishnan (1996). Modeling, measurement and evaluation of sequencing flexibility in manufacturing systems. *International Journal of Production Research* 34, 1195–1220.

Bernardo, J.J., A. Mohamed (1992). The measurement use of operational flexibility in loading flexible manufacturing systems. *European Journal of Operational Research* 60, 144–155.

Bobrowski, P.M., P.S. Park (1993). An evaluation of labor assignment workers are not perfectly interchangeable. *Journal of Operations Management* 11, 257–268.

Boyer, K.K., G. Keong Leong (1996). Manufacturing flexibility at the plant level. *Omega* 24(5), 495–510.

Browne, J., D. Dubois, K. Rathmill, S.P. Sethi, K.E. Stecke (1984). Classification of flexible manufacturing systems. *The FMS Magazine* April, 114–117.

Buzacott, J.A., J.G. Shanthikumar (1993). *Stochastic models of manufacturing systems*, N.J, Prentice Hall.

Cachon, G.P., M.A. Lariviere (2001). Contracting to assure supply: how to share demand forecasts in a supply chain. *Management Science* 47(5), 629–646.

Cagliano, R., G. Spina (2000). Advanced manufacturing technologies and strategically flexible production. *Journal of Operations Management* 18, 169–190.

Carlsson, B. (1989). Flexibility and the theory of the firm. *International Journal of Industrial Organisation* 7, 179–203.

Chandra, P., M.M. Tombak (1992). Models for the evaluation of routing and machine flexibility. *European Journal of Operational Research* 60, 156–165.

Chen, I.J., R.J. Calantone, C.H. Chung (1992). The marketing in manufacturing interface and manufacturing flexibility. *Omega* 26, 431–443.

Clark, A., H. Scarf (1960). Optimal policies for a multi-echelon inventory problem. *Management Science* 6, 475–490.

Clark, K.B., S.C. Wheelwright (1993). *Managing new product and process development; text and cases*, New York, The Free Press.

Corbey, M. (1991). Measurable economic consequences of investments in flexible capacity. *International Journal of Production Economics* 23, 47–57.

Daniels, R.L., J.B. Hooper, J.B. Mazzola (1996). Scheduling parallel manufacturing cells with resource flexibility. *Management Science* 42, 1260–1276.

Darlar, H., Z.K. Weng (1997). Designing a firm's coordinated manufacturing and supply decisions with short product life cycles. *Management Science* 43(10), 1329–1344.

Das, S.K., P. Nagendra (1993). Investigations into the impact of flexibility on manufacturing performance. *International Journal of Production Research* 31, 2337–2354.

Das, S.K., P. Nagendra (1997). Selection of routes in a flexible manufacturing facility. *International Journal of Production Economics* 48, 237–247.

De Groote, X. (1994a). The flexibility of production processes: A general framework. *Management Science* 40(7), 945–993.

De Groote, X. (1994b). Flexibility and product variety in lot-sizing models. *European Journal of Operational Research* 75, 264–274.

De Meyer, A., J. Nakana, J.G. Miller, K. Ferdows (1989). Flexibility: the next competitive battle. *Strategic Management Journal* 10, 135–144.

De Toni, A., S. Tonchia (1998). Manufacturing flexibility: a literature review. *International Journal of Production Research* 30, 1587–1617.

Dixon, J.R. (1992). Measuring manufacturing flexibility: an empirical investigation. *European Journal of Operational Research* 60, 131–143.

Dorf, R.C., A. Kusiak (1994). *Handbook of Design, Manufacturing and Automation*, Chichester, Wiley-Interscience.

D'Souza, D.E., F.P. Williams (2000). Towards a taxonomy of manufacturing flexibility dimensions. *Journal of Operations Management* 18, 577–593.

Eppen, G.D., A.V. Iyer (1997). Backup agreements in fashion buying: the value of upstream flexibility. *Management Science* 43, 1469–1484.

Ettlie, J.H., J.D. Penner-Hahn (1994). Flexibility ratios and manufacturing strategy. *Management Science* 40(11), 1444–1454.

Farlow, D., G. Schmidt, A. Tsay (1995). Supplier management at Sun microsystems. Case study Graduate School of Business, Stanford University, Stanford, CA.

Felan, J.T., T.D. Fry, P.R. Philipoom (1993). Labor flexibility and staffing levels in a dual resource constrained job shop. *International Journal of Production Research* 31, 2487–2506.

Fine, C.H. (1998). *Clockspeed*, Reading, Massachusetts, Pereus Books.

Fine, C.H., R.M. Freund (1990). Optimal investment in product flexible manufacturing capacity. *Management Science* 36, 449–460.

Forrester, J.W. (1961). *Industrial Dynamics*, New York, John Wiley and Sons Inc.

Fry, T.D., H.V. Kehr, M.K. Malhotra (1995). Managing worker flexibility and attrition in dual resource constrained job shops. *International Journal of Production Research* 33, 2163–2179.

Fryer, J.S. (1973). Operating policies in multi-echelon dual-constraint job shops. *Management Science* 19(9), 1001–1012.

Fryer, J.S. (1974). Labor flexibility in multi-echelon dual-constraint job shop. *Management Science* 20(7), 1073–1080.

Fryer, J.S. (1975). Effects of shop size and labor flexibility in labor and machine limited production systems. *Management Science* 21(5), 507–515.

Gaimon, C., V. Singhal (1992). Flexibility and the choice of manufacturing flexibilities under short product life cycles. *European Journal of Operational Research* 60, 211–223.

Garavelli, C.A. (2001). Performance analysis of a batch production system with limited flexibility. *International Journal of Production Economics* 69, 39–48.

Garg, S., P. Vrat, A. Kanda (2001). Equipment flexibility vs. inventory: a simulation study of manufacturing systems. *International Journal of Production Economics* 70, 125–143.

Gerwin, D. (1987). An agenda for research on the flexibility of manufacturing processes. *International Journal of Operations and Production Management* 7, 38–49.

Gerwin, D. (1993). Manufacturing flexibility in a strategic perspective. *Management Science* 39, 395–410.

Graves, S., D. Kletter, W. Hetzel (1998). A dynamic model for requirements planning with application to supply chain optimization. *Operations Research* 46, S35–S49.

Graves, S.C., B.T. Tomlin (2001). *Process Flexibility in Supply Chains*, A.P. Sloan School of Management, MIT, Cambridge, MA.

Graves, S.C., A.H.G. Rinnooy Kan, P.H. Zipkin (1993). *Logistics of Production and Inventory, Handbooks in Operations Research and Management Science*, Vol. 4. Amsterdam, North Holland.

Gunther, R.E. (1981). Dual-resource parallel queues with server transfer and information access delays. *Decision Sciences* 12, 97–111.

Gupta, D. (1993). On measurement and valuation of manufacturing flexibility. *International Journal of Production Research* 31, 2947–2958.

Gupta, D., J.A. Buzacott (1989). A framework for understanding flexibility of manufacturing systems. *Journal of Manufacturing Systems* 8(2), 89–97.

Gupta, Y.P., S. Goyal (1989). Flexibility of manufacturing systems: concepts and measurements. *European Journal of Operational Research* 43, 119–135.

Gupta, Y., T. Somers (1992). The measurement of manufacturing flexibility. *European Journal of Operational Research* 60, 166–182.

Gupta, Y.P., T.M. Somers (1996). Business strategy, manufacturing flexibility and organizational performance relationships: a path analysis approach. *Production and Operations Management* 5(3), 204–233.

Ho, Y.C., C.L. Moody (1996). Solving all formation problems in a manufacturing environment with flexible processing and routing capabilities. *International Journal of Production Research* 34, 2901–2923.

Hogg, G.L., D.T. Philips, M.J. Maggard (1979). Parallel-channel dual-resource constrained queuing systems with heterogeneous resources. *AIIE Transactions* 9, 352–362.

Huchzermeier, A., C.H. Loch (2001). Project management under risk: using the real options approach to evaluate flexibility in R&D. *Management Science* 47, 58–101.

Hutchinson, G.K., V.A. Pflughoeft (1994). Flexible process plans: their value in flexible automation systems. *International Journal of Production Research* 32(3), 707–719.

Hyun, J.H., B.H. Ahn (1992). A unifying framework for manufacturing flexibility. *Manufacturing Review* 5, 251–260.

Iyer, A.V., M.E. Bergen (1997). Quick response in manufacturing-retailer channels. *Management Science* 43(4), 559–570.

Jensen, J.B. (2000). The impact of resource flexibility and staffing decisions on cellular and departmental shop performance. *European Journal of Operational Research* 127, 279–296.

Jensen, J.B., M.K. Malhotra, P.R. Philipoom (1996). Machine dedication and process flexibility in a group technology environment. *Operations Management* 14, 19–39.

Jordan, W.C., S.C. Graves (1995). Principles in the benefits of manufacturing process flexibility. *Management Science* 41, 577–594.

Kamien, M.I., L. Li (1990). Subcontracting, coordination, flexibility and production smoothing in aggregate planning. *Management Science* 36(11), 1352–1363.

Karmarkar, U.S. (1987). Lot sizes, lead times and in-process inventories. *Management Science* 33, 409–423.

Kher, H.V. (2000). Examination of flexibility acquisition policies in dual resource constrained job shops with simultaneous worker learning and forgetting effects. *Journal of Operational Research Society* 51, 592–601.

Kher, H.V., M.K. Malhotra, P.R. Philipoom, T.D. Fry (1999). Modeling simultaneous worker learning and forgetting in dual resource constrained system. *European Journal of Operational Research* 115, 158–172.

Koste, L.L., M.K. Malhotra (1999). A theoretical framework for analyzing the dimensions of manufacturing flexibility. *Journal of Operations Management* 18, 75–93.

Kusiak, A. (1999). *Engineering Design: Products, Processes and Systems*, London, Academic Press.

Lambrecht, M.R., P.L. Ivens, N.J. Vandaele (1998). ACLIPS: A capacity and lead time integrated procedure for scheduling. *Management Science* 44(11), 1548–1561.

Lee, H.L. (1996). Effective management of inventory and service through product and process redesign. *Operations Research* 44, 151–159.

Lee, H., P. Padmarabhan, W. Whang (1997). Information distortion in a supply chain: the bullwhip effect. *Management Science* 43, 546–558.

Li, C., P. Kouvelis (1999). Flexible and risk-sharing supply contracts under price uncertainty. *Management Science* 45(10), 1378–1398.

Malhotra, M.K., T.D. Fry, H.V. Kher, J.M. Donohue (1993). The impact of learning and labor attrition and worker flexibility in dual resource constrained job shop. *Decision Sciences* 24(3), 641–663.

Malhotra, M.K., H.V. Kher (1994). An evaluation of worker assignment policies in dual resource-constrained job shops with heterogeneous resources and worker transfer delays. *International Journal of Production Research* 32(5), 1087–1103.

Malhotra, M.K., L.P. Ritzman (1990). Resource flexibility issues in multistage manufacturing. *Decision Sciences* 21, 673–690.

Mandelbaum, M., P.H. Brill (1989). Examples of measurement of flexibility and adaptivity in manufacturing systems. *Journal of Operations Research Society* 40, 603–609.

Morris, J.S., R.J. Terinze (1994). A simulation comparison of process and cellular layouts in a dual resource constrained environment. *Computer and Industrial Engineering* 26, 733–741.

Nagurar, N. (1992). Some performance measures of flexible manufacturing systems. *International Journal of Production Research* 30, 799–809.

Nam, I. (2001). Dynamic Scheduling for a flexible processing network. *Operations Research* 49(2), 305–315.

Nandkeolyar, U., D.P. Christy (1992). An investigation of the effect of machine flexibility and number of part families on system performance. *International Journal of Production Research* 30, 513–526.

Nelson, R.T. (1967). Labor and machine limited production systems. *Management Science* 13(9), 648–671.

Nelson, R.T. (1970). A simulation of labor efficiency and central assignment in a production model. *Management Science* 17(2), B97–B106.

Park, P.S. (1991). The examination of worker cross-training in a dual resource constrained job shop. *European Journal of Operational Research* 51, 291–299.

Park, P.S., P.M. Bobrowski (1989). Job release and labor flexibility in a dual resource constrained job shop. *Journal of Operations Management* 8(3), 230–249.

Parker, R., A. Wirth (1999). Manufacturing flexibility. Measures and relationships. *European Journal of Operational Research* 118, 429–440.

Parlar, M., Z.K. Weng (1997). Designing a firm's coordinated manufacturing and supply decisions with short product life cycle. *Management Science* 43(10), 1329–1344.

Parsons, J.A. (1966). Multiproduct lot sizes determination when certain restrictions are active. *Journal of Industrial Engineering* 27, 360–365.

Pereira, J., B. Paulré (2001). Flexibility in manufacturing systems: A relational and dynamic approach. *European Journal of Operational Research* 130, 70–82.

Peters, B.A., L.F. McGinnis (2000). Modeling and analysis of the product assignment problem in single stage electronic assembly systems. *IIE Transactions* 32, 21–31.

Porteus, E.L. (1985). Investing in reduced set-ups in the EOQ model. *Management Science* 31, 998–1010.

Ramasesh, R.V., M.D. Jayakumar (1991). Measurement of manufacturing flexibility: a value based approach. *Journal of Operations Management* 10, 446–468.

Sethi, A.K., S.P. Sethi (1990). Flexibility in manufacturing in a survey. *International Journal of Flexibility Management Systems* 2, 289–328.

Shafer, S.M., J.M. Charnes (1997). Offsetting lower routing flexibility in cellular manufacturing due to machine dedication. *International Journal of Production Research* 35(2), 551–567.

Skinner, W. (1974). The focussed factory. *Harvard Business Review* 52(3), 113–121.

Slack, N. (1983). Flexibility as a manufacturing objective. *International Journal of Operations and Production Management* 3, 4–13.

Slack, N. (1987). The flexibility of manufacturing systems. *International Journal of Operations and Production Management* 7, 35–45.

Smunt, T.L., J. Meredith (2000). A comparison of direct cost savings between flexible automation and labor with learning. *Production and Operations Management* 9(2), 158–170.

Spence, A.M., E.L. Porteus (1987). Set-up reduction and increased effective capacity. *Management Science* 33, 1291–1301.

Stalk, G., T.M. Hout (1990). Competing against time, New York, the Free Press.

Suarez, F.F., M.A. Cusumano, C.H. Fine (1996). An empirical study of manufacturing flexibility in printed circuit board assembly. *Operations Research* 44(1), 223–240.

Swamidass, P.M., W.T. Newell (1987). Manufacturing strategy, environmental uncertainty and performance: a path analytic model. *Management Science* 33, 509–524.

Tanner, J.P. (1985). *Manufacturing Engineering*, New York, Marcel Dekker Inc.

Tayur, S., R. Ganeshan, M. Magazine (1999). *Quantitative Models for Supply Chain Management*, Boston, Kluwer Academic Publishers.

Treleven, M. (1989). A review of dual resource constrained research. *IIE Transactions* 21, 279–287.

Treleven, M.D., D.A. Elvers (1985). An investigation of labor assignment rules in a dual constrained job shop. *Journal of Operations Management* 6(1), 51–68.

Tsay, A.A. (1999). The quantity flexible contract and supplier–customer incentives. *Management Science* 45(10), 1339–1358.

Tsay, A.A., W.S. Lovejoy (1999). Quantity flexibility contracts and supply chain performance. *Manufacturing and Service Operations Management* 1(2), 89–111.

Upton, D.M. (1994). The management of manufacturing flexibility. *California Management Review* Winter, 72–89.

Upton, D.M. (1995). Flexibility as process mobility: the management of plant capabilities for quick response manufacturing. *Journal of Operations Management* 12, 205–224.

Upton, D.M. (1997). Process range in manufacturing: an empirical study of flexibility. *Management Science* 43, 1079–1092.

Van Mieghem, J. (1998). Investment strategies for flexible resources. *Management Science* 44(8), 1071–1078.

Van Mieghem, J. (1999). Coordinating investment, production and subcontracting. *Management Science* 45(7), 954–971.

Vander Veen, D., W. Jordan (1989). Analyzing trade-offs between machine investments and utilizations. *Management Science* 35(10), 1215–1226.

Vorkurka, R.J., S.W. O'Leary-Kelly (2000). A review of empirical research on manufacturing flexibility. *Journal of Operations Management* 18, 485–501.

Wayson, R.D. (1965). The effects of alternative machines on two priority dispatching disciplines in the general job shop, Master Thesis Cornell University, New York.

Weeks, J.K., J.S. Fryer (1976). A simulation study of operating policies in a hypothetical dual constrained job shop. *Management Science* 22, 1362–1371.

Wisner, J.D., J.N. Pearson (1993). An exploratory study of the effects of operator relearning in dual resource constrained job shops. *Production and Operations Management* 2, 55–68.

Chapter 5

Design for Postponement

Jayashankar M. Swaminathan
The Kenan-Flagler Business School, University of North Carolina, Chapel Hill, NC 27599, USA

Hau L. Lee
Graduate School of Business, Stanford University, Stanford, CA 94305, USA

1 Introduction

In this age of increasing globalization and shortening of product life cycles, companies are faced with the demand for escalating product variety to meet the diverse needs of global customers. Indeed, mass customization has become a business requirement for many high technology companies. However, the provision of product variety comes with a price. With it forecasting becomes more difficult, overhead for product support is higher, inventory control is more difficult, manufacturing complexity increases, and after-sales support is more complex. One solution that innovative companies have exploited is the power of product and process design, by integrating design with their supply chain operations to gain control of product variety proliferation.

Design has always been viewed as a key driver of manufacturing cost. Past research has indicated that as much as 80% of the manufacturing cost of the product is determined by the design of the product or the process in which the product is to be manufactured. Design can also be leveraged to address the problem of mass customization (Martin, Hausman and Ishii, 1998). By properly designing the product structure and the manufacturing and supply chain process, one can delay the point in which the final personality of the product is to be configured, thereby increasing the flexibility to handle the changing demand for the multiple products. This approach is termed *postponement*.[1] Alderson (1950) appears to be the first who coined this term, and identified it as a means of reducing marketing costs. Alderson held that 'the most general method which can be applied in promoting the efficiency

---
[1] This approach has also been termed as *delayed product differentiation* or *late customization*.

of a marketing system is the postponement of differentiation,..., postpone changes in form and identity to the latest possible point in the marketing flow; postpone change in inventory location to the latest possible point in time'. He believed that this approach could reduce the amount of uncertainty related to marketing operations. Bucklin (1965) provided arguments as to how postponement as identified by Alderson could be a useful concept but would be difficult to implement through the channel particularly in manufacturing environments predominantly operating on a 'make-to-stock' basis. He argued that some entity in the channel would have to bear the risks associated with product variety, and postponement only helped in shifting this risk to some other partner in the channel. However, as manufacturing firms started to move away from the traditional make-to-stock environment, postponement has become an attractive alternative.

Zinn and Bowersox (1988) describe different types of postponement that could be implemented. These included *labeling postponement*, *packaging postponement*, *assembly postponement*, *manufacturing postponement*, and *time postponement*. Labeling postponement is a situation where a standard product is stocked and labeled differently based on the realized demand. In packaging postponement products are not packaged into individual packs until final orders are received. Assembly and manufacturing postponement refer to situations where additional assembly or manufacturing may be performed at the assembly facility or at a warehouse before shipping the product to the customer after demand is realized. Finally, time postponement refers to the concept that products are not shipped to the retail warehouses but are held at a central warehouse and are shipped to customers directly.

Clearly, different types of postponement strategies have different costs and benefits associated with them. For example, with packaging postponement, inventory costs are reduced due to stocking of the standard product, whereas the packaging costs are higher since it is not done in one big batch thereby losing economies of scale. Similarly, in manufacturing and assembly postponement, component costs may increase, and in some cases, a more complex process may have to be used. Moreover, there are multiple ways in which postponement can be pursued, each with different cost and service performance impacts.

Fundamentally, there are three types of factors that affect the benefits and costs associated with postponement – *market* factors, *process* factors, and *product* factors. Market factors are those related to customer demand and service requirements. These parameters include demand fluctuations or variance, correlation in demand across the different products, lead time and service requirements for customization (which affect the penalty cost for stockouts or late deliveries). Process factors are those manufacturing and distribution processes under the control of the firm. These include the sequence of operations performed to customize the product, the network structure of the supply chain (manufacturing and distribution sites), whether the product is made to order or made to stock as well as how much and at which location inventories (components, subassemblies, and finished products) are stored in the supply chain. Product factors are related to the design of the product or product lines. These include the degree of standardization that is present in the components and the costs associated with standardizing components, modularity in the product design, as well as the degree to which end products can be substituted for each other's demand.

The ability of a firm to implement a successful postponement strategy depends on how well the firm can tailor its process and product characteristics to the market requirements. Primarily these relate to the changes in the design of the product or process so that implementing postponement strategies becomes easier and more cost effective. There are mainly two types of changes – those related to process design changes, termed *process postponement*, and those related to product design changes, termed *product postponement*. Process postponement usually requires (1) process standardization, i.e., making some part of the process standard so that the different product variants share that process; and (2) process resequencing, i.e., changing the sequence of customization steps in which the product attains distinct functionalities and characteristics. Product postponement often requires standardizing some key components, or introducing parts commonality in the product structure.

In this chapter, we discuss analytical models for evaluating postponement alternatives. Earlier survey articles on similar areas include Garg and Lee (1998) and Swaminathan and Tayur (1998b). The rest of this chapter is organized as follows. In Section 2, we introduce the three key postponement enablers: process standardization, process resequencing and component standardization. These three enablers and associated performance evaluation models are described in greater details in Sections 3–5. We also describe industry applications utilizing these enablers. In Section 6, we discuss other techniques for managing product variety such as modularity and downward substitution and explore the additional benefits of postponement in pricing and information processing. In Section 7, we provide our concluding remarks.

2 Postponement enablers

Postponement can be enabled through changes in the manufacturing-distribution process or the product architecture. In this section we introduce three enablers of postponement – process standardization, process resequencing and component standardization. Process standardization refers to standardizing the initial steps in the process across the product line so that products are not differentiated at these steps, and distinct personalities of the products are added at a later stage. All the products in the product line (or a subset of it) are processed through these standard steps. A complementary approach to process standardization is process resequencing. Here, the sequence is changed so that more common components are added at the beginning of the process. The components or features that create product differentiation are added later. The key benefit from both of the above approaches is that the initial stages in the process are less differentiated, leading to partially completed products at the end of these common stages. This enables the firm to pool the risk across the different product demands and to effect lower inventory requirements. Clearly the success of the above approaches depends to a great extent on how modular in structure the process is. Process modularity is the same as the product modularity concept applied to a process. If a process can be divided into separate substeps so that these substeps can be performed in either parallel or in different sequence then it is classified as a modular process. For example, the testing process of a product may require multiple tests and burn-ins. In some cases, the whole test process may have to be carried out in a continuous fashion, while in other cases, the process can be broken up into subtests. Process modularity is closely related to the flexibility of the process in that processes that are more flexible are likely to be modular processes as well. In addition to process modularity, the feasibility of process resequencing depends on common or standard components in the product line. Indeed, the third enabler to postponement is component standardization.

We should note that all three enablers could be used individually or in any combination to achieve postponement. In the next sections, we describe the models that have been developed for each of these enablers. Some notation that is used throughout the paper is given in Table 1.

3 Process standardization

In process standardization approaches, inventory may be carried at the intermediate stage after the common steps in the process (known as *the point*

Table 1
Notations used in the paper

Symbol	Description
D_i	Realized demand for product i
μ_i	Mean demand for product i
σ_i	Standard deviation of demand for product i
ρ_{ij}	Correlation of demand for product i and product j
S_i	Base stock inventory level for i
s_i	Safety stock for i
$E(x)$	Expected value of x
$Var(x)$	Variance of x
z	Safety factor
h	Per unit holding cost
U	A vanilla box configuration in terms of the components

of differentiation) as well as at the final product level. The models developed here differ in terms of consideration of single versus multiple points of differentiation.

3.1 Single point of differentiation

Lee (1996) describes the most basic version of this model where there are M products and inventories are carried in finished form. All the products are customized from the inventory available at the end of the standard steps of the process. It is assumed that negligible stock of inventory of the generic product is stored, in that upon arrival from the generic production process it is allocated for customization. The basic assumption is that the standard part of the process takes t time periods to complete and the remaining $T-t$ periods correspond to time for the customization step (see Fig. 1). This is analogous to the warehouse lead time of t periods and $T-t$ periods of transportation time from the warehouse to the retailers. All the products are assumed to have independent and normal demands with mean μ_i per period and standard deviation σ_i per period and that the system follows a periodic review policy with a period length of one, and complete backlogging for unmet demands. Eppen and Schrage (1981) assumed that the inventory allocation at the intermediate stage to the end-products follows the equal fractile allocation rule, i.e., after allocation, the inventory position for each end product should be the sum of the mean demand for that end product over $T-t$ time periods and a common safety stock factor multiplied by the standard deviation of demand for the end product over the $T-t$ time periods. It was assumed that the probability of stock imbalance, i.e., that the stocks for the different products cannot be reallocated to satisfy the equal fractile rule after allocation, is negligible. They showed that when the costs are identical at each site it is optimal to operate the end product inventory stockpiles in an order-up-to manner with a base stock level of S_i. Erkip, Hausman and Nahmias (1990) extended the analysis to allow item demands to be correlated both across warehouses and also correlated in time. Lee (1996) studied the case where demand across

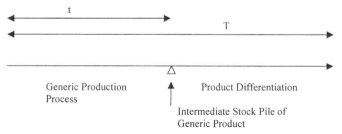

Fig. 1. Postponement with a single point of differentiation.

products (j, k) may be correlated in each period (ρ_{jk}). For such a system the steady state end of period inventory level for product i (I_i) is given by:

$$E(I_i) = A_i - R_i T \sum_j \mu_j \qquad (3.1)$$

$$Var(I_i) = R_i^2 t \left\{ \sum_j \sigma_j^2 + \sum_{j \neq k} \rho_{jk} \right\} + (T - t)\sigma_i^2 \qquad (3.2)$$

where A_i is a function of S_i and μ_i, but is independent of t and $R_i = \sigma_i / \sum_j \sigma_j$. Based on these two moments, service measures such as fill rate can be derived. The value of S_i can be determined to satisfy the target service level. Lee (1996) analyzes the above system with process standardization and addresses the impact of postponement which is reflected in the parameter t. Clearly, $E(I_i)$ is independent of t, but $Var(I_i)$ is decreasing in t for a given S_i.

$$\frac{\partial Var(I_i)}{\partial t} = R_i^2 \left\{ \sum_j \sigma_j^2 + \sum_{j \neq k} \rho_{jk} \right\} - \sigma_i^2$$

$$= \sigma_i^2 \left\{ \left(\sum_j \sigma_j^2 + \sum_{j \neq k} \rho_{jk} \right) \bigg/ \left(\sum_j \sigma_j \right)^2 - 1 \right\} \qquad (3.3)$$

For all i and j $\rho_{ij} \leq \sigma_i \sigma_j$ and therefore $(\sum_j \sigma_j^2 + \sum_{j \neq k} \rho_{jk}) \leq (\sum_j \sigma_j)^2$. Hence, the variance of the steady state end-of-period inventory for product i is decreasing in the degree of postponement. Thus, postponement will lead to reduction of inventory of finished products. Further, the reduction in inventory is greater when the end product demands are negatively correlated. For identical and independent demand for products, the expression for $Var(I_i)$ simplifies to:

$$Var(I_i) = [t/M + (T - t)]\sigma^2 \qquad (3.4)$$

Clearly one can see in this case (from the derivative with respect to t) that the reduction in variance is greater when the number of products is larger.

Lee and Whang (1998) explore this model further by assuming that demands are not IID (independent and identically distributed) over time. With non-IID demand, the value of postponement is more than just being able to make product commitments at a later point in time when realized demands have been revealed. In addition, the progression of demands may also help to improve the forecast of the future demands. Lee and Whang term these two values as the value of uncertainty resolutions and the value of forecast improvement. To illustrate these two different values, they used a random walk demand model with the characteristics that the variance of future demand increases as we look further out into the future. Hence, if we let demand for end product i after t periods from today be $D_i(t)$, then

$$D_i(t) = \mu_i + \sum_{k=1}^{t} \varepsilon_{ik} \tag{3.5}$$

where ε_{ik} is normally distributed with mean 0 and standard deviation σ_{ik}. When we have products with identical means and variances of demands, and when $\sigma_{ik} = \sigma$ for all i's and k's, then the safety stock that needs to be carried at the end product level for the original Lee (1996) model becomes:

$$s_i^*(t) = z \cdot \sigma \sqrt{\frac{M}{6}(T+1)(T+2)(2T+3) + \frac{M(M-1)}{6}(T+1-t)(T+2-t)(2T+3-2t)} \tag{3.6}$$

where z is the safety factor. In that case, the percentage savings obtained due to postponement as compared to no postponement (the case where $t=0$) is given by

$$V_i(t) = 1 - \frac{s_i^*(t)}{s_i^*(0)} = 1 - \sqrt{\frac{1}{M} + \left(1 - \frac{1}{M}\right)\frac{(T+1-t)(T+2-t)(2T+3-2t)}{(T+1)(T+2)(2T+3)}} \tag{3.7}$$

One can observe that the safety stock required is decreasing in t and the percentage savings due to postponement are increasing and convex in t. The reduction in safety stock with postponement in this case is greater than that when demand is stationary. The reason is that, with stationary demand, postponement allows the allocation decision (to the multiple end-products from the common intermediate product) be made after demand realizations of the end products have been revealed during the time when the common process was performed. When demand is a random walk process in which future demands are more variable into the future time, then there is an added value of postponement – by delaying the point when allocation to end-products has to be made, the demand variability of the end-product is reduced, since we are now closer to that future demand period than when we begin the total production process (forecast improvement). Thus, postponement with time-dependent demands may be even more valuable.

In the above models there is an implicit assumption that the production-distribution process is continuous and inventory can be stored only in finished product form. In general, manufacturing environments involve a discrete set of operations, and inventory can be stored immediately following any one of

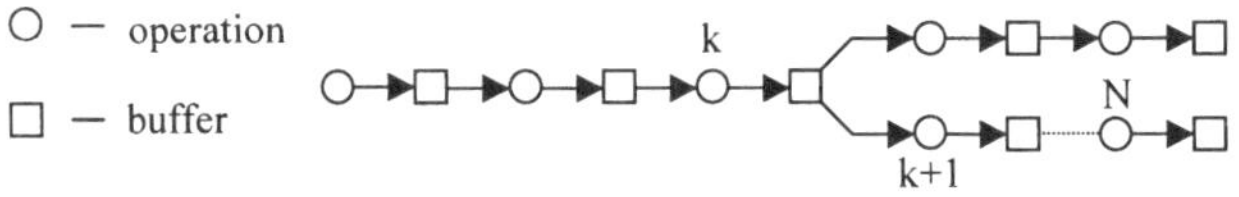

Fig. 2. Postponement with a single point of differentiation and discrete manufacturing steps.

these stages. Furthermore, the costs associated with delaying differentiation have not been considered in the above models. There is a stream of research that extended the standard postponement model by allowing for multiple manufacturing steps with intermediate buffer inventory, and by explicitly modeling the costs of resequencing the process steps.

Lee and Tang (1997) consider a model where there are two products that require N sequential tasks for completion (see Fig. 2). Inventory can be stored in a buffer after each task with the buffer after the Nth task being finished goods. The first k tasks are assumed to have been standardized, i.e., the inventory in the buffer after the kth operation can be used for customization into either products. The tasks $k+1$ to N are distinct for the two final products. Thus, the point of differentiation is right after the kth step. Under normal demand assumption for the two products and a discrete time setting, they consider the costs associated with standardizing stages. Let Z_i denote the average investment cost per period (amortized) if task i is changed into a common operation for both products. It is possible that $Z_i < 0$, e.g., when standardizing that task leads to overall reduction in costs. Let $L_i(k)$, $p_i(k)$, and $h_i(k)$ denote the lead time for task i, unit processing cost for task i and the per unit inventory holding cost for items in the buffer following task i, respectively, when the first k tasks are standardized. Further, they assume that the same safety factor z is used at all stages in the process, and a base stock policy is followed. Then, the average buffer inventory at any stage is given by

$$\frac{\mu}{2} + z\sigma \sqrt{(L+1)} \tag{3.8}$$

where μ and σ are the mean and standard deviation of demand faced at that stage. The relevant cost per period for the case when the first k operations are standardized is given by:

$$C(k) = \sum_{i=1}^{k} Z_i + \sum_{i=1}^{N} p_i(k)(\mu_1 + \mu_2) + \sum_{i=1}^{N} h_i(k)[L_i(k)(\mu_1 + \mu_2)]$$

$$+ \sum_{i=1}^{k} h_i(k) \left[\frac{(\mu_1 + \mu_2)}{2} + z\sigma_{12}\sqrt{L_i(k)+1} \right]$$

$$+ \sum_{i=k+1}^{N} h_i(k) \left[\frac{(\mu_1 + \mu_2)}{2} + z(\sigma_1 + \sigma_2)\sqrt{L_i(k)+1} \right] \tag{3.9}$$

This includes the average investments, processing costs, in-transit inventory (WIP) and buffer inventory holding costs.

Consider the special case when the lead time and holding costs at different stages are not affected by the point of differentiation, and when $p_i(k) = p_i + \beta_i$, i.e., β_i represents the additional processing cost for standardizing an operation. Then conditions under which $C(k)$ may be convex or concave can be derived. For the case when $C(k)$ is convex in k, the optimal k^* is decreasing in demand correlation among the two products, Z_i, β_i, and mean demand, but is increasing in $h_i\sqrt{L_i+1}$. As the demand correlation decreases, the resulting savings in inventory cost increase. In order to take advantage of the savings it is desirable to defer the common operation. When the cost of standardizing an operation Z_i or the incremental processing cost associated with delayed differentiation β_i increase then it is not as desirable to delay differentiation. As mean demand increases (while holding the variances constant), the resulting demand is less variable, therefore delaying differentiation is less attractive. Increase in $h_i\sqrt{L_i+1}$ leads to higher inventory savings due to delayed differentiation so the optimal delay point is further out. Furthermore, $C(k)$ is concave in k if (1) Z_i, β_i and h_i are identical; (2) S_i and β_i are proportional to h_i and L_i is constant; or (3) Z_i, β_i are identical and h_i is linear in i.

3.2 Multiple points of differentiation

So far, all the above models are restricted to only one point of differentiation. Garg and Tang (1997) consider a system with two points of differentiation, the first is the family differentiation point and the second the product differentiation point (see Fig. 3). In their system, there are three stages in the process. At the first stage all the products are in their generic form, the family differentiation point occurs at the beginning of the second stage where specific components are added to differentiate a generic product into different families. The product differentiation point occurs at the beginning of the third stage where specific components are used to customize semi-finished products into different end products of that family. Note that the points of differentiation emerge because of adding specific components. The lead times for the different stages are assumed to be T_1, T_2, and T_3, respectively. They assume that the manufacturing lead times for customizing the products into the different families T_2 are the same and that the manufacturing lead times for customizing different end products of different families T_3 are the same. Early postponement is defined as increasing T_1 to T_1+1 while reducing T_2 to T_2-1, and late postponement is defined as increasing T_2 to T_2+1 and reducing T_3 to T_3-1. For the above system, the authors consider two possible scenarios. In the first scenario, inventory is stored only in the finished goods form (called centralized system) and in the second scenario, inventory is stored at each point of differentiation as well as at the finished goods level (called decentralized system). The centralized system extends the model

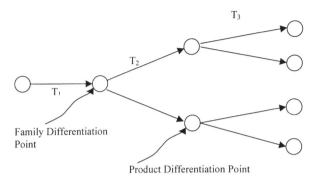

Fig. 3. Postponement with two points of differentiation.

studied by Eppen and Schrage (1981) to three stages and correlated demand. The demand for the final products are assumed to be independent normal and identical across time periods. For each time period, the demand for the final products are correlated. The assumption is that the system operates under a base stock policy and periodic review. An equal fractile allocation is assumed at the first and second stages. For the centralized system, under an equal-fractile allocation policy and identical equivalent degree of correlation of demand at the family level (defined as the ratio of variance for perfectly correlated demand and the actual variance in demand for the family), they show that both early and late postponement lead to reduction in total inventory. Further, they show that as the product demand across a family becomes more negatively correlated then late postponement becomes more preferable as compared to early postponement. In the decentralized model, they assume that inventory is stored at all locations and the service level at each of the stages is high enough that the system can be decoupled into independent single stage inventory systems. For such an environment, they analyze the inventory savings across the whole network due to early and late postponement. They show that if $T1 > T2 > T3$ then both early and later postponement are beneficial. Further, when T2 is sufficiently smaller (larger) than T1 and T3, then early (late) postponement is beneficial.

3.3 Vanilla boxes

The above papers assume that the production distribution process does not have any capacity constraints. Swaminathan and Tayur (1998a) analyze a final assembly process with production capacity where inventory is stored in the intermediate form (called vanilla boxes). In addition to the intermediate form, they allow the two extreme forms of vanilla boxes – as components and as finished products. Therefore, this model captures both assemble-to-order (where components are stocked and products assembled from the components after demand is realized) as well as make-to-stock (where inventory is carried in finished form only) as special cases. This approach allows for multiple

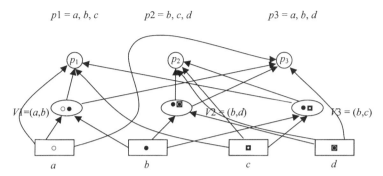

Fig. 4. Postponement with multiple points of differentiation using vanilla boxes.

points of differentiation, in that there is no restriction on the type of vanilla box that can be stored. For example, Fig. 4 shows a product line with three products $p1$, $p2$, and $p3$ made of components a, b, c, and d. Vanilla box $V1 = (a, b)$ can support products $p1$ and $p3$ while vanilla box $V3 = (b, c)$ can support $p1$ and $p2$. In general, every product i ($1, \ldots, M$) may be assembled either directly from its components, or from any vanilla box whose component set is a subset of those required by i, thus avoiding redundant components. A binary bill of material in terms of the components is assumed without loss of generality.

Demands for the final products are random but follow one of L given scenarios, each with a given likelihood. They assume that the vanilla box inventory follows a base stock policy in that every period the inventory is brought up to that level, then demand is realized, products are assembled from vanilla boxes by adding other components within the production capacity. Unsatisfied demand is lost with a penalty and remaining inventory of vanilla boxes incurs holding cost. Clearly, the main benefit of having vanilla boxes is that the amount of lead time for customization is much lower than customizing from the component level. Also, the capacity for customization may be limited which makes the problem more challenging. Under the above setting, they develop a stochastic integer program to determine the optimal types of vanilla boxes as well as their inventory levels which minimize the expected holding and penalty costs in single and multi-period settings. The first stage variables determine which components should be present in different vanilla boxes and the base stock levels for those vanilla boxes. The second stage variables determine how those vanilla boxes should be allocated to the different products on realization of product demand.

Let C denote the capacity available to assemble products from vanilla boxes or from basic components, t_{i0} and t_{ik} the per unit assembly time for product i from components or from vanilla box k ($t_{ik} = \infty$ if product i cannot be made from vanilla box k), respectively, π_i the per unit per period stock-out cost for product i, h_k the per unit per period holding cost for vanilla box k,

$S = (S_1, \ldots, S_K)$ the vector of base stock levels of vanilla boxes k $(1, \ldots, K)$, D_l a realization $(D_{1l}, \ldots, D_{Ml})$ of product demands in scenario l where $D_1, \ldots, D_M$ have a joint distribution F, and r_{ikl} the quantity of product i made using vanilla box k in scenario l ($k = 0$ implies that product i is assembled directly from components). Then the two-stage stochastic program corresponding to a vanilla box configuration U can be formulated as follows. $P1(\mathbf{U}) = \min_{\mathbf{S}} E_l Q(\mathbf{S}, \mathbf{U}, D_l)$, where

$$Q(\mathbf{S}, \mathbf{U}, D_l) = \min_{r_l} \sum_{i=1}^{M} \left(\pi_i \left(D_{il} - \sum_{k=0}^{K} r_{ikl} \right) \right) + \sum_{k=1}^{K} \left(h_k \left(S_k - \sum_{i=1}^{M} r_{ikl} \right) \right)$$ (3.10)

s.t. $\sum_{i=1}^{M} \sum_{k=0}^{K} t_{ik} r_{ikl} \leq C \quad \forall l,$ (3.11)

$\sum_{i=1}^{M} r_{ikl} \leq S_k \qquad \forall k \geq 1, \forall l,$ (3.12)

$\sum_{k=0}^{K} r_{ikl} \leq D_{il} \qquad \forall i, \forall l,$ (3.13)

$r_{ikl}, S_j \in R_+.$ (3.14)

Utilizing the above framework and through the development of an efficient simulation based algorithm, the authors explore the benefits of postponement through vanilla boxes under various settings. Among other results, they show that postponement using vanilla boxes outperforms both assemble-to-order and make-to-stock systems when the assembly capacity available is neither too slack nor too tight (representative of most real environments). Further, they find that the vanilla box approach is extremely powerful under high variance and negative correlation among product demands. Finally, they provide examples where stocking two types of vanilla boxes may be sufficient for a product family with ten products and the performance may be better than a make-to-stock approach (with all the 10 products).

Graman and Magazine (2002) consider a postponement model with capacity constraints where inventory can be stored in an intermediate form. On realization of demand all the finished goods are used first, and then the semi-finished product is used to satisfy the demand subject to a capacity constraint. This problem can be viewed as a special case of the vanilla box problem with only one type of vanilla box. For this model, they derive analytical expressions for service measure and also inventory calculations and through a numerical study show that very little postponement capacity can actually provide all the benefits related to inventory reduction.

Benjafer and Gupta (2000) present models that utilize queuing approximations to analyze a system where both make-to-stock and make-to-order environments are utilized while delaying differentiation of the product. There are two stages in the production process; the first stage produces products to stock while the second stage produces products to order. The two stages are separated by a buffer that holds semi-finished inventory. The authors utilize queuing approximations by decoupling the two stages and assume each of them will behave similar to a M/M/1 queue. For the above approximation, solutions for inventory and service are available which the authors utilize to develop an optimization problem. The objective is to minimize the total costs subject to service level constraints by changing the stocking level of the intermediate product and the degree of differentiation. Further, the authors present several computational insights and show the impact of congestion effects on the postponement decision.

3.4 *Process standardization applications*

Lee, Billington and Carter (1993) describe the process standardization efforts at Hewlett Packard DeskJet Printer business. The printer line had three distribution centers in Europe, the US, and the Far East and needed localization for the different countries in terms of power supply module with correct voltage, power cord terminators, and a manual in the appropriate language. The existing operation was one where the products were 'localized' at the US factory before being shipped to the respective distribution centers. The manufacturing in the US was done through a pull system based on the target safety levels set for the different distribution centers while taking into account the one month lead time in transit to the overseas distribution centers. As a result, high levels of safety stock are needed in the overseas distribution centers. The re-engineering of the distribution process involved resequencing the transportation and localization steps so that localization would now be done at the distribution centers. This was accomplished by making changes to the product design so that the power supply and the manuals could be added later at the distribution centers. There were also additional investments in the form of product redesign, package redesign, and enhancement to distribution center capabilities, which were offset by the inventory savings that resulted from postponement. Additional benefits included lower capital investment for in-transit inventory, lower freight costs (due to the use of bulk packaging of the generic printers as opposed to packaging finished printers) and local presence of final assembly in the overseas markets. Based on a detailed modeling and analysis, Hewlett-Packard adopted process standardization in their inkjet business, and was rewarded with huge costs savings and improvements in customer service.

Swaminathan and Tayur (1998a) analyze the final assembly stage of RS6000 server machines produced by IBM. Each model in the product line had 50–75 end products mainly differentiated by 10 main features or components. A component is defined to be a part that is directly used in final assembly, so a component may be a subassembly in itself, e.g., a planar card. Different end products across the product line showed a high degree of component commonality. Since demand for end products were highly stochastic and correlated, the existing mode of operation was to start final assembly only after a firm customer order had been received. The typical steps in final assembly involved getting components together (kitting), putting them in the right place (assembly), testing, loading software (preloading), and packing the final product. At the time of the research this process often finished later than the customer requested arrival date leading to a sizeable percentage of late orders. This order delay problem was becoming increasingly acute, as customers who were once satisfied with the delivery within a month were now demanding the products to be shipped within seven to ten days after the orders were placed. The change in customer requirements was primarily due to the competition in the industry and increase in service expectations. The vanilla assembly process based on delayed differentiation stocked vanilla boxes (semi-finished inventory). Clearly, there were additional costs in terms of redesigning the line to enable vanilla boxes, including work force training and having inventory of vanilla boxes in the process that tied up capital. However, the benefits of such an approach were that the lead time experienced by the customer was only limited to the customization time starting from the vanilla box, hence most of the orders could be satisfied on time. After a thorough analysis of the costs and benefits as well as the change process involved, the vanilla assembly process was introduced in one of the two assembly plants (which had a satellite plant that was redesigned to produce vanilla boxes).

Brown, Lee and Petrakian (2000) describe the postponement approaches at Xilinx which involved process standardization. As a leader in the field-programmable logic business, Xilinx made use of the postponement practice to achieve significant cost savings and service improvements. The manufacturing of integrated circuits consisted of two major steps: a front-end wafer fabrication at their outsourced manufacturer in Taiwan; and a backend assembly and test at their outsourced assembly sites in the Philippines and other Asian sites. The front-end process was standardized so that multiple devices share the same process. This way, the product does not have to be highly differentiated at the end of the front-end process. Fabricated wafers are then stored as intermediate inventory, known as the die bank, and they would go through the backend process that customize the products into the exact end device, only after the customer orders have come in. This way, the lead time to the customers is only the backend process time, which is much shorter than the sum of front-end and backend process times (the lead time when a totally build-to-order process is used); but the flexibility to customer orders is much greater than if the finished goods inventory is stored under a build-to-stock process is used.

Zara (one of the famous brands of Inditex) utilizes process standardization and vanilla boxes in the design phase of the product life (Harle, Pich and Hayden, 2001 and Fraiman and Singh, 2002). Zara introduces new products at a rapid rate; in fact 70% of the products change every two weeks in a typical retail outlet. In order to create large variety and quick response to customers, the firm employs several strategies including standardization of the design modules. At the beginning of each selling season, the designers create a library of models that serve as platforms for the models that will be eventually launched. Twenty designers walk the streets and go to discos in order to get a feel of the latest fashion trends. After carefully watching the latest in fashion trends, Zara designers give adaptation (or customization) to the models from the library (which are vanilla boxes) and create 5–8 new designs every day! In total about 12,000 new products and designs are created every year.

4 Process resequencing

Process resequencing is another approach for enabling postponement and making it more effective. The basic idea is that it may be possible to change the sequence of operations in a process so that products get differentiated later. However, there may be costs associated with changing the sequence of operations, and hence it is important to have models that provide insights on these costs and benefits.

4.1 Linear process sequencing

Lee and Tang (1998) consider a two-stage system where at each stage a distinct feature is introduced into the product. They consider the knitting and dyeing tasks for garments as representative of the two stages. Each feature may have multiple options, for example, garment could be knitted under different settings or dyed with different colors. They analyze the case where each feature has two alternative options (see Fig. 5). Thus there are four possible products available to a customer. Figure 5(a) represents the case where the garments are dyed first and are knitted later, while Fig. 5(b) represents the case where the garments are knitted first and dyed later.

In such a system, changing the sequence of operations (which determines the feature that should be introduced first into the product) does not affect the ends of the process (raw material and finished products) but only affects the inventory that is stored at the end of the first stage. The objective is to minimize the total variance for the two intermediate buffers since the variance influences the inventory requirements for the system. The use of the total variance as an objective function is of course a stylized assumption. The authors argued that the cost of manufacturing, such as the use of overtime or

Fig. 5. Operations reversal in the process.

expediting, is often directly linked to the variances of production requirements (which, in this case, is the same as the variances of the intermediate buffers). Indeed, as we see below, the use of a different objective function can lead to different results. The total demand in any period (across all the four products) is assumed to be a random variable with mean μ and standard deviation σ. The demands for the end products are modeled as a multi-variate normal distribution with parameters $(N, \theta_{11}, \theta_{12}, \theta_{21}, \theta_{22})$ where θ_{11} represents the fraction of customers buying the first option in both the features. They show that it is optimal to have feature A sequenced before feature B, if

$$(\mu - \sigma^2)[p(1-p) - q(1-q)] < 0 \qquad (4.1)$$

where p is the probability of a customer buying option 1 on feature A and q is the probability of a customer buying option 1 on feature B. Clearly if the variance associated with feature A, given by $p(1-p)$, is smaller than the variance associated with feature B, given by $q(1-q)$, then one expects that feature A should be sequenced first. However if $\sigma^2 > \mu$ then the reverse result is true, which is counter-intuitive. They also show that when more options are available on the two features and each of these options are equally likely, then it is better to sequence the operation with fewer options first when $\mu > \sigma^2$ and vice versa otherwise. Kapuscinski and Tayur (1999) show that if the objective is to minimize the sum of standard deviations rather than the sum of variances at the intermediate stage, then for the two feature – two option case, the counter-intuitive result corresponding to the case $\mu < \sigma^2$ vanishes.

4.2 Assembly sequence design

It is clear from the above models that the sequence of tasks could play an important role in enabling postponement and thereby reducing inventory requirements. However, the physical assembly sequence is often defined through a complex set of precedence relationships among the different tasks. The general assembly design sequence problem has been primarily studied by researchers in engineering (see Nevins and Whitney 1989). For example, Fig. 6 shows a product line with four products and six components. The assembly

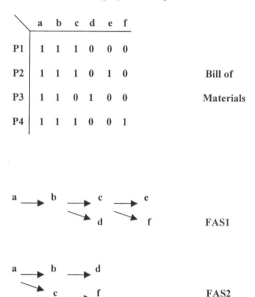

Fig. 6. Alternative feasible assembly sequences.

sequences *FAS*1 and *FAS*2 represent possible sequences for the product line. *FAS1* represents an assembly sequence where component *a* needs to be introduced first followed by *b* which can be followed by *c* or *d*. Once component *c* has been added, *e* or *f* can be added to the subassembly. *FAS2* represents another assembly sequence where component *a* still needs to be added first at which point either *b* or *c* can be added. Once *c* has been added, *e* and *f* can be added in any sequence and once *b* has been added, component *d* can be added to the subassembly. Note that in *FAS*1, component *b* needs to precede component *c* in the assembly whereas in *FAS*2 that precedence has been relaxed.

Gupta and Krishnan (1998) introduce the notion of a 'generic subassembly (GSA)' which is similar to the vanilla box concept. GSA is a subassembly that satisfies all the precedence relationships among its components and is a feasible subassembly. In the above example (*a, c, f*) is a GSA for *FAS2* but is not a GSA for *FAS1* because in *FAS1* component *b* has to be in place before component *c* can be introduced. A MGSA is a maximal generic subassembly according to criteria such as number of components in the assembly or number of final products that can be supported by it. In the above example, (*a, b, c*) which is a generic subassembly (GSA) covers three products P1, P2, and P4 and has three components. It is a MGSA in terms of number of components. On the other hand, (*a, b*) which covers P1, P2, P3, and P4 (all the four products) is a MGSA in terms of number of products covered. For a given feasible assembly sequence, Gupta and Krishnan (1998) present an algorithm that generates the MGSA for a product family with criterion of maximizing the number of end products supported. Although a useful step in the right direction, the above model does not provide a cost benefit analysis related to assembly sequence design.

Swaminathan and Tayur (1999) utilize the vanilla box model described earlier, along with an assembly sequence design model to generate useful managerial insights. In the assembly sequence design problem (ASDP) they develop a mathematical programming model that generates the best sequence taking into account costs associated with the design of components to make such an assembly sequence. Thus, they model the situation where components could be designed in a flexible manner to satisfy alternative precedence conditions during execution. In combination with the vanilla box model, they consider two approaches to task resequencing: (1) where the best vanilla boxes are determined first and the sequence design is generated to enable assembly of the vanilla boxes and finished products with minimal design costs; and (2) where the most efficient assembly sequence is determined for the set of finished products and then the best vanilla boxes are found while taking into account the assembly constraints. As opposed to earlier work, this approach integrates the assembly sequence decisions with the postponement decisions, and hence enables analysis of various 'what if' questions pertaining to process resequencing.

Their notation is as follows; products are indexed by $i = 1 \ldots M$, components by $j = 1 \ldots n$ and vanilla boxes by $k = 1 \ldots K$. Let u_{kj} denote the content of the kth vanilla box in terms of components, $\mathbf{U}$ the vanilla configuration (matrix of u_{kj}) and $\mathbf{U}_k$ the configuration of the kth vanilla box. Let a_{ij} denote the bill of materials for the products in terms of the components, g_{pq} the cost of assembling component p before component q, e_{pq} the cost of allowing independence between components p and q, and $\mathbf{Y}$ the assembly sequence defined through the Boolean variables y_{pq} (set to 1 if component p is assembled before component q and to 0 otherwise). The difference in design costs between a fixed and an independent precedence relationship is given by $c_{pq} = g_{pq} - e_{pq} \leq 0$ and the objective is to minimize the total cost incurred. For a particular vanilla box configuration $\mathbf{U}$ the problem can be formulated as given below.

$$ASDP(\mathbf{U}) : \min_{\mathbf{Y}} \sum_{p=1}^{n} \sum_{q=1}^{n} c_{pq} y_{pq}$$

s.t. $\quad 1 - y_{qp} \geq u_{kp}(1 - u_{kq}) \quad \forall p, q, k,$ \hfill (4.2)

$1 - y_{qp} \geq a_{ip}(1 - a_{iq}) \quad \forall i, p, q,$ \hfill (4.3)

$y_{pq} + y_{qp} \leq 1 \quad \forall p, q,$ \hfill (4.4)

$$y_{pq} + y_{qr} - y_{pr} \leq 1 \quad \forall p, q, r, \tag{4.5}$$

$$y_{pp} = 0 \quad \forall p, \tag{4.6}$$

$$y_{pq} \in \{0, 1\} \quad \forall p, q. \tag{4.7}$$

In the above formulation, constraint (4.2) represents that if a component q is not present in a vanilla box ($u_{kq} = 0$) then it cannot be a predecessor of any component p in that vanilla box ($u_{kp} = 1$). Constraint (4.3) represents that if a component q is not present in a product ($a_{iq} = 0$) then it cannot be a predecessor of any component p in that product ($a_{ip} = 1$). These constraints assure that all vanilla boxes and products can be assembled using the assembly sequence **Y**. Constraint (4.4) indicates that two components are either unordered in the assembly sequence ($y_{pq} = y_{qp} = 0$) or there exists a unique ordering of these components in the assembly sequence ($y_{pq} = 1$ or $y_{qp} = 1$, but not both). Constraint (4.5) maintains the transitivity relationship between components and constraint (4.6) indicates that all components of the same type are at the same level in the assembly sequence.

The authors conducted an extensive computational study which, in addition to validating earlier observations on the role of demand variance and correlation, provides additional insights on issues such as: it is better to sequence features with higher degree of variance later in the process; when the total amount of options provided across all features is kept constant, it is better to provide more options in a restricted number of features.

4.3 *Process resequencing applications*

Benetton (described in Dapiran 1992) is the earliest reported application of process resequencing that the authors are aware of. Traditionally sweaters were manufactured by first dyeing the yarn into different colors, and then knitting the garments out of the colored yarns. The garments were stored in the form of finished goods to be shipped to the retailers. Dapiran describes how Benetton interchanged the knit and dye operations when they realized that most of the demand variability was due to the uncertainty of the customers' preference of colors in a particular season. The interchange of the knit and dye operations enabled the firm to stock inventory of 'greige' (uncolored) knit garments that could be dyed once the seasonal demand was known, enabling postponement and reducing inventory. Benetton had to invest in improving the dyeing technology so that the quality of the garments would not deteriorate due to the process changes.

Swaminathan and Tayur (1999b) describe the assembly sequencing problem for US Filter, a manufacturer of reverse osmosis pumps. The sequence of operations at the final assembly was altered to enable faster response to customers. Costs related to product-process redesign as well as worker retraining at the final assembly had to be taken into account. The process sequencing approach has also been successfully applied by Garg (1999) to a large electronic manufacturer whose products are tailored for the telecommunication industry. The process involved board insertion and assembly, a station 'marrying' different modules together, and the packaging of accessories and other components to make the final product. Alternative sequences of the process would result in different inventory and waiting times for the manufacturing of the product. Garg employed a queueing network model to evaluate these alternatives. More examples of application of postponement can be found in Lee, Feitzinger and Bellington (1997) and Lee (1998).

5 Component standardization

5.1 Commonality and inventory management

Component commonality has traditionally been studied in the context of make-to-stock or assemble-to-order systems. Traditional research on component commonality in the operations management stream has been mainly focused on reduction in component inventory due to commonality. Collier (1982) introduced the notion of commonality index – a measure of degree of commonality in a product line. Gerchak and Henig (1986) showed that when components are combined (standardized), the inventory requirements for product specific components always increase. Further, they also showed that a myopic inventory policy is optimal for a dynamic multi-period inventory problem with component commonality. Baker, Magazine and Nuttle (1986) and later Gerchak, Magazine and Gamble (1988) explored the benefits of having common components in terms of reduced inventory or increased service. In particular they considered two products each with two components and analyzed the impact of standardizing one of the components. Since then, several authors including Eynan and Rosenblatt (1996) and Thonemann and Brandeau (2000) among others have explored the benefits associated with component commonality under different settings. Fisher, Ramdas and Ulrich (1999) study commonality issues in the automotive industry.

5.2 Commonality and postponement

Lee (1996) points out that in order to perform a complete analysis regarding the benefits of postponement due to component standardization, one needs a model that takes into account the following aspects: (1) inventory savings for the part; (2) increase in material costs for the common parts; (3) additional costs for the engineering change; and (4) inventory savings for the finished goods. Most of the analysis on postponement has focused on (4) and the analysis on commonality has focused on (1). Lee and Tang (1998) present a model where they consider the costs associated with standardizing a process step. In order to standardize a process step, one needs to standardize the part associated with that step and associate a cost for that. They incorporate the above cost in their model while analyzing the degree of postponement that is optimal.

The value of component standardization may be different at different stages of the product life cycle. This is due to the dynamic changes of demand uncertainty, shortage costs, inventory holding costs, and rework costs (to convert one version of end product to another). For example, the uncertainty of demand may be much higher in the product introduction and end-of-life phase than the mature phase. The shortage cost may be the highest during the product introduction phase, while the inventory holding cost may be the highest during the end-of-life phase, when left-over inventory may have to be written off as the product becomes obsolete. Consequently, models capturing these dynamics are needed to assess the value of component standardization. Lee and Sasser (1995) describe one simple model that shows that, given all the dynamics of the demand and cost characteristics, the value of standardization for postponement is high in the product introduction phase, low in the mature phase, and high in the end-of-life phase.

Swaminathan (1999) considers the problem related to optimizing the level of commonality while simultaneously considering the costs of commonality as well as the benefits due to inventory savings obtained due to higher levels of commonality. In particular, the author considers a two-product system with one common subassembly and two product specific subassemblies. The parameter to be optimized is the size of this common subassembly in terms of the degree of commonality. The author assumes that the cost of the common component increases in a convex fashion with respect to the degree of commonality while the costs of the product specific components decrease in a linear fashion. Inventory of both common and product specific components are stored in anticipation of demand. Under the above assumptions and standard inventory assumptions related to holding and penalty costs and linear costs of commonality, the author shows that the two products have either complete commonality or no commonality. Further the optimal level of commonality is lower in product lines where the costs of introducing commonality are higher. Based on a computational study, the author shows that the optimal level of commonality is always higher in postponement as compared to the optimal level of commonality when product inventories are managed independently. Moreover, the cost of commonality affects the impact of operational factors on optimal commonality and inventory under postponement. That impact is limited when the cost of providing commonality is either very high or very low.

Van Mieghem (2002) analyzes a model with two products similar to Swaminathan (1999), where each product is assembled from two components. However, Van Mieghem (2002) assumes that both common and product specific components are stocked and derives conditions under which commonality should be adopted. This condition is stated in terms of a maximal commonality threshold cost that depends on the demand forecast only through its demand correlation as well as on financial data. For high commonality cost, neither commonality nor postponement is optimal. A pure commonality strategy where each product is assembled using a common component, however, is never optimal unless complexity costs are introduced. Finally, the author shows that while the value of the commonality strategy decreases in the correlation between product demands, commonality is optimal even when the product demands move in lockstep (perfectly correlated) if there is a sufficient profit differential between the two products.

5.3 Component standardization applications

Lee (1996) provides the example of a large computer printer manufacturer which produced both mono as well as color printers. The manufacturing process for the two products are very similar except for the materials used. There are two key stages in the production, printed circuit board assembly and final assembly. At each of these stages a distinct component (print mechanism interface or head driver board) is inserted to differentiate the two types of products. The product differentiation began as soon as the head driver board was inserted at the printed circuit board stage. The demand was highly uncertain for both products and often correlated which led to high forecast errors. The firm evaluated the option of standardizing the head driver board or both the head driver board and print interface mechanism, which would lead to postponement. However, the costs of designing the additional functionality in the common components needed consideration.

Brown et al. (2000) describe the component standardization approach utilized at Xilinx where the final end product was actually designed in such a way such that customization could be done through software deployment at the customer site. The result is an integrated circuit that is field-programmable. In this case, we have an extreme case of commonality standardization, since the final product has been standardized.

Thonemann and Brandeau (2000) describe a model that determines the optimal degree of standardization of components in a multi-product environment. This model was successfully utilized by a large automobile manufacturer to determine the optimal degree of commonality of wire harnesses that go into a product family. The model provided decision support for the design of future generation of components. Hewlett-Packard Company had also used this approach to redesign their network printer (Lee, 1997). The network printer was made in Japan and used to have two distinct engines, one for 110 voltage countries (e.g., North America), and one for 220 voltage countries (e.g., Europe and Asia). The printers with different engines are thus not inter-changeable to meet the changing demands in the different continents. An alternative design called for using universal power supply and fuser, which would result in a universal printer. This way, the printers made in Japan could be used for any market. At the end of life of the product, this could be particularly beneficial, as excess inventory in one continent does not have to be written off, but instead can be trans-shipped to another continent if there are any demand/supply imbalances.

Most recently, Lucent Technologies in Spain has utilized the component standardization strategy to achieve great success (Hoyt and Lopez-Tello, 2001). In 1998, the Tres Cantos plant, which builds telecommunications-switching systems, was faced with a great sales potential in Saudi Arabia that was worth millions of dollars. However, the lead time required was much shorter than usual, due to the Saudi government's desire to have all systems implemented prior to Y2K. The specific configurations required, however, could only be known after detailed site engineering work had been performed. The complete manufacturing time for the build-to-order process was far greater than what the Saudi customers wanted. In addition, the Tres Cantos actually did not have enough capacity to meet this big order from Saudi. By redesigning the product so that they could have common building blocks, Lucent was able to: (1) prebuild the common building blocks before detailed site engineering tasks were completed; and (2) utilize the US plant in Oklahoma to help solve the capacity limitation problem. The result was that the company was able to win the contract and delivered the products on time.

6 Related strategies and other benefits

Thus far we have considered models that explore mostly the inventory-related benefits of postponement. However, postponement decisions may be tightly linked to pricing decisions and information flow decisions. We will explore models for those and other strategies related to product variety management in this section.

6.1 Postponement, information, and pricing

Although the key benefit identified with postponement has been inventory reduction, there are other issues related to postponement that are important. The main benefit related to postponement stems from the fact that one can delay the decision point for differentiation so that one can get more demand information before making a final commitment. Benefits of postponement can also be due to better forecasts generated in cases where the future forecasts are improved, as one gets closer to the period when the demand occurs. As described earlier, Lee and Whang (1998) differentiated the value of postponement as 'uncertainty resolution' and 'forecast improvement.' Anand and Mendelson (1998) study the increased flexibility and pooling benefits of delayed production in a multi-product supply chain with a noisy information system on a binary demand distribution. Aviv and Federgruen

(1999) provide a detailed characterization for the benefits related to postponement under unknown demand distributions.

Another benefit of postponement beyond risk pooling relates to more effective usage of existing capacity. Swaminathan and Tayur (1998a) show that the benefit of postponement using vanilla boxes is extremely high when the available capacity is medium (neither too high or too low). This is because, under that condition, postponement also benefits capacity utilization, which in turn affects the total costs. Gavirneni and Tayur (1999) explore this issue further by considering a two-item system where the manufacturer has the option to produce them in separate facilities or postpone the differentiation to the distribution stage. The authors explore the benefits of postponement under varying assumptions about the information at the manufacturer about the ordering policies being utilized.

Although we have only focused on the flexibilities provided by postponement strategies related to production of products, one could also envision situations where a firm tries to obtain a similar flexibility through pricing. Van Mieghem and Dada (1999) present a comparative analysis related to price, production and capacity decisions. They show that competition, uncertainty and timing of operational decisions can influence investment decisions of the firm related to capacity and inventory. Using a simple model for uncertainty in demand (captured by a random shock) they show that, in contrast to production postponement, price postponement can make the investment decisions related to capacity and inventory relatively insensitive to uncertainty.

6.2 *Postponement, modularity, and substitution*

Postponement strategy is often affected to a great extent by the product architecture decisions. One such decision relates to the degree of modularity that is present in the product architecture. Recently, several firms have started designing their products in 'families' where the individual products are distinguished by the alternative combinations that are given to the modular components. Examples include the personal computer, electronics and automobile industry. Although this is an important concept from a product design perspective, analytical models studying its relationship with postponement have been very limited.

Another concept related to postponement for managing product variety is substitution. Substitution is a strategy where the manufacturer is able to provide a customer with an alternative product when the product ordered is not available in stock, and in the process incurs some kind of goodwill cost (or a real cost related to providing a better product to the customer or providing a gift voucher of some sort). This is a powerful strategy that has been used by firms over the years in several industries. Researchers have studied several versions of this problem (see Jordan and Graves 1995, Bassok, Anupindi and Akella 1999, Rao, Swaminathan and Zhang 2002) following the initial characterization of the optimal policy by Ignall and Veinott (1969). Some firms have also utilized the substitution strategy along with the postponement strategy. Swaminathan and Kukukyavuz (2000) analyze one such environment from the biotech industry, and show the comparative benefits of these strategies. An alternative situation related to substitution where a customer automatically buys a substitute product, has also been studied (see Mahajan and Van Ryzin, 1999 for details). Swaminathan (2001) presents a managerial framework that relates product variety strategies such as postponement, substitution and commonality to the product and process modularity and identifies the most appropriate approach under different conditions.

7 Conclusions

Global markets with diverse needs and dramatic shortening of product life cycles put a great premium on effective product variety management. As a result, designing products for postponement is of high priority. The interdisciplinary and complicated nature of the problem has generated a need for models that can provide multiple perspectives on the costs and values of postponement. In this chapter, we have provided a summary of such research conducted from the operations management perspective. In particular, we focused on the three enablers of postponement; namely, process standardcization, process resequencing, and component standardization. We presented models which provided insights on their benefits as well as industrial application of these strategies.

Clearly, there is a need for research and models along two additional dimensions. First, we need models that can be incorporated into decision support systems that allow managers to benefit from model based decision support. To achieve this, more emphasis is needed on large-scale models that capture the essential characteristics of the real environment and development of algorithms to solve those models in a fast and efficient manner. Second, most of the models developed this far relate to product postponement. As a larger set of firms move towards service-oriented businesses, models that can capture postponement benefits in those environments are going to be extremely useful.

As more and more firms adopt the Internet technology to conduct their business on-line as well as have more interactions with their customers over the Internet, they are beginning to gather richer and detailed information about customer preferences. This has provided firms with an opportunity to tailor their products and services around customer preferences, i.e., mass customization. Postponement provides a powerful way for firms to pursue mass customization without incurring the usual huge operational costs associated with proliferating product variety. Indeed, as Feitzinger and Lee (1997) indicated, postponement is a strategy that allows firms to implement cost-effective mass customization. As a result, it is all the more important that new models related to postponement and other strategies for effective product variety management be studied and analyzed by future researchers.

Acknowledgement

The work of the first author was supported in part by NSF CAREER Award #0296081.

References

Alderson, W., Marketing Efficiency and the Principle of Postponement, *Cost and Profit Outlook*, September 1950.
Anand, K.,]H. Mendelson, Postponement and Information in a Supply Chain, Technical Report, Northwestern University, July 1998.
Aviv, Y., A. Federgruen (1999). The Benefits of Design for Postponement, in: S. Tayur *Quantitative Models for Supply Chain Management*, Norwell, MA, Kluwer Academic Publishers, pp. 553–586.
Baker, K.R., M.J. Magazine, H. Nuttle (1986). The Effect of Commonality on Safety Stock in a Simple Inventory Model. *Management Science* 32(8), 982–988.
Bassok, Y., R. Anupindi, R. Akella (1999). Single Period Multiproduct Inventory Model with Substitution. *Operations Research* 47(4), 632–642.
Benjafer, S., D. Gupta, Make-to-Order, Make-to-Stock or Delay Product Differentiation? – A Common Framework for Modeling and Analysis, *Working Paper*, Department of Mechanical Engineering, University of Minnesota, Minneapolis, MN, 2000.
Brown, A.O., H.L. Lee, R. Petrakian, Xilinx Improves its Semiconductor Supply Chain using Product and Process Postponement, *Interfaces*, 30(4), July–August 2000, 65–80.
Bucklin, L.P. (1965). Postponement, Speculation and the Structure of the Distribution Channel. *Journal of Marketing Research* 2, 26–31.
Collier, D.A. (1982). Aggregate Safety Stocks and Component Part Commonality. *Management Science* 28(11), 1296–1303.
Dapiran, P. (1992). Benetton – Global Logistics in Action. *International Journal of Physical Distribution and Logistics Management* 22(6), 7–11.
Eppen, G., L. Schrage, Centralized Ordering Policies in a Multi-Warehouse System with Lead Times and Random Demand, *Multi-Level Production Inventory Control Systems: Theory and Practice* (ed. L.B. Schwartz), North Holland, Amsterdam, 1981, 51–58.
Erkip, N., W. Hausman, S. Nahmias (1990). Optimal centralized ordering policies in multi-echelon inventory systems with correlated demands. *Management Science* 36(3), 381–392.
Eynan, A., M. Rosenblatt (1996). Component Commonality Effects on Inventory Costs. *IIE Transactions* 28, 93–104.
Feitzinger, E., H.L. Lee (1997). Mass Customization at Hewlett-Packard: the Power of Postponement. *Harvard Business Review* 75(1), 116–121.
Fisher, M., K. Ramdas, K. Ulrich (1999). Component Sharing in the Management of Product Variety: A Study of the Automotive Braking System. *Management Science* 45(3), 297–315.
Fraiman, N., M. Singh, "Zara," *Columbia Business School Case*, 2002.
Garg, A. (1999). An Application of Designing Products and Processes for Supply Chain Management. *IIE Transactions* 31, 417–429.
Garg, A., H.L. Lee, Effecting Postponement through Standardization and Process Sequencing, *IBM Research Report RC 20726*, IBM T.J. Watson Research Center, Yorktown Heights, NY, 1996.

Garg, A., H.L. Lee (1998). Managing Product Variety: An Operational Perspective, in: S. Tayur *Quantitative Models for Supply Chain Management*, Norwell, MA, Kluwer Academic Publishers, pp. 467–490.

Garg, A., C.S. Tang (1997). On Postponement Strategies for Product Families with Multiple Points of Differentiation. *IIE Transactions* 29, 641–650.

Gavirneni S., S. Tayur, Delayed Product Differentiation and Information Sharing, *Working Paper*, GSIA, Carnegie Mellon University, 1999.

Gerchak, Y., M. Henig (1986). An Inventory Model with Component Commonality. *Operations Research Letters* 5(3), 157–160.

Gerchak, Y., M.J. Magazine, A.B. Gamble (1988). Component Commonality with Service Level Requirements. *Management Science* 34(6), 753–760.

Graman, G.A., M.J Magazine (2002). A Numerical Analysis of Capacitated Postponement. *Production and Operations Management* 11(3), 340–357.

Gupta, S., V. Krishnan (1998). Product Family Based Assembly Design Sequencing Methodology. *IIE Transactions* 30(10), 933–945.

Harle N., M. Pich, L. Van der Hayden, Mark & Spencer and Zara: Process Competition in the Textile Apparel Industry, INSEAD case, 2001.

Hoyt D. and E. Lopez-Tello, Lucent Technologies – Provisioning and Postponement, Stanford University Case No. GS-02, 2001.

Ignall, E., A. Veinott (1969). Optimality of Myopic Inventory Policies for Several Substitute Products. *Management Science* 15, 284–304.

Jordan, W., S. Graves (1995). Principles on the Benefits of Manufacturing Process Flexibility. *Management Science* 41, 579–594.

Kapuscinski R., S. Tayur, Variance vs. Standard Deviation: Variability Reduction through Operations, *Management Science*, 45(5), 765–767.

Lee, H.L. (1993). Design for Supply Chain Management: Concepts and Examples, in: R. Sarin *Perspectives in Operations Management*, Norwell, MA, Kluwer Academic Publishers.

Lee, H.L. (1996). Effective Management of Inventory and Service Through Product and Process Redesign. *Operations Research* 44, 151–159.

Lee H.L., "Hewlett-Packard Company: Network Printer Design for Universality," *Stanford University Case*, 1997.

Lee, H.L. (1998). Postponement for Mass Customization: Satisfying Customer Demands for Tailor-made Products, in: J. Gattorna *Strategic Supply Chain Alignment*, England, Gower Publishing Ltd, pp. 77–91.

Lee, H.L., C.A. Billington, B. Carter (1993). Hewlett Packard Gains Control of Inventory and Service through Design for Localization. *Interfaces* 23(4), 1–11.

Lee, H.L., E. Feitzinger, C. Billington (1997). Getting Ahead of Your Competition Through Design for Mass Customization. *Target* 13(2), 8–17.

Lee, H.L., M. Sasser (1995). Product Universality and Design for Supply Chain Management. *Production Planning and Control: Special Issue on Supply Chain Management* 6(3), 270–277.

Lee, H.L., C.S. Tang (1997). Modeling the Costs and Benefits of Delayed Product Differentiation. *Management Science* 43(1), 40–53.

Lee, H.L., C.S. Tang (1998). Variability Reduction through Operations Reversal. *Management Science* 44(2), 162–172.

Lee, H.L., S. Whang (1998). Value of Postponement from Variability Reduction Through Uncertainty Resolution and Forecast Improvement, in: T. Ho, C. Tang (eds.), *Product Variety Management: Research Advances*, Boston, Kluwer Publishers,pp, pp. 66–84. Chapter 4.

Mahajan, S., G. Van Ryzin (1999). On the Relationship between Inventory Costs and Variety Benefits in Retail Assortments. *Management Science* 45(11), 1496–1509.

Martin M., W. Hausman, K. Ishii, Design for Variety, *Product Variety Management: Research Advances* (ed. T. Ho and C. Tang), Kluwer Academic Publishers, 1998, 103–122.

Nevins, J.L., D.E. Whitney (1989). *Concurrent Design of Products and Processes: A Strategy for the Next Generation in Manufacturing*, New York, McGraw Hill Publishing Company.

Rao U.S., J.M. Swaminathan, J. Zhang, A Multi-Product Inventory Problem with Set-up Costs and Downward Substitution, 2002, to appear, *IIE Transactions*.

Swaminathan J.M., Optimizing Commonality for Postponement, *Working Paper*, University of California, Berkeley, 1999.

Swaminathan J.M., Enabling Customization using Standardized Operations, *California Management Review*, 43(3), Spring 2001, 125–135.

Swaminathan J.M., S. Kukukyavuz, A Perishable Inventory Problem with Postponement and Substitution, *Working Paper*, University of California, Berkeley, 2000.

Swaminathan, J.M., S.R. Tayur (1998a). Managing Broader Product Lines through Delayed Differentiation using Vanilla Boxes. *Management Science* 44(12.2), S161–S172.

Swaminathan, J.M., S.R. Tayur (1998b). Managing Design of Assembly Sequences for Product Lines that Delay Product Differentiation. *IIE Transactions* 31(4), 1015–1026.

Swaminathan, J.M., S.R. Tayur (1999). Stochastic Programming Models for Managing Product Variety, in: S. Tayur *Quantitative Models for Supply Chain Management*, Norwell, MA, Kluwer Academic Publishers, pp. 485–622.

Thonemann, U., M. Brandeau (2000). Optimal Commonality in Component Design. *Operations Research* 48(1), 1–19.

Van Mieghem J., Component Commonality Strategies: Value Drivers and Equivalence with Flexible Capacity Strategies, *Working Paper*, Kellogg School of Management, Northwestern University, 2002.

Van Mieghem J., M. Dada, Price versus Product Postponement: Capacity and Competition, *Management Science*, 45(12), 1631–1649.

Zinn W., D.J. Bowersox, Planning Physical Distribution with the Principle of Postponement, *Journal of Business Logistics*, 9, 1988, 117–136. Optimal centralized ordering policies in multi-echelon inventory systems with correlated demands, *Management Science*, 36(3), 1990, 381–392.

PART II

Supply Chain Coordination

A.G. de Kok and S.C. Graves, Eds., *Handbooks in OR & MS, Vol. 11*
© 2003 Elsevier B.V. All rights reserved.

Chapter 6

Supply Chain Coordination with Contracts

Gérard P. Cachon

The Wharton School of Business, University of Pennsylvania, Philadelphia, PA 19104, USA
E-mail: cachon@wharton.upenn.edu

1 Introduction

Optimal supply chain performance requires the execution of a precise set of actions. Unfortunately, those actions are not always in the best interest of the members in the supply chain, i.e., the supply chain members are primarily concerned with optimizing their own objectives, and that self-serving focus often results in poor performance. However, optimal performance is achievable if the firms coordinate by contracting on a set of transfer payments such that each firm's objective becomes aligned with the supply chain's objective.

This chapter reviews and extends the supply chain literature on the management of incentive conflicts with contracts. Numerous supply chain models are discussed, roughly presented in order of increasing complexity. In each model the supply chain optimal actions are identified. In each case the firms could implement those actions, i.e., each firm has access to the information required to determine the optimal actions and the optimal actions are feasible for each firm.[1] However, firms lack the incentive to implement those actions. To create that incentive the firms can adjust their terms of trade via a contract that establishes a transfer payment scheme. A number of different contract types are identified and their benefits and drawbacks are illustrated.

---

[1] Even in the asymmetric information models there is an assumption that the firms can share information so that all firms are able to evaluate the optimal policies. Nevertheless, firms are not required to share information. See Anand and Mendelson (1997) for a model in which firms are unable to share information even though they have the incentive to do so.

The first model has a single supplier selling to a single retailer that faces the newsvendor problem. In that model the retailer orders a single product from the supplier well in advance of a selling season with stochastic demand. The supplier produces after receiving the retailer's order and delivers her production to the retailer at the start of the selling season.[2] The retailer has no additional replenishment opportunity. How much the retailer chooses to order depends on the terms of trade, i.e., the contract, between the retailer and the supplier.

The newsvendor model is not complex, but it is sufficiently rich to study three important questions in supply chain coordination. First, which contracts coordinate the supply chain? A contract is said to coordinate the supply chain if the set of supply chain optimal actions is a Nash equilibrium, i.e., no firm has a profitable unilateral deviation from the set of supply chain optimal actions. Ideally, the optimal actions should also be a unique Nash equilibrium; otherwise, the firms may 'coordinate' on a suboptimal set of actions. In the newsvendor model the action to coordinate is the retailer's order quantity (and in some cases, as is discussed later, the supplier's production quantity also needs coordination). Second, which contracts have sufficient flexibility (by adjusting parameters) to allow for any division of the supply chain's profit among the firms? If a coordinating contract can allocate rents arbitrarily, then there always exists a contract that Pareto dominates a noncoordinating contract, i.e., each firm's profit is no worse off and at least one firm is strictly better off with the coordinating contract. Third, which contracts are worth adopting? Although coordination and flexible rent allocation are desirable features, contracts with those properties tend to be costly to administer. As a result, the contract designer may actually prefer to offer a simple contract even if that contract does not optimize the supply chain's performance. A simple contract is particularly desirable if the contract's efficiency is high (the ratio of supply chain profit with the contract to the supply chain's optimal profit) and if the contract designer captures the lion's share of supply chain profit.

Section 3 extends the newsvendor model by allowing the retailer to choose his retail price in addition to his stocking quantity. Coordination is more complex in this setting because the incentives provided to align one action (e.g., the order quantity) might cause distortions with the other action (e.g., the price). Not surprising, it is shown that some of the contracts that coordinate the basic newsvendor model no longer coordinate in this setting, whereas others continue to do so.

Section 4 extends the newsvendor model by allowing the retailer to exert costly effort to increase demand. Coordination is challenging because the

---

[2] The author adopts the convention (first suggested to him by Martin Lariviere) that the firm offering the contract is female and the accepting firm is male. When neither firm offers the contract, then the upstream firm is female, and the downstream firm is male.

retailer's effort is noncontractible, i.e., the firms cannot write contracts based on the effort chosen (for reasons discussed later). Furthermore, as with the retail price, coordination is complicated by the fact that the incentives to align the retailer's order-quantity decision may distort the retailer's effort decision.

Section 5 studies two models, each with one supplier that sells to multiple competing retailers. Coordination requires the alignment of multiple actions implemented by multiple firms, in contrast with the price and effort models (Sections 3 and 4) that have multiple actions implemented by a single firm (the retailer). More specifically, coordination requires the tempering of downstream competition.

Section 6 has a single retailer that faces stochastic demand but two replenishment opportunities. Early production (the first replenishment) is cheaper than later production (the second replenishment), but less informative because the demand forecast is updated before the second replenishment. Coordination requires that the retailer be given the proper incentives to balance this trade-off.

Section 7 studies an infinite horizon stochastic demand model in which the retailer receives replenishments from a supplier after a constant lead time; a departure from the single-period lost-sales models of the previous sections. As in the effort model, coordination requires that the retailer chooses a 'higher action', which in this model is a larger base-stock level. The cost of this higher action is more inventory on average, but unlike in the effort model, the supplier can verify the retailer's inventory and therefore share the holding cost of carrying more inventory with the retailer.

Section 8 adds richness to the single-location base-stock model by making the supplier hold inventory, albeit at a lower holding cost than the retailer. Whereas the focus in the previous sections is primarily on coordinating the downstream actions, in this model the supplier's action also requires coordination, and that coordination is nontrivial. To be more specific, in the single-location model the only critical issue is the amount of inventory in the supply chain, but here the allocation of the supply chain's inventory between the supplier and the retailer is important as well.

Section 9 departs from the assumption that firms agree to contracts with set transfer prices. In many supply chains the firms agree to a contractual arrangement before the realization of some relevant information. The firms could specify transfer payments for every possible contingency, but those contracts are quite complex. Instead, firms could agree to set transfer prices via an internal market after the relevant information is revealed.

Section 10 endows one firm with important information that the other firm does not possess, i.e., it is private information. For example, a manufacturer may have a more accurate demand forecast for a product than the manufacturer's supplier. As in the previous models, supply chain coordination requires each firm to implement optimal actions. But since those optimal actions depend on the private information, supply chain coordination also requires the accurate sharing of information. Sharing information is challenging because there exists the incentive to provide false information in an effort to influence the actions taken, e.g., a manufacturer may wish to offer a rosy demand forecast to try to get the supplier to build more capacity.

The final section summarizes the main insights that have developed from this literature and provides some general guidance for future research.

Each section presents one or more simple models to facilitate the analysis and to highlight the potential incentive conflicts within a supply chain. The same analysis recipe is usually followed: identify the type of contracts that can coordinate the supply chain, determine for each contract type the set of parameters that achieves coordination, and evaluate for each coordinating contract type the possible range of profit allocations, i.e., what fraction of the supply chain's profit can be earned by each member in the supply chain with a coordinating contract. Implementation issues are then explored: e.g., is a contract-type compliant with legal restrictions; what are the consequences for failing to comply with the contractual terms; and what is a contract's administrative burden (e.g., what types of data need to be collected and how often must data be collected). Each section ends with a discussion of extensions and related research.

While this chapter gives a broad treatment of the supply chain contracting literature, it does not address all papers that could possibly be classified within this literature. In particular, there are (at least) six types of closely related papers that are not discussed directly. The first is the extensive literature on quantity discounts because several excellent reviews are available, see Dolan and Frey (1987) and Boyaci and Gallego (1997). The second set that is not addressed includes papers on a single firm's optimal procurement decisions given particular contractual terms. Examples include Scheller-Wolf and Tayur's (1997) study of procurement under a minimum quantity-commitment contract, Duenyas, Hopp and Bassok's (1997) study of procurement with JIT contracts, Bassok and Anupindi's (1997a) study of procurement with total minimum commitments, and Anupindi and Akella's (1993) and Moinzadeh and Nahmias' (2000) studies of procurement with standing order contracts. The third body of excluded work is research on supply chain coordination without contracts. Examples include papers on the benefit of Quick Response (Iyer & Bergen, 1997), Accurate Response (Fisher & Raman, 1996), collaborative planning and forecasting (Aviv, 2001), Vendor Managed Inventory (Aviv & Federgruen, 1998) and information sharing within a supply chain (Gavirneni, Kapuscinski & Tayur, 1999). Fourth, papers on decentralized supply chain operations which do not explicitly consider coordination are excluded: e.g., Cachon and Lariviere (1997, 1999), Corbett and Karmarkar (2001), Erhun, Keskinocak and Tayur (2000), Ha, Li and Ng (2000) and Majumder and Groenevelt (2001). Fifth, the broad literature on franchising is not directly discussed, primarily because that literature generally avoids operational detail (see Lafontaine & Slade, 2001 for a recent review of that literature) Finally, papers on vertical restraints vis-a-vis social welfare and antitrust issues are not considered, see Katz (1989).

For earlier overviews on supply chain coordination with contracts, see Whang (1995) and the three chapters in Tayur, Ganeshan and Magazine (1998) that focus on the topic: Cachon (1998), Lariviere (1998) and Tsay, Nahmias and Agrawal (1998).

2 Coordinating the newsvendor

This section studies coordination in a supply chain with one supplier and one retailer. There is one selling season with stochastic demand and a single opportunity for the retailer to order inventory from the supplier before the selling season begins. With the standard wholesale-price contract, it is shown that the retailer does not order enough inventory to maximize the supply chain's total profit because the retailer ignores the impact of his action on the supplier's profit. Hence, coordination requires that the retailer be given an incentive to increase his order.

Several different contract types are shown to coordinate this supply chain and arbitrarily divide its profit: buyback contracts, revenue-sharing contracts, quantity-flexibility contracts, sales-rebate contracts and quantity-discount contracts.

The concept of a compliance regime is introduced. The compliance regime determines the consequences for failing to adhere to a contract. For example, it is assumed that the supplier cannot force the retailer to accept more product than the retailer orders, i.e., the retailers could clearly use the courts to prevent any attempt to do so. However, it is debatable whether the supplier is required to deliver the retailer's entire order. The compliance regime matters because it influences the kinds of contracts that coordinate the supply chain: there exist contracts that coordinate with one compliance regime, but not another.

2.1 Model and analysis

In this model there are two firms, a supplier and a retailer. The retailer faces the newsvendor's problem: the retailer must choose an order quantity before the start of a single selling season that has stochastic demand. Let $D > 0$ be demand during the selling season. Let F be the distribution function of demand and f its density function: F is differentiable, strictly increasing and $F(0) = 0$. Let $\overline{F}(x) = 1 - F(x)$ and $\mu = E[D]$. The retail price is p. The supplier's production cost per unit is c_s and the retailer's marginal cost per unit is c_r, $c_s + c_r < p$. The retailer's marginal cost is incurred upon procuring a unit (rather than upon selling a unit). For each demand the retailer does not satisfy the retailer incurs a goodwill penalty cost g_r and the analogous cost for the supplier is g_s. For notational convenience, let $c = c_s + c_r$ and $g = g_s + g_r$. The retailer earns $v < c$ per unit unsold at the end of season, where v is net of any salvage expenses. Assume the supplier's net salvage value is no greater than v, so it is optimal for the supply chain to salvage leftover inventory at the retailer. The qualitative insights from the subsequent analysis do not depend on whether it is optimal for the retailer or the supplier to salvage leftover inventory. [The supply chain contracting literature generally avoids this issue by assuming the net salvage value of a unit is the same at either firm. Tsay (2001) is an exception.] For more extensive treatment of the newsvendor model, see Silver, Pyke and Peterson (1998) or Nahmias (1993).

The following sequence of events occurs in this game: the supplier offers the retailer a contract; the retailer accepts or rejects the contract; assuming the retailer accepts the contract, the retailer submits an order quantity, q, to the supplier; the supplier produces and delivers to the retailer before the selling season; season demand occurs; and finally transfer payments are made between the firms based upon the agreed contract. If the retailer rejects the contract, the game ends and each firm earns a default payoff.

The supplier is assigned to make the contract offer, rather than the retailer, only for expositional convenience, i.e., it has no impact on the subsequent analysis. The firm that offers the contract does not matter because we seek to identify the set of contracts that coordinate the supply chain and arbitrarily allocate its profit. If one firm were indeed assigned to make the only offer, then it would offer the most favorable contract in that set which the other firm will accept. Furthermore, it is unlikely in practice that either firm makes a single offer which is regarded as the final offer. Instead, firms are likely to make many offers and counter offers before they settle on some agreement. The details of this negotiation process are generally not considered in the supply chain literature, nor are they explored here.

The contract that is actually adopted at the end of the negotiation process depends on the firms' relative bargaining power, which is a concept that is easy to understand but difficult to quantify. Power, like beauty, can be in 'the eye of the beholder', or it can be more concrete. A standard approach to model power is to assume one of the firms has an exogenous reservation profit level, i.e., the firm accepts only a contract that yields that reservation level: the higher the reservation level, the higher the firm's power.[3] Ertogral and Wu (2001) are even more explicit with their bargaining process: bargaining occurs in rounds in which either firm may

---

[3] Webster and Weng (2000) impose a stronger condition. They require that both firms are at least as well off with the adoption, a contract as they would be with a default contract for all realization of demand.

make an offer, but if at the end of a round an offer is not accepted there is a fixed probability the negotiations fail, i.e., the firms are left with their reservation profit. However, the reservation level approach is not entirely satisfying: it is quite likely a firm's opportunity outside of the relationship being studied is not independent of the firm's opportunity within the relationship. Nor should it be expected that the value of a firm's outside opportunity is known with certainty a priori (van Mieghem, 1999; Rochet & Stole, 2002). Aside from the reservation level approach, some researchers adjust power by changing which firm makes the contract offer or by changing when actions are chosen. In general, a firm has more power when she makes the first offer, assuming it is a 'take-it or leave-it' offer, or when she chooses her actions first, assuming she is committed to her action. These choices matter when one wants to predict with precision the particular outcome of a negotiation process, which is not done here. Additional research is surely needed on this issue.

To continue with the description of the model, each firm is risk neutral, so each firm maximizes expected profit. There is full information, which means that both firms have the same information at the start of the game, i.e., each firm knows all costs, parameters and rules. Game theorists have been also concerned with higher levels of common knowledge: e.g., does firm A know that firm B knows all information and does firm B know that firm A knows that firm B knows all information, etc. The supply chain contracting literature has not explored this issue. See Rubinstein (1989) for a model with counterintuitive implications for less than complete common knowledge.

It is quite reasonable to assume the supplier cannot force the retailer to pay for units delivered in excess of the retailer's order quantity. But can the supplier deliver *less* than the amount the retailer orders? A failure to deliver the retailer's full order may occur for a number of reasons beyond the supplier's control: e.g., unforeseen production difficulties or supply shortages for key components. The shortage may also be due to self-interest. In recognition of that motivation, the retailer could assume the supplier operates under *voluntary compliance*, which means the supplier delivers the amount (not to exceed the retailer's order) that maximizes her profit given the terms of the contract. Alternatively, the retailer could believe the supplier never chooses to deliver less than the retailer's order because the consequences for doing so are sufficiently great, e.g., court action or a loss of reputation. Call that regime *forced compliance*.

The compliance regime in reality almost surely falls somewhere between those two extremes. However, in any regime other than forced compliance the supplier can be expected to fall somewhat short on her delivery *if* the terms of the contract give the supplier an incentive to do so. In other words, any contract that coordinates the supply chain with voluntary compliance surely coordinates with forced compliance, but the reverse is not true (because the contract may fail to coordinate the supplier's action). Hence, voluntary compliance is the more conservative assumption (albeit maybe too conservative).[4]

The approach taken in this section is to assume forced compliance but to check if the supplier has an incentive to deviate from the proposed contractual terms. This seemingly contradictory stance is adopted to simplify notation: voluntary compliance requires notation to keep track of two actions, the retailer's order quantity and the supplier's production quantity, whereas forced compliance requires notation only for one action. See Cachon and Lariviere (2001) for additional discussion on compliance regimes.[5]

Let $S(q)$ be expected sales, $\min(q, D)$,c

$$S(q) = q(1 - F(q)) + \int_0^q y f(y)\,dy$$
$$= q - \int_0^q F(y)\,dy$$

(The above follows from integration by parts.) Let $I(q)$ be the expected leftover inventory, $I(q) = (q - D)^+ = q - S(q)$. Let $L(q)$ be the lost-sales function, $L(q) = (D - q)^+ = \mu - S(q)$. Let T be the expected transfer payment from the retailer to the supplier. That function may depend on a number of observations (e.g., order quantity, leftover inventory), as is seen later.

The retailer's profit function is

$$\pi_r(q) = pS(q) + vI(q) - g_r L(q) - c_r q - T$$
$$= (p - v + g_r)S(q) - (c_r - v)q - g_r \mu - T,$$

the supplier's profit function is

$$\pi_s(q) = g_s S(q) - c_s q - g_s \mu + T,$$

---

[4] This chapter assumes the wholesale price operates with forced compliance whereas the quantity to deliver may operate with voluntary compliance. Hence, the parameters in a contract can operate under different compliance regimes, which can be justified by the differences in ease by which the courts can verify different terms. As suggested by Fangruo Chen, it is also possible to view all contracts as iron clad contracts (i.e., everything operates with forced compliance), but the kinds of contractual terms may be limited. For example, suppose the contract were written such that the retailer's order quantity is an upper bound on the supplier's delivery quantity, i.e., forced compliance of an upper bound is analogous to our voluntary compliance with a specific quantity. Additional research is needed to determine if the distinctions in these interpretations matter.

[5] See Krasa and Villamil (2000) for a model in which the contracting parties endogenously set the compliance regime. Milner and Pinker (2001) do not explicitly define a compliance regime, but it does impact their results. They show supply chain coordination is possible when one firm is able to identify any deviation by the other firm and follow through with substantial penalties. When deviations cannot be identified for sure, supply chain coordination is no longer possible. Baiman, Fischer and Rajan (2000) focus on how the compliance regime impacts a supplier's incentive to improve quality and a buyer's incentive to inspect.

and the supply chain's profit function is

$$\Pi(q) = \pi_r(q) + \pi_s(q) = (p - v + g)S(q) - (c - v)q - g\mu. \tag{2.1}$$

Given the above profit functions, it is possible to normalize some of the variables. For example, let $\hat{p} = p - v + g_r$ be the adjusted price, let $\hat{c} = c - v$ be the adjusted production cost, let $\hat{T} = T + (c_r - v)q$ be the adjusted transfer payment, let $\hat{\pi}_r(q) = \pi_r(q) + g_r\mu$ be the retailer's adjusted profit function and let $\hat{\pi}_s(q) = \pi_s(q) + g_s\mu$ be the supplier's adjusted profit function. Those adjusted profit functions simplify to $\hat{\pi}_r(q) = \hat{p}S(q) - \hat{T}$ and $\hat{\pi}_s(q) = g_s S(q) - \hat{c}q + \hat{T}$. While those functions have cleaner notation, caution is required when defining the transfer payment for a given contract because the contract's terms (e.g., the wholesale price, the buyback rate, etc.) must be given in terms of the adjusted parameters. Unfortunately, the notational clarity gained by these adjustments is often lost when the adjusted contract terms are included. As a result, this chapter works with the unadjusted profit functions.

Let q^o be a supply chain optimal order quantity, i.e., $q^o = \arg\max \Pi(q)$. To avoid uninteresting situations, assume $\Pi(q^o) > 0$. Since F is strictly increasing, Π is strictly concave and the optimal order quantity is unique. Further, q^o satisfies

$$S'(q^o) = \overline{F}(q^o) = \frac{c - v}{p - v + g}. \tag{2.2}$$

Let q_r^* be the retailer's optimal order quantity, i.e., $q_r^* = \arg\max \pi_r(q)$. The retailer's order clearly depends on the chosen transfer payment scheme, T.

A number of contract types have been applied to this model. The simplest is the wholesale-price contract: the supplier merely charges the retailer a fixed wholesale price per unit ordered. Section 2.2 studies that contract. It is shown that the wholesale-price contract generally does not coordinate the supply chain. Hence, the analysis concentrates on two questions: what is the efficiency of the wholesale-price contract (the ratio of supply chain profit to optimal profit) and what is the supplier's share of the supply chain's profit.

More complex contracts include a wholesale price plus some adjustment that typically depends on realized demand (the quantity-discount contract is an exception). As mentioned in Section 1, the analysis recipe for all of those contracts is the same: determine the set of contract parameters that coordinate the retailer's action; then evaluate the possible range of profit allocations between the firms; and then check whether the contract coordinates under voluntary compliance, i.e., whether the supplier has an incentive to deliver less than the retailer's order quantity.

2.2 The wholesale-price contract

With a wholesale-price contract the supplier charges the retailer w per unit purchased: $T_w(q, w) = wq$. See Lariviere and Porteus (2001) for a more complete analysis of this contract in the context of the newsvendor problem. Bresnahan and Reiss (1985) study the wholesale-price contract with deterministic demand.

$\pi_r(q, w)$ is strictly concave in q, so the retailer's unique optimal order quantity satisfies

$$(p - v + g_r)S'(q_r^*) - (w + c_r - v) = 0. \tag{2.3}$$

Since $S'(q)$ is decreasing, $q_r^* = q^o$ only when

$$w = \left(\frac{p - v + g_r}{p - v + g}\right)(c - v) - (c_r - v).$$

It is straightforward to confirm that $w \leq c_s$, i.e., the wholesale-price contract coordinates the channel only if the supplier earns a nonpositive profit. So the supplier clearly prefers a higher wholesale price. As a result, the wholesale-price contract is generally not considered a coordinating contract. [As discussed in Cho and Gerchak (2001) and Bernstein, Chen and Federgruen (2002), marginal cost pricing does not necessarily lead to zero profit for the supplier when the marginal cost is not constant.] Spengler (1950) was the first to identify the problem of 'double marginalization'; in this serial supply chain there is coordination failure because there are two margins and neither firm considers the entire supply chain's margin when making a decision.

Even though the wholesale-price contract does not coordinate the supply chain, the wholesale-price contract is worth studying because it is commonly observed in practice. That fact alone suggests it has redeeming qualities. For instance, the wholesale-price contract is simple to administer. As a result, a supplier may prefer the wholesale-price contract over a coordinating contract if the additional administrative burden associated with the coordinating contract exceeds the supplier's potential profit increase.

From Eq. (2.3) the retailer's optimal order quantity satisfies

$$F(q_r^*) = 1 - \frac{w + c_r - v}{p - v + g_r}.$$

Since F is strictly increasing and continuous there is a one-for-one mapping between w and q_r^*. Hence, let $w(q)$ be the unique wholesale price that induces the retailer to order q_r^* units,

$$w(q) = (p - v + g_r)\overline{F}(q) - (c_r - v).$$

The supplier's profit function can now be written as

$$\pi_s(q, w(q)) = g_s S(q) + (w(q) - c_s)q - g_s \mu. \tag{2.4}$$

It is immediately apparent that the compliance regime does not matter with this contract: for a fixed wholesale price no less than c_s the supplier's profit is nondecreasing in q, so the supplier surely produces and delivers whatever quantity the retailer orders.

The supplier's marginal profit is

$$\frac{\partial \pi_s(q, w(q))}{\partial q} = g_s S'(q) + w(q) - c_s + w'(q)q$$

$$= (p - v + g_r)\overline{F}(q)\left(1 + \frac{g_s}{p - v + g_r} - \frac{qf(q)}{\overline{F}(q)}\right) - (c - v).$$

The supplier's profit function is unimodal if the above is decreasing. $\overline{F}(q)$ is decreasing, so $\pi_s(q, w(q))$ is decreasing in q if $qf(q)/\overline{F}(q)$ is increasing. Demand distributions with that property are called increasing generalized failure rate (IGFR) distributions.[6] Fortunately, many of the commonly applied demand distributions are IGFR: the normal, the exponential, the Weibull, the gamma and the power distribution. Thus, with an IGFR demand distribution there is a unique sales quantity, q_s^*, that maximizes the supplier's profit. (Actually, the supplier sets the wholesale price to $w(q_s^*)$ knowing quite well the retailer then orders q_s^* units.)

While $\pi_s(q_s^*, w(q_s^*))$ is the best the supplier can hope for, the retailer may actually insist on more than $\pi_r(q_s^*, w(q_s^*))$. For example, the retailer may earn more by selling some other product in his store, i.e., his opportunity cost is greater than $\pi_r(q_s^*, w(q_s^*))$. In that case the supplier needs to offer the retailer more generous terms to get the retailer to carry the product. With a wholesale-price contract the retailer's profit is increasing in q,

$$\frac{\partial \pi_r(q, w(q))}{\partial q} = -w'(q)q = (p - v + g_r)f(q)q > 0,$$

so the supplier can increase the retailer's profit by reducing her wholesale price (which should surprise no one). As long as the retailer insists on less than the supply chain optimal profit, the retailer's minimum profit requirement actually increases the total supply chain profit: the supply chain's profit is increasing in q for $q \in [q_s^*, q^o]$ and so is the retailer's profit. Hence, an

---

[6] The failure rate of a demand distribution is $f(x)/\overline{F}(x)$. Any demand distribution with an increasing failure rate (IFR) is clearly also IGFR. However, there are IGFR distributions that are not IFR. See Lariviere and Porteus (2001) for additional discussion.

increase in retail power can actually improve supply chain performance (which is somewhat surprising, and controversial). However, that improvement comes about at the supplier's expense. See Messinger and Narasimhan (1995), Ailawadi (2001) and Bloom and Perry (2001) for additional discussion and empirical evidence on how power has changed in several retail markets.

The two performance measures applied to the wholesale-price contract are the efficiency of the contract, $\Pi(q_s^*)/\Pi(q^o)$ and the supplier's profit share, $\pi_s(q_s^*, w(q_s^*))/\Pi(q_s^*)$. From the supplier's perspective the wholesale-price contract is an attractive option if both of those measures are high: the product of those ratios is the supplier's share of the supply chain's optimal profit,

$$\frac{\pi_s(q_s^*, w(q_s^*))}{\Pi(q^o)} = \left(\frac{\pi_s(q_s^*, w(q_s^*))}{\Pi(q_s^*)}\right)\left(\frac{\Pi(q_s^*)}{\Pi(q^o)}\right)$$

To illustrate when that is likely, suppose $g_r = g_s = 0$ and demand follows the power distribution: $F(q) = g^k$ for $k > 0$ and $q \in [0, 1]$. In that case the efficiency of the wholesale-price contract is $(k+1)^{-(1+1/k)}(k+2)$, the supplier's profit share is $(k+1)/(k+2)$ and the coefficient of variation is $(k(k+2))^{-1/2}$. (See Lariviere & Porteus, 2001 for details.) Note that the coefficient of variation is decreasing in k but both measures are increasing in k. In fact, as $k \to \infty$ the coefficient of variation approaches zero and both measures approach 1. Nevertheless, Table 1 demonstrates the supplier's share of supply chain profit increases more quickly than supply chain efficiency.

One explanation for this pattern is that the retailer's profit represents compensation for bearing risk: with the wholesale-price contract there is no variation in the supplier's profit, but the retailer's profit varies with the realization of demand. As the coefficient of variation decreases the retailer faces less demand risk and therefore his compensation is reduced. However, the retailer is not compensated due to risk aversion. (See Tsay, 2002 for a model with risk aversion.) If the retailer were risk averse, the supplier would have to provide for yet more compensation. Instead, the retailer is being compensated for the risk that demand and supply do not match. Lariviere and Porteus (2001) demonstrate this argument holds for a broad set of demand distributions.

Anupindi and Bassok (1999) study an interesting extension to this model. Suppose the supplier sells to a retailer that faces an infinite succession of identical selling seasons.[7] There is a holding cost on leftover inventory at the end of a season but inventory can be carried over to the next season. The

---

[7] In fact, their model has two retailers. But in one version of their model the retailers face independent demands, and so that model is qualitatively identical to the single retailer model.

Table 1
Wholesale-price contract performance when demand follows a power distribution with parameter k

Demand distribution parameter, k	0.2	0.4	0.8	1.6	3.2
Efficiency, $\Pi(q_s^*)/\Pi(q^o)$	73.7%	73.9%	74.6%	76.2%	79.1%
Supplier's share, $\pi_s(q_s^*, w(q_s^*))/\Pi(q_s^*)$	54.5%	58.3%	64.3%	72.2%	80.8%
Coefficient of variation	1.51	1.02	0.67	0.42	0.25

retailer submits orders between seasons and the supplier is able to replenish immediately. Within each season the retailer faces a newsvendor problem that makes the trade-off between lost sales and inventory holding costs. Hence, the retailer's optimal inventory policy is to order up to a fixed level that is the solution to a newsvendor problem. But since inventory carries over from season to season, the supplier's average sales per season equals the retailer's average sales per season, i.e., the supplier's profit function is $(w(q) - c)S(q)$. The analysis of the supplier's optimal wholesale price is more complex in this setting because the supplier's profit is now proportional to the retailer's sales, $S(q)$, rather than to his order quantity, q. Nevertheless, since $S(q)/S'(q) > q$, the supplier's optimal wholesale price is lower than in the single season model. (The expression $S(q)/S'(q) > q$ simplifies to $qF(q) > \int_0^q F(y)\,dy$, which clearly holds.) Thus, the efficiency of the wholesale-price contract is even better than in the single season model.

Debo (1999) studies the repeated version of the single-shot newsvendor model without inventory carrying from period to period. He demonstrates that supply chain coordination is possible with just a wholesale-price contract if the firm's discount rate is not too high, i.e., the firms care about future profit. Cooperation is achieved via the use of trigger strategies that punish a defector.

The infinite horizon extensions to the model do not have an end-of-horizon effect, i.e., inventory is not salvaged. Cachon (2002) studies a two-period version of the model which has excess inventory and demand updating. The retailer can submit an order well in advance of the selling season and pay the supplier w_1 for each unit in that order. The supplier then produces and delivers the retailer's first order before the selling season starts. During the selling season the retailer can order from the supplier additional units. If the supplier has inventory available, then the supplier delivers the units and charges the retailer w_2 per unit, $w_2 \geq w_1$. The supply chain can operate in one of three modes. The first matches the single-period model studied by Lariviere and Porteus (2001): only the initial order before the season starts is allowed. This mode of operations is called 'push', because all inventory risk is pushed upon the retailer (i.e., the retailer bears the cost of disposing leftover inventory). The other extreme is called 'pull': the retailer orders only during the selling season, so now the supplier bears all inventory risk. A combination of push and pull is created by the use of an advanced purchase discount, $w_1 < w_2$: the retailer submits an initial order to take advantage of the advanced purchase discount and the supplier produces more than the initial order in anticipation of the retailer's orders during the selling season. It is shown that supply chain efficiency is substantially higher if the firms consider both push and pull contracts rather than just push or just pull contracts. Furthermore, there exist conditions in which advanced purchase discounts coordinate the supply chain and arbitrarily allocate its profit. See Ferguson, DeCroix and Zipkin (2002), Taylor (2002b) and Yüksel and Lee (2002) for additional work on the timing of the retailer's orders. See Section 6 for a model that studies demand updating with coordinating contracts.

Dong and Rudi (2001) study the wholesale-price contract with two newsvendors and transshipment of inventory between them. They find that the supplier is generally able to capture most of the benefits of transshipments and the retailers are worse off with transshipment. This is consistent (but probably not identical) to Lariviere and Porteus' (2001) finding that the supplier is better off and the retailer worse off with less variable demand. Related to transshipment, Chod and Rudi (2002) study a supplier selling a single resource to a downstream firm that can use that resource to produce multiple products.

Gilbert and Cvsa (2000) study the wholesale-price contract with demand uncertainty and costly investment to reduce production costs. They demonstrate that a trade-off exists between the beneficial flexibility of allowing the wholesale price to adjust to market demand and the need to provide incentives to reduce production costs. Additional detail on this paper is provided in Section 4.

2.3 The buyback contract

With a buyback contract the supplier charges the retailer w per unit purchased, but pays the retailer b per unit remaining at the end of the season:

$$T_b(q, w, b) = wq - bI(q) = bS(q) + (w - b)q.$$

A retailer should not profit from leftover inventory, so assume $b \leq w$. See Pasternack (1985) for a detailed analysis of buyback contracts in the context of the newsvendor problem.

Buyback contracts are also called returns policies, but, unfortunately, both names are somewhat misleading since they both imply the units remaining at the end of the season are physically returned to the supplier. That does occur if the supplier's net salvage value is greater than the retailer's net salvage value. However, if the retailer's salvage value is higher, the retailer salvages the units and the supplier credits the retailer for those units, which is sometime referred to as 'markdown money' (see Tsay, 2001). An important implicit assumption is that the supplier is able to verify the number of remaining units and the cost of such monitoring does not negate the benefits created by the contract.

With a buyback contract the retailer's profit is

$$\pi_r(q, w_b, b) = (p - v + g_r - b)S(q) - (w_b - b + c_r - v)q - g_r\mu.$$

Now consider the set of buyback parameters $\{w_b, b\}$ such that for $\lambda \geq 0$,

$$p - v + g_r - b = \lambda(p - v + g) \quad (2.5)$$

$$w_b - b + c_r - v = \lambda(c - v) \quad (2.6)$$

A comparison with Eq. (2.1) reveals the retailer's profit function with that set of contracts is

$$\begin{aligned}\pi_r(q, w_b, b) &= \lambda(p - v + g)S(q) - \lambda(c - v)q - g_r\mu \\ &= \lambda\Pi(q) + \mu(\lambda g - g_r).\end{aligned} \quad (2.7)$$

It follows immediately that $q_r^* = q^o$ is optimal for the retailer. The supplier's profit function is

$$\pi_s(q, w_b, b) = \Pi(q) - \pi_r(q, w_b, b) = (1 - \lambda)\Pi(q) - \mu(\lambda g_s - (1 - \lambda)g_r).$$

So the buyback contract coordinates with voluntary compliance as long as $\lambda \leq 1$. Some ambiguity arises with $\lambda = 1$ (or $\lambda = 0$) because then q^o is optimal for the supplier (or retailer), but so is every other quantity. Hence, coordination is possible, but the optimal solution is no longer the unique Nash equilibrium.

Interestingly, voluntary compliance actually increases the robustness of the supply chain. Suppose the retailer is not rational and orders $q > q^o$. Since the supplier is allowed to deliver less than the retailer's order quantity, the supplier corrects the retailer's mistake by delivering only q^o units. However, because the retailer can refuse to accept more than he orders, the supplier cannot correct the retailer's mistake if he orders less than q^o. See Chen (1999a), Porteus (2000) and Watson (2002) for further discussion on the robustness of a coordination scheme to irrational ordering.[8]

The retailer's profit is increasing in λ and the supplier's profit is decreasing in λ so the λ parameter acts to allocate the supply chain's profit

---

[8] While this is an intriguing idea, it is difficult to construct a theory based on irrational behavior. Maybe a better interpretation is that shocks occur in the system that are only observable to one member. For example, while q^o is the steady-state optimal order quantity, the supplier may learn some information that reveals in fact $q' < q^o$ is indeed optimal. Thus, the retailer's apparently irrational excessive order is really due to a lack of information. Some interesting research must be able to follow from these ideas. Perhaps inspiration could come from Stidham (1992). He considers the regulation of a queue when a manager sets her actions for a defined time period but actual expected demand during that period may deviate from what the manager expects. He shows there may exist unstable equilibria, i.e., a small shock to the system sends the system away from the equilibrium rather than back to it.

between the two firms. The retailer earns the entire supply chain profit, $\pi_r(q^\circ, w_b, b) = \Pi(q^\circ)$, when

$$\lambda = \frac{\Pi(q^\circ) + \mu g_r}{\Pi(q^\circ) + \mu g} \leq 1 \tag{2.8}$$

and the supplier earns the entire supply chain profit, $\pi_s(q^\circ, w_b, b) = \Pi(q^\circ)$, when

$$0 \leq \lambda = \frac{\mu g_r}{\Pi(q^\circ) + \mu g}. \tag{2.9}$$

So every possible profit allocation is feasible with this set of coordinating contracts, assuming $\lambda = 0$ and $\lambda = 1$ are considered feasible.

It is interesting to note that coordination of the supply chain requires the simultaneous adjustment of both the wholesale price and the buyback rate. This has implications for the bargaining process. For example, suppose the firms agree to a particular wholesale price. Given any fixed wholesale price, the coordinating buyback rate is not the buyback rate that maximizes the supplier's or the retailer's profit. In other words, both players would have an incentive to argue for a non-Pareto optimal (i.e., noncoordinating) contract. It would be a shame if the players then agreed upon an non-Pareto optimal contract because then, by definition, there would exist some coordinating contract that could make both players better off. However, that coordinating contract would have a different wholesale price. The key lesson for managers is that they should never negotiate these parameters sequentially (i.e., agree to one parameter and then consider the second parameter). Instead, negotiations should always allow simultaneous changes to both the wholesale price and the buyback rate.

From Eq. (2.7), the parameter λ can loosely be interpreted as the retailer's share of the supply chain's profit; it is precisely the retailer's share when $g_r = g_s = 0$. Note that the λ parameter is not actually part of the buyback contract. It is introduced for expositional clarity. Most of the supply chain contracting literature does not explicitly define a comparable parameter. Instead, it is more common to present one contract parameter in terms of the other, e.g.,

$$w_b(b) = b + c_s - (c - v)\left(\frac{b + g_s}{p - v + g}\right).$$

Furthermore, coordinating parameters are often identified from first-order conditions. The approach taken above is preferred because it is more general. For example, the strategy space does not need to be continuous, there does not need to be a unique optimum and the supply chain cost function does not have to be continuous, even though all of those conditions are satisfied in this model (which is why the first-order condition approach works).

In general, a contract coordinates the retailer's and the supplier's action whenever each firm's profit is an affine function of the supply chain's profit. In effect the firms end up with something that resembles a profit-sharing arrangement. Jeuland and Shugan (1983) note that profit sharing can coordinate a supply chain, but they do not offer a specific contract for achieving profit sharing. Caldentey and Wein (1999) show that profit sharing occurs when each firm receives a fixed fraction of every other firm's utility. With that approach each firm transacts with every other firm, which may lead to an administrative burden if the number of firms is large.

There is a substantial literature on buyback contracts. Padmanabhan and Png (1995) describe several motivations for return policies that are not included in the newsvendor model. A supplier may wish to offer a return policy to prevent the retailer from discounting leftover items, thereby weakening the supplier's brand image. For instance, suppliers of fashion apparel have large marketing budgets to enhance the popularity of their clothes. It is difficult to convince consumers that your clothes are popular if they can be found in the discount rack at the end of the season. Alternatively, a supplier may wish to accept returns to rebalance inventory among retailers. There are a number of papers that consider stock rebalancing in a centralized system (Lee, 1987; Tagaras & Cohen, 1992). Rudi, Kapur and Pyke (2001) and Anupindi, Bassok and Zemel (2001) consider inventory rebalancing in decentralized systems.

In Padmanabhan and Png (1997) a supplier uses a buyback contract to manipulate the competition between retailers (see Section 5.2) Emmons and Gilbert (1998) study buyback contracts with a retail price-setting newsvendor (see Section 3). Taylor (2000a) incorporates a buyback contract with a sales-rebate contract to coordinate the newsvendor with effort-dependent demand (see Section 4). Donohue (2000) studies buyback contracts in a model with multiple production opportunities and improving demand forecasts (see Section 6). Anupindi and Bassok (1999) demonstrate buyback contracts can coordinate a two-retailer supply chain in which consumers search among the retailers to find inventory.[9] Lee, Padmanabhan, Taylor and Whang (2000) model price protection policies in a way that closely resembles a buyback.[10] However, in Taylor (2001) price protection is distinct from

---

[9] In their model the supplier subsidizes the holding cost of leftover inventory, which is analogous to a buyback.

[10] In their first model a retailer makes a single purchase decision even though demand occurs over two periods. Price protection is modeled as a credit for each unit remaining at the end of the first period, which resembles a buyback. In their second model the retailer may purchase at the start of each period. Again, price protection is modeled as a credit for each unit not sold at the end of the first period; as with a buyback, the price protection reduces the retailer's overage cost.

buybacks. He demonstrates coordination with arbitrary allocation of profit requires price protection in addition to buybacks when the retail price declines with time.[11]

In the context of capital intensive industry, Wu, Kleindorfer and Zhang (2002) study contracts that are similar to buyback contracts. There is one supplier and one buyer. The buyer reserves Q units of capacity for a fee, s, and pays another fee for each unit of capacity utilized, g. This is analogous to a buyback contract with a wholesale price $w = s + g$ and a buyback rate $b = g$: the buyer pays $s + g$ for each unit of capacity that is reserved and receives g for each unit of capacity *not* utilized. The buyer's demand depends on the contract parameters and the uncertain spot price for additional capacity: if the spot is less than g then the buyer satisfies his demand via the spot market exclusively, but with higher spot prices the buyer uses an optimal mixture of the reserved capacity and the spot market.

2.4 The revenue-sharing contract

With a revenue-sharing contract the supplier charges w_r per unit purchased plus the retailer gives the supplier a percentage of his revenue. Assume all revenue is shared, i.e., salvage revenue is also shared between the firms. (It is also possible to design coordinating revenue-sharing contracts in which only regular revenue is shared.) Let ϕ be the fraction of supply chain revenue the retailer keeps, so $(1 - \phi)$ is the fraction the supplier earns. Revenue-sharing contracts have been applied recently in the video cassette rental industry with much success. Cachon and Lariviere (2000) provide an analysis of these contracts in a more general setting.

The transfer payment with revenue sharing is

$$T_r(q, w_r, \phi) = (w_r + (1 - \phi)v)q + (1 - \phi)(p - v)S(q).$$

The retailer's profit function is

$$\pi_r(q, w_r, \phi) = (\phi(p - v) + g_r)S(q) - (w_r + c_r - \phi v)q - g_r\mu.$$

Now consider the set of revenue-sharing contracts, $\{w_r, \phi\}$, such that $\lambda \geq 0$ and

$$\phi(p - v) + g_r = \lambda(p - v + g)$$
$$w_r + c_r - \phi v = \lambda(c - v).$$

---

[11] In this model the retailer can either order additional units at the end of the first period or return units to the supplier. Price protection is now a credit for each unit retained. Therefore, price protection is a subsidy for retaining inventory whereas the buyback is a subsidy for disposing inventory.

With those terms the retailer's profit function is

$$\pi_r(q, w_r, \phi) = \lambda \Pi(q) + \mu(\lambda g - g_r). \tag{2.10}$$

Hence, q^o is the retailer's optimal order quantity. The supplier's profit is

$$\pi_s(q, w_r, \phi) = \Pi(q) - \pi_r(q, w_r, \phi) = (1 - \lambda)\Pi(q) - \mu(\lambda g - g_r),$$

so q^o is the supplier's optimal production quantity as long as $\lambda \leq 1$. The retailer's profit is increasing in λ and the supplier's is decreasing in λ. It is easy to confirm that Eqs. (2.8) and (2.9) provide the parameter values for λ such that the retailer's profit equals the supply chain's profit with the former and the supplier's profit equals the supply chain's profit with the latter. Hence, those revenue-sharing contracts coordinate the supply chain and arbitrarily allocate its profit.

The similarity between Eqs. (2.10) and (2.7) suggests a close connection between revenue-sharing and buyback contracts. In fact, in this setting they are equivalent. Consider a coordinating buyback contract, $\{w_b, b\}$. With that contract the retailer pays $w_b - b$ for each unit purchased and an additional b per unit sold. (The common description for a buyback contract has the retailer paying w_b per unit purchased and receiving a credit of b per unit not sold, which is the same as paying $w_b - b$ for each unit purchased no matter the demand realization and an additional b per unit sold.) With revenue sharing the retailer pays $w_r + (1 - \phi)v$ for each unit purchased and $(1 - \phi)(p - v)$ for each unit sold. Therefore, revenue-sharing and a buyback contract are equivalent when

$$w_b - b = w_r + (1 - \phi)v$$
$$b = (1 - \phi)(p - v)$$

In other words, the revenue-sharing contract $\{w_r, \phi\}$ generates the same profits for the two firms for any realization of demand as the following buyback contract,

$$w_b = w_r + (1 - \phi)p$$
$$b = (1 - \phi)(p - v)$$

While these contracts are equivalent in this setting, Sections 3 and 5.2 demonstrate that their paths diverge in more complex settings.

There are several other papers that investigate revenue-sharing contracts. Mortimer (2000) provides a detailed econometric study of the impact of revenue-sharing contracts in the video rental industry. She finds that the adoption of these contracts increased supply chain profits by 7%. Dana and Spier (2001) study these contracts in the context of a perfectly competitive retail market. Pasternack (1999) studies a single retailer newsvendor model in which the retailer can purchase some units with revenue sharing and other units with a wholesale-price contract. He does not consider supply chain coordination in his model. Gerchak, Cho and Ray (2001) consider a video retailer that decides how many tapes to purchase and how much time to keep them. Revenue sharing coordinates their supply chain, but only provides one division of profit. They redistribute profits with the addition of a licensing fee. Wang, Jiang and Shen (2001) consider revenue sharing with consignment (i.e., $w_r = 0$).

2.5 The quantity-flexibility contract

With a quantity-flexibility contract the supplier charges w_q per unit purchased but then compensates the retailer for his losses on unsold units. To be specific, the retailer receives a credit from the supplier at the end of the season equal to $(w_q + c_r - v)\min(I, \delta q)$, where I is the amount of leftover inventory, q is the number of units purchased and $\delta \in [0, 1]$ is a contract parameter. [See Yüksel and Lee (2002) for a model in which the return threshold is an absolute quantity instead of a percentage of the retailer's order.] Hence, the quantity-flexibility contract fully protects the retailer on a portion of the retailer's order whereas the buyback contract gives partial protection on the retailer's entire order. (The retailer continues to salvage leftover inventory, which is why the salvage value is not included in each unit's credit.) If the supplier did not compensate the retailer for the c_r cost per unit then the retailer would receive only partial compensation on a limited number of units, which is called a backup agreement. Those contracts are studied by Pasternack (1985) and Eppen and Iyer (1997) and Barnes-Schuster, Bassok and Anupindi (1998).[12]

Tsay (1999) studies supply chain coordination with quantity-flexibility contracts in a model that resembles this one. In Tsay (1999) the retailer receives an imperfect demand signal before submitting his final order (i.e., just before deciding how much to return), whereas in this model the retailer receives a perfect signal, i.e., the retailer observes demand. Nevertheless, since production is done before any demand information is learned, the centralized solution in Tsay (1999) is also a newsvendor problem. The demand signal does not matter to the analysis or to the outcome if the retailer returns units only at the end of the season: by then the demand signal is no longer relevant. However, if the retailer is able to return units after observing the demand signal and before the selling

---

[12] Eppen and Iyer (1997) do not consider channel coordination. Instead, they consider the retailer's order quantity decision. However, their model is more complex: e.g., it includes demand updates, holding costs and customer returns.

season starts, then the demand signal does matter. Because the inventory that is produced is sunk, the supply chain optimal solution is to keep all inventory at the retailer no matter what signal is received. Allowing the retailer to return inventory (alternatively, allowing the retailer to cancel a portion of the initial order) creates a 'stranded inventory problem': inventory could be stranded at the supplier, unable to be used to satisfy demand. In that situation, as shown in Tsay (1999), a quantity-flexibility contract may actually prevent supply chain coordination. On another issue, Tsay (1999) assumes forced compliance, which does have some significance.[13]

With the quantity-flexibility contract the transfer payment is $T_q(q, w_q, \delta)$

$$T_q(q, w_q, \delta) = w_q q - (w + c_r - v) \int_{(1-\delta)q}^{q} F(y)\,dy,$$

where the last term is the retailer's compensation for unsold units, up to the limit of δq units. The retailer's profit function is

$$\begin{aligned}\pi_r(q, w_q, \delta) &= (p - v + g_r)S(q) - (c_r - v)q - T_q(q, w_q, \delta) - \mu g_r \\ &= (p - v + g_r)S(q) - (w_q + c_r - v)q \\ &\quad + (w_q + c_r - v)\int_{(1-\delta)q}^{q} F(y)\,dy - \mu g_r\end{aligned}$$

To achieve supply chain coordination it is necessary (but not sufficient) that the retailer's first-order condition holds at q^o:

$$(p - v + g_r)S'(q^o) - (w_q + c_r - v)(1 - F(q^o) + (1 - \delta)F((1 - \delta)q^o)) = 0 \tag{2.11}$$

Let $w_q(\delta)$ be the wholesale price that satisfies Eq. (2.11):

$$w_q(\delta) = \frac{(p - v + g_r)(1 - F(q^o))}{1 - F(q^o) + (1 - \delta)F((1 - \delta)q^o)} - c_r + v.$$

---

[13] There are some other minor differences that do not appear to be important qualitatively. He assumes demand is normally distributed. In addition, the retailer's final order must be in the range $[q(1+\alpha), q(1-w)]$, where q is the initial forecast and α and w are contract parameters. In this model the retailer's final order must be in the range $[\delta q, q]$, where q is the initial order and δ is a contract parameter. He does not include a supplier goodwill cost, nor a retailer marginal cost, c_r.

$w_q(\delta)$ is indeed a coordinating wholesale price if the retailer's profit function is concave:

$$\frac{\partial^2 \pi_r(q, w_q(\delta), \delta)}{\partial q^2} = -(p + g_r - w_q(\delta) - c_r)f(q) - (w_q(\delta) + c_r - v)(1 + (1-\delta)^2 f((1-\delta)q))$$

$$\leq 0$$

which holds when $v - c_r \leq w_q(\delta) \leq p + g_r - c_r$. That range is satisfied with $\delta \in [0, 1]$ because

$$w_q(0) = (p - v + g_r)\overline{F}(q^\circ) + v - c_r,$$
$$w_q(1) = p + g_r - c_r,$$

and $w_q(\delta)$ is increasing in δ.

For supply chain coordination the supplier must also wish to deliver q° to the retailer. The supplier's profit function is

$$\pi_s(q, w_q(\delta), \delta) = g_s S(q) + (w_q(\delta) - c_s)q - (w_q(\delta) + c_r - v)\int_{(1-\delta)q}^{q} F(y)\,dy - \mu g_s$$

and

$$\frac{\partial \pi_s(q, w_q(\delta), \delta)}{\partial q} = g_s(1 - F(q)) + (w_q(\delta) - c_s) - (w_q(\delta) + c_r - v)(F(q)$$
$$- (1-\delta)F((1-\delta)q))$$
$$= g_s(1 - F(q)) - c + v + (w_q(\delta) + c_r - v)(1 - F(q)$$
$$+ (1-\delta)F((1-\delta)q))$$

The supplier's first-order condition at q° is satisfied:

$$\frac{\partial \pi_s(q^\circ, w_q(\delta), \delta)}{\partial q} = g_s(1 - F(q^\circ)) - c + v + (p - v + g_r)(1 - F(q^\circ)) = 0$$

However, the sign of the second-order condition at q° is ambiguous,

$$\frac{\partial^2 \pi_s(q, w_q(\delta), \delta)}{\partial q^2} = -w_q(\delta)(f(q) - (1-\delta)^2 f((1-\delta)q)) - g_s f(q).$$

In fact, q° may be a local minimum (i.e., the above is positive). That occurs when $g_s = 0$ and $(1-\delta)^2 f((1-\delta)q^\circ)$ is greater than $f(q^\circ)$, which is possible when δ is small, $f((1-\delta)q^\circ)$ is large and $f(q^\circ)$ is small. The second condition occurs when $(1-\delta)q^\circ \approx \mu$ and there is little variation in demand (i.e., so most of the density function is concentrated near the mean). The third condition occurs when $f(q^\circ)$ is in the tail of the distribution, i.e., when the critical fractile is large. For example, q° is a local minimum for the following parameters: D is normally distributed, $\mu = 10$, $\sigma = 1$, $p = 10$, $c_s = 1$, $c_r = 0$, $g_r = g_s = v = 0$ and $\delta = 0.1$. Hence, supply chain coordination under voluntary compliance is not assured with a quantity-flexibility contract even if the wholesale price is $w_q(\delta)$. Channel coordination is achieved with forced compliance since then the supplier's action is not relevant.

Assuming a $(w_q(\delta), \delta)$ quantity-flexibility contract coordinates the channel, now consider how it allocates profit. When $\delta = 0$, the retailer earns at least the supply chain optimal profit:

$$\pi_r(q, w_q(0), 0) = (p - v + g_r)S(q) - \left(\frac{p - v + g_r}{p - v + g}\right)(c - v)q^\circ - \mu g_r$$
$$= \Pi(q^\circ) + g_s(\mu - S(q^\circ) + \overline{F}(q^\circ)q^\circ)$$
$$\geq \Pi(q^\circ)$$

When $\delta = 1$, the supplier earns at least the supply chain's optimal profit:

$$\pi_s(q, w_q(1), 1) = g_s S(q^\circ) + (p + g_r - c)q^\circ - (p + g_r - v)\int_0^q F(y)\,dy - \mu g_s$$
$$= \Pi(q^\circ) + \mu g_r$$
$$\geq \Pi(q^\circ)$$

Given that the profit functions are continuous in δ, it follows that all possible allocations of $\Pi(q^\circ)$ are possible.

There are a number of other papers that study the quantity-flexibility contract, or a closely related contract. Tsay and Lovejoy (1999) study quantity-flexibility contracts in a more complex setting than the one considered here: they have multiple locations, multiple demand periods, lead times and demand forecast updates. Bassok and Anupindi (1997b) provide an in-depth analysis of these contracts for a single-stage system with more general assumptions than in Tsay and Lovejoy (1999). (They refer to their contract as a rolling horizon-flexibility contract.) In multiple-period models it is observed that these contracts dampen supply chain order variability, which is a potentially beneficial feature that the single-period model does not capture.

Cachon and Lariviere (2001) and Lariviere (2002) study the interaction between quantity-flexibility contracts and forecast sharing. In Cachon and Lariviere (2001) a downstream firm has a better demand forecast than the upstream supplier, but needs to convince the upstream supplier that her forecast is genuine. The minimum commitment in a quantity-flexibility contract is a very effective solution for this problem (see Section 10). In Lariviere (2002) the upstream firm wishes to encourage the downstream firm to exert the proper amount of effort to improve his demand forecast.

Plambeck and Taylor (2002) study quantity-flexibility contracts with more than one downstream firm and ex-post renegotiation. With multiple retailers it is possible that one retailer needs more than its initial order, q, and the other retailer needs less than its minimum commitment, δq. This creates an opportunity to renegotiate the contracts, which influences the initial contracts signed and actions taken.

2.6 The sales-rebate contract

With a sales-rebate contract the supplier charges w_s per unit purchased but then gives the retailer an r rebate per unit sold above a threshold t. This contract form is studied by Taylor (2002a) and Krishnan, Kapuscinski and Butz (2001), where the latter refers to it as a 'markdown allowance'. Both models are more complex than the one considered here. In particular, both papers allow the retail to exert effort to increase demand: in Taylor (2002a) effort is chosen simultaneously with the order quantity, whereas Krishnan et al. (2001) focus on the case in which the retailer chooses an order quantity, a signal of demand is observed and then effort is exerted. Hence, if the demand signal is strong relative to the order quantity, then the retailer does not need to exert much effort. See Section 4 for additional discussion of coordination in the presence of retail effort.

The transfer payment with the sales-rebate contract is

$$T_s(q, w_s, r, t) = \begin{cases} w_s q & q < t \\ (w_s - r)q + r\left(t + \int_t^q F(y)\,dy\right) & q \geq t \end{cases}:$$

when $q \geq t$ the retailer pays $w_s - r$ for every unit purchased, an additional r per unit for the first t units purchased and an additional r per unit for the units *not* sold above the t threshold. The retailer's profit function is then

$$\pi_r(q, w_s, r, t) = (p - v + g_r)S(q) - (c_r - v)q - g_r\mu - T_s(q, w_s, r, t)$$

For this contract to achieve supply chain coordination, q^o must at least be a local maximum:

$$\frac{\partial \pi_r(q^o, w_s, r, t)}{\partial q} = (p - v + g_r)\overline{F}(q^o) - (c_r - v) - \frac{\partial T_s(q^o, w_s, r, t)}{\partial q} = 0.$$

(2.12)

Ch. 6. Supply Chain Coordination with Contracts

If $t \geq q^o$ the above condition is only satisfied with $w_s = c_s - g_s \overline{F}(q^o)$, which is clearly not acceptable to the supplier. But this contract is interesting only if it achieves supply chain coordination for $t < q^o$. So assume $t < q^o$.

Define $w_s(r)$ as the wholesale price that satisfies Eq. (2.12):

$$w_s(r) = (p - v + g_r + r)\overline{F}(q^o) - c_r + v \qquad (2.13)$$

Given that wholesale price, the second-order condition confirms $\pi_r(q, w_s(r), r, t)$ is strictly concave in q for $q > t$. So q^o is a local maximum. But $\pi_r(q, w_s(r), r, t)$ is strictly concave in q for $q \leq t$ and, due to a 'kink' at $q = t$, $\pi_r(q, w_s(r), r, t)$ need not be unimodal in q. Let $\overline{q} = \arg\max_{q \leq t} \pi_r(q, w_s(r), r, t)$. Hence, it is necessary to demonstrate that there exist coordinating contracts such that q^o is preferred by the retailer over $\overline{q}$. Substitute $w_s(r)$ into the retailer's profit function:

$$\pi_r(q, w_s(r), r, t) = \Pi(q) + g_s(\mu - S(q) + q\overline{F}(q^o)) - rq\overline{F}(q^o)$$
$$+ \begin{cases} 0 & q < t \\ rq - r(t + \int_t^q F(y)\,dy) & q \geq t \end{cases}$$

and

$$\pi_r(q^o, w_s(r), r, t) = \Pi(q^o) + g_s(\mu - S(q^o) + q^o \overline{F}(q^o))$$
$$+ r\left(q^o F(q^o) - t - \int_t^{q^o} F(y)\,dy\right).$$

With $t = 0$ the retailer earns more than $\Pi(q^o)$, so q^o is surely optimal. With $t = q^o$, the retailer's profit function is decreasing for $t \geq q^o$; $\overline{q}$ is at least as good for the retailer as q^o. Given that $\pi_r(q^o, w_s(r), r, t)$ is decreasing in t, there must exist some t in the range $[0, q^o]$ such that $\pi_r(q^o, w_s(r), r, t) = \pi_r(\overline{q}, w_s(r), r, t)$.

Now consider the allocation of profit. We have already established that with $t = 0$ the retailer earns more than $\Pi(q^o)$. Hence, there must be a t such that $\pi_r(q^o, w_s(r), r, t) = \Pi(q^o)$, i.e., the retailer earns the supply chain's profit. When $t = q^o$, the retailer earns $\pi_r(\overline{q}, w_s(r), r, t)$, and with a sufficiently large r such that profit is zero; the supplier earns the supply chain's profit. In fact, there is generally a set of contracts that generate any profit allocation because the sales-rebate contract is parameter rich: these three parameters are more than sufficient to coordinate one action and to redistribute rents.

Now consider the supplier's production decision. The supplier's profit function given a coordinating sales-rebate contract is

$$\pi_s(q, w_s(r), r, t) = -g_s(\mu - S(q)) - c_s q + T_s(q, w_s(r), r, t).$$

For $q > t$,

$$\frac{\partial \pi_s(q, w_s(r), r, t)}{\partial q} = g_s\overline{F}(q) - c_s + w_s(r) - r + rF(q)$$

$$= (r - g_s)(F(q) - F(q^\circ))$$

The above is positive for $q \leq q^\circ$ only if $r < g_s$. But if $r \leq g_s$, then $w_s(r) \leq c_s$; the supplier cannot earn a positive profit with $r < g_s$. As a result, it must be that $r > g_s$, which implies the supplier loses money on each unit delivered to the retailer above t: the retailer effectively pays the supplier $w_s(r) - r$ for each unit sold above the threshold t and from Eq. (2.13),

$$w_s(r) - r = c_s - v - g_s\overline{F}(q^\circ) - rF(q^\circ) < c_s.$$

So the sales-rebate contract does not coordinate the supply chain with voluntary compliance.

2.7 The quantity-discount contract

There are many types of quantity discounts.[14] This section considers an 'all unit' quantity discount, i.e., the transfer payment is $T_d(q) = w_d(q)q$, where $w_d(q)$ is the per unit wholesale price that is decreasing in q. The retailer's profit function is then

$$\pi_r(q, w_d(q)) = (p - v + g_r)S(q) - (w_d(q) + c_r - v)q - g_r\mu.$$

One technique to obtain coordination is to choose the payment schedule such that the retailer's profit equals a constant fraction of the supply chain's profit. To be specific, let

$$w_d(q) = ((1 - \lambda)(p - v + g) - g_s)\left(\frac{S(q)}{q}\right) + \lambda(c - v) - c_r + v.$$

The above is decreasing in q if $\lambda \leq \overline{\lambda}$, where

$$\overline{\lambda} = \frac{p - v + g_r}{p - v + g},$$

---

[14] Roughly speaking, the quantity-discount contract achieves coordination by manipulating the retailer's marginal cost curve, while leaving the retailer's marginal revenue curve untouched. Coordination is achieved if the marginal revenue and marginal cost curves intersect at the optimal quantity. Hence, there is an infinite number of marginal cost curves that intersect the marginal revenue curve at a single point. See Moorthy (1987) for a more detailed explanation for why many coordinating quantity discount schedules exist. See Kolay and Shaffer (2002) for a discussion on different types of quantity discounts. See Wilson (1993) for a much broader discussion of nonlinear pricing.

since $S(q)/q$, i.e., sales per unit ordered, is decreasing in q. The retailer's profit function is now

$$\pi_r(q, w_d(q)) = \lambda(p - v + g)S(q) - \lambda(c - v)q - g_r\mu$$
$$= \lambda(\Pi(q) + g\mu) - g_r\mu.$$

Hence, q^o is optimal for the retailer and the supplier. As with the buyback and revenue-sharing contracts, the parameter λ acts to allocate the supply chain's profit between the two firms. However, the upper bound on λ prevents too much profit from being allocated to the retailer with a quantity discount. Technically, the $w_d(q)$ schedule continues to coordinate even if $\lambda > \bar{\lambda}$, but then $w_d(q)$ is increasing in q. In that case the retailer pays a quantity premium. See Tomlin (2000) for a model with both quantity-discount and quantity-premium contracts.[15]

2.8 Discussion

This section studies five contracts, two of which are equivalent (revenue-sharing and buyback contracts), to coordinate the newsvendor and to divide the supply chain's profit. Each contract coordinates by inducing the retailer to order more than he would with just a wholesale-price contract. Revenue-sharing and quantity-flexibility contracts do this by giving the retailer some downside protection: if demand is lower than q, the retailer gets some refund. The sales-rebate contract does this by giving the retailer upside incentive: if demand is greater than t, the retailer effectively purchases the units sold above t for less than their cost of production. The quantity discount coordinates by adjusting the retailer's marginal cost curve so that the supplier earns progressively less on each unit. However, an argument has not yet been made for why one contract form should be observed over another.

The various coordinating contracts may not be equally costly to administer. The wholesale-price contract is easy to describe and requires a single transaction between the firms. The quantity discount also requires only a single transaction, but it is more complex to describe. The other coordinating contracts are more costly to administer: the supplier must monitor the number of units the retailer has left at the end of the season, or the remaining units must be transported back to the supplier, depending on where the units are salvaged. Hence, the administrative cost argument does not explain the selection among buyback, revenue-sharing and quantity-flexibility contracts, but may explain the selection of a quantity-discount or a wholesale-price contract.

---

[15] Tomlin (2000) studies a supplier–manufacturer supply chain in which both firms incur costs to install capacity and both firms incur costs to convert capacity into units.

The risk neutrality assumption notwithstanding, the contracts do differ with respect to risk. With the exception of the quantity-discount contract, each of the coordinating contracts shifts risk between the two firms: as the retailer's share of profit decreases, his risk decreases and the supplier's risk increases. Hence, these contracts could provide some insurance to a risk averse retailer, but would be costly to a risk averse supplier. See Eeckhoudt, Gollier and Schlesinger (1995), Schweitzer and Cachon (2000) and Chen and Federgruen (2000) for a discussion of risk in the single-firm newsvendor model. Agrawal and Seshadri (2000) do study the influence of risk aversion in supply chain contracting. They argue that risk aversion among retailers provides an incentive for a distributor to provide risk intermediation services. In their model the distributor offers a contract with a fixed fee, a wholesale price, a return rate and premium fee for units ordered on an emergency basis to cover demand in excess of the retailer's order quantity. Finally, Plambeck and Zenios (2000) provide a principle-agent model that does incorporate risk aversion.

The supplier's exposure to demand uncertainty with some of the coordinating contracts could matter to the supplier if the retailer chose an order quantity other than q^o. For example, if the supplier offers a generous buyback to the retailer, then the supplier does not want the retailer to order too much products. Under voluntary compliance the supplier can avoid this excessive ordering error by shipping only q^o. But with forced compliance the supplier bears the full risk of an irrational retailer, a risk that even a risk neutral supplier may choose to avoid. However, with voluntary compliance the supplier may ship less than the retailer's order even if everyone is quite rational: revenue sharing and quantity discounts always coordinate the supplier's action with voluntary compliance, quantity-flexibility contracts generally, but not always, coordinate the supplier's action and sales-rebate contracts never do.

Now consider the application of these contracts in a setting with heterogenous retailers that do not compete, i.e., the action of one retailer has no impact on any other retailer, probably because of geographic dispersion. In general, suppliers are legally obligated to offer the same contractual terms to their retailers; hence, it is desirable for the supplier to offer the same contract to all of her retailers, or at the very least, the same menu of contracts.[16] If only one contract is offered, then it coordinates all of the retailers as long as the set of coordinating contracts does not depend on something that varies across the retailers. For example, the coordinating revenue-sharing contracts do not depend on the demand distribution, but do depend on the retailer's marginal cost. Hence, a single revenue-sharing contract can coordinate retailers with heterogenous demands, but not necessarily retailers with different marginal costs. However, in some cases heterogeneity can be accommodated with a single contract. Consider the

---

[16] Actually, a supplier can offer different contracts to retailers that do not compete.

quantity-flexibility contract, which depends on the demand distribution, and two retailers that have demands that differ by a scale factor; let retailer i's demand distribution be $F_i(x \mid \theta_i) = F(x/\theta_i)$, where θ_i is the scale parameter. Hence, the same wholesale price coordinates different retailers, $w_q(\delta \mid \theta_i) = w_q(\delta \mid \theta_j)$.

The independence of a contract to some parameter is also advantageous if the supplier lacks information regarding that parameter. For example, a supplier does not need to know a retailer's demand distribution to coordinate the supply chain with a revenue-sharing contract, but would need to know the retailer's demand distribution with a quantity-flexibility, sales-rebate or quantity-discount contract.

However, there may also be situations in which the supplier wishes to divide the retailers by offering a menu of contracts. For example, Lariviere (2002) studies a model with one supplier selling to a retailer that may exert effort to improve his demand forecast. He considers whether it is useful to offer two types of contracts, one for a retailer that exerts effort and one for a retailer that does not. Since coordinating buyback contracts are independent of the demand distribution, this separation requires the supplier to offer noncoordinating buyback contracts, i.e., supply chain efficiency must be sacrificed to induce forecasting. Quantity-flexibility contracts do depend on the demand distribution, so a menu can be constructed with two coordinating quantity-flexibility contracts, i.e., supply chain efficiency need not be sacrificed. Surprisingly, unless forecasting is very expensive, the supplier is still better off using the menu of buyback contracts even though this sacrifices some efficiency.

To summarize, the set of coordinating contracts is quite large and it is even quite likely that there exist other types of coordinating contracts. While it is possible to identify some differences among the contracts (e.g., different administrative costs, different risk exposures, etc.) none of them is sufficiently compelling to explain why one form should be adopted over another. More theory probably will not provide the answer. We now need some data and empirical analysis.

3 Coordinating the newsvendor with price-dependent demand

In the newsvendor model the retailer impacts sales only through his stocking decision, but in reality a retailer may influence sales through many different actions. Probably the most influential one is the retailer's pricing action. This section studies coordination in the newsvendor model with price-dependent retail demand. A key question is whether the contracts that coordinate the retailer's order quantity also coordinate the retailer's pricing. It is shown that buybacks, quantity-flexibility and sales-rebate contracts do not coordinate in this setting. Those contracts run into trouble because the incentive they provide to coordinate the retailer's quantity action distorts the retailer's pricing decision. Revenue sharing coordinates if there are no goodwill penalties, $g_s = g_r = 0$. With goodwill penalties there exists a single coordinating revenue-sharing contract that provides only a single allocation of the supply chain's profit. The quantity discount does better: it coordinates and allocates profit even if $g_r \geq 0$, but $g_s = 0$ is required. Another contract is introduced, the price-discount contract, which is shown to coordinate and arbitrarily allocate profit. It is essentially a buyback contract with price contingent parameters, i.e., it is a buyback contract with parameters that are set only after the retailer chooses his price. The idea of contingent contracts can also be applied with revenue-sharing contracts when there are goodwill penalties.

3.1 Model and analysis

This model is identical to the one in Section 2 except now the retailer chooses his price in addition to his order quantity. Let $F(q|p)$ be the distribution function of demand, where p is the retail price. It is natural to assume demand decreases stochastically in price, i.e., $\partial F(q|p)/\partial p > 0$. In a realistic model the retailer would be able to adjust his price throughout the season, possibly for a fee for each adjustment. Such a dynamic pricing strategy would allow the retailer to adjust his price to reflect demand conditions: e.g., if demand were less than expected the retailer could accelerate price discounts. This dynamic pricing problem is quite complex even when supply chain coordination is not considered. Hence, to obtain initial insights, assume the retailer sets his price at the same time as his stocking decision and the price is fixed throughout the season.[17]

The integrated channel's profit is

$$\Pi(q,p) = (p - v + g)S(q,p) - (c - v)q - g\mu$$

where $S(q,p)$ is expected sales given the stocking quantity q and the price p,

$$S(q,p) = q - \int_0^q F(y|p)\,dy.$$

The integrated channel profit function need not be concave nor unimodal (see Petruzzi & Dada, 1999). Assume there exists a finite (but not necessarily unique) optimal quantity-price pair, $\{q^\circ, p^\circ\}$.

---

[17] A hybrid model may be more tractable. For example, suppose the retailer chooses q, then observes a demand signal and then chooses price. van Mieghem and Dada (1999) study a related model in the context of a single firm. The multiretailer model in Section 5.2 is also closely related: order quantities are chosen first, then demand is observed and then price is set to clear the market, i.e., price is variable but not a decision the firms have direct control over.

Ch. 6. Supply Chain Coordination with Contracts

Let $p^\circ(q)$ be the supply chain optimal price for a given q. The following first-order condition is necessary for coordination (but not sufficient),

$$\frac{\partial \Pi(q, p^\circ(q))}{\partial p} = S(q, p^\circ(q)) + (p^\circ(q) - v + g)\frac{\partial S(q, p^\circ(q))}{\partial p} = 0. \tag{3.1}$$

A contract fails to coordinate if it is unable to satisfy the first-order condition at $p^\circ(q)$, or it is able to satisfy the first-order condition at $p^\circ(q)$ only with parameters that fail to coordinate the quantity decision.

Consider the quantity-flexibility contract. The retailer's profit function is

$$\pi_r(q, p, w_q, \delta) = (p - v + g_r)S(q, p) - (w_q + c_r - v)q$$
$$+ (w_q + c_r - v)\int_{(1-\delta)q}^{q} F(y \mid p)\,dy - \mu g_r$$

For price coordination the first-order condition must hold,

$$\frac{\partial \pi_r(q, p^\circ(q), w_q, \delta)}{\partial p} = S(q, p^\circ(q)) + (p^\circ(q) - v + g_r)\frac{\partial S(q, p^\circ(q))}{\partial p}$$
$$+ (w_q + c_r - v)\int_{(1-\delta)q}^{q} \frac{\partial F(y \mid p^\circ(q))}{\partial p}\,dy$$
$$= 0. \tag{3.2}$$

The second term in Eq. (3.2) is no smaller than the second term in Eq. (3.1), so the above holds only if the third term is nonpositive. But the third term is nonnegative with a coordinating w_q, so coordination can only occur if $g_s = 0$ and either $w_q = v - c_r$ or $\delta = 0$. Neither is desirable. With $w_q = v - c_r$ the supplier earns a negative profit ($w_q < c_s$), so the supplier certainly cannot be better off with that coordinating contract. With $\delta = 0$ the quantity-flexibility contract is just a wholesale-price contract, so the retailer's quantity action is not optimal (assuming the supplier desires a positive profit). Hence, the quantity-flexibility contract does not coordinate the newsvendor with price-dependent demand.

The sales-rebate contract does not fare better in this setting. With that contract

$$\frac{\partial \pi_r(q, p^\circ(q), w_s, r, t)}{\partial p} = S(q, p^\circ(q)) + (p^\circ(q) - v + g_r)\frac{\partial S(q, p^\circ(q))}{\partial p}$$
$$- r\int_{t}^{q} \frac{\partial F(y \mid p^\circ(q))}{\partial p}\,dy.$$

Since the last term is negative when $r>0$ and $t<q$, the retailer prices below the optimal price. Coordination might be achieved if something is added to the sales-rebate contract to induce the retailer to a higher price. A buyback could provide that counterbalance: a buyback reduces the cost of leftover inventory, so a retailer need not price as aggressively to generate sales.

Now consider a buyback contract on its own. The retailer's profit function is

$$\pi_r(q,p,w_b,b) = (p - v + g_r - b)S(q,p) - (w_b - b + c_r - v)q - g_r\mu.$$

For coordination the supply chain optimal price must satisfy the first-order condition,

$$\frac{\partial \pi_r(q,p^o(q),w_b,b)}{\partial p} = S(q,p^o(q)) + (p^o(q) - v + g_r - b)\frac{\partial S(q,p^o(q))}{\partial p}$$
$$= 0. \qquad (3.3)$$

But a comparison of Eqs. (3.3) and (3.1) reveals Eq. (3.3) holds only if $b = -g_s$, which may violate the $b \geq 0$ constraint. In addition, with $b = -g_s$ the coordinating wholesale price is not acceptable to the supplier, $w_b(-g_s) = c_s - g_s$. Therefore, a buyback contract does not coordinate the newsvendor with price-dependent demand.[18] That result is also demonstrated by Marvel and Peck (1995) and Bernstein and Federgruen (2000). While Emmons and Gilbert (1998) recognize that the buyback contract does not coordinate this model, they nevertheless demonstrate a buyback contract with $b > 0$ may still perform better than a wholesale-price contract.

The buyback contract fails to coordinate in this setting because the parameters of the coordinating contracts depend on the price: from Eqs. (2.5) and (2.6), the coordinating parameters are

$$b = (1 - \lambda)(p - v + g) - g_s,$$
$$w_b = \lambda c_s + (1 - \lambda)(p + g - c_r) - g_s.$$

For a fixed λ, the coordinating buyback rate and wholesale price are linear in p. Hence, the buyback contract coordinates the newsvendor with price-dependent demand if b and w_b are made contingent on the retail price chosen, or if b and w_b are chosen after the retailer commits to a price (but before the retailer chooses q). This is the price-discount-sharing contract studied by Bernstein and Federgruen (2000), which is also called a 'bill back' in practice. To understand the name for the contract, notice that the retailer

---

[18] If $g_s = 0$, then there is one buyback contract that coordinates, $w_b = c_s$ and $b_s = 0$. But that contract does not leave the supplier with a positive profit.

gets a lower wholesale price if the retailer reduces his price, i.e., the supplier shares in the cost of a price discount with the retailer. To confirm that this contract coordinates the supply chain, substitute the above contract parameters into the retailer profit function:

$$\pi_r(q,p,w_b,b) = \lambda(p-v+g)S(q,p) - \lambda(c-v)q - g_r\mu$$
$$= \lambda(\Pi(q,p) + g\mu) - g_r\mu$$

Hence, for the retailer as well as the supplier, $\{q^\circ, p^\circ\}$ is optimal for $\lambda \in [0,1]$.

Now consider the revenue-sharing contract. With revenue sharing the retailer's profit is

$$\pi_r(q,p,w_r,\phi) = (\phi(p-v) + g_r)S(q,p) - (w_r + c_r - \phi v)q - g_r\mu.$$

Coordination requires

$$\frac{\partial \pi_r(q, p^\circ(q), w_r, \phi)}{\partial p} = S(q, p^\circ(q)) + (p^\circ(q) - v + g_r/\phi)\frac{\partial S(q, p^\circ(q))}{\partial p}$$
$$= 0. \qquad (3.4)$$

There are two important cases to consider: the first has $g_r = g_s = 0$, and the second has at least one positive goodwill cost. Begin with the first case, $g_r = g_s = 0$. In this situation,

$$\frac{\partial \pi_r(q,p,w_r,\phi)}{\partial p} = \frac{\partial \Pi(q,p)}{\partial p}$$

with any revenue-sharing contract. Thus, the retailer chooses $p^\circ(q)$ no matter which revenue-sharing contract is chosen. With full freedom to choose the ϕ and w_r parameters, revenue sharing is able to coordinate the retailer's quantity decision with precisely the same set of contracts used when the retailer price is fixed.

Recall that with the fixed price newsvendor revenue sharing and buybacks are equivalent: for every coordinating revenue-sharing contract there exists a buyback contract that generates the same profit allocation for all realizations of demand. Here, the contracts produce different outcomes. The difference occurs because with a buyback the retailer's share of revenue $(1-b/p)$ depends on the price, whereas with revenue sharing it is independent of the price, by definition. However, the price contingent buyback contract (which is also known as the price-discount contract) is again equivalent to revenue sharing: if $g_r = g_s = 0$, the coordinating revenue-sharing contracts yield

$$\pi_r(q,p,w_r,\phi) = \lambda\Pi(q,p)$$

and the price contingent buyback contracts yield the same profit for any quantity and price,

$$\pi_r(q, p, b(p), w_b(p)) = \lambda \Pi(q, p).$$

With the second case (either $g_r > 0$ or $g_s > 0$) revenue sharing is less successful. Now, according to Eq. (3.4), coordination is achieved only if $\phi = g_r/g$. With only one coordinating contract, revenue sharing is able to provide only one profit allocation, albeit both firms may enjoy a positive profit with this outcome, which contrasts with the single coordination outcome of the buyback contract. Again, the difficulty with coordination occurs because the coordinating parameters generally depend on the retail price

$$\phi = \lambda + \frac{\lambda g - g_r}{p - v},$$
$$w_r = \lambda(c - v) - c_r + \phi v.$$

The dependence on the retail price is eliminated only in the special case $\phi = \lambda = g_r/g$.

Coordination for all profit allocations is restored even in this case if, like with the buyback contract, the parameters of the revenue-sharing contract are made contingent on the retailer's price. In that case revenue sharing is again equivalent to the price-discount contract: price discounts are contingent buybacks and contingent buybacks are equivalent to contingent revenue sharing.

The final contract to investigate is the quantity discount. With the quantity discount the retailer keeps all revenue, so only the retailer's marginal cost curve is adjusted. As a result, the quantity discount does not distort the retailer's pricing decision. In many cases, this is ideal. To explain, the retailer's profit function with a quantity discount is

$$\pi_r(q, w_d(q), p) = (p - v + g_r)S(q, p) - (w_d(q) + c_r - v)q - g_r\mu.$$

If $g_s = 0$, then

$$\frac{\partial \pi_r(q, w_d(q), p)}{\partial p} = \frac{\partial S(q, p)}{\partial p} + (p - v + g_r)S(q, p) = \frac{\partial \Pi(q, p)}{\partial p}$$

and so $p^o(q)$ is optimal for the retailer. On the other hand, if $g_s > 0$, then the retailer's pricing decision needs to be distorted for coordination, which the quantity discount does not do. (It is possible that a quantity discount could be designed to correct this distortion when $g_s > 0$, but more careful analysis is required which is left to future research.)

Assuming $g_s = 0$, it remains to ensure that the quantity decision is coordinated. The same schedule can be used as with a fixed retail price, but now the schedule is designed assuming the optimal price is chosen:

$$w_d(q) = ((1-\lambda)(p^o - v + g) - g_s)\left(\frac{S(q,p^o)}{q}\right) + \lambda(c-v) - c_r + v,$$

where $p^o = p^o(q^o)$. It follows that

$$\pi_r(q, w_d(q), p) = (p - v + g_r)S(q,p) - \lambda(c-v)q - g_r\mu$$
$$- ((1-\lambda)(p^o - v + g) - g_s)S(q,p^o)$$

and so p^o is optimal for the retailer,

$$\frac{\partial \pi_r(q, w_d(q), p)}{\partial p} = \frac{\partial \Pi(q,p)}{\partial p}.$$

Given p^o is chosen,

$$\pi_r(q, w_d(q), p^o) = \lambda(p^o - v + g)S(q,p^o) - \lambda(c-v)q - g_r\mu$$
$$= (\Pi(q,p^o) + g\mu) - g_r\mu$$

and so q^o is optimal for the retailer and the supplier. Coordination occurs because the retailer's pricing decision is not distorted, and the retailer's quantity decision is adjusted contingent that p^o is chosen.

3.2 Discussion

There are surely many situations in which a retailer has some control over his pricing. However, incentives to coordinate the retailer's quantity decision may distort the retailer's price decision. This occurs with the buyback, quantity-flexibility and the sales-rebate contracts. Since the quantity discount leaves all revenue with the retailer, it does not create such a distortion, which is an asset when the retailer's pricing decision should not be distorted, i.e., when $g_s = 0$. Revenue sharing does not distort the retailer's pricing decision when $g_r = 0 = g_s$. In those situations the set of revenue-sharing contracts to coordinate the quantity decision with a fixed price continue to coordinate the quantity decision with a variable price. However, when there are goodwill costs, then the coordinating revenue-sharing parameters generally depend on the retail price. The dependence is removed with only a single revenue-sharing contract; hence coordination is only achieved with a single profit allocation. Coordination is restored with arbitrary profit allocation by making the parameters contingent on the retail price chosen, e.g., a menu of revenue-sharing contracts is offered that depend on the price selected. This technique also applies to the buyback contract: the price contingent buyback contract, which is also called a price-discount-sharing contract, coordinates the price-setting newsvendor. In fact, just as buybacks and revenue sharing are equivalent with a fixed retail price, the price contingent buyback and revenue sharing are equivalent when there are no goodwill costs. When there are goodwill costs then the price contingent buyback is equivalent to the price contingent revenue-sharing contract.

4 Coordinating the newsvendor with effort-dependent demand

A retailer can increase a product's demand by lowering his price, but the retailer can take other actions to spur demand: the retailer can hire more sales people, improve their training, increase advertising, better maintain the attractiveness of the product's display, enhance the ambiance of the store interior (e.g., richer materials, wider aisles) and he can give the product a better stocking location within the store. All of those activities are costly. As a result, a conflict exists between the supplier and the retailer: no matter what level of effort the retailer dedicates toward those activities, the supplier prefers that the retailer exert even more effort. The problem is that those activities benefit both firms, but are costly to only one.

Sharing the cost of effort is one solution to the effort coordination problem. For example, the supplier could pay some of the retailer's advertising expenses, or she could compensate the retailer for a portion of his training cost. Several conditions are needed for cost sharing to be an effective strategy: the supplier must be able to observe (without much hassle) that the retailer actually engaged in the costly activity (so the supplier knows how much to compensate the retailer), the retailer's effort must be verifiable to the courts (so that any cost sharing is enforceable) and the activity must directly benefit the supplier.[19] In many cases those conditions are met. For example, the supplier generally can observe and verify whether or not a retailer purchased advertising in a local newspaper. Furthermore, if the ad primarily features the supplier's product, then the benefit of the ad is directed primarily at the supplier. Netessine and Rudi (2000a) present a coordinating contract which involves sharing advertising costs in a model that closely resembles the one in this section. In Wang and Gerchak (2001) the retailer's shelf space can be considered an effort variable. They also allow the supplier to compensate the retailer for his effort, which in their model takes

---

[19] If the firms interact over a long horizon it may be sufficient that the action is observable even if it is not verifiable, i.e., enforcement can be due to the threat to leave the relationship rather than the threat of court action.

the form of an inventory subsidy. Gilbert and Cvsa (2000) study a model in which effort is observable but not verifiable.

There are also many situations in which cost sharing is not as effective. For example, a supplier probably will not pay for an ad that merely promotes the retailer's brand image. In that case the ad enhances the demand for all of the retailer's products, not just the supplier's product. Also, there are many demand-improving activities that are too costly for the supplier to observe. For example, it may be too costly to visit a store on a frequent basis to ensure the presentation of the supplier's product is maintained to the supplier's satisfaction.[20]

This section studies the challenge of coordinating an action for which there is no direct transfer payment. It is shown that most of the coordinating contracts with the standard newsvendor model no longer coordinate in this setting because the incentives they provide to coordinate the retailer's quantity decision distort the retailer's effort decision. Only the quantity-discount contract continues to coordinate the supply chain. In fact, the quantity-discount contract can coordinate a retailer that chooses quantity, price and effort.

4.1 Model and analysis

To model retail effort, suppose a single effort level, e, summarizes the retailer's activities and let $g(e)$ be the retailer's cost of exerting effort level e, where $g(0)=0$, $g'(e)>0$ and $g''(e)>0$. To help avoid confusion and to simplify the notation, assume there are no goodwill costs, $g_r = g_s = 0$, $v=0$ and $c_r = 0$. Let $F(q|e)$ be the distribution of demand given the effort level e, where demand is stochastically increasing in effort, i.e., $\partial F(q|e)/\partial e < 0$. Suppose the retailer chooses his effort level at the same time as his order quantity. Finally, assume the supplier cannot verify the retailer's effort level, which implies the retailer cannot sign a contract binding the retailer to choose a particular effort level. This approach to retail effort has been adopted in a number of marketing papers. For example, see Chu and Desai (1995), Desai and Srinivasan (1995), Desiraju and Moorthy (1997), Gallini and Lutz (1992), Lal (1990) and Lariviere and Padmanabhan (1997).

The integrated channel's profit is

$$\Pi(q, e) = pS(q, e) - cq - g(e),$$

where $S(q, e)$ is expected sales given the effort level e,

$$S(q, e) = q - \int_0^q F(y|e)\,dy.$$

---

[20] However, in some cases it is too costly *not* to visit a store. For example, in the salty snack food category is it common for suppliers to replenish their retailers' shelves.

The integrated channel's profit function need not be concave nor unimodal. For tractability, assume the integrated channel solution is well behaved, i.e., $\Pi(q, e)$ is unimodal and maximized with finite arguments. (For instance, if $S(q, e)$ increases sufficiently quickly with e and $g(e)$ is not sufficiently convex, then infinite effort could be optimal, which is rather unrealistic.) Let q^o and e^o be the optimal order quantity and effort.

The optimal effort for a given order quantity, $e^o(q)$, maximizes the supply chain's revenue net effort cost. That occurs when

$$\frac{\partial \Pi(q, e^o(q))}{\partial e} = p \frac{\partial S(q, e^o(q))}{\partial e} - g'(e^o(q)) = 0. \tag{4.1}$$

With a buyback contract the retailer's profit function is

$$\pi_r(q, e, w_b, b) = (p - b)S(q, e) - (w_b - b)q - g(e).$$

For all $b > 0$ it holds that

$$\frac{\partial \pi_r(q, e, w_b, b)}{\partial e} < \frac{\partial \Pi(q, e)}{\partial e}. \tag{4.2}$$

Thus, e^o cannot be the retailer's optimal effort level when $b > 0$. But $b > 0$ is required to coordinate the retailer's order quantity, so it follows that the buyback contract cannot coordinate in this setting.

With a quantity-flexibility contract the retailer's profit function is

$$\pi_r(q, e, w_q, \delta) = pS(q, e) - w_q\left(q - \int_{(1-\delta)q}^{q} F(y \mid e)\,dy\right) - g(e).$$

For all $\delta > 0$ (which is required to coordinate the retailer's quantity decision)

$$\frac{\partial \pi_r(q, e, w_q, \delta)}{\partial e} < \frac{\partial \Pi(q, e)}{\partial e}.$$

As a result, the retailer chooses a lower effort than optimal, i.e., the quantity-flexibility contract also does not coordinate the supply chain in this setting.

The revenue-sharing and sales-rebate contracts fare no better. It can be shown with $\phi < 1$,

$$\frac{\partial \pi_r(q, e, w_r, \phi)}{\partial e} < \frac{\partial \Pi(q, e)}{\partial e},$$

and so the retailer's optimal effort is lower than the supply chain's. With the sales-rebate contract it can be shown for $r > 0$ and $q > t$,

$$\frac{\partial \pi_r(q, e, w_s, r, t)}{\partial e} > \frac{\partial \Pi(q, e)}{\partial e},$$

which means the retailer exerts too much effort. Although the sales-rebate contract does not coordinate on its own, Taylor (2000) demonstrates it can coordinate the channel if it is combined with a buyback contract: the buyback reduces the retailer's incentive to exert effort, which counteracts the retailer's excessive incentive to exert effort with a sales rebate alone. However, four parameters make for a complex contract. Krishnan et al. (2001) also study the combination of a sales-rebate contract with buybacks. However, they allow the retailer to choose effort after observing demand.

The last contract to consider is the quantity discount. As in the price-setting model, coordination can be achieved in the effort model by letting the retailer earn the entire reward for exerting effort, which is the revenue function, because the retailer is charged the entire cost of effort. Therefore, the quantity discount should let the retailer retain the revenues but charge a marginal cost based on expected revenue conditional on the optimal effort. To explain, suppose the transfer payment is $T_d(q) = w_d(q)q$, where

$$w_d(q) = (1 - \lambda)p\left(\frac{S(q, e^o)}{q}\right) + \lambda c + (1 - \lambda)\frac{g(e^o)}{q}$$

and $\lambda \in [0, 1]$. Given that $S(q, e^o)/q$ is decreasing in q, this is indeed a quantity-discount schedule. As already mentioned, it is almost surely not the only coordinating quantity discount.

The retailer's profit function with the quantity-discount contract is

$$\pi_r(q, e) = pS(q, e) - (1 - \lambda)pS(q, e^o) - \lambda cq - g(e) + (1 - \lambda)g(e^o).$$

As in the price-dependent newsvendor, the retailer chooses the supply chain optimal effort because the retailer keeps all realized revenue. Given the optimal effort e^o, the retailer's profit function is

$$\pi_r(q, e^o) = \lambda pS(q, e^o) - \lambda cq - \lambda g(e^o) = \lambda \Pi(q, e^o),$$

and so the retailer's optimal order quantity is q^o, any allocation of profit is feasible and the supplier's optimal production is q^o.

This approach is sufficiently powerful that it is quite easy to design a quantity-discount contract that coordinates the newsvendor with demand-dependent on price and effort:

$$w_d(q) = (1-\lambda)p^\circ \left(\frac{S(q,e^\circ)}{q}\right) + \lambda c + (1-\lambda)\frac{g(e^\circ)}{q}.$$

Again, the retailer retains all revenue and so optimizes price and effort. Furthermore, the quantity decision is not distorted because the quantity-discount schedule is contingent on the optimal price and effort, and not on the chosen price and effort.

4.2 Discussion

Coordination with the effort-dependent demand model is complex when the firms are not allowed to contract on the retailer's effort level directly, i.e., any contract that specifies an effort level for the retailer is either unverifiable or unenforceable. Buybacks, revenue-sharing, quantity-flexibility and sales-rebate contracts all fail to coordinate the retailer's action because they all distort the retailer's marginal incentive to exert effort. [That distortion occurs even if the retailer chose its effort after observing a demand signal, as in Krishnan et al. (2001).] The quantity-discount contract does coordinate this system because the retailer incurs the entire cost of effort but also receives the entire benefit of effort.

A number of papers in the marketing and franchising literatures elaborate on the basic retail effort model. For example, in Chu and Desai (1995) the supplier can also exert costly effort to increase demand, e.g., brand building advertising, but the impact of effort occurs only with a lag: they have a two-period model and period 1 effort by the supplier increases only period 2 demand. They also enrich the retailer's effort model to include two types of effort, effort to increase short-term (i.e., current period) sales and long-term effort to increase long-term customer satisfaction and demand (i.e., period 2 sales). They allow the supplier to compensate the retailer by paying a portion of his effort cost and/or by paying the retailer based on the outcome of his effort, i.e., a bonus for high customer satisfaction scores. The issue is the appropriate mix between the two types of compensation. Lal (1990) also includes supplier effort, but, effort again is nonenforceable. Although revenue sharing (in the form of a royalty payment) continues to distort the retailer's effort decision, it provides a useful incentive for the supplier to exert effort: the supplier will not exert effort if the supplier's profit does not depend directly on retail sales. Lal (1990) also considers a model with multiple retailers and horizontal spillovers: the demand-enhancing effort at one retailer may increase the demand at other retailers. These spillovers can lead to free riding, i.e., one retailer enjoys higher demand due to the efforts of others without exerting his own effort. He suggests that the franchisor can control the problem of free riding by exerting costly monitoring effort and penalizing franchisees that fail to exert sufficient effort.

While the models mentioned so far have effort-increasing demand, effort can make other supply chain improvements. Two are discussed: effort to reduce hazardous material consumption and effort to improve product quality.

Corbett and DeCroix (2001) study shared-savings contracts between a supplier of a hazardous material and a manufacturer that uses the material in his output. They assume the product is an indirect material, i.e., the manufacturer's revenue is not correlated with the amount of the product used. For example, an automobile manufacturer does not earn more revenue if it increases the amount of paint used on its vehicle (assuming the increased amount of paint provides no perceived quality improvement). However, with a traditional contract the supplier's revenue does depend on the amount of material used, e.g., the paint supplier's revenue is proportional to the amount of paint the manufacturer purchases. They also assume both the manufacturer and the supplier can exert costly effort to reduce the needed amount of material to produce each unit of the manufacturer's output. The manufacturer clearly has an incentive to exert some effort, since using less material reduces his procurement cost. But the supplier certainly does not have an incentive to reduce the manufacturer's consumption if the supplier's revenue is increasing in consumption. However, it is also quite plausible the supplier's effort would reduce consumption, and further, the supplier may even be more efficient at reducing consumption than the manufacturer (i.e., the supplier's effort cost to reduce consumption by a fixed amount is lower than the manufacturer's effort cost). Thus the supply chain optimal effort levels may very well have both firms exerting effort to reduce consumption.

The Corbett and DeCroix (2001) model adds several twists to the newsvendor model with effort-dependent demand: both firms can exert effort, as opposed to just one firm; and effort hurts one firm and helps the other, whereas in the newsvendor model both firms benefit from effort. Given that structure it is no longer possible to assign all of the costs and all of the benefits of effort to one firm (as the quantity-discount contract does in the newsvendor model). Hence, they show shared-savings contracts (which are related to revenue-sharing contracts) do not coordinate their supply chain, and unfortunately, they are unable to identify which contract would coordinate their model.

Several papers study how effort in a supply chain influences quality. In Reyniers and Tapiero (1995) there is one supplier and one buyer. The supplier can choose between two production processes, one that is costly but produces high quality (in the form of a low-defect probability) and one that is inexpensive but produces low quality (a high-defect probability). The choice of production process can be taken as a proxy for effort in this model. The buyer can test each unit the supplier delivers, but testing is costly. Defective units that are discovered via testing are repaired for an additional cost incurred by the supplier, i.e., an internal failure cost. If the buyer does not test and the unit is defective, then an external failure cost is incurred by the buyer. They allow a contract that includes a wholesale-price rebate for internal failures and an external failure compensation, i.e., the supplier pays the buyer a portion of the buyer's external failure cost. Internal failures are less costly to the supplier (repair cost plus rebate cost) than external failures (compensation to the buyer), so the supplier benefits if the buyer tests a higher fraction of units.

In Baiman, Fischer and Rajan (2000) there is a supplier that can exert costly effort to improve quality and a buyer that exerts testing effort that yields an imperfect signal of quality. Both effort levels are continuous variables, as opposed to the discrete effort levels in Reyniers and Tapiero (1995). If testing suggests the product is defective, the buyer incurs an internal failure cost. If the testing suggests the product is not defective (and hence the buyer accepts the product) then an external failure cost is incurred if the product is in fact defective. They show that optimal supply chain performance is achievable when both effort levels are contractible. Optimal performance is also possible if the firms can verify the external and internal failures and therefore commit to transfer payments based on those failures. Baiman, Fischer and Rajan (2001) extend their model to include the issue of product architecture. With modular design the firms can attribute external failures to a particular firm: either the supplier made a defective component or the supplier made a good component but the buyer caused a defect by poor handling or assembly. However, with an integrated design it is not possible to attribute blame for a product's failure. Hence, the product architecture influences the contract design and supply chain performance. See Novak and Tayur (2002) for another model and empirical work on the issue of effort and attributing responsibility for quality failures.

The quality literature suggests that firms that cannot contract on effort directly can contract on a proxy for effort (the frequency of internal or external failures), which is one solution around potential observability or verifiability problems. See Holmstrom (1979) for another model with moral hazard and effort signals.

Gilbert and Cvsa (2000) study a model with costly effort that is observable but not verifiable, i.e., the firms in the supply chain can observe the amount of effort taken, but the amount of effort taken is not verifiable to the courts, and therefore not contractible. This distinction can be important, as it is in their model. They have a supplier that sets a wholesale price and a buyer that can invest to reduce his marginal cost. The investment to reduce the marginal cost is observed by both firms before the supplier chooses the wholesale price. The buyer cannot fully capture the benefit of cost reduction because the supplier will adjust her wholesale price based on the observed effort. Hence, the buyer invests in less effort to reduce cost than optimal. The supplier can do better if the supplier commits to a wholesale price before observing the buyer's cost reduction. However, demand is random in their model (and observed at the same time the buyer's cost reduction is observed) and so it is beneficial to choose the wholesale price after observing demand, which is in conflict with the incentive benefits of a committed wholesale price. They demonstrate that a hybrid solution works well: the supplier commits to a wholesale-price ceiling before observing the buyer's effort and the demand realization, and after the observations the supplier chooses a wholesale price that is not greater than her wholesale-price ceiling, i.e., there is partial wholesale-price commitment and partial flexibility.

5 Coordination with multiple newsvendors

This section considers two models with one supplier and multiple competing retailers. The first model has a fixed retail price and competition occurs by allocating demand among the retailers proportional to their inventory. It is shown the retailers are biased toward ordering more inventory than optimal because of a demand-stealing effect: each retailer fails to account for the decrease in the other retailer's demands when the retailer increases his order quantity. As a result, with just a wholesale-price contract the supplier can coordinate the supply chain and earn a positive profit. Nevertheless, there are limitations to that coordinating contract: it provides for only one division of supply chain profit; and it is not even the supplier's optimal wholesale-price contract. A buyback contract does not share those limitations: with a buyback contract the supplier can coordinate the supply chain and earn more than with the optimal wholesale-price contract.

The second model with retail competition yields qualitatively very different results. In that model the following sequence of events occurs: the retailers order inventory, market demand is observed and then the market-clearing price is set. The market-clearing price is the price at which consumers are willing to purchase all of the retailers' inventory. Hence, retailers might incur a loss on each unit when the market demand realization is low. The retailers anticipate that possibility and respond by ordering less than the optimal amount. As a result, in contrast to the quantity-allocation competition of the first model, now the supplier needs an instrument to increase retail inventory. Two are considered: resale price maintenance and buyback contracts. With either one the supplier can coordinate the supply chain and extract all of the supply chain's profit.

5.1 Competing newsvendors with a fixed retail price

Take the single retailer newsvendor model (described in Section 2) and make the following modifications: set $c_r = g_r = g_s = v = 0$, increase the number of retailers to $n > 1$; interpret D as the total retail demand, let F continue to be the distribution function for D; and let retail demand be divided between the n firms proportional to their stocking quantity, i.e., retailer i's demand, D_i, is

$$D_i = \left(\frac{q_i}{q}\right) D,$$

where $q = \sum_{i=1}^{n} q_i$ and $q_{-i} = q - q_i$. See Wang and Gerchak (2001) for a model that implements proportional allocation with deterministic demand.

Demands at the retailers are perfectly correlated with the proportional allocation model. Hence, either every firm has excess demand (when $D > q$) or every firm has excess inventory (when $D < q$). That could be a reasonable model when customers have low search costs; a customer that desires a unit finds a unit if there is a unit in the system. That search need not involve an actual physical inspection of each store by every customer. For example, information regarding availability could be exchanged among customers through incidental social interactions that naturally occur with daily activities. The model also presumes consumers do not care from which retailer they make their purchases, i.e., there are no retail brand preferences.

There are other demand-allocation models that maintain a constant retail price. Parlar (1988), Karjalainen (1992) and Anupindi and Bassok (1999) and Anupindi, Bassok and Zemel (1999) assign independent random demands to each retailer and then redistributed the retailer's excess demand. Netessine and Shumsky (2001) also redistribute each retailer's excess demand, but they add a twist. In their model the retailers are actually airlines and they have two fare classes. Lippman and McCardle (1997) adopt a more general approach to demand allocation which includes the independent random demand model. They represent aggregate demand as a single random variable and then allocate demand using a splitting rule that depends on the realization of demand (and not on the retailer's order quantities). The retailer's excess demand is then redistributed. In those models some of the redistributed demand may be lost, i.e., some customers may not be willing to continue their shopping if the first retailer they visit has no stock. As a result, total sales depend on the retailers' total inventory and how inventory is distributed among the retailers. With the proportional allocation model industry sales does not depend on the distribution of inventory among the retailers. Thus, the proportional allocation model is simpler to analyze.[21] However, the proportional allocation model is not a special case of the allocation model adopted in Lippman and McCardle (1995). Nevertheless, the qualitative insights from the models are consistent.

---

[21] The allocation of demand across multiple retailers is analogous to the allocation of demand across a set of products, which is known as the assortment problem. That problem is quite complex. See Mahajan and van Ryzin (1999) for a review of that literature.

There are also models that allocate demand dynamically. In Gans (2002) customers search among retailers without having perfect knowledge of the retailers' stocking levels.[22] As with Lippman and McCardle (1995), Gans (2002) does not consider channel coordination. van Ryzin and Mahajan (1999) assume customers may have different preferences for the retailers and choose to purchase from their most preferred retailer that has stock available.

Given the proportional allocation rule, the integrated supply chain faces a single newsvendor problem. Hence the optimal order quantity is defined by the familiar

$$F(q^o) = \frac{p-c}{p}. \tag{5.1}$$

Because the integrated solution remains a single-location newsvendor problem, the multiple retailer model with proportional allocation is a nice generalization of the single retailer model.

In the decentralized system we want to investigate retail behavior with either a wholesale-price contract or a buyback contract. (Since the retail price is fixed, in this case there exists a revenue-sharing contract that is equivalent to the buyback contract.) Retailer i's profit function with a buyback contract is

$$\pi_i(q_i, q_{-i}) = (p-w)q_i - (p-b)\left(\frac{q_i}{q}\right)\int_0^q F(x)\,dx.$$

The above also provides the retailer's profit with a wholesale-price contract (i.e., set $b = 0$). The second-order condition confirms each retailer's profit function is strictly concave in his order quantity. Hence, there exists an optimal order quantity for retailer i for each q_{-i}. In game theory parlance, retailer i has a unique optimal response to the other retailers' strategies (i.e., their order quantities). Let $q_i(q_{-i})$ be retailer i's response function, i.e., the mapping between q_{-i} and retailer i's optimal response. Since the retailers have symmetric profit functions, $q_j(q_{-j}) = q_i(q_{-i})$, $i \neq j$.

A set of order quantities, $\{q_1^*, \ldots, q_n^*\}$, is a Nash equilibrium of the decentralized system if each retailer's order quantity is a best response, i.e., for all i, $q_i^* = q_i(q_{-i}^*)$, where $q_{-i}^* = q^* - q_i^*$ and $q^* = \sum_{j=1}^n q_j^*$. There may not exist an Nash equilibrium, or there may be multiple Nash equilibria. If there is a unique Nash equilibrium then that is taken to be the predicted outcome of the decentralized game.

---

[22] Gans (2002) presents his model more generally. He has multiple suppliers of a service that compete on some dimension of quality, say their fill rate. Nevertheless, customers have less than perfect information about the suppliers' service levels and so they must develop a search strategy. The suppliers compete knowing customers have limited information.

Any Nash equilibrium must satisfy each retailer's first-order condition:

$$\frac{\partial \pi_i(q_i, q_j)}{q_i} = q^*\left(\frac{p-w}{p-b}\right) - q_i^* F(q^*) - q_{-i}^*\left(\frac{1}{q^*}\int_0^{q^*} F(x)\,dx\right) = 0.$$

Substitute $q_{-i}^* = q^* - q_i^*$ into the above equation and solve for q_i^* given a fixed q^*:

$$q_i^* = q^* \frac{\left((p-w)/(p-b) - (1/q^*)\int_0^{q^*} F(x)\,dx\right)}{F(q^*) - 1/q^* \int_0^{q^*} F(x)\,dx}. \tag{5.2}$$

The above gives each retailer's equilibrium order conditional on q^* being the equilibrium total order quantity. Hence, the above describes an equilibrium only if $q^* = n q_i^*$. Substitute Eq. (5.2) into $q^* = n q_i^*$ and simplify:

$$\frac{1}{n} F(q^*) + \left(\frac{n-1}{n}\right)\left(\frac{1}{q^*}\int_0^{q^*} F(x)\,dx\right) = \frac{p-w}{p-b}. \tag{5.3}$$

The left-hand side of Eq. (5.3) is increasing in q^* from 0 (when $q^* = 0$) to 1 (when $q^* = \infty$). Hence, when $b < w < p$, there exists a unique q^* that satisfies Eq. (5.3). In other words, in this game there exists a unique Nash equilibrium in which the total order quantity, q^*, is implicitly given by Eq. (5.3) and each retailer's order quantity equals $q_i^* = q^*/n$.

Consider how the equilibrium order quantity changes in n. The left-hand side of Eq. (5.3) is decreasing in n. Hence, q^* is increasing in n for fixed contractual terms: a single retailer that faces market demand D purchases less than multiple retailers facing the same demand (with proportional allocation). [This effect generalizes beyond just the proportional allocation model, as demonstrated by Lippman and McCardle (1995).] Competition makes the retailers order more inventory because of the demand-stealing effect: each retailer ignores the fact that ordering more means the other retailers' demands stochastically decrease. Anupindi and Bassok (1999) and Mahajan and van Ryzin (2001) also noticed this effect. However, the effect does not apply universally: Netessine and Rudi (2000b) find that competition may lead some retailers to understock when there are more than two retailers and demands are not symmetric. Furthermore, if retailers sell complements, rather than substitutes, then the demand-stealing effect is reversed: each retailer tends to understock because it ignores the additional demand it creates for other retailers.

Due to the demand-stealing effect the supplier can coordinate the supply chain and earn a positive profit with just a wholesale-price contract. To explain, let $\hat{w}(q)$ be the wholesale price that induces the retailers to order q units with a wholesale-price contract (i.e., with $b=0$). From Eq. (5.3),

$$\hat{w}(q) = p\left(1 - \left(\frac{1}{n}\right)F(q) - \left(\frac{n-1}{n}\right)\left(\frac{1}{q}\int_0^q F(x)\,\mathrm{d}x\right)\right).$$

By definition $\hat{w}(q^\circ)$ is the coordinating wholesale price. Given that $F(q^\circ) = (p-c)/c$ and

$$\frac{1}{q}\int_0^q F(x)\,\mathrm{d}x < F(q),$$

it can be shown that $\hat{w}(q^\circ) > c$ when $n > 1$. Hence, the supplier earns a positive profit with that coordinating contract. With the single retailer model channel coordination is only achieved when the supplier earns zero profit, i.e., marginal cost pricing, $\hat{w}(q^\circ) = c$.

Although the supplier can use the wholesale-price contract to coordinate the supply chain, that contract is not optimal for the supplier. The supplier's profit function with a wholesale-price contract is

$$\pi_s(q, \hat{w}(q)) = q(\hat{w}(q) - c).$$

Assuming $n > 1$, differentiate $\pi_s(q, \hat{w}(q))$ with respect to q and evaluate at $\hat{w}(q^\circ)$, the coordinating wholesale price,

$$\frac{\partial \pi_s(q^\circ, \hat{w}(q^\circ))}{\partial q} = -\frac{q^\circ p f(q^\circ)}{n} < 0.$$

Hence, rather than coordinating the supply chain with the wholesale price $\hat{w}(q^\circ)$, the supplier prefers to charge a higher wholesale price and sell less than q° when $n > 1$.

Although the supplier does not wish to use a wholesale-price contract to coordinate the supply chain, it is possible the supplier's profit with a coordinating buyback contract may exceed her profit with the optimal wholesale-price contract. Let $w_b(b)$ be the wholesale price that coordinates the supply chain given the buyback rate. Since the buyback rate provides an incentive to the retailers to increase their order quantity, it must be that $w_b(b) > \hat{w}(q^\circ)$, i.e., to coordinate the supply chain the supplier must use a wholesale price that is higher than the coordinating wholesale-price contract, which, recall, is lower than the supplier's optimal wholesale price. From Eqs. (5.1) and (5.3)

$$w_b(b) = p - (p-b)\left[\frac{1}{n}\left(\frac{p-c}{p}\right) + \left(\frac{n-1}{n}\right)\left(\frac{1}{q^o}\int_0^{q^o} F(x)\,dx\right)\right].$$

Given that $q_i^* = q^*/n$, retailer i's profit with a coordinating buyback contract is

$$\pi_i(q_i^*, q_{-i}^*) = (p - w(b))q^o/n - (p-b)\left(\frac{1}{n}\right)\int_0^{q^o} F(x)\,dx$$

$$= \left(\frac{p-b}{pn^2}\right)q^o\left[p - c - \frac{p}{q^o}\int_0^{q^o} F(x)\,dx\right]$$

$$= \left(\frac{p-b}{pn^2}\right)\Pi(q^o)$$

The supplier's profit with the coordinating contract is

$$\pi_s(q^o, w_b(b), b) = \Pi(q^o) - n\pi_i(q_i^*, q_{-i}^*)$$

$$= \left(\frac{p(n-1)+b}{pn}\right)\Pi(q^o).$$

Hence, the supplier can extract all supply chain profit with $b=p$. As shown earlier, the coordinating contract with $b=0$ provides a lower bound for the supplier's profit (because the supplier could do even better with a higher wholesale price than $w_b(0)$). The ratio of the supplier's lower bound to the supplier's maximum profit, $\Pi(q^o)$, provides a measure of how much improvement is possible by using a coordinating buyback contract:

$$\frac{\pi_s(q^o, w_b(0), 0)}{\Pi(q^o)} = \frac{n-1}{n}.$$

Hence, as n increases the supplier's potential gain decreases from using a coordinating buyback contract rather than her optimal wholesale-price contract. In fact, the supplier can use a wholesale-price contract to capture most of the supply chain's optimal profit with a relatively few number of retailers: for $n=5$ the supplier captures at least 80% of the optimal profit and for $n=10$ the supplier captures at least 90%. Mahajan and van Ryzin (2001) also observe that downstream competition can mitigate the need for coordinating contracts.

5.2 Competing newsvendors with market-clearing prices

In the previous model retail competition influences the allocation of demand. In this model, first analyzed by Deneckere, Marvel and Peck (1997), competition influences the retail price. Specifically, the market price depends on the realization of demand and the amount of inventory purchased.

Suppose industry demand can take on one of two states, high or low. Let q be the retailers' total order quantity. When demand is low, the market-clearing price is

$$p_l(q) = (1-q)^+$$

and when demand is high the market-clearing price is

$$p_h(q) = \left(1 - \frac{q}{\theta}\right)^+,$$

for $\theta > 1$. Suppose either demand state is equally likely.

There is a continuum of retailers, indexed on the interval $[0, 1]$. Retailers must order inventory from a single supplier before the realization of the demand is observed. After demand is observed the market-clearing price is determined. Perfect competition is assumed, which means the retailers continue to order inventory until their expected profit is zero. Leftover inventory has no salvage value and the supplier's production cost is zero. Deneckere et al. (1997) show the qualitative insights from this model continue to hold with a continuous demand state space, a general supplier cost function and a general demand function. In another paper, Deneckere, Marvel and Peck (1996) show the qualitative insights also hold if the retailers choose their prices before the realization of demand. (In that model all demand is allocated to the retailer with the lowest price, and any residual demand is subsequently allocated to the retailer with the second lowest price, etc.)

To set a benchmark, suppose a single monopolist controls the entire system. In this situation the monopolist can choose how much of her inventory to sell on the market after demand is observed. At that point the cost of inventory is sunk, so the monopolist maximizes revenue: in the low-demand state the monopolist sells $q = 1/2$ at price $p_l(1/2) = 1/2$; and in the high-demand state $q = \theta/2$ with the same price, $p_h(\theta/2) = 1/2$. So the inventory order should be one of those two quantities. Given the production cost is zero, ordering $\theta/2$ units is optimal.[23] Furthermore, the monopolist sells her

---

[23] The monopolist is actually indifferent between $\theta/2$ and a greater amount. A positive production cost would eliminate that result. But that is not an interesting issue. The important result is that the supply chain optimal order quantity is the greater amount.

entire stock in the high-demand state, but in the low-demand state the monopolist does not sell some of her inventory. The monopolist's expected profit is

$$\Pi^\circ = (1/2)p_l(1/2)(1/2) + (1/2)p_h(\theta/2)(\theta/2) = \frac{1+\theta}{8}.$$

Now consider the system in which the supplier sells to the perfectly competitive retailers with just a wholesale-price contract. The retailer's expected profit is

$$\frac{1}{2}p_l(q)q + \frac{1}{2}p_h(q)q - wq = \begin{cases} \frac{1}{2}q(2 - q - q/\theta) - wq & q \leq 1 \\ \frac{1}{2}q(1 - q/\theta) - wq & q > 1 \end{cases}.$$

Let $q_1(w)$ be the quantity that sets the above profit to zero when $q \leq 1$, which is the equilibrium outcome due to perfect competition:

$$q_1(w) = \frac{2\theta}{1+\theta}(1-w).$$

For $q_1(w) \leq 1$ to hold, it must be that $w \geq (1/2) - 1/(2\theta)$. Let $q_2(w)$ be the quantity that sets the above profit to zero when $q > 1$,

$$q_2(w) = \theta(1 - 2w).$$

For $q_2(w) > 1$ to hold it must be that $w < (1/2) - 1/(2\theta)$.

Let $\pi_s(w)$ be the supplier's profit. From the above results,

$$\pi_s(w) = \begin{cases} q_1(w)w & w \geq (1/2) - 1/(2\theta) \\ q_2(w)w & \text{otherwise} \end{cases}.$$

Let $w^*(\theta)$ be the supplier's optimal wholesale price:

$$w^*(\theta) = \begin{cases} \frac{1}{2} & \theta \leq 3 \\ \frac{1}{4} & \text{otherwise} \end{cases}$$

and

$$\pi_s(w^*(\theta)) = \begin{cases} \dfrac{\theta}{2(1+\theta)} & \theta \leq 3 \\ \frac{1}{8}\theta & \text{otherwise} \end{cases}.$$

So when $\theta \leq 3$ the retailers order

$$q_1(w^*(\theta)) = \frac{\theta}{1+\theta}$$

and the market-clearing prices are

$$p_l(q_1(w^*(\theta))) = \frac{1}{1+\theta}, \quad p_h(q_1(w^*(\theta))) = \frac{\theta}{1+\theta}.$$

When $\theta > 3$ the retailers order

$$q_2(w^*(\theta)) = \frac{\theta}{2}$$

and the market-clearing prices are

$$p_l(q_2(w^*(\theta))) = 0, \quad p_h(q_2(w^*(\theta))) = \frac{1}{2}.$$

No matter the value of θ $\pi_s(w^*(\theta)) < \Pi^\circ$, so the supplier does not capture the maximum possible profit with a wholesale-price contract. When $\theta \leq 3$ the supplier falls short because the retailers sell too much in the low-demand state. To mitigate those losses the retailers order less than the optimal quantity, but then they are unable to sell enough in the high-demand state. When $\theta > 3$ the supplier falls short because the retailers sell too much in the low-demand state even though they sell the optimal amount in the high-demand state. Hence, in either case the problem is that competition leads the retailers to sell too much in the low-demand state. Recall, the monopolist does not sell all of her inventory in the low-demand state, but the perfectly competitive retailers cannot be so restrained.

To earn a higher profit the supplier must devise a mechanism to prevent the low-demand state market-clearing price from falling below $1/2$. In short, the supplier must curtail the destructive competition that results from having more inventory than the system needs. Deneckere et al. (1997) propose the supplier implements resale price maintenance: the retailers may not sell below a stipulated price. [For other research on resale price maintenance, see Ippolito (1991), Shaffer (1991) and Chen (1999b).] Let $\bar{p}$ be that price. When $\bar{p}$ is above the market-clearing price the retailers have unsold inventory, so demand is allocated among the retailers. Assume demand is allocated so that each retailer sells a constant fraction of his order quantity, i.e., proportional allocation.

Given the optimal market-clearing price is always 1/2, the search for the optimal resale price maintenance contract should begin with $\bar{p} = 1/2$.[24] Let $q(t)$ be the order quantity of the tth retailer and let $\pi_r(t)$ be the tth retailer's expected profit. Assume the retailers' total order quantity equals $\theta/2$, i.e.,

$$\int_0^1 q(t)\,dt = \frac{\theta}{2}. \tag{5.4}$$

Hence, the market price in either demand state is 1/2. We later confirm that the retailers indeed order $\theta/2$ in equilibrium. Evaluate the tth retailer's expected profit:

$$\pi_r(t) = -q(t)w + \frac{1}{2}\left(\frac{1/2}{\theta/2}\, q(t)\right)\bar{p} + \frac{1}{2}q(t)\bar{p}:$$

the retailer sells $(1/2)q(t)/(\theta/2)$ in the low-demand state and sells $q(t)$ in the high-demand state. Simplify the above profit:

$$\pi_r(t) = q(t)\left(\frac{1+\theta}{4\theta} - w\right).$$

so the supplier can charge

$$\bar{w} = \frac{1+\theta}{4\theta}.$$

We must now confirm the retailers indeed order $\theta/2$ given that wholesale price. Say the retailers order $1/2 < q < \theta/2$, so the tth retailer's expected profit is

$$-q(t)w + \frac{1}{2}\left(\frac{1/2}{q}\,q(t)\right)\bar{p} + \frac{1}{2}q(t)\left(1 - \frac{q}{\theta}\right).$$

The above is decreasing in the relevant interval and equals 0 when the wholesale price is $\bar{w}$. So with the $(\bar{p}, \bar{w})$ resale price maintenance contract the retailers order $q = \theta/2$, the optimal quantity is sold in either state and the retailers' expected profit is zero.[25] Hence, the supplier earns Π^o with that contract.

---

[24] The optimal market-clearing price is independent of the demand realization because the demand model, $q = \theta(1-p)$, has a multiplicative shock, θ. The optimal market-clearing price would differ across states with an additive shock, e.g., if the demand model were $q = \theta - p$. In that case resale price maintenance could only coordinate the supply chain if the resale price were state dependent.

[25] Given $(\bar{p}, \bar{w})$, any $q(t)$ that satisfies Eq. (5.4) is an equilibrium, i.e., there are an infinite number of equilibria. The authors do not specify how a particular equilibrium would be chosen.

Resale price maintenance prevents destructive competition in the low-demand state, but there is another approach to achieve the same objective. Suppose the supplier offers a buyback contract with $b=1/2$. Since retailers can earn $b=1/2$ on each unit of inventory, the market price cannot fall below $1/2$: for the market price to fall below $1/2$ it must be that some retailers are willing to sell below $1/2$, but that is not rational if the supplier is willing to give $1/2$ on all unsold units. Therefore, the retailers sell at most $1/2$ in the low-demand state and $\theta/2$ in the high-demand state.

The retailers' profit with a buyback contract is

$$\frac{1}{2}(p_l(1/2)(1/2) + b(q-1/2)) + \frac{1}{2}(p_h(q)q) - qw,$$

which simplifies to

$$q\left(\frac{3}{4} - w - \frac{q}{2\theta}\right).$$

(That profit assumes $1/2 < q < \theta/2$.) The supplier wants the retailers to order $q = \theta/2$. From the above equation the retailers earn a zero profit with $q = \theta/2$ when $w = 1/2$. Hence, the supplier maximizes the system's profit with a buyback that offers a full refund on returns.

Although resale price maintenance and the buyback contract achieve the same objective, the supplier sets a higher wholesale price with the buyback contract, i.e., $1/2 > (1+\theta)/4\theta$: retailers do not incur the cost of excess inventory in the low-demand states with a buyback contract, but they do with resale price maintenance. A buyback contract is also not the same as a revenue-sharing contract in this situation. (Section 2 demonstrates the two contracts are equivalent in the single newsvendor model.) The buyback contract prevents the market-clearing price from falling below $1/2$ in the low-demand state, but revenue sharing does not prevent destructive competition: in the low-demand state the retailers have no alternative use for their inventory, so they still attempt to sell all of it in the market. [However, revenue sharing does prevent destructive price competition in Dana and Spier (2001) because in their model the retailers incur a marginal cost for each sale rather than each unit purchased.] Those contracts are also different in the single newsvendor model with price-dependent demand (see Section 3). However, in that model the revenue-sharing contract coordinates the supply chain and the buyback contract does not. The key distinction is that in the single newsvendor model the retailer controls the market price, whereas in this competitive model the retailers do not.

It is interesting that a buyback contract coordinates the supply chain in either competitive model even though in the first one the supplier must discourage the retailers from ordering too much and in the second one the supplier must encourage the retailers to order more. To explain this apparent contradiction, the wholesale-price component of the contract always reduces the retailer's order quantity and the buyback component always increases the retailer's order quantity. Thus, depending on the relative strength of those two components, the buyback contract can either increase or decrease the retailers' order quantities.

5.3 Discussion

Retail competition introduces several challenges for supply chain coordination. There may exist a demand-stealing effect which causes each retailer to order more than the supply chain optimal quantity because each retailer ignores how he reduces his competitors' demand. For coordination the supplier needs to *reduce* the retailers' order quantities, which can be done with just a wholesale-price contract above marginal cost. But that wholesale-price contract only provides for one division of the supply chain's profit, and it is not even the supplier's optimal wholesale-price contract. The supplier can do better with a buyback contract and coordinate the supply chain. However, the incremental improvement over the simpler wholesale-price contract decreases quickly as retail competition intensifies. In contrast to the demand-stealing effect, in the presence of the destructive competition effect the supplier needs to *increase* the retailers' order quantities. This occurs when demand is uncertain and the retail price is set to clear the market. When demand is high the retailers earn a profit, but when demand is low deep discounting to clear inventory leads to losses. The retailers anticipate this problem and respond by curtailing their inventory purchase. Both resale price maintenance and buyback contracts prevent deep discounting, and therefore alleviate the problem.

There are several other papers that study supply chain coordination with competing retailers. Padmanabhan and Png (1997) demonstrate a supplier can benefit by mitigating retail competition with a buyback contract even with deterministic demand and less than perfect retail competition. In their model two retailers first order stock and then choose prices. Retailer i sells $q_i = \alpha - \beta p_i - \gamma p_j$, where α, β and γ are constants and $\beta > \gamma$. With just a wholesale-price contract ($b=0$), the retailers price to maximize revenue, because their inventory is sunk. When the supplier offers a full returns policy ($b=w$), the retailers price to maximize profit because unsold inventory can be returned for a full refund. The retailers price more aggressively when they are maximizing revenue. They anticipate this behavior when choosing their order quantity, and so order less when they expect more intense price competition.[26] Thus, for any given wholesale price the

---

[26] When maximizing revenue the retailer chooses a price so that marginal revenue equals zero. When maximizing profit the retailer chooses a price so that marginal revenue equals marginal cost. Thus, the retailer's optimal price is higher when maximizing profit.

retailers order more with the full returns policy. Since demand is deterministic, in neither case does the supplier actually have to accept returns. So the supplier is better off with the full returns policy when demand is deterministic.[27] When demand is stochastic the supplier may not prefer the full returns policy because that policy may induce the retailers to order too much inventory. However, one suspects the supplier could benefit in that situation from a partial return credit, i.e., $b<w$.[28] See Bernstein and Federgruen (1999) for a more complex model with deterministic demand and competing retailers. See Tsay and Agrawal (2000) and Atkins and Zhao (2002) for models with two retailers that compete on price and service.

Several authors have studied coordination when retailers face oligopolistic competition, i.e., they may earn nonzero profit in equilibrium. Even though this is a different type of competition, the demand-stealing effect remains, but establishing the existence and possibly the uniqueness of equilibrium is generally more challenging. Cachon and Lariviere (2000) demonstrate that revenue sharing can coordinate retailers that compete along a single dimension, e.g., quantity-competing retailers or price-competing retailers. Revenue sharing is not successful if the retailers compete both on quantity and price (e.g., a firm's demand depends on its price and possibly its fill rate, for which quantity is taken as a proxy.) However, Dana and Spier (2001) find that revenue sharing does coordinate perfectly competitive price-setting newsvendors. Bernstein and Federgruen (2000) show that a nonlinear form of the price-discount contract coordinates price- and quantity-competing retailers.

Rudi et al. (2001) study a model with two retailers that each face a newsvendor problem. Inventory can be shipped between the retailers for a fee. Those shipments occur after demand is observed, but before demand is lost. Hence, if retailer i has excess inventory and retailer j has excess demand, then some portion of retailer i's excess inventory can be shipped to retailer j to satisfy retailer j's excess demand. At first glance it would appear the redistribution of excess demand that occurs in the Lippman and McCardle (1995) model is qualitatively equivalent to the redistribution of inventory that occurs in this model. One difference is that the firms in the demand redistribution model do not incur an explicit fee for each demand unit moved between the retailers. A second difference is that the firms in the inventory redistribution model control the redistribution process, and so demand is only lost if total demand exceeds total

---

[27] Padmanabhan and Png (1997) state that the full returns policy helps the manufacturer by *increasing* retail competition. They are referring to the competition in the ordering stage, not in the pricing stage. The retailers order more precisely because they anticipate less aggressive competition in the pricing stage.

[28] In a slightly different model of imperfect competition between two retailers, Butz (1997) demonstrates a buyback contract allows the supplier to coordinate the channel.

inventory (assuming the firms set transfer prices so that Pareto improving trades always occur).

Rudi et al. (2001) demonstrate the retailers may either order too much or too little inventory in this model, depending on the transfer prices for redistributed inventory. When the receiving firm must pay the maximum fee (so he is indifferent between receiving the transfer and incurring a lost sale), the firms order too much inventory. Each firm profits from selling his inventory to the other firm, so each firm is biased toward ordering too much. When the receiving firm pays the minimum fee (so the sending firm is indifferent between salvaging excess inventory and shipping it to the other firm), the firms order too little inventory (neither firm profits from excess inventory, but can depend on some portion of the other firm's inventory). Given these two extremes, there exists a set of intermediate transfer prices such that the firms order the optimal amount of inventory.

Rudi et al. (2001) do not include a supplier in their model. The supplier could be a facilitator of the inventory redistribution. For example, in their model the price at which a firm sells excess inventory is the same as the price at which a firm buys excess inventory. But with a supplier those prices need not be the same: the supplier could buy excess inventory at one price (a buyback) and redistribute at a different price (which could differ from the initial wholesale price). The inclusion of the supplier would also change each retailer's inventory problem. In their model each retailer expects to sell only a portion of his excess inventory, because the other retailer purchases only enough to meet his excess demand. With a buyback contract the supplier stands ready to buy all excess inventory at a fixed price. [Dong and Rudi (2001) do study transshipment with a supplier but they only consider the wholesale-price contract.]

Anupindi et al. (2001) study a general inventory redistribution game with multiple locations. They adopt a 'coopetive' analysis: some decisions are analyzed with concepts from cooperative game theory, whereas others implement noncooperative game theory.

Lee and Whang (2002) have a supplier and free inventory redistribution at an intermediate point in the selling season. [With Rudi et al. (2001) the redistribution occurs at the end of the season, i.e., after a perfect demand signal is received.] In their model the redistribution transfer price is the clearing price of a secondary market rather than a price dictated by the retailers or the supplier. They find that the spot market is advantageous to the supplier for low margin items, but not for high margin items. If the supplier cannot control the spot market, then the supplier can attempt to influence the spot market via minimum order quantity requirements or return policies. For example, a return policy will remove inventory from the spot market and thereby raise its price. This is analogous to the use of buybacks to prevent destructive price competition discussed in Section 5.2.

Gerchak and Wang (1999) and Gurnani and Gerchak (1998) consider supply chains with multiple upstream firms rather than multiple downstream firms. In these assembly systems the upstream firms are different suppliers, each producing a component for the manufacturer's product (the downstream firm). Total production is constrained by the supplier with the smallest output and the excess output of the other suppliers is wasted. Hence, as with destructive competition, the suppliers are biased toward producing too little. They study several contracts that encourage the suppliers to increase their production quantities. Bernstein and DeCroix (2002) also study coordination in assembly systems. They discuss how the organization of the assembly structure influences supply chain performance.

Bernstein et al. (2002) demonstrate that coordination of competing retailers with a wholesale-price contract is easier with Vendor Managed Inventory (the supplier controls the retailers' inventory policies and the retailers choose prices) than with standard operations, i.e., the retailers choose prices and inventory policies. This represents a different approach to coordination: instead of aligning incentive via contracts, the firms transfer decision rights. The pros and cons of this approach relative to formal contracts have not been fully explored.

Throughout this section it has been assumed the supplier is independent of all of the retailers. However, in some markets a supplier may choose to own her own retailer or to sell directly to consumers. Tsay and Agrawal (1999) explore the channel conflicts such a move creates. It has also been assumed that the supplier simultaneously offers a contract to all retailers that is observable by all retailers. However, McAfee and Schwarz (1994) argue that a supplier has an incentive to sequentially offer contracts to retailers and to keep these contractual terms secret. Retailers anticipate this behavior, respond accordingly and thereby destroy the effectiveness of some contracts. See the following papers for additional discussion on this issue: O'Brien and Shaffer (1992), Marx and Shaffer (2001a,b, 2002).

6 Coordinating the newsvendor with demand updating

With the standard newsvendor problem the retailer has only one opportunity to order inventory. However, it is reasonable that the retailer might have a second opportunity to order inventory. Furthermore, the retailer's demand forecast may improve between the ordering epochs (Fisher & Raman, 1996). Hence, all else being equal, the retailer would prefer to delay all ordering to the second epoch. But that creates a problem for the supplier. With a longer lead time the supplier may be able to procure components more cheaply and avoid overtime labor. If the supplier always has sufficient capacity to fill the retailer's order, the supplier prefers an earlier rather than later order commitment by the retailer.[29] Thus, supply chain coordination requires the firms to balance the lower cost of early production with the better information afforded by later production. It is shown that coordination is achieved and profits divided with a buyback contract as long as the supplier is committed to a wholesale price for each order epoch.

6.1 Model and analysis

Based on Donohue (2000), consider a model in which the retailer receives a forecast update only once before the start of the selling season. Let $\xi \geq 0$ be the realization of that demand signal. Let $G(\cdot)$ be its distribution function and $g(\cdot)$ its density function. Let $F(\cdot \mid \xi)$ be the distribution function of demand after observing the demand signal. Demand is stochastically increasing in the demand signal, i.e., $F(x \mid \xi_h) < F(x \mid \xi_l)$ for all $\xi_h > \xi_l$. For convenience, let period 1 be the time before the demand signal and let period 2 be the time between the demand signal and the start of the selling season.

Let q_i be the retailer's total order as of period i, i.e., q_1 is the retailer's period 1 order and $q_2 - q_1$ is the retailer's period 2 order. The retailer's period 2 order is placed after observing the demand signal. Within each period the supplier chooses her production after receiving the retailer's order. Early production is cheaper than later production, so let c_i be the supplier's per unit production cost in period i, with $c_1 < c_2$. The supplier charges the retailer w_i per unit for units ordered in period i. In addition, the supplier offers to buyback all unsold units for b per unit. The supplier offers these terms at the start of period 1 and commits to not change the terms.[30] Let p be the retail price. Normalize to zero the salvage value of leftover inventory and any indirect costs due to lost sales.

The supplier does not have a capacity constraint in either period and delivers stock to the retailer at the end of each period. The supplier operates under voluntary compliance, so the supplier may deliver less than the retailer orders.[31] However, the supplier may also produce more in period 1 than the retailer orders. For simplicity, there is no holding cost on inventory carried from period 1 to period 2.

---

[29] This preference is not due to the time value of income, i.e., the supplier prefers an early order even if the retailer pays only upon delivery.

[30] There may be some incentive to alter the terms after the demand signal is received. Suppose the news is good. In that case the supplier may prefer to leave the retailer with a smaller fraction of supply chain profit (if the retailer has a constant minimum acceptable profit) or the supplier may argue that it deserves a larger fraction of the profit as a reward for producing a good product. There is a large literature in economics on renegotiation and its impact on contract design (see Tirole 1986; Demougin 1989; Holden, 1999).

[31] Donohue (2000) assumes forced compliance. She also assumes the supplier offers a buyback contract.

All information is common knowledge. For example, both firms know $F(\cdot \mid \xi)$ as well as $G(\cdot)$. In particular, both firms observe the demand signal at the start of period 2.[32] Both firms seek to maximize expected profit.

Begin with period 2. Let $\Omega_2(q_2 \mid q_1, \xi)$ be the supply chain's expected revenue minus the period 2 production cost:

$$\Omega_2(q_2 \mid q_1, \xi) = pS(q_2 \mid \xi) - c_2 q_2 + c_2 q_1. \tag{6.1}$$

Let $q_2(q_1, \xi)$ be the supply chain's optimal q_2 given q_1 and ξ. Let $q_2(\xi) = q_2(0, \xi)$, i.e., $q_2(\xi)$ is the optimal order if the retailer has no inventory at the start of period 2. Given $\Omega_2(q_2 \mid q_1, \xi)$ is strictly concave in q_2,

$$F(q_2(\xi) \mid \xi) = \frac{p - c_2}{p}. \tag{6.2}$$

$q_2(\xi)$ is increasing in ξ, so it is possible to define the function $\xi(q_1)$ such that

$$F(q_1 \mid \xi(q_1)) = \frac{p - c_2}{p}. \tag{6.3}$$

$\xi(q_1)$ partitions the demand signals into two sets: if $\xi > \xi(q_1)$ then the optimal period 2 order is positive, otherwise it is optimal to produce nothing in period 2.

The retailer also faces in period 2 a standard newsvendor problem, with the modification that the retailer may already own some stock. Let $\pi_2(q_2 \mid q_1, \xi)$ be the retailer's expected revenue minus period 2 procurement cost,

$$\pi_2(q_2 \mid q_1, \xi) = (p - b)S(q_2 \mid \xi) - (w_2 - b)q_2 + w_2 q_1,$$

where assume the supplier delivers the retailer's order in full. (To simplify notation the contract parameters are not included in the arguments of the functions considered in this section.) To coordinate the retailer's period 2 decision, choose contract parameters with $\lambda \in [0, 1]$ and

$$p - b = \lambda p$$
$$w_2 - b = \lambda c_2$$

---

[32] The supplier does not have to observe the signals directly if the supplier knows $F(\cdot \mid \xi)$. In that case the supplier can infer ξ from the retailer's order quantity because $F(\cdot \mid \xi)$ is strictly decreasing in ξ. See Brown (1999) for a model in which the upstream firm is not able to use the downstream's order quantity to exactly infer the downstream firm's demand signal.

Not surprisingly, those parameters are analogous to the coordinating buyback parameters in the single-period newsvendor model. With any of those contracts

$$\pi_2(q_2 \mid q_1, \xi) = \lambda(\Omega_2(q_2 \mid q_1, \xi) - c_2 q_1) + w_2 q_1.$$

Thus, $q_2(q_1, \xi)$ is also the retailer's optimal order, i.e., the contract coordinates the retailer's period 2 decision.

Now consider whether the supplier indeed fills the retailer's entire period 2 order. Let x be the total inventory in the supply chain at the start of period 2, $x \geq q_1$. The supplier's inventory at the start of period 2 is $x - q_1$. Let y be the inventory at the retailer after the supplier's delivery in period 2. The supplier completely fills the retailer's order when $y = q_2$. The supplier clearly delivers the retailer's full order when $x \geq q_2$ because there is no reason to partially fill the retailer's order and have leftover inventory. If $q_2 > x$, the supplier must produce additional units to deliver the retailer's complete order. Let $\Pi_2(y \mid x, q_1, \xi)$ be the supplier's profit, where $x \leq y \leq q_2$,

$$\Pi_2(y \mid x, q_1, q_2, \xi) = bS(y \mid \xi) - by + w_2(y - q_1) - (y - x)c_2$$
$$= (1 - \lambda)(\Omega_2(y \mid q_1, \xi) - c_2 q_1) + c_2 x - w_2 q_1$$

where the above follows from the contract terms, $w_2 = \lambda c_2 + b$. Given $q_2 > x$, the supplier fills the retailer's order entirely as long as $q_2 \leq q_2(q_1, \xi)$, i.e., the supplier does not satisfy the retailer if the retailer happens to irrationally order too much. Therefore, in period 2 the retailer orders the supply chain optimal quantity and the supplier fills the order entirely, even with voluntary compliance and no matter how much inventory the supplier carries between periods.

In period 1, assuming a coordinating $\{w_2, b\}$ pair is chosen, the retailer's expected profit is

$$\pi_1(q_1) = -(w_1 - w_2 + \lambda c_2)q_1 + \lambda E[\Omega_2(q_2(q_1, \xi) \mid q_1, \xi)].$$

The supply chain's expected profit is

$$\Omega_1(q_1) = -c_1 q_1 + E[\Omega_2(q_2(q_1, \xi) \mid q_1, \xi)].$$

Choose w_1 so that

$$w_1 - w_2 + \lambda c_2 = \lambda c_1$$

because then

$$\pi_1(q_1) = \lambda \Omega_1(q_1).$$

It follows that the retailer's optimal order quantity equals the supply chain's optimal order quantity, q_1^o, and any portion of the supply chain's profit can be allocated to the retailer. Given $\Omega_1(q_1)$ is strictly concave, q_1^o satisfies:

$$\frac{\partial \Omega_1(q_1^o)}{\partial q_1} = -c_1 + c_2(1 - G(\xi(q_1^o))) + \int_0^{\xi(q_1^o)} pS'(q_1^o \mid \xi) g(\xi) \, d\xi$$
$$= 0. \tag{6.4}$$

With centralized operations it does not matter whether inventory is left at the supplier in period 1 because the supply chain moves all inventory to the retailer in period 2: inventory at the supplier has no chance of selling. With decentralized control supply chain coordination is only achieved if the supplier does not hold inventory between periods: there is no guarantee, even if the retailer orders the optimal period 2 quantity, that the retailer orders all of the supplier's inventory. However, it is quite plausible the supplier might attempt to use cheaper period 1 production to profit from a possible period 2 order.

Assuming the supplier fills the retailer's second-period order (which we earlier confirmed the supplier will do), the supplier's period 2 profit is

$$\Pi_2(x, q_1, q_2, \xi) = bS(q_2 \mid \xi) - bq_2 - (q_2 - x)^+ c_2$$
$$= (1 - \lambda)\Omega_2(q_2 \mid q_1, \xi) - w_2 q_2 + x c_2 - (x - q_2)^+ c_2.$$

Given that $q_2 \geq q_1$, the above is strictly increasing in x for $x \leq q_1$. Hence, the supplier surely produces and delivers the retailer's period 1 order (as long as $q_1 \leq q_1^o$). The supplier's period 1 expected profit is

$$\Pi_1(x \mid q_1) = -c_1 x + E[\Pi_2(x, q_1, q_2, \xi)]$$
$$= -c_1 x + E[(1-\lambda)\Omega_2(q_2 \mid q_1, \xi)] - w_2 q_2 + x c_2$$
$$- c_2 \int_0^{\xi(x)} (x - q_2(\xi)) g(\xi) \, d\xi.$$

It follows that

$$\frac{\partial \Pi_1(x \mid q_1)}{\partial x} = -c_1 + c_2(1 - G(\xi(x)))$$

and from Eq. (6.4)

$$\frac{\partial \Pi_1(q_1^o \mid q_1^o)}{\partial x} = -c_1 + c_2(1 - G(\xi(q_1^o))) < 0.$$

Hence, with a coordinating $\{w_1, w_2, b\}$ contract the supplier produces just enough inventory to cover the retailer's period 1 order. Overall, those contracts coordinate the supply chain and arbitrarily allocate profits.

Interestingly, with a coordinating contract the supplier's margin in period 2 is actually lower than in period 1:

$$w_2 - c_2 = w_1 - (\lambda c_1 + (1 - \lambda) c_2) < w_1 - c_1.$$

Intuition suggests the supplier should charge a higher margin for the later production since it offers the retailer an additional benefit over early production. Nevertheless, that intuition is incompatible with supply chain coordination (at least with a buyback contract).

6.2 Discussion

Forecast improvements present several challenges for supply chain coordination. Just as in the simpler single-period model, the retailer must be given incentives to order the correct amount of inventory given the forecast update. In addition, the supplier must correctly balance inexpensive early production against more expensive later production. Finally, the decentralized supply chain must be careful about inventory placement, since unlike with centralized operations, inventory is not necessarily moved to the optimal location in the supply chain, i.e., inventory can become 'stranded' at the supplier.

As in the single-location model, a buyback contract does coordinate this supply chain and arbitrarily allocates profit, even with voluntary compliance. Somewhat surprising, the supplier's margin with later production is smaller than her margin with early production, even though later production provides the retailer with a valuable service.

There are a number of useful extensions to this work. Consider a model with the following adjustments. Suppose the supplier at the start of period 1 picks her capacity, K, which costs c_k per unit. The supplier can produce at most K units over the two periods. Let c be the cost to convert one unit of capacity into one product. In this setting the supply chain optimal solution never produces in period 1: given that early production is no cheaper than later production, the supply chain should delay production until after it has the best demand forecast. A slight modification of the model lets the supplier produce K units in each period. In that case there is some incentive to conduct early production because then the total amount of inventory available in period 2 increases.

In a qualitatively similar model Brown and Lee (1998) study pay-to-delay contracts. With that contract the retailer reserves m units of the supplier's capacity in period 1 for a constant fee per unit. That commits the retailer to purchase at least m units in period 2. They show both firms can be better off with the pay-to-delay contract than with a contract that does not include minimum purchase agreements. However, the pay-to-delay contract cannot coordinate this supply chain. The reason is simple, minimum purchase agreements may result in more production than is optimal given the information signal: if a bad demand signal is observed it may be optimal for the supply chain to produce less than the minimum purchase agreement.

Information acquisition occurs exogenously in both the Brown and Lee (1998) and Donohue (2000), which is reasonable as long as the information is learned before the selling season starts. But suppose the firms' had a replenishment opportunity in the middle of the season. In that case early sales provides information on future sales. However, demand equals sales only if the retailer does not run out of inventory at the start of the season. See Barnes-Schuster, Bassok and Anupindi (2002) for a model in which the demand signal may be truncated due to lost sales. However, that is not the only additional complication in their model. The optimal solution has inventory held at the supplier between periods, hence coordination requires that inventory not be stranded at the supplier at the start of the second period. See Lu, Song and Regan (2002) for another model with midseason replenishment opportunity.

In Kouvelis and Gutierrez (1997) demand occurs in each period, with period 1 demand being the primary market demand and period 2 demand being the secondary market demand. Leftover inventory from period 1 can either be salvaged in the primary market or moved to the secondary market. That decision depends on the realization of the exchange rate between the two markets' currencies. Hence, the information learned between periods is not a demand signal, as in Donohue (2000), but rather the realization of period 2 effective production cost. They coordinate this supply chain (with one manager responsible for each market's decisions) using a nonlinear scheme. Kouvelis and Lariviere (2000) show an internal market can also coordinate this supply chain (see Section 10).

van Mieghem (1999) studies forecast updating with several additional twists. In his model the downstream firm is the manufacturer and the upstream firm is the subcontractor. At issue is the production of a component that is part of the manufacturer's product. The manufacturer has only one market for his product, but the subcontractor can sell her component either to the manufacturer or to an outside market. (But the subcontractor has access to the manufacturer's market only via the manufacturer, i.e., the subcontractor cannot sell directly to that market.) Both markets have random demand. Both firms choose a capacity level in period 1, where the manufacturer's capacity produces the component. At the start of period 2 the firms observe demand in their respective markets and then convert capacity into final output. Hence, like Donohue (2000), the firms receive a demand signal between their early decision (how much capacity to construct) and their later decision (how much to produce), albeit in van Mieghem (1999) it is a perfect demand signal. Unlike Donohue (2000), in van Mieghem (1999) the downstream firm has production capability, the upstream firm has a random opportunity cost for capacity not sold to the downstream firm, and buyback contracts are not considered. See Milner and Pinker (2000) for another model with early capacity decisions and later forecast adjustments.

van Mieghem (1999) and Donohue (2000) also differ on what they assume about the firms ability to commit to future actions. In Donohue (2000) the supplier commits to a period 2 wholesale price, whereas in van Mieghem (1999) the firms renegotiate their agreement between periods. Anand, Anupindi and Bassok (2001) demonstrate that a supplier's inability to commit to future prices may cause a retailer to carry inventory purely for strategic reasons. They have a two-period model with deterministic demand. The supplier's period 2 wholesale price is decreasing in the retailer's period 1 inventory, thereby providing the retailer with a motivation to carry inventory. See Gilbert and Cvsa (2000) for another model in which the ability to commit to future wholesale prices matters.

Future research should consider a model with endogenous information acquisition, i.e., the firms must exert effort to improve their demand forecasts. Should one firm exert the effort or should both firms undertake forecast improvement activities? To the best of my knowledge, that issue has not been explored. There is also the possibility the firms could have different forecasts: the firms could exert a different amount of effort toward forecasting or the firms could have different sources of forecasting information. If there are asymmetric forecasts, supply chain performance may improve via forecast sharing: see Section 10 for a discussion of that issue.

7 Coordination in the single-location base-stock model

This section considers a model with perpetual demand and many replenishment opportunities. Hence, the newsvendor model is not appropriate. Instead, the base-stock inventory policy is optimal: with a base-stock policy a firm maintains its inventory position (on-order plus in-transit plus on-hand inventory minus backorders) at a constant base-stock level. It is assumed, for tractability, that demand is backordered, i.e., there is no lost sale. As a result, expected demand is constant (i.e., it does not depend on the retailer's base-stock level). Optimal performance is now achieved by minimizing total supply chain costs: the holding cost of inventory and the backorder penalty costs. In this model the supplier incurs no holding costs, but the supplier does care about the availability of her product at the retail level. To model that preference, it is assumed that a backorder at the retailer incurs a cost on the supplier. Since the retailer does not consider that cost when choosing a base-stock level, it is shown the retailer chooses a base stock level that is lower than optimal for the supply chain, which means the retailer carries too little inventory. Coordination is achieved and costs are arbitrarily allocated by providing incentives to the retailer to carry more inventory.

This model also provides a useful building block for the two-location model considered in the next section.

7.1 Model and analysis

Suppose a supplier sells a single product to a single retailer. Let L_r be the lead time to replenish an order from the retailer. The supplier has infinite capacity, so the supplier keeps no inventory and the retailer's replenishment lead time is always L_r, no matter the retailer's order quantity. (There are two firms, but only the retailer keeps inventory, which is why this is considered a single-location model.) Let $\mu_r = E[D_r]$. Let F_r and f_r be the distribution and density functions of D_r, respectively: assume that F_r is strictly increasing, differentiable and $F_r(0) = 0$, which rules out the possibility that it is optimal to carry no inventory.

The retailer incurs inventory holding costs at rate $h_r > 0$ per unit of inventory. For analytical tractability, demand is backordered if stock is not available. [There are a few papers that consider multiple demand periods with lost sales: e.g., Moses and Seshadri (2000), Duenyas and Tsai (2001) and Anupindi and Bassok (1999).] The retailer incurs backorder penalty costs at rate $\beta_r > 0$ per unit backordered. The supplier has unlimited capacity, so the supplier need not carry inventory. However, suppose the supplier incurs backorder penalty costs at rate $\beta_s > 0$ per unit backordered at the *retailer*. In other words, the supplier incurs a cost whenever a customer wants to purchase the supplier's product from the retailer but the retailer does not have inventory. This cost reflects the supplier's preference for maintaining sufficient availability of her product at the retail level in the supply chain. Let $\beta = \beta_r + \beta_s$, so β is the total backorder cost rate incurred by the supply chain. Cachon and Zipkin (1999) adopt the same preference structure for the retailer, but in their model the supplier has limited inventory. Their model is discussed in the next section. Narayanan and Raman (1997) adopt a different preference structure: they assume a fixed fraction of consumers who experience a stockout for the supplier's product choose to purchase another product from a different supplier at the same retailer. Hence, the backorder cost for the supplier is her lost margin, whereas the backorder cost for the retailer is the difference in the margin between selling the supplier's product and selling the other product (which may actually benefit the retailer).

Sales occur at a constant rate μ_r, due to the backorder assumption, no matter how the firms manage their inventory. As a result, the firms are only concerned with their costs. Both firms are risk neutral. The retailer's objective is to minimize his average inventory holding and backorder cost per unit time. The supplier's objective is to minimize her average backorder cost per unit time.

Define the retailer's inventory level to equal inventory in-transit to the retailer plus the retailer's on-hand inventory minus the retailer's backorders. (This has also been called the effective inventory position.) The retailer's inventory position equals his inventory level plus on-order inventory (inventory ordered, but not yet shipped). Since the supplier immediately ships all orders, the retailer's inventory level and position are identical in this setting.

Let $I_r(y)$ be the retailer's expected inventory at time $t+L_r$ when the retailer's inventory level is y at time t:

$$I_r(y) = \int_0^y (y-x)f_r(x)\,dx = \int_0^y F_r(x)\,dx. \tag{7.1}$$

Let $B_r(y)$ be the analogous function that provides the retailer's expected backorders:

$$B_r(y) = \int_y^\infty (x-y)f_r(x)\,dx = \mu_r - y + I_r(y). \tag{7.2}$$

Inventory is monitored continuously, so the retailer can maintain a constant inventory position. In this environment it can be shown that a base-stock policy is optimal. With that policy the retailer continuously orders inventory so that his inventory position always equals his chosen base-stock level, s_r.

Let $c_r(s_r)$ be the retailer's average cost per unit time when the retailer implements the base-stock policy s_r:

$$c_r(s_r) = h_r I_r(s_r) + \beta_r B_r(s_r)$$
$$= \beta_r(\mu_r - s_r) + (h_r + \beta_r)I_r(s_r).$$

Given the retailer's base-stock policy, the supplier's expected cost function is

$$c_s(s_r) = \beta_s B_r(s_r)$$
$$= \beta_s(\mu_r - s_r + I_r(s_r)).$$

Let $c(s_r)$ be the supply chain's expected cost per unit time,

$$c(s_r) = c_r(s_r) + c_s(s_r)$$
$$= \beta(\mu_r - s_r) + (h_r + \beta)I_r(s_r). \tag{7.3}$$

$c(s_r)$ is strictly convex, so there is a unique supply chain optimal base-stock level, s_r^o. It satisfies the following critical ratio equation

$$I_r'(s_r^o) = F_r(s_r^o) = \frac{\beta}{h_r + \beta}. \tag{7.4}$$

Let s_r^* be the retailer's optimal base-stock level. The retailer's cost function is also strictly convex, so s_r^* satisfies

$$F_r(s_r^*) = \frac{\beta_r}{h_r + \beta_r}.$$

Given $\beta_r < \beta$, it follows from the above two expressions that $s_r^* < s_r^o$, i.e., the retailer chooses a base-stock level that is less than optimal. Hence, channel coordination requires the supplier to provide the retailer with an incentive to raise his base-stock level.

Suppose the firms agree to a contract that transfers from the supplier to the retailer at every time t

$$t_I I_r(y) + t_B B_r(y)$$

where y is the retailer's inventory level at time t and t_I and t_B are constants. Furthermore, consider the following set of contracts parameterized by $\lambda \in (0, 1]$,

$$t_I = (1 - \lambda)h_r$$
$$t_B = \beta_r - \lambda\beta.$$

(Here we choose to rule out $\lambda = 0$ since then any base-stock level is optimal.) Given one of those contracts, the retailer's expected cost function is now

$$c_r(s_r) = (\beta_r - t_B)(\mu_r - s_r) + (h_r + \beta_r - t_I - t_B)I_r(s_r). \tag{7.5}$$

The contract parameters have been chosen so that

$$\beta_r - t_B = \lambda\beta > 0$$

and

$$h_r + \beta_r - t_I - t_B = \lambda(h_r + \beta) > 0.$$

It follows from Eqs. (7.3) and (7.5) that with these contracts

$$c_r(s_r) = \lambda c(s_r). \tag{7.6}$$

Hence, s_r^o minimizes the retailer's cost, i.e., those contracts coordinate the supply chain. In addition, those contracts arbitrarily allocate costs between the firms, with the retailer's share of the cost increasing in the parameter λ. Note, the λ parameter is not explicitly incorporated into the contract, i.e., it is merely used for expositional clarity.

Now consider the sign of the t_I and the t_B parameters. Since the contract must induce the retailer to choose a higher base-stock level, it is natural to conjecture $t_I > 0$, i.e., the supplier subsidizes the retailer's inventory holding cost. In fact, that conjecture is valid when $\lambda \in (0, 1]$. It is also natural to suppose $t_B < 0$, i.e., the supplier penalizes the retailer for backorders. But $\lambda \in (0, 1]$ implies $t_B \in [-\beta_s, \beta_r)$, i.e., with some contracts the supplier subsidizes the retailer's backorders ($t_B > 0$): in those situations the supplier encourages backorders by setting $t_B > 0$ because without that encouragement the large inventory subsidy leads the retailer to $s_r^* > s_r^o$.

The above analysis is reminiscent of the analysis with the newsvendor model and buyback contracts. This is not a coincidence, because this model is qualitatively identical to the newsvendor model. To explain, begin with the retailer's profit function in the newsvendor model (assuming $c_r = g_r = g_s = v = 0$):

$$\pi_r(q) = pS(q) - wq$$
$$= (p-w)q - pI(q).$$

The retailer's profit has two terms, one that increases linearly in q and the other that depends on the demand distribution. Now let $p = h_r + \beta_r$ and $w = h_r$. In that case,

$$\pi_r(q) = \beta_r q - (h_r + \beta_r)I_r(q) = -c_r(q) + \beta_r \mu_r.$$

Hence, there is no difference between the maximization of $\pi_r(q)$ and the minimization of $c_r(s_r)$.

Now recall that the transfer payment with a buyback contract is $wq - bI(q)$, i.e., there is a parameter (i.e., w) that affects the payment linearly in the retailer's action (i.e., q), and a parameter that influences the transfer payment through a function (i.e., $I(q)$) that depends on the retailer's action and the demand distribution. In this model

$$t_I I_r(y) + t_B B_r(y) = (t_I + t_B)I_r(y) + t_B(\mu_r - y)$$

so t_B is the linear parameter and $t_I + t_B$ is the other parameter. In the buyback contract the parameters work independently. To get the same effect in the base-stock model the supplier could adopt a transfer payment that depends on the retailer's inventory position, s_r, and the retailer's inventory. That contract would yield the same results.

7.2 Discussion

Coordination with the infinite horizon base-stock model is qualitatively the same as coordination in the single-period newsvendor model. In particular, coordination via a holding cost and backorder cost transfer payment is like coordination via a buyback contract. One suspects that quantity-flexibility or sales-rebate like contracts could also coordinate in this setting. Choi, Dai and Song (2002) consider a similar model with the addition of a capacity constraint. They demonstrate that standard service level measures do not provide sufficient control over the firm managing inventory.

8 Coordination in the two-location base-stock model

The two-location base-stock model builds upon the single-location base-stock model discussed in the previous section. Now the supplier no longer has infinite capacity. Instead, she must order replenishments from her source and those replenishment always are filled within L_s time (i.e., her source has infinite capacity). So in this model the supplier enjoys reliable replenishments but the retailer's replenishment lead time depends on how the supplier manages her inventory. Only if the supplier has enough inventory to fill an order does the retailer receive that order in L_r time. Otherwise, the retailer must wait longer than L_r to receive the unfilled portion. That delay could lead to additional backorders at the retailer, which are costly to both the retailer and the supplier, or it could lead to lower inventory at the retailer, which helps the retailer.

In the single-location model the only critical issue is the amount of inventory in the supply chain. In this model the allocation of the supply chain's inventory between the supplier and the retailer is important as well. For a fixed amount of supply chain inventory the supplier always prefers that is more allocated to the retailer, because that lowers both her inventory and backorder costs. (Recall that the supplier is charged for retail backorders.) On the other hand, the retailer's preference is not so clear: less retail inventory means lower holding costs, but also higher backorder costs. There are also subtle interactions with respect to the total amount of inventory in the supply chain. The retailer is biased to carry too little inventory: the retailer bears the full cost of his inventory but only receives a portion of the benefit (i.e., he does not benefit from the reduction in the supplier's backorder cost). On the other hand, there is no clear bias for the supplier because of two effects. First, the supplier bears the cost of his inventory and does not benefit from the reduction in the retailer's backorder cost, which biases the supplier to carry too little inventory. Second, the supplier does not bear the cost of the retailer's inventory (which increases along with the supplier's inventory), which biases the supplier to carry too much inventory. Either bias can dominate, depending on the parameters of the model.

Even though it is not clear whether the decentralized supply chain will carry too much or too little inventory (however, it generally carries too little inventory), it is shown that the optimal policy is never a Nash equilibrium of the decentralized game, i.e., decentralized operation is never optimal. However, the competition penalty (the percent loss in supply chain performance due to decentralized decision making) varies considerably: in some cases the competition penalty is relatively small, e.g., less than 5%, whereas in other cases it is considerable, e.g., more than 40%. Therefore, the need for coordinating contracts is not universal.

8.1 Model

Let h_s, $0 < h_s < h_r$, be the supplier's per unit holding cost rate incurred with on-hand inventory. (When $h_s \geq h_r$ the optimal policy does not carry inventory at the supplier and when $h_s \leq 0$ the optimal policy has unlimited supplier inventory. Neither case is interesting.) The firms' operating decisions have no impact on the amount of in-transit inventory, so no holding cost is charged for either the supplier's or the retailer's pipeline inventory. Let $D_s > 0$ be demand during an interval of time with length L_s. (As in the single-location model, $D_s > 0$ ensures that the supplier carries some inventory in the optimal policy.) Let F_s and f_s be the distribution and density functions of that demand. As with the retailer, assume F_s is increasing and differentiable. Let $\mu_s = E[D_s]$. Retail orders are backordered at the supplier but there is no explicit charge for those backorders. The supplier still incurs per unit backorder costs at rate β_s for backorders at the retailer. The comparable cost for the retailer is still β_r. Even though there are no direct consequences to a supplier backorder, there are indirect consequences: lower retailer inventory and higher retailer backorders.

Both firms use base-stock policies to manage inventory. With a base-stock policy firm $i \in \{r, s\}$ orders inventory so that its inventory position remains equal to its base-stock level, s_i. (Recall that a firm's inventory level equals on-hand inventory, minus backorders plus in-transit inventory and a firm's inventory position equals the inventory level plus on-order inventory.) These base-stock policies operate only with local information, so neither firm needs to know the other firm's inventory position.

The firms choose their base-stock levels once and simultaneously. The firms attempt to minimize their average cost per unit time. (Given that one firm uses a base-stock policy, it is optimal for the other firm to use a base-stock policy.) They are both risk neutral. There exists a pair of base-stock levels, $\{s_r^o, s_s^o\}$, that minimize the supply chain's cost. [In fact, that policy is optimal among all possible policies. See Chen and Zheng (1994) for an elegant proof.] Hence, it is feasible for the firms to optimize the supply chain, but incentive conflicts may prevent them from doing so.

This model is essentially the same as the one considered by Cachon and Zipkin (1999).[33] However, the notation differs somewhat. (Caution, in some cases the notation is inconsistent.)

The first step in the analysis of this model is to evaluate each firm's average cost. The next step evaluates the Nash equilibrium base-stock levels. The third step identifies the optimal base-stock levels and compares them to the Nash equilibrium ones. The final step explores incentive structures to coordinate the supply chain. The remaining portions of this section describe alternative coordination techniques, summarize the results and discuss research in related models.

8.2 Cost functions

As in the single-location model, $c_r(y)$, $c_s(y)$ and $c(y)$ are the firms' and the supply chain's expected costs incurred at time $t + L_r$ at the retail level when the retailer's inventory level is y at time t. However, in the two-location model the retailer's inventory level does not always equal the retailer's inventory position, s_r, because the supplier may stockout. Let $c_i(s_r, s_s)$ be the average rate at which firm i incurs costs at the retail level and $c(s_r, s_s) = c_r(s_r, s_s) + c_s(s_r, s_s)$. To evaluate c_i, note that at any given time t the supplier's inventory position is s_s (because the supplier uses a base-stock policy). At time $t + L_s$ either the supplier's on-hand inventory is $(s_s - D_s)^+$ or the supplier's backorder equals $(D_s - s_s)^+$. Therefore, the retailer's inventory level at time $t + L_s$ is $s_r - (D_s - s_s)^+$. So

$$c_i(s_r, s_s) = F_s(s_s) c_i(s_r) + \int_{s_s}^{\infty} c_i(s_r + s_s - x) f_s(x) \, dx :$$

at time $t + L_s$ the supplier can raise the retailer's inventory level to s_r with probability $F_s(s_s)$, otherwise the retailer's inventory level equals $s_r + s_s - D_s$.

---

[33] Cachon and Zipkin (1999) assume periodic review, whereas this model assumes continuous review. That difference is inconsequential. In addition to the local inventory measure, they allow firms to use echelon inventory to measure their inventory position: a firm's echelon inventory position equals all inventory at the firm or lower in the supply chain plus on-order inventory minus backorders at the retail level. For the retailer there is no difference between the local and echelon measures of inventory position, but those measures are different for the supplier. They allow for either $\beta_s = 0$ or $\beta_r = 0$, but those special cases are not treated here. They include holding costs for pipeline inventory into their cost functions. Finally, they also study the Stackelberg version of this game (the firms choose sequentially instead of simultaneously).

Based on the analogous reasoning, let $I_r(s_r, s_s)$ and $B_r(s_r, s_s)$ be the retailer's average inventory and backorders given the base-stock levels:

$$I_r(s_r, s_s) = F_s(s_s)I_r(s_r) + \int_{s_s}^{\infty} I_r(s_r + s_s - x)f_s(x)\,dx,$$

$$B_r(s_r, s_s) = F_s(s_s)B_r(s_r) + \int_{s_s}^{\infty} B_r(s_r + s_s - x)f_s(x)\,dx.$$

Let $\pi_i(s_r, s_s)$ be firm i's total average cost rate. Since the retailer only incurs costs at the retail level,

$$\pi_r(s_r, s_s) = c_r(s_r, s_s).$$

Let $I_s(s_s)$ be the supplier's average inventory. Analogous to the retailer's functions (defined in the previous section)

$$I_s(y) = \int_0^y F_s(x)\,dx.$$

The supplier's average cost is

$$\pi_s(s_r, s_s) = h_s I_s(s_s) + c_s(s_r, s_s).$$

Let $\Pi(s_r, s_s)$ be the supply chain's total cost, $\Pi(s_r, s_s) = \pi_r(s_r, s_s) + \pi_s(s_r, s_s)$.

8.3 Behavior in the decentralized game

Let $s_i(s_j)$ be an optimal base-stock level for firm i given the base-stock level chosen by firm j, i.e., $s_i(s_j)$ is firm i's best response to firm j's strategy. Differentiation of each firm's cost function demonstrates that each firm's cost is strictly convex in its base-stock level, so each firm has a unique best response.

With a Nash equilibrium pair of base stocks, $\{s_r^*, s_s^*\}$, neither firm has a profitable unilateral deviation, i.e.,

$$s_r^* = s_r(s_s^*) \quad \text{and} \quad s_s^* = s_s(s_r^*).$$

Existence of a Nash equilibrium is not assured, but in this game existence of a Nash equilibrium follows from the convexity of the firm's cost functions (see Friedman, 1986). (Technically it is also required that the firms' strategy spaces have an upper bound. Imposing that bound has no impact on the analysis.) In fact, there exists a unique Nash equilibrium. To demonstrate uniqueness begin by bounding each player's feasible strategy space, i.e., the set of strategies a player may choose. For the retailer it is not difficult to show that $s_r(s_s) > \hat{s}_r > 0$, where $\hat{s}_r$ minimizes $c_r(y)$, i.e.,

$$F_r(\hat{s}_r) = \frac{\beta}{h_r + \beta}.$$

In other words, if the retailer were to receive perfectly reliable replenishments the retailer would choose $\hat{s}_r$, so the retailer certainly does not choose $s_r \leq \hat{s}_r$ if replenishments are unreliable. (In other words, $\hat{s}_r$ is optimal for the retailer in the single-location model discussed in the previous section.) For the supplier, $s_s(s_r) > 0$ because $\partial \pi_s(s_r, s_s)/\partial s_s < 0$ given $F_s(s_s) = 0$ and $c'_s(y) < 0$.

Uniqueness of the Nash equilibrium holds if for the feasible strategies, $s_r > \hat{s}_r$ and $s_s > 0$, the best reply functions are contraction mappings (see Friedman, 1986), i.e.,

$$|s'_i(s_j)| < 1. \tag{8.1}$$

From the implicit function theorem

$$s'_r(s_s) = -\frac{\int_{s_s}^{\infty} c''_r(s_r + s_s - x) f_s(x)\,dx}{F_s(s_s) c''_r(s_r) + \int_{s_s}^{\infty} c''_r(s_r + s_s - x) f_s(x)\,dx}$$

and

$$s'_s(s_r) = -\frac{\int_{s_s}^{\infty} c''_s(s_r + s_s - x) f_s(x)\,dx}{[h_s - c'_s(s_r)] f_s(s_s) + \int_{s_s}^{\infty} c''_s(s_r + s_s - x) f_s(x)\,dx}.$$

Given $s_r > \hat{s}_r$ and $s_s > 0$, it follows that $c''_r(x) > 0$, $c'_s(y) < 0$, $c''_s(y) > 0$ and $F_s(s_s) > 0$. Hence, Eq. (8.1) holds for both the supplier and the retailer.

A unique Nash equilibrium is quite convenient, since that equilibrium is then a reasonable prediction for the outcome of the decentralized game. (With multiple equilibria it is not clear the outcome of the game would even be an equilibrium, since the players may choose strategies from different equilibria.) Hence, the competition penalty is an appropriate measure of the gap between optimal performance and decentralized performance, where the competition penalty is defined to be

$$\frac{\Pi(s_s^*, s_r^*) - \Pi(s_s^o, s_r^o)}{\Pi(s_s^o, s_r^o)}.$$

In fact, there always exists a positive competition penalty, i.e., decentralized operations always leads to suboptimal performance in this game.[34] To explain, note that the retailer's marginal cost is always greater than the supply chain's

$$\frac{\partial c_r(s_r, s_s)}{\partial s_r} > \frac{\partial c(s_r, s_s)}{\partial s_r}$$

because $c'_r(s_r) > c'(s_r)$. Since both $c_r(s_r, s_s)$ and $c(s_r, s_s)$ are strictly convex, it follows that, for any s_s, the retailer's optimal base stock is always lower than the supply chain's optimal base stock. Hence, even if the supplier chooses s_s^o, the retailer does not choose s_r^o, i.e., $s_r(s_s^o) < s_r^o$.

Although the Nash equilibrium is not optimal, Cachon and Zipkin (1999) find in a numerical study that the magnitude of the competition penalty depends on the parameters of the model. When the firms' backorder penalties are the same (i.e., $\beta_r/\beta_s = 1$) the median competition penalty for their sample is 5% and the competition penalty is no greater than 8% in 95% of their observations. However, very large competition penalties are observed when either $\beta_r/\beta_s < 1/9$ or $\beta_r/\beta_s > 9$. The retailer does not have a strong concern for customer service when $\beta_r/\beta_s < 1/9$, and so the retailer tends to carry far less inventory than optimal. Since the supplier does not have direct access to customers, the supplier can do little to prevent backorders in that situation, and so the supply chain cost is substantially higher than need be. In the other extreme, $\beta_r/\beta_s > 9$, the supplier cares little about customer service, and thus does not carry enough inventory. In that situation the retailer can still prevent backorders, but to do so requires a substantial amount of inventory at the retailer to account for the supplier's long lead time. The supply chain's cost is substantially higher than optimal if the optimal policy has the supplier carry inventory to provide reliable replenishments to the retailer. However, there are situations in which the optimal policy does not require the supplier to carry much inventory: either the supplier's holding cost is nearly as high as the retailer's (in which case keeping inventory at the supplier gives little holding cost advantage) or if the supplier's lead time is short (in which case the delay due to a lack of inventory at the supplier is negligible). In those cases the competition penalty is relatively minor.

In the single-location model decentralization always leads to less inventory than optimal for the supply chain. In this setting the interactions between the firms are more complex, and so decentralization generally leads to too little inventory, but not always. Since the retailer's backorder cost rate is lower than the supplier's backorder cost rate, for a fixed s_s the retailer always carries too little inventory, which certainly contributes to a less than optimal amount of inventory in the system. However, the retailer is only a part

---

[34] However, when $\beta_s = 0$ the optimal policy is a Nash equilibrium with just the right parameters. See Cachon and Zipkin (1999) for details.

of the supply chain. In fact, from Cachon and Zipkin's (1999) numerical study, the supplier's inventory may be so large that even though the retailer carries too little inventory, the total amount of inventory in the decentralized supply chain may exceed the supply chain's optimal quantity. Suppose β_r is quite small and β_s is quite large. In that case the retailer carries very little inventory. To attempt to mitigate the build up of backorders at the retail level the supplier provides the retailer with very reliable replenishments, which requires a large amount of inventory, an amount that may lead to more inventory in the supply chain than optimal.

The main conclusion from the analysis of the decentralized game is that the competition penalty is always positive, but only in some circumstances is it very large. It is precisely in those circumstances that the firms could benefit from an incentive scheme to coordinate their actions.

8.4 Coordination with linear transfer payments

Supply chain coordination in this setting is achieved when $\{s_r^o, s_s^o\}$ is a Nash equilibrium. Cachon and Zipkin (1999) propose a set of contracts to achieve that goal, but they do not answer two important questions: do the contracts allow for an arbitrary division of the supply chain's cost and is the optimal solution a unique Nash equilibrium? This section studies their contracts and answers both of those questions.

In the single-location model the supplier coordinates the supply chain with a contract that has linear transfer payments based on the retailer's inventory and backorders. Suppose the supplier offers the same arrangement in this model with the addition of a transfer payment based on the supplier's backorders:

$$t_1 I_r(s_r, s_s) + t_B^r B_r(s_r, s_s) + t_B^s B_s(s_s),$$

where t_1, t_B^r and t_B^s are constants and $B_s(s_s)$ is the supplier's average backorder:

$$B_s(y) = \mu_s - y + I_s(y).$$

Recall that a positive value for the above expression represents a payment from the supplier to the retailer and a negative value represents a payment from the retailer to the supplier. While both firms can easily observe $B_s(s_s)$, an information system is needed for the supplier to verify the retailer's inventory and backorder.

The first step in the analysis provides some results for the optimal solution. The second step defines a set of contracts and confirms those contracts coordinate the supply chain. Then the allocation of costs is considered. Finally, it is shown that the optimal solution is a unique Nash equilibrium.

The traditional approach to obtain the optimal solution involves reallocating costs so that all costs are preserved. Base stock policies are then optimal and easily evaluated.[35] However, to facilitate the comparison of the optimal policy to the Nash equilibrium of the decentralized game, it is useful to evaluate the optimal base-stock policy without that traditional cost reallocation.[36]

Given $\Pi(s_r, s_s)$ is continuous, any optimal policy with $s_s > 0$ must set the following two marginals to zero

$$\frac{\partial \Pi(s_r, s_s)}{\partial s_r} = F_s(s_s) c'(s_r) + \int_{s_s}^{\infty} c'(s_r + s_s - x) f_s(x) \, dx \qquad (8.2)$$

and

$$\frac{\partial \Pi(s_r, s_s)}{\partial s_s} = F_s(s_s) h_s + \int_{s_s}^{\infty} c'(s_r + s_s - x) f_s(x) \, dx. \qquad (8.3)$$

Since $F_s(s_s) > 0$, there is only one possible optimal policy with $s_s > 0$, $\{\tilde{s}_r^1, \tilde{s}_s^1\}$, where $\tilde{s}_r^1$ satisfies

$$c'(\tilde{s}_r^1) = h_s, \qquad (8.4)$$

and $\tilde{s}_s^1$ satisfies $\partial \Pi(\tilde{s}_r^1, \tilde{s}_s^1)/\partial s_s = 0$. Eq. (8.4) simplifies to

$$F_r(\tilde{s}_r^1) = \frac{h_s + \beta}{h_r + \beta},$$

so it is apparent $\tilde{s}_r^1$ exists and is unique. Since $\Pi(\tilde{s}_r^1, s_s)$ is strictly convex in s_s, $\tilde{s}_s^1$ exists and is unique.

There may also exist an optimal policy with $s_s \leq 0$. In that case the candidate policies are $\{\tilde{s}_r^2, \tilde{s}_s^2\}$, where $\tilde{s}_s^2 \leq 0$, $\tilde{s}_r^2 + \tilde{s}_s^2 = \bar{s}$ and $\bar{s}$ satisfies

$$\int_0^{\infty} c'(\bar{s} - x) f_s(x) \, dx = 0. \qquad (8.5)$$

The above simplifies to

---

[35] Clark and Scarf (1960) pioneered that approach for the finite horizon problem. Federgruen and Zipkin (1984) make the extension to the infinite horizon case.

[36] This procedure does not prove base-stock policies are optimal, it merely finds the optimal base-stock policies. Furthermore, it relies on the continuity of $\Pi(s_r, s_s)$, so it does not trivially extend to discrete demand.

$$\Pr(D_r + D_s \leq \bar{s}) = \frac{\beta}{h_r + \beta},$$

so $\bar{s}$ exists and is unique.

Given that $\Pi(\tilde{s}_r^1, s_s)$ is strictly convex in s_s, $\Pi(\tilde{s}_r^1, \tilde{s}_s^1) < \Pi(\tilde{s}_r^2, \tilde{s}_s^2)$ whenever $\partial \Pi(\tilde{s}_r^1, 0)/\partial s_s < 0$. Since $\partial \Pi(s_r, 0)/\partial s_s$ is increasing in s_r, from Eq. (8.5), that condition holds when $\tilde{s}_r^1 < \bar{s}$, otherwise it does not. Therefore, $\{\tilde{s}_r^1, \tilde{s}_s^1\}$ is the unique optimal policy when $\tilde{s}_r^1 < \bar{s}$, otherwise any $\{\tilde{s}_r^2, \tilde{s}_s^2\}$ is optimal:

$$\{s_r^o, s_s^o\} = \begin{cases} \{\tilde{s}_r^1, \tilde{s}_s^1\} & \tilde{s}_r^1 < \tilde{s}_r^2 \\ \{\tilde{s}_r^2, \tilde{s}_s^2\} & \tilde{s}_r^1 \geq \tilde{s}_r^2 \end{cases}.$$

Now consider the firms' behavior with the following set of contracts parameterized by $\lambda \in (0, 1]$,

$$t_I = (1 - \lambda) h_r, \tag{8.6}$$

$$t_B^r = \beta_r - \lambda \beta, \tag{8.7}$$

$$t_B^s = \lambda h_s \left(\frac{F_s(s_s^o)}{1 - F_s(s_s^o)} \right). \tag{8.8}$$

Cachon and Zipkin (1999) also propose those contracts, but they do not include the λ parameter.

The retailer's cost function, adjusted for the above contracts is

$$c_r(y) = (h_r - t_I) I_r(y) + (\beta_r - t_B^r) B_r(y) - t_B^s B_s(s_s)$$
$$= \lambda c(y) - t_B^s B_s(s_s)$$

and so

$$\pi_r(s_r, s_s) = \lambda \Pi(s_r, s_s) - t_B^s B_s(s_s). \tag{8.9}$$

Recall $c(s_r, s_s) = c_r(s_r, s_s) + c_s(s_r, s_s)$, so

$$\pi_s(s_r, s_s) = h_s I_s(s_s) + (1 - \lambda) c(s_r, s_s) + t_B^s B_s(s_s)$$
$$= (h_s + t_B^s) I_s(s_s) + (1 - \lambda) c(s_r, s_s) + t_B^s (\mu_s - s_s). \tag{8.10}$$

There are two cases to consider: either $s_s^o > 0$ or $s_s^o = 0$. Take the first case. If $s_s^o > 0$, then Eq. (8.4) implies

$$\frac{\partial \pi_r(s_r^o, s_s^o)}{\partial s_r} = \left(\frac{\lambda}{1-\lambda}\right) \frac{\partial \pi_s(s_r^o, s_s^o)}{\partial s_s} = \lambda \frac{\partial \Pi(s_r^o, s_s^o)}{\partial s_s} = 0.$$

Further, $\pi_r(s_r, s_s)$ is strictly convex in s_r and $\pi_s(s_r^o, s_s)$ is strictly convex in s_s, so $\{s_r^o, s_s^o\}$ is indeed a Nash equilibrium. In fact, it is the unique Nash equilibrium. From the implicit function theorem, $s_r(s_s)$ is decreasing, where, as in the decentralized game without the contract, $s_i(s_j)$ is firm i's best response to firm j's strategy,c

$$\frac{\partial s_r(s_s)}{\partial s_s} = -\frac{\int_{s_s}^{\infty} c''(s_r + s_s - x) f_s(x) \, dx}{f_s(s_s) c''(s_r) + \int_{s_s}^{\infty} c''(s_r + s_s - x) f_s(x) \, dx} \leq 0.$$

Hence, for $s_r = s_r(s_s)$ and $\lambda \leq 1$ the supplier's marginal cost is increasing:

$$\frac{\partial \pi_s(s_r(s_s), s_s)}{\partial s_s} = F_s(s_s)\left(h_s - (1-\lambda) c'(s_r(s_s)) + t_B^s\right) - t_B^s.$$

Thus, there is a unique s_s that satisfies $s_s(s_r(s_s)) = s_s$, i.e., there is a unique Nash equilibrium.

Now suppose $s_s^o \leq 0$. It is straightforward to confirm all of the $\{\tilde{s}_r^2, \tilde{s}_s^2\}$ pairs satisfy the firms' first and second-order conditions. Hence, they are all Nash equilibria. Even though there is not a unique Nash equilibrium, the firms' costs are identical across the equilibria.

These contracts do allow the firms to arbitrarily allocate the retail level costs in the system, but they do not allow the firms to arbitrarily allocate all of the supply chain's costs. This limitation is due to the $\lambda \leq 1$ restriction, i.e., it is not possible with these contracts to allocate to the retailer more than the optimal retail level costs: while the retailer's cost function is well behaved even if $\lambda > 1$, the supplier's is not; with $\lambda > 1$ the supplier has a strong incentive to increase the retail level costs. Of course, fixed payments could be used to achieve those allocations if necessary. But since it is unlikely a retailer would agree to such a burden, this limitation is not too restrictive.

An interesting feature of these contracts is that the t_1 and t_B^r transfer payments are identical to the ones used in the single-location model. This is remarkable because the retailer's critical ratio differs across the models: in the single-location model the retailer picks s_r such that

$$F_r(s_r) = \frac{\beta}{\beta + h_r}$$

whereas in the two-location model the retailer picks s_r such that

$$F_r(s_r) = \frac{\beta + h_s}{\beta + h_r}.$$

8.5 Other coordination methods

Alternatives to Cachon and Zipkin's linear scheme have been studied to coordinate similar supply chains under the special case that $\beta_s = 0$, i.e., the supplier incurs no cost associated with retail backorders. That preference structure is most appropriate for an internal supply chain in which each location is operated by a separate manager. For example, instead of a supplier, suppose the second stage in the supply chain is controlled by a warehouse manager. That manager may have no direct interest in the availability of the firm's product at the retail level.

Lee and Whang (1999) base their coordination scheme on the work by Clark and Scarf (1960). (They consider a periodic review model and their firms minimize discounted costs rather than average costs.) Clark and Scarf (1960), which focuses only on system wide performance, demonstrates that base-stock policies are optimal and they can be evaluated from a series of simple single-location optimization problems after the costs in the system are reallocated among the locations. Lee and Whang (1999) take the Clark and Scarf cost reallocation and show it can be used to coordinate decentralized operations. In their arrangement the supplier subsidizes the retailer's holding cost at the rate of h_s and charges the retailer an additional backorder penalty cost per unit at rate h_s. Given those transfers let $g_r(y)$ be the retailer's expected cost at time $t + L_r$ when the retailer's inventory level is y at time t,

$$\begin{aligned} g_r(y) &= (h_r - h_s)I_r(y) + (\beta + h_s)B_r(y) \\ &= (h_r + \beta)I_r(y) + (\beta + h_s)(\mu_r - y) \end{aligned}$$

where $\beta = \beta_r$ since $\beta_s = 0$. $g_r(y)$ is strictly convex and minimized by s_r^o. However, due to shortages at the supplier, the retailer's inventory level, IL_r, may be less than his inventory position, IP_r. To penalize the supplier for those shortages, the supplier transfers to the retailer at time t, $g^p(IL_r, IP_r)$,

$$g^p(x, y) = g_r(x) - g_r(y).$$

That transfer may be negative, i.e., the retailer pays the supplier, if the retailer (for irrational reasons) orders $IP_r > s_r^o$ and the supplier does not fill that request completely, $IL_r < IP_r$. In addition, that transfer is not linear in the retailer's inventory position.

The retailer's final cost function is $\pi_r(s_r) = g_r(s_r)$, i.e., from the retailer's perspective the supplier provides perfectly reliable deliveries, since the supplier exactly compensates the retailer for any expected cost consequence of delivering less than the retailer's order. Hence, s_r^o is the retailer's optimal base-stock level.

The supplier's cost function with this arrangement is

$$c_s(s_r, s_s) = h_s[I_s(s_s) + I_r(s_r, s_s) - B_r(s_r, s_s)] + \int_{s_s}^{\infty} g^P(s_r + s_s - x, s_r) f_s(x) \, dx$$

$$= h_s[s_s + s_r - \mu_r - \mu_s] - (1 - F_s(s_s))g_r(s_r) + \int_{s_s}^{\infty} g_r(s_r + s_s - x) f_s(x) \, dx.$$

Differentiate:

$$\frac{\partial c_s(s_r, s_s)}{\partial s_s} = h_s + \int_{s_s}^{\infty} g_r'(s_r + s_s - x) f_s(x) \, dx$$

$$= -\beta + (h_s + \beta)\left[F_s(s_s) + \int_{s_s}^{\infty} F_r(s_r + s_s - x) f_s(x) \, dx\right],$$

$$\frac{\partial^2 c_s(s_r, s_s)}{\partial s_s^2} = (h_s + \beta)\left[f_s(s_s)(1 - F_r(s_r)) + \int_{s_s}^{\infty} f_r(s_r + s_s - x) f_s(x) \, dx\right] > 0.$$

So the supplier's cost function is strictly convex and if the retailer's base-stock level is s_r^o, then the supplier's optimal base-stock level is s_s^o, i.e., $\{s_r^o, s_s^o\}$ is a Nash equilibrium. It is also not difficult to show that $\{s_r^o, s_s^o\}$ is the unique Nash equilibrium (assuming $s_s^o > 0$ is optimal).[37] However, this arrangement provides for only one division of the supply chain's profit. Fixed payments could be used to reallocate costs differently.

There are several differences between this transfer payment contract and Cachon and Zipkin's contract. Lee and Whang charge the supplier a nonlinear cost for supplier backorders, whereas Cachon and Zipkin charge a linear one. In the linear contract the retailer must account for supplier shortages, i.e., he does not receive direct compensation for those shortages, whereas with the nonlinear contract the retailer need not be concerned with the supplier's inventory management decision. (In other words, with the linear contract $s_r(s_s)$ is not independent of s_s, whereas it is with the nonlinear contract.) Furthermore, they both have linear transfers associated with I_r and B_r, but those transfers are different: with the nonlinear contract $t_1 = h_s$ and $t_B^r = -h_s$, but that pair is not a member of the linear contracts ($t_1 = h_s$ requires $\lambda = 1 - h_s/h_r$ whereas $t_B^r = -h_s$ requires $\lambda = (\beta_r + h_s)/\beta$).

---

[37] The retailer's best reply function is independent of s_s. From the implicit function theorem, the supplier's reaction function is decreasing in s_r; hence there is a unique Nash equilibrium.

Chen (1999a) uses an accounting inventory approach to coordinate a serial supply chain.[38] One firm is chosen to compensate all other firms for their actual costs.[39] In this case, assume that firm is the supplier, i.e., the supplier's transfer payment rate to the retailer equals h_r per unit of inventory and β_r per backorder. That leaves the retailer with no actual cost, and so no incentive to choose s_r^o. To provide that incentive, the supplier charges the retailer t_I^a per unit of accounting inventory and t_B^a per accounting backorder, where the retailer's accounting inventory equals the inventory the retailer would have if the supplier always delivered the retailer's full order. Hence, the retailer's expected payment to the supplier per unit time is

$$t_I^a I_r(s_r) + t_B^a B_r(s_r) = (t_I^a + t_B^a) I_r(s_r) + t_B^a(\mu_r - s_r).$$

The retailer's optimal base-stock level, s_r^*, minimizes that payment,

$$F_r(s_r^*) = \frac{t_B^a}{t_I^a + t_B^a}.$$

Now set

$$t_I^a = \lambda(h_r - h_s)$$
$$t_B^a = \lambda(h_s + \beta)$$

for $\lambda > 0$. It follows that

$$\frac{t_B^a}{t_I^a + t_B^a} = \frac{h_s + \beta}{h_r + \beta},$$

and so from Eq. (8.4) the retailer chooses s_r^o. With those contracts the retailer's expected payment is

$$\lambda(h_r + \beta)[I_r(s_r) + F_r(s_r^o)(\mu_r - s_r)].$$

---

[38] There are several differences between this model and his. In his model there are order-processing delays between stages whereas this model assumes orders are received immediately. In addition, he accommodates more than two levels in the supply chain. Finally, he studies a periodic review model and allows for discrete demand.

[39] Chen also considers the possibility that there exists an owner of the entire supply chain who does not make any operating decisions. Those decisions are made by the managers. In that situation the owner can adopt all of the supply chain's costs.

Hence, the retailer's action can be coordinated and any cost can be assigned to the retailer. The supplier has no control over the transfer payment received (once the terms are set), so the supplier minimizes the costs under her control, which equal the supply chain's costs. Thus, $\{s_r^o, s_s^o\}$ is a Nash equilibrium.

Relative to the two previously discussed approaches for coordination, Chen's accounting inventory is most closely related to Lee and Whang's approach. In fact, in this setting they are essentially equivalent. In both approaches the retailer's cost function is based on the presumption that the retailer's orders are always filled immediately. Second, in Lee and Whang the retailer's effective holding cost per unit is $h_r - h_s$, and in Chen the retailer's holding cost is $\lambda(h_r - h_s)$. Similarly, the backorder penalty costs are $\beta + h_s$ and $\lambda(\beta + h_s)$. When $\lambda = 1$, the approaches are the same. By allowing for $\lambda \neq 1$, the accounting inventory approach allows for any division of profit, whereas the Lee and Whang approach does not have that flexibility (merely because it lacks that parameter). Finally, in both cases the supplier bears all remaining costs in the supply chain, and so the supplier's cost function is equivalent with either scheme.

The equivalence between Chen's accounting inventory and Lee and Whang's nonlinear contract is surprising because Chen does not appear to charge the supplier for her backorders whereas Lee and Whang do charge a nonlinear penalty function. However, accounting inventory does charge a nonlinear penalty because the supplier compensates the retailer's actual costs. Thus, accounting inventory and the Lee and Whang approach are two different ways to describe the same transfer payments. This is probably a general result, but additional research is needed to confirm that conjecture after all differences between the models are reconciled.

Porteus (2000) offers responsibility tokens to coordinate the supply chain, which is also closely related to the Lee and Whang approach.[40] As with Lee and Whang, in Porteus the transfer rate from the supplier to the retailer is $h_s I_r - h_s B_r$. However, Porteus does not include an explicit charge associated with B_s. Instead, in lieu of actual inventory, the supplier issues a responsibility token for each unit the retailer orders that the supplier is unable to fill. From the retailer's perspective that token is equivalent to inventory: the retailer incurs a holding cost of $h_r - h_s$ per token per unit time and incurs no backorder penalty cost if the token is used to 'fill' demand. If a token is used to fill demand then the supplier is charged the full backorder cost until the supplier provides a unit of actual inventory to fill that demand. Similarly, the supplier receives the retailer's holding cost on each token since the supply chain does not actually incur a holding cost on these imaginary tokens. Hence, with this system the retailer receives a perfectly reliable supply from the supplier and the supplier bears the consequence of her backorders, just as with Lee and Whang. However, in Lee and Whang the supplier pays

---

[40] He considers a periodic review, finite horizon model with multiple firms in a serial supply chain.

the expected cost consequence of a backorder whereas with responsibility tokens the supplier pays the actual cost consequence. The same holds with accounting inventory. When all players are risk neutral this distinction does not matter.

Watson (2002) considers coordination of a serial supply chain with AR(1) demand. Given this demand process the optimal order at each stage of the supply chain is not equal to demand, as it is with independent demand. Hence, schemes that use local penalties, e.g., Chen's (1999a) accounting inventory or the linear transfer payments of Cachon and Zipkin (1999), require each stage to forecast the ordering process of the subsequent stage, which is nontrivial with this demand process, especially for the highest stages. Watson proposes an alternative which is computationally friendlier. In his scheme the stage is given incentives to manage echelon inventory, and so each stage needs only to observe the demand process.

8.6 Discussion

In the two-location base-stock model decentralized operations always leads to suboptimal performance, but the extent of the performance deterioration (i.e., the competition penalty) depends on the supply chain parameters. If the firms' backorder costs are similar then the competition penalty is often reasonably small. The competition penalty can be small even if the supplier does not care about customer service because her operations role in the supply chain may not be too important (e.g., if h_s is large or if L_s is small).

To coordinate the retailer's action the firms can agree to a pair of linear transfer payments that function like the buyback contract in the single-period newsvendor model. To coordinate the supplier's action Cachon and Zipkin (1999) propose a linear transfer payment based on the supplier's backorders. With these contracts the optimal policy is the unique Nash equilibrium. Further, they allow the firms to arbitrarily divide the retail level costs. Lee and Whang (1999), Chen (1999a) and Porteus (2000) propose alternative coordination schemes for the special case in which the supplier does not care about retail level backorders.

There are a number of worthwhile extensions to the two-location base-stock model. Caldentey and Wein (2000) study a model that is the two-location model considered here, with the exception that the supplier chooses a production rate rather than an inventory policy. They demonstrate that many of the qualitative insights from Cachon and Zipkin (1999) continue to hold. Cachon (1999) obtains the same finding for a two-echelon serial supply chain with lost sales. Duenyas and Tsai (2001) also study a two-echelon serial supply chain with lost sales but they include several twists. First, they consider dynamic policies rather than just base-stock policies. Second, upstream inventory is needed by the downstream firm to satisfy its demand, but that inventory can also be used to satisfy demand in some outside market.

Given a choice between the two demands, the supply chain prefers to use the inventory to meet the needs of the downstream firm. But since demand is stochastic, it may be optimal to satisfy some outside demand. They demonstrate that while fixed wholesale-price contracts do not coordinate the supply chain, they are nevertheless quite effective, i.e., the competition penalty is small for most parameter settings. Finally, Parker and Kapuscinsky (2001) tackle the considerably harder problem of coordinating a serial supply chain with capacity constraints.

A natural next step after serial supply chains is to consider supply chains with multiple retailers. However, the analytical complexity of even centralized operations in those models poses quite a challenge. Cachon (2001) obtains results for a two-echelon model with multiple retailers and discrete stochastic demand using the theory of supermodular games (see Topkis, 1998). He demonstrates that multiple Nash equilibria may exist, and the optimal policy may even be a Nash equilibrium. Hence, decentralized operations do not necessarily lead to suboptimal performance.[41] Andersson and Marklund (2000) and Axsater (2001) consider a similar model but approach coordination differently.

Chen, Federgruen and Zheng (2001) (CFZ) study a model with one supplier, multiple noncompeting retailers and deterministic demand. [Bernstein and Federgruen (1999) study a closely related model with competing retailers.] While the centralized solution is intractable, for the case with fixed prices, and therefore fixed demand rates, Roundy (1985) provides a class of policies that is nearly optimal. Within that class CFZ find the centralized solution. They show that a single-order quantity-discount policy cannot coordinate the action of heterogenous retailers and they propose a set of transfer payments that does coordinate the supply chain. The coordination issues in this model are quite different than in the previously discussed multiechelon models with stochastic demand. For example, in the two-location model the retailer's action has no impact on the supplier's holding cost, whereas in the CFZ model the retailer's action impacts the supplier's holding and order processing costs. Furthermore, in the CFZ model, unlike with stochastic demand, all customer demands are met without backordering and the supplier never delays a shipment.

9 Coordination with internal markets

In each of the previous models the firms agree to a contract that explicitly stipulates transfer payments, e.g., the buyback rate is b or the revenue share is ϕ. However, there are situations that call for more flexibility, i.e., the transfer payment rates are contingent on the realization of some

---

[41] The true optimal policy is unknown for that supply chain, so performance is measured relative to the optimal policy within the class of reorder point policies.

9.1 Model and analysis

Suppose there is one supplier, one production manager and two retailers. The production manager is the supplier's employee and the retailers are independent firms. The following sequence of events occurs: the production manager chooses a production input level e, which yields an output of $Q = Ye$ finished units, where $Y \in [0, 1]$ is a random variable; the production manager incurs cost $c(e)$, where $c(e)$ is strictly convex and increasing; retailer i observes α_i, the realization of the random variable $A_i > 0$; each retailer submits an order to the supplier; the supplier allocates q_i units to retailer i, being sure that q_i does not exceed retailer i's order and $q_1 + q_2 \leq Q$; and finally retailer i earns revenue $q_i p_i(q_i)$, where $p_i(q_i) = \alpha_i q_i^{-1/\eta}$ and $\eta > 1$ is the constant demand elasticity. Let θ be the realization of Y, $A = \{A_1, A_2\}$ and $\alpha = \{\alpha_1, \alpha_2\}$. This model is closely related, albeit with different notation, to one developed by Kouvelis and Lariviere (2000), which in turn is a variant of the model developed by Porteus and Whang (1991).[42] See Agrawal and Tsay (2001) and Erkoc and Wu (2002b) for related models.

Before considering transfer payments between the firms, consider the supply chain optimal actions. Given that each retailer's revenue is strictly increasing in his allocation, it is always optimal to allocate the entire output to the retailers. Thus, let γ be the fraction of Q that is allocated to retailer one. Let $\pi(\gamma, \alpha, Q)$ be total retailer revenue if retailer one is allocated γQ units and retailer two is allocated $(1-\gamma)Q$ units:

$$\pi(\gamma, \alpha, Q) = (\alpha_1 \gamma^{(\eta-1)/\eta} + \alpha_2 (1-\gamma)^{(\eta-1)/\eta}) Q^{(\eta-1)/\eta}.$$

The optimal allocation of production to the two retailers depends on the demand realizations but not on the production output. Further, revenue is concave in γ, so let $\gamma^o(\alpha)$ be the optimal share to allocate to retailer one:

$$\gamma^o(\alpha) = \frac{\alpha_1^\eta}{\alpha_1^\eta + \alpha_2^\eta}. \tag{9.1}$$

Conditional on an optimal allocation, the retailers' total revenue is

$$\pi(\alpha, Q) = \pi(\gamma^o(\alpha), \alpha, Q) = (\alpha_1^\eta + \alpha_2^\eta)^{1/\eta} Q^{(\eta-1)/\eta}.$$

---

[42] Kouvelis and Lariviere (2000) do not label their players 'supplier' and 'retailers'. More importantly, they have two agents responsible for production: the output of production is $\theta q_1 q_2$, where q_i is the action taken by the ith agent. Porteus and Whang (1991) have a single production agent and N demand agents (which are the retailers in this model). Their production agent faces an additive output shock rather than a multiplicative one. Their demand agents face a newsvendor problem with effort-dependent demand.

Total expected supply chain profit, $\Pi(e, A, Y)$, is thus

$$\Pi(e, A, Y) = E[\pi(A, Ye)] - c(e).$$

Profit is concave in e, so the unique optimal production effort level, e^o, satisfies

$$\Pi'(e^o) = \left(\frac{\eta-1}{\eta}\right)(e^o)^{-1/\eta} E[(A_1^\eta + A_2^\eta)^{1/\eta} Y^{(\eta-1)/\eta}] - c'(e^o) = 0. \quad (9.2)$$

Now consider decentralized operations. To achieve channel coordination it must be that the retailers purchase all of the supplier's output and that output must be allocated to the retailers properly. It is now apparent that a fixed per unit wholesale price cannot achieve those tasks: for any fixed wholesale price there is some realization of Y such that the retailers do not purchase all of the supplier's output. In addition, with a fixed wholesale price it is possible the retailers desire more than the supplier's output, in which case the supplier must implement some allocation rule. The possibility of rationing could cause the retailers to submit strategic orders, which in turn could lead to an inefficient allocation of output (see Cachon & Lariviere, 1999). To alleviate those problems the supplier could make the transfer price contingent on the realization of A and Q. To be specific, suppose the supplier charges the retailers w per unit. Assuming retailer i receives q_i units, retailer i's profit is

$$\pi_i(q_i, w) = \alpha_i q_i^{(\eta-1)/\eta} - w q_i.$$

Retailer i's optimal quantity, q_i^*, satisfies

$$\frac{\partial \pi_i(q_i^*, w)}{\partial q_i} = 0 = \left(\frac{\eta-1}{\eta}\right) \alpha_i (q_i^*)^{-1/\eta} - w.$$

It follows that $q_i^* = \gamma^o(\alpha) Q$ when $w = w(\alpha, Q)$,

$$w(\alpha, Q) = \left(\frac{\eta-1}{\eta}\right)(\alpha_1^\eta + \alpha_2^\eta)^{1/\eta} Q^{-1/\eta}.$$

Hence, when the supplier charges $w(\alpha, Q)$ the retailers order exactly Q units in total and the allocation of inventory between them maximizes supply chain revenue. Note that $w(\alpha, Q)$ is precisely the marginal value of additional production,

$$\frac{\partial \pi(\alpha, Q)}{\partial Q} = w(\alpha, Q).$$

The supplier could offer the retailers a contract that identifies $w(\alpha, Q)$ as the wholesale price contingent on the realizations of A and Y, but that surely would be an unruly contract (in length and complexity). Furthermore, implementation of that contract requires the supplier to actually observe the realizations of A, which may not occur. Fortunately, there is a simpler alternative. The supplier merely commits at the start of the game to hold a market for output after the retailers observe their demand realizations. The unique market-clearing price is $w(\alpha, Q)$, and so the market optimizes the supply chain's profit without the supplier observing A.

It remains to determine the supplier's compensation scheme for her production manager. To complicate matters, assume the supplier is unable to observe the production manager's effort, i.e., e is noncontractible. But the supplier is able to observe the firm's final output, Ye. So suppose the supplier pays the production manager

$$\left(\frac{\eta-1}{\eta}\right)(e^o)^{-1/\eta}E[(A_1^\eta + A_2^\eta)^{1/\eta}Y^{(\eta-1)/\eta}]/E[Y] = E[Qw(A,Q)\mid e^o]/E[Q\mid e^o] \quad (9.3)$$

per unit of realized output. Hence, the supplier pays the production manager the expected shadow price of capacity and sells capacity to the retailers for the realized shadow price of capacity. Given that scheme, the production manager's expected utility is

$$u(e) = \left(\frac{\eta-1}{\eta}\right)(e^o)^{-1/\eta}E[(A_1^\eta + A_2^\eta)^{1/\eta}Y^{(\eta-1)/\eta}]e - c(e)$$

and the marginal utility is

$$u'(e) = \left(\frac{\eta-1}{\eta}\right)(e^o)^{-1/\eta}E[(A_1^\eta + A_2^\eta)^{1/\eta}Y^{(\eta-1)/\eta}] - c'(e).$$

A comparison with Eq. (9.2) reveals that the production manager's optimal effort is e^o.

The supplier earns zero profit in expectation from the internal market: $E[Qw(A,Q)\mid e^o]$ is the expected revenue from the retailers, and, from Eq. (9.3), it is also the expected payout to the production manager. To earn a positive profit the supplier can charge the production manager and/or the retailers fixed fees. In fact, in more general settings, Kouvelis and Lariviere (2000) show that supplier breaks even or loses money with the internal market approach to supply chain coordination.[43] Hence, this is a viable strategy for the supplier only if it is coupled with fixed fees.

---

[43] For example, suppose there were two production managers and output equaled Ye_1e_2, where e_i is the effort level of production manager i. In that case Kouvelis and Lariviere (2000) show that the supplier/market maker loses money on the market. See Holmstrom (1982) for additional discussion on coordination and budget balancing.

9.2 Discussion

In this model one agent in the supply chain produces a resource (production output) that has uncertain value to the other part of the supply chain (i.e., the retailers). Coordination therefore requires the proper incentive to produce the resource as well as the proper incentive to consume the resource. Production can be coordinated with a single price (the expected value of output) but its consumption requires a state-dependent price, which can be provided via a market mechanism. Interestingly, the expected revenue from selling the resource via the market may be less than the expected cost to purchase the resource. Hence, there must be a market maker that stands between the producers and consumers of the resource and the market maker must be ready to lose money in this market. So the market maker is willing to participate only if there exists some other instrument to extract rents from the participants (e.g., fixed fees). In other words, the market maker uses the market to align incentives, but does not directly profit from the market.

The Donohue (2000) model is similar to this one in the sense that the value of first-period production is uncertain: it depends on the realization of the demand signal at the start of period 2. A market was not necessary in that model because it is optimal for the retailer to purchase the supplier's entire period 1 output. Now suppose there is a holding cost for inventory and the retailer's holding cost is higher than the supplier's. In that case it may be optimal for the supply chain to produce more in period 1 than the retailer orders, i.e., the excess production is held in storage at the supplier until needed in period 2. It remains optimal to move all period 1 production to the retailer in period 2: inventory at the supplier cannot satisfy demand. But only $w_2 = 0$ can ensure the retailer orders all of the supplier's inventory. Unfortunately, with that price any period 2 order is optimal for the retailer. Hence, the buyback contract with fixed wholesale prices in each period is no longer a practical coordination scheme for the supply chain. See Barnes-Schuster et al. (1998) for a more formal treatment of this argument.

10 Asymmetric information

In all of the models considered so far the firms are blessed with full information, i.e., all firms possess the same information when making their decisions. Hence, any coordination failure is due exclusively to incentive conflicts and not due to an inability by one firm to evaluate the optimal policy. However, in practice full information is rare. Given the complexity and geographic breadth of most modern supply chains it is not surprising that at least one firm lacks some important piece of information that another firm possesses. For instance, the manufacturer of a product may have a more accurate forecast of demand than the manufacturer's supplier of a critical component. In that case optimal supply chain performance requires more than the coordination of actions. It also requires the sharing of information so that each firm in the supply chain is able to determine the precise set of optimal actions.

Sometimes sharing information is not difficult. For example, suppose the relevant information is the demand distribution of a product with stationary stochastic demand. Hence, the demand forecast can be shared by sharing past sales data. (That might be technically challenging, but the credibility of the forecast is not in doubt.) In that case the interesting research questions include how to use that information to improve supply chain performance and by how much performance improves (see Cachon & Fisher, 2000; Chen, 1998; Gavirneni et al., 1999; Lee, So & Tang, 2000).

Unfortunately, there is also the possibility of opportunistic behavior with information sharing. For example, a manufacturer may tell her supplier that demand will be quite high to get the supplier to build a substantial amount of capacity. This is particularly problematic when the demand forecast is constructed from diverse and unverifiable pools of information. Consider the demand forecast for a new product. The manufacturer's sales manager may incorporate consumer panel data into her forecast, which could be shared with a supplier, but her forecast may also include her subjective opinion based on a myriad of information gathered from her years of experience in the industry. If the sales manager knows her market well, those guesses and hunches may be quite informative, yet there really is no obvious way for her to convey that information to the supplier other than to say in her opinion expected demand is 'x'. In other words, the sales manager has important information that the supplier cannot easily verify, so the supplier may not be sure 'x' is indeed the sales manager's best forecast. Furthermore, it is even difficult for the supplier to verify the forecast *ex post*: if demand turns out to be less than 'x' the supplier cannot be sure the sale manager gave a biased forecast since a low-demand realization is possible even if 'x' is the correct forecast.

This section considers a supply chain contracting model with asymmetric demand forecasts that is based on Cachon and Lariviere (2001).[44] As before, the main issues are which contracts, if any, achieve supply chain coordination and how are rents distributed with those contracts. In this model coordination requires (1) the supplier takes the correct action and (2) an accurate demand forecast is shared.

In addition to information sharing, this model highlights the issue of contract compliance, as was first discussed in Section 2. With forced compliance (i.e., all firms must choose the actions specified in the contract)

---

[44] See Riordan (1984) for a similar model: he has asymmetric information regarding demand, asymmetric information regarding the supplier's cost and the capacity and production decisions are joined, i.e., production always equals capacity.

supply chain coordination and accurate forecast sharing are possible. However, with voluntary compliance (i.e., each firm chooses optimal actions even if they deviate from those in the contract) information sharing is possible but only if optimal supply chain performance is sacrificed.

10.1 The capacity procurement game

In the capacity procurement game a manufacturer, M, develops a new product with uncertain demand. There is a single potential supplier, S, for a critical component, i.e., even M is not able to make this component. [See Milner and Pinker (2001) for a capacity contracting model without asymmetric information in which the downstream firm is able to provide some of its own capacity but still may depend on the upstream firm's capacity when demand is high.] Let D_θ be demand, where $\theta \in \{h, l\}$. Let $F(x|\theta)$ be the distribution function of demand, where $F(x|\theta) = 0$ for all $x < 0$, $F(x|\theta) > 0$ for all $x \geq 0$, and $F(x|\theta)$ is increasing and differentiable. Furthermore, D_h stochastically dominates D_l, i.e., $F(x|h) < F(x|l)$ for all $x \geq 0$.

With full information both firms observe the θ parameter. With asymmetric information the θ parameter is observed only by the manufacturer. In that case the supplier's prior beliefs are that $\Pr(\theta = h) = \rho$ and $\Pr(\theta = l) = 1 - \rho$. The manufacturer also knows ρ, i.e., the prior is common knowledge.

The interactions between M and S are divided into two stages. In stage 1, M gives S a demand forecast and offers S a contract which includes an initial order, q_i. Assuming S accepts the contract, S then constructs k units of capacity at a cost $c_k > 0$ per unit. In stage 2, M observes D_θ and places her final order with S, q_f, where the contract specifies the set of feasible final orders. Then S produces $\min\{D_\theta, k\}$ units at a cost of $c_p > 0$ per unit and delivers those units to M. Finally, M pays S based on the agreed contract and M earns $r > c_p + c_k$ per unit of demand satisfied. The salvage value of unused units of capacity is normalized to zero. The qualitative behavior of the model is unchanged if M only observes an imperfect signal of demand in stage 2.

Like the newsvendor model studied in Section 2, this model has only one demand period. But in the newsvendor model production occurs before the demand realization is observed, whereas in this model production, constrained by the initial capacity choice, occurs after the demand realization is observed. This model is also different than the two-stage newsvendor model considered in Section 6. In that model some production can be deferred until after demand information is learned, but that production is more expensive than early production. In this model the cost of production is the same no matter in which stage it occurs. Hence, unlike in the Section 6 model, it is never optimal to produce before the demand information is learned.

10.2 Full information

To establish a benchmark, in this section assume both firms observe θ and begin the analysis with the supply chain optimal solution. The supply chain makes two decisions: how much capacity to construct, k, and how much to produce. The latter is simple, produce $\min\{k, D_\theta\}$ after observing demand. Hence, the only substantive decision is how much capacity to build. Let $S_\theta(x)$ be expected sales with x units of capacity,

$$S_\theta(x) = x - E[(x - D_\theta)^+]$$
$$= x - \int_0^x F_\theta(x)\,dx.$$

Let $\Omega_\theta(k)$ be the supply chain's expected profit with k units of capacity,

$$\Omega_\theta(k) = (r - c_p)S_\theta(k) - c_k k.$$

Given that $\Omega_\theta(k)$ is concave, the optimal capacity, k_θ^o, satisfies the newsvendor critical ratio:

$$\overline{F}_\theta(k_\theta^o) = \frac{c_k}{r - c_p},$$

where $\overline{F}_\theta(x) = 1 - F_\theta(x)$.[45] Let $\Omega_\theta^o = \Omega_\theta(k_\theta^o)$. Thus, supply chain coordination is achieved if the supplier builds k_θ^o units of capacity and defers all production until after receiving the manufacturer's final order.

Now turn to the game between M and S. There are many different types of contracts the manufacturer could offer the supplier. Consider an options contract: M purchases q_i options for w_o per option at stage 1 and then pays w_e to exercise each option at stage 2. Hence, the total expected transfer payment is

$$w_o q_i + w_e S_\theta(q_i).$$

That contract could also be described as a buyback contract: M pays $w = w_o + w_e$ at stage 1 for an order quantity of q_i and S pays $b = w_e$ per unit in stage 2 that M 'returns', i.e., does not take actual delivery. Alternatively, that contract could be described as a wholesale-price contract ($w_o + w_e$ is the wholesale price) with a termination penalty charged for each unit M cancels in stage 2 (where w_o is the termination penalty). Erkoc and Wu (2002a) study

---

[45] Given that $F_\theta(0) > 0$, it is possible that $k_\theta^o = 0$, but that case is not too interesting, so assume $k_\theta^o > 0$. Boundary conditions are also ignored in the remainder of the analysis.

reservation contracts in a capacity procurement game with convex capacity costs: with a reservation contract M reserves a particular amount of capacity before observing demand and then pays a fee to S for each unit of reserved capacity that is not utilized once demand is observed. That contract is not considered in this section.

On the assumption the supplier builds enough capacity to cover the manufacturer's options, $k = q_i$, the manufacturer's expected profit is

$$\Pi_\theta(q_i) = (r - w_e)S_\theta(q_i) - w_o q_i.$$

If contract parameters are chosen so that $(r-w_e) = \lambda(r-c_p)$ and $w_o = \lambda c_k$, where $\lambda \in [0, 1]$, then

$$\Pi_\theta(q_i) = \lambda \Omega_\theta(q_i).$$

Hence, $q_i = k_\theta^o$ is the manufacturer's optimal order, the supply chain is coordinated and the supply chain's profit can be arbitrarily allocated between the firms. Indeed, the supplier's profit is $(1-\lambda)\Omega_\theta(q_i)$, so k_θ^o also maximizes the supplier's profit, apparently confirming the initial $k = q_i$ assumption.

But there is an important caveat to the above analysis. Can the manufacturer be sure the supplier indeed builds $k = q_i$? Suppose the manufacturer is unable to verify the supplier actually builds $k = q_i$. Given the supplier's capacity depends on a number of factors that are hard for the manufacturer to verify (labor practices, production schedules, component yields, etc.), it is not surprising if there were situations in which the manufacturer would be unable to prove the supplier built less than q_i capacity. With that in mind, consider the following profit function for a supplier (assuming $k < q_i$) who believes demand is τ,

$$\pi(k, q_i, \tau) = (w_e - c_p)s_\tau(k) + w_o q_i - c_k k$$
$$= (1 - \lambda)(r - c_p)s_\tau(k) - c_k(k - \lambda q_i).$$

It follows that

$$\frac{\partial \pi(k_\theta^o, k_\theta^o, \theta)}{\partial k} < 0,$$

i.e., k_θ^o does not maximize the supplier's profit if $q_i = k_\theta^o$. The source of the problem is that the above cost function assumes the manufacturer pays w_o per option no matter what capacity is constructed. Therefore, only the w_e parameter of the contract impacts the supplier's decision on the margin, i.e., the supplier sets his capacity as if the supplier is offered just a wholesale-price contract.

Cachon and Lariviere (2001) define forced compliance to be the case when the supplier must choose $k = q_i$ and voluntary compliance to be the case when the supplier can choose $k < q_i$ even though the manufacturer pays $w_o q_i$ for q_i options. Both situations represent extreme ends of a spectrum: with forced compliance the supplier acts as if any deviation from $k < q_i$ is infinitely costly, whereas with voluntary compliance the supplier acts as if there are no consequences. Reality is somewhere in the middle. (This discussion is analogous to the one with retail effort, which in reality is neither fully contractible nor fully uncontractible.) Nevertheless, voluntary compliance is worth study because the supplier is likely to build $k < q_i$ even if there is some penalty for doing so. For example, suppose the supplier must refund the manufacturer for any option the manufacturer purchased that the supplier is unable to exercise, i.e., if $D_\theta > k$, the supplier refunds the manufacturer $(\min\{q_i, D_\theta\} - k)^+ w_o$. Even with such a penalty, the supplier has a $1 - F(k)$ chance of pocketing the fee for $(q_i - k)$ options without having built the capacity to cover those options: if $D_\theta < k$, then the manufacturer would not know the supplier was unable to cover all options because the supplier covers the options the manufacturer exercises. That incentive is enough to cause the supplier to choose $k < q_i$. Erkoc and Wu (2002a) propose an alternative approach in the context of their reservation contract: they study a game in which the supplier pays a penalty for each unit of capacity that is reserved but is not delivered. They find sufficient penalties such that compliance is achieved.

To summarize, with forced compliance the manufacturer can use a number of contracts to coordinate the supply chain and divide its profit. However, coordination with those contracts is not assured with anything less than forced compliance. As a result, voluntary compliance is a more conservative assumption (albeit, possibly too conservative).[46]

The remainder of this section studies the manufacturer's contract offer under voluntary compliance. As suggested above, with voluntary compliance the manufacturer's initial order has no impact on the supplier's capacity decision. It follows that transfer payments based on the initial order also have no impact on the supplier's capacity decision. To influence the supplier's capacity decision the manufacturer is relegated to a contract based on his final order, q_f. An obvious candidate is the wholesale-price contract.

With a wholesale-price contract the supplier's profit is

$$\pi_\theta(k) = (w - c_p)S_\theta(k) - c_k k.$$

---

[46] Intermediate compliance regimes are challenging to study because the penalty for noncompliance is not well behaved in k: in general $1 - F(k)$ is neither concave nor convex in k. See Krasa and Villamil (2000) for a model in which the compliance regime is an endogenous variable. As already mentioned, see also Erkoc and Wu (2002a) for an approach to achieve compliance.

Note that the supplier evaluates his expected profit on the assumption that θ is certainly the demand parameter. The supplier's profit is strictly concave in k, so there exists a unique wholesale price for any k such that k is optimal for the supplier. Let $w_\theta(k)$ be that wholesale price:

$$w_\theta(k) = \frac{c_k}{\overline{F}_\theta(k)} + c_p.$$

The manufacturer's profit function can now be expressed as

$$\Pi_\theta(k) = (r - w_\theta(k))S_\theta(k),$$

i.e., the manufacturer can choose a desired capacity by offering the wholesale-price contract $w_\theta(k)$. Differentiate the manufacturer's profit function:

$$\Pi'_\theta(k) = (r - w_\theta(k))S'_\theta(k) - w'_\theta(k)S_\theta(k),$$
$$\Pi''_\theta(k) = (r - w_\theta(k))S''_\theta(k) - w''_\theta(k)S_\theta(k) - 2w'_\theta(k)S'_\theta(k)$$
$$= -\left(r - c_p + \frac{c_k}{\overline{F}_\theta(k)}\right)f_\theta(k) - w''_\theta(k)S_\theta(k).$$

If $w''_\theta(k) > 0$, then $\Pi_\theta(k)$ is strictly concave. So for convenience assume $w''_\theta(k) > 0$, which holds for any demand distribution with an increasing failure rate, i.e., the hazard rate, $f_\theta(x)/(1 - F_\theta(x))$, is increasing. The normal, exponential and the uniform meet that condition, as well as the gamma and Weibull with some parameter restrictions (see Barlow & Proschan, 1965). Therefore, there is a unique optimal capacity, k^*_θ, and a unique wholesale price that induces that capacity, $w^*_\theta = w(k^*_\theta)$. It follows from $\Pi'_\theta(k^*_\theta) = 0$ that the supply chain is not coordinated, $k^*_\theta < k^o_\theta$:

$$\overline{F}_\theta(k^*_\theta) = \frac{c_k}{r - c_p}\left(1 + \frac{f_\theta(k^*_\theta)}{\overline{F}_\theta(k^*_\theta)^2} S_\theta(k^*_\theta)\right)$$
$$= \overline{F}_\theta(k^o_\theta)\left(1 + \frac{f_\theta(k^*_\theta)}{\overline{F}_\theta(k^*_\theta)^2} S_\theta(k^*_\theta)\right).$$

10.3 Forecast sharing

Now suppose the supplier does not observe θ. The supplier has a prior belief regarding θ, but to do better the supplier might ask the manufacturer for her forecast of demand, since the supplier knows the manufacturer knows θ. If the manufacturer announces demand is low, the supplier should believe the forecast: a high-demand manufacturer is unlikely to bias her forecast down. But a high-demand forecast is suspect, since there is the real possibility a low-demand manufacturer would offer a high-demand forecast to get the supplier to build more capacity: the manufacturer always wants more capacity built if the manufacturer is not paying for it. Thus, a sophisticated supplier ignores the manufacturer's verbal forecast, and instead infers the manufacturer's demand from the contract the manufacturer offers. With that understanding, the manufacturer shares her demand forecast by offering the right contract.

To continue with that reasoning, in this model forecast sharing takes place via a signaling equilibrium. With a signaling equilibrium the supplier assigns a belief to each possible contract the manufacturer could offer: either the contract is offered only by a high-demand manufacturer, offered only by a low-demand manufacturer or it could be offered by either type. There are different kinds of signaling equilibria. With a separating equilibrium the supplier divides the contracts into the former two sets: either the contract is a high-type contract or a low-type contract, but no contract is offered by both types. With a pooling equilibrium the contracts are divided into the latter two sets: either the contract is offered by a low-type or it is offered by both types. In a pooling equilibrium there is no means to share a high-demand forecast (since no contract is designated for only the high-type manufacturer) and the low-demand forecast is shared only out of equilibrium (since each type prefers the contract designated for both types, which conveys no information regarding the manufacturer's demand). Thus, forecast sharing only occurs with a separating equilibrium: a high-type M offers the best contract among those designated for high-types and a low-type M offers the best contract among those designated for low-types.

A separating equilibrium is rational if a high-type M indeed prefers to offer a high-type contract rather than to offer a low-type contract. The high-type recognizes that if she offers a low-type contract the supplier builds capacity based on the assumption demand is indeed low. Similarly, a separating equilibrium must also have that a low-type M prefers to offer a low-type contract rather than to offer a high-type contact. In other words, the low-type manufacturer must not prefer to mimic a high-type manufacturer by choosing a high-type contract. That condition is more onerous because if the low-type M offers a high-type contract then the supplier builds capacity under the assumption demand is high. See Cachon and Lariviere (2001) for a more formal description of these conditions.

First consider separating equilibria with forced compliance. Recall, the manufacturer can coordinate the supply chain with an options contract (and the parameters of those contracts do not depend on the demand distribution). The type θ manufacturer's profit with one of those contracts is $\lambda \Omega_\theta(q_i)$: with forced compliance the supplier must build $k=q_i$ if the supplier accepts the contract, so the supplier's belief regarding demand has no impact on the capacity choice given the supplier accepts the contract. The supplier's profit is

$(1-\lambda)\Omega_\theta(q_i)$. Let $\hat{\pi}$ be the supplier's minimum acceptable profit, i.e., the supplier rejects the contract if his expected profit is less than $\hat{\pi}$.[47] Clearly both types of manufacturers want a large share of the supply chain profit and for a fixed share, both manufacturers want to maximize the supply chain's profit. However, a manufacturer could prefer a larger share of suboptimal profit over a smaller share of optimal profit. Hence, the supplier must be diligent about biased forecasts.

Suppose the low-type manufacturer offers an options contract with

$$\lambda_l = 1 - \frac{\hat{\pi}}{\Omega_l^o}$$

and the high-type manufacturer offers $\min\{\lambda_h, \hat{\lambda}_h\}$, where

$$\lambda_h = 1 - \frac{\hat{\pi}}{\Omega_h^o}$$

and

$$\hat{\lambda}_h = \frac{\Omega_l^o - \hat{\pi}}{\Omega_l(k_h^o)}.$$

Since $\min\{\lambda_h, \hat{\lambda}_h\} > \lambda_l$, the high-type manufacturer has no interest in offering the low-type's contract: with that contract the manufacturer earns a smaller share (λ_l vs. $\min\{\lambda_h, \hat{\lambda}_h\}$) of a lower profit ($\Omega_h(k_l^o)$ vs. Ω_h^o). The low-type manufacturer also has no interest in offering the high-type's contract. By construction, the low-type manufacturer is indifferent between earning her low-type profit, $\Omega_l^o - \hat{\pi}$, and earning $\hat{\lambda}_h$ percent of the high-type contract profit, $\Omega_l(k_h^o)$. As long as the high-type captures no more than $\hat{\lambda}_h$ percent of the supply chain's profit, the low-type has no interest in pretending to be a high-type. Thus, the above contracts are a separating equilibrium.

With low demand the supplier earns $\hat{\pi}$. With high-demand the supplier earns the same amount if $\lambda_h < \hat{\lambda}_h$, otherwise the supplier earns more. The manufacturer would prefer to take more from the supplier when $\lambda_h > \hat{\lambda}_h$, but then the supplier cannot trust the manufacturer's forecast (because a low-type would gladly accept some suboptimal supply chain performance in exchange for a large fraction of the profit). Even though the manufacturer may be unable to drive the supplier's profit down to $\hat{\pi}$, the supply chain is coordinated in all situations. Hence, with forced compliance

---

[47] It is assumed $\hat{\pi}$ is independent of θ. But it is certainly plausible that a supplier's outside opportunity is correlated with the manufacturer's demand: if the manufacturer has high demand, then other manufacturers may have high demand, leading to a higher than average opportunity cost to the supplier. Additional research is needed to explore this issue.

information is exchanged via the parameters of the contract and not via the form of the contract.

Now consider voluntary compliance. Here the manufacturer is relegated to offering a wholesale-price contract. In a separating equilibrium the low-type manufacturer offers w_l^* and the high-type manufacturer would like to offer w_h^*. If $w_h^* > w_l^*$ then it is possible the low-type manufacturer will not mimic the high-type: mimicking gets the low-type manufacturer more capacity, but she must pay a higher price. If $w_h^* \leq w_l^*$ then mimicking certainly occurs: the low-type manufacturer gets more capacity and pays no more per unit. In that case the high-type manufacturer needs to supplement the wholesale-price contract with some additional transfer payment that only a high-type manufacturer would be willing to pay.

One suggestion is for the high-type M to also offer the supplier a fixed fee, A. The low-type manufacturer earns $\Pi_l(w_l^*, l)$ when she reveals herself to be a low-type, where $\Pi_\theta(w, \tau)$ is a type θ manufacturer's profit when the supplier believes demand is type $\tau \in \{l, h\}$. She earns $\Pi_l(w_h^*, h) - A$ when she mimics the high-type (i.e., offers the high-type contract), so she does not wish to mimic when

$$\Pi_l(w_l^*, l) \geq \Pi_l(w_h^*, h) - A.$$

The high-type prefers to offer that fixed fee rather than to offer the low-type's contract when

$$\Pi_h(w_h^*, h) - A > \Pi_h(w_l^*, l).$$

There exists a fixed fee that satisfies both conditions when

$$\Pi_h(w_h^*, h) - \Pi_h(w_l^*, l) > \Pi_l(w_h^*, h) - \Pi_l(w_l^*, l).$$

The above states that the high-type manufacturer has more to gain from the supplier believing demand is high than the low-type manufacturer.

While the fixed fee works, there may be a cheaper approach for the high-type manufacturer to signal. An ideal signal is not costly to a high-type manufacturer but very costly to a low-type manufacturer. Clearly the fixed fee is not ideal because it is equally costly to each type. A better signal is for the high-type manufacturer to offer a higher wholesale price. For the high-type manufacturer, a higher wholesale price is not costly at all initially, while it is costly for the low-type:

$$\frac{\partial \Pi_h(w_h^*, h)}{\partial w} = 0,$$

whereas, if $w_h^* \geq w_l^*$, then

$$\frac{\partial \Pi_l(w_h^*, h)}{\partial w} < 0.$$

Another option for the high-type manufacturer is to offer a firm commitment: in stage 1 the manufacturer commits to purchase at least m units in stage 2, i.e., $q_f \geq m$, with any remaining units purchased for a wholesale price w. That contract could also be called a capacity reservation contract: at stage 1 the manufacturer reserves m units of the supplier's capacity that she promises to utilize fully at stage 2. The firm commitment is not costly to the manufacturer when $D_\theta \geq m$, but is costly when $D_\theta < m$. Since $D_l < m$ is more likely than $D_h < m$, a firm commitment is more costly to the low-demand manufacturer than the high-demand manufacturer. Hence, it too is a cheaper signal than the fixed payment. Interestingly, a firm commitment is not desirable with full information because it may lead to an ex-post-inefficient action: if $D_\theta < m$ then the production of $m - D_\theta$ units is wasted. However, with asymmetric information these contracts can enhance supply chain performance by allowing for the credible communication of essential information. See Cachon and Lariviere (2001) for a more detailed analysis of these different types of signaling instruments.

10.4 Discussion

This section considers a model in which one member of the supply chain has a better forecast of demand than the other. Since supply chain coordination requires that the amount of capacity the supplier builds depend on the demand forecast, supply chain coordination is achieved only if the demand forecast is shared accurately. With forced compliance the manufacturer can use an options contract to coordinate the supplier's action and to share information. However, sharing information is more costly with voluntary compliance. Nevertheless, some techniques for credibly sharing forecasts are cheaper than others. In particular, firm commitments, which are not optimal with full information, are effective for a manufacturer that needs to convince a supplier that her high-demand forecast is genuine.

Given that forecast sharing is costly even with the best signaling instrument, the high-demand manufacturer may wish to consider options other than signaling. One option is for the manufacturer to pay the supplier for units and take delivery of them in stage 1, i.e., before the demand realization is observed. In that case the supplier's profit does not depend on the manufacturer's demand distribution, so there is no need to share information. However, this option completely disregards the benefit of deferring production until after the demand realization is observed. A second option is for the manufacturer to choose a contract associated with a pooling equilibrium. In that case the supplier evaluates the contract as if he is dealing with a representative manufacturer. Hence, the terms are not as good as the high-demand manufacturer could get with full information, but this option could be attractive if there is a substantial cost to signal her high demand. (For example, a person in excellent health might opt for a standard health group life insurance plan merely to avoid the hassle of medical exams to demonstrate her excellent health.) Integration of all of these options awaits further research.

There are several other papers on supply chain contracting with asymmetric information. Cohen, Ho, Ren and Terwiesch (2001) study the forecasting process in the semiconductor equipment supply chain. In this setting the supplier has a long lead time to complete a piece of equipment and the manufacturer's desired completion time is uncertain. The manufacturer has an internal forecast for the desired delivery date and can provide a forecast of the delivery date to the supplier. However, since the manufacturer does not want the supplier to be late with a delivery, the manufacturer is biased to forecast a delivery date that is sooner than really needed. (The manufacturer is powerful in this supply chain and so can refuse to take delivery of completed equipment until the equipment is actually needed.) The supplier is well aware of this bias, but the research question is whether the supplier acts as if the forecast is biased. For example, if the manufacturer announces that the equipment is needed in the third quarter of a year does the supplier act as if the manufacturer really needs it in the first quarter of the following year. Using data from the industry they find that the supplier indeed acts as if the forecasts are biased. Terwiesch, Ren, Ho and Cohen (2002) extend this result to demonstrate empirically that suppliers given poor delivery lead times to manufacturers that are notorious for biased forecasting. Terwiesch and Loch (2002) also study signaling, but in the context of a product designer with an unknown ability to create valuable designs to their clients.

Porteus and Whang (1999) study a model that closely resembles the model in this section except they have the supplier offering the contract rather than the manufacturer. Hence, they study screening (the party without the information designs the contract to learn information) rather than signaling (the party with the information designs the contract to communicate information). As a result, the supplier offers a menu of contracts, one designed for each type of manufacturer.

Ha (1996) also studies a screening model. He has a supplier offering a contract to a manufacturer with stochastic demand. [Corbett and Tang (1999) study a similar model with deterministic demand.] The manufacturer knows his cost, but that cost is uncertain to the supplier. After the supplier offers the contract the manufacturer orders q units and sets the retail price. Supply chain coordination is possible with full information; however, the coordinating parameters depend on the manufacturer's cost. (His coordinating contract prohibits the manufacturer from setting a price below a specified threshold, i.e., a resale price maintenance provision.) If the supplier does not know the manufacturer's cost, Ha suggests the supplier offer a menu of contracts: for each order quantity the supplier stipulates a transfer payment and a minimum resale price. Each contract is targeted to a particular manufacturer, i.e., manufacturers with different costs choose different contracts and each manufacturer chooses the contract designed for his cost. That latter property is due to the revelation principle (Myerson, 1979).[48] Unfortunately, supply chain coordination is no longer possible. See Lim (2001) for a screening model that separates producers that vary in the quality of their output.

Corbett and de Groote (2000) study a model with one buyer, one supplier, deterministic demand and fixed ordering costs for each level in the supply chain. As in Porteus and Whang (1999) and Ha (1996), the contract designer has a prior belief regarding the other firm's cost and that firm knows his cost precisely. They propose a quantity-discount schedule, which is like a contract menu: there is a unique per unit price for each quantity the buyer may choose. As in Ha (1996) supply chain coordination is not achieved if there is asymmetric information. Corbett (2001) studies coordination with one supplier, one buyer, stochastic demand, fixed ordering costs and asymmetric information with respect to either the fixed ordering cost or the backorder penalty cost. He finds that consignment stock influences incentives, sometimes in a beneficial way, sometimes in a destructive way. (Consignment assigns ownership of inventory at the downstream firm to the upstream firm.)

Brown (1999) is related to the capacity procurement model. He has one supplier, one manufacturer and a single demand period with stochastic demand. But there are some important differences. First, he considers only forced compliance. Second, in his model the manufacturer announces her demand forecast to the supplier, and then immediately places her final order, i.e., there is no intermediate step in which the supplier builds capacity between the initial and final order. Brown requires that the manufacturer's order be consistent with the announced forecast, i.e., assuming the forecast is true the manufacturer's order is optimal. This constraint is reasonable because the supplier is able to immediately verify with certainty any inconsistency between the forecast and the order quantity. In the capacity procurement model the supplier is never able to verify for sure whether a biased forecast was provided, so that constraint would be problematic. Furthermore, it is necessary for the manufacturer to provide both an order quantity and a forecast, because there is a continuum of forecasts (where a forecast includes a mean and a standard

---

[48] Kreps (1990) describes the revelation principle as "... obvious once you understand it but somewhat cumbersome to explain." (p. 691). See his book for a good entry into mechanism design. In a nutshell, the revelation principle states that if there exists an optimal mechanism and that mechanism does not completely reveal the player's types, then the outcome of that mechanism can be replicated with a mechanism that does reveal the player's types. Thus, the search for optimal mechanisms can be restricted to truth-inducing mechanisms.

deviation) that yield the same optimal order quantity. In other words, he allows for a continuum of manufacturer types.

Brown studies two related contracts, an options contract (which he refers to as a buyback contract) and an options-futures contract. The latter is a buyback contract with a maximum threshold that the manufacturer can return: the manufacturer orders z units, pays c_f for the first y units, pays $w > c_f$ for the remaining $z-y$ units, and can return up to $z-y$ units for credit $b < w$. Hence, the options-futures contract contains firm commitments.

Both types of contracts coordinate the supply chain with full information and arbitrarily allocate profit. With asymmetric information Brown assumes the supplier accepts the contract only if it yields a minimum expect profit. However, with the buyback contract this fixed profit benchmark creates an incentive for the manufacturer to announce a biased forecast. To explain, let $\hat{\pi}$ be that benchmark, let $\{\mu(\sigma), \sigma\}$ be the mean-standard deviation pairs such that the manufacturer's order quantity is optimal and let $\Omega(\mu(\sigma), \sigma)$ be the supply chain's profit with that order quantity. The manufacturer must offer a contract in which the supplier's share of supply chain profit, $1-\lambda$, is at least $\hat{\pi}/\Omega^o(\mu(\sigma), \sigma)$. In the newsvendor setting $\Omega(\mu(\sigma), \sigma)$ is decreasing in σ, which implies $1-\lambda$ is increasing in σ, and the manufacturer's share, λ, is decreasing in σ. Hence, the manufacturer's optimal forecast has $\sigma = 0$ even if in reality $\sigma > 0$. If the supplier accepts that contract then his expected profit is in fact less than $\hat{\pi}$. Brown shows this incentive is eliminated if the manufacturer offers a futures-options contract: i.e., there exists a one for one relationship between the set of options-futures contracts and the set of manufacturer types such that a manufacturer always prefers the contract designated for his type. Hence, as in capacity procurement model, firm commitments are a useful tool for conveying information. They are particularly attractive in Brown's model because they do not result in suboptimal performance.

11 Conclusion

Over the last decade the legitimacy of supply chain contracting research has been established and many research veins have been tapped. Several key conclusions have emerged. First, coordination failure is common; incentive conflicts plausibly arise in a wide range of operational situations. As a result, suboptimal supply chain performance is not necessarily due to incompetent managers or naive operating policies. Rather, poor supply chain performance may be due to conflicting incentives and these incentive conflicts can be managed. Second, in many situations there are multiple kinds of contracts that achieve coordination and arbitrarily divide profit. Hence, the contract selection process in practice must depend on criteria or objectives that have not been fully explored, i.e., there is a need for additional research that investigates the subtle, but possibly quite important, differences among the set of coordinating contracts. Third, the consequence of coordination failure is context specific: there are situations in which supply chain performance is nearly optimal with naive and simple contracts, but there are also situations in which decentralized operations without proper incentive management leads to substantially deteriorated performance. It is quite useful to have theory that can help to contrast those cases. Fourth, this body of work emphasizes that managing incentive conflicts can lead to Pareto improvements, which is often referred to as a 'win-win' situation in practice. This insight should help to break the 'zero sum game' mentality which is understandably so prevalent among supply chain managers and is a strong impediment to significant supply chain progress.[49] While vigilance is always prudent, a wise supply chain manager recognizes that not every offer is a wolf in sheep's clothing.

Unfortunately, theory has almost exclusively followed practice in this domain, i.e., practice has been used as a motivation for theoretical work, but theoretical work has not found its way into practice. This need not be so. As already mentioned, one of the surprises of this research is that coordination can be achieved with many different contractual forms. An understanding of the subtle differences among these contracts may allow a researcher to identify a particularly suitable contract form for an industry, even if that contract form has no precedence in the industry. Just as there has been documented improvements with innovations like delayed differentiation (Lee, 1996) and accurate response (Fisher & Raman, 1996), it should be possible to generate equally valuable improvements via innovations in incentive design.

As a first step toward wider implementation, this research needs to develop an empirical–theoretical feedback loop. As this chapter illustrates, the literature contains a considerable amount of theory, but an embarrassingly paltry amount of empiricism. Thus, we have little guidance on how the theory should now proceed. For example, we have identified a number of contracts that coordinate a supplier selling to a newsvendor but can we explain why certain types have been adopted in certain industries and not others? Can we explain why these contracts have not completely eliminated the Pareto inferior wholesale-price contract? A standard argument is that the wholesale-price contract is cheaper to administer, but we lack any evidence regarding the magnitude of the administrative cost of the more complex contracts. And even if the coordinating contracts are adopted, such as buybacks or revenue sharing, are coordinating parameters chosen in practice? For example, the set of revenue-sharing contracts is much larger than the set of coordinating revenue-sharing contracts. If we observe that firms choose noncoordinating contracts, then we need an explanation. Irrational or incompetent behavior on the part of managers is a convenient explanation, but it is not satisfying to build a theory on irrational behavior. A theory is

---

[49] In a zero sum game one player's payoff is decreasing in the other player's payoff, so one player can be made better off only by making the other player worse off.

interesting only if it can be refuted and irrational behavior cannot be refuted. A better approach is to challenge the assumptions and analysis of the theory. With some empiricism we should be able to identify which parts of the theory are sound and which deserve more scrutiny.

The franchise literature could provide a useful guide to researchers in supply chain contracting. An excellent starting point is Lafontaine and Slade (2001). They review and compare the extensive theoretical and empirical results on franchising. Some of the predictions from theory are indeed supported by numerous empirical studies, while others are lacking. It is clear that the give-and-take between theory and data has been enormously successful for that body of work.

On a hopeful note, some preliminary activity in the empirical domain has fortunately begun. Mortimer (2000) provides an analysis of revenue-sharing contracts in the video cassette rental industry. Cohen, Ho, Ren and Terwiesch (2002) carefully evaluate forecast sharing in the semiconductor equipment industry. Their findings are consistent with the premise in Cachon and Lariviere (2001): if forecasts are not credible, then they will be ignored and supply chain performance suffers. Follow on work of theirs demonstrates that providing poor forecasts leads to lower future credibility and lower received service from a supplier. Novak and Eppinger (2001) empirically evaluate the interaction between product complexity and the make or buy decision, and find support for the property rights theory of vertical integration. Finally, Randall, Netessine and Rudi (2002) study whether e-retailers choose to drop ship or hold their own inventory. The appropriate strategy for a retailer depends on the characteristics of its product and industry, as predicted by the theoretical work in Netessine and Rudi (2000a), and they indeed find that e-retailers that chose the appropriate strategy were less likely to bankrupt.

Even though our most rewarding efforts now lie with collecting data, it is still worthwhile to comment on areas of the theory that need on additional investigation. Current models are too dependent on single-shot contracting. Most supply chain interactions occur over long periods of time with many opportunities to renegotiate or to interact with spot markets. For some steps in this direction, see Kranton and Minehart (2001) for work on buyer–supplier networks and long-run relationships; Plambeck and Taylor (2002) for a model with renegotiation of quantity-flexibility contracts; and Wu, Kleindorfer and Zhang (2002) and Lee and Whang (2002) for the impact of spot markets on capacity contracting and inventory procurement, respectively.

More research is needed on how multiple suppliers compete for the affection of multiple retailers, i.e., additional emphasis is needed on many-to-one or many-to-many supply chain structures. In the context of auction theory, Jin and Wu (2002) and Chen (2001) study procurement in the many-to-one structure and Bernstein and Véricourt (2002) offer some initial work on a many-to-many supply chain. Forecasting and other types of information sharing require much more attention. Lariviere (2002) provides recent work in this area. Finally, more work is needed on how scarce capacity is allocated in a supply chain and how scarce capacity influences behavior in the supply chain. Recent work in this area is provided by Zhao, Deshpande and Ryan (2002) and Deshpande and Schwarz (2002).

To summarize, opportunities abound.

Acknowledgements

I would like to thank the many people who carefully read and commented on the first two drafts of this manuscript: Ravi Anupindi, Fangruo Chen, Charles Corbett, James Dana, Ananth Iyer, Ton de Kok, Yigal Gerchak, Mark Ferguson, Paul Kleindorfer, Howard Kunreuther, Marty Lariviere, Serguei Netessine, Ediel Pinker, Nils Rudi, Leroy Schwarz, Sridhar Seshadri, Greg Shaffer, Yossi Sheffi, Terry Taylor, Christian Terwiesch, Andy Tsay and Kevin Weng. I am, of course, responsible for all remaining errors.

References

Agrawal, N., A.A. Tsay (2001). Intrafirm incentives and supply chain performance, in: J.S. Song, D.D. Yao (eds.), *Supply Chain Structures: Coordination, Information and Optimization (Volume 42 of the International Series in Operations Research and Management Science)*, Norwell, MA, Kluwer Academic Publishers, pp. 44–72.

Agrawal, V., S. Seshadri (2000). Risk intermediation in supply chains. *IIE Transactions* 32, 819–831.

Ailawadi, K. (2001). The retail power-performance conundrum: what have we learned. *Journal of Retailing* 77, 299–318.

Anand, K., H. Mendelson (1997). Information and organization for horizontal multimarket coordination. *Management Science* 43, 1609–1627.

Anand, K., R. Anupindi, Y. Bassok (2001). Strategic inventories in procurement contracts. University of Pennsylvania Working Paper, Philadelphia, PA.

Andersson, J., J. Marklund (2000). Decentralized inventory control in a two-level distribution system. *European Journal of Operational Research* 127(3), 483–506.

Anupindi, R., R. Akella (1993). An inventory model with commitments. University of Michigan Working Paper.

Anupindi, R., Y. Bassok (1999). Centralization of stocks: retailers vs. manufacturer. *Management Science* 45(2), 178–191.

Anupindi, R., Y. Bassok, E. Zemel (2001). A general framework for the study of decentralized distribution systems. *Manufacturing and Service Operations Management* 3(4), 349–368.

Atkins, D., X. Zhao (2002). Supply chain structure under price and service competition. University of British Columbia Working Paper.

Aviv, Y. (2001). The effect of collaborative forecasting on supply chain performance. *Management Science* 47(10), 1326–1343.

Aviv, Y., A. Federgruen (1998). The operational benefits of information sharing and vendor managed inventory (VMI) programs. Washington University Working Paper, St. Louis, MO.

Axsater, S. (2001). A framework for decentralized multi-echelon inventory control. *IIE Transactions* 33, 91–97.

Baiman, S., P. Fischer, M. Rajan (2000). Information, contracting and quality costs. *Management Science* 46(6), 776–789.

Baiman, S., P. Fischer, M. Rajan (2001). Performance measurement and design in supply chains. *Management Science* 47(1), 173–188.

Barlow, R.E., F. Proschan (1965). *Mathematical Theory of Reliability*, John Wiley & Sons.

Barnes-Schuster, D., Y. Bassok, R. Anupindi (2002). Coordination and flexibility in supply contracts with options. *Manufacturing and Service Operations Management* 4, 171–207.

Bassok, Y., R. Anupindi (1997a). Analysis of supply contracts with total minimum commitment. *IIE Transactions* 29(5).

Bassok, Y., R. Anupindi (1997b). Analysis of supply contracts with commitments and flexibility. University of Southern California Working Paper.

Bernstein, F., G. DeCroix (2002). Decentralized pricing and capacity decisions in a multi-tier system with modular assembly. Duke University Working Paper.

Bernstein, F., A. Federgruen (1999). Pricing and replenishment strategies in a distribution system with competing retailers. Forthcoming *Operations Research*.

Bernstein, F., A. Federgruen (2000). Decentralized supply chains with competing retailers under demand uncertainty. Forthcoming *Management Science*.

Bernstein, F., F. Véricourt (2002). Allocation of supply contracts with service guarantees. Duke University Working Paper.

Bernstein, F., F. Chen, A. Federgruen (2002). Vendor managed inventories and supply chain coordination: the case with one supplier and competing retailers. Duke University Working Paper.

Bloom, P., V. Perry (2001). Retailer power and supplier welfare: the case of Wal-Mart. *Journal of Retailing* 77, 379–396.

Boyaci, T., G. Gallego (1997). Coordination issues in simple supply chains. Columbia University Working Paper, New York, NY.

Bresnahan, T., P. Reiss (1985). Dealer and manufacturer margins. *Rand Journal of Economics* 16(2), 253–268.

Brown, A. (1999). A coordinating supply contract under asymmetric demand information: guaranteeing honest information sharing. Vanderbilt University Working Paper, Nashville, TN.

Brown, A., H. Lee (1998). The win-win nature of options based capacity reservation arrangements. Vanderbilt University Working Paper, Nashville, TN.

Butz, D. (1997). Vertical price controls with uncertain demand. *Journal of Law and Economics* 40, 433–459.

Cachon, G. (1998). Competitive supply chain inventory management, in: S. Tayur, R. Ganeshan, M. Magazine (eds.), *Quantitative Models for Supply Chain Management*, Boston, Kluwer.

Cachon, G. (1999). Competitive and cooperative inventory management in a two-echelon supply chain with lost sales. University of Pennsylvania Working Paper, Philadelphia, PA.

Cachon, G. (2001). Stock wars: inventory competition in a two echelon supply chain. *Operations Research* 49(5), 658–674.

Cachon, G. (2002). The allocation of inventory risk and advanced purchase discounts in a supply chain. University of Pennsylvania Working Paper.

Cachon, G., M. Fisher (2000). Supply chain inventory management and the value of shared information. *Management Science* 46(8), 1032–1048.

Cachon, G., M. Lariviere (1997). Capacity allocation with past sales: when to turn-and-earn. *Management Science* 45(5), 685–703.

Cachon, G., M. Lariviere (1999). Capacity choice and allocation: strategic behavior and supply chain performance. *Management Science* 45(8), 1091–1108.

Cachon, G., M. Lariviere (2000). Supply chain coordination with revenue sharing: strengths and limitations. Forthcoming. *Management Science*.

Cachon, G., M. Lariviere (2001). Contracting to assure supply: how to share demand forecasts in a supply chain. *Management Science* 47(5), 629–646.

Cachon, G., P. Zipkin (1999). Competitive and cooperative inventory policies in a two-stage supply chain. *Management Science* 45(7), 936–953.

Caldentey, R., L. Wein (1999). Analysis of a production-inventory system. Massachusetts Institute of Technology Working Paper, Cambridge, MA.
Chen, F. (1998). Echelon reorder points, installation reorder points, and the value of centralized demand information. *Management Science* 44(12), S221–S234.
Chen, F. (1999). Decentralized supply chains subject to information delays. *Management Science* 45(8), 1076–1090.
Chen, Y. (1999). Oligopoly price discrimination and resale price maintenance. *Rand Journal of Economics* 30(3), 441–455.
Chen, F. (2001). Auctioning supply contracts. Columbia University Working Paper.
Chen, F., A. Federgruen (2000). Mean-variance analysis of basic inventory models. Columbia University Working Paper.
Chen, F., Y.S. Zheng (1994). Lower bounds for multi-echelon stochastic inventory system. *Management Science* 40(11), 1426–1443.
Chen, F., A. Federgruen, Y. Zheng (2001). Coordination mechanisms for decentralized distribution systems with one supplier and multiple retailers. *Management Science* 47(5), 693–708.
Cho, R., Y. Gerchak (2001). Efficiency of independent downstream firm could counteract coordination difficulties. University of Waterloo Working Paper.
Chod, J., N. Rudi (2002). Resource flexibility with responsive pricing.
Choi, K.S., J.G. Dai, J.S. Song (2002). On measuring supplier performance under vendor-managed inventory programs. University of California at Irvine Working Paper.
Chu, W., P. Desai (1995). Channel coordination mechanisms for customer satisfaction. *Marketing Science* 14(4), 343–359.
Clark, A.J., H.E. Scarf (1960). Optimal policies for a multi-echelon inventory problem. *Management Science* 6, 475–490.
Cohen, M., T. Ho, J. Ren, C. Terwiesch (2001). Measuring inputed costs in the semiconductor equipment supply chain. University of Pensylvania Working Paper.
Corbett, C. (2001). Stochastic inventory systems in a supply chain with asymmetric information: cycle stocks, safety stocks, and consignment stock. *Operations Research* 49(4), 487–500.
Corbett, C., G. DeCroix (2001). Shared savings contracts in supply chains. *Management Science* 47(7), 881–893.
Corbett, C., X. de Groote (2000). A supplier's optimal quantity discount policy under asymmetric information. *Management Science* 46(3), 444–450.
Corbett, C., U. Karmarkar (2001). Competition and structure in serial supply chains. *Management Science* 47(7), 966–978.
Corbett, C., C. Tang (1999). Designing supply contracts: contract type and information asymmetry, in: S. Tayur, R. Ganeshan, M. Magazine (eds.), *Quantitative Models for Supply Chain Management*, Boston, Kluwer Academic Publishers.
Dana, J., K. Spier (2001). Revenue sharing and vertical control in the video rental industry. *The Journal of Industrial Economics* 49(3), 223–245.
Debo, L. (1999). Repeatedly selling to an impatient newsvendor when demand fluctuates: a supergame theoretic framework for co-operation in a supply chain. Carnegie Mellon University Working Paper.
Demougin, D.M. (1989). A renegotiation-proof mechanism for a principal-agent model with moral hazard and adverse selection. *Rand Journal of Economics* 20(2), 256–267.
Deneckere, R., H. Marvel, J. Peck (1996). Demand uncertainty, inventories, and resale price maintenance. *Quarterly Journal of Economics* 111, 885–913.
Deneckere, R., H. Marvel, J. Peck (1997). Demand uncertainty and price maintenance: markdowns as destructive competition. *American Economic Review* 87(4), 619–641.
Desai, P., K. Srinivasan (1995). Demand signalling under unobservable effort in franchising. *Management Science* 41(10), 1608–1623.
Desiraju, R., S. Moorthy (1997). Managing a distribution channel under asymmetric information with performance requirements. *Management Science* 43, 1628–1644.
Deshpande, V., L. Schwarz (2002). Optimal capacity allocation in decentralized supply chains. Purdue University Working Paper.

Dolan, R., J.B. Frey (1987). Quantity discounts: managerial issues and research opportunities/commentary/reply. *Marketing Science* 6(1), 1–24.

Dong, L., N. Rudi (2001). Supply chain interaction under transshipments. Washington University Working Paper.

Donohue, K. (2000). Efficient supply contracts for fashion goods with forecast updating and two production modes. *Management Science* 46(11), 1397–1411.

Duenyas, I., C.-Y. Tsai (2001). Centralized and decentralized control of a two-stage tandem manufacturing system with demand for intermediate and end products. University of Michigan Working Paper, Ann Arbor, MI.

Duenyas, I., W. Hopp, Y. Bassok (1997). Production quotas as bounds on interplant JIT contracts. *Management Science* 43(10), 1372–1386.

Eeckhoudt, L., C. Gollier, H. Schlesinger (1995). The risk-averse (and prudent) newsboy. *Management Science* 41(5), 786–794.

Emmons, H., S. Gilbert (1998). Returns policies in pricing and inventory decisions for catalogue goods. *Management Science* 44(2), 276–283.

Eppen, G., A. Iyer (1997). Backup agreements in fashion buying – the value of upstream flexibility. *Management Science* 43(11), 1469–1484.

Erhun, F., P. Keskinocak, S. Tayur (2000). Spot markets for capacity and supply chain coordination. Carnegie Mellon Working Paper, Pittsburgh, PA.

Erkoc, M., S.D. Wu (2002a). Managing high-tech capacity expansion via reservation contracts. Lehigh University Working Paper.

Erkoc, M., S.D. Wu (2002b). Due-date coordination in an internal market via risk sharing. Lehigh University Working Paper.

Ertogral, K., D. Wu (2001). A bargaining game for supply chain contracting. Lehigh University Working Paper.

Federgruen, A., P. Zipkin (1984). Computational issues in an infinite-horizon, multiechelon inventory model. *Operations Research* 32(4), 818–836.

Ferguson, M., G. DeCroix, P. Zipkin (2002). When to commit in a multi-echelon supply chain with partial information updating. Duke University Working Paper.

Fisher, M., A. Raman (1996). Reducing the cost of demand uncertainty through accurate response to early sales. *Operations Research* 44, 87–99.

Friedman, J. (1986). *Game Theory with Applications to Economics*, Oxford, Oxford University Press.

Gallini, N., N. Lutz (1992). Dual distribution and royalty fees in franchising. *The Journal of Law, Economics, and Organization* 8, 471–501.

Gans, N. (2002). Customer loyalty and supplier quality competition. *Management Science* 48(2), 207–221.

Gavirneni, S., R. Kapuscinski, S. Tayur (1999). Value of information in capacitated supply chains. *Management Science* 45(1), 16–24.

Gerchak, Y., Y. Wang (1999). Coordination in decentralized assembly systems with random demand. University of Waterloo Working Paper, Waterloo, Ontario, Canada.

Gerchak, Y., R. Cho, S. Ray (2001). Coordination and dynamic shelf-space management of video movie rentals. University of Waterloo Working Paper, Waterloo, Ontario.

Gilbert, S., V. Cvsa (2000). Strategic supply chain contracting to stimulate downstream process innovation. University of Texas at Austin Working Paper.

Gurnani, H., Y. Gerchak (1998). Coordination in decentralized assembly systems with uncertain component yield. University of Waterloo Working Paper, Waterloo, Ontario, Canada.

Ha, A. (2001). Supplier-buyer contracting: asymmetric cost information and the cut-off level policy for buyer participation. *Naval Research Logistics* 48(1), 41–64.

Ha, A., L. Li, S.-M. Ng (2000). Price and delivery logistics competition in a supply chain. Yale School of Management Working Paper, New Haven, CT.

Holden, S. (1999). Renegotiation and the efficiency of investments. *Rand Journal of Economics* 30(1), 106–119.

Holmstrom, B. (1979). Moral hazard and observability. *Bell Journal of Economics* 10(1), 74–91.

Holmstrom, B. (1982). Moral hazard in teams. *Bell Journal of Economics* 13(2), 324–340.

Ippolito, P. (1991). Resale price maintenance: economic evidence from litigation. *Journal of Law and Economics* 34, 263–294.

Iyer, A., M. Bergen (1997). Quick response in manufacturer-retailer channels. *Management Science* 43(4), 559–570.

Jeuland, A., S. Shugan (1983). Managing channel profits. *Marketing Science* 2, 239–272.

Jin, M., S.D. Wu (2002). Procurement auctions with supplier coalitions: validity requirements and mechanism design. Lehigh University Working Paper.

Karjalainen, R. (1992). The newsboy game. University of Pennsylvania Working Paper, Philadelphia, PA.

Katz, M. (1989). Vertical contractual relations, in: R. Schmalensee, R. Willig (eds.), *Handbook of Industrial Organization*, Vol. 1. Boston, North-Holland.

Kolay, S., G. Shaffer (2002). All-unit discounts in retail contracts. University of Rochester Working Paper.

Kouvelis, P., G. Gutierrez (1997). The newsvendor problem in a global market: optimal centralized and decentralized control policies for a two-market stochastic inventory system. *Management Science* 43(5), 571–585.

Kouvelis, P., M. Lariviere (2000). Decentralizing cross-functional decisions: coordination through internal markets. *Management Science* 46(8), 1049–1058.

Kranton, R., D. Minehart (2001). A theory of buyer-seller networks. *American Economic Review* 91(3), 485–508.

Krasa, S., A. Villamil (2000). Optimal contracts when enforcement is a decision variable. *Econometrica* 68(1), 119–134.

Kreps, D.M. (1990). *A Course in Microeconomic Theory*, Princton, Princeton University Press.

Krishnan, H., R. Kapuscinski, D. Butz (2001). Coordinating contracts for decentralized supply chains with retailer promotional effort. University of Michigan Working Paper.

Lafountaine, F. (1992). Agency theory and franchising: some empirical results. *Rand Journal of Economics* 23(2), 263–283.

Lafontaine, F., M. Slade (2001). Incentive contracting and the franchise decision, in: K. Chatterjee, W. Samuelson (eds.), *Game Theory and Business Applications*, Boston, Kluwer Academic Publishing.

Lal, R. (1990). Improving channel coordination through franchising. *Marketing Science* 9, 299–318.

Lariviere, M. (1998). Supply chain contracting and co-ordination with stochastic demand, in: S. Tayur, R. Ganeshan, M. Magazine (eds.), *Quantitative Models for Supply Chain Management*, Boston, Kluwer.

Lariviere, M. (2002). Inducing forecast revelation through restricted returns. Northwestern University Working Paper.

Lariviere, M., V. Padmanabhan (1997). Slotting allowances and new product introductions. *Marketing Science* 16, 112–128.

Lariviere, M., E. Porteus (2001). Selling to the newsvendor: an analysis of price-only contracts. *Manufacturing and Service Operations Management* 3(4), 293–305.

Lee, H. (1987). A multi-echelon inventory model for repairable items with emergency lateral transshipments. *Management Science* 33(10), 1302–1316.

Lee, H. (1996). Effective inventory and service management through product and process redesign. *Operations Research* 44(1), 151–159.

Lee, H., S. Whang (1999). Decentralized multi-echelon supply chains: incentives and information. *Management Science* 45(5), 633–640.

Lee, H., S. Whang (2002). The impact of the secondary market on the supply chain. *Management Science* 48, 719–731.

Lee, H., V. Padmanabhan, T. Taylor, S. Whang (2000). Price protection in the personal computer industry. *Management Science* 46(4), 467–482.

Lee, H., K.C. So, C. Tang (2000). The value of information sharing in a two-level supply chain. *Management Science* 46, 626–643.

Lim, W. (2001). Producer-supplier contracts with incomplete information. *Management Science* 47(5), 709–715.
Lippman, S., K. McCardle (1997). The competitive newsboy. *Operations Research* 45, 54–65.
Lu, X., J.S. Song, A. Regan (2002). Rebate, returns and price protection policies in supply chain coordination. University of California at Irvine Working Paper.
Mahajan, S., G. van Ryzin (1999). Retail inventories and consumer choice, in: S. Tayur, R. Ganeshan, M. Magazine (eds.), *Quantitative Models for Supply Chain Management*, Boston, Kluwer.
Mahajan, S., G. van Ryzin (2001). Inventory competition under dynamic consumer choice. *Operations Research* 49(5), 646–657.
Majumder, P., H. Groenevelt (2001). Competition in remanufacturing. *Production and Operations Management* 10(2), 125–141.
Marvel, H., J. Peck (1995). Demand uncertainty and returns policies. *International Economic Review* 36(3), 691–714.
Marx, L., G. Shaffer (2001a). Bargaining power in sequential contracting. University of Rochester Working Paper.
Marx, L., G. Shaffer (2001b). Rent shifting and efficiency in sequential contracting. University of Rochester Working Paper.
Marx, L., G. Shaffer (2002). Base contracts in games with ex-post observability. University of Rochester Working Paper.
Messinger, P., C. Narasimhan (1995). Has power shifted in the grocery channel. *Marketing Science* 14(2), 189–223.
Milner, J., E. Pinker (2001). Contingent labor contracting under demand and supply uncertainty. *Management Science* 47(8), 1046–1062.
Moinzadeh, K., S. Nahmias (2000). Adjustment strategies for a fixed delivery contract. *Operations Research* 48(3), 408–423.
Moorthy, K.S. (1987). Managing channel profits: comment. *Marketing Science* 6, 375–379.
Mortimer, J.H. (2000). The effects of revenue-sharing contracts on welfare in vertically separated markets: evidence from the video rental industry. University of California at Los Angeles Working Paper, Los Angeles, CA.
Moses, M., S. Seshadri (2000). Policy mechanisms for supply chain coordination. *IIE Transactions* 32, 245–262.
Myerson, R. (1979). Incentive compatibility and the bargaining problem. *Econometrica* 47, 399–404.
Nahmias, S. (1993). *Production and Operations Analysis*, Boston, Irwin.
Narayanan, V., A. Raman (1997). Contracting for inventory in a distribution channel with stochastic demand and substitute products. Harvard University Working Paper.
Netessine, S., N. Rudi (2000). Supply chain structures on the internet: marketing-operations coordination. University of Pennsylvania Working Paper, Philadelphia, PA.
Netessine, S., N. Rudi (2000b). Centralized and competitive inventory models with demand substitution. Forthcoming, *Operations Research*.
Netessine, S., R. Shumsky (2001). Revenue management games. University of Pennsylvania Working Paper, Philadelphia, PA.
Novak, S., S. Eppinger (2001). Sourcing by design: product complexity and the supply chain. *Management Science* 47(1), 189–204.
O'Brien, D., G. Shaffer (1992). Vertical control with bilateral contracts. *Rand Journal of Economics* 23, 299–308.
Padmanabhan, V., I.P.L. Png (1995). Returns policies: make money by making good. *Sloan Management Review* Fall, 65–72.
Padmanabhan, V., I.P.L. Png (1997). Manufacturer's returns policy and retail competition. *Marketing Science* 16(1), 81–94.
Parker, R., R. Kapuscinsky (2001). Managing a non-cooperative supply chain with limited capacity. University of Michigan Working Paper.
Parlar, M. (1988). Game theoretic analysis of the substitutable product inventory problem with random demands. *Naval Research Logistics Quarterly* 35, 397–409.

Pasternack, B. (1985). Optimal pricing and returns policies for perishable commodities. *Marketing Science* 4(2), 166–176.

Pasternack, B. (1999). Using revenue sharing to achieve channel coordination for a newsboy type inventory model. CSU Fullerton Working Paper, Fullerton.

Petruzzi, N., M. Dada (1999). Pricing and the newsvendor problem: a review with extensions. *Operations Research* 47, 183–194.

Plambeck, E., T. Taylor (2002). Sell the plant? The impact of contract manufacturing on innovation, capacity and profitability. Stanford University Working Paper.

Plambeck, E., S. Zenios (2000). Performance-based incentives in a dynamic principal-agent model. *Manufacturing and Service Operations Management* 2, 240–263.

Porteus, E. (2000). Responsibility tokens in supply chain management. *Manufacturing and Service Operations Management* 2(2), 203–219.

Porteus, E., S. Whang (1991). On manufacturing/marketing incentives. *Management Science* 37(9), 1166–1181.

Porteus, E., S. Whang (1999). Supply chain contracting: non-recurring engineering charge, minimum order quantity, and boilerplate contracts. Stanford University Working Paper.

Randall, R., S. Netessine, N. Rudi (2002). Inventory structure and internet retailing: an empirical examination of the role of inventory ownership. University of Utah Working Paper.

Reyniers, D., C. Tapiero (1995). The delivery and control of quality in supplier-producer contracts. *Management Science* 41(10), 1581–1589.

Riordan, M. (1984). Uncertainty, asymmetric information and bilateral contracts. *Review of Economic Studies* 51, 83–93.

Rochet, J., L. Stole (2002). Nonlinear pricing with random participation. *Review of Economic Studies* 69, 277–311.

Roundy, R. (1985). 98%-effective integer ratio lot-sizing for one-warehouse multi-retailer systems. *Management Science* 31, 1416–1430.

Rubinstein, A. (1989). The electronic mail game: strategic behavior under 'almost common knowledge'. *American Economic Review* 79(3), 385–391.

Rudi, N., S. Kapur, D. Pyke (2001). A two-location inventory model with transhipment and local decision making. *Management Science* 47(12), 1668–1680.

Scheller-Wolf, A., S. Tayur (1997). Reducing international risk through quantity contracts. Carnegie Mellon Working Paper, Pittsburgh, PA.

Schweitzer, M., C. Cachon (2000). Decision bias in the newsvendor problem: experimental evidence. *Management Science* 46(3), 404–420.

Shaffer, G. (1991). Slotting allowances and resale price maintenance: a comparison of facilitating practices. *Rand Journal of Economics* 22, 120–135.

Silver, E., D. Pyke, R. Peterson (1998). *Inventory Management and Production Planning and Scheduling*, New York, John Wiley and Sons.

Spengler, J. (1950). Vertical integration and antitrust policy. *Journal of Political Economy*, 347–352.

Stidham, S. (1992). Pricing and capacity decisions for a service facility: stability and multiple local optima. *Management Science* 38(8), 1121–1139.

Tagaras, G., M.A. Cohen (1992). Pooling in two-location inventory systems with non-negligible replenishment lead times. *Management Science* 38(8), 1067–1083.

Taylor, T. (2001). Channel coordination under price protection, midlife returns and end-of-life returns in dynamic markets. *Management Science* 47(9), 1220–1234.

Taylor, T. (2002). Coordination under channel rebates with sales effort effect. *Management Science* 48(8), 992–1007.

Taylor, T. (2002b). Sale timing in a supply chain: when to sell to the retailer. Columbia University Working Paper.

Tayur, S., Ganeshan, R., Magazine, M. (eds.) (1998). *Quantitative Models for Supply Chain Management*, Boston, Kluwer.

Terwiesch, C., C. Loch (2002). Collaborative prototyping and the pricing of custom designed products. University of Pennsylvania Working Paper.

Terwiesch, C., J. Ren, T. Ho, M. Cohen (2002). An empirical analysis of forecast sharing in the semiconductor equipment supply chain. University of Pennsylvania Working Paper.

Tirole, J. (1986). Procurement and renegotiation. *Journal of Political Economy* 94(2), 235–259.

Tomlin, B. (2000). Capacity investments in supply chain: sharing-the-gain rather than sharing-the-pain. University of North Carolina Working Paper.

Topkis, D. (1998). *Supermodularity and Complementarity*, Princeton, Princeton University Press.

Tsay, A. (1999). Quantity-flexibility contract and supplier-customer incentives. *Management Science* 45(10), 1339–1358.

Tsay, A. (2001). Managing retail channel overstock: markdown money and return policies. *Journal of Retailing* 77, 457–492.

Tsay, A. (2002). Risk sensitivity in distribution channel partnership: implications for manufacturer return policies. *Journal of Retailing* 78, 147–160.

Tsay, A., N. Agrawal (1999). Channel conflict and coordination: an investigation of supply chain design. Santa Clara University Working Paper, Santa Clara, CA.

Tsay, A., N. Agrawal (2000). Channel dynamics under price and service competition. *Manufacturing and Service Operations Management*. 2(4), 372–391.

Tsay, A., W.S. Lovejoy (1999). Quantity-flexibility contracts and supply chain performance. *Manufacturing and Service Operations Management* 1(2), 89–111.

Tsay, A., S. Nahmias, N. Agrawal (1998). Modeling supply chain contracts: a review, in: S. Tayur, R. Ganeshan, M. Magazine (eds.), *Quantitative Models for Supply Chain Management*, Boston, Kluwer.

van Mieghem, J. (1999). Coordinating investment, production and subcontracting. *Management Science* 45(7), 954–971.

van Mieghem, J., M. Dada (1999). Price vs production postponement: capacity and competition. *Management Science* 45(12), 1631–1649.

van Ryzin, G., S. Mahajan (1999). Supply chain coordination under horizontal competition. Columbia University Working Paper, New York, NY.

Wang, Y., Y. Gerchak (2001). Supply chain coordination when demand is shelf-space dependent. *Manufacturing and Service Operations Management* 3(1), 82–87.

Wang, Y., L. Jiang, Z.J. Shen (2001). Consignment sales, price-protection decisions and channel performances. Case Western Reserve University Working Paper.

Watson, Noel (2002). Execution in supply chain management: dynamics, mis-steps and mitigation strategies. University of Pennsylvania Dissertation.

Webster, S., S.K. Weng (2000). A risk-free perishable item returns policy. *Manufacturing and Service Operations Management* 2(1), 100–106.

Whang, S. (1995). Coordination in operations: a taxonomy. *Journal of Operations Management* 12, 413–422.

Wilson, R. (1993). *Nonlinear Pricing*, Oxford, Oxford University Press.

Wu, D.J., P. Kleindorfer, J. Zhang (2002). Optimal bidding and contracting strategies for capital-intensive goods. *European Journal of Operational Research* 137, 657–676.

Yüksel, O., H. Lee (2002). Sharing inventory risks for customized components. Stanford University Working Paper.

Zhao, H., V. Deshpande, J. Ryan (2002). Inventory sharing and rationing in decentralized dealer networks. Purdue University Working Paper.

A.G. de Kok and S.C. Graves, Eds., *Handbooks in OR & MS, Vol. 11*
© 2003 Elsevier B.V. All rights reserved.

Chapter 7

Information Sharing and Supply Chain Coordination

Fangruo Chen

Graduate School of Business, Columbia University
New York, NY 10027, USA

1 Introduction

The performance of a supply chain depends critically on how its members coordinate their decisions. And it is hard to imagine coordination without some form of information sharing. A significant part of supply chain management research is devoted to understanding the role of information in achieving supply chain coordination. It is the purpose of this chapter to review this literature.[1]

The first part of the chapter focuses on papers that have contributed to our understanding of the value of shared information. We first consider information pertaining to the downstream part of the supply chain. The next is upstream information. Finally, we discuss papers that study the consequences of imperfect transmission of information. All the papers here adopt the perspective of a central planner whose goal is to optimize the performance of the entire supply chain.

The chapter then proceeds to discuss papers that address incentive issues in information sharing. Here it is made explicit and prominent that supply chains are composed of independent firms with private information. The goal is to understand whether or not incentives for sharing information exist, and if not, how they can be created. This section is divided into three parts. When one firm has superior information, it may hide it to gain strategic advantage or reveal it to gain cooperation. The former, the less-informed party may try to provide incentives for the informed to release

---

[1] For a summary of recent industry initiatives to improve supply chain information flows, see Lee and Whang (2000).

its information. This is called screening, and it constitutes the first part of the section. If the informed tries to convey its private information, it is often the case that he has to 'put his money where his mouth is' in order to be credible, i.e., signaling. This is the second part of the section. The last part of the section deals with situations where it is difficult to say if a supply chain member has more or less information; they simply have different information about something they all care about (e.g., the potential market size of a product). Here a common question is if information sharing will emerge as an equilibrium outcome in some noncooperative game.

The chapter ends with some thoughts on future research directions.

The structure of the chapter provides an implicit taxonomy for thinking about research on supply chain information sharing. Specifically, the sections and subsections provide categories so that (hopefully) every piece of relevant research finds its home. It is important to mention that the unnumbered, boldfaced headings are meant to represent examples within a category, and the examples may not be exhaustive. For example, Section 2.2 deals with the value of upstream information, and within this subsection are several examples (i.e., cost, lead time, capacity information). This should not be taken to mean either that these are the only types of upstream information or that they can only come from upstream. For example, an upstream supply chain member (e.g., a manufacturer) may have some private information about demand that the (downstream) retailers do not have, so demand information is a possible type of upstream information. On the other hand, a seller may not know a buyer's cost structure, so cost information can also come from downstream. In other words, the headings without section/subsection numbers are not part of the taxonomy anymore.

A few words on how we choose the papers and what we are going to do with them. The emphasis here is modeling, not analysis. Therefore, we prefer papers with modeling novelties, and we want as much variety as possible given limited space. So if there are several papers that are close to each other in the 'novelty space', we will just take one with a simple reference to the others. Within a collection of papers, if there is no clear logical progression, we will simply review them in chronological sequence. We will often present a model without stating all the assumptions. The mentioning of results is often brief and is meant to whet your appetite so that the original paper becomes irresistable. We sometimes purposely 'trivialize' a model by further specializing it (e.g., by making assumptions that a top journal referee would be hard pressed to swallow). The reader should be assured that this is done for ease of presentation and for crystallizing the main ideas without getting bogged down on details. In terms of notation, we try to be consistent with the original paper. The risk of this is that the reader may see different symbols for, say, the wholesale price. But the chapter is sufficiently modularized that we hope, the reader can easily see which symbol belongs where. Needless to say, the papers presented in this chapter reflect the author's knowledge and taste at a certain point in time, the former of which is inevitably incomplete while the latter is constantly evolving.

2 Value of information

The perspective taken by this part of the literature is often that of a central planner, who determines decision rules to optimize the performance of the entire supply chain. The decision rules reflect the information available to the managers who implement the rules. For example, the inventory manager at a supply chain stage only has access to local inventory information, and so the decision rule (determined by the central planner) for this manager must be based on the local information. Clearly, if we increase the information available to the manager (e.g., by providing access to inventory information at other supply chain locations), the set of feasible decision rules is enlarged and the supply chain's performance may improve. The resulting improvement is then the value of the additional information. This section reviews papers that try to quantify the value of information in different supply chain settings.

2.1 Downstream information

A significant part of the literature is interested in the value of information pertaining to the downstream part of the supply chain (i.e., the part that is closer to the end customers). We first consider papers that deal with information sharing within a supply chain. A typical setup here is one where the members of a supply chain share their information about the end customer demand, in the form of realized demand or updated demand forecasts, although other types of information are also discussed. We then review models where information sharing takes place at a supply chain's boundary, e.g., when the supply chain's customers provide advance warnings of their demands. While most researchers use the supply chain-wide costs as the performance measure, there is a stream of papers that use the 'bullwhip effect' (i.e., the amplification of the order variance up the supply chain) as a surrogate performance measure. These papers are considered at the end of this subsection.

Information sharing within a supply chain
Here are some papers that study the value of giving the upstream members of a supply chain access to downstream information, which can be the point-of-sale data or information about the control rule used by a downstream member. The customer demand process can be stationary or nonstationary, and the structure of the supply chain can be serial or distributional.

Chen (1998a) studies the value of demand/inventory information in a serial supply chain. The model consists of N stages. Stage 1 orders from stage 2, 2 from 3,..., and stage N orders from an outside supplier with unlimited stock. The lead times from one stage to the next are constant and represent delays in production or transportation. The customer demand process is compound Poisson. When stage 1 runs out of stock, demand is backlogged. The system incurs linear holding costs at every stage, and linear backorder costs at stage 1. The objective is to minimize the long-run average total cost in the system.

The replenishment policy is of the (R, nQ) type. Each stage replenishes a stage-specific inventory position according to a stage-specific (R, nQ) policy: when the inventory position falls to or below a reorder point R, the stage orders a minimum integer multiple of Q (base quantity) from its upstream stage to increase the inventory position to above R. In case the upstream stage does not have sufficient on-hand inventory to satisfy this order, a partial shipment is sent with the remainder backlogged at the upstream stage. The base quantities are fixed and the reorder points are the only decision variables. Moreover, the base quantities, which are denoted by Q_i for stage i, $= 1,\ldots,N$, satisfy the following integer-ratio constraint:

$$Q_{i+1} = n_i Q_i, \quad i = 1,\ldots,N-1$$

where n_i is a positive integer. This assumption is made to simplify analysis, but it also reflects some practical considerations aimed at simplifying material handling such as packaging and bulk breaking. Moreover, there is evidence that the system-wide costs are insensitive to the choice of base quantities.[2]

Two variants of the above (R, nQ) policy are considered. One is based on *echelon stock*: each stage replenishes its echelon stock with an echelon reorder point. A stage's echelon stock is the inventory position of the subsystem consisting of the stage itself as well as all the downstream stages, which includes the outstanding orders of the stage, either in transit or backlogged at the (immediate) upstream stage, plus the inventories in the subsystem, on hand or in transit, minus the customer backorders at stage 1. Let R_i be the echelon reorder point at stage i, $i = 1,\ldots,N$. Therefore, under an echelon-stock (R, nQ) policy, stage i orders a multiple of Q_i from stage $i+1$ every time its echelon stock falls to or below R_i.

---

[2] Such insensitivity results have been established for single-location models, see Zheng (1992) Zheng and Chen (1992). This property is likely to carry over to multistage models. Also, the optimality of (R, nQ) policies has been established by Chen (2000a) for systems where the base order quantities are exogenously given.

Alternatively, replenishment can be based on *installation stock*: each stage controls its installation stock with an installation reorder point. A stage's installation stock refers to its local inventory position, i.e., its outstanding orders (in transit or backlogged at the upstream stage) plus its on-hand inventory minus backlogged orders from its (immediate) downstream stage. Let r_i be the installation reorder point at stage i, $i = 1, \ldots, N$. Therefore, under an installation-stock (R, nQ) policy, stage i orders a multiple of Q_i from stage $i+1$ every time its installation stock falls to or below r_i. Note that echelon-stock (R, nQ) policies require centralized demand information, while installation-stock (R, nQ) policies only require local 'demand' information, i.e., orders from the immediate downstream stage. When every customer demands exactly one unit, i.e., the demand process is simple Poisson, each order by stage i is exactly of size Q_i, $i = 1, \ldots, N$. In this case, the (R, nQ) policy reduces to the (R, Q) policy.

From Axsäter and Rosling (1993), we know that installation stock (R, nQ) policies are special cases of echelon stock (R, nQ) policies. The two policies coincide when

$$R_1 = r_1, \text{ and } R_i = R_{i-1} + Q_{i-1} + r_i, \quad i = 2, \ldots, N.$$

(Note that r_i is an integer multiple of Q_{i-1} for $i \geq 2$.) To see the intuition behind this result, suppose the demand process is simple Poisson so that (R, nQ) policies reduce to (R, Q) policies. Under the installation stock (R, Q) policy, orders are 'nested' in the sense that every order epoch at stage i coincides with an order epoch at stages $i-1, i-2, \ldots, 1$. The installation stock at stage j after each order is $r_j + Q_j$ for all j. Consequently, just before stage i places an order, its echelon stock, which is the sum of the installation stocks at stages 1 to i, is $\sum_{j=1}^{i-1}(r_j + Q_j) + r_i$. Let this echelon stock level be R_i, $i = 1, \ldots, N$. It is easy to verify that the echelon reorder points so defined satisfy the above equalities and the resulting echelon-stock policy is identical to the installation-stock policy.

Echelon stock (R, nQ) policies have very nice properties. As a result, the optimal echelon reorder points can be determined sequentially in a bottom-up fashion starting with stage 1. Essentially, after a proper transformation, the batch-transfer model reduces to a base-stock model of the Clark and Scarf (1960) type. On the other hand, installation stock (R, nQ) policies are not as nice; a heuristic algorithm is available for determining the optimal installation-stock reorder points, based on several easy-to-compute bounds.

As mentioned earlier, echelon-stock policies require centralized demand information, while installation stock policies only require local information. The relative cost difference between the two is a measure of the value of centralized demand information. An extensive numerical study (with 1,536 examples) shows that the value of information ranges from 0 to 9% with an average of 1.75%.

Gavirneni, Kapuscinski and Tayur (1999) study different patterns of information flow between a retailer and a supplier. The retailer faces independent and identically distributed (i.i.d.) demands and replenishes his inventory by following an (s, S) policy. At the beginning of each period, the retailer reviews his inventory level (on-hand inventory minus customer backorders), and if it is below s, he places an order with the supplier to raise the inventory level to S. The supplier satisfies this order as much as possible. In the event the supplier does not have sufficient on-hand inventory to satisfy the retailer order, a partial shipment is made to the retailer, and the retailer obtains the unfilled part of the order from an external source. There is no delivery lead time with both sources of supply. Customer demand arises at the retailer during the period, with complete backlogging. The focus of the analysis is the supplier, who after satisfying (partially or fully) a retailer order at the beginning of each period, decides how much to produce for the period. Production takes one period and is subject to a capacity constraint. The supplier incurs linear inventory holding costs and linear penalty costs for lost retailer orders. The objective is to determine a production strategy to minimize the supplier's costs, under various scenarios that differ in terms of the supplier's information about the downstream part of the supply chain.

The first scenario assumes that the supplier has no information about the retailer except for the orders the retailer has placed in the past. Moreover, the supplier is rather naive in assuming that the orders from the retailer are i.i.d. Under this assumption, the best the supplier can do is to follow the modified base-stock policy with the same order-up-to level in every period.[3]

The second model assumes that at the beginning of each period, the supplier knows the number of periods i since the last retailer order. In addition, the supplier knows the demand distribution at the retailer, the fact that the retailer follows an (s, S) policy, and the specific policy parameters used by the retailer. Given this information, the supplier is able to determine the probability that the retailer is going to place an order in the coming period and the distribution of the order size. This influences the current production decision. It is shown that the optimal policy for the supplier in this case is again a modified base-stock policy with state-dependent order-up-to level z_i.

---

[3] A modified base-stock policy with order-up-to level z works like this: if the inventory position (on-hand inventory plus work-in-process minus backorders) is less than z, produce as much as possible under the capacity constraint to increase it to z; if the inventory position is above z, produce nothing. In the context of Gavirneni et al., there are no backorders at the supplier, only lost sales, and there is no work-in-process at any review time. The optimality of such a policy has been established by Federgruen and Zipkin (1986a,b).

The third and final model assumes that the supplier has access to all the information available to her in the second model. In addition, at the beginning of each period, the supplier knows the value of j, the number of units sold by the retailer since the last retailer order. Again, a modified base-stock policy with state-dependent order-up-to level z_j is optimal.[4]

A numerical study has been conducted by Gavirneni et al. to understand the differences between the above three models. From the first model to the second, the percentage decrease in supplier costs varies from 10 to 90%; and the savings increase with capacity. From the second model to the third model, the savings range from 1 to 35%. They also report sensitivity results on the cost savings as a function of the supplier capacity, the supplier cost parameters, the retailer demand distribution, and the retailer's policy parameters. The key observations are: (1) when the retailer demand variance is high, or the value of $S-s$ is either very high or very low, information tends to have low values, and (2) if the retailer demand variance is moderate, and the value of $S-s$ is not extreme, information can be very beneficial.

Lee, So and Tang (2000) study the value of sharing demand information in a supply chain model with a nonstationary demand process. The supply chain consists of two firms, one retailer and one manufacturer. The customer demand process faced by the retailer is an AR(1) process:

$$D_t = d + \rho D_{t-1} + \varepsilon_t$$

where $d > 0$, $-1 < \rho < 1$, and ε_t are independent random variables with a common normal distribution with mean zero and variance σ^2. Both firms know the values of the parameters of the demand process, i.e., d, ρ and σ. The retailer sees the realization of demand in each period, while the manufacturer's information depends on, well, what the retailer provides.

The retailer reviews its inventory at the end of each period. Take period t. The retailer satisfies D_t, the demand for period t, from its on-hand inventory with complete backlogging. At the end of the period, the retailer places an order for Y_t units with the manufacturer. The manufacturer

---

[4] One can imagine that the retailer transmits his demand data to the supplier in every period via some electronic medium. The supplier can then determine the value of j and use that information in her production decisions (through the state-dependent policy). A supplier's optimal use of timely demand information from a retailer has been addressed in other papers. For example, Gallego, Huang, Katircioglu and Leung (2000) address this issue in a continuous-time model without capacity constraints. They also show that it is not always in the retailer's interests to share demand information with the supplier. Another reference is Bourland, Powell and Pyke (1996) who study a supply chain model with a component plant (seller) and a final assembly plant (buyer). The production cycles of the two factories do not coincide. Traditionally, information sharing occurs only when the buyer places an order. They study the impact of real-time communication of the buyer's demand data.

satisfies this order from its own on-hand inventory, also with complete backlogging.[5] At the beginning of the next period (period $t+1$) the manufacturer places an order with an outside supplier with ample stock to replenish its own inventory. For easy exposition, we assume that the lead times at both sites are zero, i.e., transportation from the outside supplier to the manufacturer or from the manufacturer to the retailer is instantaneous. (Note that if part of Y_t is backlogged at the manufacturer, that backlog will remain there until the end of period $t+1$, even though the manufacturer's replenishment lead time is zero. This is because the manufacturer only fills the retailer's orders at the ends of periods. The case with a different sequence of events can be analyzed similarly.)

We begin with the retailer's ordering decisions. Suppose we are now at the end of period t. What is the ideal inventory level for the retailer going into period $t+1$? Since the delivery lead time is zero, the retailer can be myopic, i.e., to minimize its expected holding and backorder costs incurred in period $t+1$. The demand in period $t+1$ is $D_{t+1} = d + \rho D_t + \varepsilon_{t+1}$, which is normally distributed with mean $d + \rho D_t$ and standard deviation σ. (D_t has been realized by the end of period t.) Therefore, the ideal inventory level going into period $t+1$ is

$$S_t = d + \rho D_t + k\sigma$$

where k is a constant depending on the holding and backorder cost parameters at the retailer. (This is a well-known formula for the newsvendor model with normal demand.) To derive the order quantity Y_t, suppose the retailer's inventory at the beginning of period t is S_{t-1}. Thus

$$S_{t-1} - D_t + Y_t = S_t$$

or

$$Y_t = D_t + S_t - S_{t-1}.$$

This gives us the demand process facing the manufacturer. (The value of Y_t can be negative, an unpleasant scenario, which should indicate to you the potential suboptimality of the myopic policy. But let us confine ourselves to cases where this rarely happens.)

---

[5] The original assumption made in Lee et al. is that if the manufacturer's on-hand inventory is insufficient to satisfy a retailer order, the manufacturer will make up the shortfall from an external source. The analytical benefit of this assumption is that the retailer always gets its orders filled in full. But this actually complicates the demand process at the manufacturer, who is effectively operating under a lost-sales regime. It is well known that when we combine lost sales with a positive replenishment lead time, it is very difficult to characterize the distribution of the total demand (or satisfied retailer orders in this case) over a lead time. This problem is not addressed in Lee et al. The same comment applies to Raghunathan (2001), to be reviewed shortly.

Now consider the manufacturer's ordering decisions. Suppose we are at the beginning of period $t+1$, having just received and satisfied (completely or partially) the retailer order Y_t. What is the manufacturer's ideal inventory level for the beginning of period $t+1$? Since the manufacturer's replenishment lead time is zero and the outside supplier has ample stock, the manufacturer can also be myopic, minimizing its expected holding and backorder costs in period $t+1$ alone. (This argument is again not water-tight because the manufacturer's myopic inventory level in one period may prevent it from reaching its myopic inventory level in the next period, i.e., there can be too much inventory. If so, the myopic policy is not optimal. Let us not worry about this here.) In period $t+1$, the retailer order is Y_{t+1}, which can be expressed as

$$Y_{t+1} = D_{t+1} + S_{t+1} - S_t = D_{t+1} + \rho(D_{t+1} - D_t).$$

Since $D_{t+1} = d + \rho D_t + \varepsilon_{t+1}$, we have

$$Y_{t+1} = (1+\rho)d + \rho^2 D_t + (1+\rho)\varepsilon_{t+1}. \tag{2.1}$$

The manufacturer's ideal inventory for period $t+1$ can be written as

$$T_t = E[Y_{t+1}] + K\,Std[Y_{t+1}]$$

where K is a constant depending on the manufacturer's holding and backorder costs. Moreover, the manufacturer's minimum expected (one-period) cost is proportional to $Std[Y_{t+1}]$, the value of which depends on what the manufacturer knows about the retailer's demand process at the beginning of period $t+1$.

As mentioned earlier, the manufacturer knows the value of Y_t in any case. If there is no sharing of demand information between the retailer and the manufacturer, the latter does not see D_t. Since $Y_t = D_t + \rho(D_t - D_{t-1})$ and $D_t = d + \rho D_{t-1} + \varepsilon_t$, we have

$$D_t = \frac{Y_t - d - \varepsilon_t}{\rho}.$$

Plugging this into (2.1), we have

$$Y_{t+1} = d + \rho Y_t - \rho \varepsilon_t + (1+\rho)\varepsilon_{t+1}.$$

Therefore

$$Std[\,Y_{t+1}|\text{ no sharing }] = \sigma\sqrt{\rho^2 + (1+\rho)^2}. \tag{2.2}$$

On the other hand, if demand information is shared with the manufacturer, the latter sees the value of D_t, then we have from (2.1)

$$Std[\,Y_{t+1}|\text{ sharing }] = \sigma(1+\rho). \tag{2.3}$$

From (2.2) to (2.3), we see the reduction in the manufacturer's costs as a result of information sharing. (Recall that the manufacturer's costs are proportional to the standard deviation of its lead time demand, i.e., Y_{t+1}.) The savings can be significant, as Lee et al. have shown by analytical and numerical results.

In a note commenting on Lee et al., Raghunathan (2001) argues that the manufacturer can do much better in the case without information sharing. The idea is that the manufacturer can use its information about the retailer's order history to greatly sharpen its demand forecast. Let us see how this works. From $Y_t = D_t + \rho(D_t - D_{t-1})$, we have

$$D_t = \frac{1}{1+\rho}Y_t + \frac{\rho}{1+\rho}D_{t-1}.$$

Applying the above equation repeatedly, we have

$$D_t = \frac{1}{1+\rho}\sum_{i=1}^{t-1}\left(\frac{\rho}{1+\rho}\right)^i Y_{t+1-i} + \left(\frac{\rho}{1+\rho}\right)^t D_0$$

where it is assumed $D_0 = d + \varepsilon_0$. Plugging the above into (2.1),

$$Y_{t+1} = (1+\rho)d + \frac{\rho^2}{1+\rho}\sum_{i=1}^{t-1}\left(\frac{\rho}{1+\rho}\right)^i Y_{t+1-i} + \frac{\rho^{t+2}}{(1+\rho)^t}D_0 + (1+\rho)\varepsilon_{t+1}$$

with

$$Std[Y_{t+1}|\text{ no sharing }] = \sigma\sqrt{\frac{\rho^{2t+4}}{(1+\rho)^{2t}} + (1+\rho)^2}.$$

It can be shown that the above expression is less than (2.2), suggesting that the value of information is less than what is reported in Lee et al. Moreover, as $t \to \infty$, the benefits effectively disappear.

Cachon and Fisher (2000) provide a model to quantify the value of downstream inventory information in a one-warehouse multiretailer system. When the warehouse has access to real-time inventory status at the retailers as opposed to just retailer orders, it can make better ordering and allocation decisions. The supply chain benefits, but by how much? (We choose not to refer to the warehouse as a supplier, for the view taken here is that of a central planner.)[6]

The model consists of one warehouse and multiple, identical retailers. The periodic customer demands at the retailers are i.i.d., both across retailers and across time periods. The retailers replenish their inventories by ordering from the warehouse, who in turn orders from an external source with unlimited inventory. Complete backlogging is assumed at both the retail and the warehouse level. The replenishment lead time at the warehouse is constant, so are the transportation lead times from the warehouse to the retailers. Linear inventory holding costs are incurred at the warehouse and the retailers, and linear penalty costs are incurred at the retailers for customer backorders. The objective is to minimize the long-run average system-wide holding and backorder costs (i.e., the central planner's view).

Inventory transfers from the warehouse to the retailers are restricted to be integer multiples of Q_r, an exogenously given base quantity. Similarly, orders by the warehouse (to the external source) must be integer multiples of $Q_s Q_r$, for some positive integer Q_s, another given parameter. The decisions for the retailers are when to place an order with the warehouse, and how many batches (each of size Q_r) to order, and the decisions for the warehouse are when to place an order with the outside source, and how many sets of batches to order (each set consists of Q_s batches, each of size Q_r).

In the scenario with traditional information sharing, the warehouse only observes the retailers' orders. Thus a replenishment policy can only be based on local information. Specifically, each retailer follows an (R_r, nQ_r) policy, i.e., whenever its inventory position (its outstanding orders, in transit or backlogged at the warehouse, plus its on-hand inventory minus its customer backorders) falls to R_r or below, order a minimum integer multiple of Q_r to increase its inventory position to above R_r. Similarly, the warehouse

---

[6] There is a large body of literature on one-warehouse multiretailer systems. One way to categorize this literature is by looking at whether or not there are economies of scale in transferring inventory from one location to another. If the answer is no, then the focus is on the so-called one-for-one replenishment policies. The key references in this area are: for continuous-time models, Sherbrooke (1968), Simon (1971), Graves (1985), Axsäter (1990), Svoronos and Zipkin (1991), Forsberg (1995) and Graves (1996); and for discrete-time models, Eppen and Schrage (1981), Federgruen and Zipkin (1984a,b), Jackson (1988) and Diks and de Kok (1998). If there are economies of scale, then a batch-transfer policy makes more sense. The key references here are: for continuous-time models, Deuermeyer and Schwarz (1981), Moinzadeh and Lee (1986), Lee and Moinzadeh (1987a,b), Svoronos and Zipkin (1988), Axsäter (1993b, 1997, 1998, 2000) and Chen and Zheng (1997); and for discrete-time models, Aviv and Federgruen (1998), Chen and Samroengraja (1999, 2000a) and Cachon (2001). For comprehensive reviews on the above literature, see Axsäter (1993a) and Federgruen (1993), and Chapter 10 of this volume by Sven Axsäter. Most of the replenishment policies studied are based on local inventory information, and only a couple use centralized demand/inventory information. The objectives of these papers are typically to show how to determine the system-wide costs for a given class of policies. The desire to understand the value of demand/inventory information appeared only recently.

follows an (R_s, nQ_s) policy: whenever its inventory position (orders in transit plus on-hand inventory minus backlogged retailer orders) falls to $R_s Q_r$ or below, order an integer multiple of $Q_s Q_r$ units. The decision variables are the reorder points R_r and R_s.

When the warehouse is unable to satisfy every retailer's order in a period, it follows an allocation policy. It is called a batch priority policy that works as follows. Suppose a retailer orders b batches in a period, then, the first batch in the order is assigned priority b, the second batch is assigned priority $b - 1$, etc. All batches ordered in a period (by all retailers) are placed in a shipment queue, with the batch having the highest priority enters the queue first. (The rationale for this allocation policy is that a retailer ordering the most batches in a period is, naturally, considered to have the highest need for inventory.) When multiple batches have the same priority, they enter the queue in a random sequence. A shipment queue is maintained for each period. The retailers' orders are satisfied in the order in which the shipment queues are created and within each shipment queue, on a first-in first-out basis. Notice that the warehouse's stock allocation is based on the retailers' 'needs' at the time the orders are placed.

The second scenario, called full information sharing, is where the warehouse has access to the retailers' inventory status on a real-time basis. In this case, the retailers continue to use the (R_r, nQ_r) policy described earlier. The warehouse, however, uses more sophisticated rules for ordering and allocation. The exact policy is complicated. The idea behind the new ordering policy is that the warehouse should perform some sort of cost-benefit analysis for each set of batches added to an order, with the cost being additional holding cost at the warehouse and the benefit being less delay for the retailers' orders. On the other hand, with immediate access to retailers' inventory status, the warehouse can allocate inventory (to satisfy backlogged retailer orders) based on the retailers' needs at the time of shipment.

By comparing the system-wide costs under traditional and full information sharing, one obtains a measure of the value of downstream inventory information. In a numerical study with 768 examples, it is found that information sharing reduces supply chain costs by 2.2% on average, with the maximum at 12%.[7]

Aviv and Federgruen (1998) consider a supply chain model consisting of a supplier and multiple retailers. The members of the supply chain are independent firms. In this decentralized setting, they attempt to characterize the value of sales information, which is defined to be the reduction in supply

---

[7] Cachon and Fisher also provide a lower bound on the system-wide costs under full information, and compare that with the costs under traditional information sharing. This does not change the picture on the value of information in any significant way, meaning the proposed full-information policy is near-optimal. Moreover, they show, again via numerical examples, that significant savings can be had if the lead times or batch sizes (due to fixed ordering costs) are reduced, which may be expected from better information-linkup. Similar findings have been reported in Chen (1998b).

chain-wide costs if the supplier has access to real-time sales data at the retail level.[8] They then proceed to consider the impact of a vendor managed inventory (VMI) program, which comes with real-time information sharing and puts the supplier in the position of a central planner for the supply chain. Their main conclusions are based on three models: a decentralized model without information sharing, a decentralized model with information sharing, and a centralized model with information sharing.

We begin with the base model. There are J retailers. Customer demands are stochastic and occur at the retail sites only. The retailers monitor their inventories periodically. The demand processes at the retailers are independent. Demands in different periods at the same retailer are i.i.d. according to a retailer-specific distribution. When demand exceeds on-hand inventory at a retailer, the excess demand is backlogged. The retailers replenish their inventories from the supplier, who in turn replenishes its own inventory through production. The transportation lead times from the supplier to the retailers are constant but may be retailer-specific. The production lead time at the supplier is also constant. The production quantity that the supplier can initiate in a period is subject to a constant capacity constraint. Each retailer incurs a fixed cost for each order it places with the supplier, linear inventory holding costs, and linear penalty costs for customer backorders. These cost parameters are stationary over time, but they can be retailer-specific. The supplier incurs linear holding costs for its on-hand inventory and linear penalty costs for backorders of retailer orders. This latter cost component is a contract parameter, which is given exogenously and represents a revenue for the retailers. There are no fixed costs for initiating a production run at the supplier.

The replenishment policies at the retailers are of the (m, β) type, whereby the retailer reviews its inventory position every m periods and places an order to increase it to β. The values of the policy parameters can be retailer-specific, with (m_j, β_j) for retailer j, $j = 1, \ldots, J$. Let M be the least common multiple of $m_1, \ldots, m_J$. Therefore, the joint order process of the retailers regenerates after a grand replenishment cycle of M periods. In general, the replenishment cycles of the retailers are not coordinated. Aviv and Federgruen consider two extreme arrangements in this regard. One is called 'peaked', in which all retailers order at the beginning of a grand replenishment cycle, and the other 'staggered', in which the retailer cycles are spread out to achieve a smooth order process for the supplier. (The staggered pattern is clearly defined if the retailers are identical. Otherwise, one needs to spell out what is meant by 'smooth'. Staggered policies have been proposed and studied by Chen and Samroengraja (2000a) in one-warehouse multiretailer systems.)

---

[8] This paper therefore deviates from the mainstream approach of quantifying the value of information in centralized models.

Given that the demand process at the supplier is cyclical, it is reasonable to expect that the supplier's production policy is also cyclical. Aviv and Federgruen assume that the supplier follows a modified base-stock policy with cyclical order-up-to levels, whereby the supplier initiates a production run, subject to the capacity constraint, in period t to increase its inventory position to β^m, if period t is the mth period in the grand cycle, $m = 0, 1, \ldots, M - 1$.[9] In the event that the supplier cannot satisfy all retailers' orders in a period, an allocation mechanism is given that is based on some measure of expected needs of the retailers. A shipment can be sent to a retailer even though it is not the retailer's ordering period. (Recall that the retailers order intermittently.) Of course, this occurs only when the supplier backlogs an order (or part of it) from the retailer and fills it in a subsequent period.

The firms minimize their own long-run average costs in a noncooperative fashion. Ideally, the solution to this noncooperative game can be obtained by using some established equilibrium concept. Since this is intractable, Aviv and Federgruen take a two-step approach: first, the retailers optimize assuming the supplier has ample stock, and then, given the retailers' decisions, the supplier optimizes. This completes our description of the base model.

The second model retains the above decentralized structure but assumes that the supplier observes the realized demands at the retail sites immediately. This information allows the supplier to better anticipate the orders that the retailers are going to place in future periods. As a result, the supplier can use a state-dependent, modified base-stock policy, where the state now includes not only where in a grand cycle the current period is but also a summary of relevant sales information from the retail sites.

Finally, the third model assumes that a VMI program is in place, which provides the supplier with immediate access to the sales data at the retail sites and gives the supplier the rights to decide when and how much to ship to each retailer. It is assumed that the VMI contract is such that it is in the supplier's interests to minimize the total costs in the supply chain. Aviv and Federgruen propose a heuristic method to solve this centralized planning problem.

A numerical study shows that the average improvement in supply chain costs from the first to the second model is around 2%, with a range from 0 to 5%. Most of these savings accrue to the supplier. From the second model to the third, the average improvement is 4.7%, with a range from 0.4 to 9.5%. It is also found that the value of information sharing and VMI increases, as the degree of heterogeneity among the retailers increases, the lead times become longer, or the capacity becomes tighter. Finally, the system tends to perform better with staggered retailer cycles rather than the peaked pattern.

---

[9] They also consider a policy with a constant order-up-to level. For proofs of the optimality of the cyclical base-stock policies in single-location settings, see Aviv and Federgruen (1997) and Kapuscinski and Tayur (1998).

One of the key drivers for production-inventory planning decisions is demand forecast. In any given period, the firm determines a set of predictions for the demands in future periods based on its information about the operating environment and planned activities. As time progresses and new information becomes available, the firm revises its demand forecast. From the standpoint of production-inventory planning, an important question is how to integrate the evolving demand forecast into the planning decisions. Below, we summarize several papers that address this question.

Gullu (1997) studies a two-echelon supply chain consisting of a central depot and N retailers. The depot serves as a transshipment center where an order arriving at the depot from an outside supplier is immediately allocated among the retailers. (Thus the depot does not hold inventory.) Customer demand arises only at the retailers, with unsatisfied demand fully backlogged. The objective is to determine a depot replenishment/allocation policy that minimizes the system-wide costs. A unique feature of the model is that each retailer maintains a vector of demand forecasts for a number of future periods, and this vector is updated from one period to the next. Gullu considers two models, one that utilizes the demand forecasts in the replenishment/allocation decision and the other that ignores the forecasts. By comparing the two models, one sees the value of demand information (contained in the forecasts).

The evolution of demand forecasts is described by the martingale model of forecast evolution (MMFE).[10] Let D_t^j be retailer j's demand forecasts for periods $t, t+1, \ldots$ at the end of period, t, $j = 1, \ldots, N$. That is,

$$D_t^j = (d_{t,t}^j, d_{t,t+1}^j, \ldots)$$

where $d_{t,t}^j$ is the realized demand for period t (thus not really a forecast) and $d_{t,t+k}^j$, $k \geq 1$, is the retailer's forecast, made at the end of period t, for the demand in period $t+k$. In the additive model, D_t^j is obtained by adding an error term (or adjustment) to each relevant component of D_{t-1}^j, i.e.,

$$d_{t,t}^j = d_{t-1,t}^j + \varepsilon_{t,1}^j$$

$$d_{t,t+1}^j = d_{t-1,t+1}^j + \varepsilon_{t,2}^j$$

$$\vdots$$

Let $\bar{\varepsilon}_t^j = (\varepsilon_{t,1}^j, \varepsilon_{t,2}^j, \ldots)$ and $\bar{\varepsilon}_t = (\bar{\varepsilon}_t^1, \ldots, \bar{\varepsilon}_t^N)$. Gullu assumes that $\varepsilon_{t,k}^j = 0$ for all $k > M$ for some positive integer M, for all t and all j. In other words, the new

---

[10] For the development of the MMFE model, see Hausman (1969), Graves, Meal, Dasu and Qiu (1986), Graves, Kletter and Hetzel (1998), Heath and Jackson (1994). Hausman (1969) and Heath and Jackson (1994) also consider a multiplicative model.

information collected during period t only affects the demand forecasts in M periods (i.e., the current period and the next $M-1$ periods). Moreover, it is assumed that $\bar{\varepsilon}_t$, for all t, are independent and have the same multivariate normal distribution with zero mean. It is, however, possible that for a given t, the different components of $\bar{\varepsilon}_t$ are correlated. This allows for modeling of demand correlation over time and across retailers. Finally, the initial forecast for the demand in a period that is more than M periods away is a constant, i.e.

$$d^j_{t,t+k} = \mu^j, \quad \forall t \quad \text{and} \quad \forall k \geq M.$$

(Thus μ^j is the mean demand per period at retailer j.) Consequently,

$$d^j_{t,t} = d^j_{t-1,t} + \varepsilon^j_{t,1} = \cdots = \mu^j + \sum_{i=1}^{M} \varepsilon^j_{t-i+1,i}$$

Letting $\sigma^2_{j,i}$ be the variance of $\varepsilon^j_{t,i}$, we have

$$Var[d^j_{t,t}] = \sum_{i=1}^{M} \sigma^2_{j,i}$$

Note that at the end of period $t-1$, the conditional variance of $d^j_{t,t}$ given $d^j_{t-1,t}$ is only $\sigma^2_{j,1}$. In other words, as the demand forecast for a fixed period is successively updated, the variance for the demand in that period is successively reduced. This reduction in demand uncertainty in turn leads to improvement in supply chain performance.[11]

Under the above demand model, Gullu considers two scenarios depending on whether or not the demand forecasts are used in the depot's allocation decision. The depot's replenishment policy is the order-up-to S policy, i.e., in each period, the depot places an order with the outside supplier to increase the system-wide inventory position to the constant level S.[12] The key analytical results are that the use of demand forecasts leads to lower system-wide costs and if and only if the backorder penalty cost rate is higher than the holding cost rate (identical cost rates are assumed across retailers), a lower system-wide inventory position. (This latter result is well known for the newsvendor model with normal demand.) There are also various asymptotic results for some special cases, which we omit.

---

[11] Updates of demand forecasts do not always make them more accurate, see Cattani and Hausman (2000) for both empirical and theoretical evidence.

[12] It is possible that the supply chain's performance can be improved if the depot's replenishment decision takes into account the demand forecasts at the retail level. This should be investigated. If you are familiar with Eppen and Schrage (1981), then you can see that the Gullu model is basically the Eppen–Schrage model with forecast evolution. The analysis is also similar to Eppen-Schrage's.

Toktay and Wein (2001) consider a single-item, single-stage, capacitated production system with an MMFE demand process. Random demand for the item arises in each period. Demand is satisfied from the finished-goods inventory, which is replenished by a production system with capacity C_t in period t. That is, the system can produce up to C_t units of the product in period t, with C_t in different periods being i.i.d. normal random variables with mean μ and variance σ_C^2. If demand exceeds the finished-goods inventory, the excess demand is fully backlogged. Let I_t be the (finished goods) inventory level at the end of period t. The system incurs holding and backorder costs equal to $(hI_t^+ + bI_t^-)$ in period t, where h and b are the holding and backorder cost rates. Let P_t be the production quantity in period t. Thus $P_t = \min\{Q_{t-1}, C_t\}$, where Q_{t-1} is the number of production orders waiting to be processed at the end of period $t-1$. At the end of each period t, R_t, new orders are released to the production system. Thus $Q_t = Q_{t-1} - P_t + R_t$. The objective is to find a release policy $\{R_t\}$ to minimize the expected steady-state holding and backorder costs.[13]

Let D_t be the demand in period t. The demand process is stationary with $E[D_t] = \lambda$. (Thus we need $\mu > \lambda$ for stability.) Let $D_{t,t+i}$ be the forecast for D_{t+i} determined at the end of period t, $i \geq 0$. Thus $D_{t,t}$ is the realized demand in period t. It is assumed that nontrivial forecasts are available only for the next H periods, i.e., $D_{t,t+i} = \lambda$ for all $i > H$. Define $\varepsilon_{t,t+i} = D_{t,t+i} - D_{t-1,t+i}$, $i \geq 0$. It is then clear that $\varepsilon_{t,t+i} = 0$ for al $i > H$. Thus $\varepsilon_t = (\varepsilon_{t,t}, \varepsilon_{t,t+1}, \ldots, \varepsilon_{t,t+H})$ is the forecast update vector whose value is observed at the end of period t. The forecast update vectors (for different periods) are assumed to be i.i.d. normal random variables with zero mean.

Toktay and Wein consider two classes of release policies: one ignores the forecast information, and the other utilizes it. In the former case, $R_t = D_t$. Under the initial condition that $Q_0 = 0$ and $I_0 = s_m$, the proposed release policy leads to $Q_t + I_t = s_m$ for all t.[14] To integrate the demand forecasts into the release policy, consider

$$R_t = \sum_{i=0}^{H} D_{t,t+i} - \sum_{i=0}^{H-1} D_{t-1,t+i} = \sum_{i=0}^{H-1} \varepsilon_{t,t+i} + D_{t,t+H} = \sum_{i=0}^{H} \varepsilon_{t,t+i} + \lambda.$$

Under this release policy and with proper initial conditions, one can show that for all t

$$Q_t + I_t - \sum_{i=1}^{H} D_{t,t+i} = s_H$$

---

[13] What queuing folks call a release policy is called a replenishment policy by inventory folks.
[14] In traditional inventory lingo, Q_t is the outstanding orders, while I_t is the inventory level. The sum of the two is the inventory position. The proposed policy is thus a base-stock policy whereby the inventory position is maintained at a constant level.

for some constant s_H which can be controlled by setting the initial inventory level.[15] A key finding of the paper is the observation that excess production capacity, as measured by $\mu - \lambda$, and demand information, as contained in the demand forecasts, are substitutes. (Other studies on capacitated problems with forecast evolution include Gullu (1996) and Gallego and Toktay (1999).)

Aviv (2001) considers a supply chain model with one retailer and one supplier. Customer demand arises at the retailer, who replenishes its inventory from the supplier, who in turn orders from an outside source with ample stock. The two members of the supply chain independently forecast the customer demands in future periods and periodically adjust their forecasts as more information becomes available. The retailer and the supplier are modeled as a team in the sense that they share a common objective to minimize the system-wide costs, but they do not necessarily share their demand forecasts. Aviv studies the following three scenarios. In scenario one, the two members neither share their demand forecasts nor use their own demand forecasts in making replenishment decisions. In scenario two, they still do not share their demand forecasts, but now they each integrate their own forecasts in their replenishment decisions. In scenario three, they share their demand forecasts and use the shared information in their replenishment decisions.

Aviv uses an MMFE demand model. The demand in period t, d_t, is the sum of a constant and a stream of random variables representing adjustments to the forecast of d_t made at different times leading up to period t. Specifically,

$$d_t = \mu + \varepsilon_t + \sum_{i=0}^{\infty} (\varepsilon^r_{t,i} + \varepsilon^s_{t,i})$$

where μ is a constant, $\{\varepsilon_t\}_{t\geq 1}$ are i.i.d. normal random variables, the components of the vector $\{(\varepsilon^r_{t,i}, \varepsilon^s_{t,i})\}_{i=0}^{\infty}$ are independent and each a bi-variate normal and the vectors (for different t) are i.i.d., and $\{\varepsilon_t\}_{t\geq 1}$ are independent of $\{(\varepsilon^r_{t,i}, \varepsilon^s_{t,i})\}_{t\geq 1, i\geq 0}$. All the random variables have zero mean. (Different notation is used in Aviv.) As a result, $\{d_t\}$ is a sequence of i.i.d. normal random variables with mean μ. At the beginning of period τ, for any τ, the retailer privately observes the vector $\{\varepsilon^r_{t',t'-\tau}\}_{t'\geq \tau}$, and the supplier privately

---

[15] Again one can draw some connection to inventory theory here. As noted above $Q_t + I_t$ is the inventory position at the end of period t. So the second release policy corresponds to a modified base-stock policy that is based on an 'adjusted inventory position.' The adjustment is the total forecasted demand in the next H periods. From inventory theory, the 'optimal' adjustment should be based on the demand during the replenishment 'lead time.' The problem is that there is no lead time here, only capacity. The proposed release policy seems to draw an equivalence between 'lead time' and the forecast horizon in a capacitated production system. If capacity is tight, then the 'lead time' should be long, and the opposite holds if capacity is ample. But that has little to do with the forecast horizon.

observes the vector $\{\varepsilon^s_{t',t'-\tau}\}_{t'\geq\tau}$. Therefore, by the beginning of period $t-k$, $k\geq 0$, the retailer has observed the value of $\sum_{i=k}^{\infty}\varepsilon^r_{t,i}$ whereas the supplier has observed the value of $\sum_{i=k}^{\infty}\varepsilon^s_{t,i}$. It is easy to see that as they get closer and closer to period t, i.e., as k decreases, the supply chain members have less and less uncertainty about d_t (or better and better forecast for d_t). In scenario three, the supply chain members share their private information. This enables them to improve (by unifying) their demand forecasts.

In a numerical study, Aviv found that integrating the forecast updates in the replenishment decisions reduces, on average, the supply chain costs by 11%, and information sharing between the retailer and the supplier brings in an additional reduction of 10%.

It is worthwhile to note that demand forecasts can take other forms with different patterns of evolution. For example, a parameter of the demand distribution may be unknown. Beginning with a prior distribution for the unknown parameter, one can sharpen the estimate of the parameter after each observation of demand. The production/inventory decisions can be made to dynamically reflect the new information that becomes available as time progresses. Alternatively, demands in different periods may be correlated and the data on early sales can be used to update the forecasts for the later sales. The following are additional papers that incorporate adjustments of demand forecasts: Scarf (1959, 1960), Iglehart (1964), Murray and Silver (1966), Hausman and Peterson (1972), Johnson and Thompson (1975), Azoury and Miller (1984), Azoury (1985), Bitran, Haas and Matsuo (1986), Miller (1986), Bradford and Sugrue (1990), Lovejoy (1990, 1992), Matsuo (1990), Fisher and Raman (1996), Eppen and Iyer (1997a,b), Sobel (1997), Barnes-Schuster, Bassok and Anupindi (1998), Brown and Lee (1998), Larivicre and Porteus (1999), Dong and Lee (2000), Donohue (2000), Milner and Kouvelis (2001), and Ding, Puterman and Bisi (2002). On the other hand, the demand process can be modulated by an exogenous Markov chain; the state of the exogenous Markov chain determines the current period's demand distribution. For inventory models with Markov-modulated demands, see Song and Zipkin (1992, 1993, 1996a), Sethi and Cheng (1997), Chen and Song (2001), and Muharremoglu and Tsitsiklis (2001).

Advance warnings of customer demands

When members of a supply chain share information, no new information is created; only existing information moves from one place to another. In some situations, however, customers can, and are willing to, provide advance warnings of their demands. These warnings represent new information for the supply chain. And the question is how to exploit such information.

Hariharan and Zipkin (1995) study inventory models where customers provide advance warnings of their demands. Customer orders arise randomly. Each order comes with a due date, a future time when the customer wishes to receive the goods ordered. They call the time from a customer's order to its due date the *demand lead time*. The customer does not want to receive delivery before the due date. Deliveries after due dates are possible, but undesirable. They show that demand lead times are the opposite of supply lead times in terms of their impact on the system performance.

A simple model illustrates the basic idea. Suppose customer orders arrive according to a simple Poisson process. Each customer orders a single unit. The demand lead time is a constant l, i.e., an order at time t calls for a 'demand' at time $t+l$. Demand is satisfied from on-hand inventory, with complete backlogging. Inventory is replenished from an outside source with ample stock, and the supply lead time is a constant L. There are no economies of scale in ordering.

If $l \geq L$, then one can satisfy all customer demands perfectly without holding any inventory. Here is how to achieve that. Whenever a customer order arrives, wait $(l - L)$ units of time before placing an order for one unit with the outside supplier. This replenishment unit will arrive just in time to satisfy the customer's demand (at time $t+l$).

Now suppose $l < L$. From basic inventory theory, we know that the inventory position at time t should 'cover' the total demand during the supply lead time (i.e., the lead time demand). Note that at time t, the total demand in the interval $(t, t+l)$ is already known due to advance ordering, whereas the demand in $(t+l, t+L)$ remains unknown (it is a Poisson random variable with mean $\lambda(L-l)$ where λ is the arrival rate of customer orders). Let d_t be the known part, and D_t the unknown portion. The lead time demand is $D_t^L = d_t + D_t$. Therefore, the inventory position at time t should consist of d_t and a buffer inventory S for protection against the uncertain part of the lead time demand D_t. The inventory level at time $t+L$ is $(d_t + S) - D_t^L = S - D_t$. Therefore, the expected holding and backorder cost rate at time $t+L$ can be written as

$$E[h(S - D_t)^+ + b(S - D_t)^-]$$

where h and b are the holding and backorder cost rates. Let S^* be the S value that minimizes this cost expression. If we set the inventory position at time t to $d_t + S^*$, then we know that the expected holding and backorder costs one supply lead time later are minimized. If this inventory position can be achieved for all t, then the system's long-run average costs are minimized and we have an optimal policy. Here is a proof. Assume at $t = 0$, the inventory position is $d_0 + S^*$. (If it is lower than this target level, order enough to make up the shortfall; otherwise, just wait until the inventory position at some time τ equals $d_\tau + S^*$.) Then, whenever a customer order arrives, order one unit from the outside source. This is just like the one-for-one replenishment policy used in the conventional system without advance ordering, with a caveat that replenishment orders are based on customer orders not customer demands.

It is a simple matter to see that the inventory position at any time coincides with the ideal target. The key point of the above exercise is that a system with demand lead time l and supply lead time L is essentially the same as the conventional system with supply lead time $(L - l)$. Thus, demand lead time is the opposite of supply lead time, an elegant characterization of the value of (one type of) demand information.

Hariharan and Zipkin also study advance ordering in other models, where the supply lead time is stochastic or where the replenishment process consists of multiple stages. We omit the details.

One limitation of the Hariharan–Zipkin construct is that all customers come with the same demand lead time. This assumption is relaxed in Chen (2001a) the customer population is divided into M segments. The customers from segment m are homogeneous and provide a common demand lead time l_m, $m = 1, \ldots, M$. In this multisegment case, it is still rather straightforward to characterize the value of advance ordering, even in a multistage serial inventory system. But the main objective of Chen (2001a) is to study the incentives required by the customers in order for them to willingly offer advance warnings of their demands and how these incentives can be traded off against the benefits of demand information embodied in the advance orders. We will review Chen (2001a) in greater detail in Section 3.1.

Gallego and Özer (2001) provide a discrete-time version of the above multisegment model of advance ordering. Time is divided into periods. In each period t, a demand vector is observed: $\vec{D}_t = (D_{t,t}, \ldots, D_{t,t+N})$, where $D_{t,s}$ is orders placed by customers in period t for deliveries in period s and N is a constant (positive integer) and is referred to as the information horizon.[16] (Therefore, the customer population effectively consists of $N + 1$ segments.) For this demand process, Gallego and Özer prove optimal policies in a single-location model with or without fixed ordering costs. They consider multiple scenarios where the planning horizon can be finite or infinite and the cost parameters can be nonstationary. The main result is that if there are fixed order costs, the optimal policy is a state-dependent (s, S) policy; otherwise, the optimal policy is a state-dependent base-stock policy. But what is the state? Define for any $s \geq t$

$$O_{t,s} = \sum_{\tau=s-N}^{t-1} D_{\tau,s}$$

---

[16] The reader may notice that this demand model, where customers place orders in advance of their requirements, resembles the MMFE model considered earlier. In fact, strictly speaking, the demand model with advance orders can be considered a special case of the forecast evolution model. The only, perhaps superficial, distinction is that the updates in the advance-orders model represent actual customer orders, whereas the updates in the MMFE don't have to be. Moreover, the advance-orders model assumes no order cancellation (i.e., the updates are always nonnegative). No such assumption has been detected under the MMFE framework.

which represents what we know at the beginning of period t about the demand in period s. As in any inventory model, we care about the total demand during the supply lead time, which is assumed to be a constant L. Define

$$O_t^L = \sum_{s=t}^{t+L} O_{t,s}$$

which is what we, standing at the beginning of period t, know about the future demands in periods $t, t+1, \ldots, t+L$. (Therefore, the lead time demand is total demand over $L+1$ periods; the extra period is simply due to the convention that orders are placed at the beginning of a period and costs are assessed at the end of a period.) The modified inventory position at the beginning of period t is simply the inventory level (on-hand minus backorders) plus outstanding orders minus O_t^L. (Therefore, the known part of the lead time demand has been taken out of the inventory position. This is just for control purposes, of course, as we will see.) The state of the inventory system consists of the above modified inventory position plus what we know about the demands beyond the supply lead time, i.e.

$$\vec{O}_t = (O_{t,t+L+1}, \ldots, O_{t,t+N-1}).$$

The optimal (s, S) policy has control parameters that are dependent on $\vec{O}_t$ and operates based on the modified inventory position. That is, at the beginning of period t, if the modified inventory position is at or below $s(\vec{O}_t)$, order to increase it to $S(\vec{O}_t)$; otherwise, do nothing. When there are no fixed order costs, the action is simply ordering to increase the modified inventory position up to $S(\vec{O}_t)$ every period. For this latter case, and when the problem is stationary, the base-stock level no longer depends on $\vec{O}_t$. This makes intuitive sense.

Other related studies include Gallego and Özer (2000), Özer (2000), Karaesmen, Buzacott and Dallery (2001) and Özer and Wei (2001). These papers show how advance demand information can be used to improve performance in various production/distribution systems with or without capacity constraints.

A mirror image of customers providing advance demand information is the decision maker postponing a decision until after customers have placed their orders. This is, e.g., the case when a firm switches from a make-to-stock regime to a make-to-order regime. The postponement reduces the uncertainty confronting the decision-maker, improving the quality of the decision and thus performance. For more on the impact of the postponement of operations decisions, see the cases of Benetton (by Signorelli and Heskett, 1984) and Hewlett-Packard (by Kopczak and Lee,

1994) and the papers by Lee and Tang (1998) and Van Mieghem and Dada (1999) and the references therein. Refer to Chapter 5 of this volume by Hau L. Lee and Jayashankar M. Swaminathan for extensive discussions on postponement strategies.

The bullwhip phenomenon

The bullwhip effect refers to a phenomenon where the replenishment orders generated by a stage in a supply chain exhibit more volatility than the demand the stage faces. Recently there has been a flurry of activities on the bullwhip effect. We review this part of the literature here mainly because information sharing (e.g., sharing of customer demand information) is often suggested to combat the undesirable effect.

Many economists have studied the bullwhip phenomenon; they are interested in it because empirical observations refute a conventional wisdom that inventory smoothes production. A firm carries inventory, the conventional wisdom goes, which serves as a buffer to smooth out the peaks and valleys of demand. This in turn generates a relatively stable environment for production. So production should be smoother than demand. Unfortunately, industry data point to the other way. Why? Possible explanations include: the use of (s, S) type of replenishment policies, the presence of positive serial correlation in demand, etc. See Blinder (1982, 1986), Blanchard (1983), Caplin (1985) and Kahn (1987). Other explanations call for industrial dynamics and organizational behavior (Forrester 1961) and irrational behavior on the part of decision-makers (Sterman 1989).

We focus on the operations management literature, which has provided some new insights into the bullwhip phenomenon. The general approach in this literature is to first specify the environment (e.g., revenue/cost structure, characteristics of the demand process, etc.) in which a supply chain member operates, and then show that when the supply chain member optimizes its own performance, it generates orders that are more volatile than the demand process it faces. The implicit message here is that the supply chain member should not be blamed for the bullwhip effect; it is the environment that has created the observed behavior. Sometimes, one can change the environment, with potential benefits for some members of the supply chain or the supply chain as a whole.

Lee, Padmanabhan and Whang (1997a,b), Lee, Padmanabhan and Whang, (1997b) exemplify the above approach. They have identified four causes for the bullwhip effect. The first is the demand characteristics. In a single-location inventory model with a positively correlated demand process, they show that the optimal policy that minimizes, say, the retailer's costs leads to variance amplification. There is an intuitive explanation. When the retailer observes a low demand, he takes that as a signal of low future demands as well and places an order that reflects that lowered forecast. Conversely, a high demand suggests to him that the demands in the future periods are likely to be high as well. He then places a large order based on that outlook. In sum, due to the positive correlation between the demands in different periods, the orders placed by the retailer exhibit larger swings than the demands he observes. The second cause is the possibility of supply shortage. To see the intuition, consider a supply chain with one supplier, whose capacity fluctuates over time, and multiple retailers. In periods when the supplier's capacity is likely to be insufficient, the retailers – engaged in a rationing game to secure an adequate supply for themselves – place large orders (larger than what they would order if there were no capacity shortage). Suppose a retailer faces a deterministic demand stream, say, 4 units per period. So he would order 4 units when no capacity shortage is expected, and order more otherwise. Clearly, when the capacity fluctuates over time, the order stream has a larger variance than the demand stream (which has zero variance). The third cause is economies of scale in placing orders. When there is a fixed cost in placing an order, it makes sense to order once every few periods. This order batching leads to the bullwhip effect. The fourth cause is fluctuating-purchase costs. In periods when the supply is cheap, you want to buy a lot and stockpile, whereas in periods when supply is expensive, you wait. It is easy to see that the high–low prices encourage extreme orders. In order to dampen the bullwhip effect, one has to attack the root causes. Lee et al. have described several industry initiatives that do just that.

Below, we describe a supply chain model that has been used to show that a quasi-optimal operating policy amplifies the order variance.[17] Graves (1999) considers a supply chain-model with a nonstationary demand process. The demand process is an autoregressive integrated moving average (ARIMA) process:

$$d_1 = \mu + \varepsilon_1$$
$$d_t = d_{t-1} - (1-\alpha)\varepsilon_{t-1} + \varepsilon_t, \qquad t = 2, 3, \ldots$$

where d_t is the demand in period t, α and μ are known constants, and ε_t are i.i.d. normal random variables with mean zero and variance σ^2. It is assumed that $0 \leq \alpha \leq 1$.[18] From the above description of the demand process, one can write

$$d_t = \varepsilon_t + \alpha\varepsilon_{t-1} + \cdots + \alpha\varepsilon_1 + \mu.$$

Note that each period, there is a shift in the mean of the demand process: the random shock ε_t shifts the mean of the demand process by $\alpha\varepsilon_t$, starting from period $t+1$. Therefore, each random shock has a permanent effect on the

---

[17] For other such studies, see, e.g., Chen, Drezner, Ryan and Simchi-Levi (2000), Ryan (1997) and Watson and Zheng (2001).

[18] The above demand process is also known as an integrated moving average (IMA) process of order (0,1,1), see Box, Jenkins and Reinsel (1994).

demand process. Note that $\alpha = 0$ corresponds to an i.i.d. demand process, whereas $\alpha = 1$ is a random walk. In general, a larger α means the process depends more on the most recent demand realization.

For the above demand process, a first-order exponential-weighted moving average provides a minimum mean square forecast. Define

$$F_1 = \mu$$
$$F_{t+1} = \alpha d_t + (1-\alpha)F_t, \qquad t = 1, 2, \ldots.$$

where F_{t+1} is the forecast for the demand in period $t+1$, after observing the demand in period t. It is easy to verify that

$$d_t - F_t = \varepsilon_t, \qquad t = 1, 2, \ldots.$$

Therefore, the exponential-weighted moving average is an unbiased forecast with a minimum mean square error. Note that

$$F_{t+1} = d_{t+1} - \varepsilon_{t+1} = \alpha\varepsilon_t + \alpha\varepsilon_{t-1} + \cdots + \alpha\varepsilon_1 + \mu.$$

At the end of period t (after the realization of d_t or ε_t), the forecast for d_{t+i} for any $i \geq 1$ is equal to F_{t+1}.

Consider, for a moment, a single-location, single-item inventory system with the above demand process. Assume that when demand exceeds on-hand inventory, the excess demand is completely backlogged. Moreover, the replenishment lead time is L periods, where L is a known integer. The events in each period are sequenced as follows: demand is realized, an order is placed, the order from L periods ago is received, and demand and backorders (if any) are filled from inventory. Consider a base-stock policy, with the order-up-to level for period t being

$$S_t = S_0 + LF_{t+1}$$

where S_0 is some constant. Note that LF_{t+1} is the forecast for the lead time demand from period $t+1$ to period $t+L$. (And recall that the order-up-to level should cover the lead time demand.) There is no optimality proof for this policy; but if orders are allowed to be negative, the policy is optimal. Let us assume that orders can be negative, and thus the order-up-to level for each period is reached exactly. We have the order quantity in period t,

$$q_t = d_t + (S_t - S_{t-1}) = d_t + L(F_{t+1} - F_t)$$

Note that the order quantity q_t reflects the most recent demand d_t as well as an update on the forecast for the total demand during the next L periods.[19]

Let x_t be the inventory level (on-hand inventory minus backorders) at the end of period t. Graves shows that under the above policy,

$$E[x_t] = S_0 + \mu, \quad Std[x_t] = \sigma \sqrt{\sum_{i=0}^{L-1}(1 + i\alpha)^2}.$$

Note that when the lead time demand is normally distributed, the minimum costs of the system are proportional to $Std[x_t]$, which can sometimes, especially when α is large, be a convex function of L. This is in sharp contrast with the traditional setting with i.i.d. demands, where the minimum costs are proportional to the square root of L. Our intuition is challenged, and it is because of the nonstationary demand process!

Another observation is that the variance of q_t is larger than the variance of d_t, given F_t. To see this, first recall that $d_t = F_t + \varepsilon_t$. Thus, $Var[d_t|F_t] = \sigma^2$. On the other hand, since $F_{t+1} - F_t = \alpha\varepsilon_t$, we have

$$q_t = d_t + L(F_{t+1} - F_t) = (F_t + \varepsilon_t) + L\alpha\varepsilon_t = F_t + (1 + L\alpha)\varepsilon_t.$$

Therefore, $Var[q_t|F_t] = (1 + L\alpha)^2\sigma^2$. This shows that the variance of the order process exceeds the variance of the demand process, and this amplification increases with lead time and α (larger α means a less stable demand process). Graves shows that the order process $\{q_t\}$ has the same characteristics as the demand process $\{d_t\}$. Thus one can easily extend the analysis to a multistage serial system and show that the order variance is further amplified upstream.

Although attention to the bullwhip effect can sometimes help us identify opportunities to improve supply chain performance, it is dangerous if we take as our goal the reduction or elimination of the bullwhip effect. This point is illustrated in a paper by Chen and Samroengraja (1999). They consider a supply chain model with one supplier and N identical retailers. The perspective is that of a central planner whose goal is to minimize the total cost in the supply chain. The supplier's production facility is subject to a capacity constraint, and transportation from the supplier to the retailers incurs fixed costs as well as variable costs. They

---

[19] We are not going to make a big deal out of negative orders here. If you continue to feel that negative orders are annoying, first consult Graves (1999) for further discussions on this and if that is still not enough, then you have a challenging, and potentially rewarding, task ahead of you.

consider two classes of replenishment strategies at the retail level. One is the staggered policy, whereby each retailer places an order to increase its inventory position to a constant base-stock level Y every T periods, and the reorder intervals of different retailers are staggered so as to smooth the aggregate demand process at the supply site. The other strategy is the (R, Q) policy, whereby each retailer orders Q units from the supplier as soon as its inventory position decreases to R. These two types of replenishment strategies are commonly used in practice when there are fixed ordering costs. The supplier replenishes its inventory through production; the production policy is a base-stock policy modified by a capacity constraint. Numerical examples show that although the (T, Y) policy gives a smoother demand process at the supply site, the (R, Q) policy often provides a lower system-wide cost.[20]

It is also interesting to note that discussions on the bullwhip effect can sometimes become confusing and pointless. Consider a supply chain with a manufacturer and a retailer. The retailer agrees to share its point-of-sale information with the manufacturer. For some unexplained reasons (historical?), the manufacturer has a quantity discount policy in place that charges a lower per-unit price for a larger order. Finally, the manufacturer can ship a retailer order in any way it desires so long as a certain service level is achieved at the retail site. In this decentralized model with information sharing and a specific contractual relationship, the manufacturer can plan its production based on the true demand information at the retail site. From the standpoint of the supply chain, what matters is the manufacturer's production quantities and the shipment quantities to the retailer. The retailer's orders do not matter very much; they exist largely for accounting purposes. There is nothing to worry about even if the retailer's orders are more volatile than the customer demands.

In summary, the existence of the bullwhip effect is only a characteristic of an operating policy, which reflects the economic forces underlying the supply

---

[20] Cachon (1999) also studies the impact of staggered ordering policies, which he calls scheduled or balanced ordering policies, on the supply chain performance. The setup is still the one-warehouse N-identical-retailer supply chain. The class of policies considered is that of (T, R, Q) policies: each retailer orders every T periods according to an (R, nQ) policy based on its own inventory position, and the reorder intervals of different retailers are staggered. Cachon provides an exact method to evaluate the supply chain costs under a (T, R, Q) policy as well as numerical examples that illustrate how the supply chain costs respond to changes in the parameters T and Q. Primary conclusions are that the staggering of retailer reorder intervals generally reduces the demand variance at the warehouse and that the combination of increasing T and decreasing Q is an effective way to decrease the total supply chain costs in systems with a small number of retailers and low customer demand variability. Although the general objective of Cachon (1999) coincides with that of Chen and Samroengraja (1999), i.e., to study the impact of variance-reduction policies on supply chain performance, the models are different (whether or not there is a capacity constraint at the warehouse), so are the approaches (the former focuses on a sensitivity analysis whereas the latter compares the optimal solutions from two classes of policies that offer different degrees of variance reduction).

chain and the experience and knowledge of the people who manage it. It is a symptom, not a problem.[21]

2.2 Upstream information

So far, our discussions have been confined to the sharing of information coming from the demand side, i.e., an upstream supply chain member's access to downstream information. We now turn to supply-side information. Interestingly, upstream information has received much less attention in the literature.

Cost information

Chen (2001b) considers a procurement problem facing an industrial buyer. Given Q units of input, the buyer can generate profits $R(Q)$, an increasing and concave function. The buyer's net profit is therefore $R(Q)$ minus the purchase cost incurred for the input. For convenience, let us call $R(\cdot)$ the buyer's revenue function. The buyer seeks a procurement strategy to maximize its expected (net) profit.

There are $n \,(>1)$ potential suppliers for the buyer's input. For supplier i, $i = 1,\ldots,n$, the cost of producing Q units of the buyer's input is $c_i Q$, for any Q. It is common knowledge that the suppliers' unit costs, c_i's, are independent draws from a common probability distribution $F(\cdot)$ over $[c, \bar{c}]$. Supplier i privately observes the value of c_i, but not the costs of other suppliers, $i = 1,\ldots,n$.

Here is an optimal solution to the buyer's procurement problem. The buyer announces a quantity-payment schedule, $P(\cdot)$, which is basically a commitment that says that the buyer will pay $P(Q)$ for Q units of input, for

---

[21] Any discussion of the bullwhip effect would be incomplete without mentioning the beer game, which is described in Sterman (1989) and some of the references therein. The game simulates a four-stage supply chain, consisting of a manufacturer, a distributor, a wholesaler, and a retailer. The demand at the retail site is 4 kegs of beer per period for the first several periods, and then jumps to 8 kegs per period for the rest of the game. The players, who manage the four supply chain stages, do not know the demand process a priori. For several decades, the beer game has been a very effective tool to illustrate the bullwhip effect to an uncountable number of students in many countries. But it has a shortcoming; it merely demonstrates a phenomenon without offering any solutions. How should we play the game? Nobody knows the answer, a quite awkward situation especially in a classroom setting. It is easy to say what we should have done in hindsight, but that is not helpful to the supply chain's managers. In fact, it is quite possible that most strategies could be explained with a belief system that uses the past to predict the future in a particular way. Frankly, there is little we can teach our students about how to manage a supply chain that resembles the beer game setup (at least, not yet). Interestingly, if we replace the 4–8 demand stream with a stream of i.i.d. random variables, and suppose the players all know the demand distribution, then we know how the game should be played. (This game will be discussed in Section 2.3 of this chapter). Under the optimal strategy, the bullwhip effect does not exist. But it may still occur (and it has) depending on the strategies used by the managers. So here is a game that can be used to illustrate the bullwhip effect, which we can say with confidence is bad. For a description of the i.i.d. version of the beer game and some teaching experience with it, see Chen and Samroengraja (2000b).

any Q. A supplier, if chosen by the buyer, is free to choose any quantity to deliver to the buyer and be paid according to the preannounced plan. Therefore, the buyer has effectively proposed a business proposition to the potential suppliers. Of course, different suppliers will value this business deal differently, with the lowest-cost supplier deriving the highest value. In an English auction, the suppliers openly bid up the price they are willing to pay for the buyer's proposed contract, with the winner being the supplier willing to pay the highest price.[22] With a little bit of thinking, the lowest-cost supplier always wins the contract and pays a price equal to the value that the second-lowest cost supplier derives from the contract.

To better understand the above solution, suppose the buyer is a retailer, who buys a product from a supplier and resells it to customers. The selling price to the customers is p per unit, which is exogenously given. The total customer demand is D, a random variable with cumulative distribution function $G(\cdot)$. If demand exceeds supply, the excess demand is lost. Otherwise, the excess supply is useless and can be disposed of at no cost. The total quantity sold to customers is thus $\min\{Q, D\}$. The buyer's expected revenue is

$$R(Q) = pE[\min\{Q, D\}] = pE[Q - (Q - D)^+] = pQ - p\int_0^Q G(y)dy.$$

Note that this revenue function is concave and increasing in Q. To make things even simpler, suppose the suppliers' costs are drawn from the uniform distribution over $[0, 1]$. Under this condition, the optimal quantity-payment schedule is

$$P(Q) = \frac{1}{2}R(Q).$$

Note that this payment schedule is independent of the number of potential suppliers. Moreover, it is a revenue-sharing contract: the business deal that the buyer proposes calls for a 50–50 split of the buyer's revenue. It is also a returns contract, which says that the buyer pays the winning supplier a wholesale price of $w = p/2$ for each unit of input delivered (before demand realization), and in case there is excess supply after demand is realized, the buyer can return the excess inventory to the supplier for a full refund. Under this contract, and assuming the returned inventory has no value to any supplier, a supplier, if he wins, earns the following expected revenue (as a function of the production quantity Q):

$$E\left[\frac{p}{2}Q - \frac{p}{2}(Q - D)^+\right] = P(Q).$$

---

[22] The theory of auctions is huge and well developed. Vickrey (1961) is seminal. Myerson (1981) and Riley and Samuelson (1981) are important milestones for their contributions to optimal auction design. McAfee and McMillan (1987) and Klemperer (1999) provide comprehensive reviews.

To put our hands around the inefficiencies caused by the asymmetric cost information, let us further assume that $G(x) = x$ for $x \in [0, 1]$ and $p = 2$. In this case, $R(Q) = 2Q - Q^2$, and the optimal quantity-payment schedule becomes $P(Q) = Q - Q^2/2$. As mentioned earlier, the lowest-cost supplier wins the contract. Let C_1 be the cost of the winning supplier. (Thus $C_1 = \min\{c_1, \ldots, c_n\}$.) The quantity delivered by the winning supplier solves the following problem

$$Q(C_1) = \mathrm{argmax}_Q P(Q) - C_1 Q = \mathrm{argmax}_Q (1 - C_1)Q - Q^2/2.$$

Therefore $Q(C_1) = 1 - C_1$. The total profit for the supply chain (the buyer plus the winning supplier) is $R(Q(C_1)) - C_1 Q(C_1) = 1 - C_1$, with an expected value

$$\pi = 1 - E[C_1].$$

On the other hand, the efficient input quantity, one that maximizes the supply chain profit, is

$$Q^*(C_1) = \mathrm{argmax}_Q R(Q) - C_1 Q = \mathrm{argmax}_Q (2 - C_1)Q - Q^2.$$

Therefore, $Q^*(C_1) = (2 - C_1)/2$. Note that $Q^*(C_1) > Q(C_1)$; asymmetric cost information reduces the input quantity. The maximum expected supply chain profit under full information is

$$\pi^* = \pi + \frac{1}{4} E[C_1^2].$$

Therefore the supply chain inefficiency due to asymmetric information is

$$\pi^* - \pi = \frac{1}{4} E[C_1^2] = \frac{1}{2(n+1)(n+2)}$$

which is decreasing in n. This is the value created if the suppliers disclose their cost information. But why should they?

Lead time information

Another important piece of information coming from the supply side is the status of a replenishment order. Chen and Yu (2001a) address the value of lead time information in the following inventory model. A retailer buys a single product from an outside supplier, stores it in a single location, and sells it to her customers. Customer demand arises periodically, with demands in different periods being i.i.d. random variables. If demand exceeds the on-hand inventory in a period, the excess demand is backlogged. On-hand inventories incur holding costs, and customer backorders incur penalty costs. The analysis is done from the retailer's standpoint; how to make replenishment decisions so as to minimize the retailer's long-run average holding and backorder costs.

Here is the supply process. Let L_t be the lead time for an order placed in period t. And $\{L_t\}$ is a Markov chain with a finite state space. The one-step transition matrix of the Markov chain is chosen so as to prevent order crossovers (so orders are received in the sequence in which they were placed). The supply process is exogenous, i.e., the evolution of the Markov chain is independent of the operations of the retailer's inventory system.[23] The supplier observes the state of the Markov chain $\{L_t\}$, and he may or may not share this information with the retailer.

Two scenarios are considered. First, suppose the retailer knows the value of L_t for each period t before her replenishment decision. In this case, the optimal policy is to place an order in period t so as to increase the retailer inventory position up to a base-stock level that is a function of L_t, for all t. That is, a state-dependent, base-stock policy is optimal. On the other hand, suppose now that the supplier does not share with the retailer the lead time information. In this case, the retailer has to rely on the history of order arrivals to infer something about the current lead time and make her replenishment decisions accordingly. By comparing these two solutions, one sees the value of lead time information. Numerical evidence indicates that the value of lead time information is small for small-volume items, but significant for high-volume items where the percentage cost savings due to lead time information can be as high as 41%.

Capacity information

Our third example deals with the value of capacity information. Chen and Yu (2001b) consider a model with one retailer and one supplier. There is a single selling season. The retailer has two opportunities to place orders with the supplier before the season starts, one at time 0 and one at time 1. At time 0, the supplier has unlimited capacity, i.e., whatever the retailer orders will be ready for the selling season. At time 1, the supplier's capacity is uncertain, and it can be written as $C - \varepsilon$, where C is the 'forward capacity', i.e., the supplier's capacity at time 1 perceived at time 0, and ε, which can be positive or negative, is an external random shock reflecting uncertainties between time 0 and time 1. At time 0, there are two possible states, high or low, for the total demand in the selling season. A cumulative distribution is given for each demand state. At time 1, the true demand-state is revealed; the retailer now has better demand information. This suggests that there is a benefit for the retailer to postpone the ordering decision to time 1. But the cost

---

[23] Song and Zipkin (1996b) have provided several concrete examples to motivate such a lead time process. Other models of random lead times have been provided by Kaplan (1970), Nahmias (1979), Ehrhardt (1984) and Zipkin (1986).

of doing this is that the retailer may not get what he orders (at time 1), due to the supplier's capacity constraint. Let q_0 be the quantity ordered by the retailer at time 0, and q_1^s the quantity ordered at time 1 if the demand state is s, s = high or low. The optimal values of these quantities balance the benefit from demand information with the cost of capacity risk.

Before time 0, it is common knowledge that C comes from a given probability distribution. At time 0, before the retailer decides on the value of q_0, the supplier privately observes the realized value of C. The retailer then offers a menu of contracts: a mapping from the supplier's reported capacity (which can be different from the true value of C) to q_0. The supplier then reports a value of C, effectively choosing a value for q_0. (This is the screening idea, to be discussed in detail in Section 3.1.) The solution to this asymmetric information case is then compared with the full-information scenario where the retailer also sees the value of C (before deciding q_0). The comparison gives the value to the retailer of knowing the supplier's forward capacity.

2.3 Information transmission

Chen (1999a) considers a supply chain model where information transmission is subject to delays. A firm has N divisions arranged in series. Customer demand arises at division 1, division 1 replenishes its inventory from division 2, 2 from 3, etc., and division N orders from an outside supplier. The demands in different periods are independent draws from the same probability distribution. Each division is managed by a division manger. Information in the form of replenishment orders flows from downstream to upstream, triggering material flow in the opposite direction. Both flows are subject to delays.[24]

An important feature of the model is that the division managers only have access to local inventory information. That is, each manager knows (1) his on-hand inventory, (2) the orders he has placed with the upstream division, (3) the shipments he has received from the upstream division, (4) the orders he has received from the downstream division, and (5) the shipments he has sent to the downstream division. However, he does not exactly know the shipments that are in transit from the upstream division that may be unreliable, and neither does he know the orders from the downstream division that are currently being processed. The decisions made by each manager can only be based on what he knows.

The first model considered by Chen assumes that the division managers behave as a *team*, i.e., they have a common goal to minimize the system-wide costs.[25] This is reasonable when, for e.g., the owner of the firm has

---

[24] This may remind you of the beer game, which is described in Sterman (1989). A key difference is the i.i.d. demand process assumed here.
[25] For the economic theory of teams, see Marschak and Radner (1972).

implemented a cost-sharing plan whereby each manager's objective function is a fixed, positive proportion of the overall cost of the system. It is shown that the optimal decision rule for each division manager is to follow an installation, base-stock policy. Division i's installation stock is equal to its net inventory (on-hand inventory minus backorders) plus its outstanding orders. Recall that manager i knows the orders he has placed (with the upstream division) as well as the shipments he has received (from the upstream division). The difference between the two is the outstanding orders. Therefore, installation stock is *local information*. The optimal decision rule for each division manager is to place an order in each period to restore the division's installation stock to a constant target level, which may be division-specific.

The solution to the team model reveals the role played by the information lead times (delays in the information flow). In terms of division i's safety stock, the information lead time from division i to $i+1$ plays exactly the same role as the production/transportation lead time from division $i+1$ to i; the safety stock level only depends on the sum of the two lead times. (This is intuitive at first glance. But the optimality proof requires some finessing.)

An alternative to the team model is the cost-centers model, where each manager is evaluated based on his division's performance. But how should local performance be determined? Chen suggests using the so-called accounting inventory level. The accounting inventory level at a division is its net inventory under the hypothetical scenario where no orders by the division will ever be backlogged at the upstream division. Note that the accounting inventory level may differ from the actual inventory level, because the upstream division is not always reliable. A division is charged a holding cost if its accounting inventory level is positive and a penalty cost otherwise.[26] It has been shown that the owner of the firm can choose the cost parameters so that when the individual division managers minimize their own (accounting) costs, the system-wide costs are also minimized.[27]

Firms decentralize the control of their operations for many reasons. One key reason is that the local managers are better informed about the local operating environments than the owner is. Therefore, it makes sense to let the local managers make local decisions. In the above supply chain model, let us suppose the division managers all know the true demand distribution, but the owner of the firm does not. Consider the following two scenarios. In one, the owner solves the team model based on her (erroneous) knowledge of the

---

[26] The accounting and management literature advocates that individuals should only be evaluated on controllable performance, see, e.g., Horngren and Foster (1991). For this reason, the actual local inventory level (on-hand inventory minus backorders) at a division is inadequate as a basis for measuring local performance, since it is also affected by decisions made at the other divisions. The accounting inventory level removes the impact of the upstream division, but is still affected by the downstream division's orders.

[27] For other coordination mechanisms for serial inventory systems, see Lee and Whang (1999) and Porteus (2000). Gérard P. Cachon discusses these papers in Chapter 6 of this volume.

demand distribution and tells her employees to implement the installation, base-stock policies she found. (The owner only provides the decision rule, leaving the division managers to implement it. Since the division managers only have access to local information, the decision rule must be based on local information.) Call this the dictator scenario. In the other scenario, the owner organizes the divisions as cost centers. After the owner has specified a measurement scheme, each division manager chooses a replenishment strategy to minimize his accounting costs by using the true demand distribution. In both cases, the system-wide performance will be suboptimal, because the owner's inaccurate knowledge about the demand distribution has been used in one way or another. Numerical examples provided by Chen show that the system-wide performance under cost centers is nearly optimal, whereas the dictator scenario can be far from optimal. The benefit of decentralization is clear. Moreover, the measurement scheme for the cost centers is rather robust with respect to shifts in the demand distribution; a scheme based on an outdated demand distribution works very well for a new demand distribution so long as the division managers update their replenishment strategies based on the new information.

Finally, what if managers make mistakes? To explore this issue, consider the following specific irrational behavior. Manager i strives to maintain his *net inventory* at a constant level Y: if it is below Y, order the difference; otherwise, do nothing. It is a mistake because the decision-maker forgets about the outstanding orders. (Recall that the optimal strategy is to maintain the installation stock at a constant level.) This mistake corresponds to the 'misperceptions of feedback' Sterman (1989) found in the beer game. A simulation study shows that such mistakes can be very costly, especially those committed at the downstream part of the supply chain.[28]

When a downstream manager follows an erroneous strategy, the upstream managers receive distorted (and delayed) demand information. This is at the heart of the problem. Now consider the following alternative design of information flow in the supply chain. When division 1 places an order, he is also required to report the demand in the previous period. This demand information is then relayed to the upstream managers along with the orders. Assume that the rational managers place their orders according to the accurate demand information, whereas the irrational ones follow the above 'forgetful' strategy. In this way, a downstream ordering mistake can no longer corrupt the upstream order decisions. Simulation results indicate that by making the accurate demand information accessible to the upstream members of the supply chain, the system becomes much more robust. This is another reason for sharing demand information.

Most of the supply chain models on information sharing assume that the transmission of information is instantaneous and reliable. (We just saw one

---

[28] Watson and Zheng (2001) provide a more recent attempt to address supply chain mismanagement due to irrational managerial behavior.

exception.) Moreover, they assume (implicitly) that information/knowledge is always transmittable. (We will see an exception soon.) However, managers are sometimes endowed with knowledge that is so specific to the local operating environment that it is very difficult to share such knowledge. This is perhaps what people mean by 'experience', the sharing (or rather the acquisition) of which may take years of apprenticeship. So a more realistic view of organizations is that there are two kinds of knowledge: one can be readily shared (e.g., sales data) and the other is difficult to share, i.e., information sharing takes time and effort, is imperfect with noise, or is just impossible. When the local, specific knowledge plays a dominant role, it is important to give the manager possessing the knowledge the authority to make decisions that have the most use of the knowledge. In other words, decision rights should reflect the dispersion of knowledge in an organization. This is, however, not the only challenge in designing an organization because the distribution of knowledge is, to some extent, manageable. This points to another aspect of organizational design, i.e., an organization's information structure (or 'who knows what'). We refer the reader to Hayek (1945), Jensen and Meckling (1976, 1992) for further discussions on specific knowledge and the design of organizations. Below, we review a paper from the operations literature that studies the above issues in a supply chain context.

Anand and Mendelson (1997) consider a firm that produces and sells a product in n markets. Production takes place in one location, and the total cost of producing Q units is assumed to be

$$TC(Q) = cQ + \frac{1}{2}\gamma Q^2.$$

A common reason cited for assuming increasing marginal costs is capacity constraints, e.g., overtime is used when production exceeds a certain threshold level and the overtime wage is higher than the regular wage. The n markets each face an independent, linear demand curve. Consider market i, $i = 1, \ldots, n$. There are only two possible market states, high or low. If the market is high, the (inverse) demand curve is $P(q_i) = a_H - bq_i$, where q_i is the quantity of the product allocated to market i, and $P(q_i)$ is the corresponding market clearing price. On the other hand, if the market is low, the demand curve is $P(q_i) = a_L - bq_i$, with $a_L < a_H$. (No transshipments are allowed among the markets after the initial allocation.) The market state is denoted by a binary random variable, s_i: if $s_i = 1$ (0) the market is high (low).

A key feature of the model is that the n markets are managed by branch managers who possess two types of information: one is specific knowledge that is not transmittable to anyone else, and the other is transferable data. This is modeled by assuming that $s_i = x_i y_i$, where both x_i and y_i are binary random variables, with x_i representing transferable market-i data and y_i the unobservable market-i condition, $i = 1, \ldots, n$. It is assumed that

$\{x_i, y_i, i = 1, \ldots, n\}$ are independent random variables with $Pr(x_i = 1) = t$ and $Pr(x_i = 0) = 1 - t, 0 < t < 1$, and $Pr(y_i = 0) = Pr(y_i = 1) = 1/2$, $i = 1, \ldots, n$. The value of t is common knowledge. The branch manager at market i observes the value of x_i as well as a binary signal L_i that may contain information about y_i with $y_i = L_i$ with probability $1 - \alpha$ and $y_i = 1 - L_i$ with probability α. If $\alpha = 0$ or 1, then the signal is perfect; if $\alpha = 1/2$, then the signal does not provide any new information beyond the prior on y_i. It is clear that we can restrict to $0 \leq \alpha \leq 1/2$ without loss of generality. Under this restriction, the value of $(1 - \alpha)$ represents the precision of the signal. The local signal L_i is branch manager i's specific knowledge that is not transferable to anyone else, $i = 1, \ldots, n$. (Refer to the original paper for motivating stories behind this elaborate design. An alternative view one may take is that information is always transmittable, but some kinds of information are very costly to transmit.)

The decision variables are the supply quantities q_i. The objective of any decision maker (to be specified below) is to maximize the firm's expected profits, which are equal to the revenues generated by the branches minus the production cost that depends on the total quantity.

Anand and Mendelson then consider three different organizational designs, depending on where decision rights reside and how information is distributed (through the design of the firm's information system, e.g.). The first design is a centralized one, where a 'center' makes all the decisions by using all the transferable data but none of the specific knowledge. The firm thus has in place an information system that allows the branches to report their transferable data, x_i for $i = 1, \ldots, n$, to the center. The second design is decentralized, where each branch manager i makes his own quantity decision q_i based on his own specific knowledge (L_i) and transferable data (x_i). Therefore, in this case, there is no information sharing so that all local knowledge (transferable or not) remains local. The third design is in between the previous two, with the branches making their own quantity decisions based on their specific knowledge and all the transferable data (again, enabled by an intra-firm information system). That is, branch manager i determines the value of q_i with knowledge of $(x_1, x_2, \ldots, x_n)$ and L_i, $i = 1, \ldots, n$. This design is referred to as the 'distributed' structure. The analysis of the second and the third organizational structures follows that of a team model, where the team members (i.e., branch managers) share a common goal but have access to different sets of information. (The team model thus assumes away all potential incentive problems. Anand and Mendelson also consider transfer-pricing schemes when incentive issues cannot be ignored.)

It is intuitive (and true) that the distributed design dominates the decentralized design in terms of the firm's expected profits. The difference represents the value of information sharing, which is shown to increase in the number of branches at first and then decrease. In other words, the distributed structure adds more value to firms that operate in a moderate number of markets. On the other hand, the difference between the centralized and decentralized systems captures the tradeoff between coordination, information sharing, and local knowledge. The centralized system benefits from better coordination of quantity decisions (due to centralized decision-making) and the pooling of all the transferable data. However, the decentralized system sometimes performs better than the centralized system, indicating the usefulness of the local knowledge.

In their concluding remarks, Anand and Mendelson said that 'the design of organizations requires an analysis of what kinds of information the firm needs to acquire, alternative ways of distributing this informational endowment and ways of structuring the organization (i.e., allocation of decision rights) to match its information structure'. (The parenthetical explanation is added.) What they have done in their paper is to treat the allocation of information and decision rights as design variables to be jointly determined, leaving out information acquisition.

3 Incentives for sharing information

Information sharing in supply chains with independent players is tricky. When a player has superior information, two things may happen. He may withhold it to gain strategic advantage, or he may reveal it to gain cooperation from others. If the former, the other (less informed) players try to provide incentives for him to reveal his private information; this is called *screening*. If the latter, we have *signaling*, i.e., revealing information in a credible way. Sometimes it is impossible to say who has more or less information; players simply have different information about something they all care about. For example, different retailers may obtain different signals about the market demand for a product. In this case, a player's willingness to share his information depends on if the others are going to share their information and how the revealed information will be put to use. This section reviews papers that deal with information exchanges in decentralized supply chains.[29]

---

[29] A branch of economics (sometimes called information economics) addresses issues arising from various information asymmetries. One type of information asymmetry is often studied under the heading 'moral hazard,' which refers to situations where one party (called agent) performs a task on behalf of another (called principal), and the agent's effort level is unobservable to the principal. A conflict arises because the principal prefers the agent to work hard while the agent dislikes exerting effort. The solution is an incentive contract that pays the agent for his output. The principal-agent theory, originating from economics, has been used/developed in the accounting, marketing, and lately operations literatures. Kreps (1990) provides an excellent introduction to the principal-agent theory; some of the seminal papers in this area are cited later in this chapter. For principal-agent models in the operations literature, see, e.g., Porteus and Whang (1991), Chen (2000b), Plambeck and Zenios (2000, 2002). We choose not to review this part of the literature here because the goal of providing incentives in principal-agent models is to induce a certain level (or pattern) of effort by an agent (not to facilitate information sharing). By the way, many of the ideas behind the papers reviewed in this section originated from economics often under rubrics such as adverse selection, mechanism design, or signaling.

3.1 Screening

This subsection presents several examples where a firm tries to 'smoke out' either consumer preferences or private information held by an employee or by a supply chain partner.[30]

A simple example

A product line usually refers to a range of goods of the same generic type, but differentiated along some attributes. For example, Dell offers two lines of notebook computers, Inspiron and Latitude, and each product line consists of models with different speeds, storage capacities, etc. An important question is how a firm can optimally design and price a product line.

Suppose a monopolist is offering a line of goods differentiated along a quality dimension. There are two consumers, with different demand intensities for quality. (This is thus a toy problem. But the basic ideas are here.) Consumer i values the variety with quality level q at $\theta_i q$, $i = 1, 2$, with $0 < \theta_1 < \theta_2$. Therefore, both consumers prefer more quality, but differ in their willingness to pay for any given quality level. Each consumer buys one unit of the good, or nothing at all. There are constant marginal costs of production at any given quality level, and the marginal cost of producing variety q is q^2.

Consider first the case of a perfectly discriminating monopolist who is able to sell to each consumer individually (i.e., deny one consumer access to the product offered to the other consumer) and prevent any resales. The monopolist will thus charge each consumer his reservation price and the only remaining problem is what variety to offer to each consumer. To solve this problem, simply maximize $\theta_i q - q^2$ over q for each i. Therefore, the optimal strategy is to offer quality level $\theta_i/2$ to consumer i and charge him $\theta_i^2/2$ for it. Both consumers will buy the products offered them and derive zero surplus. The firm may be able to achieve this ideal solution in some cases. For example, a telephone company is routinely charging different rates to business users and residential users (it is relatively easy for the company to verify if a user is business or residential and it is very difficult to trade phone calls between business and residential users). In other cases, it is impossible to deny one consumer access to the products offered to other consumers, for technical or legal reasons. In other words, a product line, whatever it may be, must be made available to all types of consumers. What should the monopolist do then?

First, note that the above solution will not work when consumers self-select. When given the above two variety-price combinations: $(\theta_i/2, \theta_i^2/2)$

---

[30] There is a large body of research on screening in queuing contexts. For example, a service provider can charge different prices for different priority levels. An arriving customer decides which priority class to join based on his/her (private) cost of waiting. For a comprehensive survey of this literature, see Hassin and Haviv (2001).

for $i = 1, 2$, consumer 1 will continue to choose $(\theta_1/2, \theta_1^2/2)$, but consumer 2 will switch from $(\theta_2/2, \theta_2^2/2)$ to $(\theta_1/2, \theta_1^2/2)$ and earns a positive surplus (before, he earned zero surplus). In order to prevent consumer 2 from switching, the firm must lower the price for variety $\theta_2/2$, if everything else stays unchanged. Finding the monopolist's optimal strategy involves a systematic tradeoff among multiple dimensions. Let (q_i, p_i), $i = 1, 2$, be the quality-price pairs offered to the market (i.e., the two consumers). Suppose consumer i chooses (q_i, p_i). (If the two pairs are identical, then the product line consists of only one good.) We first consider the case where both consumers are served (i.e., they each buy a unit). The monopolist's problem can be written as:

$$\max_{q_1, p_1, q_2, p_2} (p_1 - q_1^2) + (p_2 - q_2^2)$$

s.t. $\theta_1 q_1 \geq p_1$ \hfill (P1)

$\theta_2 q_2 \geq p_2$ \hfill (P2)

$\theta_1 q_1 - p_1 \geq \theta_1 q_2 - p_2$ \hfill (SL1)

$\theta_2 q_2 - p_2 \geq \theta_2 q_1 - p_1$ \hfill (SL2)

where the objective function represents the firm's total profits, the first two constraints ((P1) and (P2)) are necessary in order for the consumers to participate, and the last two constraints ((SL1) and (SL2)) ensure that the consumers choose the right bundle. This problem is easy. First note that $\theta_2 q_1 - p_1 \geq \theta_1 q_1 - p_1$ and the right side is nonnegative from (P1). This, together with (SL2), implies that $\theta_2 q_2 \geq p_2$. Therefore, (P2) is redundant and is thus deleted. Moreover, (P1) must bind, for otherwise, one can simultaneously increase p_1 and p_2 by the same amount without violating any constraints. Note that (SL1) and (SL2) can be combined to produce:

$$\theta_2(q_2 - q_1) \geq p_2 - p_1 \geq \theta_1(q_2 - q_1).$$

It then follows that $q_2 \geq q_1$, which then implies $p_2 \geq p_1$. On the other hand, (SL2) must bind, because if not, one can increase p_2 to get a better solution for the monopolist. Therefore, $p_2 - p_1 = \theta_2(q_2 - q_1) \geq \theta_1(q_2 - q_1)$ because $q_2 \geq q_1$. Consequently, (SL1) is implied by the binding version of (SL2) plus $q_2 \geq q_1$. In sum, the above optimization is equivalent to

$$\max_{q_1, p_1, q_2, p_2} (p_1 - q_1^2) + (p_2 - q_2^2)$$

s.t. $p_1 = \theta_1 q_1$

$\theta_2 q_2 - p_2 = \theta_2 q_1 - p_1$

$q_2 \geq q_1.$

This has a closed-form solution:

$$q_1^* = \max\left\{\theta_1 - \frac{\theta_2}{2}, 0\right\}, \quad q_2^* = \frac{\theta_2}{2}.$$

Note that compared with the previous solution without consumer self-selection, the lower demand-intensity consumer (i.e., consumer 1) gets a lower quality product, a lower price, and the same (zero) surplus, whereas the higher demand-intensity consumer gets the same quality product, a lower price, and a positive surplus. Moreover, if we interpret $q_2^* - q_1^*$ as the breadth of the product line, self-selection leads to a broader range of goods. Finally, if $\theta_1 \leq \theta_2/2$, consumer 1 is not served.

The above simple example captures the basic idea of screening. For more sophisticated models involving consumer self-selection, see, for e.g., Mussa and Rosen (1978), Maskin and Riley (1984) and Moorthy (1984).

Quality, broadly interpreted, represents product attributes for which the consumer preference is of the more-is-better kind. Other product attributes include color, size, taste, etc., and each consumer is likely to have a unique ideal point in the attribute space.[31] It is possible to differentiate products along these dimensions as well. For this part of the literature, we refer the reader to Shocker and Srinivasan (1979), Green and Krieger (1985), Lancaster (1979, 1990), de Groote (1994), Dobson and Kalish (1988, 1993), Nanda (1995), Chen, Eliashberg and Zipkin (1998), and Yano and Dobson (1998). A recent book edited by Ho and Tang (1998) contains further references on this topic.

Market segmentation and product delivery

Different customers may exhibit different degrees of aversion to waiting: some want to have their orders delivered right away, while others can tolerate a delay. Therefore, the delivery schedule of a product can be a useful tool for segmenting the market. One benefit of such a segmentation strategy is that when a customer places an order that does not have to be shipped immediately, the firm obtains advance demand information that can be used for better production-distribution planning. A potential cost of this strategy occurs when a price discount must be offered in order for a customer to accept a delay. How can a firm design an optimal price-delivery schedule?

Chen (2001a) provides a model to address the above question. A firm sells a single product to consumers. The firm announces a price-delay schedule $\{(p_k, \tau_k)\}_{k=0}^{K}$, for some nonnegative integer K, where p_k is the price a customer

---

[31] The *ideal-point model* is often used to describe consumer preferences along dimensions that are not quality-like. For example, a consumer's utility of buying a product with level x of certain attribute can be written as $A - (x - a)^2$, where A is the maximum possible utility level and a is the ideal attribute level for the consumer. Different consumers can have different ideal attribute levels.

pays if he agrees to have his order shipped τ_k units of time after order placement, with $0 = \tau_0 < \tau_1 < \cdots < \tau_K$ and $p_0 > p_1 > \cdots > p_K$. The maximum price p_0 is paid only if a consumer wants immediate shipment.

The consumers divide into M segments. Let $u_m(\tau)$ be the maximum (or reservation) price that the customers in segment m are willing to pay for one unit of the product if their orders are shipped τ units of time after order placement, $m = 1, \ldots, M$. It is assumed that $u_m(\cdot)$ is decreasing, differentiable, convex with $u'_m(\tau) < u'_{m+1}(\tau) \, (<0)$ for all τ.[32] Thus, segment 1 is uniformly more sensitive to waiting than segment 2, segment 2 more so than segment 3, and so on. The surplus that a type-m customer derives from (p_k, τ_k) is $u_m(\tau_k) - p_k$. His objective is to choose a pair from the schedule that maximizes his surplus, i.e., consumer self-selection.

The product is replenished by an N-stage supply chain. Stage 1 is the final stocking point from which product is shipped to customers, stage 1 is replenished by stage 2, stage 2 by stage 3, etc., and stage N by an outside supplier with ample stock. The transit time from one stage to the next is constant. Customer orders are satisfied first-come, first-served according to the sequence of their shipping dates, not the dates in which the orders are placed. If the firm cannot ship an order on the date chosen by the customer because the product is out of stock (at stage 1), the order is backlogged. The backlogged orders are shipped as soon as inventory becomes available. In this case, the firm incurs a goodwill loss (or backorder cost). In addition, the firm incurs holding costs for inventories held in the supply chain and variable costs for every unit sold.

Determining an optimal price-delivery schedule turns out to be a hard problem. Below is a brief description of the solution.

The optimal schedule always has the segments bundled in a sequential manner, with lower segments choosing higher prices and shorter delays. For example, suppose there are five segments in the market, and the firm offers $\{(p_0, 0), (p_1, \tau_1), (p_2, \tau_2)\}$. Sequential bundling means something like the following. Suppose segments 1 and 2 choose $(p_0, 0)$, segment 3 chooses (p_1, τ_1), and segments 4 and 5 choose (p_2, τ_2). Given this, a property of the optimal schedule is that segment 3 is indifferent between (p_1, τ_1) and $(p_0, 0)$, and segment 4 is indifferent between (p_2, τ_2) and (p_1, τ_1). (It is assumed that when indifferent, a consumer will choose the lower price.) These indifference relationships imply that a price-delay schedule is fully specified by the delays and the marginal segments, i.e., (τ_1, τ_2) and segments 3 and 4 in the above example.

---

[32] Note that $u_m(0) - u_m(\tau)$ is the cost of waiting for a segment-m customer. The assumption that $u_m(\cdot)$ is convex implies that the cost of waiting is concave, i.e., the marginal cost of waiting is decreasing. This is true if, for example, the excitement about the product decreases over time after the order is placed. In this case, the marginal cost of waiting is very high in the first few days, while the excitement still lingers, and decreases as time goes by.

Now fix the price-delay schedule and consider the firm's supply chain. Suppose segment m chooses a shipping delay equal to l_m, $m = 1, \ldots, M$. Therefore, if a segment-m customer places an order at time t, the corresponding demand occurs at time $t + l_m$. Suppose $l_m > 0$ for some m. Therefore, some orders serve as warnings of future demand, and the question is how this information can be incorporated into the firm's replenishment strategy. The optimal strategy, and the corresponding minimum supply chain costs, can be obtained by carefully separating the known demand information from the unknown and following the approach of Chen and Zheng (1994). The optimal policy is an echelon base-stock policy with floating order-up-to levels (one for each stage).

To gain some intuition on how the shipping delays translate into cost savings, suppose $N = 2$. Let the lead times at stages 1 and 2 be 4 and 2 periods respectively. To make matters really simple, assume there is only one segment choosing a shipping delay of l periods. It is easy to see that if $l = 6$, the supply chain faces no demand uncertainty at all and as a result, no inventories need to be carried at any stage. Now suppose $l = 5$. In this case, there is no need to carry any inventory at stage 1. Consider the problem facing stage 2. Say a customer order arrives at time t (at stage 1). This order needs to be shipped out of stage 1 at time $t + l = t + 5$. This means that the order needs to be shipped out of stage 2 at time $t + 1$. Therefore, stage 2's problem is basically a single-location inventory problem where customers choose a shipping delay of 1 period. This, together with the fact that the lead time at stage 2 is 2 periods, implies that the demand uncertainty facing stage 2 is just one-period worth of demand, not the usual two-periods worth of demand. Finally, if, say, $l = 3$, then both stages will need to carry some safety stock. With some thinking, the reader will see that the demand uncertainty facing stage 1 is one-period $(= 4 - 3)$ worth of demand and the demand uncertainty facing stage 2 remains to be two-periods worth of demand.

An optimal price-delay schedule can be obtained by solving an optimization problem that captures both the costs and benefits of the segmentation strategy. Numerical results show that the net benefit of this strategy can be substantial.

Screening and moral hazard

Sometimes the party being screened also takes hidden actions. Chen (2000c) studies such a model. Suppose a firm sells a single product through a single sales agent. The market demand is the sum of the agent's selling effort (a), the market condition (θ), and a random shock (ε), i.e.

$$X = a + \theta + \varepsilon$$

where θ and ε are independent random variables, $Pr(\theta = \theta_H) = \rho$ and $Pr(\theta = \theta_L) = 1 - \rho$ for $0 < \rho < 1$ and $\theta > \theta_L > 0$, and $\varepsilon \sim N(0, \sigma^2)$. The agent privately observes the value of θ, and the agent's effort level is not observable to the firm. The firm's decisions are how to compensate the agent for his work (i.e., a wage contract) and how much to produce before demand realization. Given a contract, the agent decides whether or not to accept it and if so, how much effort to expend.[33]

The model assumes the following sequence of events: (1) the firm (or principal) offers a menu of wage contracts (for screening); (2) the agent privately observes the value of θ; (3) the agent decides whether or not to participate (work for the firm) and if so, which contract to sign; (4) under a signed contract, the firm determines the production quantity, and the agent makes the effort decision; and (5) ε is realized.

Consider the agent's decisions when offered a menu of contracts. First, he considers each contract on the menu and determines the maximum expected utility that can be obtained under the contract. Suppose $s(\cdot)$ is the contract being considered, i.e., $s(x)$ is the wage paid to the agent if the total sale is x. Assume the agent's utility for net income z is $U(z) = -e^{-rz}$ with $r > 0$.[34] Note that $U(\cdot)$ is increasing and concave, implying that the agent is risk averse. The net income is the wage received, $S(X)$, minus the cost of effort, $V(a) = a^2/2$.[35] To determine the maximum expected utility achievable under $s(\cdot)$, the agent solves the following optimization problem

$$\max_a E\left[-e^{-r(s(X)-V(a))}\right].$$

Recall that the agent has already observed the value of θ when evaluating the contract. Therefore, the above expectation is with respect to ε given the observed value of θ. If the maximum expected utility is greater than or equal to $-U_0$, $U_0 > 0$, the agent's reservation utility representing the best outside opportunity for the agent, then s is said to be acceptable to the

---

[33] Coughlan (1993) reviews the salesforce compensation literature. A common assumption is that the total sales is a function of selling effort and a random shock and that effort is unobservable to the firm. This is the moral hazard problem, which has been widely studied in the economics/agency-theory literature, see, e.g., Shavell (1979), Harris and Raviv (1978, 1979), Holmstrom (1979, 1982) and Grossman and Hart (1983). If, in addition to the moral hazard problem, the firm is in an informational disadvantage in terms of the sales environment, i.e., the sales people have superior information about the sales response function (the productivity of selling effort, the sensitivity of customers to price changes, the sales prospects, etc.), then the firm also faces an adverse selection problem. The typical solution is a menu of contracts. The salesforce compensation literature in marketing includes Basu, Lal, Srinivasan and Staelin (1985), Lal (1986), Lal and Staelin (1986), Rao (1990) and Raju and Srinivasan (1996).

[34] The negative exponential utility function is widely used in agency models.

[35] The quadratic form is not critical for the analysis. An often-assumed feature of the cost-of-effort function is increasing marginal cost of effort.

agent.[36] Among all the contracts on the menu, the agent chooses the one with the highest achievable expected utility and participates if this utility level exceeds $-U_0$.

We now turn to the principal's problem. Recall that the firm must make its production decision before observing the total sales. This is reasonable when the customers demand fast delivery of their orders and the production lead time is relatively long. (It is thus impossible to follow make-to-order.) Let Q be the production quantity. Let c be the cost per unit produced. If $X \leq Q$, the excess supply is salvaged at p per unit. On the other hand, if $X > Q$, the excess demand must be satisfied by a special production run at a cost of c' per unit. Let the unit selling price be $1 + c$ (the profit margin is thus normalized to 1). To avoid trivial cases, assume $p < c < c' < 1 + c$. As mentioned before, the firm makes contracting as well as production decisions with the objective of maximizing its expected profit (the principal is thus risk neutral). If $s(\cdot)$ is the contract signed by the agent, the firm's profit is

$$(1+c)X - s(X) - cQ + p(Q-X)^+ - c'(Q-X)^-$$
$$= X - s(X) - [(c-p)(Q-X)^+ + (c'-c)(Q-X)^-]$$

where $w^+ = \max\{w, 0\}$ and $w^- = \max\{-w, 0\}$. Note that the optimal production quantity minimizes

$$E[(c-p)(Q-X)^+ + (c'-c)(Q-X)^-]$$

where the expectation is with respect to X given the principal's knowledge about the market condition and the agent's selling effort (inferred, not observed) after a contract is signed.

Since there are only two possible market conditions, the firm needs to offer at most two contracts. Let $s_H(\cdot)$ be the contract chosen by the high-type agent, and $s_L(\cdot)$ chosen by the low type. The principal, by putting herself in the shoes of the agent, can anticipate the amount of selling effort under each type. Let a_H be the selling effort of the high-type agent, and a_L the effort of the low type. Assume that $s_H(\cdot) \neq s_L(\cdot)$. In this case, the principal discovers the market condition after observing the contract choice made by the agent. If $\theta = \theta_H$ then $X \sim N(a_H + \theta_H, \sigma^2)$; otherwise, if $\theta = \theta_L$, then $X \sim N(a_L + \theta_L, \sigma^2)$. And the principal can make her quantity decision accordingly. This is a benefit the principal obtains from screening.

One way to achieve screening is by offering a menu of (two) linear contracts. Let $s_H(x) = \alpha_H x + \beta_H$ be the contract intended for the high-type agent, and $s_L(x) = \alpha_L x + \beta_L$ the contract intended for the low-type agent, with $\alpha_H, \alpha_L \geq 0$. It can be shown that the optimal values of the contract

---

[36] It is reasonable to assume that the reservation utility does not depend on the agent's type, because what distinguishes the high type from the low type is the market condition, something unrelated to the agent's intrinsic quality.

parameters are

$$\alpha_H = \frac{1}{1+r\sigma^2}$$

$$\alpha_L = \frac{1}{1+r\sigma^2} \max\left\{1 - \frac{\rho}{1-\rho}(\theta_H - \theta_L), 0\right\}$$

$$\beta_L = -\frac{\ln U_0}{r} - \alpha_L \theta_L - \frac{1-r\sigma^2}{2}\alpha_L^2$$

$$\beta_H = -\frac{\ln U_0}{r} + \alpha_L(\theta_H - \theta_L) - \alpha_H \theta_H - \frac{1-r\sigma^2}{2}\alpha_H^2.$$

Another way to achieve screening is suggested by Gonik (1978). Under his scheme, the firm asks the salesperson to submit a forecast of the total sales. If the forecast is F, then $s(x\,|\,F)$ – a given function of the actual total sales x parameterized by F – is the compensation for the agent. Therefore, the firm is effectively offering a menu of contracts; by submitting a forecast, the agent chooses a particular contract from the menu. Gonik's original proposal uses the following functional forms: $s(x\,|\,x) = \alpha x + \beta$ for all x, and for any x and F,

$$s(x\,|\,F) = \begin{cases} s(F\,|\,F) - u(F-x) & x \leq F \\ s(F\,|\,F) + v(x-F) & x > F \end{cases}$$

where α, β, u, and v are contract parameters chosen by the firm with $u > \alpha > v > 0$. Note that $s(x\,|\,x) \geq s(x\,|\,F)$ for all F and x. Therefore, if the agent expects to sell x units, it is in his best interest to submit a forecast that is equal to x. Also, for any given F, $s(x\,|\,F)$ is increasing in x, providing the agent with incentives to generate more sales.

It can be shown that the agent's optimal effort level is $a^* = \alpha$, which is entirely determined by only one contract parameter, α, and it is independent of the agent's type. Moreover, the optimal forecast decision is $F^* = z^* + a^* + \theta = z^* + \alpha + \theta$ for some value z^*, which depends on α, u, v but is independent of β and the agent's type. Therefore, the high-type agent forecasts $F_H = z^* + \alpha + \theta_H$ and the low-type forecasts $F_L = z^* + \alpha + \theta_L$. The agent is screened!

Numerical examples comparing the menu of linear contracts with the Gonik scheme show that the former dominates the latter in terms of the firm's expected profits.

Screening in supply chains

An important type of information asymmetry in supply chains is about cost structures. A supplier may only have imperfect knowledge about a buyer's cost structure, and vice versa. Here again the less informed may try to screen the more informed with a menu of contracts. Below, we describe a few papers that deal with screening in supply chains with asymmetric cost information.[37]

Ha (2001) provides a screening model where the supplier does not know the buyer's marginal cost. The setting is that of the newsvendor model (with pricing): the buyer faces a demand that is stochastic and price-sensitive, and before demand realization, an order quantity must be determined together with the selling price. The demand model is the additive kind, i.e., $D = \mu(p) + Y$ where p is the selling price, μ is a decreasing and concave function, and Y is a random variable independent of p. Let s be the supplier's marginal cost of production, and c the buyer's marginal cost (of selling and maybe additional processing). A key feature of the model is that c is known only to the buyer, with the supplier endowed with a prior distribution of c over a finite interval. Everything else is assumed to be common knowledge. The analysis is from the standpoint of the supplier: how to offer a menu of contracts to the buyer so as to maximize the supplier's expected profit.

The contract menu is restricted to be of the form: $\{p(\hat{c}), q(\hat{c}), R(\hat{c})\}$, where $\hat{c}$ is the buyer's announced marginal cost (which can be different from the true cost c), and works as follows: if the buyer announces $\hat{c}$, then the supplier will deliver $q(\hat{c})$ units to the buyer for a total payment of $R(\hat{c})$, and the buyer is to set the selling price at $p(\hat{c})$. The functions $p(\cdot)$, $q(\cdot)$ and $R(\cdot)$ are chosen by the supplier. Given this menu, the buyer then decides whether or not to sign a contract and if so, which one (by choosing $\hat{c}$). Ha (2001) has solved this mechanism design problem.

As mentioned in Ha (2001) the above menu of contracts is a nonlinear contract with price fixing, and this may run into problem with commercial laws such as Resale Price Maintenance (RPM), see, e.g., Tirole (1988). This would not be a problem if the contract menu is changed to $\{q(\hat{c}), R(\hat{c})\}$ and the retailer is free to choose any selling price after contract signing. Ha has not solved this problem, except for a special case where the selling price is exogenously given.

Corbett and de Groote (2000) consider a model with one supplier and one retailer, where the retailer's holding cost parameter h_b is unknown to the supplier. The basic setup is a two-stage economic lot-sizing problem with deterministic demand and no backlogging, with an additional restriction that the supplier's lot size is equal to the retailer's (i.e., the lot-for-lot replenishment). The supplier, however, is endowed with a prior distribution of h_b. The problem facing the supplier can be formulated as a direct revelation game, whereby the supplier asks the retailer to announce the value of h_b: if the announced value is $\hat{h}_b$, then the lot size is $Q(\hat{h}_b)$ and the discount is $P(\hat{h}_b)$ given as a lump-sum payment per unit of time. The task is to determine the

---

[37] A careful reader would realize that some of the models discussed in Section 2.2 fall under this category.

pair of functions $Q(\cdot)$ and $P(\cdot)$ so as to minimize the supplier's expected costs subject to the incentive compatibility constraint that the retailer always wants to announce his true holding cost. Corbett and de Groote show that the optimal $Q(\cdot)$ and $P(\cdot)$ are both decreasing functions, which can thus be interpreted as a quantity discount scheme because larger quantities are associated with larger discounts.

Corbett (2001) considers a supplier–retailer model with stochastic demand. The setup is basically the same as the classic (Q, r) model: whenever the retailer inventory position falls to the reorder point r, it orders (and the supplier produces) a batch of Q units. The twist here is that the supplier makes the lot-sizing decision (i.e., the value of Q) and incurs a fixed cost for each batch produced, and that the retailer determines the reorder point r and is responsible for the holding and backorder costs incurred at the retail site.[38] The inefficiency in this supply chain is evident if there is no coordination: the supplier will set the batch size to be infinity! Corbett derives screening solutions for the following scenarios: (1) the supplier privately observes the value of the fixed cost, and (2) the retailer privately observes the backorder penalty cost. Also discussed is how consignment – the practice of giving the ownership of retailer inventory to the supplier – affects supply chain coordination. We omit the details. For other studies on supply chain models with asymmetric cost information, see Corbett, Zhou and Tang (2001) and the references therein.

Another type of information asymmetry in supply chains is about demand information. For example, a retailer, due to its proximity to the market, may possess better information about the demand than the supplier may. Cachon and Lariviere (1999) consider a one-period model with one supplier and N retailers. The retailers are local monopolists, each of which receives a private signal about its own market, which in turn determines its desired stocking level. The supplier has a finite capacity, and must determine an allocation mechanism in the event the sum of the retailer orders exceeds the capacity. (An allocation mechanism is therefore a mapping from a vector of retailer orders to a vector of capacity allocations.) They consider various allocation mechanisms and their impact on the supply chain. They found that some mechanisms induce the retailers to truthfully order their desired quantities, but the supply chain often fares better with a mechanism that induces order inflation (i.e., the retailers order more than they need hoping to get a higher allocation in the event of capacity shortage). In other words, the truthful sharing of retailer order information is not necessarily the appropriate goal for the supply chain. A recent paper by Deshpande and Schwarz (2002)

---

[38] Here lies a critical assumption: the supplier sets the retailer's order quantity. It is worth thinking about an alternative model where there are two quantity decisions: the supplier sets its production quantity, and the retailer sets its order quantity. Under the current cost structure, it is reasonable to assume that the production quantity is larger than the order quantity. Consequently, the supplier will also incur some inventory holding costs. How would supply chain coordination come about in this case? The same comment applies to Corbett and de Groote (2000).

considers a similar problem and derives an optimal mechanism from the supplier's standpoint.

3.2 Signaling

We begin with a simple example to illustrate the basics of signaling. We then review several supply chain models where an informed party likes to convey a piece of private information to the uninformed. These models are, interestingly, all set forth in the context of new product introductions.

A simple example

Consider a supply chain with one manufacturer and one retailer. The manufacturer (she) produces one product and sells it through the retailer (he). The retailer faces a linear demand function $D = a - p$. The demand intercept, a, has two possible values: $a = 8$ or $a = 4$. The manufacturer observes the value of a, while the retailer assesses a probability of ρ that $a = 8$ and a probability of $1 - \rho$ that $a = 4$, $0 < \rho < 1$. And the manufacturer is aware of this assessment by the retailer. To make numbers simple, say the manufacturer's marginal cost of production is zero. The one-period game begins with the manufacturer offering a wholesale price w. Then, the retailer sets the retail price p. Finally, the market demand is realized and profits accrue to the two players.

To start, let us consider the full information case, i.e., the value of a is also known to the retailer. In the high-type case (i.e., $a = 8$), the retailer chooses p to maximize $(p - w)(8 - p)$ when the wholesale price is w. The optimal solution is $p = (8 + w)/2$. Given this, the manufacturer maximizes $w(8 - (8 + w)/2)$. The solution is $w_H = 4$, which leads to a retail price of $p_H = 6$. The profits for the manufacturer and the retailer are $\pi_H^M = 8$ and $\pi_H^R = 4$, respectively. On the other hand, if $a = 4$, i.e., the low-type case, we have $w_L = 2$, $p_L = 3$, $\pi_L^M = 2$, and $\pi_L^R = 1$.

We can already see that the high-type manufacturer has an incentive to 'pretend' to be low type. For example, suppose, miraculously, the high-type manufacturer charges $w = 2$ and the retailer believes she is actually low type. The retailer then chooses a retail price p to maximize $(p - w)(4 - p)$, leading to $p = 3$. In this case, the high-type manufacturer's profit is $w(8 - p) = 10$, which is higher than $\pi_H^M = 8$ under full information. The intuition is clear: the manufacturer 'prefers' the retailer to think the demand intercept is low and hence, to charge a lower retail price leading to a higher demand. Interestingly, the retailer's actual profit in this case is $(p - w)(8 - p) = 5$, which is also higher than the profit under full information $\pi_H^R = 4$. On the other hand, the low-type manufacturer does not want the retailer to think otherwise. To verify this is a good exercise.

Now back to the case with asymmetric information. Suppose the manufacturer sets the wholesale price at w. Let $\mu(w)$ be the probability the retailer attributes to the event that $a = 8$. If, e.g., $\mu(w) = \rho$, then the retailer obtains no new information about the demand intercept after observing w. The other extreme is $\mu(w) = 0$ or 1, in which case the retailer learns the exact value of the demand intercept. With w and $\mu(w)$, the retailer chooses a retail price to maximize his expected profits:

$$(p - w)[8\mu(w) + 4(1 - \mu(w)) - p] = (p - w)[4 + 4\mu(w) - p].$$

Therefore, the optimal solution is

$$p(w, \mu(\cdot)) = \frac{4 + 4\mu(w) + w}{2} = 2(1 + \mu(w)) + \frac{w}{2}.$$

Given $\mu(\cdot)$, the manufacturer can anticipate what the retailer is going to do through the above equation for each possible value of w. The optimal wholesale price for the high-type manufacturer is the solution to

$$\max_w \pi_H^M(w) \stackrel{def}{=} w(8 - p(w, \mu(\cdot))) = w\left(6 - 2\mu(w) - \frac{w}{2}\right).$$

Let the solution be w_H^*. Similarly, the low-type's optimal wholesale price w_L^* solves

$$\max_w \pi_L^M(w) \stackrel{def}{=} w(4 - p(w, \mu(\cdot))) = w\left(2 - 2\mu(w) - \frac{w}{2}\right).$$

In sum, given $\mu(\cdot)$, the two players simply play a Stackelberg game with the manufacturer as the leader and the retailer as the follower. But the story does not end here. Where does $\mu(\cdot)$ come from? It must be consistent with the pricing strategies that prevail in the Stackelberg game. For example, if $w_H^* \neq w_L^*$, then a wholesale price equal to w_H^* signals to the retailer that $a = 8$, and a wholesale price of w_L^* signals $a = 4$. Therefore, to be consistent, we must have $\mu(w_H^*) = 1$ and $\mu(w_L^*) = 0$. On the other hand, if $w_H^* = w_L^*$, then the retailer is going to see only one wholesale price no matter what the manufacturer type is. In this case, consistency calls for $\mu(w_H^*) = \rho$. In the former case, we have a separating equilibrium because the retailer is able to separate the two manufacturer types; in the latter, a pooling equilibrium because the two manufacturer types do the same thing. A belief structure that is consistent is referred to as an equilibrium belief.

Let us see if there exists any equilibrium, separating or pooling, in the above game.

Suppose a pooling equilibrium exists. Let w^0 be the wholesale price chosen by both types of the manufacturer. Let $\mu^0(\cdot)$ be the retailer belief. Recall that consistency calls for $\mu^0(w^0) = \rho$. We claim that if $\rho < 3 - 2\sqrt{2} \approx 0.1716$, then the following strategy profile and belief form a pooling equilibrium: $w^0 = 2 - 2\rho$, $\mu^0(w) = \rho$ for all $w \leq 2 - 2\rho$ and $\mu^0(w) = 1$ for all other w. To verify, all we need to do is check that under the given belief, $w^0 = 2 - 2\rho$ is indeed the optimal choice for both manufacturer types. Consider first the high type. If the manufacturer offers a wholesale price w greater than w^0, then $\pi_H^M(w) = 4w - w^2/2$, which is maximized at $w = 4$ with $\pi_H^M(4) = 8$. If $w = w^0$, $\pi_H^M(w^0) = 6w^0 - (w^0)^2/2 - 2\rho w^0$. It is easy to verify that $\pi_H^M(w^0) > 8$ when $\rho < 3 - 2\sqrt{2}$. Moreover, for all $w < w^0$, $\pi_H^M(w) = 6w - w^2/2 - 2\rho w$ is increasing in w. This shows that the high-type manufacturer's optimal choice of a wholesale price is w^0. Now consider the low-type manufacturer. If the low-type manufacturer offers a wholesale price $w > w^0$, the retailer thinks the manufacturer is of high type. Under this scenario, the low-type manufacturer's profit function is $\pi_L^M(w) = -w^2/2$. Therefore, it is not in the interest of the low-type manufacturer to 'pretend' to be of high type. For all $w \leq w^0$, $\pi_L^M(w) = 2w - w^2/2 - 2\rho w$, which is maximized at $w = w^0$. This establishes that the above strategy–belief combination is a pooling equilibrium.[39]

The following is a separating equilibrium: $w_H = 4$, $w_L = 6 - \sqrt{20} \approx 1.5279$, $\mu(w) = 0$ for all $w \leq w_L$, and $\mu(w) = 1$ for all $w > w_L$. To verify, first consider the high-type manufacturer. For any wholesale price w greater than w_L, the retailer thinks $a = 8$ and the manufacturer faces the profit function $\pi_H^M(w) = 4w - w^2/2$, which is maximized at $w = 4$ with $\pi_H^M(4) = 8$. If the manufacturer wants to pretend to be of low type, he must offer a wholesale price $w \leq w_L$. Over this range, $\pi_H^M(w) = 6w - w^2/2$, which is an increasing

---

[39] Various equilibrium refinements are possible. We consider one here, i.e., the test of equilibrium domination that is also known in the literature as the 'intuitive criterion.' The reader is referred to Kreps (1990) for discussions on the intuitive criterion and references to other refinements. In the above pooling equilibrium, the high-type manufacturer obtains an expected profit $(8 - p^0)w^0$, where p^0 is the retailer's selling price in the equilibrium, i.e.

$$p^0 = 2(1 + \rho) + w^0/2 = 3 + \rho.$$

Likewise, the low-type manufacturer's equilibrium expected profit is $(4 - p^0)w^0$. The first step in the test is to identify all signals (i.e., wholesale prices) that are 'equilibrium dominated.' A wholesale price w is equilibrium dominated if the maximum achievable profit for the manufacturer under w is less than what she gets in equilibrium. Consider the high-type manufacturer. We know her expected profit in equilibrium is $(8 - p^0)w^0$. If she charges wholesale price w and the retailer sets the retail price at p, her profit is $(8 - p)w$. The maximum achievable profit is $8w$, which is obtained when $p = 0$. Thus w is equilibrium dominated at the high type if

$$8w < (8 - p^0)w^0 \quad \text{or} \quad w < (1 - p^0/8)w^0 = \frac{(5 - \rho)(1 - \rho)}{4} \stackrel{def}{=} \tilde{w}_H.$$

Similarly, w is equilibrium dominated at the low type if

$$4w < (4 - p^0)w^0 \quad \text{or} \quad w < (1 - p^0/4)w^0 = \frac{(1 - \rho)^2}{2} \stackrel{def}{=} \tilde{w}_L.$$

The central idea of the intuitive criterion is that it should be obvious what the retailer's beliefs should be at wholesale prices that are equilibrium dominated. For example, if the retailer observes $w < \tilde{w}_H$, then the signal must not come from the high-type manufacturer and thus $\mu(w) = 0$. Similarly, a signal

function in w for $w \leq w_L$. Therefore, misleading the retailer gives the high-type manufacturer a profit lower than or equal to $6w_L - w_L^2/2 = 8$, which is not better than 'truth-telling'. This verifies that the optimal choice for the high-type manufacturer is $w_H = 4$. On the other hand, recall from our earlier discussions that the low-type manufacturer will never want to pretend to be of high-type (doing so would give him a negative profit). For wholesale prices $w \leq w_L$, the low-type manufacturer's profit function is $\pi_L^M(w) = 2w - w^2/2$. It is easy to see that this function is increasing over $w \leq w_L$. Thus the low-type manufacturer's optimal wholesale price is w_L. Moreover, the choices by the two manufacturer types confirm the above retailer belief. We thus have a separating equilibrium.

You may wonder if there exists any other pooling or separating equilibrium. It is a good exercise to try to find one. To read more about signaling games, see, Kreps (1990), Fudenberg and Tirole (1992) and Kreps and Wilson (1982).

Demand signaling in new product introductions

When a manufacturer introduces a new product to the market, it often possesses some private information about the potential market demand for the product. This information is critical for a retailer who is deciding whether or not to carry the product because the retailer may not be able to recoup the overhead for a low-demand product. Similarly, the information is valuable to a supplier who is considering how much capacity to build for the manufacturer's product; building a lot of capacity for a low-demand product is wasteful. In both cases, the manufacturer has an incentive to report

---

$w < \tilde{w}_L$ tells the retailer that the manufacturer cannot be of low type. But can she be of high type in this case? No, because $\tilde{w}_L < \tilde{w}_H$ and thus $w < \tilde{w}_H$. We are in a quandary here; a reasonable assumption is that such a wholesale price will never be observed. Under this assumption, the intuitive criterion suggests that $\mu(w) = 0$ for all $w < \tilde{w}_H$. Notice that $\mu^0(w) = \rho$ for the same range. (Check that $\tilde{w}_H < w^0$.) Let us see if this change in retailer belief will change the manufacturer's signaling strategy, assuming the retailer continues with his optimal response $p(w, \mu(\cdot))$ where $\mu(\cdot)$ is the updated belief. First, consider the high-type manufacturer. If she offers $w < \tilde{w}_H$, then $\mu(w) = 0$ and

$$\pi_H^M(w) = 6w - w^2/2 < 6\tilde{w}_H - (\tilde{w}_H)^2/2$$

which can be shown to be less than 8, which is the manufacturer's expected profit if the wholesale price is 4 (and thus the retailer thinks she is high type), which in turn is less than what she gets by charging w^0. Therefore, the high-type's choice remains intact. Now consider the low-type manufacturer. For any $w < \tilde{w}_H$, we have

$$\pi_L^M(w) = 2w - w^2/2 < 2\tilde{w}_H - (\tilde{w}_H)^2/2.$$

It can be shown that if $\rho < 0.11$ then

$$2\tilde{w}_H - (\tilde{w}_H)^2/2 < (w^0)^2/2$$

where the right-hand side is the low-type manufacturer's expected profit if she chooses w^0 as the wholesale price. Consequently, for $\rho < 0.11$, the pooling equilibrium is sustained by the intuitive criterion.

high demand, whether the actual demand is high or low. (This is different than the scenario considered in the above example, where the manufacturer benefits if the retailer thinks the market is low.) As a result, a simple announcement by the manufacturer will not be believed. To be credible, the manufacturer needs to put money where its mouth is, i.e., signaling. Below are several papers that deal with this issue.

Chu (1992) considers a distribution channel consisting of a manufacturer and a retailer. The product, produced by the manufacturer and sold by the retailer, draws a demand that depends on the market condition, the retail price P, and the manufacturer's advertising expenditure A:

$$Q^i = a - b^i P + f(A), \quad i = H, L$$

where Q^i is the demand for the product under market condition i, a is a constant, b^i is the demand sensitivity to price, and $f(\cdot)$ is a concave, increasing function. Assume $b^H < b^L$. Thus, for any given P and A, the demand is higher when the market condition is 'H'. For convenience, we say the market condition is either high ($i = H$) or low ($i = L$). The manufacturer knows the true market condition, whereas the retailer assesses a probability ρ that the market is high (and the manufacturer knows of this assessment). The manufacturer incurs a constant marginal cost of production C.

A signaling game is where the manufacturer (with superior information) moves first by offering a wholesale price P_w and spending A on advertising. Given P_w and A, the retailer updates his belief about the market condition from ρ to $\hat{\rho}$, decides whether or not to carry the manufacturer's product, and if the latter, sets the retail price. The retailer incurs a fixed cost F for carrying the product. The retailer will accept the manufacturer's offer if his expected profit (excluding the carrying cost) exceeds F, and will reject it otherwise.

Chu makes an additional assumption that as soon as the retailer accepts the manufacturer's offer, he sees the true market condition. Therefore, the retail price can be made contingent upon the value of the slope b^i.

An equilibrium for the signaling game consists of a manufacturer strategy $\{P_w^i, A^i\}$, $i = H, L$, a retailer accept/reject strategy $R(x, y)$ that is a binary function, and a retailer belief $\mu(x, y)$, which is the posterior probability that the market is high, for any possible offer $(P_w, A) = (x, y)$ from the manufacturer. Recall that the retailer's pricing decision is made after learning the market condition. Thus if the retailer accepts an offer (x, y) from the manufacturer, the optimal retail price is

$$P^i(x, y) = \operatorname{argmax}_P (P - x)(a - b^i P + f(y)), \quad i = H, L.$$

Let the retailer's profit (excluding the fixed cost F) be $\pi_R^i(x, y)$, i.e., $\pi_R^i(x, y) = (P^i(x, y) - x)(a - b^i P^i(x, y) + f(y))$. Before the accept/reject decision, the retailer's expected profit is $\pi_R(x, y) \stackrel{def}{=} \mu(x, y)\pi_R^H(x, y) + (1 - \mu(x, y))\pi_R^L(x, y)$. Therefore, if $\pi_R(x, y) \geq F$, the retailer accepts the manufacturer's offer (x, y), i.e., $R(x, y) = 1$; otherwise, if $\pi_R(x, y) < F$, the retailer rejects the offer (x, y), i.e., $R(x, y) = 0$. Anticipating all this, the manufacturer, knowing her own type, maximizes her profits:

$$(P_w^i, A^i) = \operatorname{argmax}_{P_w, A} R(P_w, A)(P_w - C)(a - b^i P^i(P_w, A) + f(A)), \quad i = H, L.$$

An equilibrium with $(P_w^H, A^H) \neq (P_w^L, A^L)$ is a separating equilibrium; otherwise, if $(P_w^H, A^H) = (P_w^L, A^L) \stackrel{def}{=} (\hat{P}_w, \hat{A})$, we have a pooling equilibrium. For a separating equilibrium, the consistency requirement for the retailer belief is $\mu(P_w^H, A^H) = 1$ and $\mu(P_w^L, A^L) = 0$. For a pooling equilibrium, consistency requires $\mu(\hat{P}_w, \hat{A}) = \rho$.

Chu has identified a separating equilibrium, where the high-type manufacturer advertises and prices above its complete information levels. He has also identified a pooling equilibrium, where both the high-type and the low-type manufacturers advertise and price at or above the complete information levels of the high-type manufacturer. (The complete information case is simply one where the retailer knows the true market condition, and the two players carry out a Stackelberg game with the manufacturer as the leader, setting the wholesale price and the advertising expenditure, and the retailer as the follower, setting the retail price, with an option to reject the manufacturer's offer.)

Chu proceeds to consider the case where the retailer moves first to screen the manufacturer. This is achieved through a slotting allowance, which is a lump-sum payment from the manufacturer to the retailer in order for the latter to carry the product. The game proceeds in the following sequence. The retailer specifies a slotting allowance, which the manufacturer can either accept or reject. If rejected, the game ends with zero profit for both parties. If the manufacturer agrees to pay the slotting allowance, she gets to set the wholesale price and an advertising expenditure. Given these, the retailer then sets the retail price after observing the market condition (as in the signaling case).

Notice that once the manufacturer has accepted to pay the slotting allowance, the rest of the game is the same as in the complete information case, because the retailer sees the market condition (due to screening) before his pricing decision. It is possible to choose a slotting allowance such that only the high-type manufacturer finds it acceptable. For example, let the slotting allowance be the high-type manufacturer's maximum profits in the complete information case. (Recall that this is what the manufacturer can achieve in the Stackelberg game with complete information, where the manufacturer moves first by announcing the wholesale price P_w and the advertising expenditure A, and the retailer follows by setting the retail price P.) It is clear that such a slotting fee is unacceptable to the low-type manufacturer. In this case, only high-type products will be carried by the retailer, who takes all the channel profits.

So, signaling or screening? This is, of course, determined by the balance of power in the channel. Clearly, the manufacturer prefers to move first, by signaling, whereas the retailer prefers to move first too, by screening. From the channel's perspective, the result depends on the effectiveness of advertising.

Consider, for a moment, the complete information case. Compared with the channel-optimal solution, the Stackelberg solution leads to a wholesale price greater than the manufacturer's marginal cost, which in turn leads to a retail price greater than the channel-optimal retail price. This is the well-known double-marginalization phenomenon.[40] On the advertising side, because the manufacturer reaps only part of the benefits from advertising (because the retailer also makes a margin), the advertising level in the Stackelberg solution tends to be lower than the channel-optimal advertising level. In sum, inefficiencies result because the wholesale price is too high and the advertising expenditure is too low.

Now consider the asymmetric information case with a high-type manufacturer. (It is reasonable to ignore the low-type manufacturer if a low-type product is not sustainable.) As mentioned earlier, the signaling game leads to a wholesale price and an advertising level both higher than those in the Stackelberg solution (with complete information). Therefore, signaling increases the wholesale-price distortion (further away from the channel optimum) but decreases the advertising distortion. When advertising has low effectiveness, the former effect dominates the latter, leading to a channel profit even lower than in the Stackelberg solution. In contrast, screening with a slotting allowance restores the channel profit to the Stackelberg-game level, because the slotting allowance, being a fixed fee, does not alter the pricing and advertising decisions in the channel. Therefore, one can say that signaling involves wasteful expenditures, whereas screening keeps the money in the channel. On the other hand, if advertising is highly effective, then the channel may be better off with signaling.

Lariviere and Padmanabhan (1997) further investigate the role of slotting allowances in new product introductions. Suppose, as before, a manufacturer introduces a new product through an independent retailer. The manufacturer begins by offering the terms of trade, consisting of a wholesale price w and a slotting allowance A. The retailer then either accepts or rejects the terms. If the former, the retailer agrees to carry the product and proceeds to set a retail price p and exert merchandising effort e. The quantity sold can be expressed as

$$D(e,p) = \tau + f(e) - \beta p$$

---

[40] Spengler (1950) is the first to discuss this phenomenon.

where τ is a market-size parameter, β measures demand sensitivity to price, and $f(\cdot)$ is an increasing, concave function. The cost of merchandising effort is assumed to be e as well, i.e., a linear effort-cost model. The retailer incurs a fixed cost K for carrying the product. He accepts the contract offered by the manufacturer if and only if his profit is nonnegative. A key feature of the model is that the two players possess asymmetric information about the market size. It is assumed that τ takes one of two possible values H and L with $H > L$. The manufacturer knows the value of τ, whereas the retailer assesses a probability θ that $\tau = H$ (and the manufacturer knows about this assessment). As in all signaling games, the retailer may infer something about the value of τ from the terms of trade offered (w, A), i.e., forming a posterior belief $\mu(w, A)$ that $\tau = H$.

Lariviere and Padmanabhan have characterized a separating equilibrium in the above signaling game. The equilibrium consists of a contract offered by the manufacturer, $(\hat{w}, \hat{A})$, and a supporting retailer belief $\mu(\cdot, \cdot)$ such that $\mu(\hat{w}, \hat{A}) = 1$ and $\mu(w, A) = 0$ for all $(w, A) \neq (\hat{w}, \hat{A})$. The parameters of the model are such that if the market is low, it is impossible for the manufacturer and the retailer to make nonnegative profits at the same time. Therefore, in any separating equilibrium where the retailer learns the true type of the manufacturer, only the high-type product may be accepted by the retailer.

Let us identify the constraints that a separating equilibrium must satisfy. Suppose the manufacturer offers (w, A) that leads the retailer to believe that the market is high. The retailer's profit function (excluding fixed costs) is thus

$$\pi_R(e, p) = (p - w)(H + f(e) - \beta p) + A - e.$$

The retailer's optimal response is thus $(\tilde{e}, \tilde{p}) \stackrel{def}{=} \mathrm{argmax}_{(e, p)} \pi_R(e, p)$. Note that the slotting allowance, since it is fixed, does not affect the retailer's pricing and effort decisions. But it certainly affects whether or not the retailer will accept the manufacturer's offer. Acceptance results only if

$$\pi_R(\tilde{e}, \tilde{p}) \geq K.$$

In order for the belief to be correct in equilibrium, it must be unprofitable for the low-type manufacturer to mimic the high-type. Suppose the low-type manufacturer offers (w, A), the contract offered by the high-type. As a result, the retailer believes $\tau = H$ and thus responds by choosing $(\tilde{e}, \tilde{A})$. The manufacturer's profit is thus:

$$\pi_M^L(w, A) = (w - c)(L + f(\tilde{e}) - \beta \tilde{p}) - A$$

where c is the manufacturer's marginal production cost. Doing so must be unprofitable for the low-type manufacturer, i.e., $\pi_M^L(w, A) \leq 0$. On the other hand, the high-type manufacturer's profit is

$$\pi_M^H(w, A) = (w - c)(H + f(\tilde{e}) - \beta\tilde{p}) - A.$$

The high-type manufacturer seeks a separating equilibrium that maximizes her profit by solving the following optimization problem:

$$\max_{(w, A)} \pi_M^H(w, A)$$
$$\text{s.t.} \quad (\tilde{e}, \tilde{p}) = \text{argmax}_{(e, p)} \pi_R(e, p)$$
$$\pi_R(\tilde{e}, \tilde{p}) \geq K$$
$$\pi_M^L(w, A) \leq 0.$$

The solution $(\hat{w}, \hat{A})$ is the contract the high-type manufacturer will offer in equilibrium.

The following results have been obtained. First, when the fixed cost K is lower than a threshold level K^*, the separating equilibrium does not involve any slotting allowance, i.e., $\hat{A} = 0$, and the wholesale price $\hat{w}$ is greater than the full information wholesale price (for high market condition).[41] Second, when $K \geq K^*$, the separating equilibrium requires a positive slotting allowance, i.e., $\hat{A} > 0$, and a wholesale price $\hat{w}$ that is still greater than the full information price. The main message here is that slotting allowances can be a useful signaling device to show to the retailers that the products the manufacturers are introducing are promising. This is in contrast with Chu (1992) who models slotting allowances only as a screening device.

Desai and Srinivasan (1995) study demand signaling in the presence of moral hazard. A principal sells a product through an independent agent. The product is new, and the principal has private information about the demand for the product.[42] The agent can also influence the demand by expending selling effort, which is not observable to the principal. Both the principal and the agent are risk neutral. The issue is the contract emerging between the two parties in this two-sided information asymmetry model.

More specifically, the demand function is either Q^H or Q^L with

$$Q^J = T^J - p + f(a) + \varepsilon, \qquad J = H, L$$

---

[41] The full information case is where the retailer also observes the value of τ. In this case, due to the assumptions made about the model parameters, the only relevant problem is what faces the high-type manufacturer. This problem is solved in a Stackelberg fashion with the manufacturer being the leader and the retailer the follower.

[42] The model can be interpreted in other ways. We choose this principal-agent, new-product story for convenience.

where $T^H > T^L$. The principal knows the true demand function, whereas the agent remains uncertain. For convenience, if the demand function is Q^H, we say the principal is of high-type; otherwise, the principal is of low-type. The agent's prior belief is that the principal is of high-type with probability ρ, with $0 < \rho < 1$. It is assumed that both types of principals have the same marginal cost c. The agent determines the selling price p and selling effort a, whose impact on sales is captured by an increasing concave function $f(\cdot)$. The cost of effort (to the agent) is $w(a)$, a convex increasing function. The demand is subject to a random shock ε with mean zero. Although the realized demand is observable and contractible, the selling effort is not.

The sequence of events is as follows. The principal offers a contract to the agent; the contract specifies a payment from the agent to the principal that can be a function of the realized demand. Given a contract, the agent then updates his belief about the principal's type, from his prior belief ρ to a posterior belief $\hat{\rho}$. Based on this updated belief, the agent chooses a selling price and an effort level to maximize his expected profit. The principal, anticipating the agent's behavior, selects a contract that maximizes her expected profit, under the constraint that the expected profit for the agent is at least $\bar{u}$, a reservation utility level reflecting the agent's outside opportunities.

Desai and Srinivasan, citing Arrow (1985) that simple linear contracts are prevalent in practice, start with a simple, two-part contract with a fixed fee F and a variable fee r, so that the agent pays the principal $F + rq$ if the realized demand is q. They then consider a nonlinear contract with a quadratic component.

First, notice that although the model involves private information on both sides, it is crucial to have the private demand information (held by the principal). To see this, let us first consider the first-best scenario where both parties have full information, i.e., the agent also knows the true demand function, and the principal can observe the agent's selling effort. In this case, the principal can demand a specific level of effort. Given a contract (r, F, a) of variable fee, fixed fee, and effort level, the agent chooses a retail price to maximize his expected profit. This is an easy problem. In the solution, the principal sets the variable fee r equal to her marginal cost c (to avoid double marginalization), the agent earns exactly his reservation utility, and the principal extracts all the surplus (via the fixed fee). It can be verified that the first-best effort level and fixed fee are higher with the high-type principal than with the low-type principal.

Now suppose the agent's selling effort is unobservable to the principal (but the agent still knows the true demand function). The solution to this problem is also well known. From agency theory, since the agent is risk neutral, the moral hazard problem can be easily overcome by making the agent a residual claimant. For either type of principal, simply set the variable fee to the marginal cost and the fixed fee to make the agent's participation constraint binding. The first-best solution prevails.

The above discussions make it clear that if the principal did not hold private demand information, the problem would be trivial. From now on, we focus on the case with asymmetric demand information. To understand the impact of moral hazard on demand signaling, Desai and Srinivasan first consider the signaling game without moral hazard, i.e., assuming the principal can observe the agent's selling effort. The signaling instruments are therefore (r, F, a). They have identified a separating equilibrium in which the high-type principal offers (r^S, F^S, a^S), whereas the low-type principal offers her corresponding first-best contract. It is shown that $r^S > r_{fb}^H$, $F^S < F_{fb}^H$, and $a^S = a_{fb}^H$, where r_{fb}^H, F_{fb}^H and a_{fb}^H are the variable fee, fixed fee, and effort level in the high-type principal's first-best solution. The somewhat unintuitive result is that it is unnecessary for the high-type principal to distort the effort level (from the first-best level) in order to credibly convey its type; a distortion in the variable fee suffices. (Since the variable fee is raised, the fixed fee must be reduced to meet the agent's participation constraint.)

If the agent's selling effort is unobservable to the principal (thus not contractible) – as in the original case with two-sided information asymmetry – the principal's signaling instruments are reduced to (r, F). A separating equilibrium for this game has also been identified, where the high-type principal offers (r^{SM}, F^{SM}), and the low-type principal continues to offer her first-best contract. It has been shown that $r_{fb}^H < r^{SM} < r^S$ and $F^S < F^{SM} < F_{fb}^H$. This leads to a main conclusion of the paper that moral hazard dampens the signaling distortions of the fixed and variable fees. Moreover, the high-type principal makes less profit when the agent's effort is unobservable.

Interestingly, when we allow nonlinear contracts, the first-best solution can be achieved for both types of principals even in the presence of moral hazard. This is established under a three-part contract where the agent's payment to the principal is $F + r_1 q + r_2 q^2$, with q being the realized demand. A separating equilibrium has been identified where both types of principals earn their first-best profits. The above conclusion is obtained under the following additional assumptions: $\bar{u} = 0$, $c = 0$, $f(a) = a$, and $w(a) = a^2$. It is worth noting that in equilibrium, the two variable fees are such that $r_1 > 0$ and $r_2 < 0$. In other words, the first best is achieved not by making the agent a residual claimant. This is in contrast with the way in which the first best is achieved when the principal does not hold private information about demand.

The above three papers on demand signaling focus on a manufacturer's task of conveying her private demand information to a retailer, a downstream supply chain partner. We next consider the problem of signaling to an upstream supply chain partner.

In the relationship between a manufacturer and her supplier, it is often the case that the manufacturer knows more about the demand (for the end product) than the supplier does. Suppose the supplier is always the bottleneck of the supply chain, i.e., production is constrained only by the supplier's capacity. From the manufacturer's standpoint, the more capacity the supplier installs the better (a higher capacity relaxes the constraint on production). Therefore, the manufacturer has an incentive to inflate her demand forecast, hoping the supplier will increase his capacity. But, of course, the supplier knows this and will not take the manufacturer's demand forecast at face value. The result is a communication breakdown leading to an efficiency loss. Is there a way for the manufacturer to reveal her demand forecast in a credible way?

Cachon and Lariviere (2001) provide a model to address the above question. A manufacturer sells a single product that has uncertain demand, D. The manufacturer relies on a single supplier for a critical component of the product. Let K be the supplier's capacity, which must be built before demand realization. Some demand may be lost due to the capacity constraint, with $\min\{D, K\}$ being the demand fulfilled and the rest lost. The supplier incurs a cost for every unit of capacity built. A key feature of the model is that the manufacturer possesses private information about the demand. It is assumed that $D = \theta X$, where θ and X are independent random variables. Moreover, θ has only two possible values, H and L with $H > L$. The parameter θ may be interpreted as an indicator of the market condition. The manufacturer observes the value of θ, whereas the supplier assesses a probability ρ that $\theta = H$ and $1 - \rho$ that $\theta = L$. (Both players have the same information about X, i.e., its distribution.) For convenience, the manufacturer observing $\theta = H$ is said to be of high-type, whereas if $\theta = L$, low-type. The manufacturer moves first by offering a contract to the supplier, the supplier builds capacity given the contract and his belief about demand, demand is then realized, and finally, production takes place (within the capacity set by the supplier). This is a signaling game because the informed party (i.e., the manufacturer) takes a more active role by designing the contract.

Let Z be the set of all admissible contracts. Both the supplier and the manufacturer agree that only contracts in Z are considered. (Cachon and Lariviere consider linear contracts with commitments and options.) The supplier's capacity decision depends on the contract offered by the manufacturer and his belief about the market condition. Let $z \in Z$ be the contract offered, and b the supplier's belief, i.e., b is a probability distribution for the values of θ. Note that b might be a function of z. Let $\pi(K, z, b)$ be the supplier's expected profit if he builds capacity K. The optimal capacity level, from the supplier's standpoint, is then $K(z, b) = \operatorname{argmax}_K \pi(K, z, b)$. Consequently, the manufacturer's expected profit is a function of her type t, the contract offered to the supplier z, and the supplier's belief b about the market condition, i.e., $\Pi_t(z, b)$, $t = H, L$.

We focus on separating equilibria, whereby the supplier learns the manufacturer's type upon seeing the contract offered by the manufacturer. That is, given a contract z, the supplier assigns probability one to either $\theta = H$ or $\theta = L$. For convenience, denote the former by $b = H$ and the latter $b = L$.

Therefore, the supplier's belief can be characterized by a partition of Z, i.e., Z_H and Z_L with $Z = Z_H \cup Z_L$ and $Z_H \cap Z_L = \emptyset$. If a contract $z \in Z_H$ is offered, $b = H$; otherwise if $z \in Z_L$ is offered, $b = L$. In equilibrium, the supplier's belief must be correct, i.e.

$$\max_{z \in Z_L} \Pi_L(z, L) \geq \max_{z \in Z_H} \Pi_L(z, H) \quad \text{and} \quad \max_{z \in Z_H} \Pi_H(z, H) \geq \max_{z \in Z_L} \Pi_H(z, L).$$

In words, given the supplier's belief (i.e., a partition of Z), a type-t manufacturer has no incentives to mislead the supplier into believing that she is of type-t', $t \neq t'$. A partition or belief with the above properties is called an equilibrium partition or equilibrium belief. Corresponding to an equilibrium partition, Z_H and Z_L, is a pair of contracts, z_H and z_L, that are offered by the high- and low-type manufacturer, respectively. Clearly,

$$z_t = \mathrm{argmax}_{z \in Z_t} \Pi_t(z, t), \quad t = H, L.$$

A separating equilibrium, then, consists of an equilibrium partition, Z_H and Z_L, and a corresponding pair of contracts, z_H and z_L.

Now assume that

$$\Pi_t(z, H) \geq \Pi_t(z, L), \quad \forall z \in Z, \quad t = H, L. \tag{3.1}$$

In words, for any contract, the manufacturer, regardless of her type, is better off if the supplier believes that the market condition is high rather than low. This seems plausible because a belief of high market condition leads the supplier to build more capacity, to the benefit of the manufacturer. Below, we characterize a separating equilibrium under the above condition.

Let z_L^* be the optimal contract for the low-type manufacturer in the full-information case, i.e.,

$$z_L^* = \mathrm{argmax}_{z \in Z} \Pi_L(z, L).$$

Assume that the above optimization problem has a unique solution; thus $\Pi_L(z_L^*, L) > \Pi_L(z, L)$ for all $z \neq z_L^*$. It is easy to establish that any equilibrium partition (Z_H, Z_L) has $z_L^* \in Z_L$. The proof is by contradiction. Suppose $z_L^* \in Z_H$. Then $\Pi_L(z_L^*, H) \geq \Pi_L(z_L^*, L) > \Pi_L(z, L)$ for all $z \in Z_L$, where the first inequality follows from (3.1). Thus the low-type wants to pretend to be the high-type. Thus (Z_H, Z_L) is not an equilibrium partition, a contradiction.

Now take any equilibrium partition (Z_H, Z_L). Let (z_H, z_L) be the corresponding pair of contracts offered by the two manufacturer types.

From the above discussion, $z_L = z_L^*$. Also, by the definition of equilibrium partitions, we have

$$\Pi_L(z_H, H) \leq \Pi_L(z_L^*, L) \tag{3.2}$$

and

$$\Pi_H(z_H, H) \geq \max_{z \in Z_L} \Pi_H(z, L). \tag{3.3}$$

Also by the definition of z_H and (3.1), for any $z \in Z_H$, $\Pi_H(z_H, H) \geq \Pi_H(z, H) \geq \Pi_H(z, L)$. This inequality, together with (3.3), implies

$$\Pi_H(z_H, H) \geq \max_{z \in Z} \Pi_H(z, L). \tag{3.4}$$

Therefore, the high-type manufacturer's expected profit, i.e., $\Pi_H(z_H, H)$, is no greater than the maximum value of the objective function in

$$\max_{z \in Z} \quad \Pi_H(z, H)$$
$$\text{s.t.} \quad \Pi_L(z, H) \leq \Pi_L(z_L^*, L)$$
$$\Pi_H(z, H) \geq \max_{z' \in Z} \Pi_H(z', L)$$

where the first and second constraints are from (3.2) and (3.4) respectively. Suppose this problem has a unique solution $z = z_H^*$. Define $Z_H^* = \{z_H^*\}$ and $Z_L^* = Z \setminus Z_H^*$. It is easy to verify that (Z_H^*, Z_L^*) is an equilibrium partition. The contract offered by the high-type (resp., low-type) manufacturer is z_H^* (resp., z_L^*) with the corresponding expected profit $\Pi_H(z_H^*, H)$ (resp., $\Pi_L(z_L^*, L)$). Thus $(Z_H^*, Z_L^*, z_H^*, z_L^*)$ is a separating equilibrium. Note that from the above arguments, both manufacturer types can do no better than this with any other equilibrium partition.

While the supplier in Cachon and Lariviere (2001) is the only source of supply for the manufacturer, Van Mieghem (1999) considers a setting with two sources of supply: the manufacturer's in-house production facility and an outside supplier (i.e., subcontractor). At the beginning of the game, the manufacturer and the subcontractor simultaneously and independently make their capacity investment decisions, i.e., the manufacturer decides how much in-house capacity K_M to build and the subcontractor decides on its own capacity K_S. The manufacturer faces market demand D_M, which can be served by in-house as well as the subcontractor's production. Moreover, the subcontractor can also sell its product to a separate market with demand D_S. At the time of the capacity decisions, only a joint probability distribution of D_M and D_S is known (to both players). After the capacity decisions, the demands are realized and the firms decide on their production/sales quantities. More specifically, the manufacturer determines how much to produce in-house (i.e., x_M) and how much to order from the subcontractor (i.e., x_t^M); the subcontractor decides how much to sell to its own market (i.e., x_S) and how much of the manufacturer's order to fill (i.e., x_t^S). Of course, $x_M \leq K_M$, $x_M + x_t^M \leq D_M$, $x_t^S \leq x_t^M$, $x_t^S + x_S \leq K_S$, and $x_S \leq D_S$. These production/sales decisions are governed by various contractual arrangements between the two firms. In one contract (called the price-only contract), a transfer price p_t is specified ex ante for each unit supplied by the subcontractor. (This price is known before the capacity decisions.) In this scenario, the two firms sequentially solve for their production/sales quantities, with the manufacturer as the first mover. Under another contractual arrangement, there are simply no ex-ante contracts and the parties negotiate the transfer quantity and price after the demands are realized. In this scenario, the firms arrive at production/sales quantities that maximize their total profits and the surplus (relative to a scenario with no transactions between the two) is split between them based on their relative bargaining power, which is captured by an exogenous index. Van Mieghem also considers state-dependent contracts: a price-only contract with p_t being a function of the installed capacities and the realized demands, or an incomplete contract/bargaining arrangement where the bargaining-power index is state-dependent. For each of these contractual arrangements, one can solve the resulting two-stage stochastic game and examine a contract's impact on the coordination of both capacity and production/sales decisions. Although information sharing is not the focus of Van Mieghem (1999) (infact, there is no information asymmetry), the paper does provide an interesting discussion on coordinating capacity and quantity decisions in a manufacturer-subcontractor supply chain. Among the key findings are (1) a higher transfer price can actually increase the manufacturer's profit and (2) only state-dependent contracts (price-only or incomplete contract) can coordinate both the quantity and capacity decisions. It remains an interesting open question as to the impact of information asymmetries (about demands, capacities, costs, etc.) on the manufacturer-subcontractor supply chain.

3.3 Information sharing in competitive environments

We begin by considering papers that deal with information sharing among horizontal competitors, e.g., competing retailers sharing market demand information. These papers are all published in economics journals in the 1980s. (Are economists tired of this problem? Extensions to supply chains may breathe new life.) Recently, there have been several attempts to generalize the horizontal information-sharing literature to vertical information sharing in supply chains, e.g., will competing retailers share

A simple example

We begin with a simple example to illustrate the incentives of sharing demand information between two competitors.[44] Consider a duopoly facing stochastic linear demand, $p = a - Q$, where p is the market price, a is a random variable, and Q is the total output of the duopolists. Assume that $a = 50, 150$ with equal probabilities. Also, the marginal production costs are zero for both firms. The two firms are engaged in a Cournot competition, i.e., making quantity decisions independently. Before their quantity decisions, firm 1 observes the true value of a, while firm 2 does not receive any private signal about a. Consider a two-stage game. In the first stage, firm 1 decides whether or not to share its information about a with firm 2. Information sharing, if it occurs, is assumed to be truthful. Firm 1 then observes the true value of a. Information transmission takes place according to the agreement reached in the first stage.[45] In the second stage, the firms make quantity decisions based on their information about a.

First, suppose firm 1 has decided to conceal its information about a. The second-stage game is thus a Bayesian game (with incomplete information). Firm 1's strategy is a decision rule that specifies a quantity for each signal it may receive. Let firm 1's output be q_1^h if it observes $a = 150$ and q_1^l if it

---

[43] A somewhat related paper is Lee and Whang (2002) who consider a supply chain with one supplier and n retailers. These supply chain members are independent firms. The retailers face a selling season with two periods. The retailers independently make their inventory decisions at the beginning of the first period by ordering from the supplier. The retailers are able to adjust their inventory levels after the first-period demands are realized by trading among themselves through a secondary market. They study the impact of the secondary market on the supply chain. The paper is related if one considers the secondary market as an institution that facilitates the sharing (more precisely, the aggregation) of the information embodied in the first-period demands (about the needs for inventories at the beginning of the second period). In this respect, Mendelson and Tunca (2001) is also related. Another paper that touches upon information sharing in a competitive environment is Anupindi and Bassok (1999), who study a decentralized supply chain with one manufacturer and two retailers. The retailers order inventories from the manufacturer, and their ordering decisions are interdependent because of consumer market search, i.e., there is a positive probability that a consumer who experiences a stockout at one retailer will look for inventory at the other retailer. It is clear that market search provides incentives for one retailer to increase its inventory (to capture the benefits of demand spillovers), given the other retailer's ordering decision; and such incentives are stronger when search becomes easier. The consequence is that the manufacturer sees an increase in total retailer order quantities as market search increases, which can be facilitated by the installation of an information system that makes the inventory status at the retailers visible to the consumers.

[44] This example was given by Gal-Or (1985).

[45] If firm 1 first observes a and then decides whether or not to share the information with firm 2, we have a very different model. It looks like a signaling problem, with retail competition as a subgame.

observes $a = 50$. Let q_2 be firm 2's output. Firm 2's best response to firm 1's strategy is obtained by solving:

$$\max_{q_2} \frac{1}{2}(50 - q_1^l - q_2)q_2 + \frac{1}{2}(150 - q_1^h - q_2)q_2.$$

That is,

$$q_2 = \frac{200 - q_1^l - q_1^h}{4}. \qquad (3.5)$$

Similarly, firm 1's best strategy against firm 2's quantity is

$$q_1^l = \frac{50 - q_2}{2}, \quad \text{and} \quad q_1^h = \frac{150 - q_2}{2}. \qquad (3.6)$$

Solving (3.5) and (3.6) gives a Bayesian Nash equilibrium: $(q_1^l, q_1^h, q_2) = (\frac{25}{3}, \frac{175}{3}, \frac{100}{3})$. The expected profits for firm 1 and firm 2 are: $(\pi_1^{ns}, \pi_2^{ns}) = (\frac{15,625}{9}, \frac{10,000}{9})$, where the superscript 'ns' stands for 'no sharing of information'.

Now suppose firm 1 has decided to share its information with firm 2. If $a = 50$, the two firms each produce $50/3$; otherwise, if $a = 150$, they each produce 50. The expected profits are $(\pi_1^s, \pi_2^s) = (\frac{12,500}{9}, \frac{12,500}{9})$.

Comparing the above two scenarios, we see that it is to the interest of the informed firm to conceal its information. Note that the total profits of the two firms are also higher with no information sharing. This is a pretty gloomy picture for information sharing.

A body of literature in economics is devoted to the investigation of information sharing among horizontal competitors. It turns out that whether or not it is optimal to share information depends on many things, including the type of competition (Cournot or Bertrand), the type of information (e.g., common demand information or private cost information), and whether the products sold by the competitors are substitutes or complements.

Duopoly with demand information: Cournot and Bertrand, substitutes and complements

Vives (1984) considers the following duopoly model. Two firms, each producing a differentiated good, face the following inverse demand functions:

$$p_i = \alpha - \beta q_i - \gamma q_j, \quad i, j = 1, 2, j \neq i$$

where q_i are the quantities of the goods and p_i their prices, with $|\gamma/\beta| \leq 1$. If $\gamma = 0$, the goods are independent and the firms are local monopolists.

If $\gamma > 0$, the goods are substitutes. If $\gamma < 0$, they are complements. Both firms have constant and equal marginal costs, which are normalized to zero. The two firms are engaged in either Cournot competition where they compete in quantities or Bertrand competition where they compete in prices.

Note that profits of firm i are given by $\pi_i = p_i q_i$. Since π_i is symmetric in p_i and q_i and the demand curves are linear, Cournot (resp., Bertrand) competition with substitutes has similar strategic properties as Bertrand (resp., Cournot) competition with complements.

The common demand intercept, α, is a normally distributed random variable with known mean and variance. It is assumed that the firms employ an 'independent testing agency' to collect samples of α. Through the agency, firm i has contracted for n_i observations: $(r_{i1}, \ldots, r_{in_i})$, where $r_{ik} = \alpha + u_{ik}$, and the u_{ik}'s are i.i.d. normal random variables with zero mean and variance σ_u^2 and they are independent of α. Moreover, firm i has instructed the agency to put the first m_i observations that it has contracted for in a common pool, available for the other firm. Therefore, firm i's best (minimum variance unbiased) estimate of α based on $n_i + m_j$, $j \neq i$, observations is

$$s_i = \alpha + \frac{\sum_{k=1}^{n_i} u_{ik} + \sum_{k=1}^{m_j} u_{jk}}{n_i + m_j}.$$

If $m_1 = m_2 = 0$, then there is no sharing of information. On the other hand, if $m_1 = n_1$ and $m_2 = n_2$, there is a complete sharing of information.

The firms play a two-stage game. First, they decide how much information to put in the common pool, i.e., choosing m_1 and m_2 independently. (n_1 and n_2 are not decision variables in the model.) The values of n_i and m_i, $i = 1, 2$, are common knowledge. The agency then collects independent observations of α and distributes the information according to the agreement reached in the first stage (i.e., transmitting some information to a firm privately and some to the common pool). At the second stage, the firms independently choose their quantities in Cournot competition or prices in Bertrand competition. Each pair of (m_1, m_2) defines a subgame with incomplete information, which can be solved by using the concept of Bayesian Nash equilibrium.

Here are some results obtained by Vives. In Cournot competition with substitutes (or Bertrand competition with complements), expected profits of firm i decrease with m_i. So no information sharing is the unique equilibrium. In Cournot competition with complements (or Bertrand competition with substitutes), expected profits of firm i increase with m_i and with m_j, $j \neq i$. So complete information sharing is the unique equilibrium. If the goods are independent, expected profits of firm i are increasing in m_j and unaffected by m_i, $j \neq i$. In this case, any pair of (m_1, m_2) is an equilibrium.

Oligopoly with demand information: Cournot, substitutes

Gal-Or (1985) provides an alternative model of information sharing in an oligopoly model with Cournot competition. There are n firms producing a common product at no cost. The industry is facing a linear demand function:

$$p = a - bQ + u, \quad a, b > 0$$

where p is price and Q is the total quantity produced. The prior distribution of u is normal with zero mean and a finite variance. Before making their quantity decisions, each firm receives a noisy signal for u. The signal observed by firm i is x_i. The following assumptions characterize the signals:

$$x_i = u_i + e_i, \quad u_i \sim N(0, \sigma), \quad e_i \sim N(0, m);$$
$$Cov(e_i, e_j) = 0, \quad i \neq j;$$
$$Cov(u_i, e_j) = 0, \quad \forall i, j;$$
$$Cov(u_i, u_j) = h, \quad i \neq j;$$
$$u = \sum_{i=1}^{n} u_i / n.$$

Furthermore, it is assumed that $h \geq 0$, a parameter that measures the (positive) level of correlation among the signals.

Before making quantity decisions, the firms choose whether or not to reveal their private signals to the other firms, and how complete this revelation will be. This is modeled by assuming that an outside agency is responsible for information transmission. The firms are required to commit themselves to a fixed amount of garbling prior to learning their signals. Upon learning its private signal, each firm i reports its private signal x_i to the agency, who then reports a message $\hat{x}_i$ to the other firms, with

$$\hat{x}_i = x_i + f_i, \quad f_i \sim N(0, s_i)$$

where the f_i's are independent of each other and of any u_j and $e_j, j = 1, \ldots, n$. The value of s_i represents the level of garbling. If this noise variance is zero for all firms, we have complete information sharing. If it is infinite for all firms, there is no sharing of information. The case with a finite noise variance represents partial information sharing.

Gal-Or characterizes a symmetric equilibrium in the following two-stage game. (A symmetric equilibrium is reasonable because the firms have symmetric cost/information structure.) At the first stage, each firm i chooses s_i independently. Once chosen, this vector of noise variances becomes common knowledge. The firms then receive their private signals, and the outside agency reports messages with levels of garbling determined in the first stage. At the second stage, the firms make their quantity decisions simultaneously. Each firm i's strategy is a decision rule that determines its output level as a function of its private signal x_i and the vector of messages reported by the outside agency. The conclusion is that no information sharing is the unique symmetric Nash equilibrium. Therefore, allowing for partial revelation and various degrees of correlation between the private signals does not change the incentives for sharing demand information. Similar results have been obtained by Clarke (1983).

Duopoly with cost information: Cournot and Bertrand, substitutes

Gal-Or (1986) shows that the incentives for sharing information depend not only on the type of competition (Cournot or Bertrand) and the relationship between the products (substitutes or complements), as Vives (1984) has shown, but also on the type of information under consideration. The private signals received by Gal-Or's firms are about unknown private costs, instead of an unknown demand intercept as in previous works. She considers a duopoly model consisting of two firms each producing a differentiated product. The demand is linear, as in Vives (1984), with an additional assumption that $\gamma > \mathbf{0}$, i.e., the products are substitutes. The production costs are linear, with c_i the unit cost of production for firm i, $i = 1, 2$. The value of c_i is a normal random variable with zero mean and a known variance, with c_1 and c_2 being independent. Each firm receives a signal for its own unit cost. Firm i receives signal z_i, where

$$z_i = c_i + e_i, \qquad i = 1, 2$$

where $e_i \sim N(0, m)$, e_i and c_j are independent for any i and j, and e_1 and e_2 are independent.

Information sharing is implemented by an outside agency. Prior to receiving their private signals, each firm commits to a level of garbling that the agency will use in reporting the private information. The reported signal is

$$\hat{z}_i = z_i + f_i, \quad f_i \sim N(0, s_i), \qquad i = 1, 2$$

where f_1 and f_2 are independent, and f_i is independent of c_j and e_j for all i and j. As in Gal-Or (1985), the values of the s_i's are determined independently by the firms at the first stage of the game, and they represent the degree of information sharing. At the second stage of the game, the firms choose their output levels (in Cournot competition) or prices (in Bertrand competition). The second-stage strategy is a decision rule based on available information, i.e., z_i and $\hat{z}_j$, $j \neq i$, for firm i.

The main finding is that complete (resp., no) information sharing is a dominant strategy in Cournot (resp., Bertrand) competition. Notice that changing the type of information (from demand to cost) reverses the incentives for information sharing. A similar set of results was obtained by Shapiro (1986).[46]

Vertical information sharing in the presence of horizontal competition

Consider a supply chain with multiple parties and dispersed information. Suppose a subset of those parties, called the insiders, have decided to share information among themselves. What is the impact of such an agreement on the insiders, the outsiders (i.e., those who do not engage in information sharing), and the supply chain as a whole? The answer to this question is complicated partly due to the spillover effect of information sharing. That is, an outsider may gain valuable information from the insiders either directly if the insiders fail to keep the shared information confidential or indirectly by observing the actions taken by the insiders. The former kind of information leakage may be prevented by the insiders through a contract protecting the confidentiality of the shared data, while the latter kind is impossible to avoid as long as the shared information affects the behavior of an insider that is observable to the outsiders. The outsiders may change what they do upon learning the information, with potentially significant impact on the insiders' profits and their decisions whether or not to share information in the first place.

Li (2002) considers a model with one manufacturer and n symmetric retailers ($n \geq 2$). The retailers sell an identical product, which is produced by the manufacturer. Both production and sales incur constant marginal costs, which are normalized to zero. The consumer market (to which the retailers sell) is characterized by the inverse demand function, $p = a + \theta - Q$, i.e., the prevailing retail price p is determined by a known constant a, a random variable θ, and the total supply Q, which is the sum of the individual quantities given by the retailers. (Thus the retailers engage in a Cournot competition.) The manufacturer determines the wholesale price P. Each retailer i receives a private signal Y_i about θ, with the joint distribution of $(\theta, Y_1, \ldots, Y_n)$ being common knowledge.

The sequence of events is as follows:

(1) Each retailer decides whether or not to share his private signal with the manufacturer. If a retailer decides to share, the information revelation is assumed to be truthful. Let K be the set of retailers who decide to share their information. Because the retailers are symmetric, we only need to know the cardinality of K, i.e., $|K| \stackrel{def}{=} k$, $k = 0, 1, \ldots, n$.
(2) Each retailer receives his private signal. Information transmission occurs according to the arrangements made in the first step.
(3) The manufacturer sets the wholesale price. The wholesale price P is thus a function of the disclosed information $\{Y_j, j \in K\}$.

---

[46] Li (1985) generalizes the above literature on the incentives for sharing demand or cost information in Cournot oligopolies by making weaker distributional assumptions about the random variables.

(4) The retailers simultaneously choose their sales quantities and place orders with the manufacturer. Retailer i's strategy thus depends on whether or not i is a member of K, his signal Y_i, the wholesale price P, and the information embedded in the wholesale price P (which is observable to all).
(5) The manufacturer produces the retailer orders.

Li shows that there is an equilibrium outcome where P is a monotone function of $\sum_{j \in K} Y_j$. Thus in equilibrium, the retailers $i \in K$ (i.e., the outsiders) can infer the value of o$\sum_{j \in K} Y_j$. Moreover, knowing this sum is as good as knowing the individual signals $Y_j, j \in K$. Therefore, the leakage of the private signals from the insiders to the outsiders is complete. In other words, even though the information sharing is between the retailers in K and the manufacturer, we could as well imagine that the retailers in K announce their private signals in public.

Suppose that k retailers have decided to share information with the manufacturer, $k = 0, 1, \ldots, n$. Let $\Pi_R^S(k)$ be the expected profits for a retailer who shares information, and $\Pi_R^N(k)$ the expected profits for a retailer who does not share information. Let $\Pi_M(k)$ be the manufacturer's expected profits. Li shows that $\Pi_M(k)$ is increasing and concave in k. Therefore, the manufacturer always benefits if a retailer decides to share information. On the other hand, $\Pi_R^N(k-1) > \Pi_R^S(k)$ for all $k = 1, \ldots, n$. In words, a retailer is always better off by switching from sharing information to not sharing. Consequently, no information sharing is the unique equilibrium outcome.

If the manufacturer's gains from information sharing exceed the losses of the retailers, the manufacturer can pay the retailers for their private information. Let $\Pi(k)$ be the supply chain's total profit when k retailers share information, $k = 0, 1, \ldots, n$. Thus $\Pi(k) = \Pi_M(k) + k\Pi_R^S(k) + (n-k)\Pi_R^N(k)$. Li shows that $\Pi(n) \geq \Pi(0)$ if and only if $(n-2)(n+1) \geq 2s$, where s is an indicator of the informativeness of the retailers' private signals, with a smaller s value meaning more informative. Consequently, there is no guarantee that the supply chain will benefit from information sharing. In cases where the supply chain does benefit from information sharing (when the number of retailers is large or the demand signals are informative), there exists a Pareto improvement if the manufacturer pays the retailers for sharing their information.

Li has also considered a case where the retailers hold private information about their costs. This is done by modifying the above model with demand uncertainty as follows. First, let $\theta \equiv 0$, thus eliminating demand uncertainty. Let C_i be retailer i's marginal cost, $i = 1, \ldots, n$. After making decisions about whether or not to share their cost information with the manufacturer but before making quantity decisions, each retailer i observes his own cost C_i. The retailers' costs are assumed to be positively correlated.

As expected, the manufacturer always benefits if a retailer decides to share his cost information. However, complete information sharing, i.e., all retailers decide to share their cost information with the manufacturer, is now always an equilibrium. And it is the unique equilibrium sometimes. Moreover, complete information sharing increases the supply chain's total profits.[47]

There are two recent extensions of Li (2002). One is Li and Zhang (2001), where the manufacturer can produce before the retailers make their quantity decisions. Production after receiving the retailers' orders is still possible, but incurs a higher cost. Moreover, the manufacturer must satisfy all retailers' orders. The manufacturer makes the early-production decision and the wholesale price decision at the same time. By and large, replacing make-to-order with make-to-stock does not alter the qualitative insights. The second extension is Zhang (2002), who focuses on the sharing of demand information but considers both Cournot and Bertrand competition with substitutes and complements. It is thus a direct generalization of Vives (1984) to a supply chain setting.

4 Future research

The role of information in achieving supply chain coordination will continue to be a fruitful research area. As in the past, research will progress in many directions.

4.1 Full information, centralized control

Here we imagine a supply chain controlled by a central planner with all relevant information. The challenge is to determine a strategy that optimizes the supply chain-wide performance. Many people say that this is the traditional way of thinking in operations management/operations research. But that should not be construed to mean that the area is unimportant. In fact, there are many important problems that are begging for solutions.

To see how difficult these problems can be and how little we understand them, simply consider a two-level supply chain with one distribution center replenishing multiple local sales offices. (This kind of supply chain structure is often under the unglamorous name of one-warehouse multiretailer systems.) The truth of the matter is nobody knows what the optimal policy is. Many have studied 'heuristic' policies; the ones that seem to make intuitive sense. One example is the control rules that are based on echelon inventory positions, i.e., the replenishment strategy at the distribution center is based on the total inventory (on-hand and in-transit) in the system. The allocation

---

[47] The fact that the retailers will share their cost information with the manufacturer is striking at first glance. This may have a lot to do with the assumption that the decision whether or not to share information is made before observing the private information. This assumption is particularly strong with the cost information; if the retailers have been in the business for a while, it seems reasonable that they have better information about their own costs than anyone else prior to making information-sharing decisions. Another scenario that may alter the result is when the parties interact repeatedly.

decision at the distribution center when it runs out of stock is often of the myopic sort or based on some, often arbitrary, priority rule. This heuristic approach sometimes puts us in a very awkward position, where a strategy based on full information is actually inferior to a strategy that is based only on local information (e.g., the so-called installation-stock policies). Information has a negative value!

To be sure, many years of work has suggested to us that the optimal strategy for the above system (or many other multiechelon systems with common cost or topological structures) is, if it exists, likely to be very complex. So for all practical purposes, we should confine ourselves to heuristic policies that are easy to implement. But the quest for better heuristics will never stop, unless we know that the heuristic we have is already very close to optimality. The most powerful statement one can make about a heuristic's closeness to optimality is the worst-case gap between the two. Since the optimal policy is unknown, the worst-case analysis must rely on bounds on the optimal performance. Discovering heuristics with small worst-case gaps is the holy grail of multiechelon inventory theory.[48]

4.2 Decentralized information, shared incentives

Now replace the central planner with local managers who are each responsible for managing part of the supply chain. These managers only have access to local information. But they share a common goal to optimize the supply chain-wide performance. This is the team-model approach (Marschak and Radner 1972). And it is appropriate when, e.g., the supply chain consists of multiple divisions of the same firm, and the divisions' incentives are aligned. It is an intermediate step to a full-blown decentralized system.

An important feature of team models is that the control rules used by the local managers can only be based on local information. To obtain a solution to a team model, it is convenient to take the view of an analyst, who optimizes the system-wide objective by restricting to strategies that can be implemented by the local managers (i.e., a strategy, when followed by a local manager using his local information, leads to an unambiguous decision). Therefore, it looks like a centralized planning model, but with an added informational

---

[48] A lower-bounding methodology has been given by Chen and Zheng (1994) for general stochastic multiechelon inventory systems. But the use of lower bounds in evaluating the optimality of heuristic policies has been sporadic and is largely numerical. It is hard to resist the temptation to mention the spectacular successes achieved for general multiechelon inventory systems with deterministic demand, for which a class of heuristic policies – the so-called power-of-two policies – have been guaranteed to be within 2% of optimality. See, e.g., Maxwell and Muckstadt (1985), Roundy (1985, 1986), and Federgruen, Queyranne and Zheng (1992). So far, unfortunately, the only comparable result for a stochastic system is the one in Chen (1999b) established for a simple two-stage serial system. In Chapter 10 of this volume, Sven Axsäter reviews studies of heuristic policies in one-warehouse multiretailer systems.

constraint. The difference between the team model and its full-information, central-planner counterpart reveals the value of information. Section 2 of this chapter has reviewed many papers in this area. This will continue to be a fruitful research direction.

4.3 Decentralized information, independent entities

A full-blown decentralized supply chain consists of independent firms with asymmetric information. Section 3 of this chapter has covered many such models. As mentioned there, a member of the supply chain may take the initiative of 'setting the stage' by either screening or signaling, or the supply chain's members come together (cooperatively or noncooperatively) to form some trading rules. This is a relatively new area for many in operations management. Below we describe several promising research directions.

One potentially fruitful research area is the integration of price discovery with a firm's internal optimization. In Section 2.2, we have already seen a procurement example with one buyer and multiple potential suppliers with private cost information. There, the solution is a marriage between an auction mechanism and a supply contract. Infusing auction theory into operations management research is exciting.

Another interesting research direction is information acquisition. In Section 3.1, we saw an example where a firm can 'buy' advance demand information from customers. The challenge was to balance the cost of information acquisition with the benefit of the acquired information. It is certainly possible to study information acquisition in other contexts, with other kinds of information and between members of a supply chain.

In Section 3.3, we have seen papers dealing with information sharing among competing firms. How about competing supply chains? Information sharing between two supply chains can happen in many different ways: (1) same-layer, cross-channel (e.g., retailer to retailer, supplier to supplier), (2) inter-layer, same-channel (retailer to supplier), and (3) inter-layer, cross-channel (retailer in supply chain A to supplier in supply chain B, and vice versa). Opportunities abound.[49]

4.4 Bounded rationality and robust supply chain design

Real firms (and people) have limitations. First of all, their data may be inaccurate. For example, a retailer may not know exactly how many units of a product are in the store. This occurs even at successful retailers who have

---

[49] For an example on competing supply chains, see Corbett and Karmarkar (2001) and the references therein.

invested large sums in information technology, mostly to track sales and automate transactions.[50] In supply chains, inaccurate data may also result from imperfect transmission of information, which can be noisy and laden with delays. On the other hand, managers like to have easy, intuitively appealing control rules. This is simply because people have limited information-processing power. The same holds for modelers/analysts/researchers: real-world problems are complex with multiple facets, and it is impossible to include all the complexities in a model. (Simply put, people – managers or not – are boundedly rational.) As a result, any 'optimal solution' obtained from a model is unlikely to be implemented as is; at best, it will inform a manager's 'insight' or 'intuition', which in turn influences the final decision. Given inaccurate data, modeling limitations, and managers' desire for simplicity, there is a pressing need to develop simple control mechanisms that are robust to such imperfections. This is virtually an uncharted territory. But it is worthwhile to ask the question.

Acknowledgements

Part of this chapter was completed during the author's visit at the Stanford University Graduate School of Business, 2000–2001. Thanks to the OIT group of the Stanford Business School for their hospitality. The chapter has benefited from the comments and suggestions made by the following colleagues: Gérard P. Cachon, Charles J. Corbett, Ton G. de Kok, Guillermo Gallego, Constantinos Maglaras, Özalp Özer, L. Beril Toktay, and Jan A. Van Mieghem. Financial support from National Science Foundation, Columbia Business School, and Stanford Global Supply Chain Management Forum is gratefully acknowledged

References

Anand, K., H. Mendelson (1997). Information and organization for horizontal multimarket coordination. *Management Science* 43(12), 1609–1627.
Anupindi, R., Y. Bassok (1999). Centralization of stocks: retailers vs. manufacturer. *Management Science* 45(2), 178–191.
Arrow, K. (1985). The economics of agency, in: J. Pratt, R. Zeckhauser (eds.), *Principals and Agents*, Cambridge, MA, Harvard Business School Press.
Aviv, Y. (2001). The effect of collaborative forecasting on supply chain performance. *Management Science* 47(10), 1326–1343.
Aviv, Y., A. Federgruen (1997). Stochastic inventory models with limited production capacity and periodically varying parameters. *Probability in the Engineering and Informational Science* 11, 107–135.

---

[50] See Raman, DeHoratius and Ton (2000) and Raman and Ton (2000) for discussions on the magnitudes and drivers of inaccurate inventory data in retail stores.

Aviv, Y., A. Federgruen. (1998). The operational benefits of information sharing and vendor managed inventory (VMI) programs. Working paper, Washington University and Columbia University.

Axsäter, S. (1990). Simple solution procedures for a class of two-echelon inventory problems. *Operations Research* 38(1), 64–69.

Axsäter, S. (1993a). Continuous review policies for multi-level inventory systems with stochastic demand, in: S. Graves, A. Rinnooy Kan, P. Zipkin (eds.), *Handbook in Operations Research and Management Science, Vol. 4, Logistics of Production and Inventory*, North Holland.

Axsäter, S. (1993b). Exact and approximate evaluation of batch ordering policies for two-level inventory systems. *Operations Research* 41(4), 777–785.

Axsäter, S. (1997). Simple evaluation of echelon stock (R, Q) policies for two-level inventory systems. *IIE Transactions* 29, 661–669.

Axsäter, S. (1998). Evaluation of installation stock based (R, Q) policies for two-level inventory systems with Poisson demand. *Operations Research* 46 (Supp. No. 3), S135–S145.

Axsäter, S. (2000). Exact analysis of continuous review (R, Q) policies in two-echelon inventory systems with Compound Poisson demand. *Operations Research* 48(5), 686–696.

Axsäter, S., K. Rosling (1993). Installation vs. echelon stock policies for multi-level inventory control. *Management Science* 39(10), 1274–1280.

Azoury, K. (1985). Bayes solution to dynamic inventory models under unknown demand distribution. *Management Science* 31, 1150–1160.

Azoury, K., B. Miller (1984). A comparison of the optimal ordering levels of Bayesian and non-Bayesian inventory models. *Management Science* 30, 993–1003.

Barnes-Schuster, D., Y. Bassok, R. Anupindi (1998). Coordination and flexibility in supply contracts with options. Working paper, University of Chicago.

Basu, A., R. Lal, V. Srinivasan, R. Staelin (1985). Salesforce-compensation plans: An agency theoretic perspective. *Marketing Science* 4(4), 267–291.

Bitran, G., E. Haas, H. Matsuo (1986). Production planning of style goods with high setup costs and forecast revisions. *Operations Research* 34, 226–236.

Blanchard, O. (1983). The production and inventory behavior of the American automobile industry. *Journal of Political Economy* 91, 365–400.

Blinder, A. (1982). Inventories and sticky prices. *American Economic Review* 72, 334–349.

Blinder, A. (1986). Can the production smoothing model of inventory behavior be saved? *Quarterly Journal of Economics* 101, 431–454.

Bourland, K., S. Powell, D. Pyke (1996). Exploiting timely demand information to reduce inventories. *European Journal of Operational Research* 92, 239–253.

Box, G., G. Jenkins, G. Reinsel (1994). *Time Series Analysis: Forecasting and Control*, 3rd Edition, San Francisco, CA, Holden-Day, pp. 110–114.

Bradford, J., P. Sugrue (1990). A Bayesian approach to the two-period style-goods inventory problem with single replenishment and heterogeneous Poisson demands. *Journal of Operational Research Society* 41, 211–218.

Brown, A., H. Lee (1998). The win-win nature of options based capacity reservation arrangements. Working paper, Vanderbilt University.

Cachon, G. (1999). Managing supply chain demand variability with scheduled ordering policies. *Management Science* 45(6), 843–856.

Cachon, G. (2001). Exact evaluation of batch-ordering inventory policies in two-echelon supply chains with periodic review. *Operations Research* 49(1), 79–98.

Cachon, G., M. Fisher (2000). Supply chain inventory management and the value of shared information. *Management Science* 46(8), 1032–1048.

Cachon, G., M. Lariviere (1999). Capacity choice and allocation: strategic behavior and supply chain performance. *Management Science* 45(8), 1091–1108.

Cachon, G. P., M. Lariviere (2001). Contracting to assure supply: how to share demand forecasts in a supply chain. *Management Science* 47(5), 629–646.

Caplin, A. (1985). The variability of aggregate demand with (s, S) inventory policies. *Econometrica* 53, 1395–1409.
Cattani, K., W. Hausman (2000). Why are forecast updates often disappointing?. *Manufacturing & Service Operations Management* 2(2), 119–127.
Chen, F. (1998a). Echelon reorder points, installation reorder points, and the value of centralized demand information, *Management Science* 44 (12, part 2), S221–S234.
Chen, F. (1998b). On (R,nQ) policies in serial inventory systems, in: S. Tayur, R. Ganeshan and M. Magazine (eds.) *Quantitative Models for Supply Chain Management*, Kluwer Academic Publishers, Boston/Dordrecht/London.
Chen, F. (1999a). Decentralized supply chains subject to information delays. *Management Science* 45(8), 1076–1090.
Chen, F. (1999b). 94%-effective policies for a two-stage serial inventory system with stochastic demand. *Management Science* 45(12), 1679–1696.
Chen, F. (2000a). Optimal policies for multi-echelon inventory problems with batch ordering. *Operations Research* 48(3), 376–389.
Chen, F. (2000b). Salesforce incentives and inventory management. *Manufacturing & Service Operations Management* 2(2), 186–202.
Chen, F. (2000c). Salesforce incentives, market information, and production/inventory planning. To appear in *Management Science*.
Chen, F. (2001a). Market segmentation, advanced demand information, and supply chain performance. *Manufacturing & Service Operations Management* 3(1), 53–67.
Chen, F. (2001b). Auctioning supply contracts. Working paper, Columbia University.
Chen, F., J. Eliashberg, P. Zipkin (1998). Customer preferences, supply-chain costs, and product-line design, in: T.-H. Ho, C. S. Tang (eds.), *Product Variety Management: Research Advances*, Kluwer Academic Publishers, Boston/Dordrecht/London.
Chen, F., R. Samroengraja (1999). Order volatility and supply chain costs. To appear in *Operations Research*.
Chen, F., R. Samroengraja (2000a). A staggered ordering policy for one-warehouse multi-retailer systems. *Operations Research* 48(2), 281–293.
Chen, F., R. Samroengraja (2000b). The stationary beer game. *Production and Operations Management* 9(1), 19–30.
Chen, F., J.-S. Song (2001). Optimal policies for multi-echelon inventory problems with Markov-modulated demand. *Operations Research* 49(2), 226–234.
Chen, F., B. Yu (2001a). Quantifying the value of leadtime information in a single-location inventory system. Working paper, Columbia Business School.
Chen, F., B. Yu (2001b). A supply chain model with asymmetric capacity information. Near completion.
Chen, F., Y.-S. Zheng (1994). Lower bounds for multi-echelon stochastic inventory systems. *Management Science* 40(11), 1426–1443.
Chen, F., Y.-S. Zheng (1997). One-warehouse multi-retailer systems with centralized stock information. *Operations Research* 45(2), 275–287.
Chen, Frank, Z. Drezner, J. Ryan, D. Simchi-Levi (2000). Quantifying the bullwhip effect in a simple supply chain: The impact of forecasting, lead times, and information. *Management Science* 46(3), 436–443.
Chu, W. (1992). Demand signaling and screening in channels of distribution. *Marketing Science* 11(4), 327–347.
Clark, A., H. Scarf (1960). Optimal policies for a multi-echelon inventory problem. *Management Science* 6, 475–490.
Clarke, R. N. (1983). Collusion and the incentives for information sharing. *Bell Journal of Economics* 14, 383–394.
Corbett, C. (2001). Stochastic inventory systems in a supply chain with asymmetric information: Cycle stocks, safety stocks, and consignment stock. *Operations Research* 49(4), 487–500.
Corbett, C., X. de Groote (2000). A supplier's optimal quantity discount policy under asymmetric information. *Management Science* 46(3), 444–450.

Corbett, C., U. Karmarkar (2001). Competition and structure in serial supply chains with deterministic demand. *Management Science* 47(7), 966–978.
Corbett, C., D. Zhou, C. Tang (2001). Designing supply contracts: Contract type and information asymmetry. Working paper, UCLA.
Coughlan, A. (1993). Salesforce compensation: a review of MS/OR advances, in: J. Eliashberg, G. Lilien (eds.), *Handbooks in Operations Research and Management Science: Marketing*, Vol. 5. North-Holland.
de Groote, X. (1994). Flexibility and product variety in lot-sizing models. *European Journal of Operations Research* 75, 264–274.
Desai, P. S., K. Srinivasan (1995). Demand signaling under unobservable effort in franchising: Linear and nonlinear price contracts. *Management Science* 41(10), 1608–1623.
Deshpande, V., L. Schwarz (2002). Optimal capacity allocation in decentralized supply chains. Working paper, Purdue University.
Deuermeyer, B., L. Schwarz (1981). A model for the analysis of system service level in warehouse/retailer distribution systems: the identical retailer case, in: L. Schwarz *Studies in the Management Sciences: The Multi-Level Production/Inventory Control Systems*, Vol. 16. Amsterdam, North-Holland Publishing Co., pp. 163–193.
Diks, E. B., A. G. de Kok (1998). Optimal control of a divergent multi-echelon inventory system. *European Journal of Operational Research* 111, 75–97.
Ding, X., M. L. Puterman, A. Bisi (2002). The censored newsvendor and the optimal acquisition of information. *Operations Research* 50(3), 517–527.
Dobson, G., S. Kalish (1988). Positioning and pricing a product line. *Marketing Science*, 2 7, 107–125.
Dobson, G., S. Kalish (1993). Heuristics for positioning and pricing a product line using conjoint and cost data. *Management Science* 39, 160–175.
Dong, L., H. Lee (2000). Optimal policies and approximations for a serial multi-echelon inventory system with time-correlated demand. Working paper, Washington University.
Donohue, K. (2000). Efficient supply contracts for fashion goods with forecast updating and two production modes. *Management Science* 46(11), 1397–1411.
Ehrhardt, R. (1984). (s, S) policies for a dynamic inventory model with stochastic leadtimes. *Operations Research* 32, 121–132.
Eppen, G., A. Iyer (1997a). Backup agreements in fashion buying-the value of upstream flexibility. *Management Science* 43(11), 1469–1484.
Eppen, G., A. Iyer (1997b). Improved fashion buying with Bayesian updates. *Operations Research* 45(6), 805–819.
Eppen, G., L. Schrage (1981). Centralized ordering policies in a multiwarehouse system with leadtimes and random demand, in: L. Schwarz (ed.) *Multi-Level Production/Inventory Control Systems: Theory and Practice*, North Holland Publishing Co., pp. 51–69.
Federgruen, A. (1993). Centralized planning models for multi-echelon inventory systems under uncertainty, in: S. Graves, A. Rinnooy Kan, P. Zipkin (eds.), *Handbook in Operations Research and Management Science, Vol. 4, Logistics of Production and Inventory*, North-Holland.
Federgruen, A., M. Queyranne, Y.-S. Zheng (1992). Simple power of two policies are close to optimal in general class of production/distribution networks with general joint setup costs. *Mathematics of Operations Research* 17, 951–963.
Federgruen, A., P. Zipkin (1984a). Approximation of dynamic, multi-location production and inventory problems. *Management Science* 30, 69–84.
Federgruen, A., P. Zipkin (1984b). Allocation policies and cost approximation for multi-location inventory systems. *Naval Research Logistics Quarterly* 31, 97–131.
Federgruen, A., P. Zipkin (1986a). An inventory model with limited production capacity and uncertain demands I. The average cost criterion. *Mathematics of Operations Research* 11, 193–207.
Federgruen, A., P. Zipkin (1986b). An inventory model with limited production capacity and uncertain demands II. The discounted-cost criterion. *Mathematics of Operations Research* 11, 208–215.

Fisher, M., A. Raman (1996). Reducing the cost of demand uncertainty through accurate response to early sales. *Operations Research* 44(1), 87–99.
Forrester, J. (1961). *Industrial Dynamics*, New York, John Wiley and Sons Inc.
Forsberg, R. (1995). Optimization of order-up-to S policies for two-level inventory systems with compound Poisson demand. *European Journal of Operational Research* 81, 143–153.
Fudenberg, D., J. Tirole (1992). *Game Theory*, Cambridge, MA, The MIT Press.
Gallego, G., Y. Huang, K. Katircioglu, Y.T. Leung (2000). When to share demand information in a simple supply chain? Working paper, Columbia University.
Gallego, G., O. Özer (2000). Optimal replenishment policies for multi-echelon inventory problems under advance demand information. Working paper, Columbia University and Stanford University.
Gallego, G., O. Özer (2001). Integrating replenishment decisions with advance demand information. *Management Science* 47(10), 1344–1360.
Gallego, G., B. Toktay (1999). All-or-nothing ordering under a capacity constraint and forecasts of stationary demand. Working paper, Columbia University.
Gal-Or, E. (1985). Information sharing in oligopoly. *Econometrica* 53(2), 329–343.
Gal-Or, E. (1986). Information transmission – Cournot and Bertrand equilibria. *Review of Economic Studies* LIII, 85–92.
Gavirneni, S., R. Kapuscinski, S. Tayur (1999). Value of information in capacitated supply chains. *Management Science* 45(1), 16–24.
Gonik, J. (1978). Tie salesmen's bonuses to their forecasts. *Harvard Business Review* May–June 1978.
Graves, S. (1985). A multi-echelon inventory model for a reparable item with one-for-one replenishment. *Management Science* 31, 1247–1256.
Graves, S. (1996). A multiechelon inventory model with fixed replenishment intervals. *Management Science* 42(1), 1–18.
Graves, S. (1999). A single-item inventory model for a nonstationary demand process. *Manufacturing & Service Operations Management* 1(1), 50–61.
Graves, S., D. Kletter, W. Hetzel (1998). A dynamic model for requirements planning with application to supply chain optimization. *Operations Research* 46, S35–S49.
Graves, S., H. Meal. S. Dasu, Y. Qiu (1986). Two-stage production planning in a dynamic environment, in: S. Axsäter, C. Schneeweiss, E. Silver (eds.), *Multi-Stage Production Planning and Control*, Lecture Notes in Economics and Mathematical Systems, Vol. 266, Springer-Verlag, Berlin, pp. 9–43.
Green, P., A. Krieger (1985). Models and heuristics for product-line selection. *Marketing Science (Winter)* 4, 1–19.
Grossman, S., O. Hart (1983). An analysis of the principal-agent problem. *Econometrica* 51(1), 7–45.
Gullu, R. (1996). On the value of information in dynamic production/inventory problems under forecast evolution. *Naval Research Logistics* 43, 289–303.
Gullu, R. (1997). A two-echelon allocation model and the value of information under correlated forecasts and demands. *European Journal of Operational Research* 99, 386–400.
Ha, A. (2001). Supplier-buyer contracting: Asymmetric cost information and cutoff level policy for buyer participation. *Naval Research Logistics* 48, 41–64.
Hariharan, R., P. Zipkin (1995). Customer order information, lead times, and inventories. *Management Science* 41(10), 1599–1607.
Harris, M., A. Raviv (1978). Some results on incentive contracts with applications to education and employment, health insurance, and law enforcement. *American Economic Review (March)* 68, 20–30.
Harris, M., A. Raviv (1979). Optimal incentive contracts with imperfect information. *Journal of Economic Theory (April)* 20, 231–259.
Hassin, R., M. Haviv (2001). *To Queue or Not to Queue: Equilibrium Behavior in Queueing Systems*, Boston/Dordrecht/London, Kluwer Academic Publishers. to be published in the International Series in Operations Research and Management Science.
Hausman, W. (1969). Sequential decision problems: A model to exploit existing forecasters. *Management Science* 16(2), B-93–B-111.

Hausman, W., R. Peterson (1972). Multiproduct production scheduling for style goods with limited capacity, forecast revisions and terminal delivery. *Management Science* 18, 370–383.

Hayek, F. (1945). The use of knowledge in society. *American Economic Review* 35(4), 519–530.

Heath, D., P. Jackson (1994). Modeling the evolution of demand forecasts with application to safety stock analysis in production/distribution systems. *IIE Transactions* 26(3), 17–30.

T. Ho, C. Tang (eds.) (1998). Product Variety Management: Research Advances, Kluwer Academic Publishers, Boston/Dordrecht/London.

Holmstrom, B. (1979). Moral hazard and observability. *Bell Journal of Economics* 10(1), 74–91.

Holmstrom, B. (1982). Moral hazard in teams. *Bell Journal of Economics* 13(2), 324–340.

Horngren, C. T., G. Foster (1991). *Cost Accounting: A Managerial Emphasis*, 7th Edition, New Jersey, Prentice Hall, Englewood Cliffs.

Iglehart, D. (1964). The dynamic inventory model with unknown demand distribution. *Management Science* 10, 429–440.

Jackson, P. (1988). Stock allocation in a two-echelon distribution system or 'what to do until your ship comes in'. *Management Science* 34, 880–895.

Jensen, M., W. Meckling (1976). Theory of the firm: managerial behavior, agency costs and ownership structure. *Journal of Financial Economics* 3, 305–360.

Jensen, M., W. Meckling (1992). Specific and general knowledge, and organizational structure, in: L. Werin, H. Hijkander (eds.), *Contract Economics*, Cambridge, MA, Basil Blackwell.

Johnson, G., H. Thompson (1975). Optimality of myopic inventory policies for certain dependent demand processes. *Management Science* 21, 1303–1307.

Kahn, J. (1987). Inventories and the volatility of production. *American Economic Review* 77(4), 667–679.

Kaplan, R. (1970). A dynamic inventory model with stochastic lead times. *Management Science* 16, 491–507.

Kapuscinski, R., S. Tayur (1998). A capacitated production-inventory model with periodic demand. *Operations Research* 46(6), 899–911.

Karaesmen, F., J. Buzacott, Y. Dallery (2001). Integrating advance order information in make-to-stock production systems. Working paper, York University, Canada.

Klemperer, P. (1999). Auction theory: a guide to the literature. *Journal of Economic Surveys* 13(3), 227–286.

Kopczak, L., H. Lee (1994). Hewlett-Packard: Deskjet Printer Supply Chain (A). Stanford University Case.

Kreps, D. (1990). *A Course in Microeconomic Theory*, Princeton, New Jersey, Princeton University Press.

Kreps, D., R. Wilson (1982). Sequential equilibrium. *Econometrica* 50, 863–894.

Lal, R. (1986). Delegating pricing responsibility to the salesforce. *Marketing Science* 5(2), 159–168.

Lal, R., R. Staelin (1986). Salesforce-compensation plans in environments with asymmetric information. *Marketing Science* 5(3), 179–198.

Lancaster, K. (1979). *Variety, Equity and Efficiency*, New York, NY, Columbia University Press.

Lancaster, K. (1990). The economics of product variety: A survey. *Marketing Science* 9(3), 189–206.

Lariviere, M. A., V. Padmanabhan (1997). Slotting allowances and new product introductions. *Marketing Science* 16(2), 112–128.

Lariviere, M. A., E. L. Porteus (1999). Stalking information: Bayesian inventory management with unobserved lost sales. *Management Science* 45(3), 346–363.

Lee, H., K. Moinzadeh (1987a). Two-parameter approximations for multi-echelon reparable inventory models with batch ordering policy. *IIE Transactions* 19, 140–149.

Lee, H., K. Moinzadeh (1987b). Operating characteristics of a two-echelon inventory system for reparable and consumable items under batch ordering and shipment policy. *Naval Research Logistics Quarterly* 34, 365–380.

Lee, H., P. Padmanabhan, S. Whang (1997a). The bullwhip effect in supply chains. *Sloan Management Review* 38, 93–102.

Lee, H., P. Padmanabhan, S. Whang (1997b). Information distortion in a supply chain: The bullwhip effect. *Management Science* 43, 546–558.
Lee, H., K. So, C. Tang (2000). The value of information sharing in a two-level supply chain. *Management Science* 46(5), 626–643.
Lee, H., C. Tang (1998). Variability reduction through operations reversals. *Management Science* 44(2), 162–172.
Lee, H., S. Whang (1999). Decentralized multi-echelon supply chains: Incentives and Information. *Management Science* 45(5), 633–640.
Lee, H., S. Whang (2000). Information sharing in a supply chain. *Int. J. Technology Management* 20(3/4), 373–387.
Lee, H., S. Whang (2002). The impact of the secondary market on the supply chain. *Management Science* 48(6), 719–731.
Li, L. (1985). Cournot oligopoly with information sharing. *Rand Journal of Economics* 16(4), 521–536.
Li, L. (2002). Information sharing in a supply chain with horizontal competition. *Management Science* 48(9), 1196–1212.
Li, L., H. Zhang (2001). Competition, inventory, demand variability, and information sharing in a supply chain. Working paper, Yale School of Management.
Lovejoy, W. (1990). Myopic policies for some inventory models with uncertain demand. *Management Science* 36, 724–738.
Lovejoy, W. (1992). Stopped myopic policies for some inventory models with uncertain demand distributions. *Management Science* 38, 688–707.
Marschak, J., R. Radner (1972). *Economic Theory of Teams*, New Haven, Yale University Press.
Maskin, E., J. Riley (1984). Monopoly with incomplete information. *Rand Journal of Economics* 15(2), 171–196.
Matsuo, H. (1990). A stochastic sequencing problem for style goods with forecast revisions and hierarchical structure. *Management Science* 36, 332–347.
Maxwell, W., J. Muckstadt (1985). Establishing consistent and realistic reorder intervals in production-distribution systems. *Operations Research* 33, 1316–1341.
McAfee, R., J. McMillan (1987). Auctions and bidding. *Journal of Economic Literature (June)* XXV, 699–738.
Mendelson, H., T. Tunca (2001). Business to business exchanges and supply chain contracting. Working paper, Stanford Business School.
Miller, B. (1986). Scarf's state reduction method, flexibility, and a dependent demand inventory model. *Operations Research* 34, 83–90.
Milner, J., P. Kouvelis (2001). More demand information or more supply chain flexibility: what does the answer depend on? Working paper, Washington University, St. Louis.
Moinzadeh, K., H. Lee (1986). Batch size and stocking levels in multi-echelon reparable systems. *Management Science* 32, 1567–1581.
Moorthy, K. (1984). Market segmentation, self-selection, and product line design. *Marketing Science* 3, 288–307.
Muharremoglu, A., J. Tsitsiklis (2001). Echelon base stock policies in uncapacitated serial inventory systems. Working paper, MIT.
Murray, G., Jr., Silver, E. (1966). A Bayesian analysis of the style goods inventory problem. *Management Science* 12, 785–797.
Mussa, M., S. Rosen (1978). Monopoly and product quality. *Journal of Economic Theory* 18, 301–317.
Myerson, R. (1981). Optimal auction design. *Mathematics of Operations Research* 6(1), 58–73.
Nahmias, S. (1979). Simple approximations for a variety of dynamic leadtime lost-sales inventory models. *Operations Research* 27, 904–924.
Nanda, D. (1995). Strategic impact of just-in-time manufacturing on product market competitiveness. Working paper, University of Rochester, Rochester, NY.
Novshek, W., H. Sonnenschein (1982). Fulfilled expectations Cournot duopoly with information acquisition and release. *Bell Journal of Economics* 13, 214–218.

Özer, O. (2000). Replenishment strategies for distribution systems under advance demand information. Working paper, Stanford University.

Özer, O., W. Wei (2001). Inventory control with limited capacity and advance demand information. Working paper, Stanford University.

Plambeck, E., S. Zenios (2000). Performance-based incentives in a dynamic principal-agent model. *Manufacturing & Service Operations Management* 2(3), 240–263.

Plambeck, E., S. Zenios (2000b). Incentive efficient control of a make-to-stock production system, *Operational Research* 51(3), 371–386.

Porteus, E. (2000). Responsibility tokens in supply chain management. *Manufacturing & Service Operations Management* 2(2), 203–219.

Porteus, E., S. Whang (1991). On manufacturing/marketing incentives. *Management Science* 37(9), 1166–1181.

Raghunathan, S. (2001). Information sharing in a supply chain: A note on its value when demand is nonstationary. *Management Science* 47(4), 605–610.

Raju, J., V. Srinivasan (1996). Quota-based compensation plans for multiterritory heterogeneous salesforces. *Management Science* 42(10), 1454–1462.

Raman, A., N. DeHoratius, Z. Ton (2000). Execution: the missing link in retail operations. Working Paper, Harvard Business School.

Raman, A., Z. Ton (2000). An empirical analysis of the magnitude and drivers of misplaced SKUs in retail stores. Working Paper, Harvard Business School.

Rao, R. (1990). Compensating heterogeneous salesforces: some explicit solutions. *Marketing Science* 9(4), 319–341.

Riley, J., W. Samuelson (1981). Optimal auctions. *American Economic Review* 71(3), 381–392.

Roundy, R. (1985). 98%-effective integer-ratio lot-sizing for one-warehouse multi-retailer systems. *Management Science* 31, 1416–1430.

Roundy, R. (1986). 98%-effective lot-sizing rule for a multi-product, multi-facility production-inventory systems. *Mathematics of Operations Research* 11, 699–727.

Ryan, J. (1997). Analysis of inventory models with limited demand information. Unpublished Ph.D. dissertation, Northwestern University.

Scarf, H. (1959). Bayes solutions of the statistical inventory problem. *Ann. Math. Statist.* 30, 490–508.

Scarf, H. (1960). Some remarks on Bayes solutions to the inventory problem. *Naval Research Logistics Quarterly* 7, 591–596.

Sethi, S., F. Cheng (1997). Optimality of (s, S) policies in inventory models with Markovian demand. *Operations Research* 45, 931–939.

Shapiro, C. (1986). Exchange of cost information in oligopoly. *Review of Economic Studies* LIII, 433–446.

Shavell, S. (1979). Risk sharing and incentives in the principal and agent relationship. *Bell Journal of Economics* 10(1), 55–73.

Sherbrooke, C. (1968). METRIC: a multi-echelon technique for recoverable item control. *Operations Research* 16, 122–141.

Shocker, A., V. Srinivasan (1979). Multiattribute approaches for product concept evaluation and generations: A critical review. *Journal of Marketing Research* XVI, 159–180.

Signorelli, S., J. Heskett (1984). Benetton (A) and (B). Harvard Business School Case (9-685-014).

Simon, R. (1971). Stationary properties of a two echelon inventory model for low demand items. *Operations Research* 19, 761–777.

Sobel, M. (1997). Lot sizes in production lines with random yield and ARMA demand. Working paper, New York State University at Stony Brook.

Song, J.-S., P. Zipkin (1992). Evaluation of base-stock policies in multi-echelon inventory systems with state-dependent demands. Part I. State-independent policies. *Naval Research Logistics* 39, 715–728.

Song, J.-S., P. Zipkin (1993). Inventory control in a fluctuating demand environment. *Operations Research* 41, 351–370.

Song, J.-S., P. Zipkin (1996). Evaluation of base-stock policies in multi-echelon inventory systems with state-dependent demands. Part II. State-dependent policies. *Naval Research Logistics* 43, 381–396.

Song, J.-S., P. Zipkin (1996). Inventory control with information about supply conditions. *Management Science* 42, 1409–1419.

Spengler, J. (1950). Vertical integration and antitrust policy. *Journal of Political Economy* 58, 347–352.

Sterman, J. (1989). Modeling managerial behavior: Misperceptions of feedback in a dynamic decision making experiment. *Management Science* 35, 321–339.

Svoronos, A., P. Zipkin (1988). Estimating the performance of multi-level inventory systems. *Operations Research* 36, 57–72.

Svoronos, A., P. Zipkin (1991). Evaluation of one-for-one replenishment policies for multi-echelon inventory systems. *Management Science* 37, 68–83.

Tirole, J. (1988). *The theory of industrial organization*, Cambridge, MA, The MIT Press.

Toktay, B., L. Wein (2001). Analysis of a forecasting-production-inventory system with stationary demand. *Management Science* 47(9), 1268–1281.

Van Mieghem, J. (1999). Coordinating investment, production, and subcontracting. *Management Science* 45(7), 954–971.

Van Mieghem, J., M. Dada (1999). Price versus production postponement: capacity and competition. *Management Science* 45(12), 1631–1649.

Vickrey, W. (1961). Counterspeculation, auctions, and competitive sealed tenders. *Journal of Finance* 16(1), 8–37.

Vives, X. (1984). Duopoly information equilibrium: Cournot and Bertrand. *Journal of Economic Theory* 34, 71–94.

Watson, N., Y.-S. Zheng (2001). Adverse effects of over-reaction to demand changes and improper forecasting. Working Paper, University of Pennsylvania.

Yano, C., G. Dobson (1998). Profit optimizing product line design, selection and pricing with manufacturing cost considerations: a survey, in: T.-H. Ho, C. S. Tang (eds.), *Product Variety Management: Research Advances*, Kluwer Academic Publisher, pp. 145–176.

Zhang, H. (2003). Vertical information exchange in a supply chain with duopoly retailers. *Production and Operations Management* 11(4), 531–546.

Zheng, Y.-S. (1992). On properties of stochastic inventory systems. *Management Science* 38, 87–103.

Zheng, Y.-S., F. Chen (1992). Inventory policies with quantized ordering. *Naval Research Logistics* 39, 285–305.

Zipkin, P. (1986). Stochastic leadtimes in continuous-time inventory models. *Naval Research Logistics Quarterly* 33, 763–774.

Chapter 8

Tactical Planning Models for Supply Chain Management

Jayashankar M. Swaminathan
The Kenan-Flagler Business School, University of North Carolina, Chapel Hill, NC 27599, USA

Sridhar R. Tayur
GSIA, Carnegie Mellon University, Pittsburgh, PA 15213, USA

1 Introduction

Supply chain management includes the implementation of efficient policies related to procurement of raw material, transforming them into semifinished and finished products and distributing them to the end customer, thereby transcending multiple business units. Poor supply chain management (more often than not) predominantly results in excessive amounts of inventory, the largest asset for many firms. Inventory is generally carried by firms to hedge against uncertainty of different types (demand, process and supply) as well as to account for economic efficiencies. The former is managed with safety stocks (either in raw material or in finished goods), and the latter through batches (lot sizes). Typically, both these types of inventories need to be considered simultaneously, as one is affected by the other. Outside the manufacturing floor, a major challenge that companies face relates to the management of safety stocks rather than with the choice of economic lot sizes. Thus, efficient coordination of the supply chain relies heavily on how well the uncertainties related to demand, process and supply are managed. Tactical planning is the setting of key-operating targets (such as safety stocks, planned lead times and batch sizes) across the different units in a coordinated manner. These key-operating targets then provide guidance as to which day-to-day operations (either in manufacturing, logistics or procurement) can be executed. While several software tools are available in the execution arena (using the more mature area of deterministic mathematical programming), effective tactical planning tools are yet to be fully developed.

Although it would be ideal from a research standpoint to develop large-scale integrated models consisting of multiple entities while trying to understand effective supply chain practices, it is often very difficult (and in most cases impossible) to get any useful insights from such large models because they are intractable. As a result, researchers in the area of operations management over the past 50 years have tried to develop insights on simpler models which could then be used as building blocks to study more complex and real supply chains. The approach adopted has been one where one decomposes a multilevel supply chain – such as an assembly or a distribution system – and analyzes individual facilities with specific characteristics under different conditions. Such models will be the main focus of this chapter. For practical applications large-scale models have been developed and implemented based on these building blocks. Some of these papers will be briefly reviewed at the end of this chapter (these also discussed in Chapters 9 and 12).

Before introducing the different parameters related to analysis of such models, we will introduce three important notions related to the modeling and analysis that differentiate supply chain models developed by researchers in the past. First is related to the time granularity of the model, second to time horizon of the study and the third related to performance measures. In terms of granularity, the model of analysis of any inventory system (for single or multiple facility) could be based either on a continuous basis or on a periodic basis. In models with continuous review (of inventory and other parameters), the assumption is that demand occurs continuously with a demand rate (units/time) that could be deterministic or stochastic, and costs that are incurred every instant of time. In a periodic review setting (or discrete-time models), the assumptions are that demand occurs every period whose granularity could be dependent on the actual environment (say a day, a week, a month or quarter). In many real environments the review process is periodic; we will focus on such models called discrete-time models in this chapter. Discrete-time models can be developed for a single period, multiple periods or for an infinite horizon. The performance measures related to the analysis of discrete-time models could be single-period costs, discounted multiperiod or infinite-horizon costs or average costs over the infinite horizon (the breakup of these costs will be explained later). Alternatively, for each horizon of analysis, we may specify a service-level requirement. Previous research-oriented books that have addressed supply chain models include Graves, Rinnoy Kan and Zipkin (1993), Tayur, Ganeshan and Magazine (1998) and Zipkin (2000).

Once a complex multistage, multiproduct supply chain has been decomposed to its basic building block, we are left with single-product, single-stage (or facility) models that interact with each other through upstream and downstream parameters. The decomposition into a single-product setting itself needs some care: for example, if multiple products share a certain common capacity, or if there are fixed costs in ordering a set of products, these

Ch. 8. Tactical Planning Models for Supply Chain Management   425

interactions across products have to be accounted for. Once decomposed, any single facility in a supply chain faces three types of parameters – downstream parameters, upstream parameters and facility parameters.

- *Downstream parameters:* Downstream parameters are those that depend on (1) actions of the facilities downstream (such as the customers); (2) the way information obtained from downstream facilities is processed; (3) the contractual relationship with downstream facilities. In developing a supply chain model for a single facility, the main downstream parameters to be considered are as follows.
    - Demand process: The demand that gets generated at any facility depends on the operations and decisions of the downstream facilities (the customers). Seldom do we find real environments where demand is deterministic because of the uncertainties in the business environment. As a result, we will concentrate on models with stochastic demand. In a discrete-time setting, the simplest stochastic demand assumption is that it is i.i.d (independent and identically distributed). This implies that in each period the demand is independent from other periods but is generated from the same distribution. Another (more realistic) related assumption is that of independent but nonidentical demands in different periods also called *nonstationary* distribution. Finally, the demand process can be modeled in a more complex way in terms of being dependent as well as different in each period (autocorrelated demands are discussed in greater detail in Chapter 7).
    - Forecasts and information: In many business environments, it is not possible to respond to the generated demand immediately (due to lead time for production, supply and distribution as well as capacity constraints). In such cases, the facility develops forecasts for demand in any period and utilizes that to produce 'enough' to match the requirements of demand. Another way to predict the demand is to gain more information about the ordering process at the downstream facility (which generates the demand) or try to predict unknown parameters in the demand distribution using the information about realized demand until then. These predictions are utilized to develop the inventory policy for the facility.
    - Contracts: Contracts with downstream facilities typically determine the costs as well service that need to be provided by the facility. Contracts dictate whether late or partial shipments will be allowed as well as the penalty cost for stocking-out or delaying the shipment. Multiperiod discrete-time models can be differentiated based on none, partial or complete backlogging of demand. In the case of no backlogging (also called *lost sales*), the firm loses all the demand that it fell short off in a given period whereas in the complete backlogging case, the firm is allowed to ship the remaining order in future periods (but has to incur the penalty). Other parameters in the contract could be level of service

(such as fill-rate constraint), quality, vendor-managed inventory, end of life returns as well as delivery flexibility. Supply chain models have been developed with the fundamental objective of optimal contracts. For details, see Lariviere (1998), Tsay, Nahmias and Agarwal (1998), Cachon (2003) and Chen (2003).

- *Upstream parameters:* Upstream parameters depend on the decisions of the suppliers upstream related to their production process.

  – Lead time: In most real environments, there are significant lead times involved before the material ordered is obtained from the supplier. Some suppliers are more reliable than others in that their lead times for fulfillment are more accurate and do not vary a whole lot from period to period. In extreme cases, suppliers may situate a hub near the facility in which case the lead time is negligible and can be ignored. Supply chain models can be developed with zero lead time, fixed deterministic lead time and stochastic lead time. As will be noted later, many results related to zero lead times can be extended to fixed deterministic lead times. It is also quite common today to have dual lead times, either because of multiple suppliers, or because there are multiple options on how to obtain material from the same supplier.
  – Yield: Yield refers to the percentage of requested order that got delivered from the supplier. Generally, yield is modeled as a random number that represents the fraction of the order that was satisfied. Additive models of yield have also been developed. Clearly, both yield and the lead time together determine the supply process. For example, a supplier may deliver the products always in 2 weeks but may falter in terms of amount of delivery. On the other hand, the supplier could be delivering the exact quantities ordered but may be delivering them with different lead times.

- *Facility parameters:* The performance of the facility among other factors depends on the capacity available for production, the number of products produced, setup costs and variable costs associated with production, randomness in the production process, and operational policies such as inventory decisions as well as sequencing and scheduling.

  – Capacity: Most real facilities have finite capacity for production in any given period which can be increased to an extent through outsourcing on a need basis. However, incorporating capacity into a supply chain model may make it more difficult to analyze. As a result, the earliest models assumed infinite capacities in the process and more recent models have incorporated finite fixed capacity in their analysis.
  – Costs: There are four types of costs associated with the production and inventory at the facility. First, there is a per unit production cost (or variable cost of production). Second, there could be a fixed cost associated with production. This cost typically reflects the costs associated

with changing the setup of machines (equivalent of setup or changeover time). Third, there is a per unit holding cost that is charged to inventory remaining at the facility at the end of the period. Finally, there may be a salvage cost (usually negative) which represents the salvage value of the inventory at the end of the horizon. These costs along with the stock-out or penalty cost described earlier comprise the total costs incurred by the facility.
- Product and process characteristics: The number of products that are produced at the facility and their interrelationships (complementary or substitutes) affect the performance of the facility. Further, the process characteristics such as yield influence the performance. As more product and process characteristics are incorporated in a single model, it becomes more difficult to obtain analytical insights. For most part of this chapter, we will discuss the base case with single product, neglecting the above characteristics. However, we will discuss results related to some of the above in Section 5 on extensions.
- Operational decisions: The fundamental decision in almost all supply chain models relates to (1) how much inventory to stock in a given period and (2) when to produce/order. All the models discussed in this chapter develop insights on these two fundamental questions. This is often called the *inventory policy* or inventory ordering policy. The inventory policy determines the operating performance of the facility. Clearly, the scheduling of different products, their lot sizes and their sequence affects the real performance in cases where there are multiple products. In most of the discrete-time models (the focus of this chapter) this aspect is typically ignored.

In Section 2, we present the notation that will be used throughout the chapter. In Sections 3 and 4, we discuss models with the stationary and nonstationary demand, respectively. In Section 5, we discuss various extensions – multilevel systems, multiproduct systems, multiple suppliers, random yield and perishability of products. We discuss industry applications in Section 6 and conclude in Section 7.

2 Notations used

In this section we will provide a list of all notations used in this chapter for ease of reference.

h	per unit holding cost incurred on inventory in a period
s	per unit salvage cost at the end of the horizon
π	per unit stock-out cost on demand not satisfied in a period
c	per unit production/ordering cost of the product
K	setup cost for production/ordering
ξ	demand realized in a period

μ	mean value of ξ
σ	standard deviation of ξ
f, F	probability density and cumulative density functions for ξ
x	beginning inventory in a period
y	inventory level after an order has been placed
$J_n(x)$	optimal cost to go with n periods remaining in the horizon with x units on hand
$G(y)$	expected one-period cost given the inventory level is y after ordering
α	discount rate $0 < \alpha \leq 1$
$\delta(x)$	threshold function where $\delta(x) = 1$, $x > 0$ and $\delta(0) = 0$
C	capacity available in a period
l	lead time for supply in a period
p_l	probability density for lead time from the supplier equal to l periods

Any of the variables above with a subscript t (such as ξ_t or x_t) represent the value of the variable for the time period t. Similarly, any of the above variables with a * in the superscript (such as y^*) represent the optimal value.

3 Stationary and independent demand

Stationary and independent demand models assume that the demand ξ in every period comes from i.i.d distribution. With the i.i.d assumption in demand, one typically need not be concerned about the particular period t one is analyzing as well as the demand history up to that period if other variables such as cost are also stationary. This simplifies the analysis and as a result we will focus on these models first. Note that an elaborate description and analysis of these models have appeared in earlier handbooks edited by Heyman and Sobel (1984) and Graves et al. (1993) (Table 1).

3.1 Single period

The single-period stochastic inventory model deals with the problem of deciding how much to order at the beginning of the period given that demand is uncertain and there are penalty costs for lost demand and holding costs for excess inventory. This problem is also called the newsvendor problem because it mimics the issue faced by a newsvendor who needs to decide how many copies of a newspaper need to be purchased at the beginning of the day given that for every copy that sells there is a profit and every copy that remains at the end of the day there is a loss. Note that in the following discussions we assume cost minimization to be the objective (by assigning a penalty cost for lost demand); however, these problems can also be studied as profit-maximization problems.

To begin we will assume the simplest model where there are no salvage and setup costs ($s = 0$, $K = 0$) as well as the lead time for delivery is zero. Let x be

Table 1
Papers on the base case: single-product single-stage stationary and independent demand

Year	Reference	Year	Reference
1951	Arrow, Harris and Marschak	1971	Morton
1958	Karlin	1972	Wijngaard
1958	Karlin and Scarf	1979	Ehrhardt
1960	Scarf	1979	Nahmias
1963	Iglehart	1986a,b	Federgruen and Zipkin
1965b	Veinott	1989	De Kok
1965	Veinott and Wagner	1991	Zheng
1966b	Veinott	1991	Zheng and Federgruen
1970	Kaplan	1993	Tayur
1971	Porteus	1996	Van Donselaar, De Kok and Rutten

the starting inventory at the facility, then the objective is to minimize the expected costs during the period by producing/ordering enough to bring the inventory level to $y \geq x$ after ordering. We further assume that $-h < c < \pi$. Then the single-period expected cost $L(y, x)$ given x is

$$L(y, x) = c(y - x) + \pi \int_y^\infty (\xi - y) \, dF(\xi) + h \int_0^y (y - \xi) \, dF(\xi) \quad (3.1)$$

The first term represents the ordering cost while the second and third terms represent the penalty and holding costs, respectively.

Let $G(y)$ be defined as follows

$$G(y) = cy + \pi \int_y^\infty (\xi - y) \, dF(\xi) + h \int_0^y (y - \xi) \, dF(\xi)$$

$$= L(y, x) + cx \quad (3.2)$$

It is clear that $G(y)$ is convex in y. As a result, $L(y, x)$ is convex in y. The optimal value of the cost L^* is obtained by setting the first derivative equal to 0 which gives

$$y^* = F^{-1}\left(\frac{\pi - c}{\pi + h}\right) = F^{-1}\left(\frac{\pi - c}{(\pi - c) + (c + h)}\right) \quad (3.3)$$

The ratio $(\pi - c)/(\pi + h)$ is called the *critical fractile* and the value of y^* is called the *base-stock* level. Since it is optimal to order or produce up to an inventory level of y^* at the beginning of each period, such a policy is also called an *order-up-to* policy. An order-up-to policy orders up to y^* if $x < y^*$ and does not order anything if $x \geq y^*$. An important thing to notice about this policy is that the order-up-to level y^* is independent of the initial inventory.

Another interesting point about (3.3) is that the critical fractile can be written as a fraction of *underage costs* and *overage + underage costs* where overage cost is the cost of having one additional unit than demanded $(c+h)$ and the underage cost is the cost having one less unit than demanded $(\pi-c)$. The above value of y^* can also be computed by equating marginal benefits to marginal costs as given in (3.4).

$$(\pi - c)(1 - F(y)) = (c + h)F(y) \qquad (3.4)$$

The difference between y^* and μ (the average demand) is referred to as the *buffer stock*. The earliest reference of this term (as indicated by Arrow, Karlin & Scarf, 1958) appears in Edgeworth (1888) within a banking context where the probability of running out was prespecified. Arrow, Harris and Marschak (1951) provide the first reference of this model with underage and overage costs.

If the per unit salvage cost s is included in the model, then the overage cost is equal to $c+h+s$ and the critical fractile is adjusted accordingly. If the facility has a production capacity of C units in the period, then the optimal policy is as follows.

$$y^* = \begin{cases} F^{-1}\left(\dfrac{\pi-c}{\pi+h}\right) & \text{if } x \leq F^{-1}\left(\dfrac{\pi-c}{\pi+h}\right) \leq x+C \\ x+C & \text{if } x+C \leq F^{-1}\left(\dfrac{\pi-c}{\pi+h}\right) \\ x & \text{if } x \geq F^{-1}\left(\dfrac{\pi-c}{\pi+h}\right) \end{cases} \qquad (3.5)$$

The policy given above is also termed as a *modified base-stock* policy since the policy tries to get as close to the base-stock level when there is capacity constraint.

3.1.1 Setup costs

In the above model one could include a setup cost K each time an order is placed (or production initiated). The corresponding cost function is given by (for $y \geq x$)

$$L(y,x) = c(y-x) + K\delta(y-x) + \pi \int_y^\infty (\xi-y)\,dF(\xi) + h\int_0^y (y-\xi)\,dF(\xi)$$

$$= G(y) + K\delta(y-x) - cx \qquad (3.6)$$

We know that given that an order is going to be placed ($y > x$), the function $L(y,x)$ is convex in y since $G(y)$ is convex in y and the value is minimized at $y^* = F^{-1}((\pi-c)/(\pi+h))$. The associated cost is equal to $G(y^*)+K-cx$.

Clearly, if $G(x) - cx \leq G(y^*) + K - cx$ or $G(x) \leq G(y^*) + K$ then it is not optimal to place an order. However, if $G(x) > G(y^*) + K$ then it is optimal to order to reach y^*. Thus, with the introduction of the setup cost the optimal inventory policy has two parameters often referred to as (s, S) policy where if $x < s$ then the inventory level is brought to S and if $x \geq s$ then no orders are placed. In the above policy $S = y^*$ and $s \leq S$ is chosen in such a way so that $G(s) = G(S) + K$.

3.1.2 General cost assumptions

Thus far we have assumed that ordering, holding and shortage costs are linear. One can also have other types of functions for these costs which are nonlinear and maybe concave or convex. Let us denote the holding, penalty and ordering cost functions as $\hat{h}$, $\hat{\pi}$ and $\hat{c}$, respectively (for now we will neglect the salvage costs). Clearly, if $\hat{h}$ and $\hat{\pi}$ are convex then $G(y)$ as defined in (3.2) is convex in y so one can find the optimal y^*. The optimal y^* will not be given by a simple critical fractile anymore. As indicated in Porteus (1991), the case of quadratic holding and stock-out costs (defined on the positive values of the argument) leads to an interesting result in that the optimal inventory level is equal to mean μ when overage and underage costs are identical under other standard assumptions. Note that this is different from the linear case where we stock the median value when the underage and overage costs are equal. If the holding and stock-out costs are nonlinear and nonconvex even then under certain conditions the optimal y^* can be found. Those conditions are that if $G(y)$ can be written as $G(y) = A + \int_0^\infty g(y - \xi) f(\xi) \, d(\xi)$ where A is constant and g is quasi convex and $f(\xi)$ is a polya frequency function (P.F.F) of order 3 [Karlin, 1958].

The model and results get somewhat changed if we have convex ordering/production costs but linear penalty and holding costs. Karlin (1958) shows that if $\hat{c}(x)$ is convex in x and $\lim_{x \to 0} \hat{c}(x) = 0$ then $y^*(x)$ is increasing in x but $y^*(x) - x$ (the amount ordered) is decreasing in x. This is called a *generalized base-stock* policy. The special case of piecewise linear convex costs leads to a generalized policy with finite number of distinct base-stock levels. For example, the case with two piecewise linear function often results in real life when there are two suppliers and the less-expensive supplier may have limited capacity.

If the production costs are concave, Karlin (1958) presented conditions under which a *generalized* (s, S) policy is optimal. This policy is represented by two parameters s and S as well as the optimal inventory level $y(x)$, where no orders are placed if $x \geq s$ and y is such that $y(u) \geq y(x) \geq S \geq s$ when $u \leq x \leq s$. Basically, this policy tends to place large orders when the inventory level is lower, thereby benefiting from economies of scale associated with the concave costs. Porteus (1971) considered the special case where the costs are concave and piecewise linear (consider a case where there is a setup cost) and showed that there exist finite breakpoints $s_1 \leq \cdots \leq s_n \leq S_n \leq \cdots \leq S_1$

such that the ordering policy is to order up to S_1 if $x < s_1$; order up to S_2 if $s_1 \leq x \leq s_2$ and so on, and do not order if $x \geq s_n$. Porteus (1991) provides detailed examples and explanations for the piecewise linear convex and concave costs.

3.2 Multiple-period dynamic model

The single-period models studied in the previous section are applicable only in very limited settings such as products that sell in one season, perishable products and products at the end of their life cycle. Most of the other settings require analysis of the inventory system over multiple periods. There are two types of models that are studied in this context – finite horizon and time horizon. In the time horizon models, there are a fixed number of periods and the objective is to minimize the discounted total expected costs over the horizon. In the infinite-horizon models, the objective could be either to minimize the discounted expected costs or the average expected costs over the infinite horizon. Another source of differentiation is related to whether unsatisfied demand is lost or backlogged.

The demand in every period is assumed to be independent and identically distributed. In addition, the costs are assumed linear and stationary in order to obtain nice structures on the optimal policy. The sequence of events in every period is similar to the single-period model in that orders are placed at the beginning of the period; demand is observed during the period; maximum demand is satisfied at the end of the period and resulting costs are incurred. The assumption is that the lead time is negligible so that the orders placed are available at the end of the period. The finite-horizon problem with backlogging can be formulated as follows. Let $J_n(x)$ be the optimal cost to go given that there are n periods remaining in the horizon and x is the on-hand inventory.

$$J_n(x) = \min_{y \geq x}\left\{c(-x) + G_n(y) + \alpha \int_0^\infty J_{n-1}(y - \xi)\,dF(\xi)\right\} \quad (3.7)$$

Note that in the above formulation, we assume that the terminal costs are zero but one could add salvage costs at the end of the horizon. Since the single-period cost function $G(y)$ is convex in y, one can use recursion to show that the objective is convex in y and hence there exists an optimal base-stock level y^* in each period. In the lost-sales case the cost-to-go recursion is given as follows and a similar analysis can be done.

$$J_n(x) = \min_{y \geq x}\left\{c(-x) + G_n(y) + \alpha\left(\int_0^y J_{n-1}(y - \xi)\,dF(\xi) + \int_y^\infty J_{n-1}(0)\,dF(\xi)\right)\right\}$$
$$(3.8)$$

A more compact representation of the above problem can be presented by creating a function $v(y, \xi)$ given by

$$v(y, \xi) = \begin{cases} a(y - \xi) & \xi \leq y \\ b(y - \xi) & \xi \geq y \end{cases} \tag{3.9}$$

and utilizing that in the cost-to-go function.

$$J_n(x) = \min_{y \geq x} \left\{ c(-x) + G_n(y) + \alpha \left(\int_0^\infty J_{n-1}(v(y, \xi)) \, dF(\xi) \right) \right\} \tag{3.10}$$

Clearly, if $a = 1$, $b = 1$ it represents backlogging; if $a = 1$, $b = 0$ it represents lost sales and if $a = 0$, $b = 0$ we have an example of perishable goods. Now define $M(y) = G(y) - \int_0^\infty \alpha c v(y, \xi) \, dF(\xi)$. Let y_n be the optimal base stock for period n. Then

$$J_n(x) = c(-x) + G_n(y_n) + \alpha \left(\int_0^\infty J_{n-1}(v(y_n, \xi)) \, dF(\xi) \right)$$

$$= -cx + \sum_{i=1}^n \alpha^{n-i} M_n(y_n) \tag{3.11}$$

Veinott (1965b) showed that the above transformation enables one to simplify the problem associated with finding the optimal base-stock inventory levels for each period because they are the y_n values that minimize $M_n(y_n)$ which depends only on the parameters and expected cost when n periods are remaining. This solution is also called the *myopic* solution since we need to solve only for the current period. Veinott (1965b) presented general conditions under which the myopic solution is optimal for the finite- and infinite- ($\alpha < 1$)-horizon discounted cost problem. The optimal myopic solution depends on whether demand is backlogged or not. In the case of backlogging the optimal inventory is given by

$$y^* = F^{-1}\left(\frac{\pi - c + \alpha c}{\pi + h}\right) \tag{3.12}$$

3.2.1 Lead time

In a dynamic inventory model, the concept of lead time becomes important. Karlin and Scarf (1958) showed that if the lead time from the supplier is fixed L and all demands are backlogged then that problem could be converted into an equivalent single-period problem with some adjustments. The assumption of complete backlogging is critical here because the approach relies on keeping track of the system stock (rather than just the on-hand inventory). The basic idea is that one keeps track of inventory on-hand plus all orders that have been placed (but not received yet) minus any backlogged demand. The effect of orders placed in period t is felt in period $L+t$, so the approach is to consider the total demand in the next $L+1$ periods and bring the system stock to that level. Note the critical fractile still remains the same as given in (3.12), what changes is the cumulative probability density function F (which is now a convolution of $L+1$ demand distributions). This is easy to compute for stationary and independent normal distributions since the resulting distribution is also normal with $\hat{\mu} = (L+1)\mu$ and $\hat{\sigma} = \sqrt{L+1}\,\sigma$. The reason that backlogging assumption is important because the state space for the dynamic program can be collapsed into a single state (that represents the system stock) rather than having a vector of $L+1$ variables which represent how much delivery is expected in the next L periods in addition to the current inventory level.

The case of stochastic lead time has been studied by several researchers. The problem arises in this case because it is difficult to compute the system stock as in the deterministic demand case because knowing when an order was placed does not help in terms of identifying when the order will arrive. Also it is statistically possible that orders that were placed later are delivered before earlier orders. Kaplan (1970) was the first one to show that under two simple assumptions, it is possible to replicate results corresponding to the deterministic case. The two assumptions that are required are: (1) later orders are not delivered before earlier orders; (2) the lead-time distribution does not change due to outstanding backorders. Nahmias (1979) showed that the above assumptions are equivalent to a delivery process generated by a sequence $\{A_t\}$ of independent and identically distributed random variables such that if $A_t = k$ then all the orders placed k or more periods before would be delivered in the current period. This transformation allows one to mimic the optimality of the myopic policy. Further, for the average-cost analysis, it suffices to assume identically distributed lead times and independence is not always necessary.

The lost-sales model even with deterministic lead times is an open problem in terms of determining the optimal policy. Morton (1971) presents bounds for the optimal ordering policy as well as the discounted cost function for the stationary problem. The heuristics presented are myopic (or near myopic) in nature and are not necessarily base-stock policies. Through a limited computational study, the author provides evidence that such heuristics may be very close to optimal. More recently, van Donselaar, de Kok and Rutten (1996) compare the performance of a base-stock policy with another myopic heuristic and show empirically that their dynamic myopic heuristic outperforms the stationary base-stock policy in a significant manner.

3.2.2 Setup costs

Scarf (1960) showed that under general conditions on the cost function (such as K-convexity) of the one-period expected cost, the n-period dynamic inventory problem has an optimal (s, S) policy. Iglehart (1963) considers the discounted infinite-horizon problem and shows the optimality of the (s, S) policy by giving bounds on sequences $\{s_n\}$ and $\{S_n\}$ and establishing their limiting behavior. He also extends this result to the case with fixed lead time. Veinott (1966b) showed the above results for a different set of conditions such as the negative of the one-period expected costs are unimodal and that the absolute minima of the one-period costs are rising over time. Zheng (1991) presents a simple proof for the optimality of the (s, S) policy for the discounted and average infinite-horizon problems by constructing an (s, S) or a variant solution to the optimality equation. Although the (s, S) parameters are computable for the single-period problem, they are more difficult to compute for the dynamic case. Veinott and Wagner (1965) give an optimal algorithm for computing these parameters. Ehrhardt (1979) presented a heuristic for computing these parameters using a power approximation. This approximation has been shown to be very accurate under a wide variety of settings. However, these approximations suffer when the variance of the demand is very small (if setup cost is also large). Zheng and Federgruen (1991) provide an efficient algorithm to compute these values.

3.2.3 Capacity constraint

In most realistic environments, it is not possible for the firm to produce (order) as much as required because of production or storage capacity. This poses fundamental problems in the analysis. Wijngaard (1972) introduces the notion of a *modified base-stock* policy where a firm tries to produce as much as possible (if unable to reach the base stock). Federgruen and Zipkin (1986a,b), Wijngaard (1972) address the optimality and (non)optimality of this policy for finite- and infinite-horizon problems under restrictive assumptions. In a series of two papers, Federgruen and Zipkin (1986a,b) show that a *modified base-stock* policy is optimal for the discounted multiple period and discounted and average cases in the infinite horizon under general conditions such as when the expected single-period cost is convex and a discrete-demand distribution. Although the optimality of such a policy was established, it was very hard to compute the actual parameters efficiently. de Kok (1989) notes that the modified base-stock policy can be computed using the fact that the inventory position at the start of a period equals $S-X$, where X is the waiting time in a D/G/1 queue. The author also provides a simple algorithm to compute the first two moments of X and a heuristic for computing S. Tayur (1993) introduced the parallel between the dam model and the inventory dynamics equation and used the notion of a *shortfall* – representing the cumulative amount of falling short of the optimal base-stock level due to capacity constraint. This allows one to construct a sequence of uncapacitated infinite-horizon problems that converge to the capacitated solution under consideration. Then the optimal base-stock level can be easily computed. For capacitated inventory systems, infinitesimal perturbation analysis has become an efficient approach to compute (through stimulation) the optimal parameter values.

4 Alternative demand assumptions

In all the discussions thus far we assumed that the demand distributions in the different periods were identical and known (hence a stationary distribution). However, in several real environments the demand distributions may be different in different periods. In this section, we highlight the key developments in those areas.

4.1 Nonstationary demand

In the nonstationary demand case, the demand is assumed to be from nonidentical distributions in each period. Karlin (1960a) studied the nonstationary inventory problem with zero lead time and showed that the time-specific base-stock policy is optimal. That is, based on the distributions of demand there is an optimal base-stock level in each period. Further, he showed that if the demand distributions are stochastically increasing in different periods, then the optimal base-stock levels are also increasing. Veinott (1965b) showed that if the optimal base-stock levels are such that in each period one needed to place an order to get to the base-stock level then a myopic policy is still optimal for the nonstationary case. Thus, a *myopic* policy is optimal for the case when $y_t^* \leq y_{t+1}^*$ for all t. Further, the stationary distribution case is a special case where the identical base-stock levels across the different periods imply that one would necessarily have to place an order given that the initial inventory is less than the base-stock level, leading to the optimality of the myopic solution. In general, when the myopic policy is not optimal it is difficult to obtain the exact optimal parameters. Morton (1978) provided a sequence of upper and lower bounds for the optimal base-stock levels such that the nth bound requires the knowledge about the first n demand distributions giving planning horizon results for the infinite-horizon case. Lovejoy (1992) considers the nonstationary inventory problem and provides conditions and stopping rules for utilizing myopic policies under very general settings. Morton and Pentico (1995) provide myopic solutions that may be ε close to the optimal solution and hence, term it as near-myopic solution and test their efficacy through a detailed computational study. Gavirneni and Tayur (2001) provide a quick method to compute the base-stock levels using Direct Derivative Estimation (DDE). Bollapragada and Morton (1999) study the nonstationary inventory problem with setup costs and provide a very effective myopic heuristic to

Table 2
Papers on single-product single-stage nonstationary and dependent demands

Year	Reference	Year	Reference
1959	Scarf	1998	Graves, Kletter and Hetzel
1960a	Scarf	1998	Kapuscinski and Tayur
1960a,b	Karlin	1999	Bollapragada and Morton
1964	Iglehart	1999	Lariviere and Porteus
1965b	Veinott	1999	Gavirneni and Tayur
1972	Hausman and Peterson	1999	Gavirneni, Kapuscinski and Tayur
1975	Johnson and Thompson	1999	Cheng and Sethi
1978	Morton	2000	Lee, So and Tang
1985	Azoury	2000	Scheller-Wolf and Tayur
1989	Zipkin	2001	Gavirneni and Tayur
1992	Lovejoy	2001	Gallego and Ozer
1994	Heath and Jackson	2001	Kaminsky and Swaminathan
1995	Morton and Pentico	2001	Toktay and Wein
1997	Sethi and Cheng	2001	Huang, Scheller-Wolf and Tayur
1997	Aviv and Federgruen	2002	Aviv

the problem by approximating the future problem in each period by a stationary problem and obtaining the solution for that problem (Table 2).

4.1.1 Cyclic demand schedule

Many firms encounter a demand pattern where the demand follows a cyclic pattern in that the cycles repeat themselves after a while. For example, one could look at the demand during the four quarters in traditional industries. Karlin (1960a,b) discusses the optimality of the *periodic base-stock policy* which implies that there are different base-stock levels for each of the periods, under stationary costs and cyclic nonstationary demand for the discounted infinite-horizon case, and also provides an optimal algorithm. Zipkin (1989) extends the above results to the infinite-horizon average-cost case with both nonstationary cyclic demand and nonstationary costs. More recently, Kapuscinski and Tayur (1998) consider the capacitated version of the problem and prove the optimality of the *modified periodic base-stock policy* for dynamic multiperiod and infinite-horizon (discounted and average-cost) cases. They also provide an algorithm to find the optimal base-stock levels using infinitesimal perturbation analysis. Independently, Aviv and Federgruen (1997) also proved the optimality of the *modified periodic base-stock policy*. Scheller-Wolf and Tayur (2000) extend the above to include minimum-order quantities and lead times.

4.2 Bayesian demand updates

In many cases, the demand distribution may not be completely known but as more information is obtained (with demand realization) the estimate of the demand distribution can be refined. Scarf (1959, 1960a) introduces the Bayesian demand updates where the distribution of the demand is supposed to depend on one or more parameters and those parameters are refined using Bayesian updates as more information related to the demand process is obtained. In particular, he assumes that the demand distribution is gamma with an unknown scale parameter and shows the optimality of the base-stock policy. Karlin (1960b) and Iglehart (1964) extended the analysis to the case where the demand distribution is of the range type between 0 and ω where ω is unknown. The fundamental assumption in this approach is that the prior for the parameter and the demand distribution are from the same conjugate family so that the posterior distribution has an easily workable form. Azoury (1985) extends these results to other types of distributions such as Weibull and Normal (with known and unknown mean and variance). Further, she explains how the optimal base-stock levels can be determined easily. Basically, a single-normalized base-stock level is computed in advance and then the optimal base-stock level is obtained by scaling this value. The scale factor depends on a function of the sufficient statistic of the unknown parameter that is generated based on past demand. Recently, Lariviere and Porteus (1999) extend the above observations to other environments and provide conditions under which a firm would invest in additional inventory to learn more about the demand as well as cases where despite poor sales, the product is stocked in order to obtain better information. Huang, Scheller-Wolf and Tayur (2001) use a Hidden Markov Model (HMM) to update the state of the unknown demand of a new product.

4.3 evolution

Another important reality that needs to be incorporated in inventory models is the forecasting process utilized. Hausman and Peterson (1972) develop a multiperiod model with terminal demand where the forecast errors get refined in a lognormal process. In a capacitated environment they show that optimal policy is of threshold-type and present heuristics to solve the problem. More recently, Kaminsky and Swaminathan (2001) consider a forecast generation process which depends on forecast bands (and the demand is expected to be uniformly spread in the interval) that get refined over time. In a terminal demand-capacitated setting, they show the optimality of the threshold-type policy and provide very efficient algorithm for cases with and without holding costs. Researchers have also studied the Martingale model for forecast evolution along with production-inventory decisions. Heath and Jackson (1994) and Graves, Kletter and Hetzel (1998) independently introduce these models and develop heuristical methods to solve the problem. Recently, Toktay and Wein (2001) study this problem and use heavy traffic approximations to prove the optimality of the base-stock policy under those assumptions. More recently, Aviv (2002) presents a supply chain model where different members observe subsets of the underlying demand information, and adapt their forecasting and replenishment policies accordingly. For each member, he identifies an associated demand evolution model, for which he proposes an adaptive inventory replenishment policy that utilizes the Kalman Filter technique. He provides a simple methodology for assessing the benefits of various types of information-sharing agreements between members of the supply chain.

4.4 Demand dependencies

Demand across the different periods could be related to each other in some environments. Veinott (1965a,b) through his work on nonstationary demands provides general conditions under which even a dependent demand process may have *myopic* solutions. Johnson and Thompson (1975) utilize those results to show that even when the demand is described by a ARMA (autoregressive moving average) model with an additive shock, a myopic policy remains optimal under mild conditions. Sobel (1988) provided general conditions under which a myopic solution remains optimal.

Another way to represent the dependencies is to assume that the demand gets generated from a Markov process, so that the state in the current period affects the demand in the next period. Karlin and Fabens (1960) introduced a Markovian demand model and postulated that a state-dependent (s, S) policy would be optimal. However, they restrict themselves to stationary (s, S) policy due to complexity. Sethi and Cheng (1997) prove the optimality of the state-dependent (s, S) policy for Markovian demand for both finite- and infinite-horizon problems. They can also extend the model to capture periods with no orders as well as capacity and service constraint. Gavirneni and Tayur (1999) consider a modified version of the Markovian process which may be generated due to a fixed ordering schedule ('Target Reverting') at the customer end. They prove the optimality of a modified base-stock policy and provide computational results. Gallego and Ozer (2001) study a situation where customers may place orders in advance (more common in a make to order environment) which provides the firm with advance demand information. They show that state-dependent (s, S) and base-stock policies are optimal for stochastic inventory systems with and without fixed costs. The state of the system reflects the knowledge of advance demand information. They also determine conditions under which advance demand information has no operational value.

Another reason for demand dependencies is that the firm may be receiving orders from another firm that may be following some optimal policy such as (s, S). Gavirneni, Kapuscinski and Tayur (1999) study the value of this additional demand information in a capacitated multiperiod setting. Lee, So and Tang (2000) study the value of information in a two-level supply chain with nonstationary end demand and show that the value of information could be very high, particularly in cases where the demand may be correlated over time. Demand is also affected by the pricing decisions used by firms. Cheng and Sethi (1999) consider a general model where the customer demand is generated by a Markov process whose state is dependent on the promotion decisions. They assume a fixed cost for promotion and that the demand moves to a stochastically higher state the next period. The firm tries to find the optimal promotion schedule in terms of which periods to promote and what inventory levels to stock. For a finite-horizon problem they show that there exists a threshold level P such that if the initial inventory is greater than P then it is optimal to promote, and also show that a base-stock policy is optimal for the linear-cost case.

5 Generalizations

There have been several generalizations of the single-firm single-product form that has been discussed so far. In this section, we will briefly explore these generalizations.

5.1 Multiechelon

The natural extension of a single-firm model in a supply chain setting relates to considering multiple echelons under a single firm. Clark and Scarf (1960) study a serial system with each facility (representing an echelon) supplying the downstream facility within a deterministic nonzero lead time. The echelon stock is defined as the stock at that facility plus the stock at all the facilities downstream. The holding and stock-out costs are assessed independently, taking into account the echelon stock of each facility. Under the above assumptions, they show that the problem can be decomposed into independent problems each for one facility and that the base-stock policy still remains optimal in that case where each facility tries to bring the inventory as close as possible to the optimal echelon base-stock level. They also provide a mechanism by which the optimal base stocks can be computed by sequentially going from the last facility moving backwards (Table 3).

Federgruen and Zipkin (1984b) provide a simple method to streamline the computations in the infinite-horizon case with normal demands. Chen and Song (2001) consider a multistage serial system with Markov-modulated demand in that the demand distribution in each period is determined by the current state of an exogenous Markov chain. They show the optimality of a state-dependent echelon base-stock policy for the long-run average-costs case. They also provide an algorithm for determining the optimal base-stock levels and extend their results to serial systems in which there is a fixed ordering cost at the last stage and to assembly systems with linear ordering costs. In a more recent work, Muharremoglu and Tsitsiklis (2001) employ a novel approach based on decomposition of the problem into a series of single-item single-customer problems that enables them to provide a simpler

Table 3
Papers on generalizations of the base case

Year	Reference	Year	Reference
Multiple echelons			
1960	Clark and Scarf	1962	Clark and Scarf
1981	Eppen and Schrage	1984b	Federgruen and Zipkin
1985	Roundy	1985	Schmidt and Nahmias
1989	Rosling	1994, 1995	Glasserman and Tayur
1994, 1998	Chen and Zheng	1999	Chen
2000	Parker and Kapuscinski	2001	Chen and Song
2001	Muharremoglu and Tsitsiklis		
Multiple products			
1963	Hadley and Whitin	1969	Ignall and Veinott
1969	Ignall	1981	Silver
1984	Federgruen, Gronevelt and Tijms	1985a,b	Karmarkar, Kekre and Kekre
1987	Karmarkar, Kekre and Kekre	1988	Atkins and Iyogun
1990	Gallego	1993	Karmarkar
1993	Lee and Billington	1996	Federgruen and Catalan
1996	Lambrecht et al.	1997	Lee and Tang
1998	Swaminathan and Tayur	1998	Eynan and Kropp
1998	Anupindi and Tayur	1999	Bollapragada and Rao
1999	Bassok, Anupindi and Akella	2001	Rajagopalan and Swaminathan
2001	Bispo and Tayur	2002	Swaminathan and Lee
2002	Rao, Swaminathan and Zhang	2002	Van Mieghem and Rudi
Multiple suppliers			
1964	Fukuda	1966a	Veinott
1969	Wright	1993	Anupindi and Akella
1999	Swaminathan and Shanthikumar	2000	Scheller-Wolf and Tayur
Process randomness			
1958	Karlin	1990	Henig and Gerchak
1991	Bassok and Akella	1994	Ciarallo, Akella and Morton
1995	Lee and Yano		

proof for the optimality of echelon base-stock policies. This approach enables them to extend their results to several variants of the problem including stochastic lead time and Markovian demand processes.

One variant of the serial system is the assembly system where more than one facility may be involved in an upstream echelon to provide parts for the downstream assembly operation. Schmidt and Nahmias (1985) study the case with two echelons where the upstream echelon has two suppliers with different deterministic lead times and a fixed assembly lead time. The optimal policy has an interesting structure in that the assembly level has a base-stock policy while the policy at the upstream echelon is such that it tries to balance the echelon stock of the two components taking into account the difference in lead times of the two suppliers. Rosling (1989) showed that under certain initial conditions, an assembly system can be reduced to a serial system with modified lead times so that results due to Clark and Scarf (1960) may apply to the modified system. Another variant of the serial system is the distribution system where one facility at an upstream level supplies to multiple facilities downstream. The results of the serial system do not carry forward easily to the distribution network. Eppen and Schrage (1981) analyze a one warehouse and multiple retailer network and explore the trade-off involved therein.

The addition of capacity restrictions on the above generalization leads to several complications. Firstly, the simple base-stock policies may no longer be optimal, and secondly, even under restricted set of base-stock policies the computation of these parameters is challenging. Glasserman and Tayur (1994, 1995) develop a solution for finding the optimal base-stock levels at the different echelons under a modified base-stock policy and utilize a simulation-based optimization procedure using infinitesimal perturbation analysis to develop an efficient solution methodology for finding the optimal parameter values. They also extend their approach to the assembly and distribution networks under certain conditions. Parker and Kapuscinski (2000) demonstrate that a modified echelon base-stock policy is optimal in a two-stage system for a capacity-dominating condition. They show that this holds for both stationary and nonstationary stochastic customer demand for finite and infinite horizons under discounted and average-cost criteria. There have been numerous attempts to develop a better grasp of the case with setup costs at both stages even for a serial system starting with Clark and Scarf (1962). Recently, Chen (1999) utilized the nested policy ideas developed by Roundy (1985) for deterministic systems, in a two-stage serial system with Poisson demand and setup costs to develop 94% optimal policies for the problem.

5.2 Multiple products

Another dimension of extension of these models is along the number of products. If all the products have independent demand and there are no capacity or production restrictions then naturally the problem can be decoupled into independent single-product problems. However, in reality, a firm produces multiple products (which may be similar in functionality) and may have a common capacity to utilize for those products. Hadley and Whitin (1963) consider the capacitated newsvendor problem when there is a common capacity constraint of the form

$$\sum_i a_i y_i \leq b$$

They solve the problem by relaxing the constraint and obtain an explicit expression for the optimal quantities in terms of the Lagrangian multiplier λ

as follows. If the optimal base-stock levels are not feasible then

$$y_i^*(\lambda) = F^{-1}\left\{\frac{\pi_i - c_i - a_i\lambda}{\pi_i + h_i}\right\} \quad (5.1)$$

and λ is chosen so that

$$\sum_i a_i y_i^*(\lambda) \leq b.$$

Recently, Bispo and Tayur (2001) study base-stock policies under various scenarios of capacity sharing across products in a single stage, serial and reentrant systems.

The notion of similarity among the different products and the fact that they could be substituted for each others demand was explored by Ignall and Veinott (1969). They showed that in fact the base-stock policies are optimal for the problem with nested downward substitution (where product 1 can substitute for product $2, 3, \ldots, n$; product 2 can substitute for product $3, 4, \ldots, n$ and so on) in a multiple-period infinite-horizon setting. Bassok, Anupindi and Akella (1999) provide an alternative proof for the same result. The problem related to downward substitution as well as setup costs is extremely hard to obtain theoretical insights on. Rao, Swaminathan and Zhang (2002) provide a highly efficient algorithm for finding the optimal production/substitution strategy for that problem using a combination of dynamic programming and simulation-based optimization.

Another important concept with multiple products relates to postponing the point of differentiation of the products in order to reduce inventory as a result of *risk pooling* – storing inventory of semifinished products reduces the risk associated with stocking that inventory. Lee and Billington (1993) and Lee and Tang (1997) study postponement in the context of distribution through the channel. Swaminathan and Tayur (1998) study the postponement issue within the context of a capacitated final assembly facility and term the semifinished products as *vanilla boxes*. For more details on research conducted on postponement strategies, see Swaminathan and Lee (2003).

Another complexity with multiple products is studied in the Joint Replenishment Planning (JRP) context, where there is a major setup cost at each order (across products), and a minor setup cost that may be product dependent. The stochastic version is particularly important as it is a very common problem in practice. A reasonable ordering policy that has been studied extensively are the *can-order* or (s, c, S) policies. In such a policy, when any item i inventory drops below its reorder point s_i, a reorder is scheduled and all other item j whose inventory is below their can-order limit c_j are also included in the order. Ignall (1969) showed that *can-order* policies are not optimal in general. However, Silver (1981) and Federgruen, Groenevelt and

Tijms (1984) have empirically shown that can-order policies perform well. Atkins and Iyogun (1988) provide a lower bound for the joint replenishment problem above and show that can-order policies are not very efficient when the joint setup cost is high. They show that a heuristic based on a periodic ordering can outperform *can-order* policies in a significant manner. Eynan and Kropp (1998) propose yet another periodic heuristic and show that is close to the optimal solution through a computational study.

Another important problem is the stochastic economic lot-sizing problem (ELSP), where several items need to be produced in a common facility with limited capacity, under significant uncertainty-regarding demands, production times or combinations thereof. Gallego (1990) considers the problem of scheduling the production of several items in a single facility that can produce only one item at a time. He assumes that demands are random with constant-expected rates, and allows backorders and charge holding and backlogging costs at linear time-weighted rates. Items are produced at continuous constant rate, and setup times and setup costs are item-dependent constants. He proposes a real time-scheduling system that utilizes the expected demand to create the initial schedules and adjusts them for the randomness. Federgruen and Catalan (1996) propose cyclical base-stock policies for the problem. Under this scheme, when the facility is assigned to a given item, production continues until either a specific target-inventory level is reached or a specific production batch has been completed; different items are produced in a given sequence or rotation cycle, possibly with idle times inserted between the completion of an item's production batch and the setup for the next item. Optimal policies within this class which minimize holding, backlogging and setup costs are effectively determined and evaluated. Bollapragada and Rao (1999) as well as Anupindi and Tayur (1998) are more recent contributions to ELSP and cyclic schedule problems.

Earlier research related to manufacturing lead times, order release and capacity releases is summarized in Karmarkar (1993). Among that Karmarkar, Kekre and Kekre (1985a,b, 1987) in a series of papers studied lot sizing in multi-item multimachine job shops and cellular environments. Lambrecht, Chen and Vandaele (1996) introduce the notion of safety lead times in queuing models of a make-to-order manufacturing environment. They show that there is a convex relationship of expected waiting time, variance of the waiting time and the quoted lead time as a function of the lot size, and a concave relationship of the service level as a function of the lot size. This allows them to accurately quantify the safety time and to compute the associate service levels. Although sizing and capacity expansion are very closely related they have not been studied extensively in integrated models. In a recent work, Rajagopalan and Swaminathan (2001) study the capacity expansion and lot sizing in a multiproduct environment with deterministic known demand and present effective heuristics and bounds for the problem. More recently, Van Mieghem and Rudi (2002) develop a framework to study multiperiod multiproduct problems of stochastic capacity investment and inventory management. The optimal capacity and inventory decisions balance overages with underage costs. The optimal balancing conditions are interpreted as specifying multiple 'critical fractiles' of the multivariate demand distribution; they also suggest appropriate measures for and trade-offs between product service levels. They establish dynamic optimality of inventory and capacity policies for the lost-sales case.

5.3 Multiple suppliers

Several firms have more than one supplier for any particular product in order to hedge against the uncertainty in the delivery process as well as to avoid being held captive by the supplier. Fukuda (1964), Veinott (1966a) and Wright (1969) study the case where rush orders from a reliable supplier could be obtained one period earlier (at an additional cost) than the normal lead time in an emergency situation. They showed that there are two base-stock levels – an emergency base-stock level and a normal base-stock level. If the inventory level is lower than the emergency stock level then orders are placed with the more reliable supplier to get to the emergency stock level. Then additional orders are placed with the normal supplier to reach the normal base-stock level. Anupindi and Akella (1993) study a different version of the problem where the lead times and their unit costs are different. They show that the optimal policy has two-parameter base-stock levels. If the inventory is higher than the larger base-stock level then no orders are placed. If it is in between the two base-stock levels then orders are placed only with the less-expensive supplier and else orders are placed with both. In particular, they note that orders are never placed with the more-expensive supplier alone. Scheller-Wolf and Tayur (2000) extend these results into a more general model. Swaminathan and Shanthikumar (1999) showed that the above structure is driven by the continuity assumption in demand and need not hold in general for discrete-demand distributions. There are several other papers which deal with supply contracts, see Anupindi and Bassok (1998), Lariviere (1998) and Corbett and Tang (1998), Tsay et al. (1998) for reviews on this topic.

5.4 Randomness in process

Another important generalization of the traditional inventory models relates to the randomness in production process or also called random yield. Several industries particularly semiconductors face a critical problem related to managing the yield of the process and simultaneously determining the inventory levels. Karlin (1958) explores the notion of randomness in supply by assuming a probability distribution for the receipt of harvest from the producer. The decision is whether to order given a set of probability of possible harvests. Bassok and Akella (1991) study the joint production and ordering decisions in an environment where demand is random and amount received is a random fraction of that ordered. Henig and Gerchak (1990) consider the case where the amount received is a fraction of amount ordered and characterize the optimal policy. For the multiperiod case they show the convexity of the cost function and show the existence of the optimal order point. Ciarallo, Akella and Morton (1994) consider a problem where the total capacity itself is random and show that there are optimal order points in that case as well although the convexity of the cost function is lost in multiperiod and infinite-horizon problems. They also present the notion of extended myopic policies in the infinite-horizon case. Lee and Yano (1995) present a comprehensive literature review on random yield research.

5.5 Approximations

Given the difficulty of solving complex inventory problems exactly, many approximate methods for solving these problems have been proposed. One such approximation is the large-deviation approximation to study capacitated systems, both in discrete-time and continuous-time models. The common characteristics in such systems is that inventory is held in part to compensate for the capacity restriction. The basic idea in this approximation is that if the tail of the distribution of demand is exponentially bounded, then the tail of the distribution of inventory shortfall is approximately exponential. Further, the exponent in this approximation can be easily computed. This approximation is useful because the performance of the inventory system with respect to service level directly depends on the tail distribution. Glasserman (1998) provides a detailed overview of this approach.

Another approximation that has been used by researchers is the one related to approximately characterize the optimal policy for a multiechelon inventory system with economies of scale. Chen and Zheng (1994, 1998) discuss near-optimal policies (in a continuous setting) for multiechelon inventory systems. Chen (1998) provides a detailed description of various approximations that have been considered by researchers in this area.

Finally, approximations related to results under heavy-traffic assumptions have been used in inventory models as well. These approximations are useful when the load of the system is very high and utilization is close to 100% [see Toktay & Wein, 2001].

6 Applications

Several applications in the past years have been developed within the context of supply chain management. These applications can be clearly classified into two categories – one that builds large-integrated models of a multitiered supply chain primarily based on deterministic assumptions about demand, supply and process, and the other based on decomposition of

Table 4
Papers on applications

Year	Reference
1988	Cohen and Lee
1990	Cohen, Kamesam, Kleindorfer, Lee and Tekerian
1993	Lee and Billington
1998	Swaminathan, Smith and Sadeh
2000	Ettl, Feigin, Lin and Yao
2000	Rao, Scheller-Wolf and Tayur
2000	Graves and Willems

the large-scale supply chains and approximation of the behavior through the development of detailed tactical supply chain models discussed in this chapter at each of those nodes. We will focus on the latter (Table 4).

One of the first large-scale model framework that linked decision and performance throughout the material–production–distribution supply chain was developed by Cohen and Lee (1988). The model structure could be used to predict the performance of a firm with respect to the cost of its products, the level of service provided to its customers and the responsiveness and flexibility of the production/distribution system. The analysis took into account the nature of the product produced, the process technologies used to manufacture the products, the structure of the facility network used to manage the material flow and the competitive environment in which the firm operates. It differed from earlier work in that a decentralized control was assumed and combined the performance of the single nodes to create the combined supply chain effect. A series of linked, approximate submodels and a heuristic optimization procedure were developed. Each submodel in the model framework used tractable stochastic models. A software package to support the structure was also introduced. Another large-scale implementation was the development of Optimizer, a decision support for IBM's multiechelon inventory system for managing spare-parts inventory [see Cohen, Kamesam, Kleindorfer, Lee & Tekerian, 1990]. The model and analysis in this work relied heavily on decoupling the multiechelon inventory system into several single-level inventory systems and determining the optimal (or near-optimal) parameters for those single-echelon systems. Starting from the echelon closest to the customer, the parameters are found in an iterative manner, by coupling the demands at the higher echelons of the supply chain with decisions at lower echelons regarding the inventory parameter decisions, namely the (s, S) values.

Lee and Billington (1993) consider a model of the supply chain with a periodic review policy and stochastic demand that has decentralized control. The idea is similar, to analyze the performance of individual entities and then create their combined effect. This model provided various insights for supply chain planning at HP. Ettl, Feigin, Lin and Yao (2000) adopt a queuing approximation to determine the service-level implications in a multitiered network to find the optimal inventory levels to stock at different points in the supply chain. In their approach, each of the sites follows a base-stock policy and they assume that there is a nominal lead time for production and transportation at each of the echelons. The actual lead times may be greater due to shortages. They model each inventory buffer as an infinite-server $M/G/\infty$ queue following a base-stock policy (as in Buzzacott and Shanthikumar, 1993). They assume a Poisson arrival process for demand. Using the above assumptions, they couple the service measures across the supply chain with the base-stock levels chosen. Using a heuristic approach they find the optimal base-stock levels. This model and approximation are validated and refined by supply chain managers through detailed simulation analysis based on an enhanced version of the supply chain library developed in Swaminathan, Smith and Sadeh (1998). The above model and implementation led to an estimated $750 million reduction in inventory at IBM and was awarded the Franz Edelman Prize in 1999.

More recently, Rao, Scheller-Wolf and Tayur (2000) describe the successful implementation of a dynamic supply chain model at Caterpillar. They analyze alternative supply chain configurations for a new product line incorporating expedited deliveries, partial backlogging of orders and sales that were responsive to service provided. Utilizing a combination of models from network flow theory, inventory management and simulation, they analyze alternative choices for the supply chain configuration. Graves and Willems (2000) develop a framework for strategic inventory placement in a supply chain that is subject to demand or forecast uncertainty. They model the supply chain as a network where each entity operates according to a base-stock policy, faces bounded demand and has a guaranteed delivery lead time between the echelons. They utilize the spanning-tree concept and formulate the problem as a deterministic optimization problem to obtain the safety stock. This model was utilized by product flow teams at Eastman Kodak. A more detailed description of approximations for multistage multi-item models appears in de Kok and Fransoo (2003), Chapter 12 of this handbook.

7 Conclusions and future directions

In the Internet age as firms try to completely integrate their operations, tactical planning models for supply chain integration are becoming extremely relevant. In this chapter, we have provided an overview of several streams of research on this broad topic that have been conducted by researchers in the past. Clearly the above stream of research has had tremendous impact on both academic research as well as on practice. However, there are certain changes that are taking place with the advent of the Internet which have opened rich topics for new research (see Swaminathan and Tayur, 2003 for details). Firstly, more and more firms are trying to integrate their production decisions with their pricing decisions and this opens up several topics of research which coordinate supply chain and pricing decisions under cooperative and competitive settings. Cachon (2003) explores some of the models developed therein. Secondly, the focus of operations in many firms is changing from a single-entity optimization model to a more collaborative decision-making process. Analysis needs to be conducted on models which integrate information and supply chain decisions. Chen (2003) explores such models in another chapter. Finally, there is a growing need to decision-support systems that can operate in real time to provide solutions for tactical supply chain problems.

Acknowledgements

The authors wish to thank Ton de Kok and Srinagesh Gavirneni for their detailed comments on an earlier version of this chapter.

References

Anupindi, R., R. Akella (1993). Diversification under supply uncertainty. *Management Science* 39(8), 944–963.

Anupindi, R., Y. Bassok (1998). Supply contracts with quantity commitments and stochastic demand, in: Tayur, Ganeshan, Magazine (eds.), *Quantitative Models for Supply Chain Management*, Kluwer Publishing, Norwell, MA, pp. 197–232.

Anupindi, R., S. Tayur (1998). Managing stochastic multi product systems: model, measures and analysis. *Operations Research*, S98–S111.

Arrow, K., T. Harris, J. Marschak (1951). Optimal inventory policy. *Econometrica* 19, 250–272.

Arrow, K., S. Karlin, H. Scarf (1958). *Studies in the Mathematical Theory of Inventory and Production*, Stanford, CA, Stanford University Press.

Atkins, D., P. O. Iyogun (1988). Periodic versus 'can-order' policies for coordinated multi-item inventory systems. *Management Science* 34(6), 791–796.

Aviv, Y. (2002). A time series framework for supply chain inventory management. To appear, *Operations Research*.

Aviv, Y., A. Federgruen (1997). Stochastic inventory models with limited production capacity and periodically varying parameters. *Probability of Engineering Information Sciences* 11, 107–135.

Azoury, K. (1985). Bayes solution to dynamic inventory models under unknown demand distribution. *Management Science* 31, 1150–1160.

Bassok, Y., R. Akella (1991). Ordering and production decisions with supply quantity and demand uncertainty. *Management Science* 37, 1556–1574.

Bassok, Y., R. Anupindi, R. Akella (1999). Single period multiproduct inventory models with substitution. *Operations Research* 47(4), 632–642.

Bispo, C., S. Tayur (2001). Managing simple re-entrant flow lines: theoretical foundation and experimental results. *IIE Transactions* 33(8), 609–623.

Bollapragada, S., T. E. Morton (1999). A simple heuristic for computing nonstationary (s,S) policies. *Operations Research* 47(4), 576–584. July–August.

Bollapragada, R., U. S. Rao (1999). Single stage resource allocation and economic lot sizing. *Management Science* 45(6), 889–904.

Buzzacott, J., J. G. Shanthikumar (1993). *Stochastic Models of Manufacturing Systems*, Englewood Cliffs, NJ, Prentice Hall.
Cachon, G. (2002). Supply chain coordination: pricing tactics for coordinating the supply chain, to appear, in: Graves, de Tok (eds.), *Supply Chain Management – Handbook in OR/MS*, North-Holland, Amsterdam.
Chen, F. (1998). On (R, NQ) policies serial inventory systems, in: Tayur, Ganeshan, Magazine (eds.), *Quantitative Models for Supply Chain Management*, Kluwer Publishing, Norwell, MA, pp. 71–110.
Chen, F. (1999). 94% Effective policies for two stage serial inventory system with stochastic demand. *Management Science* 12, 1679–1696.
Chen, F. (2002). Information sharing and supply chain coordination, to appear, in: Graves, de Tok (eds.), *Supply Chain Management – Handbook in OR/MS*, North-Holland, Amsterdam.
Chen, F., J. Song (2001). Optimal policies for multiechelon inventory problems with Markov-modulated demand. *Operations Research* 49(2), 226–234.
Chen, F., Y. Zheng (1994). Lower bounds for multi-echelon stochastic inventory systems. *Management Science* 40, 1426–1443.
Chen, F., Y. Zheng (1998). Near optimal echelon stock (R,Q) policies in multi-stage serial systems. *Operations Research* 46(4), 592–602.
Cheng, F., S. P. Sethi (1999). A periodic review inventory model with demand influenced by promotion decisions. *Management Science* 45, 1510–1523.
Ciarallo, F., R. Akella, T. E. Morton (1994). A periodic review production planning model with uncertain capacity. *Management Science* 40, 320–332.
Clark, A., H. Scarf (1960). Optimal policies for a multi-echelon inventory problem. *Management Science* 6, 475–490.
Clark, A., H. Scarf (1962). Approximate solutions to a simple multi-echelon inventory problem, in: Arrow, Scarf (eds.), *Studies in Applied Probability and Management Science*, Stanford University Press, Stanford, CA, pp. 88–110.
Cohen, M. A., H. L. Lee (1988). Strategic analysis of integrated production distribution systems. *Operations Research* 34(4), 482–499.
Cohen, M. A., P. V. Kamesam, P. Kleindorfer, H. L. Lee, A. Tekerian (1990). Optimizer: IBM's multi-echelon inventory system for managing service logistics. *Interfaces* 20(1), 65–82.
Corbett, C.J., C.S. Tang (1998). Designing supply contracts: contract type and information asymmetry, in: Tayur, Ganeshan, Magazine (eds.), *Quantitative Models for Supply Chain Management*, Kluwer Publishing, Norwell, MA, pp. 269–298.
Debodt, M., S. C. Graves (1985). Continuous review policies for a multi-echelon inventory problem with stochastic demand. *Management Science* 31, 1286–1299.
de Kok, T. (1989). A moment-iteration method for approximating the waiting time characteristics of the GI/G/1 queue. *Probability of Engineering and Information Sciences* 3, 273–287.
de Kok, T., J. Fransoo (2002). Supply chain operations: comparing planning concepts, to appear, in: Graves, de Tok (eds.), *Supply Chain Management – Handbook in OR/MS*, North-Holland, Amsterdam.
Edgeworth, F. (1888). The mathematical theory of banking. *Journal of Royal Statistics Society* 51, 113–127.
Ehrhardt, R. (1979). The power approximation for computing (s,S) inventory policies. *Management Science* 25, 777–786.
Ehrhardt, F. (1984). (s,S) policies for a dynamic inventory model with stochastic lead times. *Operations Research* 32, 121–132.
Eppen, G., L. Schrage (1981). Centralized ordering policies in a multi-warehouse system with lead times and random demand, in: L. B. Schwartz *Multi-Level Production Inventory Control Systems: Theory and Practice*, Amsterdam, North Holland, pp. 51–58.
Ettl, M., G. Feigin, G. Lin, D. D. Yao (2000). A supply network model with base-stock control and service requirements. *Operations Research* 48(2), 216–232.
Eynan, A., D. Kropp (1998). Periodic review and joint replenishment in stochastic demand environments. *IIE Transactions* 30(11), 1025–1033.

Federgruen, A., Z. Catalan (1996). Stochastic economic lot scheduling problem: cyclical base stock policies with idle times. *Management Science* 42(6), 783–796.
Federgruen, A., P. Zipkin (1984). An efficient algorithm for computing optimal (s,S) policies. *Operations Research* 32, 818–832.
Federgruen, A., P. Zipkin (1984). Approximation of dynamic multi-location production inventory problem. *Management Science* 30, 69–84.
Federgruen, A., P. Zipkin (1986). An inventory model with limited production capacity and uncertain demands I: the average cost criterion. *Mathematics of Operations Research* 11, 193–207.
Federgruen, A., P. Zipkin (1986). An inventory model with limited production capacity and uncertain demands II: the discounted cost criterion. *Mathematics of Operations Research* 11, 208–215.
Federgruen, A., H. Groenevelt, H. C. Tijms (1984). Coordinated replenishment in a multi-item inventory system with compound Poisson demand. *Management Science* 30, 344–357.
Fukuda, Y. (1964). Optimal policies for inventory problems with negotiable lead time. *Management Science* 10, 690–708.
Gallego, G. (1990). Scheduling the production of several items with random demands in a single facility. *Management Science* 36, 1579–1592.
Gallego, G., O. Ozer (2001). Integrating replenishment decisions with advance demand information. *Management Science* 47, 1344–1360.
Gavirneni, S., S. Tayur (1999). Managing a customer following a target reverting policy. *Manufacturing and Service Operations Management* 1, 157–173.
Gavirneni, S., S. Tayur (2001). An efficient procedure for non-stationary inventory control. *IIE Transactions* 33(2), 83–89.
Gavirneni, S., R. Kapuscinski, S. Tayur (1999). Value of information in capacitated supply chains. *Management Science* 45, 16–24.
Glasserman, P. (1998). Service levels and tail probabilities in multistage capacitated production inventory systems, in: Tayur, Ganeshan, Magazine (eds.), *Quantitative Models for Supply Chain Management*, Kluwer Publishing, Norwell, MA, pp. 41–70.
Glasserman, P., S. Tayur (1994). The stability of a capacitated multi-echelon production inventory system under a base stock policy. *Operations Research* 42, 913–925.
Glasserman, P., S. Tayur (1995). Sensitivity analysis for base stock levels in multi-echelon production inventory system. *Management Science* 41, 263–281.
Graves, S., S. Willems (2000). Optimizing strategic safety stock placement in supply chains. *Manufacturing and Service Operations Management* 2(1).
Graves, S., A. H. G. Rinnoy Kan, P. Zipkin (1993). *Handbook in OR/MS on Logistics of Production and Inventory*, Vol. 4. Amsterdam, Netherlands, North-Holland.
Graves, S., D. B. Kletter, W. B. Hetzel (1998). A dynamic model for requirements planning with application to supply chain optimization. *Operations Research* 46, S35–S49.
Hadley, G., T. Whitin (1963). *Analysis of Inventory Systems*, Englewood Cliffs, NJ, Prentice-Hall.
Hausman, W. H. (1969). Sequential decision problems: a model to exploit existing forecasters. *Management Science* 16(2), B93–B110.
Hausman, W. H., R. Peterson (1972). Multiproduct production scheduling for style goods with limited capacity, forecast revisions and terminal delivery. *Management Science* 18(7), 370–383.
Heath, D. C., P. L. Jackson (1994). Modeling the evolution of demand forecasts with application to safety stock analysis in production/distribution systems. *IIE Transactions* 26(3), 17–30.
Henig, M., Y. Gerchak (1990). The structure of periodic review policies in the presence of random yield. *Operations Research* 38, 634–643.
Heyman, D., M. Sobel (1984). *Stochastic Models in Operations Research*, Vol. II. New York, McGraw-Hill.
Huang, P., A. Scheller-Wolf, S. Tayur (2001). Dynamic capacity partitioning during new product introduction. GSIA Working Paper, Carnegie Mellon University.
Iglehart, D. (1963). Optimality of (s,S) policies in the infinite horizon dynamic inventory problem. *Management Science* 9, 259–267.

Iglehart, D. (1964). The dynamic inventory model with unknown demand distribution. *Management Science* 10, 429–440.
Ignall, E. (1969). Optimal continuous review policies for two product inventory systems with joint setup costs. *Management Science* 15, 277–279.
Ignall, E., A. Veinott (1969). Optimality of myopic inventory policies for several substitute products. *Management Science* 15, 284–304.
Johnson, G., H. Thompson (1975). Optimality of myopic inventory policies for certain dependent demand process. *Management Science* 21, 103–1307.
Kaminsky, P., J. M. Swaminathan (2001). Utilizing forecast band refinement for capacitated production planning. *Manufacturing and Service Operations Management* 3(1), 68–81.
Kaplan, R. (1970). A dynamic inventory model with stochastic lead times. *Management Science* 16, 491–507.
Kapuscinski, R., S. Tayur (1998). A capacitated production inventory model with periodic demand. *Operations Research*.
Karlin, S. (1958). One-stage inventory models with uncertainty, in: K. Arrow, S. Karlin, H. Scarf (eds.), *Studies in the Mathematical Theory of Inventory and Production*, Stanford, CA, Stanford University Press.
Karlin, S. (1960). Dynamic inventory policy with varying stochastic demands. *Management Science* 6, 231–258.
Karlin, S. (1960). Optimal policy for dynamic inventory process with stochastic demands subject to seasonal variations. *Journal of Society of Industrial Applied Mathematics* 8, 611–629.
Karlin, S., A. Fabens (1960). The (s,S) inventory model under Markovian demand process, in: Arrow, Karlin, Suppes (eds.), *Mathematical Methods in the Social Sciences*, Stanford University Press, Stanford, CA, pp. 159–175.
Karlin, S., H. Scarf (1958). Inventory models of the Aroow-Harris-Marschak type with time lag, in: Arrow, Karlin, Scarf (eds.), *Studies in the Mathematical Theory of Inventory and Production*, Stanford University Press, Stanford, CA.
Karmarkar, U.S. (1993). Manufacturing lead times, in: S. Graves, A.H.G. Rinnoy Kan, P.H. Zipkin (eds.), *Handbook in OR/MS on Logistics of Production and Inventory*, pp. 287–321.
Karmarkar, U. S., S. Kekre, S. Kekre (1985). Lot sizing in multi-item multi-machine job shops. *IIE Transactions* 17, 290–292.
Karmarkar, U. S., S. Kekre, S. Kekre (1985). Lot sizing and lead time performance in a manufacturing cell. *Interfaces* 15, 1–9.
Karmarkar, U. S., S. Kekre, S. Kekre (1987). The dynamic lot sizing problem with startup and reservation costs. *Operations Research* 35(3), 389–398.
Lambrecht, S. Chen, N.J. Vandaele (1996). A lot sizing model with queuing delays: the issue of safety time. *European Journal of Operational Research* 89, 269–276.
Lariviere, M.A. (1998). Supply chain contracting and coordination with stochastic demand, in: Tayur, Ganeshan, Magazine (eds.), *Quantitative Models for Supply Chain Management*, Kluwer Publishing, Norwell, MA, pp. 233–268.
Lariviere, M. A., E. Porteus (1999). Stalking information: Bayesian inventory management with unobserved lost sales. *Management Science* 45(3), 346–353.
Lee, H. L., C. Billington (1993). Materials management in decentralized supply chains. *Operations Research* 41(5), 835–847.
Lee, H. L., C. Tang (1997). Modelling the costs and benefits of delayed product differentiation. *Management Science* 43(1), 40–53.
Lee, H. L., R. So, C. S. Tang (2000). The value of information in a two level supply chain. *Management Science* 46(5), 626–643.
Lee, H. L., C. A. Yano (1995). Lot sizing with random yields: a review. *Operations Research* 43(2), 311–334.
Lovejoy, W. (1992). Stopped myopic policies in some inventory models with generalized demand processes. *Management Science* 38, 688–707.

Morton, T. (1971). The near-myopic nature of the lagged proportional cost inventory problem with lost sales. *Operations Research* 19, 1708–1716.

Morton, T. (1978). The non-stationary infinite horizon inventory problem. *Management Science* 24, 1474–1482.

Morton, T., D. Pentico (1995). The finite horizon non-stationary stochastic inventory problem. *Management Science* 41, 334–343.

Muharremoglu, A., J. Tsitsiklis (2001). Echelon base stock policies in uncapacitated serial inventory systems. Working Paper, MIT.

Nahmias, S. (1979). Simple approximations for a variety of dynamic lead time lost sales inventory models. *Operations Research* 27, 904–924.

Parker, R., R. Kapuscinski (2000). Optimal inventory policies for a capacitated two echelon system. Working Paper, University of Michigan.

Porteus, E. (1971). On the optimality of generalized (s,S) policies. *Management Science* 17, 411–427.

Porteus, E. (1972). The optimality of generalized (s,S) policies under uniform demand densities. *Management Science* 18, 644–646.

Porteus, E. (1991). In: Heyman, Sobel (eds.), *Stochastic Inventory Theory, Stochastic Models – Handbook in OR/MS*, North-Holland, Amsterdam, pp. 605–652.

Rajagopalan, S., J. M. Swaminathan (2001). A coordinated production planning model with capacity expansion and inventory management. *Management Science* 47(11), 1562–1580.

Rao, U. S., A. Scheller-Wolf, S. Tayur (2000). Development of a rapid response supply chain at caterpillar. *Operations Research* 48(2), 189–204.

Rao, U.S., J.M. Swaminathan, J. Zhang (2002). Multi-product inventory planning with downward substitution, stochastic demand and setup costs. To appear, *IIE Transactions*.

Rosling, K. (1989). Optimal inventory policies for assembly systems under random demands. *Operations Research* 37, 565–579.

Roundy, R. (1985). 98% effective integer ratio lot sizing for one warehouse multi retailer systems. *Management Science* 31, 1416–1430.

Scarf, H. (1959). Bayes solution of the statistical inventory problem. *Annals of Mathematical Statistics* 30, 490–508.

Scarf, H. (1960). The optimality of (s,S) policies in dynamic inventory problem, in: K. Arrow, S. Karlin, P. Suppes (eds.), *Mathematical Methods in the Social Sciences*, Stanford, CA, Stanford University Press.

Scarf, H. (1960). Some remarks on the Bayes solution to the inventory problem. *Naval Research Logistics Quarterly* 7, 591–596.

Scheller-Wolf A., S. Tayur (2000). A Markovian dual source production-inventory system with order bands. GSIA Working Paper, Carnegie Mellon University.

Schmidt, C., S. Nahmias (1985). Optimal policy for a two stage assembly system under random demand. *Operations Research* 33, 1130–1145.

Sethi, S. P., F. Cheng (1997). Optimality of (s,S) policies in inventory models with Markovian demand. *Operations Research* 45, 931–939.

Silver, E. (1981). Establishing reorder points in the (S,c,s) coordinated control system under compound Poisson demand. *International Journal of Production Research* 9, 743–750.

Sobel M. (1988). Dynamic affine logistics models. Technical Report, SUNY, Stony Brook.

Swaminathan, J.M., H.L. Lee (2003). Design for postponement, to appear, in: Graves, de Kok (eds.), *Supply Chain Management – Handbook in OR/MS*, North-Holland, Amsterdam.

Swaminathan, J. M., J. G. Shanthikumar (1999). Supplier diversification: the effect of discrete demand. *Operations Research Letters* 24(5), 213–221.

Swaminathan, J. M., S. Tayur (1998). Managing broader product lines through delayed differentiation using vanilla boxes. *Management Science* 44, S161–S172.

Swaminathan, J.M., S. Tayur (2003). Models for supply chains in e-business. Working Paper, The Kenan-Flagler Business School, University of North Carolina, Chapel Hill, To appear, *Management Science*.

Swaminathan, J. M., S. F. Smith, N. Sadeh (1998). Modelling supply chain dynamics: a multi-agent approach. *Decision Sciences* 29(3), 607–632.

Tayur, S. (1993). Computing the optimal policy in capacitated inventory models. *Stochastic Models* 9, 585–598.

Tayur, S., R. Ganeshan, M. Magazine (1998). *Quantitative Models for Supply Chain Management*, Norwell, MA, Kluwer Academic Publishers.

Toktay, L. B., L. M. Wein (2001). Analysis of a forecasting production inventory system with stationary demand. *Management Science* 47(9), 1268–1281.

Tsay, A.A., S. Nahmias, N. Agarwal (1998). Modeling supply chain contracts: a review, in: Tayur, Ganeshan, Magazine (eds.), *Quantitative Models for Supply Chain Management*, Kluwer Publishing, Norwell MA, pp. 299–336.

van Donselaar, K., T. de Kok, W. Rutten (1996). Two replenishment strategies for the lost sales inventory model: a comparison. *International Journal of Production Economics* 46–47, 285–295.

Van Mieghem, J., N. Rudi (2002). Newsvendor networks: inventory management and capacity investments with discretionary activities. To appear, *Manufacturing and Service Operations Management*.

Veinott, A., Jr. (1965). Optimal policy for a multi-product dynamic non-stationary inventory problem. *Management Science* 12, 206–222.

Veinott, A., Jr. (1965). Optimal policy in a dynamic single product non-stationary inventory model with several demand classes. *Operations Research* 13, 776–778.

Veinott, A., Jr. (1966). The status of mathematical inventory theory. *Management Science* 12, 745–777.

Veinott, A., Jr. (1966). On the optimality of the (s,S) inventory policies: new conditions and a new proof. *SIAM Journal of Applied Mathematics* 14, 1067–1083.

Veinott, A., Jr., Wagner, H. (1965). Computing optimal (s,S) inventory policies. *Management Science* 11, 525–552.

Wijngaard, J. (1972). An inventory problem with constrained order capacity. TH-Report 72-WSK-63, Eindhoven University of Technology.

Wright, G. (1969). Optimal ordering policies for inventories with emergency ordering. *Operations Research Quarterly* 20, 111–123.

Zheng, Y. (1991). A simple proof for optimality of (s,S) policies in infinite horizon inventory systems. *Journal of Applied Probability* 28, 802–810.

Zheng, Y., A. Federgruen (1991). Finding optimal (s, S) policies is about as simple as evaluating a single policy. *Operations Research* 39(4), 654–665.

Zipkin, P. (1989). Critical number policies for inventory models with periodic data. *Management Science* 35, 71–80.

Zipkin, P. (2000). *Foundations of Inventory Management*, Boston, McGraw Hill.

PART III

Supply Chain Operations

A.G. de Kok and S.C. Graves, Eds., *Handbooks in OR & MS, Vol. 11*
© 2003 Elsevier B.V. All rights reserved.

Chapter 9

Planning Hierarchy, Modeling and Advanced Planning Systems

Bernhard Fleischmann and Herbert Meyr
Lehrstuhl für Produktion und Logistik, Universität Augsburg, Universitätsstr. 16,
D-86135 Augsburg, Germany

Along a supply chain, various decisions have to be made continuously, from the rather simple choice, which job to be processed next on a certain machine, to the serious question, whether to build a new factory or to close down an existing one. Within this chapter '*Supply chain planning*' is used as a generic term for the whole range of those decisions on the design of the supply chain, on the mid-term coordination and on the short-term scheduling of the processes in the supply chain. This definition also applies to the traditional notion of *Logistics*. However, in the scope of the recent development of supply chain management, the focus is on the following two aspects of planning:

- *Integral planning* of the entire supply chain: The planning process should consider the supply chain of an enterprise, at least from its suppliers up to its customers, as a whole and take into account the interdependencies of the various activities.
- *True optimization of decisions*: The planning process should be based on a proper definition of alternatives, objectives and constraints and use (exact or heuristic) optimization algorithms.

While the last aspect is quite familiar to Operations Researchers, it has not been the common view in practice for a long time: The wide-spread Enterprise Resource Planning (ERP) software and the included Material Requirements Planning (MRP) logic, in spite of their name, do not provide planning functions in the above sense [Drexl, Fleischmann, Günther, Stadtler, & Tempelmeier, 1994].

The postulates of integral planning and of true optimization are scarcely compatible. A practicable compromise between both postulates is the use of *Hierarchical Planning* (HP) concepts allowing to decompose the overall task into partial planning tasks and to still consider their interdependencies and to coordinate their solution. Hierarchical Planning has also a long tradition in Operations Research, starting with the work of Hax and Meal in

1975, and many planning algorithms have been developed for the various partial planning tasks.

What then is '*Advanced Planning*'? This notion has been introduced by software providers for a new type of planning software, the *Advanced Planning Systems* (APS), which are based on the above ideas. It is true that neither the HP concept underlying the APS architecture nor the algorithms used in the single modules are particularly advanced, but the real advance is the *implementation* of these concepts in standard software, enabling the dissemination of reasonable planning concepts and OR based algorithms in practice. This is indeed a great progress as compared to the traditional ERP systems. Moreover, the APS architecture is open to include new modules and new algorithms. This is necessary, because there are various partial problems in supply chain planning that still miss practicable solution algorithms. Thus, APS provide an appropriate framework for putting OR developments into practice.

However, there seems to be some confusion in both science and practice what APS are, which modules and solution methods they include, which purposes they can be used for and how they should be used. Therefore, the aim of this chapter is to analyze first the various supply chain planning tasks and their hierarchical relationship and then the architecture and functionality of APS.

Fisher (1997) already has shown how important it is to identify different types of supply chains in order to derive fitting management strategies. Thus, planning systems, which try to implement these strategies operatively, also have to be tailored to the particular requirements of the type of supply chain under consideration. To support the processes of describing and analyzing different types of supply chains, Section 1 first introduces a *typology* of supply chains, which is illustrated by means of two contrasting examples: consumer goods manufacturing and computer assembly. Section 2 then provides a general framework for deriving the corresponding planning tasks of the respective supply chain type identified. This framework again is applied to the two examples. As it is shown in Fleischmann, Meyr, and Wagner (2002, Chapter 4.3), planning concepts that fit these planning requirements can be designed by means of HP. Therefore, at the end of Section 2, HP concepts, including recent developments, are reviewed.

Section 3 shows that a common modular architecture, which is along the lines of HP, is underlying all APS and discusses the functions of the typical modules. Finally, Section 4 presents a snapshot (state: January 2003) of five particular APS, reveals – as far as possible – the OR methods applied within their modules and reviews case studies of actual APS implementations which have been published in the literature.

1 Types of supply chains

Experience with Production Planning and Control systems has shown that a single production planning concept like the MRP II–concept cannot cover the large variety of planning problems that arise in practice for different production layouts and market requirements [Drexl et al., 1994]. Different types of production processes, e.g., job shop, batch flow, assembly or continuous processes, imply particular requirements for planning. Thus planning concepts have to be tailored to these special requirements [Silver, Pyke, & Peterson, 1998, p. 36]. What is true for the (relatively small) production area is even more valid for the supply chain as a whole.

But one can define certain types of supply chains having substantial features in common and thus sharing similar planning tasks. In order to identify such *supply chain types* (*SC-types*), a typology may be helpful that lists and categorizes the most important attributes characterizing a given supply chain. Examples for such typologies are given in Meyr, Rohde, and Stadtler (2002b) and Silver et al. (1998, Chapter 3.5.3).

1.1 Supply chain attributes

We will now briefly introduce the typology of Meyr et al. (2002b). Since typologies can never be comprehensive, we will concentrate on attributes that are relevant for planning functions. These SC-types will in Section 2.2 be used to demonstrate how the specific planning requirements of a certain SC-type can be derived from the type's properties.

Meyr et al. distinguish between two types of attributes, functional and structural ones. *Functional attributes* can be assigned to *each individual* member (partner, entity) of a supply chain, e.g., a manufacturer, carrier or wholesaler. *Structural attributes* ought to describe the relations *between all* members of a supply chain. All attributes are grouped into six categories. Thereof structural attributes comprise the two categories 'topography of an SC' and 'integration and coordination.' According to the supply chain processes of each member, the functional attributes are classified into the categories 'procurement type,' 'production type,' 'distribution type' and 'sales type.' An overview over the categories and attributes considered here is given in Table 1.

Note that the SCOR model [Meyr et al., 2002b, Chapter 3.1] uses a quite similar approach for supply chain *analysis* by differentiating (beside others) between the processes *source*, *make* and *deliver*. As can be seen in Fig. 1, just like functional attributes these processes have to be defined for each member of an SC separately. However, whereas the SCOR-model is a well suited tool for analyzing SCs and revealing redundancies and weaknesses, the typology proposed in the following aims at identifying planning tasks common for a specific type of SC, deriving requirements for planning and supporting the design of planning concepts fitting these particular requirements [Meyr et al., 2002b, Chapter 3.2].

The category *procurement type* contains all attributes that characterize the inflow of goods (like raw materials or parts for a manufacturer) to the respective member of the supply chain. For example, the attribute *type of*

Table 1
Overview of the categories and attributes of the SC-typology

Functional (for each member of the SC)	Structural (for the SC as a whole)
Procurement type: – type of products procured – sourcing type – supplier lead time – number or raw materials – materials' life cycle	*Topography of an SC:* – network structure – location of the decoupling point(s) – major constraints
Production type: – organization of the production process – repetition of operations – changeover characteristics – type of bottlenecks – working time flexibility	*Integration and coordination:* – legal position – distribution of power
Distribution type: – distribution structure	
Sales type: – products being sold – products' life cycle – shelf lives – bill of materials – seasonal demand patterns	

products procured may range from standard products easily available on the market to highly specific products that can only be sourced from one or a few suppliers. Further important attributes are e.g., the *sourcing type* indicating the number of alternative suppliers actually being used, the *supplier lead time* describing the time span between ordering and receiving incoming goods, the *number of* different types of *raw materials* being purchased or the length of the *life cycle of materials*.

The *production type* comprises the attributes that characterize the production process itself (which may be of minor importance for service providers like carriers). The *organization of the production process* (e.g., job shop or flow shop) defines the flow of work in process (WIP). The *repetition of operations* scales the frequency of producing the same items and may range from mass production (continuous repetition) to making one-of-a-kind products (no repetition). If none of these extremes applies, products of the same kind are processed in groups, so-called batches or lots. The effort in time and money spent in preparing machines to produce a new lot is expressed by the *changeover characteristics*. A vast amount of other attributes concerning the production type is conceivable and can be found in literature, e.g., the *type of bottlenecks* in production (shifting,

Ch. 9. Planning Hierarchy, Modeling and Advanced Planning Systems   461

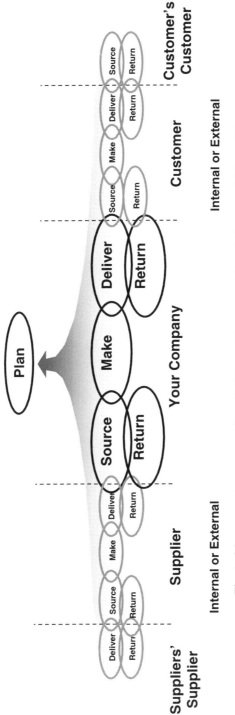

Fig. 1. Management processes of the SCOR model, version 5.0a [cf. Stephens, 2001, p. 11].

stationary, unproblematic) or the *flexibility* to extend or reduce *working time* (cf. Chapter 4).

The *distribution type* concerns outgoing material flows, i.e., the delivery of products from one member of the SC to its next downstream neighbors. For example, the *distribution structure* states the type and number of stages of this member's distribution system, e.g., direct delivery to its neighbors, indirectly via central warehouses (two stages) or via central and regional (RW) warehouses (three stages).

The *sales type* comprehends both the product and market characteristics for the member of the SC considered. The *products being sold* may range from standard products in a fixed configuration to highly specific products with an open configuration. While the *products' life cycle* spans the time a product will be present on the market, the *shelf life* of a product limits the time an item can be held in inventory until it has to be consumed. The *bill of materials* (BOM) may vary between a pure assembly structure (many raw materials or parts are put together to build one final item) and a purely divergent structure where one starting substance or material is decomposed into several successors, e.g., by a chemical process or simply by packaging in different sizes and shapes. Moreover, the *demand* for products may be influenced to a different degree by *seasonal* peaks.

Structural attributes of the SC as a whole are grouped into the two broad categories 'topography of a supply chain' and 'integration and coordination'. The category *topography of a supply chain* contains all attributes describing the characteristics of a supply chain as a whole. The *network structure* indicates whether a serial, convergent, divergent flow of material is given or a mixture of these types applies.

As will be shown later, the *location of the decoupling points* [Hoekstra & Romme, 1991, Chapters 1.5 and 4.2] has a crucial impact on planning: The (physical) processes of each member of the supply chain can be subdivided into anticipative and reactive ones. While reactive processes are triggered by an explicit order of a subsequent member of the supply chain (e.g., the final customer), anticipative ones are triggered by forecasts instead of orders, i.e., they try to anticipate an order that has not yet been placed. A decoupling point forms the interface between upstream anticipative processes that are executed '*to-stock*' and downstream reactive processes that are executed '*to-order.*' The term 'to-stock' is due to the fact that a stocking point always comes along with the decoupling point in order to hedge against forecast errors. For example, an assemble-to-order decoupling point implies that upstream processes like ordering of raw materials and manufacturing of parts are executed to stock, that final products are assembled from stock only if the respective order has arrived, and that downstream processes like delivery are also executed to order. Note that there are supply chains that consist exclusively of reactive processes (e.g., when making highly specialized one-of-a-kind products) or of anticipative processes (e.g., vendors in a Vendor Managed Inventory (VMI) setting [Meyr et al., 2002b]).

Depending on the availability and usage of resources like equipment, personnel and material one can identify capacity, labor and/or material as being the *major constraint(s)* of a supply chain.

The last category focuses on *integration and coordination* aspects between different members of a supply chain. If several organizational units of the same company form a supply chain, an intra-organizational supply chain is given. In an inter-organizational supply chain several legally separated companies, i.e., companies in different *legal positions*, are members of the supply chain. In this case, the *distribution of power* between these members is of particular importance. In a material constrained supply chain, for example, suppliers often play a dominant role.

As already mentioned, this typology is by far not comprehensive. Just a few attributes were selected which will help to identify peculiarities of different SC-types and their impact on planning.

1.2 Examples

In the following, two types of supply chains, '*consumer goods manufacturing*' and '*computer assembly*', respectively, will be shown in some detail. The description will mainly be based on two case studies (Wagner & Meyr, 2002, Kilger & Schneeweiss, 2002a) which – in our opinion – are representative for these kinds of industries. For demonstration purposes it will be sufficient to concentrate on functional attributes of *only a single member* (the manufacturer) of each supply chain. Structural attributes will give some hints on each supply chain as a whole and on relations to other members as well.

For introducing the functional attributes *procurement*, *production*, *distribution* and *sales* a somewhat 'natural' sequence was upstream to downstream, i.e., in direction of the material flow. However, because the 'products sold to the final customer' are the decisive part of any supply chain it is now more convenient to start the two examples with the *sales type* category first and to illustrate the further functional attributes – according to the flow of information – in the opposite direction downstream to upstream.

1.2.1 Consumer goods manufacturing

Sales type. In the following, 'consumer goods' will denote standard products with a low volume, weight and value per item (e.g., food, low tech electronics like light bulbs or fluorescent lamps) that are sold to the final customer via retailers in a grocery or electronics store, for example. Such products typically have a rather long life cycle (one to several years), but may have different shelf lives varying from a few days (e.g., fresh milk) to a few years (e.g., canned food, electronics). A consumer goods manufacturer often produces a few related product types, which are sold in many variants due to minor variations in the production process (e.g., applying different colors) or in sizes and shapes of packaging. In this sense, the BOM is divergent. Demand has to be estimated and is quite unstable because of seasonal influences or price promotions, for example. However, due to the long product life cycle, historical sales figures are available making reliable forecasts easier. Because of the low product differentiation, in shortage situations sometimes final items can be purchased from external suppliers and be sold under the manufacturer's own brand.

Distribution type. The distribution of consumer goods (cf. Fig. 2) needs quite a lot of care since the variety of final items has to be distributed from one or a few production sites to a bulk of downstream members of the supply chain like wholesalers, retailers or department stores before being bought by the final customers. Retailers want frequent deliveries in small quantities. Thus, usually a three stage distribution system is installed where goods are temporarily stored in fully assorted distribution centers (DCs) until they are brought to intermediate RWs or stock-less transshipment points (TPs). From there, goods are delivered in dynamic routes to the manufacturer's customers (see Section 3.2.5). Mostly, unlimited transportation capacity can be assumed because third party carriers are employed.

Production type. Consumer goods usually are produced on one or a few parallel production lines which are organized in a flow shop. Sometimes the same products can alternatively be produced at several production sites.

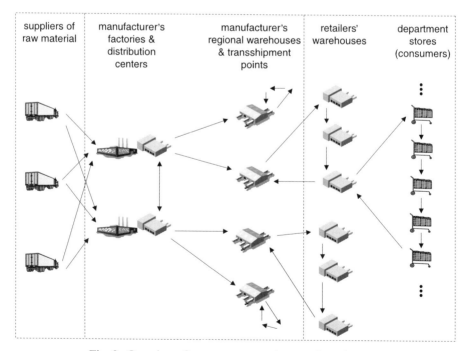

Fig. 2. Overview of a consumer goods manufacturing SC.

Because goods are made in high volume, production speed has to be very high and thus always a high level of automation is predominant. As typical for flow shops, WIP inventory is low and production lead times are short. Mostly, there are only two to three stages of production (e.g., manufacturing and packaging) with one of them being a well-known stationary bottleneck. Changeovers between different product types often have to be done manually causing significant setup times and costs. Not seldom setup times and costs are sequence dependent, i.e., the amount of time and costs depends on the product (type) produced before on the same line (e.g., changes from white to black color are easier than changes from black to white). Not every product is produced every day, but lot-sizes cover the demand of several days up to months, causing finished product inventory. Due to the highly skilled workers needed to operate the production lines, shift patterns have to be determined on a mid-term time range. The capital intensive equipment entails a high utilization.

Procurement type. Since the BOM is rather flat and simple, procurement usually is unproblematic. A few standard raw materials with a low value are sourced from a handful of suppliers. Often mid- and long-term contracts ensure a stable relationship between manufacturer and suppliers. Supplier lead times are quite short and reliable.

Topography of the SC. The supply chain as a whole consists of a network of some raw material suppliers, one or a few production sites with DCs being associated, several RWs or TPs and a bulk of downstream members selling the goods to the final customers. For this reason, usually a network structure of the mixture type is present. Decoupling points typically (except for VMI settings) are located at the department stores of the retailers ('orders' of the final customers) and at the DCs of the manufacturer (deliver to orders of the retailers). The capacity of the production lines is mostly the major constraint within the supply chain.

Integration and coordination. Thus intra-organizational relations between the different planning units of the consumer goods manufacturer play an important role and their coordination is difficult. Nevertheless, the inter-organizational integration and information flows between consumer goods manufacturers and retailers was paid attention to in recent years by concepts like Efficient Consumer Response, Continuous Replenishment and Vendor Managed Inventory. A main reason for this was the discussion about the Bullwhip effect [Lee, Padmanabhan, & Whang, 1997] and the insight that stocks can be reduced when improving the information flow (especially concerning final customers' demand) between these two members of the SC.

1.2.2 Computer assembly

A computer assembly supply chain (see Fig. 3 for an overview), on the other hand, shows quite different characteristics.

Sales type. Computers have a typical assembly structure. An incoming order of a computer manufacturer usually consists of several order lines for different product families (like consumer and professional PCs or servers) and for external units like printers, monitors etc. Whereas the computers are produced by the manufacturer itself, the external units are purchased from external suppliers, but delivered to the customers in 'complete order' shipments. A computer consists of the system unit (housing, system board, processor, disk drive...) and accessories like keyboard, mouse, software, manual, and so on. Altogether a complex BOM with several stages is given. As far as the manufacturer offers *fixed configurations*, this BOM is predefined, i.e., the BOM cannot be customized by the customer. However, the manufacturer usually is free to substitute some parts with equivalent ones in case of shortage situations, e.g., comparable hard disks of several suppliers may alternatively form the same final product. Product differentiation is very low. Due to technological improvements and price war, the product life cycle is very short (at the most a few months). For this reason and because of seasonal influences (e.g., Christmas business), demand for final products is hard to predict.

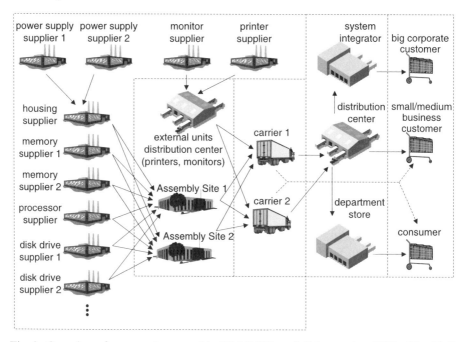

Fig. 3. Overview of a computer assembly SC [cf. Kilger & Schneeweiss, 2002a, Fig. 20.1].

Distribution type. Typical customers of a computer manufacturer are system integrators providing complete solutions for big corporate customers like banks or insurance companies, small and medium business customers, and department stores. Only some manufacturers also directly sell to private customers. Often a two-stage distribution system is used where computers and external units are merged in an intermediate distribution center.

Production type. The production process mainly comprises the stages 'assembly of the system unit', 'loading of software', 'testing', and 'packing'. If the manufacturer also assembles the system board, a further upstream stage is necessary which may be located at another production site. Dependent on the sizes of the production lots, both job shop (for small lots) and flow shop (for larger lots) environments arise. Production speed is (even in job shops) quite high so that short production lead times can be achieved (one to a few days). Changeovers do not cause noticeable setup times or costs. Altogether, there are no serious bottlenecks in production.

Procurement type. Since an assembly structure is given, incoming material flows dominate the supply chain. Altogether, several hundreds to thousands of components (electronic and mechanical parts, external units and accessories) have to be purchased. However, the main part of value procured is provided from a few suppliers (20–40) only. Quite often the same components can be and are procured from several alternative suppliers (e.g., hard disks and CD-ROMs). Nevertheless, some crucial components (e.g., processors) are offered by just one or two suppliers. In this case the distribution of power is shifted towards the supplier(s). Supplier lead times are quite inhomogeneous, long, and unreliable. They vary from one week to several months. For many components (like processors, hard disks etc.) the life cycle is very short because of technological progress. Thus, there is a high risk of obsolete stocks.

Topography of the SC. The supply chain is of a mixture type and consists of the five (to six) stages suppliers, (system board manufacturer), computer manufacturer, logistic service providers, resellers and final customers. For fixed configurations an assemble-to-order decoupling point is the normal case. In order to improve customer service, however, deliver-to-order may be applied, thus shifting stocks from the component to the finished goods level. As opposite to the consumer goods SC-type, the major constraints of a computer assembly supply chain are materials. Thus more parties have an influence on the performance of the supply chain.

The computer assembly scenario described above is specific for computers sold in predefined, fixed configurations. The character of the supply chain changes somewhat if *open configurations* are offered. In this case the customer is able to customize his system with respect to his particular preferences. Thus no static BOM exists, but a customer request has to be checked according to some (rather loose) configuration rules, and a customer-specific BOM has to be generated. The decoupling point moves from assemble-to-order to configure-to-order. Because of these configuration checks the order lead time increases. Lot-sizes and the frequency of repetition decrease so that job shops will be better utilized.

The most important attributes (according to our typology) of both SC-types are summarized in Table 2 in order to stress their different characteristics. Of course, a large variety of further SC-types occurs. For example, the chemical industry is related to consumer goods production, but has to handle more complex production processes (several stages of production with restrictive interconnections, chemical batch processes in reactors, stochastic output quantities and quality) and divergent BOMs due to joint production. The automotive industry is again a type of supply chain where assembly processes are dominant. But the distribution of power is clearly shifted towards the automobile manufacturer. A major focus of production is the balancing of the model mix, i.e., one tries to counterbalance the different production speeds of various car variants at the successive assembly stations by an appropriate mix of the variants. Most of the transport activities are on the procurement side where the just-in-time-concept or regional carriers are used to synchronize the incoming material flow. At least in the high price segment, mass customization and (online) order promising will be of increasing importance in the future.

2 Supply chain planning

Before deriving the particular planning requirements of the above two supply chain examples from their SC-attributes, some general planning tasks – to various extent occurring in any type of supply chain – are introduced and the term 'supply chain *planning*' is specified in some more detail. Section 2.3 finally reviews the principles of Hierarchical Planning, which provide a practicable and useful way to integrate these planning tasks by means of APS.

2.1 Planning tasks in the supply chain

Since the fundamental work of Anthony (1965), usually three levels of managerial decision making are referred to (see e.g., Bitran & Tirupati, 1993, Miller, 2001, Silver et al., 1998). They mainly differ with respect to the time during which the decisions will have an impact on the future development of a supply chain or company. According to this categorization and the *planning horizon* they comprise, planning tasks are commonly assigned to one of the three planning levels 'long-term,' 'mid-term' and 'short-term planning.'

Long-term planning prepares decisions whose implications on the supply chain can be felt for several years. These decisions essentially determine the

Table 2
Supply chain types *consumer goods production* and *computer assembly*

Category	Attributes	Consumer goods	Computer assembly fixed/configurable
Functional			
Procurement type	Products procured	Standard	Standard and specific
	Sourcing type	Multiple	Multiple
	Supplier lead time	Short and reliable	Long and unreliable
	Number of raw materials	Few	Many
	Materials' life cycle	Several years	Few months
Production type	Organization of the production process	Flow shop	Flow shop/job shop
	Repetition of operations	Frequent	Frequent/no repetition
	Changeover characteristics	Sequence dependent setup times and costs	Irrelevant
	Bottlenecks in production	Known bottlenecks/high influence	Low influence
	Working time flexibility	Low	Low
Distribution type	Distribution structure	Three stages	Two stages
Sales type	Products being sold	Standard	Standard/customized
	Products' life cycle	Several years	Few months
	Shelf lives	Perishable or stable	Stable
	Bill of materials	Divergent	Assembly
	Seasonal demand patterns	Highly seasonal	Weakly seasonal
Structural			
Topography of a SC	Network structure	Mixture	Mixture
	Location of the decoupling point(s)	Deliver-to-order	Assemble-/configure-to-order
	Major constraints	Capacity	Material
Integration and coordination	Legal position	Intra-organizational	Inter- and intra-organizational
	Control over suppliers	High	Low

physical structure of a supply chain and should directly reflect a company's business strategies. *Mid-term planning* has to effectively use and act within the infrastructure set by the long-term 'strategic' planning. The validity of a mid-term plan ranges from half a year to two years [Silver et al., 1998]. The planning horizon of *short-term planning* is restricted to a few weeks or at most a few months. Short-term planning has to put into practice the guidelines given by the upper two levels and to prepare detailed instructions for immediate execution and control of the operations. According to Fleischmann et al. (2002), mid- and short-term planning will also be denominated 'operational planning' in the following.

The supply chain planning matrix (SCP-matrix [Fleischmann et al., 2002], see Fig. 4) makes use of the supply chain processes *procurement, production, distribution and sales* – already helpful in the supply chain typology of Section 1.1 – to further classify the planning tasks typically emerging for each member of a supply chain. The structure of the SCP-matrix will also be used to characterize the role of APS in supply chain planning in Section 3. Some typical planning tasks will now briefly be introduced. Note that this selection is by far not comprehensive. For further investigation the reader is referred to Miller (2001, Chapter 1.1).

Long term planning on the procurement side concerns questions like 'What material should be purchased from which suppliers?', 'Are strategic cooperations useful?' or 'What type of cooperation should be chosen?'. Decisions about the location and sizes of plants, the organization of the production processes and the capacity of the production system are further typical long-term planning tasks. On the distribution side, the structure of the distribution system and the locations of warehouses or transshipment points have to be determined. Since a supply chain is mainly characterized by the products it sells, the planning of the product program and strategic sales planning concerning questions like 'Which product to place on what markets?' need particular care.

On the *mid-term* planning level, decisions to be made concern

- rough quantities of material and components to be obtained from (alternative) suppliers,
- workforce requirements and the degree of external purchasing of final products,
- the assignment of production quantities and seasonal stock to different plants and warehouses, and
- the use of different distribution channels,

for example. These decisions usually have to be based on mid-term sales forecasts for product groups and sales regions. Furthermore, contracts with suppliers (and customers) usually are thought over after one or two years. In case of SCs consisting of multiple legally separated entities these contracts also comprise agreed lead times between upstream and downstream members, which have to be determined carefully.

Short-term procurement calculates the schedule and the quantities of materials actually to be ordered from suppliers. Also the short-term deployment of personnel has to be considered. On the production side, appropriate lot-sizes have to be determined, the lots are scheduled on the shop floor and the progress of production must be controlled in order to meet due dates and to quickly react to unforeseen events like machine breakdowns. Production output has to be assigned to warehouses and customers, transportation means must be chosen and vehicle routes are to be determined. Short-term sales planning deals with different aspects of demand fulfillment, e.g., the promising of delivery dates and the allocation of make-to-stock quantities to actual customer orders.

The SCP-matrix gives a general overview over the planning tasks arising in any possible supply chain. However, according to the particular SC-type considered, the importance of the single planning tasks is quite different. Furthermore, the assignment of planning tasks to planning levels and supply chain processes in Fig. 4 is somewhat fuzzy because the positioning may also vary with respect to the SC-type considered. For example, Bertrand, Wortmann, and Wijngaard (1990, Chapter 8.5.3.1) present a case study from electronic component manufacturing where lot-sizing decisions already have to be made in the medium term.

2.2 Examples

In the following we will demonstrate how the planning requirements of the consumer goods and computer assembly SC-types can be derived from their

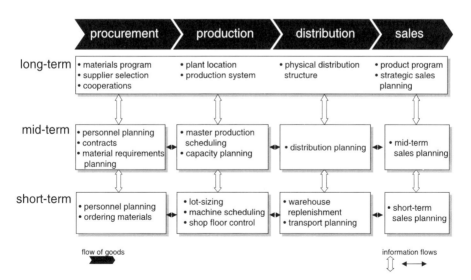

Fig. 4. Planning tasks according to the SCP-matrix [cf. Fleischmann et al., 2002, Fig. 4.3].

respective attributes (see Table 3 for an overview). This presentation is certainly not exhaustive, but concentrates on the most important tasks.

2.2.1 Consumer goods manufacturing

As soon as *multiple sourcing* from several suppliers is practiced, the share of each supplier and the quantities procured have to be determined in the medium term.

In a make-to-stock environment 'lot-sizing,' i.e., defining the sizes of production lots, has to be done simply because orders are not available at time of production (planning). Since there are significant setup times and costs in consumer goods supply chains, this expense has to be balanced against the inventory costs caused by lot-sizes exceeding (forecasted) demand of the near future.

If setup costs or times are sequence dependent, the sizes *and* the sequence of the lots have to be determined jointly, i.e., simultaneous lot-sizing and scheduling is necessary. This has to be done with respect to the limited availability of production lines which are the main bottlenecks in production. The low working time flexibility prevents from a short-term extension of capacity by means of overtime.

Because of this low flexibility and the limited capacity of production and because of the seasonal demand pattern, an integrated mid-term planning (master planning, MP) of production quantities, seasonal stock and working time is necessary respecting all costs (and revenues) arising. Only demand forecasts can drive this planning. They may be derived from historical sales data which usually are available because of the long product life cycles. As soon as several plants producing the same product are involved, the allocation to plants has to be considered, too. The three-stage distribution system allows to serve a single customer via several distribution paths (DC, RW, TP). Therefore, the use of the distribution system and the resulting transportation costs has to be included into MP, as well. But the main focus of mid-term planning is to ensure feasibility with respect to the limited production capacity available.

The location of the decoupling point [Hoekstra & Romme, 1991] (deliver-to-order, i.e., make-to-stock) has the strongest impact on planning. Incoming orders have to be served immediately from stock of final items. Except for the deployment, i.e., the assignment of incoming orders to stock of final items (and the subsequent delivery), all planning tasks are driven by demand estimates. In order to hedge against inevitable forecast errors, safety stocks of final items have to be introduced. Not only the quantity, but also the allocation of this safety stock to the DCs and RWs of the distribution system has to be planned.

Note that the degree of freedom for building up the different types of stock (lot-size, seasonal, safety) is restricted by the shelf life of final items and may be very tight in case of perishable goods.

Table 3
Impact of the SC-type on planning

Attributes	Impact on planning	
	Consumer goods	Computer assembly
Supplier lead time		Mid-term master planning is basis for purchasing and order promising
Changeover characteristics Type of bottlenecks Working time flexibility	Simultaneous lot-sizing and scheduling Explicit capacity planning Mid-term planning of working time	Rough capacity planning
Distribution structure	Choice of distribution channels Allocation of safety stock	
Products being sold		Open config.: config. check BOM generation
Shelf lives	Perishable: limited inventory	
Loc. of decoupling points	Safety stock of final items Deployment	Safety stock of components Order promising Allocation planning
Major constraints	Forecasts for final items Main focus of master planning: ...feasibility w.r.t. capacity	Forecasts for components ...synchronize materials

2.2.2 Computer assembly

The long supplier lead times and the limited availability of material enforce purchasing of material to be planned on a mid-term basis. In order to receive supply of material as an output of mid-term (master) planning, component demand is needed as an input. To obtain forecasts for components, either directly or indirectly from sales forecasts for final items or product groups, is a hard problem because of missing historical data, the possibility of material or component substitution, and customized BOMs. Thus a forecast accuracy of only 65% can be found in computer assembly [Kilger & Schneeweiss, 2002a], while 90% are attainable in consumer goods production [Fisher, 1997]. Mid-term master planning should also take into account aggregate machine capacity, but bottlenecks in production do not need to be paid as much attention as is necessary in consumer goods supply chains.

If an assemble-to order decoupling point [Hoekstra & Romme, 1991] is given, safety stocks have to be held for components instead of final items. Safety stock planning (SSP) is a difficult task because of the long and unreliable supplier lead times, the high risk of obsolete stocks due to short material life cycles and the low forecast accuracy.

Since product differentiation is low and since all production/assembly processes have to be executed during the order lead time, '*order promising*,' i.e., the estimation of reliable (and preferably soon) customer delivery dates, is of very high importance. Order promising is even more difficult if additional configuration checks are necessary because of a configure-to-order decoupling point (open configurations). Due to the long order lead times (as compared to consumer goods supply chains), order promising less relies on actual, but mainly on *planned* supply. Thus the mid-term master plan must build a basis for order promising. Since order promising in computer assembly usually is an online task, there is a high risk that – when processing a query for an order with low revenue – later on a more lucrative order will arrive. In a material constrained supply chain, the latter one cannot be confirmed as requested if material has been assigned to the less preferable order very early. So there may be a need to reserve planned stock for different order classes (ATP allocation [Kilger & Schneeweiss, 2002b]).

In addition to order promising, further short-term matching of demand and supply, i.e., of orders and stock of material or components on hand, is necessary at every stage of the assembly structure (e.g., assembly of the system unit, assembly of the computer, delivery of the complete order including external units). This is a similar task as deployment in consumer goods supply chains. It is only problematic if demand exceeds supply so that shortages occur. In this case, one may try to accelerate the supply process manually (e.g., by negotiating with some critical suppliers). If this fails, the unlucky orders that have to be delayed (thus decreasing the delivery-on-time performance) must be selected. This problem is also known as *shortage gaming* or *rationing game* (see Chapter 4). Note that for intermediate products and fixed configurations the assignment to a respective customer order is of preliminary character and could be changed in the short run (even if this is not desirable).

These two contrary examples show that different types of SCs have quite different requirements for (short- and mid-term operational) planning. Thus, planning concepts have to be tailored to the particular requirements of the SC-type under consideration. The next section will show that Hierarchical Planning is valuable in designing such planning concepts and that APS call for HP, as well. Presenting HP-concepts for the two SC-types *consumer goods manufacturing* and *computer assembly* would go beyond the scope of this chapter. However, the interested reader can be referred to Fleischmann et al. (2002, Chapter 4.3), where HP-concepts for both SC-types are discussed. Furthermore, in [Meyr, Rohde, Schneeweiss, & Wagner, 2002a, Chapter 17.2.2] a workflow of the APS-provider J.D. Edwards is sketched showing how commercial APS modules can be used to implement such an HP-concept of the consumer goods type.

2.3 Hierarchical planning

Thinking about the SCP-matrix, the idea could arise to tackle all planning tasks with one comprehensive, overall planning model simultaneously. Clearly such an approach will never work for reasons of mathematical complexity. But independent of the power of solution procedures and OR methods, such an approach would not be useful, anyway, for the following reasons [cf. e.g., Meal, 1984]:

1. The longer the planning horizon is, the higher will be the *uncertainty*. Thus operational planning models can approximate reality by far closer than strategic models do. What-if-analyses and risk scenarios only play a dominant role in strategic planning.
2. Planning horizons of various lengths also imply different *frequencies* of planning. While strategic decisions have to be made only once or have to be thought over very seldom, short-term operational planning iterates weekly or even daily.

    Mainly because of (1.) and (2.) *rolling horizon planning* [Silver et al., 1998, Chapter 14.3] is very popular. Here the planning horizon is split into several time buckets, but only the first time bucket is put into practice. After this 'frozen horizon' is elapsed, a replanning is done considering new and probably more reliable information. The shorter the planning cycle is, the better decisions can be postponed until they really *have* to be made. At an extreme, replanning is not bound to a specific time structure, but is reacting to critical 'events' like machine breakdowns or significant changes of demand estimates. The progress being made in communication technology in the last few years supports this kind of *event-driven planning*. In this case, however, 'frozen actions' (rather than frozen periods) must be carefully and consistently defined in order to prevent from serious nervousness of planning. It is still an open question how to find the best trade-off between information reliability and nervousness of planning, i.e., how to define the best replanning frequency or share of frozen actions. Note that the usefulness of rolling horizon planning – as compared to planning concepts based on stochastic models – is discussed in Chapter 12.

3. Planning tasks on different planning levels need a different *degree of aggregation* in terms of

- time (e.g., hourly, daily, weekly, monthly or even yearly time buckets),
- place (e.g., individual customer, zip code, country, sales region),
- products (e.g., final product, product family i.e., groups of final items sharing the same setup behavior), and
- resources (e.g., individual machines, groups of alternative machines, plant as a whole).

A strategic decision, comprising a time horizon of several years, cannot be based on the same detailed information as an operational decision does and – on the other hand – cannot produce such detailed information.

4. Decisions are of different importance. Thus they are made by decision-makers with more or less responsibility and influence. Generally one can say, the longer the impact of a decision can be noticed, the higher the decision maker is ranked within the company's organizational hierarchy and the more likely the decision is prepared and/or made by a central instead of a decentral planning unit:

- Long-term strategic decisions are made by top management, but often prepared by central, multi-functional (production, distribution) planning units of middle management.
- However, top management is no longer involved in (at most informed about) mid-term decisions coordinating several decentral planning units of a supply chain. Here middle management is also responsible for the various decisions to be made.
- 'Routine' decisions with short-term operational character like machine scheduling are made decentrally, for example by the production planning unit of a plant or even by machine control people.

Due to these reasons '*planning modules*' have to be built which pool all decisions that are within the responsibility of the same planning unit, share a similar planning horizon (i.e., planning level according to Anthony's framework), and should be made simultaneously because of their strong interdependence.

These planning modules have to be directly derived from (the SC-type's and) the company's specific planning tasks. For example, the need for simultaneous lot-sizing and scheduling that has been identified as a characteristic planning task of the consumer goods industry (cf. Section 2.2.1) implies a short-term, decentral planning module 'lot-sizing and scheduling'.

On the other hand, planning modules themselves interact. Information and guidelines are exchanged between planning modules in all possible directions (cf. Fig. 5):

Top down: Upper planning levels set limits to lower levels, e.g., strategic planning prepares a framework where operational decisions have to be made in.

Bottom up: Feedback information of a lower level drives decision making on an upper level, e.g., setup times determined by short-term shop floor scheduling are input data for mid-term capacity planning.

Downstream: Early (with respect to the material flow) supply chain processes set a frame for subsequent ones, e.g., short-term production outcome limits the amount of final products that can be delivered to customers in the short run. However, this is not only valid for decisions made at the same level of planning. Also mid- and long-term decisions of downstream members of an SC are affected (or even caused) by short-term decisions of upstream members.

Upstream: Demand information – quite often in terms of orders – which is propagated upstream (i.e., contrary to the material flow of a supply chain) drives earlier planning processes.

Hierarchical planning seeks to coordinate planning modules such that the right degree of integration can be achieved (for a given SC-type). In HP *at least* two levels of planning covered by several planning modules exist. One or several planning modules of a lower level, the base level, are coordinated by a single planning module of an upper level, the top level, by means of instructions (see the discussion of Schneeweiss' framework below). The other way around – and opposite to simple successive planning – feedback information of the lower level guides the planning and instructions of the upper level (cf. gray area in Fig. 5).

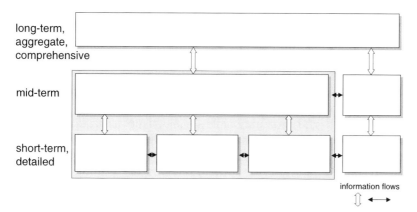

Fig. 5. Interaction between planning modules [cf. Fleischmann et al., 2002].

Further characteristics of HP are an increasing level of detail, decreasing planning horizon, and increasing planning frequency the lower a planning module is settled within this hierarchy. As can easily be seen, these characteristics perfectly fit the requirements of supply chain planning stated in (1.)–(4.) above.

Although HP has been known for several decades (see Chapter 12 and Bitran & Tirupati (1993) for a comprehensive overview), most practical applications concentrate on the production part of a supply chain (known as 'Hierarchical Production Planning,' HPP). The progress being made in information and communication technology encourages and enforces to extend HP over all SC processes. First attempts are shown in Miller (2001) and Stadtler and Kilger (2002). As we will see in the following, APS are well suited to such an approach. Schneeweiss (1999) even goes one step ahead and investigates all types of hierarchies arising in distributed decision making, including mathematical decomposition methods, principal agent relationships and negotiation processes. We will now briefly introduce the framework of Schneeweiss, but restricted to the ideas helping to point out how HP can be implemented by use of APS.

Schneeweiss differs several classes of hierarchies [Schneeweiss, 1999, p. 9]. Two of them are of particular interest in the context of this chapter:

Constructional hierarchies decompose a complex system into simplified subsystems only for reasons of complexity, e.g., if no solution methods exist to solve a monolithic model of the complex system in a single step. The decision has to be made by a single decision-maker (having all, i.e., symmetric information) at a single point in time.

Organizational hierarchies are characterized by an asymmetric information status: either a single person decides on the top and the base level at two different points in time (decision time hierarchy), e.g., a mid-term decision is made first and some short-term decisions are – based on updated demand information – made afterwards. Or several persons at different planning levels having different status of information are (even at the same point in time) involved in the decision process, e.g., a company's central SC planning unit and some representatives of the (decentral) plants or regional sales offices have to agree about a mid-term decision.

The general framework of Fig. 6 shows the possible interrelations between a top and a base level: The top level (1) makes some decision implying an (in its opinion optimal) instruction IN^* (6) that is given to the base level (7). The base level may react (8) to this instruction, e.g., in case of serious problems or expected suboptimalities, so that a replanning of the top level is kicked off. After some further rounds a final decision IN^{**} (9) is implemented in real world causing some consequences to the object system (10) that may influence the next decision of the top level (expost feedback (11) because it can only be observed after the decision has been implemented).

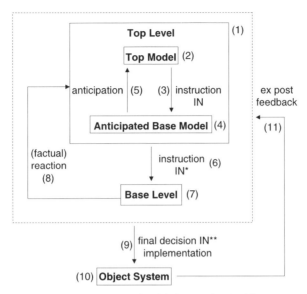

Fig. 6. Framework of Schneeweiss (1999).

In order to shorten the factual planning cycle (1), (6), (7), (8) and to integrate both planning levels more closely, the top level may try to anticipate the base levels behavior by means of an anticipated base model (4). Thus the top level (1) tries to estimate and simulate this planning cycle in advance through its own top (2) and base models (4) and the anticipated instructions (3) and reactions (5).

The type of anticipation may vary substantially [Schneeweiss, 1999, p. 42]. In a *nonreactive anticipation* an explicit reaction ((5), (8)) of the base level is not taken into account; only a feedback influence (11) is possible, e.g., by means of setup times having been observed in the past. Otherwise, the factual reaction (8) of the base level may be anticipated exactly, approximately or implicitly (*reactive anticipation*). Further discussion on anticipation issues can be found in Chapter 12.

A further important characterization of HP is the degree to which the base level's objective function is represented within the top model. If the objective function of the top model does not consider the base level's objectives at all, a *top-down hierarchy* is given. *Tactical-operational hierarchies*, however, additively consider both the top and the base levels' objectives when solving the top model and determining the instruction IN^*. For constructional hierarchies it also may be useful that the top model exclusively optimizes the base model's objective function.

Finally we refer to the production control framework of Bertrand et al. (1990) which also presents basic principles to meet the above requirements (1.)–(4.). By distinguishing between *self-controlling production units* and a (central) *instance controlling the goods flow between these units* entire production networks can be modeled. Since additionally the coordination of production and sales may be considered, the scope can be extended to the planning of complete supply chains. The further distinction between *detailed, item-oriented control* and *aggregate, capacity-oriented control* shows the obvious similarities of this framework to HP concepts.

3 Advanced Planning Systems: General structure

More than 20 years ago companies have started to introduce *Enterprise Resource Planning* (*ERP*) systems (e.g., Baan or SAP/R3) which integrate data of all major business units like controlling, finance, human resources, production or sales. Despite their name ERP systems are rather transactional than planning systems, i.e., their main focus is to provide and exchange consistent data for and between business units.

Production Planning and Control components of ERP systems support planning only in a limited way [Drexl et al., 1994]. This lack of functionality, the recent progress in communication and information technology (like Internet technology, Gigabytes of main memory etc.) and the wide spreading of ERP systems promoted the genesis of *Advanced Planning Systems* (APS).

3.1 Common architecture

APS do not replace ERP systems. They can be seen as add-ons to plan and optimize the supply chain. They often are part of a larger software suite containing ERP or e-Business and SCP software of vendors like i2 Technologies, J.D. Edwards or SAP.

APS extract data from ERP systems, support decision making (through preparing 'optimized' proposals which still have to be controlled, possibly revised, and eventually released by human decision makers) and send the decisions back to the ERP system for final execution. Even though a lot of vendors are on the market, most APS have a common structure. As Fig. 7 [cf. Meyr, Wagner, & Rohde, 2002c] shows, they are comprised of several software modules covering all segments of the SCP-matrix (Fig. 4) introduced in Section 2.1. In the line of Meyr et al. (2002c), the names of the software modules have been chosen independently of the respective software suppliers. Section 4 will give some examples for actual modules of a few widely known APS suppliers.

The software modules themselves contain planning methods such as OR methods, forecasting procedures or simulation tools. However, not all planning tasks of the SCP-matrix are supported by APS. Thus in the following sections for each software module it will be shown in detail which planning tasks are tackled, how the respective sections of the SCP-matrix should be modeled and which potential solution methods are and

Ch. 9. Planning Hierarchy, Modeling and Advanced Planning Systems

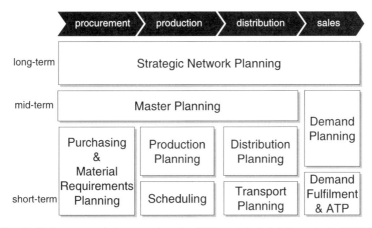

Fig. 7. Software modules covering the SCP-matrix [cf. Meyr et al., 2002c].

should be available in APS. Here only a brief overview over the software modules is given:

Strategic Network Planning (SNP) covers the quantitative part of strategic planning. Questions of network design like plant location, dimension of stocking or production capacities, and the choice of procurement and distribution channels are answered from a quantitative view. Quite often linear (LP) and mixed integer programming (MIP) methods simultaneously evaluate the optimal material flow within different prespecified supply chain scenarios.

Demand Planning (DP) mainly cares about forecasting future demand in a make-to-stock environment on both a mid-term aggregate and short-term detailed basis. Conventional statistical forecasting methods as well as additional features, e.g., to incorporate causal factors, are almost always offered. Quite seldom, however, support for safety stock setting can be found.

Master Planning (MP) coordinates the material flow of the supply chain as a whole for a mid-term planning horizon, mostly by means of LP or MIP. Additionally, rough capacity and material planning is possible with respect to mid-term shortages and seasonalities. The information about customer demand being necessary is usually received from the DP module.

The Production Planning and *Scheduling* (PP&S) modules deal with lot-sizing, machine assignment, scheduling and sequencing. These short-term tasks are strictly dependent on the SC-type. Thus it is not surprising that some vendors offer alternative modules especially designed to satisfy the particular needs of a specific line of business. For the same reason sometimes all of the above tasks are tackled by a single software module.

The same is true for Distribution and Transport Planning. *Distribution Planning* concerns the mid-term tactical constraints within the distribution system, such as the regular transport links, the delivery areas of warehouses, the allocation of customers to sources, and the use of service providers. This module may even overlap with SNP in so far as it supports the detailed design of the distribution network. *Transport Planning* deals with short-term dispatching of the shipments in the distribution and – as far as controlled by the receiver – also on the procurement side.

Demand Fulfillment and Available to Promise (ATP) takes care of the arriving customer orders. It comprises the tasks of order promising, which includes checking the availability of materials and due date setting, and of measures in case of shortage.

Despite its importance short- and mid-term procurement is only seldom supported by an APS module *Purchasing & Material Requirements Planning*. The main tasks of this section, *bill of materials explosion* and *ordering of material*, are often left open for the respective software components of the ERP system or an additional e-Business solution (which usually do not have the functionality that would actually be needed). Because of the rather seldom occurrence and since material release and coordination issues are also discussed in Chapter 12, this software module will not further be investigated in the following.

Commonly, ERP and AP systems of different software suppliers can be integrated by means of standardized interfaces. Since some data like penalty costs or aggregate data are generated for planning purposes only, APS need to carry an own data base, anyway [Shapiro, 1999].

Usually only some of the software modules of Fig. 7 will be installed in a respective company. Due to their central and decentral character, however, these modules may be settled at different locations. Thus further software components – not especially shown in Fig. 7 – are offered to integrate software modules and to collaborate over the Internet. These also contain an alert management system which is able to communicate serious problems within the supply chain.

As briefly indicated, alternative software modules are increasingly launched, especially designed to satisfy the particular needs of some lines of business. Thus APS implicitly orient on the particular planning tasks of a respective SC-type as already claimed in Section 2.1. So the general structure of APS is well suited to the SCP-matrix. As this also holds valid for HP, we will now discuss how APS and HP coincide (in terms of the framework of Schneeweiss, see Section 2.3).

Having identified a planning module and its respective planning tasks, a planning model has to be built and an appropriate solution method has to be found. In other words, a tool of the APS's respective software module has to be selected that solves the planning model (at least approximately). Quite often however, the planning problem will be too complex with respect to the power of the solution methods being offered. In this case a decomposition approach may be appropriate. Thus a *constructional hierarchy* is given where one decision-maker (with symmetric information) at a single point in time is only interested in a final solution to his respective planning problem. The Dantzig–Wolfe scheme [Dantzig & Wolfe, 1961] is a sophisticated example for such a type of hierarchical decomposition. Linear Programming, the solution method used by Dantzig–Wolfe, is applied in most Master Planning modules of APS.

On the other hand, several planning modules covering distinct functional areas on different levels of the SCP-matrix have to be linked and coordinated in a hierarchical manner. In this case, an *organizational hierarchy* is given where usually several persons/planning units make decisions at different points in time (decision time hierarchy, e.g., monthly mid-term and daily short-term decisions) on basis of their – more or less – private knowledge because of the different planning frequencies. The framework of Schneeweiss and Fig. 6 will help to demonstrate in four scenarios how HP can be implemented by means of APS:

1. Most HPP systems like the one of Hax and Meal (1975) employ a nonreactive anticipation function with a simple top down criterion, i.e., a binding instruction of the top level (process (6) according to Fig. 6) is sent to the base level and will directly be implemented. The base level cannot kick off a replanning of the top level if serious problems are identified in the short term. It rather has to solve these problems by itself employing short-term fire fighting actions. The top level does not explicitly consider the base level's objectives. There is only *ex post* feedback given back from the base to the top level in a rolling horizon context, e.g., setup times having been observed during the last planning period are input for the top level's next regular planning round.
2. This planning cycle (1), (6), (7), (9), (10), (11) can be implemented with APS, but is not typical for APS. If the base level uses a decision support system, too, it is able to propagate the top level's instruction and to simulate its consequences as soon as the instruction arrives. When serious problems are detected, an alert is generated which is given back to the top level (8) by the alert management system of the APS. The top level evaluates the alert and changes its own plan when necessary.
3. If this reactive process can be executed not only after a planning activity of the top level, but after each planning activity of the base level, a real event-driven planning supplements the traditional rolling horizon time scheme.
4. However, (2.) and (3.) still use top down criteria not directly taking into account the base level's objectives. With (2.) an online reaction to the top level's decision is possible, but a lot of (physical) communication has to be done by these two levels. In order to reduce this and to accelerate the planning process, an explicit reactive anticipation can be

installed, i.e., the top level uses the base level's planning system (software module) in order to simulate the behavior of the base level by itself ((2), (3), (4), (5)). In this case, the top level is also able to evaluate to what extent the base level's objectives are met. Thus a tactical-operational hierarchy can be installed. This bivalent use of the base level's planning system – as simulation tool on the top level and as operational tool on the base level – has also been suggested in Miller (2001) and can be supported by APS. Because of the otherwise high expense in time and the additional effort to be made the scenarios (2.) and (3.) usually are preferred.

3.2 Typical modules

In this section, the single modules of an APS, as shown in Fig. 7, are considered in more detail. For every module, the tasks and their links to other modules are discussed, which may depend on the SC-type. Models and solution methods available for the respective tasks are briefly reviewed. This presentation is mainly restricted to those models and methods occurring in some APS, but also includes some important models that are recommended, but not implemented in APS so far.

3.2.1 Strategic network planning

The task of SNP is to decide on the major facility locations in the supply chain, i.e., the plants, suppliers and distribution centers (see also Fig. 8).

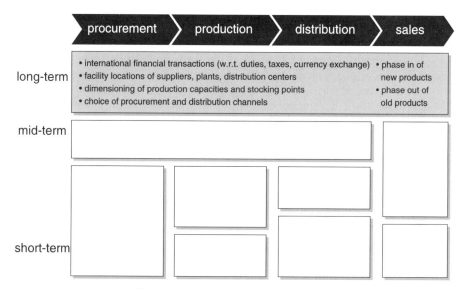

Fig. 8. Planning tasks of SNP modules.

These locational decisions require simultaneous decisions on the flows between the facilities and to the customers, i.e., the quantities to be supplied, produced and shipped. Possible objectives are to minimize the variable flow costs and fixed facility costs for satisfying expected market demands or to maximize the profit or the net present value. However, besides these quantitative financial values, there are important qualitative 'locational factors' which influence the decisions, such as the political stability, the infrastructure and the economic policies of the governments and regional authorities.

The basic model for this type of problems is a network flow problem where the nodes represent existing and potential facilities with associated binary decision variables. The requirements for refinements of this model depend on the planning situation. The most comprehensive problem is the design of a *global supply network*, a highly important issue in the fast changing global economy. In that planning situation, the adequate consideration of the international financial transactions is essential, whereas details of the physical flows are less important. However, in the design of the national part of a supply chain, e.g., a distribution network, the focus may be on the flows and operations implied by the locational decisions. In the following we discuss extensions of the basic model suggested in the literature. A recent survey of models can be found in Vidal and Goetschalckx (1997) and Goetschalckx (2002).

Several products. The distinction of groups of products and materials is very important in most SNP situations, because the pure decision on a plant location makes little sense without the decision on what to produce there. This has a great impact on the production costs, which usually include fixed costs for the allocation of a product to a certain plant. In addition, the BOM allows to consider the dependencies between the suppliers and plants and defines the flows in a multi-stage production network. Note that the decision on the product/plant allocation requires additional binary variables [see Arntzen, Brown, Harrison, & Trafton, 1995; Cohen & Moon, 1991].

Financial variables. In a global network, costs and revenues are affected by duties, exchange rates, national taxes and, within a multi-national company, by the transfer prices. The objective function should be based on the profit after tax in every country, which may also involve the depreciation due to the investment decisions and additional investment incentives [see Canel & Khumawala, 1997, Popp, 1983, Vidal & Goetschalckx, 2001].

Several periods. The subdivision of the strategic planning horizon into several periods, typically years, is very important for three reasons: First, it permits to model the development from the present supply network into the future one, i.e., every locational decision is assigned to a certain period.

Second, the implied depreciation expands over several periods depending on the year of the investment. Surprisingly, this effect is not considered in recent supply network models, but clearly elaborated by Popp (1983). Third, a multi-period model allows to consider an important strategic objective function, the net present value of the cash flow plus the final value of the facilities. However, the consideration of inventory carried from period to period, as it is usual in MP, has no importance in multi-period SNP, contrary to the suggestion of Arntzen et al. (1995) and Goetschalckx (2002); because, the end-of-period inventory in a multi-period model is a seasonal inventory protecting against temporary under-capacity, and hence its development over yearly periods is not a strategic issue.

Lead time aspects. Lead times in a supply chain, which are composed of the duration of the processes in the various arcs of the network, are important factors for the evaluation of a supply chain. Arntzen et al. (1995) include the weighted average operation time, in particular the shipment times, into the objective function. Vidal and Goetschalckx (2000, 2001) express the inventory in transit, the cycle stock due to transportation, and the safety stock as functions of the shipment times and frequencies and consider the corresponding holding cost in the objective function.

Dealing with uncertainties. The development of market demand, prices, exchange rates and cost factors in a long-term planning horizon is highly uncertain. But the inclusion of these aspects in a stochastic optimization model also requires rather uncertain assumptions on the probability distribution of those values. The usefulness of an 'optimal solution' based on such assumptions can be doubted. Instead, the use of several scenarios helps to evaluate a certain network configuration with respect to robustness and flexibility. However, stochastic variation of operational factors such as lead times can be taken into account by including the resulting costs, in particular for safety stocks, in the objective function. For instance, the above mentioned safety stock calculation [Vidal & Goetschalckx, 2000] is based on deterministic (aggregate) flows and stochastic travel times. The necessary safety stock is then obtained by multiplying the flow variable with a safety stock factor, which depends on the travel time distribution and is a characteristic of the transport mode. Another way of taking short-term stochastic variability into account is chance constraint modeling. Vidal and Goetschalckx (2000) restrict the selection of suppliers by the condition that, for every product in every plant, temporal availability of all components has a minimum probability, where each supplier has a known probability for delivering on time.

Economies of scale. A major source of economies of scale is expressed by the fixed costs per facility and per product/facility combination, as explained above. More detailed modeling of concave cost curves for production, handling and transport might be of interest, if the focus is on operational details, in particular in the design of national distribution systems [see Chap. 2 and Fleischmann, 1998a]. But for global SNP, this will mostly be of little importance, in particular transportation is modeled in a highly aggregated form, where the end nodes of the transport flows just represent a country or a regional market.

All these extensions can be easily modeled in MIP form. The number of binary variables equals the number of periods times the number of potential product/plant combinations. Nonlinear cost curves are usually piecewise linearized, which requires additional binary variables. Primary solution methods are the standard MIP techniques, based on Branch and Bound, as implemented in MIP software. However, only models with a limited number of binary variables can be solved to optimality in reasonable computation time by these methods. For more complicated models, either more specific optimization techniques, such as Benders decomposition and Lagrangian relaxation [see Nemhauser & Wolsey, 1988], or heuristics, such as iterative LP rounding schemes or local search, are required.

Most APS provide an SNP module with an LP/MIP solver. The MIP model is not formulated directly by the user, but by means of a problem-oriented modeling interface, which makes modeling easy even for nonexperts, but, on the other hand, reduces the flexibility of modeling. Thus, it depends on the particular interface, whether the above discussed SNP features can be implemented in the system. As to solution methods, APS provide standard MIP solvers and problem-oriented heuristics, but rarely special optimization techniques for facility location. But the notion of 'optimal facilities', stressed by many authors, is, at least for global SNP, misleading. In a decision problem with many, partly qualitative, criteria and highly uncertain data, there is no optimal solution. The task of an SNP software is rather to evaluate given network configurations with financial criteria, a valuable information for the decision maker. This evaluation, too, requires the optimization of the global operations, and the power of SNP software is its ability to provide fast evaluations by global flow optimization. Arntzen et al. (1995), in their detailed report on the use of an SNP model, state that the model is 'typically executed several hundred times during a major study.'

3.2.2 Demand planning

The central role of decoupling points for planning has already been emphasized in Section 2.1. All processes downstream of a decoupling point are based on orders whereas all processes upstream are based on forecasts. In order to hedge against inevitable forecast errors, safety stocks have to be built at the decoupling point. The forecast error is one key parameter when determining the optimal safety stock level. This close relationship between forecasting and SSP justifies that methods for both tasks (see also Fig. 9) are put into a single module 'Demand Planning' (DP). The DP framework of Wagner (2002) supplements these two planning tasks by a

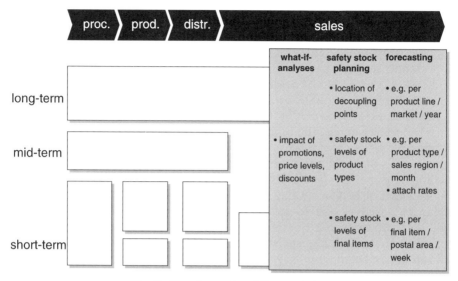

Fig. 9. Planning tasks of DP modules.

third one called 'Simulation/What-if-Analysis.' These methods are valuable when planning to actively influence and guide customer demand instead of simply estimating it.

In the following we will concentrate on forecasting since this is the *major* planning task being tackled by APS. Note that DP is closely related to Demand Fulfillment and ATP (see Section 3.2.6). It is there that the transformation from forecasted ATP quantities to real customer orders takes place.

Forecasting. The task of forecasting is to predict some unknown future events. The question '*What to predict?*' depends upon the planning tasks the forecast ought to give input to. For example, when thinking about product design, yearly sales forecasts of product lines for the company as a whole (measured in monetary units) may be appropriate. Concerning APS, long-term SNP requires estimated sales quantities per product line, year and market as an input for the scenario-based design of SCs (see Section 3.2.1). For mid-term MP, however, monthly estimates per product type and sales region are the desired level of detail. In deliver-to-order SCs, furthermore weekly or daily sales forecasts for final items per postal area are needed for short-term production and distribution planning. Additionally, one has to consider whether final products, parts or raw materials need to be forecasted. As we have seen in Section 2.2, this mainly depends on the locations of the decoupling points and the type of SC considered.

In order to support these tasks, hierarchies have to be defined which at least are comprised of a product (line, type, item), geographic (market, region, postal area), and time (year, month, week) dimension. But also alternative aggregations may be useful, basing, for example, on the company or sales organization [Miller, 2001, Chapter 4.6].

There are several ways to forecast such a hierarchy. For instance, each level of a hierarchy can be estimated independently of each other with respect to past observations on the respective level. In this case, however, forecasts for different levels may not be consistent – even if made at the same point in time. In order to avoid this phenomenon, bottom-up, top-down or middle-out approaches exist [Miller, 2001, Chapter 6.4] which guarantee the consistency within a hierarchy. To check and improve forecasting, sometimes several of these approaches are separately applied. If results differ too much, some sort of agreement (for example in a collaborative way or by using weights) has to be made [see e.g., Vollmann, Berry, & Whybark, 1997, Chapter 8, Miller, 2001, Chapter 6.4].

Furthermore, the product structure may help to generate forecasts. If a steady BOM exists, it can be used to directly derive component forecasts from forecasts for final products. Even if a steady BOM is missing, 'mappings' may be defined and 'attach rates' may be estimated that substitute this linkage [Kilger & Schneeweiss, 2002a].

Independent forecasts for a *single level* can be made on various ways. We will give a brief (and of course incomplete) review of these models and methods. For this purpose, we use a framework (see Fig. 10) of Wagner (2002) and Silver et al. (1998, Fig. 4.1) showing the successive steps that should be executed when establishing a forecast:

1. *Statistical*[1] *(quantitative) forecasting: Time series models* only use past observations of the phenomenon (in this case 'demand') that is to be predicted. Moving average methods and exponential smoothing try to estimate systematic pattern of demand (like a steady level, trend or seasonality) from a history of observations. While these methods assume statistical independency of demand observations within different time periods, Box-Jenkins and ARIMA models [Hanke & Reitsch, 1995, Chapter 10] are more general to the debit of a higher complexity and experienced users being required.

    Causal (deterministic, explanatory) models use data from other sources than the history of the time series predicted. They search for systematic dependencies between one or several leading factors (so-called 'independent variables' like the temperature or promotions) and a dependent variable to be forecasted (e.g., demand). With respect to the number of leading factors considered, simple and multiple regression are distinguished.

    Literature for quantitative forecasting, for example, can be found in

---

[1] Note that the term 'statistical forecasting' is not uniquely used in the literature. While Hanke and Reitsch (1995) exclude causal models, Wagner (2002) and Silver et al. (1998, Chapter 4) use statistical forecasting as a synonym for quantitative forecasting methods in general.

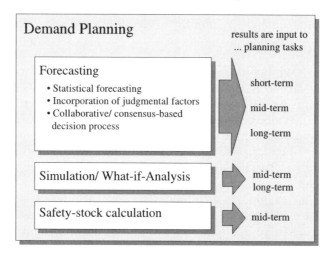

Fig. 10. Methods for demand planning [cf. Wagner, 2002].

Hanke and Reitsch (1995), Makridakis, Wheelwright, and Hyndman (1998), and Nahmias (2001, Chapter 2).

2. *Judgment:* Statistical forecasting as a stand-alone method often is not sufficient because exceptional actions like promotions and price discounts take place. These are only known to the planner and have to be included into the forecasting process in a structured way [Wagner, 2002, Chapter 7.3]. One approach to do this is known as 'Bayesian forecasting' [see Silver et al., 1998, Chapter 4.8.1, and Pole, West, & Harrison, 1994, for further references].

   Judgment is especially important for long-term forecasting where further aspects like technological progress or global economical and political trends have to be considered, too. For this purpose and when historical data are missing at all, scenario writing and similar qualitative forecast methods, which exclusively base upon human knowledge (e.g., the Delphi-method), may be appropriate [Hanke & Reitsch, 1995, Chapter 11].

3. *Collaboration:* These latter methods profit from human knowledge of different sources which are put together in an, if possible, unbiased way. Collaboration processes – not only in forecasting, but also in general – get increasing importance due to new communication technology and the general availability of the Internet. Focusing on forecasting and APS, one can distinguish between intra- and inter-company collaboration.

   'A small example ought to show the need for *intra-company* collaboration. Usually – due to their close contact to the final customer – regional sales offices have the best information of what the customer really wants. On the other hand, price discounts and promotions are coordinated by central headquarter sales. A good forecast has to consider both sources of information what requires an active collaboration between both parties.

One can also profit from bringing together different functional units of a company like sales, production and procurement. However, in this case it should carefully be thought about the question '*Which piece of information can be contributed best by a respective unit?*' When focusing on *forecasting*, such a team has to decide about input data of planning processes and not about results of planning. For example, to support MP (see Section 3.2.3) such a team has to *estimate* future demand and potential limits on future production capacity or material supply. The actual production quantities and stock levels to be aspired to are output of the MP process and nevertheless can be checked by a collaborative team, too. This check, however, is not a task of forecasting and DP, anymore. Quite often, and mainly in *Sales and Operations Planning* [Miller, 2001, Chapter 6.5], the planning tasks DP and MP are not neatly separated from each other. This leads to such confusing and misleading terms like 'constrained demand forecast' often met in practice.

The discussion about the bullwhip effect and the benefits of sharing information [Lee et al., 1997] also supported *inter-organizational collaboration* between legally separated companies. Forecasting probably is the most promising and in practice implemented collaboration function because demand forecasts are less critical to exchange than resource utilization or cost data, for instance. Nevertheless, since an upstream partner typically supplies competitive customers as well, confidentiality of demand forecasts should be ensured as far as possible. Since there never can be a 100% guarantee of information privacy, long-term relations and trust have to be the basic prerequisites for supply chain management and any inter-company collaboration. The best-known example for collaborative forecasting is the integration of manufacturers and retailers according to the standardized processes of the CPFR initiative [CPFR, 2003]. These processes are also supported by some APS.

Concluding the discussion on forecasting, one can say that time-series and causal models are mainly applied to short- and mid-term planning tasks while qualitative methods are appropriate for long-term planning. For more detailed information, the interested reader is referred to Hanke and Reitsch (1995 Tables 4.6 and 11.3). APS usually provide most of the methods mentioned.

What-if-analysis/simulation. Forecasting tries to predict phenomena, but not to actively influence them. Causal models, for example, can be used to estimate the impact certain actions – like promotions, discounts or new products' introduction – have on a product's customer demand. Also the impact on sales of other, correlated products may be predicted. The mid-term planning tasks '*what product and when to promote*', '*what a price level to choose*' or '*when to introduce a new product*' are *decision problems* that go one step ahead.

Analogously to SNP, scenario techniques, what-if-analyses, and simulation may be applied to answer these questions and to decide how to actively influence demand. Again, Master Planning models and methods may help to evaluate such scenarios and to predict the impact of such actions not only on sales (the input of Master Planning), but also on stock levels and production or transportation quantities (the results of Master Planning).

Safety stocks. The relationship between DP and SSP has already been stressed above. Long-term planning tasks in SSP comprise the questions '*where to put stocking points within an SC*' and especially '*where to place decoupling points*'. In Master Planning aggregate safety stocks of product types are required as lower bounds on end-of-period inventories (see Section 3.2.3). Thus the sizes of these stock levels have to be determined in advance with respect to the structure of the SC, the reliability of the forecasting system (forecast errors), the lead times between stocking points, the inventory management system going to be used (review and reorder policies), and the desired customer satisfaction. This planning task also arises in short-term planning for individual items. Note that consistency between mid- and short-term safety stocks has to be ensured, for example by using a bottom-up approach.

APS often do not employ standard order-point or order-up-to-level policies as supposed in the literature [see e.g., Silver et al., 1998, Chapter 7.4] which automatically consider short-term refilling of safety stocks since such policies do not respect limited production capacities. In this case, both safety stock review and refilling policies have to be determined.

Unfortunately, there is only poor methodical support for the above planning tasks (in the literature and in APS as well) because common safety stock models do not respect limited capacities and quite often make critical assumptions with regard to some of the requirements mentioned. The current state of strategic SSP is reviewed in Chapter 3 and in Minner (2000), short- and mid-term safety stock setting is dealt with in Chapters 8, 10, 11 and 12.

3.2.3 Master planning

Master planning (MP) has to synchronize the flow of materials in the *complete supply chain* on a mid-term time horizon [see Section 3.1 and Rohde & Wagner, 2002]. It has to act within the limitations strategic planning sets (e.g., a given network structure, technical properties of machines, etc.) and to effectively utilize the infrastructure that has been established by strategic planning. Thus maximization of net revenues or minimization of total SC costs are typical objectives of MP. On the other hand, MP has to create targets for short-term operational planning that allow a well-integrated coordination of decentral, locally operating planning units (like procurement, production, distribution). Along the lines of Anthony (1965), MP also is denoted as '*Tactical Planning*' [Miller, 2001, Section 1.1.2] or '*Tactical Optimization Modeling*' [Shapiro, 1999].

Ch. 9. Planning Hierarchy, Modeling and Advanced Planning Systems

According to the nature of mid-term planning (see Section 2.1) and the objective to optimize the SC as a whole, a high level of *aggregation* is necessary in MP. Regarding the entities of an SC, only key suppliers, aggregate plants (or at most bottleneck production segments or bottleneck production lines), DCs (TPs), and sales regions are modeled. The planning horizon comprises at least one seasonal cycle, typically one year. This time span usually is subdivided into weekly, monthly or even quarterly time buckets. Products are also aggregated into product groups. For such a mid-term planning problem, Hax and Meal (1975) propose product types, i.e., families of products which share similar (seasonal) demand pattern, inventory holding costs and production rates.

In order to give an idea of the problems being tackled (see also Fig. 11), a small 'basic model' will be formulated as an LP (see also Table 4; for a more general, yet still simple model the reader is referred to Chapter 12). Here p, w, and s denote some production plants, warehouses and sales regions in a two-stage distribution system. t denotes monthly time buckets within a planning horizon of one year. The production quantities x_{pwt} of a single product type in plant p that are pushed to warehouse w in period t are to be determined. Further decision variables are the end of period stocks I_{wt} and the transportation (from warehouse w to sales region s) and sales quantities y_{wst} in months t.

$$\text{Max} \sum_{w,s,t} r_{st} y_{wst} - \sum_{p,w,t} c^p_{pw} x_{pwt} - \sum_{w,t} c^h_w I_{wt} - \sum_{w,s,t} c^d_{w,s} y_{wst} \qquad (3.1)$$

subject to

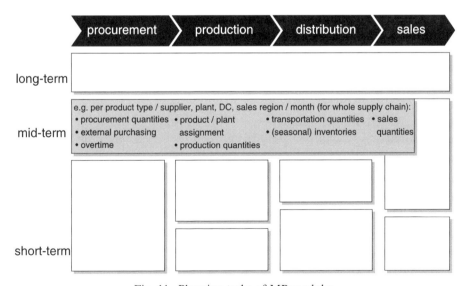

Fig. 11. Planning tasks of MP modules.

Table 4
Data of the basic model

r_{st}	per unit revenue in sales region s (varying over time)
c^p_{pw}	aggregate production (plant p) and transportation costs (to warehouse w)
c^h_w	inventory holding costs (per unit and month)
c^d_{ws}	distribution costs (from warehouse w to sales region s)
I_{w0}	initial inventory of warehouse w
a_p	time needed to produce one unit of the product group
K_{pt}	capacity of plant p in month t (hours)
$\text{Min}_{st}, \text{Max}_{st}$	minimum and maximum sales quantities (regional forecasts per month)

$$I_{wt} = I_{w,t-1} + \sum_p x_{pwt} - \sum_s y_{wst} \quad \forall w, t \tag{3.2}$$

$$a_p \sum_w x_{pwt} \leq K_{pt} \quad \forall p, t \tag{3.3}$$

$$\text{Min}_{st} \leq \sum_w y_{wst} \leq \text{Max}_{st} \quad \forall s, t \tag{3.4}$$

$$x_{pwt}, y_{wst}, I_{wt} \geq 0 \quad \forall p, w, s, t \tag{3.5}$$

The objective is to maximize the total profit of the SC (3.1). The inventory holding constraints (3.2) balance the inflow and outflow of the warehouses and ensure a correct booking of stocks. Limited capacity of the production plants is considered by (3.3). Finally, lower and upper bounds for sales quantities being forecasted by the sales regions give the freedom to put emphasis to the markets with the highest revenues (3.4). Note that, if $\text{Min}_{st} = \text{Max}_{st}$ for all s and t, a predefined demand (forecast) has to be met and cost minimization is pursued instead of profit maximization.

This model can also be reformulated as a simple network flow model. However, this pleasant property gets lost as soon as the basic model is extended and more realistic features are introduced. Quite easy extensions which retain the character of a linear program (with continuous variables) are the following:

Indices: Several types of (final) products, intermediates or raw materials can be introduced by means of further indices. Also additional stages of the production and distribution system and different suppliers/supply locations can easily be formulated. If a higher level of detail is necessary, a plant's production segments or (parallel) production lines may be distinguished, different modes of transport can be introduced, and even alternative modes of production can be considered.

Decisions: Besides production, inventory (especially seasonal stock), transport and sales quantities already included in the base model, backlog, supply quantities, and stock of raw materials or parts can be determined. Further important decisions concerning hiring and firing, external purchasing of final products, and overtime can be made.

Constraints: Similarly to (3.3) and (3.4) capacity limitations of supply, resources and transport can be modeled. Minimum purchase quantities (due to mid-term contracts), minimum inventory levels (safety and lot-size stock), and maximum inventory levels (shelf life restrictions or limited stocking space) can be respected. When formulating a linear program, lead times usually are (but need not to be [Hackman & Leachman, 1989]) considered as multiples of uniform time buckets.

Unfortunately, often further properties have to be modeled which enforce the use of binary and integer variables or nonlinear constraints and thus usually necessitate heuristic solution techniques: Although lot-sizing should be tackled in short-term planning, sometimes setup times or costs, minimum lot-sizes or batch production quantities have already to be respected in mid-term planning (see Vercellis, 1999, Wagner & Meyr, 2002). Also assigning product types to plants usually entails fixed costs for switching the plant to the new products [see e.g., Hax & Meal, 1975]. If production capacity can only be extended by additional full shifts, integer variables are required. On the distribution side, the need for full truck loads [Özdamar & Yazgac, 1999] or single sourcing make things difficult.

The peculiarities of *global SCs* – already introduced in Section 3.2.1 in a strategic context – may also have a strong impact on mid-term decisions like production or transport quantities. Mohamed (1999) demonstrates this for the example of varying exchange rates. A further planning task that has a mid-term character is the determination of transfer prices [Vidal & Goetschalckx, 2001].

The importance of the above features for MP again depends on the SC-type considered. In a material constrained SC like computer assembly, the emphasis is put on the supply side [Kilger & Schneeweiss, 2002a] and plant or transport capacities need not to be modeled in detail. In consumer goods industries on the other hand, production lines are the main bottlenecks and have to be modeled very carefully [see Vercellis, 1999, Wagner & Meyr, 2002].

However, the 'art of modeling' lies in the linkage to the short-term planning processes. The outcome of MP should synchronize different decentral planning units, but also leave them a sufficient degree of freedom in order to hedge against the significant uncertainty due to the long planning horizon. If possible, decisions should be postponed until the point in time they really *have* to be made [de Kok, 1990, Zijm, 1992]. For example, lower bounds for seasonal inventories rather than absolute production quantities are targets for short-term production planning [Fleischmann, 1998b]. So the decentral plants are still able to quickly react to short-term fluctuations of demand.

In practice, MP is very often done by simple spreadsheet calculations without considering capacity limitations. However, also practitioners become more and more aware of the need for a simultaneous consideration of all major constraints of a SC. Thus it is not surprising that most of the major suppliers of APS offer MP modules that base on mathematical optimization methods like LP or MIP. Even if commercial optimizers like CPLEX [ILOG, 2003] have become very powerful in the meantime, (constructional) hierarchies in MP still do exist and problem decomposition is scarcely supported. It remains a challenge to choose the right degree of aggregation and to establish proper links between different planning units.

No clear demarcation between MP and short-term operational or strategic planning can be found in the literature. Thus, for the following brief review, we use the loose definition that MP comprises a supply *network*, i.e., more than a single production plant, and that a decision being made in MP must have a mid-term character. Note that often further plant-wide, mid-term aggregate production plans are used [Silver et al., 1998, Chapter 14] which will be dealt with in Section 3.2.4.

Whereas the term 'Master Planning' is rather seldom used in the literature, the *planning task* MP has been well-known in the context of hierarchical planning for a long time. Even the early papers of Hax and Meal (1975), Glover, Jones, Karney, Klingman, and Mote (1979), and Liberatore and Miller (1985) engage a tactical planning level that constitutes combined production and distribution planning on a mid-term, aggregate basis. Miller (2001) gives an excellent overview over these papers and the respective planning tasks.

Examples for MP can not only be found in hierarchical planning, but also in the literature about (mid-term) integrated production and distribution planning. For instance, Özdamar and Yazgac (1999) describe a practical case of detergents' production and distribution in Turkey. Wagner and Meyr (2002) and Vercellis (1999) present two similar case studies of the consumer goods industry where rough lot-sizing decisions also have to be made on a mid-term level. In Barbarosoglu and Özgür (1999) the close relationship between organizational structure and mathematical solution methods is stressed using integrated production and distribution as an example. Zuo, Kuo, and McRoberts (1991) show that MP is not only relevant in industrial SCs, but also necessary in agricultural systems. Mohamed (1999) and Vidal and Goetschalckx (2001) apply MP in global SCs.

Although Thomas and Griffin's review about coordinated SCM [Thomas & Griffin, 1996] contains a section about operational planning, concerning the separate links '*buyer–vendor coordination*', '*production and distribution coordination*' and '*inventory–distribution coordination*', MP in this sense is not explicitly considered. Shapiro (1999, 2001, p. 45) claims for a '*Tactical Optimization Modeling*' system equaling MP, but also states that such models and methods are not used in practice. The case study [Wagner & Meyr, 2002] pointed out in Section 4.6, for example, disproves this opinion. A quite comprehensive description of MP and its implementation within APS is given by Rohde and Wagner (2002).

3.2.4 Production planning and scheduling

As opposite to MP, the *Production Planning and Scheduling* (PP&S) modules mainly concentrate on a single plant. The overall objective is to establish a detailed, daily or even minutely schedule for each resource of the plant. However, this usually cannot be achieved in a single step. Thus traditionally at least two levels of planning are distinguished, aggregate production planning and detailed scheduling.

Aggregate *production planning* essentially is very similar to MP. Also mid-term planning is aimed at. However, not a whole SC, but only the production processes of a single plant are considered. In return, a higher degree of detail is possible regarding products, production processes and time. Here also time buckets are used, but the sizes of the buckets may be smaller. The planning tasks (see also Fig. 12) tackled are, for example, the allocation of production quantities (of product groups) to production segments or resource groups, production smoothing (by means of overtime, subcontracting, seasonal inventory, back-logging or external purchasing), and aggregate lot-sizing for groups of final items. Because of the close relationship to MP, essentially the same solution methods can be applied. Very often LP and MIP are proposed. Thomas and McClain (1993) give an excellent overview over aggregate PP and examples for LP/MIP formulations being applied. Further literature concerning aggregate PP can be found in Silver et al. (1998, Chapter 14) and Vollmann et al. (1997, Chapters 6, 7, 14,

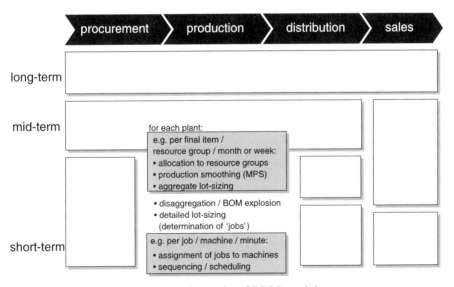

Fig. 12. Planning tasks of PP&S modules.

15), for example. Note that aggregate PP itself can comprise several planning levels with varying degree of detail.

Short-term *scheduling*, on the other extreme, deals with the final assignment of already defined 'jobs' (production lots of final items) to machines, and sequencing and scheduling of these jobs on the respective machines. At this detailed level usually no monetary objectives are pursued anymore, but meeting due dates or achieving a high utilization of resources are aimed at. Time buckets are no more precise enough, thus time-continuous schedules are made. There is a rich literature on (production) scheduling, see e.g., Blazewicz, Ecker, Pesch, Schmidt, and Weglarz (2001), Brucker, 1995, Lawler, Lenstra, Rinnooy Kan, and Shmoys (1993), Morton and Pentico, 1993, Silver et al. (1998, Chapter 17), Vollman et al. (1997, Chapter 13) to name only a few. Note that quick and reliable online *rescheduling* [Morton & Pentico, 1993, Chapter 1.3.6, Smith, 1995] of already existing plans (due to short-term breakdowns of machines, for example) becomes more and more important as PP&S and 'Capable To Promise' (CTP, see Section 3.2.6) modules of APS are increasingly used.

Important planning tasks like disaggregation, BOM explosion or detailed lot-sizing for individual final items and components, which have to be done in between the Production Planning and the Scheduling level, have not been mentioned so far. Indeed, there is much literature on how to link these levels. Overviews of production planning and control concepts are for example given by Zäpfel and Missbauer (1993) or Zijm (2000). Frameworks for PP&S are – among others – proposed in Drexl et al. (1994), Orlicky (1975), Silver et al. (1998, Chapter 13), Vollman et al. (1997, Chapter 1) and in a broader, SC-wide context by Miller (2001) and Shapiro (1999). Some of them will briefly be reviewed.

The probably best known concepts for production planning and control are the *MRP* concept [Orlicky, 1975] and its extension *manufacturing resources planning* (MRP-II) [Wight, 1981]. MRP mainly consists of the processes *master production scheduling* (generating a bucket-oriented production plan for final items) and *BOM explosion* (computing internal demand for parts and raw materials with respect to predefined lead times). Since capacities are not considered in MRP, it has been extended to MRP-II. Here further aggregate planning functions (*business planning and aggregate production planning*), capacity checks (*resource requirements planning, rough cut capacity planning, and capacity requirements planning*) and *short-term scheduling* have been added together with closed loop feedback mechanisms. However, capacity and material (see Chapter 12) violations can only be detected, but not be resolved automatically. So actually no finite capacity loading is given and a real integration of these planning functions thus cannot be achieved. Voss and Woodruff (2000, 2003) illustrate the relationship between MRP and MRP-II and the latter mentioned deficiencies of MRP-II by formulating mathematical optimization models.

Aggregate PP and detailed scheduling also are part of all HPP implementations. Already the early concept of Hax and Meal contained a seasonal planning model and family/item scheduling subsystems, hierarchically integrated and respecting limited capacities. There are a lot of case studies applying such (plant-wide) HPP models in different types of industries, for example by Günther (1986) (washing powder), Stadtler (1986) (food manufacturing), Fleischmann (1998b) (luminescent lamps) and Negenman (2000) (motivated by consumer electronics industry). Miller (2001, Chapter 3) focuses exclusively on HPP and scheduling at the plant level.

The deficiencies of MRP(-II) inspired Drexl et al. (1994) to propose a hierarchical planning framework that is capacity oriented at all levels of planning, especially respecting the peculiarities and particular planning requirements of different production segments within a plant. These production segments differ due to their organizational structure. The concept is comprised of the stages '*SC-wide MP*,' '*plant-specific capacitated master production scheduling*' (MPS), '*detailed lot-sizing and resources allocation*' (DLRA) and '*segment specific shop floor scheduling and control*' (SFSC). MPS has to coordinate all production segments of a plant with respect to limited capacity. Short-term planning has to consider both the targets set by MPS and the particular requirements of each production segment. For instance, for a production segment making one-of-a-kind products, methods of resource constrained project scheduling are proposed [see e.g., Kolisch, 1995]. For a job shop, the two separate planning levels DLRA and SFSC, applying methods of multi-stage capacitated lot-sizing [see e.g., Tempelmeier & Derstroff, 1996] and job shop scheduling, have to be hierarchically integrated. In a production segment consisting of parallel flow lines, however, lot-sizing and scheduling have to be done simultaneously in a single step of planning [see e.g., Drexl & Kimms, 1997, Meyr, 2000], in order to adapt lot-sizes to the tight line capacities and to consider sequence dependent setups.

The framework of Zijm (2000) stresses aspects of technological planning and the importance of safety stocks. Zijm also emphasizes the need to respect the peculiarities of different product/market and organizational structures. PP&S appear in the three modules '*demand management and aggregate capacity planning*,' '*job planning and resource group loading*,' and '*shop floor scheduling and shop floor control*'.

APS usually designate one to two modules to the tasks PP&S. If only a single module is available, this often is intended for scheduling. The aggregate PP then has to be supported by a further MP module. As we have seen, this can easily be done because the solution methods LP and MIP – commonly implemented in MP modules – can be applied to aggregate PP, too.

3.2.5 Distribution and transport planning

This section considers the planning of the external transport in a supply chain. Transport planning mainly occurs as part of the *distribution* function. The transports of materials from an external supplier or from an own remote factory to a production site are usually controlled by the supplier in the scope of his distribution task. But there are important exceptions, where the receiver controls the supply transports, e.g., in the automotive industry. In this case transport planning occurs in the *procurement* function. These two cases will be distinguished in the following.

In an APS, distribution and transport planning is positioned below the MP level and is therefore provided with information and restrictions from SNP, such as the locations in the distribution or procurement network and the potential paths and transport modes, and from MP, such as aggregate quantities to be shipped on every transport link and the increase or decrease of seasonal stocks in the warehouses. In addition, this module uses information from DP, such as the customers to be delivered, the demand forecast and necessary safety stock at every DC. If the APS has separate modules for distribution planning and for transport planning, the latter concerns the short-term (usually daily) dispatch of the transports, whereas the distribution planning module deals with longer-term decisions, which may even overlap with SNP decisions, e.g., the detailed structure of the distribution network.

The appropriate structure of a transport system mainly depends on the size of the shipments. Large shipments can go directly from the source to the destination in full transport units, such as truck loads or containers, whereas small shipments have to be consolidated in a network. In a typical distribution network [see Fleischmann, 1998a, p. 56 ff.], the products from different factories are first brought together to one or several *distribution centers*, then the transports are bundled over long distances up to *RWs* or *TPs*, where the deliveries of small orders to the customers start from. In a typical procurement network the materials from all suppliers in one region are first collected, consolidated at a TP and then shipped to the factory. Another concept uses a warehouse close to the factory, which has to be replenished by the suppliers, as starting point for JIT supply.

Transports as well as TPs and warehouses are often operated by *logistics service providers* (LSPs), who can bundle the flows of various supply chains.

Short-term planning tasks. Short-term transport planning is usually carried out daily with a horizon of a few days. This task, also called *deployment*, has to make the following decisions (see also Fig. 13):

The *quantities to be shipped* have to be determined in a *distribution system* for the replenishment of every DC and every product, in a *procurement system* for the supply of every material item. For the distribution transports to customers the shipment quantity is fixed by the customer orders, except for the case of a customer that is supplied in a VMI concept. The latter case can be treated in exactly the same way as the replenishment of a DC. The shipment quantities can be influenced by the mid-term decisions on shipment frequencies, as discussed below.

The *paths*, along which the transports are performed, have to be selected, in a distribution system among direct delivery from factory, from DC or

Ch. 9. Planning Hierarchy, Modeling and Advanced Planning Systems 501

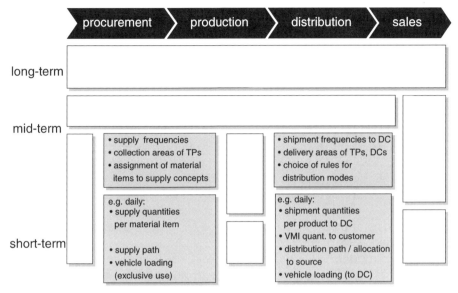

Fig. 13. Planning tasks of Distribution and Transport Planning modules.

shipment via TP, in a procurement system among direct supply, supply via regional TP or from a warehouse. In case of multiple sources for the same product or material, shipments have to be allocated to sources, considering aggregate allocations from MP. All these decisions usually follow simple rules, mainly single source allocations depending on product and shipment size, which are fixed by longer-term planning (see below). Only in case of shortage, deviations from these rules are considered.

For products made to stock, planning deliveries to customers requires matching the current customer orders with the available stock. This is part of the *ATP* function considered in Section 3.2.6.

The task of *vehicle loading* is to adjust the sum of the shipment quantities of the various items on the same transport link to a full vehicle load or a multiple thereof. It is relevant for warehouse replenishment in distribution and for the supply of materials, if the vehicle, as usual in these cases, is used exclusively for the concerned supply chain.

The task of *vehicle routing* occurs mainly for the tours delivering smaller orders from TP to customers and for the tours collecting materials from suppliers for TP consolidation. However, such tours are mostly operated by an LSP, who uses the vehicles simultaneously for flows outside the supply chain under consideration, in order to increase the efficiency of the transport processes. Therefore, vehicle routing must be the task of the LSP, whereas the integration of vehicle routing in an APS, though offered by several APS providers, makes little sense in most cases. In view of this fact and the rich literature on vehicle routing (see Chapters 4 and 13), it will not be considered in the following.

Mid-term planning tasks. For regular transports on the same relation, the *frequency* of the shipments is a key cost factor. It is a mid-term decision variable for the warehouse replenishment on the distribution side and for the supply of materials on the procurement side, setting a target value for the short-term decisions on shipment quantities.

The short-term planning of the shipment paths can be guided by the following mid-term decisions, some of which could even be made on the SNP level:

Delivery areas (or collection areas) of TPs in distribution (or procurement, respectively) and of DCs are to be formed by single-source allocation of locations (e.g., postal codes) to TPs and of TPs to DCs.

Rules for the distribution mode are usually based on limits for the shipment quantity, e.g., up to 30 kg by a parcel service, up to 1000 kg from DC via TP, up to 3000 kg directly from DC and larger orders directly from factory [Fleischmann, 2002, p. 199].

The *assignment of material items to the supply concepts* – directly, via regional TP or via LSP warehouse – is closely related with the supply frequencies.

Models and methods

Shipment frequencies: Planning the frequency of regular transports from a source to a destination, say from a factory to a DC or from a supplier to a factory, is a lot-sizing problem involving the cost of transport and the cycle stock at both locations. Note that the average stock in transit is independent on the frequency [see Fleischmann, 2002, p. 201 f.]. The following *single link model* is a basic model:

Several products i are supplied at the source and needed at the destination with the same steady rate d_i and have holding cost h_i at both locations. The cost per shipment with a vehicle of fixed capacity Q is F. Then it is optimal to ship every t time units all products together with quantities $q_i = d_i t$ [Fleischmann, 1999]. The optimal cycle time t^* is given by the slightly modified EOQ formula [cf. Blumenfeld, Burns, & Daganzo, 1991]:

$$t^* = \min\left\{\frac{Q}{\sum_i d_i};\ \sqrt{\frac{F}{\sum_i h_i d_i}}\right\}.$$

In general, production of several products at the source does not take place steadily, but in consecutive lots of different products, and there are several destinations. *Synchronization* of production lots and shipment quantities reduces the cycle stock [Blumenfeld et al., 1991], but may become very difficult or impractical for a large number of items. However, if transportation planning is done independently from production planning, it decomposes into the above single link cases.

A more complicated interdependence between frequencies, transport cost and cycle stock exists in the case, where the transports are performed in tours, in particular for the collection of materials from suppliers by a regional LSP and for deliveries to several VMI customers: The frequencies, which may be different for every supplier, influence not only the cycle stock, but also the locations to be visited at a certain day and hence the cost of routing. Models for combined vehicle routing and inventory planning have been developed only recently (see the review of Baita, Ukovich, Pesenti, & Favaretto, 1998).

However, in current APS, the mid-term optimization of shipment frequencies is not supported, but frequencies are input to the short-term transportation modules.

TP and DC areas can be planned by means of single-source network flow models or general MIP models, which are provided by most distribution planning modules in APS. But these modules mostly assume linear transportation costs, although economies of scale are particularly important for the shipments via TP. Single source models with nonlinear costs are considered by Simchi-Levi (Chapter 2). This type of models can also be used to set the quantity limits for the distribution mode by parametric variation. Experience shows that the quantity limits have a rather low cost sensitivity [Fleischmann, 1998a].

Shipment quantities: The short-term decision on the shipment quantity of a certain product to a certain DC (or from a certain supplier) is based on the transport frequency, which implies the *transport cycle*, the forecast for the next transport cycle, the safety stock, and the current stock at the destination. For the distribution side, the recent inventory control theory provides optimal replenishment policies in a so-called 'one-warehouse multi-retailer system' (see Chapter 10), which can be interpreted for the manufacturer's supply chain, as one-factory multi-DC system. However, these are single-product models, whereas production mostly is done for several products consecutively in cycles. Therefore, periodic review models, with the review period equal to the production cycle, have to be used at the factory level.

In APS, only simpler rules are used. In a *pull policy*, the general form is

shipment quantity = demand forecast at the destination for the
                    transport cycle
                  + safety stock at the destination
                  ÷ available stock at the destination.

The quantity can be modified by a vehicle loading procedure, as explained below. If the stock at the source is not sufficient for all destinations, it is allocated to the destinations using a 'Fair Share' rule, which takes into account the stock situation at every destination. In a *push policy*, every production lot being produced at the factory is immediately distributed to the DCs according to the fair shares. The critical determination of the safety stocks and of the fair shares, however, is not yet supported sufficiently in APS (see Chapters 10–12 for further details).

Vehicle loading has to consider all products i to be shipped from a certain source to all destinations k and comprises the following steps:

- round up or down the target quantity of every product and every destination to whole transport units.
- adjust the size of the joint shipment, i.e., the sum of the single product quantities, for every destination possibly to a full vehicle capacity Q (or a multiple thereof).

Both steps are constrained by minimum quantities $q_{ik}^{\min}$ and the available stock s_i at the source. The problem can be expressed by an LP with the integer variables

q_{ik} = number of transport units of product i to destination k

as follows:

$$\text{Max} \sum_{i,k} q_{ik} \tag{3.6}$$

subject to

$$\sum_i q_{ik} \leq \left\lceil \sum_i q_{ik}^{\min} / Q \right\rceil Q \quad \forall k \tag{3.7}$$

$$\sum_k q_{ik} \leq s_i \quad \forall i \tag{3.8}$$

$$q_{ik} \geq q_{ik}^{\min}, \quad \text{integer} \quad \forall i, k. \tag{3.9}$$

The objective function can be extended by adding a goal programming term representing the deviation of q_{ik} from the target quantity. However, simple rounding heuristics are used in APS.

3.2.6 Demand fulfillment and available to promise

Demand fulfillment deals with the arriving customer orders. It has to decide on the acceptance of orders and to set the delivery dates that are promised to the customers. This task of *order promising* requires, as a first step, to check the availability of materials, finished products and capacity, the *ATP* check. In a proper sense, ATP considers only available stock and released orders for production and/or purchasing, whereas the check for potential further supply gives the quantities *capable to promise* (*CTP*). But often, ATP is used as the general term. If the ATP quantities are not sufficient for the current customer orders, a *shortage planning* for the open orders or even a *repromising* of promised orders is necessary.

Order promising is a very critical task, as it has a strong impact on the customer service: Setting promised dates too late generates unnecessarily long *order lead times*, and setting them too early makes them unrealistic and generates a poor *delivery reliability*.

Before discussing the tasks (see also Fig. 14) in more detail, we first have to make a distinction w.r.t. the decoupling point [Hoekstra & Romme, 1991], because this is the point in the supply chain where the available stock is kept (see Section 1.1).

In a *deliver-to-order* (make-to-stock) system, the customers expect an immediate fulfillment of the orders, i.e., an order lead time equal to the delivery lead time, typically between 24 and 72 hr. In this case, ATP is concerned with *finished product stocks*, the order promising is rather a yes-or-no decision and the most important step is the shortage planning, i.e., the allocation of tight stocks to the current orders. This step overlaps with the deployment task of short-term transport (see Section 3.2.5) as the quantities allocated from certain stock locations to the customer orders are just the shipment quantities.

In an *assemble-to-order* system ATP is concerned with *stock of components*, and the normal order lead time comprises the lead time for assembly and delivery, typically 5–15 working days, whereas the lead times for supply of the components may be much longer. Then, there is a danger for a pure ATP consideration without CTP: In case of a short component stock, all orders which are not covered by the ATP quantity are shifted to the day after the supply lead time, i.e., to the earliest time when a new supply order will arrive. This may be quite unrealistic, if the assembly capacity (or the supply capacity) is not sufficient to handle so many orders the same day. Hence, both ATP and CTP are important in that case.

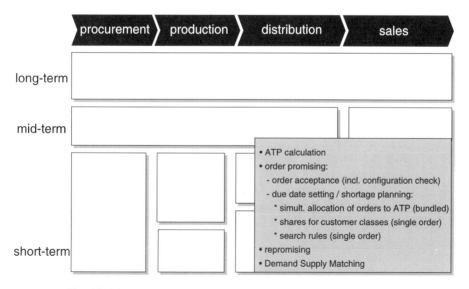

Fig. 14. Planning tasks of Demand Fulfillment and ATP modules.

In a *make-to-order* situation, with a more complex 'make' process, the focus of order promising is on the availability of capacity rather than of materials. In this case, order promising has to be integrated into the PP&S function. It will not be considered in the following.

Tasks. The ATP *calculation* has to provide information for a quick availability check, i.e., the quantities available of a certain product or component at a certain time. For times within the supply lead time, it can be based on available stock and released orders, but for a longer horizon, additional information is required. In APS, usually the master plan is used for that purpose [Kilger & Schneeweiss, 2002b]. As it reflects capacity restrictions, it ensures feasible CTP quantities. However, if the master plan does not exhaust the available capacity and if the current demand is higher than expected, this may lead to CTP quantities which are more restrictive than necessary. A calculation against the available capacity would be preferable. Kilger and Schneeweiss (2002b), instead, suggest a frequent, normally weekly, adjustment of the master plan to the current demand, together with an ATP update.

The *order promising* uses the ATP information in order to decide on current orders, whether they can be confirmed for the customer's requested date or have to be delayed or refused. An important factor for that task is the response time that is allowed after the arrival of an order: In many cases, the answer is expected immediately, for instance for telephone sales and, of course, for the rapidly growing Internet sales. As a consequence, the order promising process has to be performed separately for every single order arrival, without knowledge of future orders. Then, naive use of the ATP quantities would lead to a *first-come-first-served* priority, which is usually undesired. In order to consider specific priorities for different customer classes, shares for every class have to be defined and rules for using up these shares are required. If a longer response time is feasible, arriving orders can be collected and promised in intervals, e.g., once a day. In this case, a bundle of orders is confirmed simultaneously, or, in case of shortage, is subject to allocation of ATP quantities, as explained below.

The *shortage planning* for a single order consists in searching for alternatives, if the ATP quantity for the desired date is not sufficient. Potential alternatives are ATP quantities

- of an earlier date, causing inventory,
- of a later date, causing delay,
- at an alternative source, causing extra cost,
- of a substitute product, if accepted by the customer, usually by downgrading a more expensive product.

The most undesired alternative

- reducing the customer order, delaying or refusing it

cannot be completely avoided in general.

If orders are bundled for order promising, the *allocation* of those alternatives should be done simultaneously to the set of orders. This same task occurs, if there are changes in the expected supplies or in the master plan, which reduce some ATP quantities to negative values. Then, the already promised orders cannot all be delivered on time, hence all orders or a subset of orders have to be subject to a reallocation of the reduced ATP quantities, leading to repromising some orders. Also, a similar allocation task occurs in the assemble-to-order case: As some time passes between order promising and the start of the corresponding production order, an additional short-term allocation of available component stock to orders, before the daily release of production orders, is recommended, called *Demand Supply Matching* [Kilger & Schneeweiss, 2002a]. The various planning tasks of Demand Fulfillment are further discussed in Fleischmann and Meyr (2003). In this paper also LP and MIP models for order promising and order repromising are proposed.

Models and methods. The *ATP quantities* of a certain item ATP_t are calculated for a planning horizon consisting of the periods $t = 1,\ldots,T$ (weeks or days) from the data

I_0 initial inventory on hand
S_t projected supply in period t (released supply orders and master plan)
C_t promised customer orders for period t,

where all quantities and times refer to the decoupling point level. The projected stock on hand is

$$I_t = I_{t-1} + S_t - C_t \quad (t = 1, \ldots, T),$$

where $I_t < 0$ for some t indicates a shortage requiring shortage planning as explained below. If $I_t \geq 0$ $(t = 1, \ldots, T)$, the ATP quantities can be calculated backwards as follows:

$$\left.\begin{aligned} I_T^* &= I_T \\ I_{t-1}^* &= \min\{I_{t-1}, I_t^*\} \\ ATP_t &= I_t^* - I_{t-1}^* \end{aligned}\right\} t = T, \ldots, 1.$$

Note that ATP_t is the maximum amount that can be subtracted from S_t without causing a negative stock on hand in the future and that $ATP_t = 0$ if $C_t \geq S_t$. Alternatively, one may use the cumulative ATP defined as

$$CATP_t = \sum_{s=1}^{t} ATP_s,$$

which shows the total amount available for order promising for period t, but which does not show a potential time lag between supply and consumption. Unfortunately, the ATP calculation is neglected in most textbooks and incorrect in others.

For the *order promising*, the traditional ATP logic in ERP systems uses these ATP quantities on a first-come-first-served basis for the arriving orders whereas APS provide more sophisticated rules: First, every ATP_t or a sum over some time interval can be decomposed into shares for customer groups differing by sales channel, sales volume, region, etc. Kilger and Schneeweiss (2002b) suggest a hierarchical system of customer groups and a top-down allocation of shares following certain rules. In addition, rules for using up these shares may allow some high prioritized customer group to access shares of lower prioritized groups, if the own share is exhausted, but not vice versa [Fischer, 2001]. Setting these shares must be based on forecasts for the groups.

Also, for the *shortage planning* for a single order, APS use simple rules, which check the above mentioned alternatives in a certain sequence [see Kilger & Schneeweiss, 2002b, p. 172 f.]. For the shortage allocation to a set of orders, the rules for the customer group shares are used again. In the latter case, allocations proportional to the order sizes should be avoided, because they generate the 'rationing game' effect, known as one of the causes of the bullwhip effect [Lee et al., 1997].

Both the order promising and shortage planning for a set of orders can be modeled and optimized by LP [Fischer, 2001; Fleischmann & Meyr, 2003]: Given the open orders i with quantities q_i and desired dates d_i and the ATP alternatives j with available quantities a_j and dates t_j, the variables are

x_{ij} the amount of order i covered by alternative j.

One alternative, say j_0, stands for reducing or cancelling orders, with a sufficiently large availability. The model is of the network flow type

$$\text{Min} \sum_{ij} c_{ij} x_{ij} \qquad (3.10)$$

subject to

$$\sum_i x_{ij} \leq a_j \quad \forall j \qquad (3.11)$$

$$\sum_j x_{ij} = q_i \quad \forall i \qquad (3.12)$$

$$x_{ij} \geq 0 \qquad \forall i,j \qquad (3.13)$$

where the major difficulty consists in the definition of the objective coefficients c_{ij}: They may be composed of the costs caused by the alternative

j, a penalty cost for the deviation d_i-t_j depending on the priority of order i and lost sales for $j=j_0$. It is also possible to distinguish between customers accepting delays and quantity reductions and those who do not. Orders i of the latter can only be allocated to alternatives j such that $t_j \leq d_i$ and to j_0 only completely. This condition requires one binary variable for every such order. Forbidding the split of an order i requires one binary variable for every combination (i,j) for such orders i. In the assemble-to-order case, for demand supply matching with component commonality, the model has to be extended, using the BOM structure, so as to include simultaneously all components and open orders.

Models of that type are available in some APS for the deployment function, where the alternatives are primarily thought as different source locations.

4 Advanced planning systems: Particular systems

In this section the peculiarities of the advanced planning solutions of some selected suppliers are shown with respect to the general structure of APS introduced in Section 3.1. Furthermore, mathematical methods applied in the respective software modules are described – as far as this is possible. Finally, (the few) case reports that can be found in the literature up to now are reviewed.

One has to keep in mind that this section only can give a snapshot of the situation up to January 2003. We are aware of the fact that descriptions of APS (Sections 4.1 to 4.5) may be outdated before these pages are printed. The names of software modules and their placement within the overall product range of an APS-vendor change quite often (sometimes semi-annually). However, we think that this section contains valuable information for validating the general structure of APS (because even an aged illustration is better than none). Reasons for such a rapid change of APS are, for example, acquisitions of further APS modules, mergers of software companies, new marketing strategies or the extension of the product spectrum in order to supply tools for collaboration, e-Business etc. In the following, however, we will exclusively concentrate on 'advanced planning' modules supporting the planning tasks introduced so far.

Sections 4.1, 4.2, 4.3, 4.4 and 4.5 are based on firsthand product descriptions of the software suppliers (e.g., Internet publications up to January 2003), on secondhand reports given by users or published in the literature [e.g., Stadtler & Kilger, 2002] and on the authors' own experience with some of the software modules. Please note that the APS and software modules presented are only a small selection of the overall offering – chosen according to the level of information that was accessible to the authors.

When reading APS brochures be aware of the fact that the word 'optimization' very often is misused for marketing purposes because neither a formal mathematical model nor a planning objective are defined. But even if real optimization takes place, software vendors show only slight interest to reveal their planning and optimization methods although this would be desirable from a customer's point of view. Therefore, statements about mathematical methods applied in the following often cannot go into detail and further investigation is troublesome.

4.1 Baan

Baan (2003) originally is a supplier of ERP systems who also offers in the meantime an APS called *iBaan for Supply Chain Management*. Its software modules are shown in Fig. 15. The modules in bold letters came along by the merger with *CAPS Logistics* [CAPS, 2003]. Since then, *iBaan for SCM* has a clear emphasis on transport and distribution processes.

SC Designer and *Coordinator* both apply LP and MIP methods (CPLEX, ILOG, 2003) and further proprietary algorithms tackling the respective long- and mid-term planning problems. Two software modules, *Coordinator* and *SC Planner*, can be used for master planning. While *Coordinator* has a close integration to the *SC Designer* and is recommended as a tool for tactical planning, the *SC Planner* is said to address shorter-term, operational planning issues and order fulfillment. *TransPro* supports freight consolidation, mode/carrier selection and pooling. It is one of the rare tools that allows a comparison of the alternatives 'private/dedicated fleets' and 'third-party carriers'.

The *RoutePro* suite is comprised of several components covering strategic (*RoutePro Designer*) to operational (*RoutePro Dispatcher*) routing decisions, thus suitable for companies with private or dedicated fleets. All modules acquired from *CAPS Logistics* are based on the *iBaan Logistics Toolkit*, a development environment for building customized optimization models of supply chains using a layered architecture [Ratliff & Nulty, 1997].

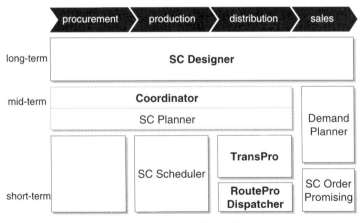

Fig. 15. iBaan for Supply Chain Management.

4.2 i2 Technologies

i2 Technologies (2003a) is market leader in the Advanced Planning segment and propagates the vision to '... add $ 75 Billion of value, in growth and savings, for our customers by the year 2005' (with $ 29.9 Billion being achieved until June 2001 [Miller-Williams, Inc. and i2 Technologies, 2000, Miller-Williams, Inc., 2003]). In Fig. 16 the most important advanced planning components of *i2 Technologies* are shown. They are assigned to the categories *i2 Supplier Relationship Management, i2 Supply Chain Management* and *i2 Demand Chain Management* which again are part of i2's overall software suite *Five.Two* [i2 Technologies, 2003c]. So the advanced planning components are only a small subset of i2's total product range and not promoted as a stand-alone product as Fig. 16 might imply.

Supply Chain Strategist uses 'mathematical optimization methods' (without saying which ones) to propose a network design whose behavior under probabilistic conditions can be validated subsequently via simulation using *Supply Chain Strategist Simulator* (the former IBM tool *Supply Chain Analyzer*).

Supply Chain Planner (also sometimes called '*Master Planner*') tries to generate master plans, feasible with respect to limited material and capacity. With i2's meta-heuristic *Strategy Driven Planning* different types of 'problems' a current plan may have (like lateness, negative inventories etc.) can be defined. Furthermore 'strategies', meant for locally resolving a single problem, have to be determined (mostly simple rules like 'moving a lot backward', but complex algorithms like LP, genetic algorithms or further optimization heuristics could also be incorporated). Each strategy tries to pursue a particular objective, e.g., minimization of lateness, of cost, of the number of problems identified or just feasibility with respect to the respective problem being tackled. For a given plan, a list of problems is gathered and

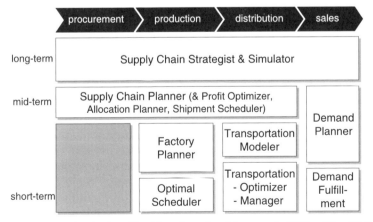

Fig. 16. i2 Technologies: Five.Two [see also Meyr et al., 2002a, Fig. 17.1].

the user is free to apply these strategies to selected problems manually or to define a sequence in which strategies are automatically worked off.

In order to include revenue or cost aspects in a more detailed manner, the *Profit Optimizer* extension can be applied to a Supply Chain Planner model using hierarchical optimization in combination with LP (CPLEX). *Allocation Planner* disaggregates the results of master planning into more detailed shares for order promising. For this, it applies allocation rules as described in Kilger and Schneeweiss (2002b, Chapter 9.2.2).

The *Factory Planner* can be used to match supply and demand with respect to material and capacity constraints and to build a production schedule. In a first step, *Factory Planner* neglects capacity and calculates a (probably infeasible) plan by simply scheduling backward from the requested due dates, checking availability of material and again scheduling forward from the date where material becomes available. In a second step, capacity violations are detected which can be resolved manually by the planner or automatically by use of i2's proprietary *Constraint Anchored Optimization* (*CAO*). CAO performs a rule-based optimization by iteratively prioritizing (and eliminating) capacity violations according to their impact on further resources. Changes of the plan are propagated to all resources affected. The rules to resolve capacity problems aim at short-term objectives like maximal utilization or minimal WIP. If unresolved problems remain, the planner is forced to add capacity (e.g., by additional shifts) or to accelerate material supply (e.g., by negotiation with suppliers). In a final scheduling step, the plan – up to now only defined within buckets of time – can further be refined into a time-continuous plan using the priority rules that are well-known from the job shop scheduling literature.

Additionally to the scheduling tool integrated in the *Factory Planner*, a stand-alone solution called *Optimal Scheduler* is offered which is based on genetic algorithms. *Demand Fulfillment* supports order promising by rule-based consumption of ATP and CTP quantities [see Kilger & Schneeweiss, 2002b, Chapter 9.3].

The modules *Supply Chain Strategist* and *Transportation Modeler*, *Optimizer* and *Manager* go back to the merger of i2 with *InterTrans* in 1998. *Transportation Modeler* and *Optimizer* use proprietary heuristics and MIP solvers.

4.3 J.D. Edwards

J.D. Edwards (2003), another major supplier of ERP-systems, in 1999 acquisited the APS of Numetrix, Ltd. This system, complemented by an order promising module, is now offered as *J.D. Edwards 5 Supply Chain Planning* (cf. Fig. 17). Originally designed for continuous production processes, it now addresses discrete parts production, too, in particular by an additional production scheduling module.

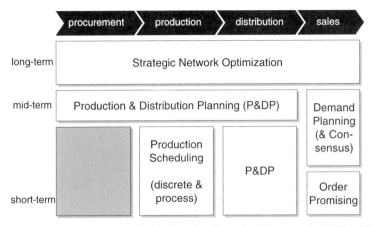

Fig. 17. J.D. Edwards 5 Supply Chain Planning [see also Meyr et al., 2002a, Fig. 17.2].

Strategic Network Optimization (SNO, the former *Linx* and *Enterprise Planning*) engages LP and MIP (CPLEX) in combination with special purpose heuristics to determine a network structure and to check its cost simulatively. The special purpose heuristics tackle problem features that require binary variables like the opening or closing of facilities (capital asset management) and minimum or batch lot-sizes. They perform quite good if only one of these features is given, but may cause problems if several have to be combined. The most striking feature of SNO is its visualization. A supply chain model can completely be built graphically (and not only be depicted graphically; cf. e.g., Günther, Blömer, & Grunow (1998), Wagner & Meyr (2002)). So SNO is easy to use even for practitioners without OR background. Of course, such an approach cannot offer the same flexibility as modeling languages like AMPL [Fourer, Gay, & Kernighan, 1993] or ILOG OPL Studio [ILOG, 2003] do. Since no APS with an interface to such modeling languages is known to the authors, this is not a particular problem of SNO. All major planning tasks of Master Planning can be modeled with SNO. Thus, it is not surprising that SNO is very often used for this purpose, too.

However, because of its multi-user architecture and its communication and alert management features, *Production and Distribution Planning* (P&DP) is recommended for master (and distribution) planning. Four solvers support these planning tasks: *Linear Programming* can be extended by *rounding heuristics* in order to ensure full truck loads. Furthermore, SC-wide demand/supply matching and assignment of customer orders are supported by a rule-based heuristic ('*connect*') which is also able to respect single sourcing. An additional *heuristic* supplements a DRP-like upstream propagation of demand with a subsequent downstream propagation of inventory on hand, following fair share rules. In this case, only networks without cycles can be considered.

Production Scheduling Process (PSP) is applied for short-term planning of continuous (parallel) production lines. It was originally designed for a one- to two-stage production system, but has been extended to several stages of production in the meantime. The solution algorithm is related to neighborhood search with descent algorithms. Basic operations like 'moving lots' or 'increasing inventory' can be applied manually or are put together to predefined algorithms [Kolisch, Brandenburg, & Krüger, 2000]. The (penalty) costs of these operations are evaluated and a plan is changed if a cost improvement has been achieved. With help of the scripting language TCL [Ousterhout, 1994], quite flexible algorithms can be formed respecting peculiarities of practical problems.

4.4 Manugistics

The software suite *Supply Chain Management* (cf. Fig. 18) is part of the NetWORKS software of Manugistics [Manugistics, 2003], besides i2 the second traditional vendor of APS not offering an ERP solution. Only rare information is available about the Manugistics components and methods.

Strategy also relies on LP and MIP techniques. *Sequencing* essentially uses a two phase approach for scheduling jobs on resources. First, an initial solution is quickly created which is subsequently reoptimized and improved. Dependent on the objectives defined by the planner (like minimization of cycle times or maximization of utilization), in both steps specialized algorithms are automatically chosen from a predefined set of heuristic methods.

Early in 2001 Manugistics acquired *STG Holding*, a successor of *Creative Output* which was a production planning software company founded by Eli Goldratt. Out of this acquisition the two software modules

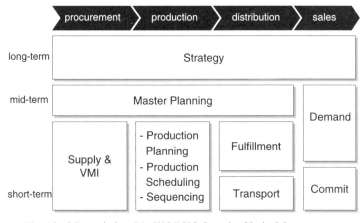

Fig. 18. Manugistics: NetWORKS-Supply Chain Management.

Production Planning and *Production Scheduling* resulted, at the time of STG known as *Advanced Planner Opt2l* and *ST Point Planner and Scheduler*. Both build on the *Optimized Production Technology* system (OPT, [Silver et al., 1998, Chapter 16.2]), which has its origins in the Theory of Constraints and the Drum-Buffer-Rope scheduling concept [see e.g., Fogarty, Blackstone, & Hoffmann, 1991, Chapter 19]. These principles have become very popular – first in the mid of the 1980s and again in the context of APS – by two books of Goldratt (and Cox and Fox, respectively) called '*The Goal*' [Goldratt & Cox, 1986] and '*The Race*' [Goldratt & Fox, 1986]. So all in all three different modules for short-term production planning and scheduling are offered by Manugistics in the meantime.

The *Demand* module is based on FORSYS [Lewandowski, 1982] and applies the Lewandowski method (OPS) in order to automatically determine optimal parameter settings for forecasting [Lewandowski, 1969].

4.5 SAP

The *SAP Advanced Planner and Optimizer* (APO) is part of the *mySAP Supply Chain Management* suite. SAP AG (2003), a major supplier of ERP software, offers its advanced planning solution since 1998. Because of this late entry into the APS market, there is a high chance to incorporate state of the art optimization methods. The wide spreading of the ERP-software *SAP/R3* probably will support the acceptance of the APO. However, development is still ongoing. An overview of the APO software modules (release 3.1) is given in Fig. 19.

Network Design contains both continuous and discrete models for facility location. Voronoi diagrams and Weber problems are used to analyze an existing distribution system and to propose candidate locations for the opening of new facilities. These can further be investigated by discrete MIP models [Kalcsics, Melo, Nickel, & Schmid-Lutz, 2000]. As solver, again, the CPLEX optimization library is used.

The *Supply Network Planning* (SNP) module supports Master Planning mainly by means of linear and mixed integer programming (Branch & Bound, CPLEX). Additionally, constraint based programming (also from ILOG, 2003) can be applied to further reduce the solution space. Also a heuristic is offered which generates master plans in a DRP/MRP like manner: customer demand is propagated upstream in order to calculate net demand of intermediate products at different locations. Thus, an infinite plan is built, limited capacity is checked, and capacity and net demand have to be balanced manually according to priority rules for different types of demand (orders, forecast, safety stock refilling etc.). Furthermore, the setting of mid-term safety stocks based on lead times, forecasting accuracy and aspired service level is supported.

For large models, such basic optimization methods can be embedded in meta-heuristics which (to the debit of solution quality) reduce the complexity by means of aggregation (of periods, products or priorities assigned to

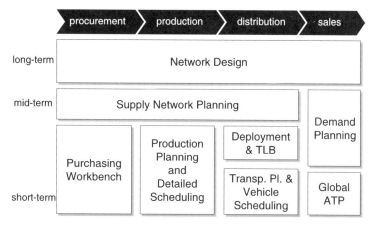

Fig. 19. SAP: Advanced Planner and Optimizer [see also Meyr et al., 2002a, Fig. 17.3].

'subproblems') and decomposition (sequential solving per time window, product or priority group). *Capable to match*, the demand/supply matching engine of SNP is said to apply constraint propagation techniques and goal programming.

For the *Deployment* of production quantities to DC's, simple push and pull rules are used. In case of shortages, fair share rules like 'delivery of stock proportional to the requests of the DCs' are implemented. Additionally to the *Transport Load Builder* (*TLB*) a further, rather short-term *Transport Planning & Vehicle Scheduling* module has been launched which is driven by ILOG components and applies a proprietary genetic algorithm and further heuristics [Meyr et al., 2002a].

Production Planning and *Detailed Scheduling* (PP/DS) also uses a set of basic solution methods like genetic algorithms, constraint propagation (ILOG) and a campaign optimizer (ILOG) which can be embedded into meta-heuristics in order to solve complex problem instances. In a multi-stage production system, the 'bottleneck heuristic', for example, may schedule a known bottleneck resource with help of the campaign optimizer first, and then propagate the results to upstream and downstream resources. Another meta-heuristic performs multi-criterion optimization by means of multi-agent strategies where several agents are coordinated, each focusing on a separate objective and each maybe using an individual optimization strategy.

The Purchasing Workbench tackles supplier selection and purchase order sizing by a two-phase descent heuristic based on local search [Tempelmeier, 2000].

4.6 Case studies

There is a vast amount of 'success stories' (comprising 1–2 pages) available in the Internet which promote successful installations and practical applications of software modules of APS. They euphoricly describe the benefits of APS implementations. Very often DP modules (including collaborative planning) are mentioned, but the type of software module preferably sold depends on the respective APS vendor. However, detailed descriptions how APS are used in practice and how to model with APS are missing.

The following examples that can be found in the literature give at least a little more information than the Internet. In Hoffman (2000), two projects between *SynQuest* (yet another APS vendor [SynQuest Inc., 2003]) and *Ford* are presented concerning the Automotive industry. In the first one, SynQuest's Strategic Network Design tool, the *Supply Chain Designer*, has been applied to design the inbound network of Ford's North American assembly plants. The network comprises 21 assembly plants, 1500 suppliers and 4600 different inbound vehicle parts and components. Original data have been aggregated so that 194 supplier regions remained. Forty-five candidates for cross docking points had to be considered. Forty to fifty different scenarios have been tested which brought up 15 cross docking points to be the best configuration. This solution is reported to contribute 'significant savings' as compared to the configuration proposed originally. Secondly, SynQuest and Ford together developed a tool for the short- to mid-term planning of inbound logistics called *Inbound Planning Engine*. It is used to route and schedule incoming trucks at the receiving docks of the plants. By doing this, the frequencies of part deliveries to assembly plants could be increased significantly without incurring higher overall costs.

In Tappe and Mussäus (1999) and Rodens-Friedrich and Friedrich (2002) a continuous replenishment and VMI project between the consumer goods manufacturer *Reckitt & Colman* and the drugstores' chain *dm* in Germany is described. Two APS modules of Manugistics support DP and ordering processes, but this is rather done by collaborative planning than by use of mathematical (forecasting) methods. So benefits are an increased flexibility and lower stocks due to a shortened planning and ordering time, and decreased replenishment cycles.

Only seldom, the *modeling* with help of APS is presented. Zoryk-Schalla (2001) presents an implementation of the i2 modules *Demand Planner*, *Master Planner* and *Factory Planner* in an European multi-site aluminium manufacturing company. Zoryk-Schalla also emphasizes the hierarchical nature of the i2 system and points out how hierarchical anticipation [Schneeweiss, 1999] could be brought into APS [Zoryk-Schalla, 2001, Chapter 4].

Henrich (2002) demonstrates how J.D. Edwards' *Strategic Network Optimization* (*SNO*) supports the global supply chain planning of the German car manufacturer BMW. SNO is used to simulate and optimize the material flow within BMW's world-wide supply chain when deciding about the assignment of new car models to assembly plants. For this strategic planning task, a planning horizon of 12 years, subdivided into yearly buckets, is considered. The aggregate global supply chain consists of 16 suppliers,

9 assembly plants, 7 customer markets and 42 products. Several scenarios, varying with respect to likely demand figures, future currency exchange rates and potential capacity extensions, are evaluated. The resulting LP models (about 350,000 continuous variables and 230,000 constraints) maximize revenues minus sourcing, production and transportation costs with respect to supplier, plant and transport capacities, customs regulations and currency conversion. If investments were necessary for a certain test scenario, the investment profitability is checked *after* solving the LP by comparing investment costs with the cash flows of the optimal LP solution.

Most of the other case studies concerning modeling (as known to the authors) are given in Stadtler and Kilger (2002). Richter and Stockrahm (2002) describe modeling with SAP APO's PP/DS by means of a practical case in the process industry. Production of synthetic granulate has to be scheduled on a four-step-hybrid flow shop. There are significant sequence-dependent setup times and setup costs. Transport containers and personnel have to be explicitly considered because they are potential bottlenecks. The product spectrum altogether is comprised of 2000 different products and rapidly changing. About 240 jobs are active in the plan. The introduction led to reduced planning times (from several days to one hour), a shorter fixed horizon (one week to two days), a decreased makespan and reduced buffer times.

In Schneeweiss and Wetterauer (2002), a practical application of i2 software in the semiconductor industry is shown. The overall project tackles the planning tasks demand planning, long-term production and distribution planning, mid-term master planning and short-term production scheduling. For the first task, i2's *Demand Planner* and for the latter tasks i2's *Supply Chain Planner* are used. The case study tests whether an alternative modeling of the short-term production scheduling with *Factory Planner* was useful.

Kilger and Schneeweiss (2002a) and Kilger and Stahuber (2002) present an implementation of several i2 modules at a large international computer manufacturer (see SC-type computer assembly in Sections 1.2.2 and 2.2.2). The planning tasks demand planning, master planning, demand supply matching and demand fulfillment are supported by the software modules *Demand Planner*, *Supply Chain Planner*, *Factory Planner* and *Demand Fulfillment*. As it is also the case for Zoryk-Schalla (2001), the scope of these papers is rather the collaboration and integration of the modules than the detailed modeling with one particular software module. The overall project was not finished at the time the paper was written, but the expected benefits of the APS introduction are an increase of forecast accuracy, on time delivery, and inventory turns by 10 to 20%.

In Wagner and Meyr (2002), the implementation of some J.D. Edwards' APS modules at the food and beverages department of a large European consumer goods manufacturer is considered (see Sections 1.2.1 and 2.2.1). The modules SNO and *PSP* have been introduced to support the processes long-term production and distribution planning, master planning (both SNO) and short-term production scheduling (PSP). For DP, the Manugistics module

DP/EE (now NetWORKS Demand) has successfully been in use for several years and needed not to be replaced. The focus of the paper is the (graphical) modeling of the master planning process by means of SNO. The supply chain model comprises three plants (producing up to 20 final items per plant), three DCs, two stages of production that may be potential bottlenecks, and four different product types. Because of minimum lot-sizes, each production line has to be considered separately. The planning horizon of half a year is subdivided into 26 weekly buckets. Decisions to be made are the weekly transportation quantities, material requirements and inventory levels at the DCs, the assignment of products to production lines, and the necessary overtime. The planning objective is the minimization of all relevant costs. Reduced planning time (30% decrease), inventory levels, overtime, and less emergency transports are reported to be the major benefits of this APS implementation.

References

Anthony, R.N. (1965). *Planning and Control Systems: A Framework for Analysis*, Division of Research, Harvard Business School, Boston, Mass.

Arntzen, B. C., G. G. Brown, T. P. Harrison, L. L. Trafton (1995). Global supply chain management at Digital Equipment Corporation. *Interfaces* 25(1), 69–93.

Baan. Homepage. URL: http://www.baan.com, Jan. 2003.

Baita, F., W. Ukovich, R. Pesenti, D. Favaretto (1998). Dynamic routing-and-inventory problems: A review. *Transportation Research A* 32(8), 585–598.

Barbarosoglu, G., D. Özgür (1999). Hierarchical design of an integrated production and 2-echelon distribution system. *European Journal of Operational Research* 118, 464–484.

Bertrand, J. W. M., J. C. Wortmann, J. Wijngaard (1990). *Production Control: A Structural and Design Oriented Approach*, Elsevier, Amsterdam.

Bitran, G.R., D. Tirupati (1993). Hierarchical production planning, in: S.C. Graves, A.H.G. Rinnooy Kan (eds.), *Logistics of Production and Inventory*, volume 4 of *Handbooks in OR and MS*, North-Holland, Amsterdam, pp. 523–568, Chapter 10.

Blazewicz, J., K. H. Ecker, E. Pesch, G. Schmidt, J. Weglarz (2001). *Scheduling Computer and Manufacturing Processes*, 2nd edn., Springer, Berlin.

Blumenfeld, D. E., L. D. Burns, C. F. Daganzo (1991). Synchronizing production and transportation schedules. *Transportation Research B* 25B(1), 23–37.

Brucker, P. (1995). *Scheduling Algorithms*, Springer, Berlin.

Canel, C., B. M. Khumawala (1997). Multi-period international facilities location: an algorithm and application. *International Journal of Production Research* 35(7), 1891–1910.

CAPS. Homepage. URL: http://www.caps.com, Jan. 2003.

Cohen, M. A., S. Moon (1991). An integrated plant loading model with economies of scale and scope. *European Journal of Operational Research* 50, 266–279.

CPFR (Collaborative Planning Forecasting and Replenishment Committee). Homepage. URL: http://www.cpfr.org, Jan. 2003.

Dantzig, G. B., P. Wolfe (1961). The decomposition algorithm for linear programs. *Econometrica* 29, 767–778.

De Kok, A. G. (1990). Hierarchical production planning for consumer goods. *European Journal of Operational Research* 45, 55–69.

Drexl, A., B. Fleischmann, H.-O. Günther, H. Stadtler, H. Tempelmeier (1994). Konzeptionelle Grundlagen kapazitätsorientierter PPS-Systeme. *Zeitschrift für betriebswirtschaftliche Forschung* 46(12), 1022–1045.

Drexl, A., A. Kimms (1997). Lot sizing and scheduling – survey and extensions. *European Journal of Operational Research* 99, 221–235.

Fischer, M.E. (2001). *"Available-to-Promise": Aufgaben und Verfahren im Rahmen des Supply Chain Management*, S. Roderer Verlag, Regensburg.

Fisher, M. L. (1997). What is the right supply chain for your product?. *Harvard Business Review* Mar.–Apr., 105–116.

Fleischmann, B. (1998a). Design of freight traffic networks, in: B. Fleischmann, J.A.E.E. van Nunen, M.G. Speranza, P. Stahly (eds.), *Advances in Distribution Logistics*, volume 460 of *Lecture Notes in Economics and Mathematical Systems*, Springer, Berlin, pp. 55–81.

Fleischmann, B. (1998b). Produktionsplanung bei kontinuierlicher Fließfertigung, in: H. Wildemann *Innovationen in der Produktionswirtschaft – Produkte, Prozesse, Planung und Steuerung*, Kommission Produktionswirtschaft, TCW Transfer-Centrum-Verlag, München, pp. 217–245.

Fleischmann, B. (1999). Transport and inventory planning with discrete shipment times, in: B. Fleischmann, J.A.E.E. van Nunen, M.G. Speranza, and P. Stähly (eds.), *New Trends in Distribution Logistics*, volume 480 of *Lecture Notes in Economics and Mathematical Systems*, Springer, Berlin, pp. 159–178.

Fleischmann, B. (2002). Distribution and transport planning, in: H. Stadtler, C. Kilger (eds.), *Supply Chain Management and Advanced Planning*, 2nd edn., Springer, Berlin, pp. 195–210. Chapter 11.

Fleischmann, B., H. Meyr (2003). Customer orientation in Advanced Planning Systems. To appear in H. Dyckhoff, R. Lackes, J. Reese (eds.), *Supply Chain Management and Reverse Logistics*, Springer, Berlin.

Fleischmann, B., H. Meyr, M. Wagner (2002). Advanced planning, in: H. Stadtler, C. Kilger (eds.), *Supply Chain Management and Advanced Planning*, 2nd edn., Springer, Berlin, pp. 71–95. Chapter 4.

Fogarty, D. W., J. H., Jr., Hoffmann, T. R. Blackstone (1991). *Production and Inventory Management*, 2nd edn., South-Western Publishing Co., Cincinnati Ohio.

Fourer, R., D. M. Gay, B. W. Kernighan (1993). *AMPL – A Modeling Language for Mathematical Programming*, Boyd & Fraser Publishing Company, Danvers, MA.

Glover, F., G. Jones, D. Karney, D. Klingman, J. Mote (1979). An integrated production, distribution, and inventory planning system. *Interfaces* 9(5), 21–35.

Goetschalckx, M. (2002). Strategic network planning, in: H. Stadtler, C. Kilger (eds.), *Supply Chain Management and Advanced Planning*, 2nd edn., Springer, Berlin, pp. 105–121. Chapter 6.

Goldratt, E.M., J. Cox (1986). *The Goal: A Process of Ongoing Improvement*, North River Press, Croton-on-Hudson, N.Y., rev. ed.

Goldratt, E. M., R. E. Fox (1986). *The Race*, North River Press, Croton-on-Hudson, N.Y.

Günther, H.O. (1986). The design of an hierarchical model for production planning and scheduling, in: S. Axsäter, C. Schneeweiss, E. Silver (eds.), *Multi-Stage Production Planning and Inventory Control*, volume 266 of *Lecture Notes in Economics and Mathematical Systems*, Springer, Berlin, pp. 227–260.

Günther, H. O., F. Blömer, M. Grunow (1998). Moderne Softwaretools für das Supply Chain Management. *Zeitschrift für wirtschaftlichen Fabrikbetrieb* 93, 7–8.

Hackman, S. T., R. C. Leachman (1989). A general framework for modeling production. *Management Science* 35(4), 478–495.

Hanke, J. E., A. G. Reitsch (1995). *Business Forecasting*, 5th edn., Prentice Hall, Englewood Cliffs, NJ.

Hax, A.C., H.C. Meal (1975). Hierarchical integration of production planning and scheduling, in: M.A. Geisler (eds.), *Logistics*, volume 1 of *TIMS Studies in Management Science*, North-Holland, Amsterdam, pp. 53–69.

Henrich, P. (2002). *Strategische Gestaltung von Produktionssystemen in der Automobilindustrie*, Shaker Verlag, Aachen.

Hoekstra, S., J. Romme (eds.) (1991). *Integral Logistic Structures: Developing Customer-oriented Goods Flow*, Industrial Press Inc., New York.
Hoffman, K.C. (2000). Ford develops different kind of engine – one that powers the supply chain. *Global Logistics & Supply Chain Strategies*, pp. 42–50, Nov.
i2 Technologies. Homepage. URL: http://www.i2.com, Jan. 2003a.
i2 Technologies. i2 five.two: The complete platform for dynamic value chain management. URL: http://www.i2.com/web505/media/B6636301-3318-4DEA-8B7E6EBDD0CBE681.pdf, Jan. 2003c.
ILOG. Homepage. URL: http://www.ilog.com, Jan. 2003.
J.D. Edwards, Homepage. URL: http://www.jdedwards.com, Jan. 2003.
Kalcsics, J., T. Melo, S. Nickel, V. Schmid-Lutz (2000). Facility location decisions in supply chain management: *Operations Research Proceedings 1999*, Springer, Berlin, pp. 467–472.
Kilger, C., L. Schneeweiss (2002a). Computer assembly, in: H. Stadtler, C. Kilger (eds.), *Supply Chain Management and Advanced Planning*, 2nd edn., Springer, Berlin, pp. 335–352. Chapter 20.
Kilger, C., L. Schneeweiss (2002b). Demand fulfilment and ATP, in: H. Stadtler, C. Kilger (eds.), *Supply Chain Management and Advanced Planning*, 2nd edn., Springer, Berlin, pp. 161–175. Chapter 9.
Kilger, C., A. Stahuber (2002). Integrierte Logistiknetzwerke in der High Tech Industrie – Case Study i2 Technologies, in: H. Baumgarten, H. Stabenau, J. Weber, J. Zentes (eds.), *Management integrierter logistischer Netzwerke*, Verlag Paul Haupt, Bern, pp. 477–505.
Kolisch, R. (1995). *Project Scheduling under Resource Constraints*, Physica-Verlag, Heidelberg.
Kolisch, R., M. Brandenburg, C. Krüger (2000). Numetrix/3 Production Scheduling. *OR Spektrum* 22, 307–312.
Lawler, E.L., J.K. Lenstra, A.H.G. Rinnooy Kan, D.B. Shmoys (1993). Sequencing and scheduling: Algorithms and complexity, in: S.C. Graves, A.H.G. Rinnooy Kan (eds.), *Logistics of Production and Inventory*, volume 4 of *Handbooks in OR and MS*. North-Holland, Amsterdam, pp. 445–522, Chapter 9.
Lee, H., V. Padmanabhan, S. Whang (1997). Information distortion in a supply chain: The bullwhip effect. *Management Science* 43(4), 546–558.
Lewandowski, R. (1969). *Ein voll adaptionsfähiges Modell zur kurzfristigen Prognose*, AKOR-Tagung, Aachen.
Lewandowski, R. (1982). Sales forecasting by FORSYS. *Journal of Forecasting* 1, 205–214.
Liberatore, M. J., T. C. Miller (1985). A hierarchical production planning system. *Interfaces* 15(4), 1–11.
Makridakis, S., S. C. Wheelwright, R. Hyndman (1998). *Forecasting*, 3rd edn., John Wiley and Sons, New York.
Manugistics. Homepage. URL: http://www.manugistics.com, Jan. 2003.
Meal, H. C. (1984). Putting production decisions where they belong. *Harvard Business Review* 2(Mar.–Apr.), 102–111.
Meyr, H. (2000). Simultaneous lotsizing and scheduling by combining local search with dual reoptimization. *European Journal of Operational Research* 120(2), 311–326.
Meyr, H., J. Rohde, L. Schneeweiss, M. Wagner (2002a). Architecture of selected APS, in: H. Stadtler, C. Kilger (eds.), *Supply Chain Management and Advanced Planning*, 2nd edn., Springer, Berlin, pp. 293–304. Chapter 17.
Meyr, H., J. Rohde, H. Stadtler (2002b). Basics for modelling, in: H. Stadtler, C. Kilger (eds.), *Supply Chain Management and Advanced Planning*, 2nd edn., Springer, Berlin, pp. 45–70. Chapter 3.
Meyr, H., M. Wagner, J. Rohde (2002c). Structure of Advanced Planning Systems, in: H. Stadtler, C. Kilger (eds.), *Supply Chain Management and Advanced Planning*, 2nd edn., Springer, Berlin, pp. 99–104. Chapter 5.
Miller, T. C. (2001). *Hierarchical Operations and Supply Chain Planning*, Springer, Berlin.
Miller-Williams, Inc. Miller-Williams Estimates i2 Customers Have Received $ 29.9B in Value. URL: http://www.millwill.com/news/pr-201338.htm, Jan. 2003.
Miller-Williams, Inc., i2 Technologies. i2 Technologies customer value report. Technical report, Oct. 2000.

Minner, S. (2000). *Strategic Safety Stocks in Supply Chains*. Volume 490 of *Lecture Notes in Economics and Mathematical Systems*, Springer, Berlin.

Mohamed, Z. M. (1999). An integrated production–distribution model for a multi-national company operating under varying exchange rates. *International Journal of Production Economics* 58, 81–92.

Morton, T.E., D.W. Pentico (1993). *Heuristic Scheduling Systems*. Wiley series in engineering and technology management. John Wiley and Sons, New York.

Nahmias, S. (2001). *Production and Operations Analysis*, 4th edn., McGraw-Hill/Irwin, Boston.

Negenman, E.G. (2000) *Material Coordination under Capacity Constraints*. PhD thesis, Technische Universiteit Eindhoven, Eindhoven, Netherlands.

Nemhauser, G.L., L.A. Wolsey (1988). *Integer and Combinatorial Optimization*. Wiley-Interscience series in discrete mathematics and optimization. John Wiley and Sons, New York.

Orlicky, J. (1975). *Material Requirements Planning: The New Way of Life in Production and Inventory Management*, McGraw-Hill, Hamburg.

Ousterhout, J. K. (1994). *TCL and the TK Toolkit*, Addison-Wesley, Bonn.

Özdamar, L., T. Yazgac (1999). A hierarchical planning approach for a production–distribution system. *International Journal of Production Research* 37(16), 3759–3772.

Pole, A., M. West, J. Harrison (1994). *Applied Bayesian Forecasting and Time Series Analysis*, Chapman and Hall, New York.

Popp, W. (1983). Strategische Planung für eine multinationale Unternehmung mit gemischtganzzahliger Programmierung. Eine Fallstudie. *OR Spektrum* 5(1), 45–57.

Ratliff, H. D., W. G. Nulty (1997). Logistics composite modeling, in: A. Artiba, S. E. Elmaghraby (eds.), *The Planning and Scheduling of Production Systems*, Chapman and Hall, London, pp. 10–53.

Richter, M., V. Stockrahm (2002). Scheduling of synthetic granulate, in: H. Stadtler, C. Kilger (eds.), *Supply Chain Management and Advanced Planning*, 2nd edn., Springer, Berlin, pp. 305–319. Chapter 18.

Rodens-Friedrich, B., S. A. Friedrich (2002). dm-drogerie markt: Vendor Managed Inventory, in: D. Corsten, C. Gabriel (eds.), *Supply Chain Management erfolgreich umsetzen*, Springer, Berlin, pp. 165–185.

Rohde, J., M. Wagner (2002). Master planning, in: H. Stadtler, C. Kilger (eds.), *Supply Chain Management and Advanced Planning*, 2nd edn., Springer, Berlin, pp. 143–160. Chapter 8.

SAP AG. Homepage. URL: http://www.sap.com, Jan. 2003.

Schneeweiss, C. (1999). *Hierarchies in Distributed Decision Making*, Springer, Berlin.

Schneeweiss, L., U. Wetterauer (2002). Semiconductor manufacturing, in: H. Stadtler, C. Kilger (eds.), *Supply Chain Management and Advanced Planning*, 2nd edn., Springer, Berlin, pp. 321–334. Chapter 19.

Shapiro, J. F. (1999). Bottom-up vs. top-down approaches to supply chain modeling, in: S. R. Tayur, R. Ganeshan (eds.), *Quantitative Models for Supply Chain Management*, Kluwer Academic Publishers, Boston, MA, pp. 737–759. Chapter 7.

Shapiro, J. F. (2001). *Modeling the Supply Chain*, Duxbury, Pacific Grove CA.

Silver, E. A., D. F. Pyke, R. Peterson (1998). *Inventory Management and Production Planning and Scheduling*, 3rd edn., John Wiley and Sons, New York.

Smith, S. F. (1995). Reactive scheduling systems, in: D. E. Brown, W. T. Scherer (eds.), *Intelligent Scheduling Systems*, Kluwer Academic Publishers, Boston, MA, pp. 155–192.

Stadtler, H. (1986). Hierarchical production planning: Tuning aggregate planning with sequencing and scheduling, in: S. Axsäter, C. Schneeweiss, E. Silver (eds.), *Multi-Stage Production Planning and Inventory Control*, volume 266 of *Lecture Notes in Economics and Mathematical Systems*, Springer, Berlin, pp. 197–226.

Stadtler, H., C. Kilger (eds.) (2002). *Supply Chain Management and Advanced Planning*, 2nd edn., Springer, Berlin.

Stephens, S. (2001). Supply chain council and supply chain operations reference (SCOR) model overview. Supply Chain Council, 303 Freeport Road, Pittsburgh, PA 15215, USA, URL: http://www.supplychainworld.org/WebCast/SCOR50_overview.ppt, Apr. 2001.

SynQuest Inc. Homepage. URL: http://www.synquest.com, Jan. 2003.

Tappe, D., K. Mussäus (1999). Efficient Consumer Response als Baustein im Supply Chain Management, in: S. Meinhardt, K. Hildebrand (eds.), *Supply Chain Management*, HMD – Praxis der Wirtschaftsinformatik 36/207, Hüthig, pp. 47–57.
Tempelmeier, H. (2000). A simple heuristic for dynamic order sizing and supplier selection with time-varying data. Technical report, Department of Production Management, University of Cologne, Germany (to appear in *Production and Operations Management*).
Tempelmeier, H., M. Derstroff (1996). A lagrangean-based heuristic for dynamic mulitlevel multiitem constrained lotsizing with setup times. *Management Science* 42(5), 738–757.
Thomas, D. J., P. M. Griffin (1996). Coordinated supply chain management. *European Journal of Operational Research* 94, 1–15.
Thomas, L.J., J.O. McClain (1993). An overview of production planning, in: S.C. Graves, A.H.G. Rinnooy Kan (eds.), *Logistics of Production and Inventory*, volume 4 of *Handbooks in OR and MS*, North-Holland, Amsterdam, pp. 333–370, Chapter 7.
Vercellis, C. (1999). Multi-plant production planning in capacitated self-configuring two-stage serial systems. *European Journal of Operational Research* 119, 451–460.
Vidal, C. J., M. Goetschalckx (1997). Strategic production–distribution models: A critical review with emphasis on global supply chain models. *European Journal of Operational Research* 98(1), 1–18.
Vidal, C. J., M. Goetschalckx (2000). Modeling the effect of uncertainties on global logistics systems. *Journal of Business Logistics* 21(1), 95–120.
Vidal, C. J., M. Goetschalckx (2001). A globl supply chain model with transfer pricing and transportation cost allocation. *European Journal of Operational Research* 129(1), 134–158.
Vollmann, T. E., W. L. Berry, D. C. Whybark (1997). *Manufacturing Planning and Control Systems*, 4th edn., McGraw-Hill/Irwin, Boston.
Voss, S., D. L. Woodruff (2000). Supply chain planning: Is mrp a good starting point?, in: H. Wildemann (ed.), *Supply Chain Management*, TCW Transfer-Centrum-Verlag, München, pp. 177–203.
Voss, S., D. L. Woodruff (2003). *Introduction to Computational Optimization Models for Production*, Springer, Berlin.
Wagner, M. (2002). Demand planning, in: H. Stadtler, C. Kilger (eds.), *Supply Chain Management and Advanced Planning*, 2nd edn., Springer, Berlin, pp. 123–141. Chapter 7.
Wagner, M., H. Meyr (2002). Food and beverages, in: H. Stadtler, C. Kilger (eds.), *Supply Chain Management and Advanced Planning*, 2nd edn., Springer, Berlin, pp. 353–370. Chapter 21.
Wight, O. (1981). *MRPII: Unlocking America's Productivity Potential*, CBI Publishing, Boston.
Zäpfel, G., H. Missbauer (1993). New concepts for production planning and control. *European Journal of Operational Research* 67, 297–320.
Zijm, W. H. M. (1992). Hierarchical production planning and multi-echelon inventory management. *International Journal of Production Economics* 26, 257–264.
Zijm, W. H. M. (2000). Towards intelligent manufacturing planning and control systems. *OR Spektrum* 22, 313–345.
Zoryk-Schalla, A.J. (2001). *Modeling of Decision Making Processes in Supply Chain Planning Software*. PhD thesis, Eindhoven University of Technology, Eindhoven, NL.
Zuo, M., W. Kuo, K. L. McRoberts (1991). Application of mathematical programming to a large-scale agricultural production and distribution system. *Journal of the Operational Research Society* 42(8), 639–648.

Chapter 10

Supply Chain Operations: Serial and Distribution Inventory Systems

Sven Axsäter

Lund University, Sweden

1 Introduction

1.1 Multi-echelon inventory systems

Multi-echelon inventory systems are common in supply chains, in both distribution and production. In distribution we meet such systems when products are distributed over large geographical areas. To provide good service, local stocking points close to the customers in different areas are needed. These local sites may be replenished from a central warehouse close to the production facility. In production, stocks of raw materials, components and finished products are coupled to each other in a similar way.

The management of multi-echelon inventory systems is a crucial part of supply chain operations. The overall goal is, in general, to minimize the costs for ordering, for capital tied up in the supply chain, and for not providing an adequate customer service. This chapter deals with various techniques that can be used for that purpose. Such techniques are not only useful in operative supply chain management. They are also needed when evaluating the effectiveness of alternative supply chain structures.

The possibilities for efficient control of multi-echelon inventory systems have increased substantially during the last two decades. One reason is the progress in research, which has resulted in new techniques that are both more general and more efficient. Another reason is the development of new information technologies, which have dramatically increased the technical possibilities for supply chain coordination. See also Chapters 9 and 12. A supply chain is not always part of a single company. Therefore, it may be necessary for different companies to work together in order to improve the material flow.

1.2 Different system structures

The main focus of this chapter is on *divergent* or *distribution* inventory systems with stochastic demand. Such systems are most common in distribution. Figure 1 illustrates a two-level distribution system with a central warehouse and a number of retailers. The system in Fig. 1 may also occur in production. The central warehouse can then correspond to the stock of a subassembly, which is used when producing a number of different final products. The retailer stocks in Fig. 1 correspond to the stocks of these final products.

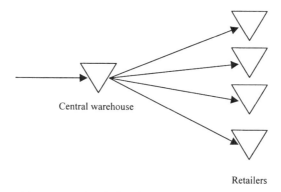

Fig. 1. Two-level distribution inventory system.

In a divergent system each installation has at most one predecessor. In the opposite case where each installation has at most one successor we have an *assembly* system. Such systems, which are dealt with in Chapter 11, are common in production. A *serial* system (Fig. 2) is obviously a special case both of a distribution system and of an assembly system.

Fig. 2. A serial system.

In practice we also quite often meet *general* systems, where some installations have multiple predecessors as well as multiple successors. Such systems are very difficult to handle by scientific methods. See also Chapter 12.

1.3 Objective

The purpose of this chapter is to present and discuss models and techniques for analysis of serial and distribution inventory systems with stochastic demand. The main focus is on policy evaluation, but we will also discuss optimization of reorder points and compare different common ordering systems. Throughout the chapter batch quantities are assumed to be given. They may, for example have been predetermined in a deterministic model. Results for single-echelon models indicate that this gives a very good approximation. See Zheng (1992) and Axsäter (1996). Efficient techniques for deterministic lot sizing are presented in Roundy (1985, 1986) and Muckstadt and Roundy (1993). Furthermore, we assume that there is a single decision maker who wants to optimize the whole supply chain.

This is not the first review dealing with serial and distribution inventory systems. The reader is also referred to related reviews by Axsäter (1993b) Federgruen (1993), Diks, de Kok and Lagodimos (1996), and Van Houtum, Inderfurh and Zijm (1996). The topic is also treated in some recent textbooks. See Sherbrooke (1992b), Silver, Pyke and Peterson (1998), Axsäter (2000a) and Zipkin (2000). Other approaches for placing and setting safety stocks are treated in Chapters 3, 8, and 12.

1.4 Consumable and repairable items

In this chapter we will deal with consumable items. It is interesting to note, however, that we can handle repairable items in much the same way. Some of the early important work in the area (e.g., Sherbrooke, 1968) was focused on systems with repairable items, which are especially common in many military applications.

To understand the relationship between consumable and repairable items consider again the distribution inventory system in Fig. 1. Assume that the stocks contain spare items, which can replace items that fail. When an item fails it is replaced by another item at a local site (retailer). If the item cannot be replaced immediately it is backordered. At the same time the local site is replenishing a new item from the central site. The failed unit is sent to the central site for repair. It is then stocked at the central site. Consider also a corresponding system with consumable items, where all installations apply one-for-one ordering policies. It turns out that the two systems are equivalent if the transportation time from a local site to the central site plus the repair time is the same as the warehouse lead-time in the system with consumable items. For a more extensive treatment of repairable items we refer to Nahmias (1981) and Sherbrooke (1992b).

1.5 Overview of chapter

In Section 2 we compare and discuss different ordering policies that are common in connection with multi-echelon inventory control. After that we consider evaluation and optimization of serial inventory systems in Section 3. Models for evaluation and optimization of order-up-to-S policies in distribution systems are treated in Section 4. Thereafter, in Section 5, we discuss general batch-ordering policies in distribution systems. In conclusion, we give some remarks in Section 6.

2 Different ordering policies

We shall first discuss advantages and disadvantages with continuous and periodic review in Section 2.1. In Section 2.2 we comment on what is known about the structure of optimal policies. Installation stock and echelon stock reorder point policies are compared in Section 2.3, and finally we discuss Material Requirements Planning and KANBAN policies in Section 2.4. See also the discussion of ordering policies in Chapter 12.

2.1 Continuous or periodic review

If the inventory system is monitored continuously and control actions can be taken at any time we have a continuous review inventory system. Another alternative is to check the status of the inventory system periodically and limit control actions to these discrete inspections. This is denoted as periodic review. A continuous review system is, in general, slightly more efficient but also more expensive to operate than a periodic review system. In case of low demand, there are relatively few orders and a continuous review system is economically feasible and often suitable to use. For items with high demand it is generally more practical with periodic review. It is evident that a periodic review system with a very short review period will essentially duplicate the corresponding continuous review system.

When dealing with multi-echelon inventory models it is, in general, possible to transform a continuous review model to a corresponding periodic review model and the other way around. In this chapter we shall deal with both types of models. Our choice of continuous or periodic review will generally reflect what is most common in the literature for each type of model.

We have pointed out that periodic review systems are less expensive to operate than continuous review systems. However, it should also be mentioned that periodic review may be advantageous in certain situations also for other reasons. For example, a periodic review system can be suitable if we want to coordinate orders for different items in order to get a smooth capacity utilization. We can then choose the review times such that orders for items produced in the same machine are not triggered at the same times. Furthermore, while using the simplest type of ordering system, a so-called base stock or order-up-to-S policy, a continuous review system will lead to replenishments that are identical to the customer demands. By using a periodic review system with a relatively long review period we can force the system to order in 'batches' while still avoiding to use a more complex batch-ordering policy. In case of continuous review it is often natural to apply a first come-first served policy at upstream sites. When implementing other allocation rules a periodic review setting may be easier to handle.

2.2 Optimal policies

Assume we have specified some performance criterion (e.g., minimization of holding and backorder costs) for a multi-echelon inventory system with stochastic demand. What does the optimal policy look like? For a single-echelon system we know that a so-called (s, S) policy is optimal under very general conditions, i.e., when the inventory position (inventory on hand, plus outstanding orders, and minus backorders) declines to or below s, we order up to S. There exist some corresponding optimality results for serial multi-echelon inventory systems, see Clark and Scarf (1960), Chen (2000) and Muharremoglu and Tsitsiklis (2001). In general, however, we can expect an optimal control policy to be quite complex. An optimal decision to send a batch from one site to another may depend on the inventory status at all sites. To simplify, it is therefore common to restrict the structure of the control policy. It is, for example, common to let all installations apply simple reorder point policies. An optimization usually means that we are trying to coordinate such simple decision rules in the best possible way. In this chapter we shall assume that such an optimization can be carried out in a centralized decision model. When the installations in a supply chain belong to different companies we may want the decisions to be taken by local decentralized units, while we still get an overall performance that is close to optimum. Such problems are dealt with, in Chapters 6 and 7.

2.3 Installation stock policies versus echelon stock policies

A very common control policy in connection with single-echelon inventory control is a so-called (R, Q) policy, where R is the reorder point and Q is the batch quantity. When the inventory position declines to or below R we order a number of batches of size Q (normally one batch) so that the resulting inventory position is strictly larger than R and less or equal to $R + Q$. In case of continuous review and continuous or unit demand we will always hit the reorder point exactly when ordering, and this policy is then equivalent to an (s, S) policy with $s = R$ and $S = R + Q$. A base stock policy, or order-up-to-S policy, or S policy, means that we always order up to the inventory position S. In case of discrete integral demand such a policy is a special case both of an (R, Q) policy ($R = S-1, Q = 1$), and of an (s, S) policy ($s = S-1$). This policy is also often denoted $(S-1, S)$ policy.

(R, Q) policies are common in connection with multi-echelon inventory control meaning that each installation applies such a policy. The reorder points and the batch quantities are typically different for the installations, though. When dealing with multi-echelon inventory systems the policy is usually denoted as an *installation stock* (R, Q) policy, and R is the installation stock reorder point. The reason is that we want to distinguish the policy from a related type of policy, a so-called *echelon stock* (R, Q) policy.

The echelon inventory position of an installation is the installation inventory position plus the sum of the installation inventory positions of all downstream installations. Consider again Figs. 1 and 2. For installations that have no downstream installations (the retailers in Fig. 1 and installation 1 in Fig. 2), the echelon inventory position is equivalent to the installation inventory position. The echelon inventory position of the warehouse in Fig. 1 is the sum of the warehouse installation stock position and all retailer installation stock positions. Similarly, the echelon stock inventory position of installation n in Fig. 2 is the sum of the installation stock inventory positions for installations $1, 2, \ldots, n$.

An echelon stock (R, Q) policy works exactly as an installation stock (R, Q) policy except that we consider the echelon stock inventory position instead of the installation inventory position. Normally the echelon stock inventory position is larger than the corresponding installation stock inventory position. This implies that an echelon stock reorder point is generally larger than an installation stock reorder point. The basic idea of an echelon stock reorder point policy is that the need to replenish at an installation is reflected by not only the local stock at the installation. If the downstream installations have low stocks levels, they will order in the near future and we need more stock at the considered installation.

While dealing with multi-echelon reorder point policies we assume that an order is always triggered at an installation when the inventory position declines to or below the reorder point. This is the case even if the upstream installation does not have stock on hand. The order is then backordered at the upstream installation. In the literature we sometimes see another equivalent convention, whereby orders are delayed until they can be satisfied by the upstream installation. This will affect the inventory position but not system control.

We shall now compare installation and echelon stock policies in more detail. Consider the serial inventory system in Fig. 2. Installation 1 faces customer demand. We shall analyze the relationship between installation stock (R, Q) policies and echelon stock (R, Q) policies. Assume that the batch quantities are given,

$Q_n =$ batch quantity at installation n.

We assume that the batch quantity at installation n is an integer multiple of the batch quantity at installation $n-1$,

$$Q_n = j_n Q_{n-1}, \tag{2.1}$$

where j_n is a positive integer. This assumption, which is common in the literature, is especially natural if the rationing policy at the installations is to satisfy all or nothing of an order. The installation stock at an installation should then always be an integral number of the lot size at the next downstream installation.

Let us now introduce some additional notation:

IP_n^i installation inventory position at installation n,
IP_n^e $IP_n^i + IP_{n-1}^i + \ldots + IP_1^i =$ echelon stock inventory position at installation n,
R_n^i installation stock reorder point at installation n,
R_n^e echelon stock reorder point at installation n.

We shall assume that the system starts with initial inventory positions IP_n^{i0} and IP_n^{e0} satisfying

$$R_n^i < IP_n^{i0} \leq R_n^i + Q_n, \qquad R_n^e < IP_n^{e0} \leq R_n^e + Q_n. \tag{2.2}$$

It is easy to see that these conditions are always satisfied as soon as each installation has ordered at least once.

Let us now make some initial observations. Assume continuous review or periodic review with the same period for all sites. Consider first the installation stock policy. An order by installation $n > 0$ is always triggered by an order from installation $n-1$. Consequently, if installation n orders, also installations $n-1, n-2, \ldots, 1$ must order at the same time. We say that the policy is *nested*. Consider then the echelon stock policy. Assume that installation $n-1$ orders. This means that the installation inventory position of installation $n-1$ will increase with the amount ordered, but at the same time the installation inventory position of installation n will decrease with the same amount. This means that the echelon inventory position of installation n is unchanged. The echelon inventory position will only decrease due to the final demand at installation 1. We can conclude that the echelon stock policy is not necessarily nested.

Let us now, for simplicity, consider continuous review and unit demand. The results that we shall derive are, however, valid also for periodic review and any type of demand. See Axsäter and Rosling (1993). We shall assume that the installation stock reorder points are chosen such that $IP_n^o - R_n^i$ is an integer multiple of Q_{n-1}. All demands at installation n are for Q_{n-1}, and all replenishments are multiples of Q_{n-1} due to (2.1). Therefore this assumption simply means that we will hit the reorder point exactly when ordering. An alternative reorder point $R_n^i + y$, where $1 \leq y < Q_{n-1}$, will trigger orders at same times and inventory positions. The only difference is that the inventory position will be y units below the reorder point when ordering, because the reorder point is y units higher. Therefore the assumption does not mean any lack of generality.

Let us assume that an installation stock reorder point policy is given. We shall prove

Proposition 1. *A given installation stock reorder point policy can be replaced by an equivalent echelon stock reorder point policy.*

Proof. Consider installation n. Recall that the installation stock policy is nested. Furthermore, due to our convention that $IP_n^{i0} - R_n^i$ is an integer multiple of Q_{n-1}, we hit the reorder point exactly when an order is triggered. Assume that installation n orders. Just after the order we must have the following echelon stock inventory position

$$IP_n^e = \sum_{k=1}^{n} (R_k^i + Q_k). \tag{2.3}$$

Just before the order at installation n the echelon inventory position is Q_n units lower. It is then clear that we can replace the installation stock policy at installation n by an echelon stock policy where

$$R_n^e = R_n^i + \sum_{k=1}^{n-1} (R_k^i + Q_k). \tag{2.4}$$

This is true for any installation and completes the proof.

We shall illustrate how Proposition 1 can be applied by a simple example.

Example 1. Consider $N = 2$ installations and the batch quantities $Q_1 = 10$, and $Q_2 = 20$. Note that (2.1) is satisfied. Assume that the installation stock reorder points are $R_1^i = 22$, and $R_2^i = 35$, and that the initial installation stock inventory positions are $IP_1^{i0} = 27$, and $IP_2^{i0} = 50$. We note that $IP_2^{i0} - R_2^i = 50 - 35 = 15$ is not a multiple of Q_1. According to our convention we therefore change to $R_2^i = 30$, which will give exactly the same control. We can then apply (2.4) to obtain the equivalent echelon stock policy: $R_1^e = R_1^i = 22$, and $R_2^e = R_2^i + R_1^i + Q_1 = 30 + 22 + 10 = 62$. Using that $IP_1^{e0} = 27$, and $IP_2^{e0} = 50 + 27 = 77$, it is easy to verify that the obtained echelon stock reorder point policy will trigger the same orders as the installation stock policy. Recall that the control does not change if we increase R_2^i slightly, for example from 30 to 35. However, if we change R_2^e from 62 to 67 the orders at installation 2 will occur earlier and the policy is no longer nested.

Note that Proposition 1 concerns only the inventory positions. Consequently, the result is not affected by the lead-times. Furthermore the result is true for any type of demand, deterministic or stochastic.

Example 1 illustrates that there exist nonnested echelon stock reorder point policies. Such policies can evidently not be replaced by equivalent installation stock policies. However, it is possible to prove: (Again we prove the result for continuous review and unit demand, although the result is valid in a more general setting.)

Proposition 2. *A nested echelon stock reorder point policy can be replaced by an equivalent installation stock reorder point policy.*

Proof. Let a nested echelon stock policy be given. Note first that if we use $R_1^i = R_1^e$ orders at installation 1 will be triggered at the same times as with the echelon stock policy. Consider an order at installation $n > 1$ with the echelon stock policy. Due to the nestedness we have just after the order:

$$IP_n^i = IP_n^e - IP_{n-1}^e = R_n^e + Q_n - R_{n-1}^e - Q_{n-1}. \tag{2.5}$$

Consequently, an installation stock policy with reorder points

$$R_1^i = R_1^e, \quad R_n^i = R_n^e - R_{n-1}^e - Q_{n-1} \quad n > 1, \tag{2.6}$$

will trigger orders at the same times as the echelon stock policy. This completes the proof.

Let us now make some observations. Propositions 1–2 simply mean that the class of installation stock (R, Q) policies is a subset of the larger class of echelon stock (R, Q) policies. Consequently, given a certain performance criterion, an optimization over the class of echelon stock reorder point policies will always give a result that is at least as good as what can be obtained by an optimization over all installation stock reorder point policies. To illustrate that the difference can be significant let us consider a simple example from Axsäter (2000a).

Example 2 We consider a serial system (Fig. 2) with $N = 2$ installations and batch quantities $Q_1 = 50$ and $Q_2 = 100$. The final demand at installation 1 is constant and continuous, 50 units per unit time. No backorders are allowed at installation 1, and the holding costs at installation 1 are higher than the holding costs at installation 2. The lead-time at installation 1 (excluding possible delays at installation 2) is one time unit and at installation 2 it is 0.5 time units.

It is obvious that the optimal control of the system is illustrated in Fig. 3. The delivery of a batch to installation 2 takes place just-in-time. Immediately after the delivery 50 units are sent to installation 1. But this can never be achieved by an installation stock (R, Q) policy because it is nested. Installation 2 can only order when installation 1 orders, i.e., at times $0, 1, 2, \ldots$, but we would like to have the orders at installation 2 triggered at times $0.5, 2.5, \ldots$, because the lead-time is 0.5. There are then two possibilities if we are using an

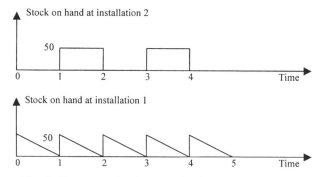

Fig. 3. Inventory development in the optimal solution.

installation stock policy. Either we order too early or too late at installation 2. If we order too late we need to raise the reorder point at installation 1 to avoid backorders. Any way this will lead to additional costs. (See Axsäter and Juntti, 1996.) The optimal control can be obtained by an echelon stock policy with $R_1^e = 50$ and $R_2^e = 75$. Just before time 0.5 the installation inventory position is zero at installation 2 and 75 at installation 1, since one batch is on its way to installation 1.

If all $Q_n = 1$, the considered (R, Q) policy degenerates to an S policy. But an echelon stock S policy, as well as an installation stock S policy, is always nested. As a consequence of Propositions 1 and 2, echelon stock S policies and installation stock S policies that satisfy the condition

$$S_n^e = \sum_{j=1}^{n} S_j^i \qquad (2.7)$$

are equivalent. It can be shown that this equivalence carries over to general multi-echelon inventory systems, for example to the distribution system in Fig. 1.

Propositions 1 and 2 are true also for assembly systems but not for distribution systems. For distribution systems the best echelon stock reorder point policy may outperform the best installation stock reorder point policy, but it may also be the other way around. See Axsäter and Juntti (1996) and Axsäter (1997a) for details.

2.4 Other ordering policies

Reorder point policies based on the installation or echelon stock are most common when dealing with a relatively stationary stochastic demand. In production contexts where demand is lumpy, it is also common to use *Material Requirements Planning (MRP)*. When this type of ordering system is used in distribution it is often denoted *Distribution Requirements Planning (DRP)*. MRP is normally used in a periodic review setting. Let us first summarize the standard MRP procedure.

The planning is carried out in a rolling horizon setting and is based on the following parameters:

T planning horizon, in periods,
L_n constant lead-time of installation n, in periods,
SS_n safety stock at installation n,
Q_n order quantity at installation n,
$D_{n,t}$ external requirements at installation n in period t.

The external requirements are forecasts that normally differ from the real stochastic demands. The real demand in a period will affect the initial inventory situation in the next period. We get the production plan by a level-by-level approach where we start with the installations, which are only facing demand by the final customers. We say that these installations belong to Level 0. A so-called Master Production Schedule gives the requirements at Level 0. A production order in a period is triggered if the installation stock minus the lead-time requirements is not covering the safety stock. Using this we obtain a schedule for all orders over the planning horizon, although only orders in the present period are implemented at this stage. Next we consider Level 1, i.e., installations facing direct requirements from Level 0. These installations may also face external demand. An order at Level 0 means a corresponding requirement at Level 1 a lead-time earlier. After planning the production at Level 1 we can go on with the remaining levels in the same way.

Material Requirements Planning is a quite general ordering system. Axsäter and Rosling (1994) note the following dominance result for a general system (see Section 1.2):

Proposition 3. *For general inventory systems any installation stock (R_n^i, Q_n) policy is duplicated by an MRP system with $L_n = 0$ and $SS_n = R_n^i + 1$.*

To see that Proposition 3 is true note that when $L_n = 0$ there are no lead-time requirements, and an order is triggered in a period if the installation inventory position is less than the safety stock, i.e., if $IP_n^i < SS_n = R_n^i + 1$, or equivalently, if $IP_n^i \leq R_n^i$.

Although simple, Proposition 3 has some interesting implications, for example concerning the concepts of 'Push' and 'Pull' policies. An installation stock reorder point policy is generally regarded as a 'Pull' policy and MRP as a 'Push' policy. It is quite often suggested that 'Pull' policies are advantageous compared to 'Push' policies when dealing with multi-echelon systems. The parameter setting in Proposition 3 is not 'normal'. For example, a system lead-time is, in general, chosen to be close to the average experienced lead-time. Still the considered parameters are feasible and possible, and with these parameters the 'Push' policy will behave exactly as the 'Pull' policy. The conclusion must be that the concepts of 'Push' and 'Pull' are misleading and not sufficiently precise when comparing ordering policies.

It is also possible to show the following result concerning echelon stock reorder point policies and MRP.

Proposition 4. *For serial and assembly systems any echelon stock reorder point policy can be replaced by an equivalent MRP system.*

The proof of Proposition 5 and the corresponding choice of parameters in the MRP system are much more complex in this case and we refer to Axsäter and Rosling (1994) for details.

An ordering policy that is very similar to an installation stock (R, Q) policy is a so-called KANBAN policy. When using a KANBAN policy at an installation we have N containers, each containing Q units and with a card, the KANBAN, at the bottom. When the container is empty the card is used for ordering a new container from the upstream installation. A KANBAN policy is therefore very similar to an installation stock policy with $R = (N-1)Q$. However, if there are already N orders outstanding, i.e., no stock on hand, there are no KANBANs available and no more orders can be triggered. A KANBAN policy can consequently be interpreted as an installation stock reorder point policy with a constraint on the number of outstanding orders. Such a constraint may sometimes be an advantage since it limits work-in-process. See Veatch and Wein (1994), and Spearman, Woodruff and Hopp (1990). Different multi-stage ordering policies with KANBAN type constraints are compared by Axsäter and Rosling (1999).

As we have pointed out in Section 2.3 echelon stock reorder point policies may sometimes outperform installation stock reorder point policies for distribution systems (Fig. 1), but it may also be the other way around. A conclusion may be that using an echelon stock policy is not the best way to include information concerning the inventory situations at the retailers. Policies that use this information in a different way have been evaluated by Marklund (2002) and Cachon and Fisher (2000).

3 Serial systems

In Section 3.1 we consider an infinite horizon version of the well-known Clark-Scarf model. Thereafter the generalization to batch-ordering policies is discussed in Section 3.2.

3.1 The Clark-Scarf model

A well-known early result in multi-echelon inventory theory is the decomposition technique by Clark and Scarf (1960). This was also the origin

Ch. 10. Serial and Distribution Inventory Systems 537

of the echelon-inventory measure. We shall describe this technique for a two-level serial system ($N=2$ in Fig. 2). Installation 1, which faces external demand, replenishes from installation 2, and installation 2 replenishes from an outside supplier with infinite supply. We shall show, that under certain conditions, we can derive optimal echelon-stock order-up-to-levels sequentially. We start with installation 1. Given the optimal base-stock policy for installation 1, we can optimize the policy for installation 2 in a similar way. Clark and Scarf (1960) considered a finite time horizon. We shall, however, instead assume an infinite horizon as in Federgruen and Zipkin (1984b).

Consider a periodic review system where both installations use echelon stock order-up-to-S policies. We assume that the lead-times (transportation times) are integral numbers of periods. The continuous period demand at installation 1 is assumed to be independent across periods. Demand that cannot be met directly from stock is backordered.

It is assumed that all events take place at the beginning of a period in the following order:

1. Installation 2 orders.
2. The period delivery from the outside supplier arrives at installation 2.
3. Installation 1 orders from installation 2.
4. The period delivery from installation 2 arrives at installation 1.
5. The stochastic period demand takes place at installation 1.
6. Evaluation of holding and shortage costs.

Let us introduce the following notation:

L_j lead-time (transportation time) for replenishments at installation j,
$D(n)$ stochastic demand at installation 1 over n periods,
$f^n(v)$ probability density function for the demand at installation 1 over n periods,
$F^n(v)$ cumulative distribution function for the demand at installation 1 over n periods,
μ average period demand at installation 1,
IL_j^e stochastic echelon stock inventory level at installation j,
IL_j^i stochastic installation stock inventory level at installation j,
e_j echelon holding cost per unit and period at installation j, $e_j \geq 0$,
h_j holding cost per unit and period at installation j, $h_1 = e_1 + e_2$, $h_2 = e_2$,
b_1 backorder cost per unit and period,
S_j^e echelon stock order-up-to position at installation j.

(Recall that $IL_1^i = IL_1^e$, i.e., for installation 1 there is no difference between installation and echelon stock. Similarly $S_1^e = S_1^i$ = installation stock order-up-to position at installation 1.)

Our purpose is to minimize the total expected holding and backorder costs. We consider the holding costs for stock on hand at the installations, i.e., $h_1 E(IL_1^i)^+ + h_2 E(IL_2^i)^+$. (We use the notation $x^+ = \max(x, 0)$, $x^- = \max(-x, 0)$, and $x^+ - x^- = x$). An alternative is to use the echelon holding costs: $e_1 E(IL_1^e)^+ + e_2 E(IL_2^e)^+ = e_1 E(IL_1^i)^+ + e_2(E(IL_2^i)^+ + E(IL_1^i)^+ + \mu L_1) = h_1 E(IL_1^i)^+ + h_2(E(IL_2^i)^+ + \mu L_1)$. The constant difference, $h_2 \mu L_1$, is the holding costs for units in transportation from installation 2 to installation 1. (Note that $e_1 = h_1 - h_2$ can be interpreted as the holding cost on the value added when going from installation 2 to installation 1.)

The echelon stock inventory position at installation 2 after ordering in an arbitrary period, t, is always S_2^e. Since the outside supplier has infinite supply, all that is ordered is also immediately sent to installation 2. Consider now the echelon stock inventory level at installation 2 in period $t + L_2$ just after the delivery, IL_2^e. Using a standard argument we can express IL_2^e as $IL_2^e = S_2^e - D(L_2)$, i.e., S_2^e minus the demand during the L periods $t, t+1, \ldots, t+L_2-1$. The inventory position of installation 1 is raised to S_1^e when ordering. It is possible, though, that a part of the order has to be backordered at installation 2 due to insufficient supply. Let us denote

$\hat{S}_1$ realized inventory position after ordering.

We obtain the installation stock inventory level at installation 2 just after the order from installation 1 as $IL_2^i = IL_2^e - S_1^e = S_2^e - D(L_2) - S_1^e$. This is then the installation inventory level at installation 2 during the whole period.

Next we consider installation 1. After ordering in period $t + L_2$, the inventory position is S_1^e, and the installation stock backorder level at installation 2 is $(IL_2^i)^- = (S_2^e - D(L_2) - S_1^e)^-$. The difference, i.e., what is on its way to or has already arrived at installation 1, is the realized inventory position

$$\hat{S}_1 = S_1^e - (S_2^e - D(L_2) - S_1^e)^- = \min(S_1^e, S_2^e - D(L_2)). \tag{3.1}$$

The average holding costs per period at installation 2 can be expressed as

$$\begin{aligned} C_2(S_1^e, S_2^e) &= h_2 E(S_2^e - D(L_2) - S_1^e)^+ \\ &= h_2(S_2^e - \mu_2' - S_1^e) + h_2 E(S_2^e - D(L_2) - S_1^e)^- \\ &= h_2(S_2^e - \mu_2') - h_2 E(\hat{S}_1), \end{aligned} \tag{3.2}$$

where $\mu_2' = L_2 \mu$. The inventory level at installation 1 after the demand in period $t + L_1 + L_2$, $IL_1^i(t + L_1 + L_2)$ is obtained as $\hat{S}_1$ minus the demands in periods $t + L_2, t + L_2 + 1, \ldots, t + L_2 + L_1$, i.e., the demand during $L_1 + 1$ periods, $D(L_1 + 1)$. Let $\mu_1'' = (L_1 + 1)\mu$. Note that $\hat{S}_1$ is independent of this demand. The average period costs at installation 1 are now obtained as

$$\begin{aligned} C_1(S_1^e, S_2^e) &= h_1 E(\hat{S}_1 - D(L_1 + 1))^+ + b_1 E(\hat{S}_1 - D(L_1 + 1))^- \\ &= h_1(E(\hat{S}_1) - \mu_1'') + (h_1 + b_1) E(\hat{S}_1 - D(L_1 + 1))^-. \end{aligned} \tag{3.3}$$

Let us now reallocate the costs slightly. We move the last term in (3.2), $-h_2 E(\hat{S}_1)$ to stage 1 since it depends exclusively on $\hat{S}_1$, and obtain

$$\tilde{C}_2(S_2^e) = h_2(S_2^e - \mu_2'), \tag{3.4}$$

and

$$\tilde{C}_1(S_1^e, S_2^e) = e_1 E(\hat{S}_1) - h_1 \mu_1'' + (h_1 + b_1) E((\hat{S}_1 - D(L_1 + 1))^-. \tag{3.5}$$

This reallocation does not, of course, affect the total costs. Note now that (3.4) is independent of S_1^e. The costs (3.5) at level 1 depend on S_2^e through $\hat{S}_1$, which is a stochastic variable, but it turns out that the optimal S_1^e is independent of S_2^e. To see this interesting fact, let us for a moment forget that the realized inventory position $\hat{S}_1$ is stochastic and depends on S_1^e and S_2^e. We simply assume that we can choose any value of $\hat{S}_1$. This means that we replace (3.5) by

$$\hat{C}_1(\hat{S}_1) = e_1 \hat{S}_1 - h_1 \mu_1'' + (h_1 + b_1) E((\hat{S}_1 - D(L_1 + 1))^-$$
$$= e_1 \hat{S}_1 - h_1 \mu_1'' + (h_1 + b_1) \int_{\hat{S}_1}^{\infty} (u - \hat{S}_1) f^{L_1+1}(u) \, du. \tag{3.6}$$

It is easy to verify that (3.6) is convex and that we can determine the best $\hat{S}_1$ from the first order newsboy type condition

$$F^{L_1+1}(\hat{S}_1^*) = \frac{h_2 + b_1}{h_1 + b_1}. \tag{3.7}$$

Consider now (3.1). If $S_2^e - D(L_2) \geq \hat{S}_1^*$ we obtain the optimal solution if we have $S_1^e = \hat{S}_1^*$. But if $S_2^e - D(L_2) < \hat{S}_1^*$ the best possible value of $\hat{S}_1$ is $\hat{S}_1 = S_2^e - D(L_2)$ due to the convexity of (3.6). Still $S_1^e = \hat{S}_1^*$ gives the optimal solution. We can consequently conclude that $S_1^{e*} = \hat{S}_1^*$ is the optimal order-up-to-level at installation 1, and that this solution is easy to obtain from (3.7). Since this is true for any S_2^e, it is true for any policy at installation 2, also for a batch-ordering policy.

Note that if $e_1 = 0$, or equivalently, if $h_1 = h_2$, (3.7) implies that $S_1^{e*} = \hat{S}_1^* \to \infty$. This means that installation 2 will never carry any stock. This is because in a serial system the whole consumption takes place at installation 1. Consequently, if there is no difference in holding costs, we can just as well move all stock to installation 1 as soon as possible. In a distribution system this is no longer the case. It can be advantageous to keep stock at an upstream installation because it is then still possible to allocate the stock to alternative lower level sites.

We shall now determine the optimal S_2^e. We consider the total costs assuming that we use the optimal S_1^{e*}, but since these costs only depend on S_2^e we denote the costs $\hat{C}_2(S_2^e)$. We obtain the costs from (3.1), (3.4), and (3.6). Since $\hat{S}_1 = S_1^{e*}$ if $D(L_2) \leq S_2^e - S_1^{e*}$, and $\hat{S}_1 = S_2^e - D(L_2)$ otherwise, we get

$$\hat{C}_2(S_2^e) = h_2(S_2^e - \mu_2') + \hat{C}_1(S_1^{e*}) + \int_{S_2^e - S_1^{e*}}^{\infty} \left[\hat{C}_1(S_2^e - u) - \hat{C}_1(S_1^{e*})\right] f^{L_2}(u) \, du. \tag{3.8}$$

The last term in (3.8) can be interpreted as the shortage costs at installation 2 induced by its inability to deliver on time to installation 1. It is easy to verify that $\hat{C}_2(S_2^e)$ is convex, which means that we only have to look for a local minimum when determining S_2^{e*}. Since there are no ordering costs the obtained policy is optimal.

The described technique is easy to generalize to more than two echelons. The additional costs at installation 2 due to insufficient supply are then used as the shortage costs at installation 3, etc. Note that the considered approach means that, starting with the most downstream installation, we can optimize one order-up-to inventory position at a time, i.e., each optimization is for a single variable.

Clark and Scarf introduced the echelon stock concept in their original model. We have also assumed echelon stock policies since this simplifies the derivation. Otherwise this is not important, though. As shown in Section 2.3 we can replace echelon stock S policies by equivalent installation stock S policies. The corresponding optimal installation stock inventory positions are $S_2^{i*} = S_2^{e*} - S_1^{e*}$, and $S_1^{i*} = S_1^{e*}$. However, if we start with initial echelon stock inventory positions that are above the optimal order-up-to positions, the echelon stock policy is still optimal while the installation stock policy may fail.

For two levels and normally distributed demand the exact solution is relatively easy to obtain, see Federgruen and Zipkin (1984a,b). However, in a general case the computations can be very time consuming. Van Houtum and Zijm (1991) have developed approximate techniques. One approximation is exact for demand distributions of mixed Erlang type. Another approximation uses a two-moment fit suggested in Seidel and de Kok (1990). A different type of approximation based on separate single-stage problems has been suggested by Shang and Song (2001). It is easy to generalize to a batch-ordering policy at the most upstream installation (installation 2 in our case). See Section 5.1. It can also be shown that the Clark-Scarf approach can be used for assembly systems. Rosling (1989) has demonstrated that an assembly system can be replaced by an equivalent serial system when carrying out the computations. Chen and Song (2001) have generalized the Clark-Scarf model to demand processes where the demands in different time periods may be correlated.

Svoronos and Zipkin (1991) have demonstrated how exogenous stochastic lead-times can be handled.

3.2 Batch-ordering policies

If all installations order in batches, i.e., not only the most upstream installation, the considered model becomes more complex. Still it is possible to handle also batch-ordering policies very efficiently for serial systems.

Assume that all orders at an installation have to be multiples of a given batch quantity. Chen (2000) has shown that under quite general conditions it is then optimal to apply a multi-stage echelon stock (R, Q) policy when the objective is to minimize holding and backorder costs. Recall from Section 2.3 that the class of echelon stock (R, Q) policies contains the class of installation stock (R, Q) policies as a subset. The best installation stock (R, Q) policy is, in general, not optimal.

Chen and Zheng (1994) have shown how echelon stock (R, Q) policies can be evaluated efficiently. They provide a recursive procedure for determination of the exact probability distributions of the echelon stock inventory levels. This procedure starts at the most upstream installation and proceeds downstream, i.e., not in the same direction as when applying the Clark-Scarf technique. Other approaches for exact policy evaluation in serial systems with batch-ordering have been presented in Axsäter (1993a,b).

For given batch quantities an optimization of the echelon stock reorder points can be carried out sequentially like the optimization of the order-up-to levels in Section 3.1, see Chen (2000).

4 Order-up-to-S policies in distribution systems

In Section 4.1 we show how the Clark-Scarf approach can be used also for approximate optimization of distribution inventory systems. In Sections 4.2, 4.3 and 4.4 we consider models with continuous review and Poisson demand. The classical METRIC approximation is presented in Section 4.2, and two exact techniques in Sections 4.3 and 4.4.

4.1 The Clark-Scarf approach for distribution systems

We shall now consider the distribution system in Fig. 1 and demonstrate how the approach in Section 3.1 can be used as an approximation by using a so-called 'balance' assumption. This technique was sketched in the original paper by Clark and Scarf (1960). Eppen and Schrage (1981) used the approach in a model with identical retailers and normally distributed demand where the central warehouse was not allowed to carry any stock. Federgruen and Zipkin (1984a,c) extended the Eppen and Schrage model in a number of ways including non-identical retailers and stock at the warehouse. See also Federgruen (1993) and Diks and de Kok (1998).

Let us introduce the following notation:

N number of retailers,
L_0 lead-time (integral number of periods) for an order generated by the warehouse,
L_j transportation time (integral number of periods) for a delivery from the warehouse to retailer j,
$D_j(n)$ stochastic demand at installation j over n periods,
$f_j^n(v)$ probability density function for the demand at retailer j over n periods,
$F_j^n(v)$ cumulative distribution function for the demand at retailer j over n periods,
$f_0^n(v)$ probability density function for the total system demand over n periods (the convolution of $f_j^n(v)$ for $j = 1, 2, \ldots,$ N),
μ_j average demand per period at retailer j,
e_j echelon holding cost per unit and period at installation j, $e_j \geq 0$,
h_j holding cost per unit and period at installation j, i.e., $h_0 = e_0$, and $h_j = e_0 + e_j$ for $j > 0$,
b_j backorder cost per unit and period at retailer j.
S_j^e echelon stock order-up-to position at installation j.

A major difference compared to Section 3.1 is that we now have a number of parallel installations at the downstream level. We still assume periodic review and that all events take place in the same order as in Section 3.1. Again we consider an arbitrary period t. The echelon stock level at the warehouse at time $t + L_0$ just after the delivery from the warehouse is obtained as S_0^e minus the total system demand during L_0 periods, $D_0(L_0)$. Since we have N parallel retailers it is now more complicated to allocate stock to the retailers. Assume that

$\hat{S}_j$ realized inventory position at retailer j after ordering in period $t + L_0$.

We consider the costs at retailer j in period $t + L_0 + L_j$. Using the same cost allocation technique as in Section 3.1, we obtain the costs at the warehouse and at retailer j in complete analogy with (3.4) and (3.6).

$$\tilde{C}_0(S_0^e) = h_0(S_0^e - \mu_0'), \tag{4.1}$$

and

$$\hat{C}_j(\hat{S}_j) = e_j \hat{S}_j - h_j \mu''_j + (h_j + b_j) E(\hat{S}_j - D_j(L_j + 1))^-. \tag{4.2}$$

In (4.1) $\mu_0' = L_0 \sum_{j=1}^{N} \mu_j$, and in (4.2) $\mu''_j = (L_j + 1)\mu_j$. We can also determine the value of $\hat{S}_j$ that minimizes (4.2) in analogy with (3.7)

$$F_j^{L_j+1}(\hat{S}_j^*) = \frac{h_0 + b_j}{h_j + b_j}. \tag{4.3}$$

It may now seem reasonable to allocate $\hat{S}_j^*$ to each retailer provided that the sum of all $\hat{S}_j^*$ does not exceed the echelon stock inventory level at the warehouse. This is also what we will do, but it is important to understand that this is just an approximation.

Assume, for example, that we have only two identical retailers and that we in a certain period are able to allocate $\hat{S}_1^* = \hat{S}_2^*$ to each of them. After this allocation we get a large period demand at retailer 2 and no demand at all at retailer 1. This means that we would like to allocate up to $\hat{S}_2^*$ at retailer 2 in the next period. But this may not be possible due to insufficient supply at the warehouse. In that case we will get unequal inventory positions at the two retailers. However, it is rather obvious that it would be better to distribute the inventory positions equally. This might have been possible if we had saved some more stock at the warehouse in the preceding period, i.e., if we had not allocated $\hat{S}_1^* = \hat{S}_2^*$ to both retailers. Due to this 'balance' problem the decision rule that was optimal in the serial case is now only approximate.

The considered approximation is often denoted the 'balance' approximation since it means that we disregard the possibility of unbalanced retailer inventory positions. Another interpretation is that we allow also negative allocations from the warehouse to the retailers. An implication is that in a two-echelon system, given this approximation and if the holding costs are the same, then there is no reason to keep stock at the warehouse.

Let us now accept the 'balance' approximation and also take into account that we cannot always allocate $\hat{S}_j^*$ to each retailer. Assume that the available amount is $v \leq S_r^{e*} = \sum_{j=1}^N \hat{S}_j^*$. Given that negative allocations are possible it is 'optimal' to solve the following myopic allocation problem:

$$\hat{C}_r(v) = \min_{\sum_{j=1}^N \hat{S}_j \leq v} \sum_{j=1}^N \hat{C}_j(\hat{S}_j), \tag{4.4}$$

and $\hat{C}_r(v)$ provides the corresponding retailer costs. If we relax the constraint $\sum_{j=1}^N \hat{S}_j \leq v$ by a Lagrange multiplier $\lambda \geq 0$, the solution of (4.4) can be obtained from

$$F_j^{L_j+1}(\hat{S}_j^*) = \frac{h_0 + b_j - \lambda}{h_j + b_j}, \tag{4.5}$$

which is a slight variation of (4.3).

Finally we obtain S_0^{e*} by minimizing (Compare to (3.8).)

$$\hat{C}_0(S_0^e) = h_0(S_0^e - \mu_0') + \hat{C}_r(S_r^{e*}) + \int_{S_0^e - S_r^{e*}}^\infty \left[\hat{C}_r(S_0^e - u) - \hat{C}_r(S_r^{e*})\right] f_0^{L_0}(u)\, du. \tag{4.6}$$

The 'balance' assumption means that we assume that the warehouse stock can be used more efficiently than what is possible in reality. Therefore $\hat{C}_0(S_0^{e*})$ is a lower bound for the real costs, and S_0^{e*} is lower than the optimal S_0^e. When implementing the solution we cannot allocate negative quantities to the retailers. Therefore the actual allocations are determined by solving a modified version of the myopic allocation problem (4.4). Let x_j denote the inventory position at retailer j just before the allocation.

$$\hat{C}_r'(v) = \min_{\substack{\sum_{j=1}^{N} \hat{S}_j \leq v \\ \hat{S}_j \geq x_j}} \sum_{j=1}^{N} \hat{C}_j(\hat{S}_j). \qquad (4.7)$$

Note that $\hat{C}_r'(v) \geq \hat{C}_r(v)$.

The 'balance' assumption has been used extensively in the inventory literature and has been shown to produce solutions of very good quality in many different situations, see for example Eppen and Schrage (1981), Federgruen and Zipkin (1984a,b,c), de Kok (1996), van der Federgruen (1993), Van Houtum, Inderfur and Zijm (1996), Verrijdt and de Heijden, Diks and de Kok (1997), Diks and de Kok (1998, 1999) and references therein. Still the 'balance' assumption may be less appropriate in situations with large differences between the retailers in terms of service requirements and demand characteristics. Other allocation approaches that are not based on the 'balance' assumption have been suggested by e.g., Erkip (1984), Jackson (1988), Jackson and Muckstadt (1989), McGavin, Schwarz and Ward (1993), Graves (1996) and Axsäter, Marklund and Silver (2002).

4.2 The METRIC approach

We shall now turn to another early and very well known technique of modeling a distribution inventory system with stochastic demand. Also this technique is approximate. In contrast to Section 4.1 we consider continuous review and independent discrete Poisson demand processes at the retailers. Due to the one-for-one ordering policies the demand at the warehouse is also Poisson. Warehouse backorders are filled on a first come-first serve basis. We shall choose to express the solution in terms of installation stock order-up-to S policies, or equivalently $(S-1, S)$ policies, which is more common in connection with continuous review and Poisson demand. Our assumptions of Poisson demand, continuous review, and one-for-one replenishments are often reasonable for items with relatively low demand and high holding costs. Such items are, for example, common among spare parts. The original contribution was presented by Sherbrooke (1968).

We introduce the following additional notation:

λ_j Poisson demand intensity at retailer j,
λ_0 $\sum_{j=1}^{N} \lambda_j =$ Poisson demand intensity at the warehouse,
S_j^i installation stock order-up-to position at installation j,
IL_j^i stochastic installation stock inventory level at installation j in steady state,
W_0 stochastic delay at the warehouse due to stock-outs in steady state.

When applying the METRIC approximation we start with the warehouse. It is easy to determine the distribution of the warehouse inventory level exactly. The inventory position is always S_0^i. We use the standard approach and consider an arbitrary time, t. At time $t+L_0$, all that is included in the inventory position at time t has reached the warehouse, and we can express the inventory level as S_0^i minus the demand in the interval $(t, t+L_0]$

$$IL_0^i(t + L_0) = S_0^i - D_0(L_0). \tag{4.8}$$

The demand in $(t, t+L_0]$ has a Poisson distribution with mean $\lambda_0 L_0$, so we obtain the exact steady state distribution as

$$P(IL_0^i = k) = P(D_0(L_0) = S_0^i - k) = \frac{(\lambda_0 L_0)^{S_0^i - k}}{(S_0^i - k)!} e^{-\lambda_0 L_0}, \quad k \leq S_0^i. \tag{4.9}$$

The average inventory on hand can now be determined as

$$E(IL_0^i)^+ = \sum_{k=1}^{S_0^i} k \cdot P(IL_0^i = k), \tag{4.10}$$

and the average number of backorders as

$$E(IL_0^i)^- = E(IL_0^i)^+ - E(IL_0^i) = E(IL_0^i)^+ - (S_0^i - \lambda_0 L_0). \tag{4.11}$$

The warehouse can be interpreted as an $M/D/\infty$ queuing system, where the backorders are the waiting customers. This means that we can get the average delay at the warehouse according to Little's formula

$$E(W_0) = E(IL_0^i)^- / \lambda_0. \tag{4.12}$$

Because of the Poisson demand at the warehouse, and the first come-first served assumption, the average delay is the same for all retailers.

Next we consider retailer j. The retailer lead-time is stochastic due to the delays at the warehouse. However, we know the average lead-time $\bar{L}_j$, i.e., the transportation time plus the average delay at the warehouse

$$\bar{L}_j = L_j + E(W_0). \tag{4.13}$$

The METRIC approximation simply means that the real stochastic lead-time is replaced by its mean as given by (4.13). Given this approximation, we have known constant lead-times for the retailers and can determine the distribution of the retailer inventory levels precisely as we did for the warehouse in (4.9).

$$P(IL_j^i = k) = P(D_j(\bar{L}_j) = S_j^i - k) = \frac{(\lambda_j \bar{L}_j)^{S_j^i - k}}{(S_j^i - k)!} e^{-\lambda_j \bar{L}_j}, \quad k \leq S_j^i. \tag{4.14}$$

Thus the approximation enables us to decompose the problem so that all installations can be handled as single-echelon systems. Given (4.14) it is easy to determine the average stock on hand $E(IL_j^i)^+$ and the average number of backorders $E(IL_j^i)^-$ for each retailer in complete analogy with (4.10) and (4.11). Consequently, we can also obtain the average holding and shortage costs. Let

$C_0(S_0^i) = h_0 E(IL_0^i)^+$ average holding costs per time unit at the warehouse,

$C_j(S_0^i, S_j^i) = h_j E(IL_j^i)^+ + b_j E(IL_j^i)^-$ average holding and backorder costs per time unit at retailer,

$C = C_0(S_0^i) + \sum_{j=1}^{N} C_j(S_0^i, S_j^i)$ average system costs per time unit.

The holding costs at the warehouse are not affected by the retailer inventory positions, and the costs at retailer j depend only on S_0^i and S_j^i.

It can be shown that the optimal order-up-to-levels must be nonnegative. Furthermore it is easy to show that $C_j(S_0^i, S_j^i)$ is convex in S_j^i. It is also easy to find lower and upper bounds for the optimal S_j^i. A lower bound, $S_j^{i\ell}$, can be found by optimizing S_j^i for the shortest possible deterministic lead-time, L, and an upper bound, S_j^{iu}, by optimizing S_j^i for the longest possible lead-time, $L_0 + L_j$.

Together this means that, for a given S_0^i, we can find the corresponding optimal $S_j^{i*}(S_0^i)$ by a simple local search where, starting with $S_j^i = S_j^{i\ell}$, we successively increase S_j^i by one unit at a time until we find a local minimum of $C_j(S_0^i, S_j^i)$. Note that this optimization can be done separately for each retailer. The optimization of C with respect to S_0^i is not that simple, though, since

$$C(S_0^i) = C_0(S_0^i) + \sum_{j=1}^{N} C_j(S_0^i, S_j^{i*}(S_0^i)), \qquad (4.15)$$

is not necessarily convex in S_0^i. We can, however, find a lower bound, $S_0^{i\ell}$, by optimizing with respect to S_0^i for $S_j^i = S_j^{iu}$, and similarly, an upper bound, S_0^{iu}, by optimizing S_0^i for $S_j^i = S_j^{i\ell}$. To find the optimal S_0^i we then finally need to evaluate the costs (4.15) for all values of S_0^i within these bounds.

The METRIC approximation has many advantages. First of all it is simple and computationally efficient. It is also easy to generalize to more than two echelons and to compound Poisson demand. At this stage it seems to be the multi-echelon technique that has been applied most widely in practice. There are especially many military applications involving high-value repairable items. The original METRIC approach has also stimulated the development of extensions in various directions. METRIC related approaches to batch-ordering policies are treated in Section 5.2.

Another extension by Muckstadt (1973), MOD-METRIC, concerns hierarchical or indentured parts structures. Consider, for example, an assembly together with a number of modules that are used when repairing the assembly. It is assumed that the assembly is repaired by exchanging exactly one of the modules. Using the MOD-METRIC technique we can include waiting times for modules in the repair time for the assembly. This means that we can analyze how different inventory policies for the assembly and the modules will affect the availability for the assembly.

There are also several papers dealing with lateral transshipments in inventory systems, which have been inspired by the METRIC framework [See e.g., Lee (1987), Axsäter (1990b), Sherbrooke (1992a), Alfredsson and Verrijdt (1999) and Grahovac and Chakravarty (2001)].

4.3 Disaggregation of warehouse backorders

We shall in this and the following section consider the same problem as in Section 4.2 and demonstrate two ways to get the exact solution. We are still assuming a first come-first serve policy at the warehouse, so an optimal policy is only optimal under this constraint. The first derivation of exact steady state distributions was carried out by Simon (1971). Our derivation in this section is essentially according to Graves (1985).

From (4.9) we have for $k > 0$

$$P((IL_0^i)^- = k) = P(IL_0^i = -k) = P(D_0(L_0) = S_0^i + k)$$
$$= \frac{(\lambda_0 L_0)^{S_0^i + k}}{(S_0^i + k)!} e^{-\lambda_0 L_0}. \qquad (4.16)$$

Let

B_j number of backorders from retailer j at the warehouse in steady state, stochastic variable.

It is clear that $\sum_{j=1}^{N} B_j = (IL_0^i)^-$. When a new backorder occurs, the probability that it emanates from retailer j is always λ_j/λ_0 due to the Poisson demand. Consequently, the conditional distribution of B_j for a given $(IL_0^i)^-$ is binomial and we are able to disaggregate the warehouse backorders in the following way:

$$P(B_j = n) = \sum_{k=n}^{\infty} P((IL_0^i)^- = k) \binom{k}{n} \left(\frac{\lambda_j}{\lambda_0}\right)^n \left(\frac{\lambda_0 - \lambda_j}{\lambda_0}\right)^{k-n}, \quad n > 0,$$

$$P(B_j = 0) = 1 - \sum_{n=1}^{\infty} P(B_j = n). \tag{4.17}$$

For the retailers we can now use an approach that is very similar to the standard approach for single-echelon inventory systems. Consider retailer j at some arbitrary time t when the system is in steady state. $S_j^i - B_j$ units are on their way to, or already at retailer j. At time $t + L_j$ all these units have reached retailer j, while ordered units backordered at the warehouse at time t and all orders that have been triggered after time t have still not reached the retailer. Consequently, we have

$$IL_j^i(t + L_j) = S_j^i - B_j - D_j(L_j), \tag{4.18}$$

or in other words, to get the inventory level at time $t + L_j$ we subtract the demand at retailer j in the interval $(t, t + L_j]$ from the amount on route to, or already at, retailer j at time t. Note that the demand $D_j(L_j)$ is independent of B_j. We obtain the exact distribution of the inventory level at retailer j as

$$P(IL_j^i = n) = P(B_j + D_j(L_j) = S_j^i - n)$$

$$= \sum_{k=0}^{S_j^i - n} P(B_j = k) \frac{(\lambda_j L_j)^{S_j^i - n - k}}{(S_j^i - n - k)!} e^{-\lambda_j L_j}, \quad n \leq S_j^i. \tag{4.19}$$

Given the distributions of the inventory levels we can evaluate the costs and optimize the policy like in Section 4.2.

Graves (1985) also suggests an approximate procedure that is more accurate than the METRIC approximation. He determines the mean and variance of the number of outstanding orders at a retailer. Using these two parameters he fits a negative binomial distribution. A related periodic review technique is presented in Graves (1996).

4.4 A recursive procedure

We shall now describe an alternative procedure, which also provides an exact solution of the problem considered in the two previous sections. It is a special case of a more general technique for installation stock (R, Q) policies and independent compound Poisson processes at the retailers, see Axsäter (2000b).

Consider retailer j and the stochastic lead-time for an ordered unit. Let

J demand at retailer j during the retailer lead-time, stochastic variable.

The stochastic lead-time for an ordered unit is independent of the stochastic demand after the order is triggered. Furthermore, orders cannot cross in time due to the first come-first served assumption. This means that we can use the relationship

$$P(IL_j^i = n) = P(J = S_j^i - n). \qquad (4.20)$$

Consequently, it only remains to determine the distribution of J.

Recall that each system demand triggers a retailer order for a unit that, in turn, triggers a warehouse order for a unit. Consider a warehouse order for a unit at some time t. Because of the first come-first served assumption, the ordered unit will fill the S_0^i-th retailer order for a unit at the warehouse after the order. (If $S_0^i = 0$ the ordered unit will fill the retailer order that triggered the warehouse order.) This means that the considered ordered unit will be assigned to the retailer order triggered by the S_0^i-th system demand after the warehouse order. Assume, for example, that $S_0^i = 2$ and consider a warehouse order. In addition to the ordered unit, there is then one more previously ordered unit that has not yet been assigned to a retailer. This other unit is either in transportation to the warehouse or in stock at the warehouse. The next system demand will trigger an order for this other unit. The second system demand will trigger a retailer order for the considered unit that has just been ordered by the warehouse. We are interested in the lead-time demand for an order from retailer j. Therefore we consider a situation where the S_0^i-th system demand occurs at retailer j. This system demand triggers the retailer order that we are studying.

The S_0^i-th system demand after time t, when the warehouse order is triggered, can occur either before or after the ordered unit has reached the warehouse. If the ordered unit has reached the warehouse the retailer lead-time is L_j, otherwise it is longer. Let

$u(S_0^i, k)$ $P(S_0^i$-th system demand occurs *before* the order has reached the warehouse and $J \leq k)$,

$v(S_0^i, k)$ $P(S_0^i$-th system demand occurs *after* the order has reached the warehouse and $J \leq k)$.

We can express the distribution function for J as

$$P(J \leq k) = u(S_0^i, k) + v(S_0^i, k) \qquad (4.21)$$

Let us now derive the probabilities $u(S_0^i, k)$ and $v(S_0^i, k)$. It is easy to determine the probability $v(S_0^i, k)$. The corresponding event occurs if there are first at most $S_0^i - 1$ system demands during the time L_0, and later at most k demands at retailer j during the time L_j. These probabilities are independent. Consequently,

$$v(S_0^i, k) = \left(\sum_{n=0}^{S_0^i - 1} \frac{(\lambda_0 L_0)^n}{n!} e^{-\lambda_0 L_0} \right) \left(\sum_{n=0}^{k} \frac{(\lambda_j L_j)^n}{n!} e^{-\lambda_j L_j} \right). \qquad (4.22)$$

Next we show how to derive $u(S_0^i, k)$ recursively. We shall divide the event that the S_0^i-th demand arrives before the time $t + L_0$ and $J \leq k$ into three disjoint subevents. In all three cases the unit ordered by retailer j will arrive at the retailer at time $t + L_0 + L_j$, i.e., it will not stop at the warehouse. One possibility is that there are exactly S_0^i system demands during L_0, and at most k demands at retailer j during the time L_j. This probability is given by the first term in (4.23) below. The second possibility is that there are at least $S_0^i + 1$ system demands during L_0, that the $(S_0^i + 1)$-th system demand is also at retailer j, and that there are at most $k-1$ additional demands at retailer j before the time $t + L_0 + L_j$. The probability for this event is $(\lambda_j/\lambda_0) u(S_0^i + 1, k-1)$. The third possibility is that there are at least $S_0^i + 1$ system demands during L_0, the $(S_0^i + 1)$-th system demand is not at retailer j, and that there are at most k additional demands at retailer j before time $t + L_0 + L_j$. This probability is obtained as $((\lambda_0 - \lambda_j)/\lambda_0) u(S_0^i + 1, k)$. Consequently, we have for $k > 0$

$$u(S_0^i, k) = \left(\frac{(\lambda_0 L_0)^{S_0^i}}{S_0^i!} e^{-\lambda_0 L_0} \right) \left(\sum_{n=0}^{k} \frac{(\lambda_j L_j)^n}{n!} e^{-\lambda_j L_j} \right)$$
$$+ \frac{\lambda_j}{\lambda_0} u(S_0^i + 1, k - 1) + \frac{\lambda_0 - \lambda_j}{\lambda_0} u(S_0^i + 1, k). \qquad (4.23)$$

For $k = 0$ (4.23) degenerates to

$$u(S_0^i, 0) = \left(\frac{(\lambda_0 L_0)^{S_0^i}}{S_0^i!} e^{-\lambda_0 L_0} \right) e^{-\lambda_j L_j} + \frac{\lambda_0 - \lambda_j}{\lambda_0} u(S_0^i + 1, 0). \qquad (4.24)$$

Ch. 10. Serial and Distribution Inventory Systems   551

Note that $u(S_0^i, k) \to 0$ as $S_0^i \to \infty$ for any value of k. We can therefore first set $u(S'+1, k) = 0$ for some large S' and all values of k. Thereafter we can first apply (4.24) for $k = 0$, and then (4.23) for $k > 0$, recursively for $S_0^i = S'$, $S'-1, \ldots, 0$. Since the coefficients λ_j/λ_0 and $(\lambda_0 - \lambda_j)/\lambda_0$ in (4.23) and (4.24) are both strictly smaller than 1, the suggested numerical procedure is always stable.

Given $u(S_0^i, k)$ and $v(S_0^i, k)$, we can determine the distribution of the inventory level from (4.20) and (4.21). We can then optimize the policy as in Section 4.2.

The general idea in the described procedure is to keep track of each supply unit as it moves through the system. A related iterative procedure for the costs in case of Poisson demand and $(S-1, S)$ policies is given in Axsäter (1990a). Forsberg (1995) and Axsäter and Zhang (1996) generalize this procedure to compound Poisson demand. Axsäter (1993c) deals with a corresponding periodic review model.

5 Batch-ordering in distribution systems

Some basic results for batch-ordering reorder point policies in distribution systems are presented in Section 5.1. Thereafter we discuss in Sections 5.2, 5.3 and 5.4 how different approaches for one-for-one ordering policies can be extended to batch-ordering policies.

5.1 Basic facts

In Section 4 we have described different techniques for evaluation and optimization of order-up-to-S policies in distribution inventory systems. Batch-ordering policies are, in general, much more difficult to evaluate. There are some exceptions, though. One exception is batch-ordering in serial systems, see Section 3.2. Another is when we only have batches at the most upstream level. Consider again the distribution system in Fig. 1. Assume that all retailers apply S-policies. This means that their inventory positions are kept constant, and there is no difference between an echelon stock (R, Q) policy and an installation stock (R, Q) policy at the warehouse. Although we shall consider installation stock policies, the following is therefore also true for echelon stock policies.

Consider first the case when also the warehouse applies an installation stock S policy. Let

$C(S_0^i, S_1^i, \ldots, S_N^i)$ total system costs per time unit for a given installation stock S policy.

Assume that these costs can be evaluated for different policies using the techniques in Section 4. Consider then instead an (R_0^i, Q_0) policy at the warehouse. For discrete integral demand this means that the warehouse inventory position is uniform on $[R_0^i + 1, R_0^i + 2, \ldots, R_0^i + Q_0]$ and we obtain the total costs for the (R_0^i, Q_0) policy at the warehouse by simply averaging over these inventory positions.

$$C(R_0^i, Q_0, S_1^i, \ldots, S_N^i) = \frac{1}{Q_0} \sum_{k=R_0^i+1}^{R_0^i+Q_0} C(k, S_1^i, \ldots, S_N^i). \tag{5.1}$$

Furthermore, as we have discussed in Section 2.1 we can also force the system to order in 'batches' while still using a simple S policy if we use periodic review with a suitably long review period. A longer review period will mean that we need more safety stock though.

Although multi-echelon inventory systems with general batch-ordering policies are difficult to analyze, there are some important properties of single-echelon inventory systems that carry over to the multi-echelon case. One such property concerns the uniform distribution of the inventory position when using (R, Q) policies. Consider a distribution system with discrete demand where all installations apply installation stock (R, Q) policies. Let q be the largest common factor of all batch quantities $Q_0, Q_1, \ldots, Q_N$. It is evident that all replenishments and all demands at the warehouse are multiples of q. Therefore it is natural to assume that also the warehouse reorder point R_0^i as well as the initial inventory position at the warehouse are multiples of q. Let us furthermore assume that not all customer demands are multiples of some unit larger than one. Under these assumptions we have (see e.g., Axsäter, 1998).

Proposition 5 *In steady state the installation stock inventory positions are independent. The warehouse installation stock inventory position is uniform on $[R_0^i + q, R_0^i + 2q, \ldots, R_0^i + Q_0]$, and the retailer installation stock inventory positions are uniform on $[R_j^i + 1, R_j^i + 2, \ldots, R_j^i + Q_j]$.*

In case of echelon stock (R, Q) policies, all installations have their echelon inventory positions uniformly distributed on $[R_j^e + 1, R_j^e + 2, \ldots, R_j^e + Q_j]$, i.e., also the warehouse. The retailer inventory positions are still independent, but the warehouse echelon stock inventory position is correlated to the retailer inventory positions. The coupling will depend on the initial stock at the warehouse (see Axsäter (1997b) and Chen and Zheng (1997)).

Recall that when dealing with batch-ordering policies a standard approach is to determine the batch quantities from a deterministic model in an initial step. The remaining main difficulty is the policy evaluation. Given a suitable evaluation method we can normally optimize the reorder points in the same way as the order-up-to inventory positions when dealing with S policies, see Section 4.2.

In the rest of this section we shall give an overview of how different researchers have approached the policy evaluation problem for distribution systems with general batch-ordering policies. To avoid getting involved in too many technical details we shall only sketch the basic ideas behind different approaches.

5.2 METRIC type approximations

A common approach when evaluating batch-ordering policies in distribution systems is to use decomposition techniques, which are similar in spirit to the METRIC approach for one-for-one ordering policies that we dealt with in Section 4.2. Recall that the METRIC approach means that we evaluate the average delay for retailer orders due to shortages at the warehouse. This average delay is added to the retailer transportation times to get exact average lead-times for the retailers. When evaluating the costs at the retailers, these averages are, as an approximation, used instead of the real stochastic lead-times.

When dealing with different batch-ordering retailers it is still easy to determine the average backorder level at the warehouse, and to obtain the average delay at the warehouse by applying Little's formula. This delay, however, is an average over all retailers. A problem in connection with this approach is that the average delays may vary substantially between the retailers. Of course, the average delays are the same in the special case of identical retailers.

The first METRIC type approach for batch-ordering retailers was by Deuermeyer and Schwarz (1981). They first approximate the mean and variance of the warehouse lead-time demand and fit a normal distribution to these parameters. Next they estimate holding costs and the average delay at the warehouse. The retailer lead-times are, as in METRIC, obtained as the transportation times plus the average delay at the warehouse. Svoronos and Zipkin (1988) use a different type of second moment approximation. Except for the average warehouse delay, they also derive an approximate estimate of the variance of the delay. Using these parameters they then fit a negative binomial distribution to the retailer lead-time demand. Another related technique is suggested in Axsäter (2002).

5.3 Disaggregation of warehouse backorders

Another possibility is to use the approach in Section 4.3. This means that we need to disaggregate the total number of backorders to obtain the distribution of the backorders that emanate from a certain retailer. In case of Poisson demand and one-for-one replenishments a binomial disaggregation is exact, and it is relatively easy to evaluate a policy exactly. In case of batch-ordering retailers an exact disaggregation is much more complicated. Chen and Zheng (1997) have still been able to derive an exact solution this way for echelon stock (R, Q) policies and Poisson demand. Lee and Moinzadeh (1987a,b) and Moinzadeh and Lee (1986) approximate the demand process at the warehouse by a Poisson process. Using this approximation they determine the distribution of the total number of outstanding orders at all retailers. Next the mean and variance of the number of outstanding orders at a single retailer are obtained by an approximate disaggregation procedure. They then use different two-moment approximations of the distribution of outstanding orders at the retailer. Axsäter (1995) is another related approach.

5.4 Following supply units through the system

A third possible approach is to try to keep track of each supply unit as it moves through the system. A batch is then seen as a 'package' of individual units. This approach is related to the technique for one-for-one ordering policies described in Section 4.4. Using this technique Axsäter (2000b) derives the exact probability distributions for the inventory levels for a quite general continuous review multi-echelon inventory system with batch-ordering retailers facing compound Poisson demand. Related papers, which also deal with batch-ordering policies, are Axsäter (1993a,b, 1997b, 1998), Forsberg (1997a,b), Andersson (1997), Marklund (2002), Cachon (2001) and Cachon and Fisher (2000).

Using this type of techniques exact policy evaluations are possible as long as we deal with relatively small problems characterized by low demand and corresponding small batch quantities. For larger problems the computational effort grows quite rapidly and it is more realistic to rely on approximate techniques.

6 Conclusions

Initially we discuss in Section 6.1 how the present knowledge can be implemented in industrial supply chains. Thereafter in Section 6.2 we comment on possible future research topics.

6.1 What does an optimal solution look like and how can it be implemented?

There is no doubt that the research concerning evaluation and optimization of distribution inventory systems with stochastic demand has made a lot of progress during the past two decades. We can handle a wider range of systems exactly, and we have better approximate techniques for systems, which cannot be analyzed exactly. Numerical experimentation with various test problems in many research papers has also illustrated what a typical optimal solution looks like (see e.g., Muckstadt and Thomas (1980), Axsäter (1993b, 2000c), Hausman and Erkip (1994), Andersson (1997), and Gallego and Zipkin

(1999)). The optimal solution of a particular problem will, of course, depend on problem parameters like holding costs, backorder costs (or service constraints), transportation times, and demand processes. Even if it is dangerous to generalize, it is, however, striking that the optimal solutions very often look the same in a certain sense. What is typical is that upstream installations should have very low stocks compared to downstream installations. Consider, for example, a distribution inventory system with a central warehouse and about five retailers. Assume that the retailers are required to provide a fill rate of 95% towards the final customers. It is then common that the optimal solution means that we should have a negative safety stock at the warehouse, and that the warehouse fill rate for orders from the retailers should be just slightly above 50%. A very first important question is how this knowledge can be used for more efficient control of practical supply chains.

In practice it is still most common to handle different stocks in a supply chain by single-echelon techniques. Quite often practitioners handle upstream and downstream stock points in a similar way. This means generally that the distribution of the total stock between upstream and downstream installations is far from optimal. One way to achieve better control is, of course, to implement exact or approximate techniques for multi-echelon inventory systems. However, it may be difficult to replace an existing simple control system by a relatively advanced multi-echelon technique. The computational effort will also grow considerably. At least in the short run it may be more realistic to implement the new knowledge in the following way: (i) Analyze a small number of representative items by multi-echelon techniques, and (ii) Adjust the service levels in the existing control system so that they roughly correspond to the optimal solution. (See also the discussion in Chapter 12.)

6.2 Future research directions

In Section 6.1 we have discussed how the present knowledge can be implemented in practice. Another question is what kind of additional knowledge that is most needed in the present state of the development process. There is no doubt that the techniques that have been discussed in this chapter can and will be improved. Such research is both welcome and needed. However, there is also a need for research in different directions.

Some other research directions concerning multi-echelon inventory systems have already attracted many researchers. One important example is the coordination of different companies that are part of the same supply chain. Not too many years ago the knowledge in this area was very limited, but at present this is a very vital research area (see e.g., Chapters 6 and 7).

However, there are also other important research areas where we definitely still need more research. For example, most of the present results are completely based on the traditional hierarchical flow pattern, from upstream installations to downstream installations. In practice it is common to have a flow of material also between installations at the same hierarchical level, e.g., between adjacent retailers. Although there exist quite a few results concerning the effect of such lateral transshipments in multi-echelon inventory systems, these results are generally considerably weaker than corresponding results for traditional hierarchical flow patterns. Another very open research area concerns the utilization of new and improved information structures. Today more or less any type of information concerning the material flow is, or can be, available at a low cost. For example, we do not only know when outstanding orders were triggered, we also know exactly where in the supply process they are situated at any time. It is reasonable to believe that such additional information should to a larger extent affect the control of multi-echelon inventory systems.

Acknowledgements

The author is indebted to Steve Graves, Ton de Kok, Jing-Sheng Song, and Paul Zipkin for their valuable comments and suggestions.

References

Alfredsson, P., J. Verrijdt (1999). Modeling emergency supply flexibility in a two-echelon inventory system. *Management Science* 45, 1416–1431.

Andersson, J. (1997). Exact evaluation of general performance measures in multi-echelon inventory systems, Lund University.

Axsäter, S. (1990a). Simple solution procedures for a class of two-echelon inventory problems. *Operations Research* 38, 64–69.

Axsäter, S. (1990b). Modelling emergency lateral transsshipments in inventory systems. *Management Science* 36, 1329–1338.

Axsäter, S. (1993a). Exact and approximate evaluation of batch-ordering policies for two-level inventory systems. *Operations Research* 41, 777–785.

Axsäter, S. (1993b). Continuous review policies for multi-level inventory systems with stochastic demand, in S. C. Graves et al. (eds.). *Handbooks in OR & MS Vol. 4*, North Holland Amsterdam, pp. 175–197.

Axsäter, S. (1993c). Optimization of order-up-to-S policies in 2-echelon inventory systems with periodic review. *Naval Research Logistics* 40, 245–253.

Axsäter, S. (1995). Approximate evaluation of batch-ordering policies for a one-warehouse, N nonidentical retailer system under compound Poisson demand. *Naval Research Logistics* 42, 807–819.

Axsäter, S. (1996). Using the deterministic EOQ formula in stochastic inventory control. *Management Science* 42, 830–834.

Axsäter, S. (1997a). On deficiencies of common ordering policies for multi-level inventory control. *OR Spektrum* 19, 109–110.

Axsäter, S. (1997b). Simple evaluation of echelon stock (R, Q) policies for two-level inventory systems. *IIE Transactions* 29, 661–669.

Axsäter, S. (1998). Evaluation of installation stock based (R, Q)-policies for two-level inventory systems with Poisson demand. *Operations Research* 46, S135–S145.

Axsäter, S. (2000a). *Inventory Control*, Boston, Kluwer Academic Publishers.

Axsäter, S. (2000b). Exact analysis of continuous review (R, Q) policies in two-echelon inventory systems with compound Poisson demand. *Operations Research* 48, 686–696.

Axsäter, S. (2002). Approximate optimization of a two-level distribution inventory system, *International Journal of Production Economics* 81-2, 545–553.

Axsäter, S., L. Juntti (1996). Comparison of echelon stock and installation stock policies for two-level inventory systems. *International Journal of Production Economics* 45, 303–310.

Axsäter, S., J. Marklund, E.A. Silver (2002). Heuristic methods for centralized control of one-warehouse N-retailer inventory systems. *Manufacturing & Service Operations Management* 4, 75–97.

Axsäter, S., K. Rosling (1993). Installation vs. echelon stock policies for multilevel inventory control. *Management Science* 39, 1274–1280.

Axsäter, S., K. Rosling (1994). Multi-level production-inventory control: Material requirements planning or reorder point policies?. *EJOR* 75, 405–412.

Axsäter, S., K. Rosling (1999). Ranking of generalised multi-stage KANBAN policies. *EJOR* 113, 560–567.

Axsäter, S., W.F. Zhang (1996). Recursive evaluation of order-up-to-S policies for two-echelon inventory systems with compound Poisson demand. *Naval Research Logistics* 43, 151–157.

Cachon, G.P. (2001). Exact evaluation of batch-ordering inventory policies in two-echelon supply chains with periodic review. *Operations Research* 49, 79–98.

Cachon, G.P., M. Fisher (2000). Supply chain inventory management and the value of shared information. *Management Science* 46, 1032–1048.

Chen, F. (2000). Optimal policies for multi-echelon inventory problems with batch ordering. *Operations Research* 48, 376–389.

Chen, F., J. Song (2001). Optimal policies for multi-echelon inventory problems with Markov-modulated demand. *Operations Research* 49, 226–234.

Chen, F., Y.S. Zheng (1994). Evaluating echelon stock (R, nQ) policies in serial production/inventory systems with stochastic demand. *Management Science* 40, 1262–1275.

Chen, F., Y.S. Zheng (1997). One-warehouse multi-retailer systems with centralized stock information. *Operations Research* 45, 275–287.

Clark, A.J., H. Scarf (1960). Optimal policies for a multi-echelon inventory problem. *Management Science* 5, 475–490.

Deuermeyer, B., L. B. Schwarz (1981). A model for the analysis of system service level in warehouse/retailer distribution systems: The identical retailer case, in: L. B. Schwarz. (ed.). *Multi-Level Production/Inventory Control Systems: Theory and Practice*, North Holland Amsterdam, 163–193.

Diks, E.B., A.G. de Kok (1998). Optimal control of a divergent multi-echelon inventory system. *EJOR* 111, 75–97.

Diks, E.B., A.G. de Kok (1999). Computational results for the control of a divergent N-echelon Inventory System. *International Journal of Production Economics* 59, 327–336.

Diks, E.B., A.G. de Kok, A.G. Lagodimos (1996). Multi-echelon systems: A service measure perspective. *EJOR* 95, 241–263.

Eppen, G. D., L. Schrage (1981). Centralized ordering policies in a multi-warehouse system with leadtimes and random demand, in: L. B. Schwarz. (ed.). *Multi-Level Production/Inventory Control Systems: Theory and Practice*, North Holland Amsterdam, 51–69.

Erkip N. (1984). Approximate Policies in Multi-Echelon Inventory Systems, unpublished Ph.D. dissertation, Department of Industrial Engineering and Engineering Management, Stanford University.

Federgruen, A. (1993). Centralized planning models for multi-echelon inventory systems under uncertainty, in: S. C. Graves et al. (eds.). *Handbooks in OR & MS Vol. 4*, North Holland Amsterdam, 133–173.

Federgruen, A., P.H. Zipkin (1984a). Approximations of dynamic multilocation production and inventory problems. *Management Science* 30, 69–84.

Federgruen, A., P.H. Zipkin (1984b). Computational issues in an infinite-horizon multiechelon inventory model. *Operations Research* 32, 818–836.

Federgruen, A., P.H. Zipkin (1984c). Allocation policies and cost approximations for multilocation inventory systems. *Naval Research Logistics Quarterly* 31, 97–129.

Forsberg, R. (1995). Optimization of order-up-to-S policies for two-level inventory systems with compound Poisson demand. *EJOR* 81, 143–153.

Forsberg, R. (1997a). Exact evaluation of (R, Q)-policies for two-level inventory systems with Poisson demand. *EJOR* 96, 130–138.

Forsberg, R. (1997b). Evaluation of (R, Q)-policies for two-level inventory systems with generally distributed customer inter-arrival times. *EJOR* 99, 401–411.

Gallego, G., P.H. Zipkin (1999). Stock positioning and performance estimation in serial production-transportation systems. *Manufacturing & Service Operations Management* 1, 77–88.

Grahovac, J., A. Chakravarty (2001). Sharing and lateral transshipment of inventory in a supply chain with expensive, low-demand items. *Management Science* 47, 579–594.

Graves, S.C. (1985). A multi-echelon inventory model for a repairable item with one-for-one replenishment. *Management Science* 31, 1247–1256.

Graves, S.C. (1996). A multiechelon inventory model with fixed replenishment intervals. *Management Science* 42, 1–18.

Hausman, W.H., N.K. Erkip (1994). Multi-echelon vs. single-echelon inventory control policies for low-demand items. *Management Science* 40, 597–602.

Jackson, P.L. (1988). Stock allocation in a two-echelon distribution system or what to do until your ship comes in. *Management Science* 34, 880–895.

Jackson, P.L., J.A. Muckstadt (1989). Risk pooling in a two-period, two-echelon inventory stocking and allocation problem. *Naval Research Logistics* 36, 1–26.

Lee, H.L. (1987). A multi-echelon inventory model for repairable items with emergency lateral transshipments. *Management Science* 33, 1302–1316.

Lee, H.L., K. Moinzadeh (1987a). Two-parameter approximations for multi-echelon repairable inventory models with batch ordering policy. *IIE Transactions* 19, 140–149.

Lee, H.L., K. Moinzadeh (1987b). Operating characteristics of a two-echelon inventory system for repairable and consumable items under batch ordering and shipment policy. *Naval Research Logistics Quarterly* 34, 365–380.

Marklund, J. (2002). Centralized inventory control in a two-level distribution system with Poisson demand. *Naval Research Logistics* 49, 798–822.

McGavin, E.J., L.B. Schwarz, J.E. Ward (1993). Two-interval inventory-allocation policies in a one-warehouse N-identical-retailer distribution system. *Management Science* 39, 1092–1107.

Moinzadeh, K., H.L. Lee (1986). Batch size and stocking levels in multi-echelon repairable systems. *Management Science* 32, 1567–1581.

Muckstadt, J.A. (1973). A model for a multi-item, multi-echelon, multi-indenture inventory system. *Management Science* 20, 472–481.

Muckstadt, J.A., R. Roundy (1993). Analysis of multistage production systems, in: S. C. Graves et al. (eds.). *Handbooks in OR & MS Vol. 4*, North Holland Amsterdam, 59–131.

Muckstadt, J.A., L.J. Thomas (1980). Are multi-echelon inventory models worth implementing in systems with low-demand-rate items?. *Management Science* 26, 483–494.

Muharremoglu, A., J. N. Tsitsiklis (2001). Echelon base stock policies in uncapacitated serial inventory systems, Operations Research Center, MIT.

Nahmias, S. (1981). Managing repairable item inventory systems: A review, in: L. B. Schwarz (ed.). *Multi-Level Production/Inventory Control Systems: Theory and Practice*, North Holland Amsterdam, 253–277.

Rosling, K. (1989). Optimal inventory policies for assembly systems under random demands. *Operations Research* 37, 565–579.

Roundy, R. (1985). 98%-effective integer-ratio lot-sizing for one-warehouse multi-retailer systems. *Management Science* 31, 1416–1430.

Roundy, R. (1986). 98%-effective lot-sizing rule for a multi-product multi-stage production/inventory system. *Mathematics of Operations Research* 11, 699–729.

Seidel, H.P., A.G. de Kok (1990). Analysis of stock allocation in a 2-echelon distribution system, Technical Report 098, CQM, Philips Electronics.

Shang, K. H., J. Song (2001). Newsvendor bounds and heuristic for optimal policies in serial supply chains, Graduate School of Management, University of California, Irvine.

Sherbrooke, C.C. (1968). METRIC: A multi-echelon technique for recoverable item control. *Operations Research* 16, 122–141.

Sherbrooke, C.C. (1992a). Multiechelon inventory systems with lateral supply. *Naval Research Logistics* 39, 29–40.

Sherbrooke, C.C. (1992b). *Optimal Inventory Modeling of Systems, Multi-Echelon Techniques*, New York, Wiley.

Silver, E.A., D.F. Pyke, R. Peterson (1998). *Inventory Management and Production Planning and Scheduling*, 3rd edition, New York, Wiley.

Simon, R.M. (1971). Stationary properties of a two-echelon inventory model for low demand items. *Operations Research* 19, 761–777.

Spearman, M.L., D.L. Woodruff, D.L. Hopp (1990). CONWIP: A pull alternative to KANBAN. *International Journal of Production Research* 28, 879–894.

Svoronos, A., P.H. Zipkin (1988). Estimating the performance of multi-level inventory systems. *Operations Research* 36, 57–72.

Svoronos, A., P.H. Zipkin (1991). Evaluation of one-for-one replenishment policies for multiechelon inventory systems. *Management Science* 37, 68–83.

Van der Heijden, M.C., E.B. Diks, A.G. de Kok (1997). Stock allocation in general multi-echelon distribution systems with (R, S) order-up-to-policies. *International Journal of Production Economics* 49, 157–174.

Van Houtum, G.J., W.H.M. Zijm (1991). Computational procedures for stochastic multi-echelon production systems. *International Journal of Production Economics* 23, 223–237.

Van Houtum, G.J., K. Inderfurth, W.H.M. Zijm (1996). Materials coordination in stochastic multi-echelon systems. *EJOR* 95, 1–23.

Veatch, M.H., L.M. Wein (1994). Optimal control of a two-station tandem production/inventory system. *Operations Research* 42, 337–350.

Verrijdt, J.H.C.M., A.G. de Kok (1996). Distribution planning for divergent depotless two-echelon network under service constraints. *EJOR* 89, 341–354.

Zheng, Y.S. (1992). On properties of stochastic inventory systems. *Management Science* 38, 87–103.

Zipkin, P.H. (2000). *Foundations of Inventory Management*, New York, McGraw-Hill.

A.G. de Kok and S.C. Graves, Eds., *Handbooks in OR & MS, Vol. 11*
© 2003 Elsevier B.V. All rights reserved.

Chapter 11

Supply Chain Operations: Assemble-to-Order Systems

Jing-Sheng Song

Graduate School of Management, University of California, Irvine, CA 92697, USA
E-mail: jssong@uci.edu

Paul Zipkin

Fuqua School of Business, Duke University, Durham, NC 27708, USA
E-mail: paul.zipkin@duke.edu

1 Introduction

An *assemble-to-order* (or *ATO*) *system* includes several *components* and several *products*. Demands occur only for products, but the system keeps inventory only of components. To make each product requires a particular selection of components, comprising only a subset of them, but possibly several units of certain ones. Some or all components are shared by several products. The time to assemble a product from its components is negligible. The time to acquire or produce a component, however, is substantial. A product is assembled only in response to demand. See Fig. 1.

A *configure-to-order* (or *CTO*) *system* is a special case. The components are partitioned into subsets, and the customer *selects* components from those subsets. A computer, for example, is configured by selecting a processor from several options, a monitor from several options, etc. The difference between a CTO system and an ATO system is important at the demand-elicitation level. At the operational level, however, the differences are minor. Our discussion focuses on general ATO systems.

Such systems have been employed for some time in various industries, but lately their popularity has soared. An ATO system is an efficient way to deliver a high level of product variety to customers, while maintaining reasonable response times and costs.

One well-known ATO system (actually, a CTO system) is Dell Computer's. Dell lets the customer select among several processors, monitors, disk drives, etc. – these are the components. Thus, the number of products (combinations of options) is huge. This approach has been so successful that most other

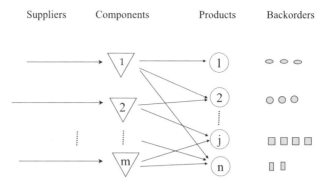

Fig. 1. Assemble-to-order system.

makers of personal computers are adopting similar systems. Indeed, the ATO approach has become widespread throughout the electronics industry. The major U.S. automobile companies are studying ambitious ATO systems for the assembly of cars [Kerwin, 2000]. (In a real system, product assembly may take some time, though not more than customers are willing to wait.)

Certain other types of systems have the same structure. Consider a mail-order or e-commerce retailer, which maintains inventories of the items in its catalogue. The items correspond to components, and a product is any combination of them. The assembly of a product entails picking out the items in the customer's order and packaging them. Also, consider the problem of stocking spare parts for the repair of equipment. The parts are the components, and a product is a particular type of repair job, requiring particular parts. The parts may be located at a central point, where equipment needing service arrives (e.g., vehicles), or the parts may travel to stationary equipment (as in field service of computers, copiers, and factory machines). In either case, the part requirements of a job are usually unknown in advance.

This chapter reviews the research to date on ATO systems. It covers modeling issues and analytical methods, and also summarizes managerial insights gained from the research. At the end, it identifies some directions for future research. (See Song and Yao (2001) for other related articles.)

Two special cases are worth identifying. An *assembly system* has just one product, and a *distribution system* has just one component. The key issue in an assembly system is the *coordination* of the components, while the key issue in a distribution system is the *allocation* of the component among the products. (This assembly system is a special one, due to the negligible assembly time, which implies that there is no reason to assemble the product in advance of demand. The distribution system is special in the same way.) An ATO system combines the elements of assembly and distribution, and so must resolve both

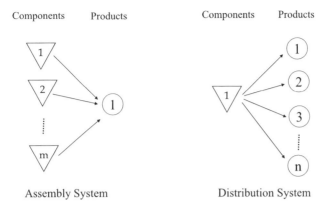

Fig. 2. Special cases.

coordination and allocation issues. This is what makes ATO systems difficult to analyze, design and manage. See Fig. 2.

Research in this area has two major goals. One is efficient operation. *Given* a particular system design, i.e., a line of products and a set of components, this work aims to evaluate its performance under various conditions, including inventory levels of the components. It also seeks to find good operating policies, including inventory levels that balance cost and customer service. The second research goal is to understand the impacts of alternative system designs, for example, the effects of designing several products to share common components.

In Section 2, we discuss one-period models. Section 3 focuses on multi-period, discrete-time models, while Section 4 presents continuous-time models. Section 5 summarizes research on system design. Section 6 points out some future directions.

2 One-period models

This section focuses on one-period models. Some systems can be understood fruitfully in a static framework. Either each time period really can be treated in isolation, or the model can serve as a myopic heuristic for more complex scenarios. As in the classic newsvendor model (one component and one product), there is no need to distinguish between backorders and lost sales, and we suppress procurement leadtimes. We present a fairly general formulation of the problem and then discuss specific works in the literature.

The sequence of events within the period is as follows: (1) Components are produced or acquired. (2) Demands for the products are realized. (3) Components are allocated to products, and costs are assessed on the ending situation. The basic approach is stochastic linear programming with simple recourse. We consider these events in reverse order.

The problem for stage (3) can be formulated as follows: Define

m	total number of components
n	total number of products
i	index for components (subscript)
j	index for products (superscript)
a_i^j	units of component i required to make one unit of product j, $A = (a_i^j)$ (matrix)
d^j	demand for product j, $\mathbf{d} = (d^j)$ (vector)
y_i	supply of component i, $\mathbf{y} = (y_i)$ (vector)
p^j	penalty cost for unit shortage of product j, $\mathbf{p} = (p^j)$ (vector)
h_i	cost for unit excess of component i, $\mathbf{h} = (h_i)$ (vector)
z^j	production of product j, $\mathbf{z} = (z^j)$ (vector)
w^j	shortage of product j, $\mathbf{w} = (w^j)$ (vector)
x_i	excess of component i, $\mathbf{x} = (x_i)$ (vector)

(The parameters a_i^j, d^j, and y_i are nonnegative real numbers, the p^j are positive real numbers, and the h_i are any real numbers. A negative h_i represents a salvage value.) The problem is then

(P3) $\quad \hat{G}(\mathbf{y}, \mathbf{d}) = $ minimize $\mathbf{hx} + \mathbf{pw}$

subject to

$$A\mathbf{z} + \mathbf{x} = \mathbf{y}$$
$$\mathbf{z} + \mathbf{w} = \mathbf{d}$$
$$\mathbf{w}, \mathbf{x}, \mathbf{z} \geq 0.$$

This is a linear program. (It can be quite large if there is a large number of products.) The minimal cost function $\hat{G}(\mathbf{y}, \mathbf{d})$ is convex.

Several modeling issues are worth noting: This formulation treats demands, supplies, etc. as continuous. In some situations these quantities may actually be discrete. If the model is revised accordingly, it becomes an integer linear program. This model is of course much harder to solve than the original.

Also, this formulation assumes that the stockout penalty cost for each product is linear in the shortage. An alternative formulation replaces the term $\mathbf{pw}$ in the objective with $\mathbf{p1(w)}$, where $\mathbf{1(w)} = (\mathbf{1}(w^j))$ is the vector of 0-1 variables indicating which of the w^j are positive. Here, a cost p^j is incurred if there is *any* shortage of product j, regardless of how much. This model too is harder to solve than the original. Yet another formulation has a term $p \max_j[\mathbf{1(w)}]$ in the objective. Here, a cost p is incurred if there is *any* shortage of *any* product. Alternatively, any of these service measures can be constrained instead of penalized. (In our opinion, the original linear formulation better represents service to customers, in addition to being more tractable computationally.)

Lastly, the formulation assumes that a demand for a product must be filled entirely or not at all. This makes sense when the product is really a product, i.e., incomplete with any of its components missing. In a retailing setting, however, the situation is less clear. The customer gets *some* benefit from partial fulfillment.

Now let us examine two special cases. First, consider a distribution system (one component, so we can omit the index i). Here, the model reduces to a continuous knapsack problem, which is easy to solve: for each product, compute the ratio p^j/a^j, the shortage cost per unit of the component. Select the product with the largest ratio, and satisfy its demand as much as possible, i.e., $z^j = \min\{y/a^j, d^j\}$. If there is any of the component left, satisfy as much as possible the demand for the product with the *next* largest ratio. Continue in this manner until the component is exhausted, or all demands are filled. (This assumes $h \geq 0$. If $h < 0$, i.e., there is a positive salvage value $-h$, omit any product with $p^j/a^j < -h$. Here, the solution can have both unfilled demand and remaining component.) Thus, in this case, the best allocation of the component is determined by a priority ranking of the products, which depends on the cost and usage data only, not demand conditions.

Next, consider an assembly system (one product, so we can omit the index j). Here too, the model is easy to solve: Set the production level z to fill all demand, if possible, or else to use up the most limiting component. i.e., $z = \min\{\min\{y_i/a_i\}, d\}$.

Now let us move to stage (1). Here, **d** is a random variable. Let $\mathbf{x}_0$ be the initial component inventory vector. The expected cost at stage (3) given **y** is $G(\mathbf{y}) = E_\mathbf{d}[\hat{G}(\mathbf{y}, \mathbf{d})]$. Let $c(\mathbf{y} - \mathbf{x}_0)$ denote the cost of acquiring the components. The problem, then, is

(P1) minimize $c(\mathbf{y} - \mathbf{x}_0) + G(\mathbf{y})$
subject to $\mathbf{y} \geq \mathbf{x}_0$

The function $G(\mathbf{y})$ is convex. If $c(\cdot)$ too is convex (e.g., linear), then the overall objective function is convex, and so the model is relatively easy to solve.

In particular, assume $c(\cdot)$ is linear with cost-coefficient vector **c**. Let $\mathbf{y}^*$ be the global minimizer of $\mathbf{cy} + G(\mathbf{y})$. Then the optimal ordering policy is a base-stock policy with base-stock level $\mathbf{y}^*$. That is, if $\mathbf{x}_0 \leq \mathbf{y}^*$, then order up to $\mathbf{y}^*$. Standard stochastic linear programming techniques can be employed to compute G and $\mathbf{y}^*$.

In the alternative formulations mentioned above, where the objective of (P3) includes **p1(w)** or something similar, the corresponding G includes terms representing stockout probabilities. Except in special cases, the function G is not convex.

Some elements of this formulation are due to Gerchak and Henig (1986). Assuming $\mathbf{x}_0 = \mathbf{0}$, they compare the ATO system with a make-to-stock (MTS) system in which the assembly is performed before demands are realized. In the latter, there is no need to solve problem (P3), and $\mathbf{y} = A\mathbf{z}$. The optimal value of $\mathbf{z}$ is the solution of problem (P1) with $A\mathbf{z}$ replacing $\mathbf{y}$. Problem (P1) now reduces to n separate newsvendor problems. Let $\mathbf{z}^S$, $\mathbf{y}^S$, and C^S denote the optimal solution and cost for this MTS setting, and let $\mathbf{z}^O$, $\mathbf{y}^O$ and C^O be their counterparts in the original ATO setting (by solving both (P3) and (P1)). Then,

(a) $C^S \geq C^O$. (The cost is lower when assembly is postponed until demand is realized.)
(b) $\mathbf{z}^O \geq \mathbf{z}^S$. (More demand is fulfilled in the ATO system.)
(c) If component i is product-specific (used by just one product), then $y_i^O \geq y_i^S$.
(d) On the other hand, one cannot predict the ordering of y_i^O and y_i^S if i is a common component (used by more than one product).

The optimal stock of a common component may be higher or lower than the combined optimal stocks of the specialized components it replaces. Thus, moving to an ATO system need not reduce inventories. It does improve overall performance, but the improvement may show up instead in reduced stockouts.

The study of Gerchak and Henig (1986) builds on earlier work comparing simple systems with and without common components among the products. See Baker, Magazine, and Nuttle (1986), Gerchak, Magazine, and Gamble (1988), and references therein. They assume a single constraint on the overall stockout probability. In our notation, the problem is

$$\text{minimize} \quad \mathbf{cy}$$
$$\text{subject to} \quad \mathbf{P}(A\mathbf{d} \leq \mathbf{y}) \geq \beta$$

Here, A is a 0-1 matrix, and β is a prespecified service level. For the sytem without common components, each column of A has only one non-zero entry. In the system with common components, A has fewer columns, and some columns have several nonzero entries. This is obtained by adding some columns of the original A, i.e., combining some components into common components.

It is shown that the total inventory investment required to meet the constraint is lower with a common component than without it. Certain additional qualitative properties of the solution derived in the first paper are, however, shown in the second paper to hold only in special cases. Thus, while commonality is certainly a good thing, all else being equal, its detailed effects are hard to predict. See also Bagchi and Gutierrez (1992), Eynan (1996) and Eynan and Rosenblatt (1996) for variations of these studies and similar conclusions. Eynan (1996) provides a summary of this line of research and additional references.

The repair-kit problem also is related. The repair kit carried by a repairman is a multi-item inventory. A demand is a repair job, which may require several different items in the kit. Typically, each job requires either one or zero unit of each item, so the matrix A contains only entries 0 and 1. The problem is to determine an optimal kit that minimizes either the expected inventory and penalty cost, or the inventory cost subject to a constraint on the job-completion rate. Some models assume that the kit can be restocked after each job, therefore the decision variables are binary; we only need to decide whether to carry an item or not. See, e.g., Smith, Chambers, and Shlifer (1980) and Graves (1982). Other models consider multiple units of each item in stock; the performance measure is the expected number of jobs, which can be completed before stock out or the probability of serving all jobs arriving in a fixed period of time. There is no inventory replenishment during the period. The difference between this model and the one-period model (P1) and (P3) is that, within the period, the allocation is dynamic as demand occurs, and it follows the FCFS rule. Network-flow and combinatorial techniques are developed to solve the optimization problem. See, e.g., Mamer and Smith (1982, 1985) and Brumelle and Granot (1993). Mamer and Smith (2001) review this literature.

Swaminathan and Tayur (1999) consider an extension of the basic model above. Between components and products is another layer of *sub-assemblies*. A sub-assembly can be made from components in stage (1), before demand is realized. The model explicitly includes a production resource (e.g., capacity or time) that is used in the creation of products from components and/or sub-assemblies in stage (3). To make a product from sub-assemblies consumes less of this resource than making it directly from components. See Chapter 8 of this handbook for a detailed discussion.

3 Multi-period, discrete-time models

This section focuses on discrete-time, multi-period models. Within a single period, the decisions and the sequence of events are the same as in the one-period problem. However, additional complications arise when we link different periods, as the ending state in one period becomes the beginning state in the next. One complication is due to leadtimes for component replenishments. When the leadtimes for different components are different, the replenishment decision in one period will affect the inventory levels in different periods in the future. The problem becomes even more complex if the leadtimes are uncertain. Another complication is how to deal with shortages – whether the unsatisfied demand in one period is backlogged or lost. In the backlogging case, there is also a partial-shipment or inventory-commitment issue. That is, if we have only part of the components a demand requests, should we ship or put aside the available components as committed inventory to this customer while waiting for those unavailable components to come? If we keep committed inventory, then we pay both the backorder costs for the unsatisfied demand and the holding cost for the committed inventory. Yet another complication is the relative priority of backlogged demand and current demand. This complication does not arise in the two special cases – one product or one component. It is problematic only when there are different products with overlapping component sets.

We first summarize results on the characterization of optimal policies and then results on performance evaluation and optimization techniques for a given type of policy.

3.1 Characterization of optimal policies

With all the complications mentioned above, the state space becomes quite large: We need to track not only the component inventory vector (**x**), but also the backordered products (**w**) and the outstanding-order vector for each component. As a result, the literature includes only partial results for special cases: zero leadtimes, or positive leadtimes but only one product.

Consider the case with no component replenishment leadtimes and no committed inventory. Backlogged demand is merged with current demand, i.e., no priority is given to the backordered demand. For simplicity, assume stationary cost factors and component usages. Let T be the time horizon and t the time index, $t=0,\ldots,T$. Use an additional subscript t to index the variables. The sequence of events within each period is the same as in the one-period model above.

Then the problem is

(P) minimize $\mathbf{E}\left\{\sum_{t=0}^{T}[c(\mathbf{y}_t - \mathbf{x}_t) + \mathbf{h}\mathbf{x}_{t+1} + \mathbf{p}\mathbf{w}_{t+1}]\right\}$

subject to

$\mathbf{x}_{t+1} = \mathbf{y}_t - A\mathbf{z}_t$

$\mathbf{w}_{t+1} = \mathbf{w}_t + \mathbf{d}_t - \mathbf{z}_t$

$\mathbf{w}_t, \mathbf{x}_t, \mathbf{z}_t \geq 0, \quad \mathbf{y}_t \geq \mathbf{x}_t, \quad t=0,\ldots,T.$

For linear $c(\cdot)$ the optimal-cost function for each period is convex, and again a base-stock policy is optimal. That is, there are vectors $\mathbf{y}_t^*(\mathbf{w})$ such that, if $\mathbf{x}_t \leq \mathbf{y}_t^*(\mathbf{w}_t)$, then order up to $\mathbf{y}_t^*(\mathbf{w}_t)$. Otherwise, we know of no results characterizing the optimal policy for the general case. (However, there are some rsults for variations of the moel. See Veinott (1965).)

Because exact solution of (P) is difficult, various computational and heuristic approaches have been developed. One approach is myopic: In each period, solve the embedded allocation problem, after the demand for that period is realized (i.e., solve for $\mathbf{z}_t$ after observing $\mathbf{d}_t$, ignoring future periods), as in (P3). Stochastic programming techniques are then used to determine the optimal order quantities for the entire horizon, as in (P1). The problem can be reformulated as a nested stochastic program. See Swaminathan and Tayur (1999).

Gerchak and Henig (1989) study a lost-sales model with stationary data and linear order cost. (The paper states that the results hold for the backlog case too, but the formulation does not completely cover that case. There are no state variables or costs for backlogs.) In particular, $\mathbf{p}=\mathbf{0}$, and $\mathbf{d}_t$ has the same distribution across t. Let r^j be the unit revenue of product j, and denote $\mathbf{r}=(r^j)$. Then, the problem can be expressed as follows:

$$\text{(Lost Sales)} \quad \text{minimize} \quad \mathbf{E}\left\{\sum_{t=0}^{T}[c(\mathbf{y}_t - \mathbf{x}_t) + \mathbf{h}\mathbf{x}_{t+1} - \mathbf{r}\mathbf{z}_t]\right\}$$

subject to

$$\mathbf{x}_{t+1} = \mathbf{y}_t - A\mathbf{z}_t$$

$$\mathbf{z}_t \leq \mathbf{d}_t$$

$$\mathbf{x}_t, \mathbf{z}_t \geq \mathbf{0}, \quad \mathbf{y}_t \geq \mathbf{x}_t, \quad t = 0, \ldots, T.$$

They too adopt the myopic-allocation policy for this problem. That is, in each period, after the demand $\mathbf{d}_t$ is realized, solve a linear program to find the that $\mathbf{z}_t$ maximizes $\mathbf{r}\mathbf{z}_t$ while satisfying the constraints. They show that, because of the stationary data, this myopic policy is optimal, so the multi-period solution is identical to the single-period solution. In other words, a base-stock order policy and the myopic-allocation policy are optimal. Van Mieghem and Rudi (2001) obtain some related results and explain in detail why the myopic result does not extend to the backlog case.

Hillier (2000) studies a model in which a common component is shared by all the products, and each product has a unique, product-specific component. Let $n+1$ index the common component. Thus, the matrix A has $n+1$ rows and n columns, with $a_i^i = 1$, $a_i^j = 0$ for $j \neq i$ and $i, j = 1, \ldots, n$, and $a_{n+1}^j = 1$ for $j = 1, \ldots, n$. The myopic-allocation policy above is employed. Under certain special assumptions about purchase, holding and backorder costs, as well as zero leadtimes, the paper concludes that commonality may not be beneficial if the common component is more expensive than the components it would replace.

3.1.1 Distribution systems

Next, consider the special case of a distribution system, that is, a single component and multiple products (or demand classes). Topkis (1968) analyzes a model in which the order decisions can be made only at certain fixed times. The number of periods between such decisions, called a stocking cycle, equals the leadtime. Thus, at any time there is only one outstanding order, and the order is received only at the end of the stocking cycle. Also, all previous backlogs are cleared at each reorder point. Thus, the problem is essentially a single-cycle problem. It is shown that, under certain conditions, a base-stock policy is optimal for ordering, and a rationing policy (described below) is optimal for allocation in each period within a cycle.

Let us take a closer look at a cycle. Suppose the cycle length is $L+1$ periods. The problem within the cycle is a special case of (P), in which $T=L$, y_0 is fixed, and $y_t=0$, $t=1,\ldots,L$. We can formulate this problem as a dynamic program. There are two state variables. One is x_t, the inventory level at the beginning of period t. The other is $\mathbf{u}_t = \mathbf{w}_t + \mathbf{d}_t$, the vector of outstanding product demands (previous backorders plus demand in the period). Define $V_t(x, \mathbf{u})$ to be the minimal expected cost in periods $t, t+1, \ldots, L$, assuming period t begins with inventory level $x_t = x$ and outstanding demand $\mathbf{u}_t = \mathbf{u}$. Then,

$$V_t(x, \mathbf{u}) = \min\{h_t x_{t+1} + \mathbf{p}_t \mathbf{u}_{t+1} + \mathbf{E}[V_{t+1}(x_{t+1}, \mathbf{u}_{t+1})]\}$$

subject to

$$x_{t+1} = x - \mathbf{A}\mathbf{z}_t$$
$$\mathbf{u}_{t+1} = \mathbf{u} + \mathbf{d}_{t+1} - \mathbf{z}_t$$
$$\mathbf{u}_{t+1}, x_{t+1}, \mathbf{z}_t \geq \mathbf{0}, \quad t = 0, \ldots, T.$$

Assume $a^j = 1$ for all j. Renumber the products so that $p_t^1 \leq p_t^2 \leq \cdots \leq p_t^n$. Topkis shows that the optimal allocation policy is determined by nonnegative rationing limits $\{\tilde{z}_t^j, j=1,\ldots,n\}$ with $\tilde{z}_t^1 \geq \tilde{z}_t^2 \geq \cdots \geq \tilde{z}_t^n$. The rule works as follows: Start with product n. Allocate as much as possible to it, as long as the stock level does not drop below the rationing limit $\tilde{z}_t^n$. If any demand remains unfilled, stop. Otherwise, apply the same rule to product $n-1$, then $n-2$, and so forth. This rule is easy to implement, but the computation of the rationing levels is difficult. (Note that the solution to the single-period problem discussed in the last section corresponds to $\tilde{z}^j = 0$ for all j. This is because, in that case, we do not need to consider future demands.)

Sobel and Zhang (2001) study a model with no leadtime and two demand sources, one deterministic ($\bar{d}_t$) and the other stochastic ($\hat{d}_t$). Think of these as two distinct products. The deterministic demand in each period must be satisfied immediately, and the stochastic demand can be backordered. So, the allocation policy in each period is fixed: Satisfy all the deterministic demand, and then satisfy as much of the stochastic demand as possible, including backorders carried from previous periods. The order cost in each period is a fixed cost k plus a linear cost with rate c. Problem (P) now becomes

$$\text{minimize} \quad E\left\{\sum_{t=0}^{T}[k\delta(y_t - x_t) + c(y_t - x_t) + hx_{t+1} + pw_{t+1}]\right\}$$

subject to

$$x_{t+1} = y_t - \bar{d}_t - z_t$$
$$w_{t+1} = w_t + \hat{d}_t - z_t$$
$$x_t, w_t \geq 0, \quad y_t \geq \max\{x_t, \bar{d}_t\}, \quad t = 0, \ldots, T,$$

where $\delta(x) = 1$ if $x > 0$ and $\delta(x) = 0$ if $x = 0$.

Assuming positive h and p, clearly, in the optimal solution, $z_t = \min\{y_t - \bar{d}_t, w_t + \hat{d}_t\}$. Also, x_{t+1} and w_{t+1} cannot be both positive. Let $\tilde{x}_t$ be the net inventory at beginning of period t, i.e., $\tilde{x}_t = x_t - w_t$, and $\tilde{y}_t$ the net inventory after ordering. The problem can be rewritten as

$$\text{minimize} \quad E\left\{\sum_{t=0}^{T}[k\delta(\tilde{y}_t - \tilde{x}_t) + c(\tilde{y}_t - \tilde{x}_t) + h(\tilde{y}_t - d_t)^+ + p(d_t - \tilde{y}_t)^+]\right\}$$

subject to

$$\tilde{y}_t \geq \tilde{x}_t + \bar{d}_t$$
$$\tilde{x}_{t+1} = \tilde{y}_t - d_t, \quad t = 0, \ldots, T,$$

where $d_t = \bar{d}_t + \hat{d}_t$. Except for the special constraint, the model is the standard one. Indeed, a modified (s, S) policy is optimal. The parameters s_t and S_t are defined as usual. The optimal policy for period t is to order up to S_t (set $y_t = S_t$) if $x_t < \max\{s_t, \bar{d}_t\}$. Otherwise, do not order. (When the leadtime is positive, the analysis breaks down.)

3.1.2 Assembly systems

Now, return to the general model (P), and consider the special case of an assembly system (one product). The optimal allocation policy is simple: Just satisfy as much backorders and demand as possible. It remains to determine the component replenishment policy. When all components have the same leadtime, then the inventories of all components (adjusted for usage) should be equal at all times, and the problem reduces to a single-item model. Different leadtimes, however, are challenging.

Building on Schmidt and Nahmias (1985), Rosling (1989) studies a multi-stage assembly system with deterministic leadtimes. He shows that, under some mild conditions on initial inventories, the assembly system is equivalent to a series system. So, following Clark and Scarf (1960), an echelon base-stock policy is optimal. Applying this result to the ATO system considered here (i.e., no subassemblies and no final assembly time), the optimal policy is a *balanced base-stock policy*. This is like a base-stock policy, but the components are coordinated as follows: Let L_i be the leadtime for component i. Without loss of generality, assume $L_m > \cdots > L_1$. Redefine units if necessary so that $a_i = 1$ for all i. Orders for component m follow a standard base-stock policy, as if component m were the only one. For component $i < m$, order precisely the quantity of component m ordered $(L_m - L_i)$ periods ago. Thus, the same amounts of all components arrive at each time. See Zhang (1995).

In the following, we relate this problem (with different leadtimes) to the formulation (P) and provide a proof of the optimal policy. Note that, with positive leadtimes, the order decision for component i at t does not influence the system cost until $t + L_i$. Thus, the variable y_{it} in (P) can no longer serve as a decision variable. Instead, a directly controllable variable is the inventory position, the sum of net inventory and inventory on order. Let

$\hat{x}_{it}$ inventory position of component i at the beginning of period t before ordering, $\hat{\mathbf{x}}_t = (\hat{x}_{it})$ (vector)

$\hat{y}_{it}$ inventory position of component i at the beginning of period t after ordering, $\hat{\mathbf{y}}_t = (\hat{y}_{it})$ (vector)

$\tilde{x}_{it}$ net inventory of component i at the beginning of period t

$d[t,s)$ cumulative demand in periods $t, t+1, \ldots, s-1$, for $s > t$

$d[t,s]$ cumulative demand in periods $t, t+1, \ldots, s$, for $s > t$.

Then $x_{it} = [\tilde{x}_{it}]^+$ and $w_t = \max_i [\tilde{x}_{it}]^-$. So, the inventory-backorder cost can be expressed in terms of the $\tilde{x}_{it}$. It is well known that

$$\tilde{x}_{i,t+1} = \hat{y}_{i,t-L_i} - d[t - L_i, t]. \tag{3.1}$$

Since there is only one product, an optimal policy must guarantee that, after some initial periods, the net inventories of all components at the end of each period are equal. Applying (3.1), we have

$$\hat{y}_{i,t-L_i} - d[t - L_i, t] = \hat{y}_{m,t-L_m} - d[t - L_m, t] \tag{3.2}$$

for all $i < m$. This is equivalent to

$$\hat{y}_{i,t-L_i} = \hat{y}_{m,t-L_m} - d[t - L_m, t - L_i) \tag{3.3}$$

for all $i < m$. Thus, the optimal policy is the balanced one described above, and the problem reduces to a single-item problem with decision variable $\hat{y}_{m,t}$. In particular, the optimal policy for component m is base-stock with base-stock level s_m^*, the solution to

$$F_m(s) = p/(p + h_m),$$

where $F_m(\cdot)$ is the cumulative distribution function of $d[1, L_m]$. Each time we order component m, we order the same amount for component i, $i < m$, but at $(L_m - L_i)$ periods later. (Zhang (1995) achieves the same result by interpreting Rosling's result in the ATO setting.)

When the leadtimes are stochastic, the above approach does not work. It is no longer possible to define m so that the difference $L_m - L_i$ is always nonnegative. Song, Yano, and Lerssrisuriya (2000) consider the special case of a one-time stochastic demand. The problem is to determine when and how much of each component to order, to minimize the total expected cost. The optimal order quantity for each component is generally less than in the standard newsvendor model with no assembly structure and no leadtime uncertainty. Several simple but reliable heuristic procedures are developed. Numerical studies indicate that, in this setting, leadtime variability often has a larger impact than demand variability. Moreover, it is better to approximate leadtime uncertainty than to ignore it.

3.2 Performance evaluation

The optimal policy for the general system with positive leadtimes is unknown. One attractive heuristic is a base-stock order policy along with some allocation rule. Several authors focus on performance evaluation and optimization techniques for such policies, assuming that demand in each period has a multivariate normal distribution. The biggest challenge here is the numerical evaluation of such distributions.

Hausman, Lee, and Zhang (1998) assume the FCFS allocation policy, so that backlogged demands are filled in order of arrival. This implies that all available inventory is committed to the earliest backlogged demands, even if those units can be used to fill later demands. The service measure of interest is the fill rate with time-window τ, the probability of filling a demand within time τ, where τ is a given nonnegative integer. This measure is hard to compute exactly. The paper focuses on a lower bound, namely, the probability of filling *all* demand in a period within time τ, denoted R_τ.

For each component i, let s_i be the base-stock level and $d_i(L_i - \tau + 1)$ the demand over $L_i - \tau + 1$ time periods. It is shown that

$$R_\tau = P(d_i(L_i - \tau + 1) \leq s_i, \quad \forall i). \tag{3.4}$$

It turns out that $R_\tau = \tilde{R}_0$, where $\tilde{R}_0$ indicates R_0 in a revised system with truncated leadtimes $\tilde{L}_i = [L_i - \tau]^+$. The paper examines the problem of maximizing the R_0, which is a multivariate normal probability, subject to a linear budget constraint. Even this objective is hard to evaluate in general, and so the paper develops heuristic methods. The best of these seems to be an *equal fractile heuristic*, which selects the base-stock levels to equalize the components' fill rates.

The model of Agrawal and Cohen (2001) minimizes the total expected component inventory costs, subject to constraints on the order fill rates. Without the constraints, the problem separates by components. The allocation policy is the following:

1. Partial FCFS: Assign the available stock of components to specific finished-product orders, and release units for delivery, even when the entire subset of components is not available. (However, the product order is considered complete only when the entire kit has been delivered.)
2. Fair-share allocation: In case of a component shortage, the available stock is allocated to the orders, based on the actual demand in the period. Specifically, for each component i, product j receives a fraction of the stock, the ratio of its demand for the component to the total component demand.

The paper develops an expression for the order fill rate under this policy. Again, this requires the evaluation of multivariate normal distributions. The paper shows that the objective function is convex and the constraints are quasi-convex, so the globally optimum solution for the optimization problem exists and is characterized by the Kuhn-Tucker conditions. This approach leads to quite different solutions from the equal-fractile heuristic of Hausman et al.

Zhang (1997) studies a similar problem using a different allocation rule. He assumes that demands in different periods are filled according to the FCFS rule. However, for demands within the same period, a product priority rule is followed. Also, the following stock-commitment policy is used: Once component units are allocated to a product as above, these units remain committed to the product, even if the demand cannot be filled due to inadequate stock of other components. Two easy-to-compute lower bounds on the order fill rate are proposed, based on properties of the multivariate normal distribution, and their performances compared through numerical experiments. The results indicate that neither bound dominates the other.

Cheng, Ettl, Lin, and Yao (2002) assume i.i.d. replenishment leadtimes and FCFS allocation rule. They study the problem of minimizing average component inventory holding cost subject to product-family dependent fill rate constraints. They use an approximation for the fill rate in each product family, so that the constraint functions are linear functions of the item fill rates. An exact algorithm and a greedy heuristic algorithm are developed.

Using the solution techniques and real data from applications in IBM, the paper further conducted numerical experiments to highlight several key benefits of the ATO operation in terms of risk pooling. Their numerical examples also show that, compared to the Make-to-Stock model, ATO can lead to substantial inventory savings and improved demand forecast accuracy.

de Kok and Visschers (1999) propose a modified base-stock policy that adapts Rosling's approach for pure assembly systems to more general ATO systems. This approach essentially fixes the allocation of components among products, before the components actually arrive in inventory. The result, instead of a series system, is a multi-level distribution system. See Chapter 12 of this handbook for details.

To summarize, the research to date has developed several plausible, reasonably effective heuristic methods. These methods are quite different, however, in spirit as well as in detail. One cannot yet draw broad conclusions about which approaches are most promising in practice.

4 Continuous-time models

This section reviews continuous-time models. Again, we discuss first the characterization of optimal policies and then results on performance evaluation and optimization of a given type of policy.

4.1 Characterization of optimal policies

Again, little is known about optimal polices in the general case, and so our discussion focuses on special cases.

For assembly systems, Chen and Zheng (1994) extend the results of Rosling (1989) discussed above in several directions. In particular, they show that the results hold for continuous as well as discrete time. So, a continuous-time, single-product ATO system again reduces to a single-item one.

Ha (1997) studies a single-item $M/M/1$ make-to-stock queue with several demand classes and lost sales. (In our terms, this is a type of distribution system.) His results have the same flavor as those of Topkis (1968). Each demand class has a rationing limit. When the inventory is at or below the limit, it is optimal to reject demands of this class in anticipation of higher-priority demands. de Véricourt, Karaesmen, and Dallery (1999) study a similar problem with backlogs. The optimal policy has the same form.

4.2 Performance evaluation

For continuous-review, multi-product systems, all research to date assumes independent compound-Poisson processes for product demands. This implies that demand for each component too is compound-Poisson. Otherwise, the works differ in the modeling of the component supply and in analytical approaches.

Before summarizing individual works, we first introduce a formulation of the demand process, the stock allocation policy, and the inventory control policy, which is largely due to Song (1998).

Let $\mathcal{I} = \{1, 2, \ldots, m\}$ denote the set of component indices. Product-demand epochs form a stationary Poisson process, denoted $\{A(t), t \geq 0\}$, with rate λ. Each demand may require several components in different amounts simultaneously. For any subset of components $K \subseteq \mathcal{I}$, we say a demand is of type K if it requests $Z_i^K > 0$ units of component $i \in K$, and 0 units in $\mathcal{I} \setminus K$. The random variable $Z^K = (Z_i^K, i \in K)$ has a known discrete probability distribution ψ^K. Assume that each order's type is independent of the other orders' types and of all other events. Also, there is a fixed probability q^K that an order is of type K, $\sum_K q^K = 1$. Thus, the type-K order stream forms a Poisson process with rate $\lambda^K = q^K \lambda$. (An order type is thus a bit different from a product, as the word is used above. A product corresponds to a fixed recipe of components. An order type has a fixed set of components, but the quantities are random.)

Let $\mathcal{K}$ be the set of all demand types, that is, $\mathcal{K} = \{K \subset \mathcal{I} : q^K > 0\}$. Note that $\mathcal{K}$ is not necessarily the set of all possible subsets of $\mathcal{I}$. For each component i, let $\mathcal{K}_i$ denote the family of subsets of $\mathcal{K}$ that contain i. The demand process for component i forms a compound Poisson process with rate $\lambda_i = \sum_{K \in \mathcal{K}_i} \lambda^K = q_i \lambda$ and batch size Z_i, the mixture of Z_i^K for all $K \in \mathcal{K}_i$.

In general, the demand model does not impose any restrictions on the batches Z_i^K among $i \in K$. It is, however, worth mentioning three important special cases:

(1) *Unit demand:* $Z_i^K = 1$ for all $i \in K$. That is, a demand of type-K requires one and only one unit of each component in K. Such a demand process is especially common when the items are relatively expensive or durable. For instance, a customer of a bookstore may buy several books but only one copy each. In the consumer market of the mail-order personal computer business, a demand typically requires one motherboard, one keyboard, one monitor, and at most one video card. Here, ψ^K concentrates on a single point.
(2) *Assembly of multiple products:* $Z_i^K = a_i^K \xi^K$, where the a_i^K are constant positive integers and ξ^K is a positive-integer random variable. In this case, a type-K demand requests a random number ξ^K of units of product K, which has the fixed bill-of-material a_i^K, $i \in K$. Here, ψ^K is in effect a one-dimensional distribution – that of ξ^K.
(3) *Pick and pack:* Z_i^K are independent across $i \in K$. This is a reasonable approximation for demands in distribution systems, such as in mail-order retailing, especially when the items in K are not too closely related, e.g., women's sweaters and men's slacks. Here, ψ^K is the product of the marginal distributions ψ_i^K of Z_i^K.

Demands are filled on a first-come-first-served (FCFS) basis. Demands that cannot be filled immediately are backlogged. When a demand arrives and some of its required components are in stock but others are not, we either ship the in-stock components or put them aside as committed inventory. However, a demand is considered backlogged until it is satisfied completely.

The inventory of each component is controlled by a base-stock policy, with

$s_i :=$ the base-stock level for component i.

Let $t \geq 0$ be the continuous time variable, and for each t denote

$IN_i(t)$	net inventory of item i,
$A^K(t)$	number of type-K demands by time t,
$D_i(t)$	cumulative demand for i by time t
$B^K(t)$	type-K backorders at t
	number of type-K orders that are not yet completely satisfied by t,
$B_i(t)$	number of backorders for item i at t.

Let D_i stand for the steady-state limit of $D_i(t - L_i, t) = D_i(t) - D_i(t - L_i)$, the lead-time demand of item i. Let IN_i be the steady-state limit of $IN_i(t)$, and define B^K and B_i similarly. Also, define

W^K = steady-state waiting time of a type-K backorder.

The performance measures of interest are, for any demand type K,

$f^{K,w}$	type-K order fill rate with time window w
	probability of satisfying a type-K order within a time window w
	$P[W^K \leq w]$
f^K	fill rate of type-K demand $= f^{K,0}$
$E[B^K]$	average number of type-K backorders.

With these order-based performance measures, one can easily obtain the following system performance measures:

f	average (over all demand types) off-shelf fill rate $= \sum_K q^K f^K$.
$E[B]$	total average order-based backorders $= \sum_K E[B^K]$.

It is also interesting to relate the order-based performance measures to the component-based ones:

f_i	off-shelf fill rate of component i,
$E[B_i]$	average number of backorders of component i.

4.3 Constant leadtimes

Let L_i be the leadtime for component i, a constant. Then,

$$IN_i = s_i - D_i.$$

Performance evaluation thus involves the joint distribution of the leadtime demands $(D_1, \ldots, D_m)$. For example,

$$f^K = P(D_i + Z_i^K \leq s_i, \quad i \in K).$$

Let $f^{K,w}(\mathbf{s}|\mathbf{L})$ be $f^{K,w}$ in a system with base-stock levels $\mathbf{s} = (s_i)_i$ and leadtimes $\mathbf{L} = (L_i)_i$. When $L_i = L$ for all i, we write $f^K(\mathbf{s}|\mathbf{L})$ as $f^K(\mathbf{s}|L)$. It is shown in Song (1998) that, for any fixed K and $0 \leq w < \max_{i \in K}\{L_i\}$,

$$f^{K,w}(\mathbf{s}|\mathbf{L}) = f^{K,0}(\mathbf{s}|(L_1 - w)^+, \ldots, (L_n - w)^+). \tag{4.1}$$

Thus, $f^{K,w}$ equals the immediate fill rate f^K in a transformed system, where the leadtimes are truncated by w. So, we only need to focus on f^K henceforth. (This result is similar to the one for discrete-time systems discussed above.)

Song (1998) observes that there are several independent random variables shared by the elements of the leadtime-demand vector $(D_i)_i$, each of which is a univariate Poisson random variable. Thus, the dimension of the distribution of the vector can be reduced by conditioning on these common elements. As a result, the order fill rates can be obtained through convolutions of one-dimensional distributions. The paper develops an algorithm that sequences the conditioning steps. This procedure makes the calculation much simpler and faster than the direct approach using the joint distribution of the net inventories.

To illustrate the result, consider a 2-component, unit-demand system. Here, there are three types of demand: A type-1 customer requires one unit of component 1 only; type-2 requires one unit of component 2 only; and type-12 (short notation for type-{1,2}) asks for one unit of each component. In this unit-demand case, D_i, which has the same distribution as $D_i(L_i)$, has a Poisson distribution with parameter $\lambda_i L_i$. For convenience, let $p(\cdot|a)$, $P(\cdot|a)$, and $P^c(\cdot|a)$ denote the probability mass function, cdf and complementary cdf, respectively, of the Poisson distribution with parameter a.

The type-i fill rate is exactly component i's fill rate

$$f_i = P(IN_i > 0) = P(D_i < s_i) = P(s_i - 1|\lambda_i L_i), \quad i = 1, 2.$$

The type-12 fill rate is

$$f^{12} = P(IN_1 > 0, IN_2 > 0) = P(D_1 < s_1, D_2 < s_2).$$

Assume $L_1 = L_2 = L$. Then,

$$D_i = D_i(L) = D^i(L) + D^{12}(L), \quad i = 1, 2.$$

Here, $D^K(L)$ has the Poisson distribution with parameter $\lambda^K L$. Moreover, the $D^K(L)$ are independent. By conditioning on $D^{12}(L)$ and then deconditioning, we obtain

$$f^{12}(\mathbf{s}|L) = \sum_{k=0}^{\min\{s_1,s_2\}-1} p(k|\lambda^{12}L)P(s_1-k-1|\lambda^1 L)P(s_2-k-1|\lambda^2 L).$$

In the case $L_1 \neq L_2$, a few extra calculations yield a similar result.

Now, let us return to the general problem. Song (1998) also develops simpler bounds on the order fill rate, which require lower-dimensional joint distributions or merely the marginal distributions. More specifically, it is shown that, for any K,

$$f^K \geq \prod_{\ell=1}^{k} f^{S_\ell}, \qquad (4.2)$$

where $\{S_1,\ldots,S_k\}$ is any partition of K. In particular,

$$f^K \geq \prod_{i \in K} f_i. \qquad (4.3)$$

Also,

$$f^K \leq \min_{i \in K} f_i.$$

It turns out that f^{ij}, the fill rate for type-$\{i,j\}$ demand, is quite easy to obtain for any i and j; it has the same formula as f^{12} given above. So, (4.2) yields a better lower bound on f^K than $\prod_{i \in K} f_i$ without too much computational effort. Finding the best *pair partition* so that this lower bound is maximized is equivalent to the *nonbipartite weighted matching problem*, which can be solved by existing algorithms in the combinatorial-optimization literature.

Song (2002) studies the evaluation of order-based backorders for the same model. Let $B_i^K(t)$ be the number of backorders for item i at time t that are due to demand type K, where $K \in \mathcal{K}_i$. Let B_i^K be the steady-state limit of $B_i^K(t)$. Then

$$E[B^K] = E\left[\max_{i \in K} B_i^K\right]. \qquad (4.4)$$

Given $B_i = n$, B_i^K is a binomial random variable with n trials and success probability λ^K/λ_i in each trial. However, since the B_i are correlated random variables, computing its joint distribution alone is difficult, not to mention the conditional binomial distributions and the max operation within the expectation. Song presents a much simpler approach. To illustrate the results, consider the 2-item, unit-demand system with equal leadtimes discussed above. Let $E[B^K(\mathbf{s}|L)]$ denote the expected type-K backorders with base-stock levels s_i and common leadtime L. First, notice that a request for item i is due to a type-i order with probability λ^i/λ_i, so the average type-i backorders equals

$$\overline{B}^i(s_i|L) = \frac{\lambda^i}{\lambda_i} E[B_i(s_i|L)],$$

where

$$E[B_i(s_i|L)] = \lambda_i L - \sum_{k=0}^{s_i-1} P^c(k|\lambda_i L).$$

Also,

$$E[B^{12}(\mathbf{s}|L)] = \lambda^{12} L - q^{12} \sum_{\ell=0}^{\min\{s_1,s_2\}-1} \sum_{m=0}^{s_1-\ell-1} \sum_{j=0}^{s_2-\ell-1} \frac{(\ell+m+j)!}{\ell! m! j!}$$
$$\times (q^{12})^\ell (q^1)^m (q^2)^j P^c(\ell+m+j|\lambda L).$$

Using the formulas, the paper discusses a few examples to gain managerial insights. It is shown, for example, that for a given level of inventory investment, using common components or fewer product configurations may *not* reduce backorders. This is in contrast to conclusions drawn from more restrictive single-period models in the literature.

Although the exact result enjoys a tremendous computational advantage over simulation, it can still be computationally demanding for large systems. The paper also develops easy-to-compute bounds. In particular,

$$LB^K := \lambda^K \max_{i \in K} \frac{E[B_i]}{\lambda_i} \leq E[B^K] \leq \lambda^K \sum_{i \in K} \frac{E[B_i]}{\lambda_i} := UB^K.$$

Summing these inequalities yields bounds on the total average order-based backorders:

$$LB := \sum_K LB^K \leq E[B] \leq \sum_K UB^K := UB.$$

It can be verified that

$$UB = \sum_{i=1}^J E[B_i] = \text{the total average item-based backorders} := E[B_I].$$

So, the total item-based backorders always dominates the total order-based backorders.

A natural approximation for $E[B^K]$ is the simple average of LB^K and UB^K. Numerical results indicate that the approximation performs extremely well.

Song (2000) extends the above results to more general policies, the (R, nQ) policies. (Here, for each component i, there is a base lot-size Q_i, a positive integer. When the inventory position of item i falls to or below the reorder point R_i, an order of size nQ_i is placed, where n is the integer such that the inventory position after ordering is between $R_i + 1$ and $R_i + Q$. When all $Q_i = 1$, the policy reduces to a base-stock policy.) Under reasonable conditions, the service measures can be computed as simple averages of their counterparts under base-stock policies.

4.4 Uncapacitated stochastic leadtimes

Next, consider the case where the supply system of each component consists of many parallel processors, so that its leadtimes are i.i.d. random variables. The constant-leadtime model is a special case. Let

$X_i(t)$ = number of outstanding orders of component i at time t.

Then (returning to base-stock policies),

$$IN_i(t) = s_i - X_i(t).$$

Thus, performance evaluation involves the distribution of the outstanding-order vector $X(t) = (X_1(t), \ldots, X_m(t))$ and its steady-state limit $X = (X_1, \ldots, X_m)$.

I.i.d. leadtimes were assumed in the earliest studies of dynamic product-based performance, in the literature on multi-indenture models of multi-echelon inventory systems. Here, an end item (product) consists of several repairable modules (components). A failure of a product is due to the failure of one component. The performance measure of interest is the number of products backordered. Cannibalization is allowed, that is, a good component in a failed product can be used to replace a failed component in another failed product. Thus, the number of products backordered is the maximum of the component backorders, i.e.,

$$B = \max_i B_i = \max_i [(X_i - s_i)^+]. \tag{4.5}$$

Suppose the product fails according to a Poisson process. Then, one can calculate the cumulative distribution of B. For other arrival processes, one can obtain the expected value of B. See Nahmias (1981) for a review.

Cheung and Hausman (1995) consider multivariate Poisson demand, so there can be simultaneous failures of several components. Again, they assume complete cannibalization, so (4.5) holds. The authors propose the following disaggregation approach for the joint distribution of X_i: Define Y^K to be the number of jobs of type-K that have one or more components outstanding. Then, according to Palm's result for $M/G/\infty$ queues, Y^K has a Poisson distribution with mean $\lambda^K E[\max_{i \in K} L_i]$, and the Y^K are independent over K.

Let X_i^K be the number of outstanding orders of component i that originated from demand type-K, so $X_i = \sum_{i \in K} X_i^K$. Then, E[$B$] can be evaluated by conditioning on $(\mathbf{Y} = \mathbf{y}) = (Y^K = y^K, \forall K)$. The conditional probability P[$X_1 \leq x_1, \ldots, X_m \leq x_m | \mathbf{Y} = \mathbf{y}$] must be obtained through computation. This probability is easy when all $y^K = 0$ or 1. For larger y^K, it becomes complicated.

Two approximations are proposed to simplify the computation of E[B] for large m. However, the approximations still employ the conditional probability mentioned above. Also, Jensen's inequality yields a simple lower bound:

$$\text{E}[B] \geq \max_i \{(\lambda_i \text{E}[L_i] - s_i)^+\}.$$

Unfortunately, as the paper shows, this bound works poorly as an approximation.

Gallien and Wein (2001) assume i.i.d. component leadtimes in a single-product assembly system. Demand is Poisson with rate λ. Inspired by Rosling's result for deterministic leadtimes, the following class of policies is considered: Start with stocks of all components at a common base-stock level, s. Every demand triggers an order for each component after a component-dependent delay $\delta_i \geq 0$. Numerical experiments show that this policy achieves nearly identical performance to the standard base-stock policy with base-stock levels $s_i = s - \lambda \delta_i$.

To keep the analysis tractable, the paper imposes a synchronization assumption, namely, that components are assembled in the same sequence they are ordered in. Thus, the time needed to replenish a complete set of components is $\max_i(L_i + \delta_i)$. Let Y be the steady-state number of replenishment orders for complete sets of components, for which at least one of the m individual component orders has not yet arrived. Then, applying Palm's result, Y has a Poisson distribution with mean $\rho = \lambda \text{E}[\max_i(L_i + \delta_i)]$. Also, $I = (s - Y)^+$ and $B = (Y - s)^+$. Using a tandem queueing network analogy, it is shown that

$$\text{E}[I_i] = \lambda(\text{E}[\max_i(L_i + \delta_i)] - \text{E}[L_i] - \delta_i).$$

Now, assume L_i has the Gumbel distribution with cdf $\exp(-\alpha_i e^{-mx})$, $m > 0$. This implies that all the L_i have the same variance σ^2. Then, E[$\max_i(L_i + \delta_i)$] can be written in closed form for any δ_i. This permits the closed-form determination of optimal values of δ_i and s that minimizes the long-run average cost

$$\sum_{i=1}^m h_i \text{E}[I_i] + \left(\sum_{i=1}^m h_i\right) \text{E}[I] + b[B],$$

where h_i and b are unit holding-cost rate and backorder cost rate, respectively. That is,

$$\delta_i^* = \max_j \left(E[L_j] - \frac{\sqrt{6}}{\pi} \sigma \ln h_j \right) - \left(E[L_i] - \frac{\sqrt{6}}{\pi} \sigma \ln h_i \right), \quad i = 1, \ldots, m$$

$$\rho^* = \frac{\lambda\sqrt{6}\sigma}{\pi} \ln \left[\sum_{i=1}^{m} \exp\left(\frac{\pi(E[L_i] + \delta_i^*)}{\sqrt{6}\sigma} \right) \right]$$

s^* is the smallest integer that satisfies $P(s^*|\rho^*) \geq \dfrac{b}{b+h}$.

Numerical results for an 11-component system indicate that this approximate solution is within 2% of the best among this class of policies (found by simulation), in all cases with a common leadtime variance. It also significantly outperforms policies that ignore either component dependence or leadtime variance. In addition, it is reasonably robust with respect to various modeling assumptions. However, the order-synchronization approximation does affect the results; it overestimates the amount of inventory required.

Song and Yao (2002) also study the single-product system with Poisson demand and i.i.d. leadtimes. They adopt the standard base-stock policy, without order synchronization. Under any base-stock policy, the outstanding-order vector $X(t)$ is precisely the numbers of jobs in m $M/G/\infty$ queues with a common arrival stream. This $X(t)$ has a steady-state limit, X. Let $G_i(\cdot)$ be the cdf of L_i and $G_i^c = 1 - G_i$. Let $N(a)$ denote a Poisson random variable with mean a. Then, X can be expressed as partial sums of 2^{m-1} independent Poisson random variables as follows:

$$X_i = \sum_{S: i \in S} N(\lambda \theta_S), \quad \text{with} \quad \theta_S = \int_0^\infty \left[\prod_{k \in S} G_k^c(x) \right] \left[\prod_{j \in \mathcal{I} \setminus S} G_j(x) \right] dx. \tag{4.6}$$

Here, for any subset S of $\mathcal{I}$, $N(\lambda \theta_S)$ is the number of jobs (in steady state) still in process in the queues $k \in S$, but completed by the other queues.

In principle, all the performance measures can be evaluated exactly using (4.6). However, there are $2^m - 1$ independent Poisson random variables. This exponential growth in the number of components means that the method is impractical for large systems.

The paper investigates the effect of leadtime variability by comparing two systems, the original system with leadtimes L_i, and another system with leadtimes $\tilde{L}_i$. Assume $E[L_i] = E[\tilde{L}_i] = \ell_i$, and that L_i is more variable than $\tilde{L}_i$ in the sense of the 'increasing convex ordering', denoted $L_i \geq_{icx} \tilde{L}_i$, i.e.,

$$\int_x^\infty G_i^c(u)\, du \geq \int_x^\infty \tilde{G}_i^c(u)\, du$$

for $x \geq 0$. (Here, $\tilde{G}_i^c$ is the complementary cdf of $\tilde{L}_i$.) Note that the above implies $\mathrm{Var}[L_i] \geq \mathrm{Var}[\tilde{L}_i]$. Let $\tilde{f}$ and $\tilde{B}$ denote the fill rate and the number of backorders in the new system. Then,

$$f \leq \tilde{f},$$
$$B \geq_{\mathrm{st}} \tilde{B}.$$

Thus, in contrast to the standard single M/G/∞ queueing system, leadtime variability degrades performance here.

Since evaluating $E[B]$ is hard, the paper develops simple upper and lower bounds on $E[B]$ and uses them as surrogate objectives in the following optimization problem:

$$\min E[B(s_1,\ldots,s_m)] \tag{4.7}$$
$$\text{s.t.} \quad c_1 s_1 + \cdots + c_m s_m \leq C.$$

Greedy algorithms are developed, and numerical results indicate that these solution techniques are fairly effective.

The paper considers another optimization problem that minimizes the average component inventory costs subject to a required fill rate. Approximating the constraint by using (4.3) yields a separable convex programming problem, which can be solved via a greedy algorithm. Numerical results show that this lower bound approach usually results in an order fill rate (in the original system) that is substantially higher than the required service level. However, the greedy algorithm has considerable advantage in computation time. Therefore, it can be used to quickly generate an initial solution, followed by a neighborhood search to find the best solution.

The extension of this analysis to multiple products turns out to be far from routine. Lu, Song, and Yao (2003a) derive the joint generating function of X:

$$\psi(z_1,\ldots,z_m) := E\left[\prod_{j=1}^m z_j^{X_j}\right],$$

$$= \exp\left[\sum_{K \in \mathcal{K}} \lambda^K \int_0^\infty (\psi^{Z^K}(G_1(u) + z_1 G_1^c(u),\ldots,G_m(u) + z_m G_m^c(u)) - 1)\, du\right].$$

In the special case of unit demands (all $Z_i \equiv 1$), the generating function takes the following form:

$$\psi(z_1,\ldots,z_m) = \exp\left[\sum_{K \in \mathcal{K}} \lambda^K \int_0^\infty \left(\prod_{j \in K} [G_j(u) + z_j G_j^c(u)] - 1\right) du\right],$$

Ch. 11. *Supply Chain Operations: Assemble-to-Order Systems*   585

which corresponds to a multivariate Poisson distribution. Thus, we obtain a generalization of (4.6):

$$X_i = \sum_{K \in \mathcal{K}_i} \sum_{S \ni i, S \subseteq K} N(\lambda^K \theta_S^K)$$

where all the Poisson variables are independent, and for any subset S of K,

$$\theta_S^K = \int_0^\infty \prod_{j \in S} G_j^c(u) \prod_{j \in K \setminus S} G_j(u) du.$$

The effort required to evaluate the performance measures is linear in the number of products. Unfortunately, it is again exponential in the number of components. On the other hand, the generating function can be used to obtain simple expressions for the means, variances and covariances. Let $\mathcal{K}_{ij}$ denote the family of subsets that contain both i and j. Then,

$$\mu_j := E[X_j] = E[L_j] \sum_{K \in \mathcal{K}_j} \lambda^K E(Z_j^K),$$

$$\sigma_j^2 := \text{Var}[X_j] = E(X_j) + \sum_{K \in \mathcal{K}_j} \lambda^K [E((Z_j^K)^2) - E(Z_j^K)] \int_0^\infty [G_j^c(u)]^2 du,$$

$$\sigma_{ij} := \text{Cov}[X_i, X_j] = \sum_{K \in \mathcal{K}_{ij}} \lambda^K E(Z_i^K Z_j^K) \int_0^\infty G_i^c(u) G_j^c(u) du \quad i \neq j,$$

This suggests using the multivariate normal distribution with these moments to approximate X. This is still a difficult calculation, and the paper develops several further approximations.

One approximation is based on an upper bound on the covariance:

$$\sigma_{ij} \leq \eta_i \eta_j,$$

where

$$\eta_i := \left[\sum_{K \in \mathcal{K}_i} \lambda^K E((Z_i^K)^2) \right]^{1/2} \left[\int_0^\infty (G_i^c(u))^2 du \right]^{1/2}.$$

It turns out that the normal calculations become easy with each σ_{ij} replaced by $\eta_i \eta_j$. This is called the *factorized* normal approximation.

A second approximation applies (4.2) to a *pairwise* partition of $\mathcal{I}$. The calculation then reduces to the evaluation of bivariate normal distributions.

A numerical study compares the factorized normal approximation, the pairwise approximation and the marginal lower bound (4.3). The pairwise approximation is the most accurate, but the other two methods require less computational effort, and their accuracy is reasonably good.

Lu, Song, and Yao (2003b) focus on the following optimization problem for the same multiproduct model:

$$\min_{\mathbf{s}} \sum_{K} w^K E[B^K(s)] \qquad (4.8)$$

$$\text{s.t.} \quad c_1 s_1 + \cdots + c_m s_m \leq C.$$

where $\mathbf{s} = (s_1, \ldots, s_m)$, and $w^K \geq 0$ is a weighting factor for the average type-K backorders. To find the optimal solution, the authors study two surrogate problems, based on upper- and lower-bound approaches to approximate the objective function. Both surrogate problems are of the same structure. Let n be the number of demand types (or products). Then the optimal base-stock levels can be determined by solving $n!$ minimizations problems of the following form:

$$\min_{\mathbf{y}} \sum_{\ell=1}^{n} v_\ell y_\ell$$

$$\text{s.t.} \quad \sum_{\ell=1}^{n} \tau_\ell (y_1 + \cdots + y_\ell) \leq C, \quad \mathbf{y} \geq \mathbf{0}.$$

Each of these problems can be solved by greedy methods. Heuristic algorithms are also developed to speed up the computation. Numerical results indicate that these solution techniques are quite effective.

Lu and Song (2002) formulate an unconstrained cost-minimization problem for the multi-product, unit-demand system. Let b^K be the backorder cost rate for each backlogged customer order of type-K, and let J_i be the steady-state number of units of item i that have been put aside and committed to demands which are backlogged due to the unavailability of other items. The expected total average cost under any base-stock policy $\mathbf{s} = (s_1, \ldots, s_m)$ is:

$$C(\mathbf{s}) = \sum_i h_i E[I_i(s_i) + J_i(\mathbf{s})] + \sum_K b^K E[B^K(\mathbf{s})]$$

$$= \sum_i h_i s_i + \sum_K \tilde{b}^K E[B^K(\mathbf{s})] - \sum_i h_i E[X_i],$$

where

$$\tilde{b}^K = b^K + \sum_{i \in K} h_i.$$

Ch. 11. Supply Chain Operations: Assemble-to-Order Systems

The paper compares the above formulation with item-based formulations – that is, treat the system as a set of independent single-item systems. Let b_i be the unit backorder cost for item i backorders. The item-based optimization problem is

$$\min_{\mathbf{s}} \sum_{i=1}^{m} (h_i E[I_i(s_i)] + b_i E[B_i(s_i)]) = \sum_{i=1}^{m} (h_i s_i + (h_i + b_i) E[B_i(s_i)]) - \sum_i h_i E[X_i].$$

This problem is separable across i. The problem for each i is a newsvendor-type problem and its solution can be obtained easily.

Lu and Song show that, if we set

$$b_i = \sum_{K \in \mathcal{K}_i} \frac{\lambda^K}{\lambda_i} \left(b^K + \sum_{j \in K, j \neq i} h_j \right),$$

the result of this item-based calculation is an upper bound on $\mathbf{s}^*$. A different choice of b_i yields a lower bound for the upper bound. Moreover, using the upper bound as a starting point, $\mathbf{s}^*$ can be obtained in a greedy fashion by employing recently developed optimization techniques for discretely convex functions.

4.5 Capacitated stochastic leadtimes

Song, Xu, and Liu (1999) consider a multi-product, unit-demand model. The supply system of each component i is modeled as single exponential processor with rate μ_i and a finite backlog buffer of capacity $b_i \geq 0$. The finite buffer works as follows: A demand for component i that cannot be filled immediately goes to the backlog queue i, provided the queue is not full. The demand will be shipped out (or put aside) as soon as a unit of item i becomes available. When a demand arrives and finds any of its items' backlog queues full, it signals the customer that a long wait is likely, and the customer decides to leave. Thus, the buffer sizes can be viewed as measures of customer impatience. (When $b_i = \infty$ for all i, unfilled demands are backlogged. When $b_i = 0$ for all i, unfilled demands are lost.)

Two blocking mechanisms are considered when an incoming demand finds the backlog queue for at least one of its components full:

- *Total order service (TOS)*: If a type K order sees at least one of its component's backlog queue is full, then the order is lost entirely. In other words, a type K order must be accepted as a whole. This model is valid for the assemble-to-order environment and also for some make-to-stock systems.

- *Partial order service* (*POS*): When a type K order arrives and the backlog queue i is full for $i \in K' \subset K$, then the order for items in K' is lost, whereas the order for items in $K - K'$ is satisfied, either immediately or in the future. This model fits many distribution systems, where customers often accept partial shipments of finished goods.

Thus, in the POS model, customer impatience is associated with individual items, while in the TOS model, it is associated with the whole order.

The paper shows that the outstanding-order vector $X(t)$ is an irreducible continuous-time Markov chain with finite state space. Its unique stationary distribution can be obtained through the *matrix-geometric* solution of a *quasi birth-and-death* (QBD) process [Neuts, 1981, Chapter 3]. The paper derives the exact expressions for $f^{K,w}$ and $E[B^K]$ in terms of this solution.

Numerical experiments indicate that the results from the POS model provide reliable estimates of their counterparts in the TOS model. So, it is sufficient to focus on one order service scheme. Also, with moderate traffic the finite-buffer model provides an accurate approximation for the infinite-buffer model.

Iravani, Luangkesorn, and Simchi-Levi (2000) employ a similar modeling framework and technique to study a system with flexible customers. Each K is partitioned into two subsets K^1 and K^2. K^1 contains the 'key' components of a type-K demand. If any components in K^1 are not available, a type-K demand is lost. On the other hand, if some 'non-key' components – those in K^2 – are not available, then a type-K customer may accept substitutions or even ignore them. Specifically, for $i \in K$, there is a probability p_{ij}^K that customers of type-K will accept component j if component i is not available. If $p_{ij}^K = 0$ for all $j \in \mathcal{S}$ \{i\}, then customer type-K accepts no substitutes for component i.

Glasserman and Wang (1998) model the supply system as a set of M/G/1 queues (G/G/1 queues for the single-product case). Assuming the fill rate $f^{K,w}$ remains high, the paper investigates the tradeoff between the delivery time window w and the total base-stock units $s = s_1 + s_2 + \cdots + s_m$. It is intuitively clear that, fixing the fill rate, s increases as w decreases. The key result of the paper is that this relationship is asymptotically linear, provided the ratios $k_i = s_i/s$ are kept constant as s increases. The two parameters of this linear relationship can be determined exactly (through analysis of the arrival and service times' cumulant generating functions) or approximately (from their moments).

Let U_i denote the random processing time of a unit of component i, and V_i be the interarrival time of demand for component i. It is shown that for large s or w, the item fill rate can be approximated by

$$1 - f_i^w \approx C_i e^{-\gamma_i w - \beta_i s_i}.$$

for some constants C_i, γ_i and β_i. For Poisson demand process (which is necessary for the multiproduct case),

$$C_i = \lambda_i^{-1}(\lambda_i + \gamma_i)(1 - \lambda_i E[Z_i]E[U_i])(\psi'_{Z_i}(\beta_i)\psi_{U_i}(\gamma_i)(\lambda_i + \gamma_i) - 1)^{-1}.$$

Recall that Z_i is the demand batch size for component i. The constants γ_i and β_i can be obtained by solving some equations involving the cumulant generating functions of the input random variables. (The cumulant generating function of a random variable Y is defined by $\psi_Y(\theta) = \log E[e^{\theta Y}]$.) They can also be approximated by the first two moments of these random variables as follows:

$$\gamma_i \approx -\frac{2E[U_i]E[Z_i] - E[V_i]}{E[Z_i]\text{Var}[U_i] + \text{Var}[Z_i](E[U_i])^2}$$

and

$$\beta_i \approx E[U_i]\gamma_i + \frac{1}{2}\text{Var}[U_i]\gamma_i^2.$$

Let

$$\gamma^K = \min_{i \in K}\{\gamma_i\} \quad \text{and} \quad \mathcal{W}^K = \{i \in K : \gamma_i = \gamma^K\}.$$

Define $\alpha_i = k_i \beta_i$ and let

$$\alpha^K = \min_{i \in K}\{\alpha_i\} \quad \text{and} \quad \mathcal{S}^K = \{i \in K : \alpha_i = \alpha^K\}.$$

Then, when the time window w is long, the order fill rate can be approximated by

$$[1 - f^{K,w}] \approx \sum_{i \in \mathcal{W}^K} C_i e^{-\gamma^K w - \alpha_i s}.$$

When the total base-stock units s is high,

$$[1 - f^{K,w}] \approx \sum_{i \in \mathcal{S}^K} C_i e^{-\gamma_i w - \alpha^K s}.$$

Assuming that all products' fill rates are the same and high, and the items' base-stock levels change in constant proportions, this approximation suggests that, when the time window w is changed, the base-stock levels should be varied according to the component-level tradeoff rule

$$\Delta s_i = -\frac{\gamma_i}{\beta_i}\Delta w.$$

to maintain the same order fill rate. Numerical experiments show that under certain conditions this linear rule provides satisfactory results.

This result, however, depends strongly on the assumption of finite capacity. (It is easy to show that, in a simple single-item system with constant leadtimes, the relationship is nonlinear.)

Wang (1999) applies this result to an optimization problem to minimize average inventory cost subject to a fill-rate constraint. The paper focuses on a single-product system and solves a surrogate problem with closed-form solution. In particular, it shows that there exists an index k, such that the following solution is effective:

$$\tilde{s}_i = \frac{1}{\beta_i} \log \frac{\theta_k C_i e^{-\gamma_i w}}{h_i/\beta_i}, \quad \text{for } i \leq k \text{ and } \tilde{s}_i = 0 \text{ for } i > k,$$

where

$$\theta_j = \frac{\sum_{i=1}^{j}(h_i/\beta_i)}{\delta - \sum_{i=j+1}^{m} C_i e^{-\gamma_1 x}}$$

and h_i is the unit holding cost of component i.

Dayanik, Song, and Xu (2001) examine several ideas scattered in diverse literature on approximations for multivariate probability distributions, and determine which approach is most effective in estimating performance in capacitated ATO systems. Tailoring different approximation ideas to the ATO setting, they derive several performance bounds, such as setwise bounds based on the dependence structure of the system, distribution-free Bonferroni-type bounds commonly used to bound multivariate distributions, Frechet-type bounds, and bounds that are combinations of these previous ones. The paper compares these bounds both analytically and numerically. The general conclusion is that the setwise bounds are most effective.

Xu (2001) summarizes several performance bounds for ATO systems based on stochastic comparison techniques.

To summarize, research on continuous-time models has made major strides in the last few years in developing robust analytical tools for design and control of ATO systems. Both exact and asymptotic results as well as bounds and approximations have been developed. The methods, however, heavily depend on the detailed model assumptions. So, in applying these methods, one should be careful about which model framework fits best in the particular application.

5 Research on system design

Another line of research aims to understand the broad issues involved in product and process design. An overview is given by Nevins and Whitney

(1989) and a review of the research literature by Krishnan and Ulrich (2001). This work, of course, considers a range of production modes, not just ATO. Here, we focus on research that explicitly treats the product- and component-variety issues posed by ATO systems.

This research tends to suppress most of the detail of the operational models discussed above. Instead, it aims to approximate the operational cost of a system by means of simple functions.

Fisher, Ramdas, and Ulrich (1999) develop a model for system design in one specific industry, automobile brakes, and test it empirically. This model represents the total operational costs (and design costs as well) by affine functions of demand. Brakes differ from each other on one critical dimension, their rotor diameters. Depending mainly on its weight, a car requires rotors of at least a certain diameter. The model aims to determine the optimal number of brakes for a given family of cars. Under certain simplifying assumptions, the optimal number of brakes is proportional to a simple index, the square root of (total demand times the range of car weights). The paper then tests the model using data from six companies, half American and half Japanese. The index accurately predicts the actual variety in brakes.

Ramdas and Sawhney (2001) develop a model to redesign a product line. First, they develop a method to estimate the impact on revenues due to product-line extensions. Second, they outline a method to estimate the operational-cost impact. This method includes terms reflecting the scale economies due to component commonality in the new products. Third and finally, they combine these methods into an integer-programming model to select the optimal line extensions. The paper reports a case study based on data from a wristwatch manufacturer.

Krishnan, Singh, and Tirupati (1999) develop another product-line design model. Although it focuses on product-development costs, it includes functions that represent part-commonality effects in operational costs. The end result, again, is an optimization model. Krishnan and Gupta (2001) use a similar model to evaluate 'product platforms', sets of components and subassemblies shared across whole families of products. They identify conditions under which such platforms may, and may not, be beneficial.

The issue of component commonality is related to the broader issue of modular design. Baldwin and Clark (2000) provide an overview of this concept. Thonemann and Brandeau (2000) develop a detailed model to optimize the level of part commonality.

These few works are, in our view, best seen as initial forays into largely uncharted territory. The design process involves many factors in addition to operational costs. Although the phrase 'design for manufacturing' represents a recognition of the importance of such costs, we do not yet understand how best to organize design resources to take them into account, along with other critical factors. Future research, we hope, will shed more light on this matter.

6 Summary and future directions

As we have seen above, recent research has made considerable progress in developing analytical methods for ATO systems. We now have tractable methods to estimate and improve performance, at least for some systems. Those methods have led to some interesting and useful managerial insights. Much work remains, nevertheless. Here we point out a few areas where further research is needed.

6.1 Optimal policies

As indicated above, little is known about the forms of optimal policies for multi-period models. The research to date mostly assumes particular policy types. It would be valuable to learn more about truly optimal policies. Even partial characterizations would be interesting. Also, better heuristic policy forms would be useful.

6.2 Tractable methods for large-scale systems

Many real ATO systems contain hundreds of components and thousands of products. A division of Hewlett-Packard, for example, uses over 100 PC components grouped into eight component families to make their computers. Such a system poses a considerable computational burden on existing models and solution methods. Even data estimation is no trivial task.

A number of approaches might improve matters. One approach is to seek model formulations with special structures that allow efficient evaluation of the performance measures. Another approach is to develop decomposition and approximation schemes allowing algorithm 'scalability' to large data sets. Sections 3 and 4 reviewed several ideas in the recent developments. Still, better methods of this sort would be most welcome.

6.3 Demand distributions

Nearly all the models in this chapter assume stationary data. However, short product life cycles imply time-varying or state-dependent demand. It is desirable that practical models in the future allow for such complex demand models.

6.4 Shifts in supply chain structures and costs

The pressure to streamline supply chain flows and to increase supply chain efficiency and reduce cost has led many manufacturers to outsource some (even all) steps of assembly operations (mostly product configuration and customization), usually to their distributors, who might in turn delegate part of the final assembly to the retailers. Hence the ATO problem might be encountered by multiple players in the same supply chain. (Chapters 6 and 7 of this volume discuss the issues involved in multi-player supply chains.)

In addition, new issues and management practices continue to emerge. Manufacturing capacity in assembly, once owned and concentrated at the manufacturer, is now shifted downstream in the supply chain and becomes distributed among the players and more flexible and less expensive to expand at the same time. Therefore, the scheduling of short-term flexible capacities to meet temporary product demand fluctuations or product mix changes, and the coordination of capacities at different stages of a system, will continue to be important research problems.

The manner in which supply-chain players share the financial risks and costs might also be changing. A particular example is the 'price protection' contract between the manufacturer and its distributor. Designed to protect the distributor from rapid price declines and shift the price decline risk to the manufacturer, the price protection policy changes the traditional definition of inventory holding cost so that much of the inventory cost might be shifted to either the supplier or the customer. Furthermore, the cost relationship between the supplier and the buyer may be made more complex by vendor-managed inventory (VMI) programs. Since the inventory holding cost is a critical parameter in ATO models, the change in the cost structure might affect the solution and the recommendation significantly. (Chapters 7 and 8 of this volume discuss these and other issues involved in multi-player supply chains.)

6.5 Product design implications

Model-based research on product design, as suggested in Section 5, is at an early stage. Unfortunately, we do not yet understand detailed operational models well enough to derive from them simple, empirically testable and usable cost models. For the time being, therefore, empirical models must rely on ad hoc cost functions with little basis in theory. We see many opportunities for future research to help bridge this gap.

Acknowledgement

We would like to thank Alex Zhang for helpful discussions on this topic.

References

Agrawal, M., M. Cohen (2001). Optimal material control and performance evaluation in an assembly environment with component commonality. *Naval Research Logistics* 48, 409–429.
Bagchi, U., G. Gutierrez (1992). Effect of increasing component commonality on service level and holding cost. *Naval Research Logistics* 39, 815–832.

Baker, K., M. Magazine, H. Nuttle (1986). The effect of commonality of safety stock in a simple inventory model. *Management Science* 32, 982–988.

Baldwin, C., K. Clark (2000). *The Power of Modularity*, Cambridge, MA, MIT Press.

Brumelle, S., D. Granot (1993). The repair kit problem revisited. *Operations Research* 41, 994–1006.

Chen, F., Y.-S. Zheng (1994). Lower bounds for multi-echelon stochastic inventory systems. *Management Science* 40, 1426–1443.

Cheng, F., M. Ettl, G. Y. Lin, D. D. Yao (2002). Inventory-service optimization in configure-to-order systems. *Manufacturing & Service Operations Management* 4, 114–132.

Cheung, K. L., W. Hausman (1995). Multiple failures in a multi-item spare inventory model. *IIE Transactions* 27, 171–180.

Dayanik, S., J.-S. Song, S.H. Xu (2001). The Effectiveness of Several Performance Bounds for Capacitated Assemble-to-Order Systems, working paper, Graduate School of Management, University of California, Irvine, CA 92697.

de Kok, A., J. Visschers (1999). Analysis of assembly systems with service level constraints. *International Journal of Production Economics* 59, 313–326.

Eynan, A. (1996). The impact of demand's correlation on the effectiveness of component commonality. *Int. J. Prod. Res.* 34, 1581–1602.

Eynan, A., M. Rosenblatt (1996). Component commonality effects on inventory costs. *IIE Transactions* 28, 93–104.

Fisher, M., K. Ramdas, K. Ulrich (1999). Component sharing in the management of product variety: A study of automotive braking systems. *Management Science* 45, 297–315.

Gallien, J., L. Wein (2001). A simple and effective component procurement policy for stochastic assembly systems. *Queueing Systems* 38, 221–248.

Gerchak, Y., M. Henig (1986). An inventory model with component commonality. *Operations Research Letters* 36, 61–68.

Gerchak, Y., M. Henig (1989). Component commonality in assemble-to-order systems: models and properties. *Naval Research Logistics* 36, 61–68.

Gerchak, Y., M. Magazine, A. Gamble (1988). Component commonality with service level requirements. *Management Science* 34, 753–760.

Glasserman, P., Y. Wang (1998). Leadtime-inventory tradeoffs in assemble-to-order systems. *Operations Research* 46, 858–871.

Graves, S. (1982). A multi-item inventory model with a job completion criterion. *Management Science* 28, 1334–1336.

Ha, A. (1997). Inventory rationing in a make-to-stock production system with several demand classes and lost sales. *Management Science* 43, 1093–1103.

Hausman, W. H., H. L. Lee, A. X. Zhang (1998). Joint demand fulfillment probability in a multi-item inventory system with independent order-up-to policies. *European Journal of Operational Research* 109, 646–659.

Hillier, M. (1999). Component commonality in a multi-period inventory model with service level constraints. *International Journal of Production Research* 37, 2665–2683.

Hillier, M. (2000). Component commonality in multi-period assemble-to-order systems. *IIE Transactions* 32, 755–766.

Iravani, S., K. Luangkesorn, D. Simchi-Levi, On assemble-to-order systems with flexible customers. Working paper, Northwestern University, 2000. Forthcoming, *IIE Transactions*.

Kerwin, K. (2000). At Ford, E-commerce Is Job 1, *Business Week*, February 28, 74–78.

Krishnan, V., S. Gupta (2001). Appropriateness and impact of platform-based product development. *Management Science* 47, 52–68.

Krishnan, V., R. Singh, D. Tirupati (1999). A model-based approach for planning and developing a set of technology-based products. *Manufacturing & Service Operations Management* 1, 132–156.

Krishnan, V., K. Ulrich (2001). Product development decisions: A review of the literature. *Management Science* 47, 1–21.

Lu, Y., J.-S. Song, (2002). Order-based cost optimization in assemble-to-order systems. Working paper, IBM Watson Research Center, Yorktown Hights, NY 10598.

Lu, Y., J.-S. Song, D.D. Yao (2003a). Order fill rate, leadtime variability, and advance demand information in an assemble-to-order system. *Operations Research* **51**, March–April.

Lu, Y., J.-S. Song, D.D. Yao (2003b). Backorder minimization in multiproduct assemble-to-order systems. Working paper, IBM Watson Research Center, Yorktown Hights, NY 10598.

Mamer, J., S. Smith (1982). Optimizing field repair kits based on job completion rate. *Management Science* 28, 1328–1333.

Mamer, J., S. Smith (1985). Job completion based inventory systems: optimal policies for repair kits and spare machines. *Management Science* 31, 703–718.

Mamer, J., S. Smith (2001). Inventories for sequences of multi-item demands, in: J. S. Song, D. D. Yao (eds.), *Supply Chain Structures: Coordination, Information and Optimization*, Norwell, MA, Kluwer Academic Publishers, pp. 415–437. Chapter 12.

Neuts, M. (1981). *Matrix Geometric Solutions in Stochastic Models*, Baltimore, MD, Johns Hopkins University Press.

Nevins, J., D. Whitney (1989). *Concurrent Design of Products and Processes*, New York, McGraw-Hill.

Ramdas, K., M. Sawhney (2001). A cross-functional approach to evaluating multiple line extensions for assembled products. *Management Science* 47, 22–36.

Rosling, K. (1989). Optimal inventory policies for assembly systems under random demands. *Operations Research* 37, 565–579.

Schmidt, C., S. Nahmias (1985). Optimal policy for a two-stage assembly system under random demand. *Operations Research* 33, 1130–1145.

Sherbrooke, C. (1992). *Optimal Inventory Modeling of Systems*, New York, Wiley.

Smith, S., J. Chambers, E. Shlifer (1980). Optimal inventories based on job completion rate for repairs requiring multiple items. *Management Science* 26, 849–852.

Song, J.-S. (1998). On the order fill rate in a multi-item, base-stock inventory system. *Operations Research* 46, 831–845.

Song, J.-S. (2000). A note on assemble-to-order systems with batch ordering. *Management Science* 46, 739–743.

Song, J.-S. (2002). Order-based backorders and their implications in multi-item inventory systems. *Management Science* 48, 499–516.

Song, J.-S., S. H. Xu, B. Liu (1999). Order fulfillment performance measures in an assembly-to-order system with stochastic leadtimes. *Operations Research* 47, 131–149.

Song, J.-S., C. Yano, P. Lerssrisuriya (2000). Contract assembly: dealing with combined supply leadtime and demand quantity uncertainty. *Manufacturing & Service Operations Management* 2, 287–296.

Song, J.-S., D.D. Yao (eds.) (2001). *Supply Chain Structures: Coordination, Information and Optimization*, Kluwer Academic Publishers, Norwell, MA.

Song, J.-S., D. D. Yao (2002). Performance analysis and optimization in assemble-to-order systems with random leadtimes. *Operations Research* 50, 889–903.

Swaminathan, J., S. Tayur (1998). Managing broader product lines through delayed differentiation using vanilla boxes. *Management Science* 44, S161–S172.

Swaminathan, J., S. Tayur (1999). Stochastic programming models for managing product variety, in: S. Tayur, R. Ganeshan, M. Magazine (eds.), *Quantitative Models for Supply Chain Management*, Boston, Kluwer. Chapter 19.

Thonemann, U., M. Brandeau (2000). Optimal commonality in component design. *Operations Research* 48, 1–19.

Topkis, D. (1968). Optimal ordering and rationing policies in a nonstationary dynamic inventory model with n demand classes. *Management Science* 15, 160–176.

Van Mieghem, J., N. Rudi (2001). Newsvendor networks: inventory management and capacity investment with discretionary activities, working paper, Northwestern University. Forthcoming, *Manufacturing & Service Operations Management*.

Veinott, A. (1965). The optimal policy for a multi-product, dynamic, nonstationary inventory problem. *Management Science* 12, 206–222.

de Véricourt, F., Karaesmen, F., Dallery, Y. (1999). Optimal stock rationing for a capacitated make-to-stock production system, working paper, Laboratoire Productique et Logistique, Ecole Centrale de Paris. Forthcoming. *Management Science*.

Wang, Y. (1999). Near-optimal base-stock policies in assemble-to-order systems under service levels requirements, working paper, MIT Sloan School.

Wemmerlov, U. (1984). Assembly-to-order manufacturing: Implications for materials planning. *Journal of Operations Management* 4, 347–368.

Xu, S.H. (2001). Dependence analysis of assemble-to-order systems, in: J. Song, D. Yao (eds.) *Supply Chain Structures: Coordination, Information and Optimization*, Kluwer Academic Publishers, Norwell, MA, pp. 359–324, Chapter 11.

Zhang, A.X. (1995) Optimal order-up-to policies in an assemble-to-order system with a single product, Working paper, School of Business Administration, University of Southern California.

Zhang, A. X. (1997). Demand fulfillment rates in an assemble-to-order system with multiple products and dependent demands. *Production and Operations Management* 6, 309–324.

Chapter 12

Planning Supply Chain Operations: Definition and Comparison of Planning Concepts

Ton G. de Kok and Jan C. Fransoo

Department of Technology Management, Technische Universiteit Eindhoven, 5600 MB Eindhoven, The Netherlands

1 Introduction

In this chapter we discuss the Supply Chain Operations Planning problem. Positioning Supply Chain Operations Planning (SCOP) in the context of Supply Chain Management (SCM), the objective of SCOP is to **coordinate the release of materials and resources** in the **supply network** under consideration such that **customer service constraints are met at minimal cost.**

We boldfaced the coordination of releases of materials and resources, since this distinguishes SCOP from other SCM decision processes discussed in this volume. First, by deciding on the releases of both materials and resources, we take into account material and resource constraints simultaneously. Second, we explicitly consider coordination of all release decisions in a multi-item, multi-period setting, i.e., operational coordination over time. Especially this operational coordination over time distinguishes SCOP from other Supply Chain Planning (SCP) activities, such as setting seasonal stock levels based on aggregate supply-demand balancing or planning availability of resources in general. The SCOP problem incorporates the outcomes of earlier planning decisions and generates material and resource release decision that are executed by the shopfloor control function. In the context of the SCP matrix presented in Stadtler and Kilger (2000), SCOP overlaps both mid-term and short-term planning, i.e., it translates mid-term planning decisions into short-term execution decisions.

Another consequence of the focus on operational coordination is the incorporation of demand uncertainty and throughput time uncertainty in the models discussed in this Chapter to the extent possible to date. In that sense our approach differs from the more common approach in the context of SCP to formulate a deterministic instance of a particular SCP problem and develop heuristics or optimization methods to solve the instance. We will discuss the difference between these two approaches in more detail in Sections 4 and 6.

We boldfaced 'supply network' since we discuss in this chapter the SCOP problem for general supply network structures. This implies that we assume that items considered in the SCOP problem may be assembled from other multiple items and may be assembled themselves into other multiple items. In that respect we extend the analysis presented in Chapters 10 and 11 of this volume, albeit that the generality of the problem posed prohibits an analysis of similar elegance.

Notice that nowadays the supply network involves multiple organizations, e.g., an OEM and 1st and 2nd tier suppliers. Furthermore, the SCOP problem involves multiple units within an organization, e.g., sales, marketing, production and purchasing, or different manufacturing or warehousing locations. This observation has important consequences for the formulation of quantitative models that formalize the SCOP problem. In our formulations we assume that there exists a centralized objective function, and that information is shared across the supply chain. Furthermore, we assume that each of the echelons in the supply chain assumes responsibility for maintaining a certain lead time, as this is not considered part of the SCOP task. We will discuss this decomposition further in the next section, where we argue that this hierarchical decomposition and ordering of planning decisions is the only possible way both from a theoretical and from a practical perspective. Note that we will thus neither consider multi-agent situations in SCP, nor gaming situations (which have only been studied until now in a supply chain contracting setting, see Chapters 6 and 7 of this volume). We would however like to point out that some form of central coordination is possible even under limited information exchange using the models discussed in this chapter (see Fransoo, Wouters, and de Kok (2001)).

We boldfaced the objective of meeting customer service level constraints at minimal cost, because our objective is to compare, where possible and by selection of appropriate case situations, the various SCOP concepts proposed in the literature to date. To our knowledge such a comparison has not yet been undertaken and we have found that it generates deeper insights into the nature of the SCOP problem. Surprisingly, the comparative study also generated further insight into the design of supply networks, in particular the positioning of inventory capital in a given supply chain structure.

The outline of this introductory section is as follows. In Section 1.1 we introduce the SCOP problem from a practical perspective in more detail. Thereafter we define the variables and notions that enable to formulate the SCOP problem as a quantitative optimization problem. These variables and notions are used throughout this chapter. In Section 1.2 we discuss the material aspect of the SCOP problem, whereas in Section 1.3 we discuss resource aspects. In Section 1.4 we briefly discuss the concept of so-called planned lead times. In Section 1.5 we present two optimization problems that are used as a basis for comparison between different SCP concepts. In Section 1.6 we deal with the Customer Order Decoupling Point (CODP) concept and its relevance for the SCOP problem.

1.1 The SCOP problem

The SCOP function is responsible for the coordination of activities along the supply chain, by making decisions on the quantities and timing of material and resource releases. In this chapter, we explicitly model the supply chain as a network, i.e., activities that transform inputs into outputs using available resources are preceded by multiple transformation activities and succeeded by multiple transformation activities. Note that a transformation activity is a general designation of any type of relationship between two items in a supply chain, and can be both referring to physical transformation activities such as manufacturing or assembly activities and to non-physical transformation activities such as transportation from one location to another. Typical activities to be considered are

1. Manufacturing activities, i.e., activities that physically transform physical inputs into physical outputs.
2. Transportation activities, i.e., activities that move physical outputs from one location to another.
3. Planning activities, i.e., all administrative activities that are required for enabling a manufacturing or transportation activity to take place, such as process planning, transportation contracting, creation of purchase orders, etc.

From the SCP point of view it is essential to identify all relevant activities and their mutual relationships. In particular planning activities can be executed parallel to manufacturing and transportation activities. On the other hand it is well possible that planning activities determine a major part of the overall supply chain throughput time, e.g., when negotiation of price is essential for economic viability or acquisition of information about future demand is of paramount interest, or when letters of credit need to be obtained before actual manufacturing can start.

In each SCOP situation, the definition of these three types of activities, including their characteristics, is the starting point for defining the SCOP problem. In this subsection we concentrate on the representation of the manufacturing activities and transportation activities in relation to the SCOP problem. Reason for this starting point is two-fold:

1. Manufacturing activities and transportation activities are usually well-defined processes of which the main characteristics like processing times, resource requirements and process yields can be easily determined relatively.
2. Supply chain planning in itself is a planning activity at a specific hierarchical level. This implies that the choice for a particular SCP concept impacts planning activities at both lower and higher hierarchical levels. We postpone a detailed discussion to Section 2.

When considering physical transformation activities and their mutual relationships, we find that two generic aspects determine these relationships:

1. One transformation activity's output is another transformation activity's input.
2. One transformation activity shares one or more resources with another transformation activity

In Section 1.2 we focus on the first aspect, defining a Bill Of Material (BOM) structure and all related variables. In Section 1.3 we discuss the second aspect, defining a Bill Of Process (BOP) structure and all related variables.

1.2 Bill of material structure

The physical supply network structure is defined by parent-child relationships between items. 'Item' is the generic term for any input into and any output from transformation activities. In the case of a manufacturing operation a set of items is transformed into one or more items. In this chapter, we restrict ourselves to the situation that a transformation process outputs only a single item, which in turn can be used in multiple other transformation processes. This assumption is valid in many situations, e.g., in discrete part manufacturing. There are situations, especially in process industries and reverse manufacturing, where the manufacturing of one item implies that another item is manufactured at the same time. Such items are called by-products. For a discussion of this phenomenon and the resulting complexities, we refer to Spengler, Püchert, Penkuhn and Rentz (1997). It is important to note that a transportation activity transforms one item into another by changzing the location of the material involved. Generally speaking an item is equivalent to a material/location combination. We omit here the time-aspect of an item; we assume the item does not change over time, e.g., due to engineering changes.

Let us consider a supply network consisting of N items. For each item i, $i = 1, 2, \ldots, N$ we define a_{ij} as the number of items i required to produce one item j ($i = 1, 2, \ldots, N, j = 1, 2, \ldots, N$).

The matrix (a_{ij}) is called the Bill Of Material (BOM). In the context of MRP-literature the BOM is usually associated with a single end-item or so-called MPS-item (MPS is Master Production Schedule) [cf. Orlicky (1975)]. Our definition of BOM comprises that definition in the sense that if our supply network would produce a single end-item for a market then both definitions of BOM would be identical.

An end-item is an item that is not used in any other item. Such an end-item is delivered to customers of the supply network. We define E as the set of end-items, i.e.,

$$E \quad \{i | a_{ij} = 0, i = 1, 2, \ldots, N, j = 1, 2, \ldots, N\}$$

We define the set I of intermediate items as

$$I \quad \{i | \exists\, 1 \leq j \leq N \text{ with } a_{ij} > 0,\, i = 1, 2, \ldots, N\}$$

For notational purposes it is convenient to introduce the following sets associated with each item,

$$V_i \quad \{j | a_{ij} > 0, \quad j = 1, 2, \ldots, N\}$$
$$W_i \quad \{j | a_{ji} > 0, \quad j = 1, 2, \ldots, N\}$$

Hence V_i is the set of successors of item i and W_i is the set of predecessors of i, $i = 1, 2, \ldots, N$. With the above definitions we have characterized completely the material structure of the supply chain.

One set of decisions that results from solving the SCOP problem is the set of *material release decisions*. In order to rigorously describe these decisions we must introduce some notation. The variables defined relate to the solution of the SCOP problem at a particular point in time. We assume that the SCOP problem is solved periodically at equidistant moments in time. Typical periods used in practice are days, weeks and months.

Let us define period t as the time interval $(t-1, t]$. At time t, $t = 0, 1, 2, \ldots$, release decisions are taken. We define for $i = 1, 2, \ldots, N$:

$D_i(t)$ independent demand for item i in period t, i.e., demand in period t for item i, that is not derived from demand for items in $I \cup E$

$G_i(t)$ dependent demand for item i in period t, i.e., demand in period t for item i, that is derived from demand for items in $I \cup E$

$p_i(t)$ quantity of item i that becomes available at the start of period t from the transformation activity generating item i

$r_i(t)$ quantity of item i released at the start of period t immediately after receipt of $p_i(t)$

$I_i(t)$ physical inventory of item i at the start of period t, immediately before receipt of $p_i(t)$

$B_i(t)$ backlog of item i at the start of period t, immediately before receipt of $p_i(t)$

$J_i(t)$ net inventory, i.e., physical inventory minus backorders, of item i a t the start of period t, immediately before receipt of $p_i(t)$

Notice that independent demand is demand generated by customers of the supply network. Such demand is usually not known beforehand and must be forecast. We define the item set P as,

$$P \quad \text{Set of items } i \text{ with } D_i(t) > 0 \quad \text{for some } t > 0.$$

Furthermore notice that $\{r_i(t)\}$ are the set of decision variables that constitute one part of the core outcome of the SCOP problem, viz. the material release decisions.

1.3 Bill Of process structure

Manufacturing and transportation activities are executed by resources. Execution of an activity at a resource requires some processing time. In general such a processing time may vary over time due to many different causes, most of which cannot be controlled. Hence in general we represent the processing time by a random variable.

We assume that the processing time at a resource depends on the item that is the output of the transformation activity. It can be easily seen that this assumption is without any loss of generality. We furthermore assume that a resource can only execute one transformation process at a time. This assumption may be a restriction, e.g., in process industries it is quite common that one transformation activity generates multiple items (e.g., by-products), which implies that multiple transformation processes are run in parallel on the same resource. In our definition, this would have to be solved by pre-allocating a specific portion of the available resource capacity to a specific item.

As stated above we assume that the SCP process is executed periodically, e.g., daily or weekly. This implies that we define resource availability as available capacity in units of time during a period. Thus we define

C_{kt} Amount of capacity available in units of time of resource k in period t, $k = 1, \ldots, K$, $t \geq 1$,

where K is the number of available resources. Let us define the following variables associated with resource usage,

U_k Set of items that can be processed on resource k
c_i Time required to process one unit of item i on its resource

For the sake of simplicity we assume that an item can be processed on one resource only. In many cases the analysis can be extended to the situation where item i can be processed on multiple resources, which implies the definition of additional variables and constraints.

The decision variables related to the release of resources at the start of an arbitrary period are given by the set $\{q_i(t)\}$, where $q_i(t)$ is defined as

$q_i(t)$ Amount of item i processed in period t, $t \geq 1$.

We note here that we assume that an amount of item i processed in period t becomes available at the start of period $t+1$. This implies that

$$p_i(t) = q_i(t-1), \quad t \geq U.$$

This implies that we do not consider the possibility of random yield. For an extensive discussion of random yield we refer to Yano and Lee (1995). Based on our experience with SCOP problems in highly volatile environments with random yields we argue that current state-of-the-art literature on this subject is not applicable to multi-item multi-echelon networks. As a quick-and-dirty solution we propose to incorporate random yield through periodic updates of the state of the supply network, taking into account actual yields, and some safety stock provisions.

1.4 The planned lead time concept

As stated above SCP coordinates release of material and resources. The coordination of material would be more or less trivial if transformation activities would require negligible time. However, transformation activities require processing times on resources. Due to various sources of uncertainty, such as demand uncertainty and random processing times, the interactions between materials and resources result into *lead times* of item orders. The phenomenon of lead times in manufacturing has been extensively studied. For an overview of this literature, which strongly relies on queuing network theory, we refer to Suri, Sanders and Kamath (1993) and Buzacott and Shantikumar (1993). From this literature we learn that a lead time consists of a processing time and a waiting time. The waiting time is typically the major part of the lead time, caused by interaction between multiple item orders, which are using the same resources for the execution of the transformation process. Waiting of an item can occur both before and after the actual transformation activity.

Thus, in order to properly coordinate the release of materials and resources, the SCOP level has to take into account lead times. For each item we thus define its lead time L_i,

L_i throughput time between time of release of an order for item i and time at which the ordered items are available for usage in other items and/or delivery to customers

Given the periodic nature of the SCP process we assume that L_i is an integer number of time units. We assume that items i released at the start of period t are available for usage at the start of period $t+L_i$, i.e., in $(t+L_i-1, t+L_i]$.

In the context of SCP we are faced with the following core issue:

Is L_i endogenous or exogenous to the SCP concept?

This issue is extensively discussed in Section 2. It is concluded there that the lead time L_i is exogenous to the SCOP problem. Given the fact that, as stated above, lead times are related to resource utilization, the actual choice of L_i should be consistent with the resource availability and resource requirements that can be derived from the BOP and the exogenous demand characteristics. Such consistency should be derived from either empirical data or by applications of the above-mentioned results from queuing (network) models.

1.5 Performance measurement as a basis for comparison

The main objective of this chapter is to provide insights into the applicability of the various SCP concepts proposed in the literature. Typically, each scientific contribution in this area selects its own case to generate managerial insights. Contrary to common practice in combinatorial optimization there is no commonly accepted set of test problems. Furthermore, it may happen that analytical results are derived based on assumptions that need not hold. A typical example of the latter is the assumption that upstream availability is guaranteed, so that a decomposition of the supply chain model results. We have seen examples where the analysis that builds on this assumption eventually yields 'optimal' solutions that strongly violate the assumption required for the analysis. Such examples are discussed in Section 5.

In order to make a proper comparison of different SCP concepts we define a cost structure and a performance criterion. We define $C(t)$ as the cost incurred at the end of period t, $t \geq 0$,

$$C(t) = \sum_{i=1}^{N} h_i I_i(t),$$

where h_i var value of item i $\forall i$

Notice that $C(t)$ is not really a cost function but represents the total supply chain inventory capital investment at the start of period t. We are interested in the long-run average value of $C(t)$,

$$\overline{C} = \lim_{t \to \infty} \frac{1}{t} \sum_{s=1}^{t} C(s).$$

We assume existence of $\overline{C}$, which holds true in the case of stationary stochastic demand. The comparisons of SCOP concepts in Sections 5 and 6 are restricted to that situation.

The long-run average supply chain inventory holding cost can be derived from multiplying $\overline{C}$ by the interest rate. Note that by taking $\overline{C}$ as a basis for comparison we circumvent discussions about proper interest rates, although a discussion about value added in the supply network cannot be circumvented.

Comparing capital investments suffices only when lot-sizing restrictions are irrelevant. More precisely, when we assume that each period for all items a positive quantity is released, we can refrain from considering fixed set-up or ordering costs. In many practical applications lot sizing is not an issue at SCOP level. Either because of sufficiently high manufacturing flexibility [cf. Bertrand (2003), Chapter 4 of this volume for an extensive discussion of flexibility in supply chain management context], or because of the time aggregation into weekly or monthly buckets, implying that lot sizing decisions are taken at the level below the SCOP level. In most cases the lot sizing restrictions have been determined at some level above the SCOP level, since lot sizing impacts the need for resources. Still, there are situations where lot sizing decisions must be taken at SCOP level. We will not deal with that situation due to the fact that to date virtually no results are available in literature on the analysis of general assembly networks under demand uncertainty and lot sizing restrictions. For an in-depth discussion of lot sizing and its position in the planning hierarchy we refer to Chapter 9, Fleischmann and Meyr (2003).

The capital investment is needed to ensure sufficient customer service. Note that we must define customer service for all items with independent demand, i.e., for all items in P. As performance criteria we choose α_i and β_i, $\forall i \in P$, defined as

$\alpha_i \quad \lim_{t \to \infty} P\{I_i(t) > 0\}, \forall i \in P$, non-stockout probability

$\beta_i \quad \lim_{t \to \infty} 1 - \dfrac{[E(I_i(t) + p_i(t) - D_i(t))^+] - E[(-I_i(t) - p_i(t))^+]}{E[D_i(t)]}, \forall i \in P$, fill rate

Likewise the case of $\overline{C}$ we assume existence of α_i and β_i. Notice that α_i is identical to the P_1-measure defined in Silver, Pyke, and Peterson (1998) and β_i is identical to the P_2-measure defined there. For each SCOP concept $\mathcal{P}$ we want to solve the following problems:

Problem (P_α)

$$\min \overline{C}(\mathcal{P})$$

s.t. $\quad \alpha_i(\mathcal{P}) \geq \alpha_i^*, i \in P$

Problem (P_β)

$$\min \overline{C}(\mathcal{P})$$

s.t. $\quad \beta_i(\mathcal{P}) \geq \beta_i^*, i \in P$

It is implicitly assumed that the SCP concept $\mathcal{P}$ satisfies the set of material and resource constraints derived in Section 3. In both problems we express the dependence of both $\overline{C}$ and α_i and β_i on $\mathcal{P}$. Hence we want to minimize capital investments subject to the service level constraints α_i^* or β_i^*, $i \in P$. An alternative to the use of service level constraints is the introduction of penalty costs. Apart from the fact that penalty costs are hard to determine in practice, one is also compelled to replace capital investments by inventory holding costs in order to make a proper trade-off. As indicated above we want to circumvent a discussion of holding costs. For results on the equivalence of penalty costs and service level constraints we refer to Silver et al. (1998), Diks, de Kok and Lagodimos (1996), Diks (1997), Janssen (1998) and Van Houtum and Zijm (2000).

Apart from a comparison of SCP concepts, we will also discuss the differences in analysis of particular SCP concepts as reported in the literature. Especially for SCP concepts that assume stochastic demand, which are discussed in Section 5, we have found different simplifying assumptions needed for tractability. We will show that some of these assumptions may yield suboptimal or erroneous results when validating the model results with simulation or an exact analysis.

We emphasize here that the comparative approach presented here is, to our knowledge, the first of its kind. Furthermore, the supply chain operations planning problem with stochastic demand is complex and to date no insight exists into the structure of optimal policies for general supply chain structures. For single period problems with two components and two end-items Rosenblatt and Eynan (1996) are able to derive the optimal policy structure. A recent paper by Hillier (2000) discusses an Assemble-To-Order situation, where the supply chain structure consists of multiple components and multiple end-items. Stocks are only held at the component level. Using Stochastic Dynamic Programming Hillier derives the optimal policy for the multi-period problem and the infinite-horizon problem. For more complex supply chain structures no results are currently available in the literature.

All this implies that this state-of-the-art comparative study identifies more questions than answers. These questions are the basis for further research, which is, as stated above, discussed in Section 7.

1.6 The customer order decoupling point concept

In this introductory Section, it is relevant to discuss a concept from the Supply Chain Management literature that is relevant for the development of an appropriate SLOP model. This concept is known as the Customer Order Decoupling Point [CODP, cf. Bertrand, Wortmann and Wijngaard (1990), Silver et al. (1998), or Hoekstra and Romme (1991), from which the Decoupling Point concept originates]. The CODP is the point that indicates how deeply the customer order penetrates into the supply chain. It is the distinction between the order-driven and forecast-driven parts of the supply chain for a particular product market combination. Items that are kept in stock at the CODP are those items for which demand must be forecast due to the fact that future demand between the moment of release of items and the moment those items are received is (partially) unknown; the lead time of supplying the item is longer than the lead time requested by the customer. Downstream of the CODP, items are not kept in stock, since future demand for these items between release moments and receipt moments is known; the lead time of supplying the item from the CODP to the customer is shorter than the lead time requested by the customer. In our models, we only consider demand upstream from the

CODP. Demand *at the CODP* then is independent demand, either originating directly from the demand forecast of end-items (Make-to-Stock), demand forecasts of modules (Assemble-to-Order), or of components (Make-to-Order). All releases *downstream of the CODP* are based on actual customer orders and thus are not planned under uncertainty of demand (cf. Final Assembly Schedule, Orlicky (1975)). All releases *upstream of the CODP* are planned based on dependent demand.

It is interesting to note here that originally the CODP concept is applied to a single organization. When taking a supply chain view, i.e., a view comprising multiple organizations, many CODPs vanish due to the fact that demand originally considered to be independent and thereby requiring forecasting, becomes dependent demand. As we will see in the sequel this observation has a major impact on the performance of the supply network. Typically, items that are stocked at the CODP are assumed to be available with high probability to satisfy independent demand. When converting these items that face independent demand into items facing dependent demand only, such a requirement is no longer relevant. Trade-offs will reveal that a supply chain view implies low availability of these items to prevent unnecessary inventory capital investments.

In this chapter, we will only consider the releases of items that are kept in stock at the CODP and the items that are part of the BOM of those CODP-items. Therefore, without loss of generality, *we denote the CODP-items as end-items* in this text. In case these end-items are not sellable products, but modules or (sets of) components, the independent demand for these end-items can be derived from forecasts of (sets of) sellable products. Alternatively, one can use historical data and market intelligence to derive forecasts of end-items directly.

1.7 Structure of the chapter

This concludes our discussion of the basic notions that enable to define the SCOP problem in the form of quantitative models. The remainder of this Chapter is structured as follows. In Section 2 we discuss the position of the SCOP problem in the context of a hierarchical planning framework comprising aggregate planning, SCOP and detailed scheduling. An important aspect of the discussion in Section 2 is the motivation for incorporating so-called planned lead times into the formulation of the SCOP problem, since in the sequel of this Chapter we restrict to supply network models with deterministic item lead times. In Section 3 we derive generic material and resource release constraints, which we denote further as 'generic SCOP constraints', which are used in later sections to test the validity of various SCOP concepts proposed in the literature. In Section 4 we use the generic SCOP constraints to derive an LP formulation for the SCOP problem without lot sizing restrictions on the material releases. The LP formulation is a benchmark in the discussion of other SCOP concepts based on deterministic exogenous demand in a rolling schedule context. The exogenous demand is derived from a forecasting or sales planning process [cf. Chapter 9, Fleischmann and Meyr (2003) of this volume]. In Section 5 we discuss SCOP concepts based on quantitative models that explicitly incorporate stochastic demand. Where possible we compare the various concepts with respect to assumptions made regarding the supply network structure and regarding the item availability assumptions required for the quantitative analysis. In Section 6 we compare the LP-based SCOP concept with a so-called synchronized base stock (SBS) policy developed by De Kok and Visschers (1999) under infinite resource availability. The reason for selecting these two concepts is that they represent two really distinct modeling concepts for the SCOP problem for general supply networks and for both concepts we can derive solutions that ensure that predefined customer service level constraints are met with a high accuracy. The LP-based concept represents the class of SCOP concepts based on deterministic models in a rolling schedule context, widely available in standard software for SCOP as discussed by Fleischmann and Meyr (2003) in Chapter 9. The SBS concept seems to be to date the only concept representing SCOP concepts explicitly incorporating stochastic demand into the SCOP model, that is able to cope with the general multi-item multi-echelon models that result from the SCOP problem for general supply networks. The comparison of these two concepts yields fundamental insights into the SCOP problem that are extensively discussed. Finally, in Section 7 we summarize our findings and discuss further SCOP research challenges.

2 The hierarchical nature of SCP

In this chapter, we position the supply chain operations planning problem in a hierarchical framework. Hierarchical planning frameworks enable us to accurately model the consecutive planning and scheduling decisions made in manufacturing organizations. The SCOP problem is only one in a series of planning and scheduling problems to be solved by (groups of) manufacturing organizations to realize their objectives in terms of customer service, turnover, profit, ROI, etc. We start with a discussion of various research perspectives that underlie the development of hierarchical planning concepts developed in the past.

2.1 Hierarchies in planning

Decisions with regard to the different components of planning of supply chain operations have traditionally been analyzed independently from one another by researchers. Research addressing the scheduling problem, the (multi-echelon) inventory problem, and the aggregate capacity planning problem have hardly been interconnected while maintaining their own characteristics. On the contrary, in the late 1960s and early 1970s attempts have been made from each of these domains to expand the scope of research and apply their available specific methods to other components of the SCP problem. In these approaches, the specific nature of each of the components has however been disregarded, and the problems have developed into conceptually monolithic models. An illustration is the work on combining lot sizing and scheduling [see, e.g., Dauzère-Pérès and Lasserre (1994)], in which two models remain to exist, but a final solution is obtained by iterating between the two models.

Managers were however still faced with this multitude of different problems in the SCP domain. They solved these issues by organizing these decisions in a hierarchical manner. Meal (1984) analyzes and describes these hierarchies and links them to the hierarchical planning hierarchies introduced by Hax and Meal (1975) and Bitran and Hax (1977).

The idea for hierarchical production planning was captured formally by Anthony (1965). He introduced three levels of hierarchical control: Strategic Planning, Management Control, and Operational Control. The principal ideas for developing this planning hierarchy into a set of formal models supporting coordinated decision making at these levels were developed by Bitran, Hax, and Meal (BHM) in the early 1970s [Winter (1989)]. In their publications, generally the following terms are used for the models supporting the three decision levels: aggregate planning, family disaggregation, and item disaggregation. The BHM hierarchy is based on capacity coordination only. Material coordination is not considered and bills-of-material are not included, which precludes the use of their methodology in SCP. Originally, the work was motivated by and the models were based on discrete parts batch manufacturing, with later applications in continuous manufacturing and job shops [see Bitran and Tirupati (1993) for a review and McKay, Safayeni and Buzacott (1995) for a historical perspective]. Note that all these environments are primarily capacity oriented [Bertrand et al., 1990]. Essentially there are two types of constraints at each of the levels of the BHM formulation:

(1) Primary process constraints: these are 'hard' constraints that are derived from physical constraints in the process, such as resources.
(2) Decision process constraints: these are 'soft' constraints that are imposed upon a level by its immediate higher level in the decision hierarchy.

At the highest level, the aggregate resource constraints form the basis of the resource hierarchy, with the decision how much time to allocate in regular time and in overtime, in line with the original HMMS model [Holt et al. (1960)], although the costs in the BHM model are linear and not quadratic as in HMMS. A distinction between the primary process constraints and the decision process constraints is not made in the model formulation, leading to the fact that, e.g., a decision to produce a certain quantity of a product family is fixed, despite possibly 'better' feasible solutions once the more detailed planning starts at a lower level.

When multiple stages are introduced into the BHM formulation, perfect aggregation becomes very difficult [see the review of the work of Axsäter by Bitran and Tirupati (1993)] and the only way in which to devise a hierarchical planning procedure is by a very loose coupling. Graves, Meal, Dasu and Qui (1986) propose a two-level model, which at the aggregate level plans capacity and at the detailed level uses a base stock approach to coordinate the various stages of production. Bitran and Tirupati (1993) and Meal, Wachter and Whybark (1987) discuss the relationship between the BHM formulation of hierarchical planning and MRP. They conclude that the hierarchical planning and MRP systems are complimentary, in that hierarchical planning focuses primarily on determining capacity levels and capacity smoothing, while MRP determines the amount of material required at various points in the manufacturing process. Due to the interactions between material release and resource release decisions as formulated in the SCOP models to be discussed in the following sections, we can argue that this argument cannot be extended to supply chain operations planning, and in fact can only be upheld if the time periods considered in BHM's hierarchical planning model are an order of magnitude longer than the time periods in the MRP model and if resources are flexible.

Decisions with regard to the planning of supply chain operations have traditionally been taken at the operational level. Meal (1984) argues that this was necessarily decentralized due to the lack of good information processing technology. In this approach, which he names the 'conventional approach', operations planning decisions were an integral part of the decision making power of the line managers in all parts and at all levels in the organization. Decisions were only coordinated marginally, and certainly not in a systematic manner. Due to the emergence of large-scale information processing technology in the 1970s, initiatives were taken to create large-scale comprehensive models of planning operations. Meal (1984) calls this the 'centralized approach', which is based on a tendency to create central decision functions which are given the power to control in detail the planning decisions of the operational process in all parts of the organization.

There are a number of difficulties associated with these centralized monolithic decision models [see also Chapter 9 of this volume, Fleischmann and Meyr (2003)]. The models tend to be very big and complex. This makes the analysis of the models and finding an optimal solution very difficult and requires a decomposition of the model in order to be able to solve this. Model decomposition is a widely used strategy in solving optimization problems. Apart from the complexity in the mathematical sense, there are also a number of organizational and people-related difficulties associated with the centralized approach. The most important difficulty is that there appears to be no owner of the monolithic model. Responsibilities within organizations tend to be dispersed over a number of people. The monolithic model assumes it is a single organizational unit deciding about a large number of details across the entire organization. If we assume that the higher-level management would actually own the model and make these decisions, a number of people and model related difficulties come about:

(1) detailed figures do not mean much to higher-level managers
(2) detailed figures give a false sense of security because they may be highly unreliable, not only if they refer to some future state of the system (e.g., forecast of exogenous data), but also if they refer to the current state of the system (data quality problems)
(3) centralized planning takes away authority from local managers further down the hierarchy and reduces their responsibility, which is not in line with the dominant management philosophy of self-contained and autonomous groups. Apart from that, it is also contradictory to a principle from control theory, which states that responsibility and decision authority should be matched with the opportunity to control. This last issue is extensively discussed by McPherson and White (1994), who state that 'Planning at superior levels must be consistent with control capabilities at subordinate levels, while planning at subordinate levels must be consistent with achieving the superior goals of the hierarchy.'
(4) A model never captures the complete richness of a situation. As a consequence, a local planner down the hierarchy will always have more information and a better representation of the actual processes than a (higher-level) model.

All this leads to the fact that a decomposition of the problem is required in order to be able to find a solution to the planning problem that can also be implemented within an organization. If a decision problem is decomposed and a hierarchy is constructed, higher levels of the hierarchy will need to aggregate the lower level models. This aggregation is necessary to overcome the difficulties just listed. Furthermore, this decomposition will lead to more or less independent units along the supply chain, that are self-contained with regard to their control within the unit, but receive objectives and constraints to be taken into account from an aggregate and centralized control function. This is in line with the idea of separating goods flow control and production unit control, as developed by Bertrand and Wortmann (1981) and further elaborated on by Bertrand et al. (1990). A consequence of this approach is that lead times of the various production units are fixed and are input to the system rather than output. These lead times are then essentially modeled in exactly the same way as in MRP [Orlicky (1975)]. We will discuss this issue further in Section 2.4. Note that the fixed lead time we are discussing here is the internal lead time of the controlled part of the supply chain that needs to be distinguished from the external lead time promised to any customers of this supply chain. The external lead time must vary to reflect the work load changes over time. As a consequence of this approach, workload control is executed over the supply chain. In summary, we can state that hierarchical decomposition of the SCP problem has two essential characteristics, namely:

- aggregation, which is necessary to construct higher level models
- fixed lead times, which are needed as a control mechanism.

In the next two subsections we will discuss the concepts of effectuation lead times and information asymmetry which underlie the notion of obtaining supply chain control by working with planned lead times. This notion of control will be further elaborated on in Section 2.4.

2.2 Effectuation lead times

Asymmetry in the decision making hierarchy and the necessity to anticipate is primarily caused by the fact that it takes time to implement a decision. We will denote this time in the remainder of this chapter as *effectuation lead time*. The effectuation lead time is the time that passes between the moment a decision is made and the moment that the consequences of this decision can be observed in the operation of the supply chain. In SCP decisions, the length of the effectuation lead times can be determined based on the product and process structure: the bill of material and the bill of resources.

An example of an effectuation lead time is the procurement time of components. If the procurement lead time for a component i is L_i, then $r_i(t)$, the quantity procured at the start of period t is supposed to be available for further assembly or sales at the start of period $t + L_i$. The immediate decisions $\{r_i, (t)\}$ are dependent on the exogenous demand forecasts $\{\hat{D}_i(t, t+s)\}$, $s \geq 0$, defined as

$\hat{D}_i(t, t+s)$ forecast of exogenous demand for item i in period $t+s$ as decided on at the start of period t, $t \geq 0$, $s \geq 0$, $\forall i$

Assuming supply is reliable and L_i is realized, we may expect that

$$\hat{p}_i(t, t + L_i) = r_i(t),$$

where we define $\hat{p}_j(t, t+s)$ as

$\hat{p}_j(t, t+s)$ forecast of quantity of item i that becomes available at the start of period $t+s$ as determined at the start of period t, $t \geq 0$, $s \geq 0$, $\forall i$

reflecting the decision of the supplier to ship as late as possible. Note that $\hat{p}_i(t, t + L_i)$ is only a planned decision from the perspective of the organization ordering the item. For the supplier this may be either a firm decision, in case the supplier has to start immediately with processing and transporting the order for item i, or a planned decision, in case the effectuation lead time incorporates some slack time.

Suppose that at time t the planned decision is taken according to the above equation. At the start of period $t+1$ we generate new forecasts $\{\hat{D}_i(t+1, t+s)\}, s \geq 1$. It is now well possible that, e.g., due to a decrease in demand for item i it is decided to change the decision made earlier, i.e.,

$$\hat{p}_i(t+1, t+L_i) \neq \hat{p}_i(t, t+L_i)$$

Following the discussion of planned versus firm decisions, we can see that dependent on the incorporation of slack time into the procurement time, it is possible or not to change the earlier decision. Flexibility can be further modelled by the choice of the (time-dependent) resource constraints.

Information asymmetry itself can be described using the following example. Consider again item i with planned lead time, i.e., the effectuation lead time of the material order release decision, L_i. Often suppliers receive forecasts about future orders in some period t multiple times in order to take consecutive decisions on e.g., buying production equipment, hiring and training people and procurement of materials. As stated above the forecasts $\{\hat{D}_i(t-s, t)\}$ for period t made at the start of an earlier period t-s differ for different s. Thus the procurement orders derived from these forecasts change over time, so that the supplier's decision to buy production equipment is based on different information then the supplier's decision to procure materials. This asymmetry in information needs to be taken into account when designing the decision hierarchy or supply chain control structure. The effectuation lead time thus leads to differences between the moments in time that certain decisions must be taken. Also, it means that decisions are often taken a substantial time before the actual action in the physical process takes place. As a consequence the decision maker in fact feeds forward in terms of control theory rather than feeds back as is often suggested in hierarchical production planning frameworks. In order to feed forward, the decision maker essentially anticipates the events over the period of time until his decision is effectuated.

It should be realized that the effectuation lead time is not only related to the bills of materials and bills of resources, but is also a characteristic of the SCP and control system. In many cases, the time buckets at higher levels of decision making are larger than at lower levels (Meal (1984)). Further, the frequency at which decisions are made, revised or processed is less at higher levels of decision making (the hierarchical structures of Gershwin (1994) are based upon this premise). This means that changes in the actual (physical) process, e.g., changes in demand, may not be observed directly. Further, if they are observed, there may be a delay in processing the consequences of this observation. This processing time due to the decreased frequency of decision making at higher levels should be included in the effectuation lead time.

2.3 Asymmetry in SCP

After constructing a hierarchical planning structure, oftentimes the resulting planning situation is characterized by asymmetry of information. Essentially having different levels of control being owned by different organizational units leads to different information statuses.

A useful framework that describes this anticipatory decision cycle, is presented by Schneeweiss (1999) and discussed by Fleischmann and Meyr (2003). In Schneeweiss's model, a decision structure within an organization can be represented as a series of decision tandems, i.e., two decision levels interacting with each other by the first of the two levels (the top level) giving an instruction to the second of the two levels (the base level), and the base level responding by giving a reaction to the top level. Before giving its instruction, the top level anticipates the base level's reaction by either implicitly or explicitly modeling the behavior of the base level in the top level's model. This is called the anticipated base model. Schalla, Fransoo and de Kok (2001) have further analyzed the various types of anticipation that may exist. In general, the anticipated base model can be constructed based on aggregating information and/or on aggregating the base level model itself.

We will now first discuss the aggregation of the base model itself. Aggregation referring to the model part explicitly deals with complexity reduction. At the top level the decision making process is represented by an aggregate and simple model in order to reduce complexity and to distribute detailed decisions to lower planning levels. We can thus distinguish the following anticipation types with regard to the model:

- *Explicit Model*: The base level model as seen at the top level is exactly the same as the original base level model.
- *Implicit Model*: The base level model as seen at the top level is different than the original base level model.

Consequently, the terms explicit and implicit with regard to anticipation refer to the fact whether the top-level base model (including the objective functions) is exactly the same as the base-level base model. If this is the case, we call this explicit anticipation; if this is not the case, we call this implicit anticipation. Explicit anticipation thus uses a detailed model of the base level, whereas implicit anticipation uses an aggregate model of the base level.

The second type of aggregation to construct an anticipation model is aggregation of information. This type of aggregation is related to uncertainty and effectuation time. In the context of the questions related to the anticipation function, special attention regarding information is paid to the concept of information asymmetry. Information asymmetry basically entails the fact that when making a decision at a higher level, the amount and quality of information may be different from when the lower level decision is made (later), and again different from when the actual execution of the decision is taking place. The fact that information asymmetry exists, leads to the necessity to *anticipate* at a higher level decision what *may* happen at the lower levels decisions. We can thus distinguish the following anticipation types with regard to information:

- *Exact Information*: The base level information as seen at the top level is exactly the same as the original base level information.
- *Approximate Information*: The base level information as seen in the top level is different than the original base level information.

Consequently, the terms exact and approximate refer to the fact whether the top level model has exact information of the base level status. Note that in most cases some time elapses between the moment at which the top level makes its decision (instruction) and the moment at which the base level makes its (final) decision. This difference in *effectuation lead times of top level decisions and base level decisions* usually entails a difference in the information status between the top level and the base level, resulting in information asymmetry and – automatically – in approximate anticipation.

Taking all combinations between anticipation types referring to the modeling part and anticipation types referring to the information part, we can distinguish three types of anticipation, based on the various types of aggregation used to construct the anticipation function as discussed above. The three types are:

- Explicit Model l Exact Information (EE)
- Explicit Model l Approximate Information (EA)
- Implicit Model l Approximate Information (IA)

Note that the combination of exact information and implicit model does not make a lot of sense, since there does not seem to be a clear reason for constructing an implicit model (i.e., more aggregate than the detailed model) if exact information is available. Further note that information asymmetry more often than not will lead to the fact that approximate information is the only information that can be used in the anticipatory model at the higher level. Given the fact that only approximate information can be used, it is not a priori clear whether the use of an explicit and detailed model is better than the use of an implicit model.

It is interesting to note that the BHM hierarchical models hardly contain any anticipation of the lower levels by the higher level, as has been noted by Schneeweiss (1999). Neither do the BHM models contain the concept of effectuation lead time, although this concept is noted as a rationale for hierarchical planning in a book largely built on the BHM models (Miller (2001), p. 8, named as 'gestation period').

2.4 The need for control

Anticipation models need to capture the base level behavior in a sufficiently accurate manner. In this sense, accurate refers to the predictive quality of the anticipation model. When designing a decision structure, two different approaches can be taken when constructing the anticipation functions. The first approach is to try and capture the base level behavior as completely as possible by enriching the anticipation function by as many details as known about the base level. The second approach is to design the decision function at the base level in such a way that the actual anticipation becomes straightforward. In this case, the objective of the base level is to realize a set of targets set by the top level (see also McPherson and White (1994), for a discussion on this matter). We will refer to this situation as a controlled situation. An example of such a design is the reliance on planned lead times maintained by workload control methods (Bertrand and Wortmann (1981), Bertrand et al. (1990), Wiendahl (1987, 1995) and Van Ooijen (1991)).

It is neither obvious nor conclusive whether working with planned lead times is a correct approach, since current research is not conclusive and there are researchers who advocate the use of variable lead times. Kanet and Sridharan (1998) demonstrate results by which the use of detailed scheduling information in material procurement reduces the inventory of components that are controlled by MRP. Tardiff (1995) and Hopp and Spearman (2000) in their concept called Capacitated MRP (or MRP-C) calculate the expected Work-in-Process and then adjust the lead times of the products accordingly, by 'building ahead' those items that would be late due to longer lead times caused by higher WIP levels. Based on work by Buzacott (1989), a research line has been developed which integrates the capacity and material planning perspectives completely using so-called 'generalized kanban systems' [Frein, Di Mascola and Dallery (1995), see also Section 5.8]. The assumptions are however very strict, such as Poisson arrivals of items orders and FIFO dispatching at resource level. These assumptions are mostly not satisfied due to the periodic review nature of the SCP process, which leads to coordinated release decisions.

Apart from the mathematical complexity of applying detailed scheduling on a supply chain wide scale, approaches that use detailed scheduling information to update the supply chain plan abstain from two basic principles that we have outlined earlier, namely the organizational hierarchical concerns and the asymmetry in information. Since the scheduling decision is generally the domain of some lower-level organizational function than the SCP decision, taking this scheduling decision at a higher level may infringe upon this organizational design [Meal (1984)]. With regard to information asymmetry, note that the actual schedule will be constructed at a later stage when more information will be available. As a consequence, the actual schedule may be very different from the projected detailed schedule constructed to make the supply chain plan. In fact the more detailed scheduling of materials will then lead to additional constraints on the operational schedule. This issue is not addressed in the paper by Kanet and Sridharan (1998) discussed above. Given only slight variations in for example the operating times, the impact on the schedule in various environments may be substantial [see e.g., Lawrence (1997), for an example of this in a job shop]. Unfortunately, this interaction between the SCP level and the detailed scheduling level has not been researched under asymmetric information conditions, so we do not know the impact of this effect on the operational performance.

With regard to dynamically adjusting the information based on aggregate status (workload) information of the shopfloor, the impact is even less clear. Much will depend upon the actual stability of the workload prediction between the moment that the SCP decision is taken and the moment the actual execution takes place, i.e., the quality of the *expected* workload as an anticipator of the *actual* lead time. If this quality is good, then the method should work fairly well in the manufacturing environments for which it was designed. In situation with multiple items and multiple resources it is however very difficult to give accurate predictions of the lead time and to adjust the supply chain plan accordingly. This will lead to very complex models and will lead to performance and accuracy problems [Hopp and Spearman (2000)]. The situation is however very different when multiple actors conduct collaborative planning actions across independent companies in the supply chain. In that case, the planned lead times act as a coordination aid between the various actors in the supply chain and planned lead times allow for independent planning of parts of the supply chain.

2.5 Positioning SCOP in the hierarchy

From the exposition above, it can be concluded that SCOP needs to be positioned hierarchically above the unit control functions that are responsible for controlling lead time in a particular unit of the supply chain [cf. Bertrand et al. (1990)]. Supply Chain Operations Planning in most industries deals with a horizon up to several months typically with weekly time buckets. In some industries, e.g., bulk chemicals, this function may have a horizon as short as a couple of weeks with daily buckets, whereas in other industries, e.g., pharmaceuticals, the horizon may be as long as a couple of years with monthly time buckets. Everything is determined by the typical effectuation lead times of the industry and the lead times that customers are willing to accept. Next to the SCOP function, an order acceptance function (often called Available-To-Promise engine in current planning software) needs to be introduced in the control loop in order to control the total amount of work accepted by the supply chain, and to externalise the portion of the customer-perceived leadtime that is due to varying demand that cannot be processed within the fixed and controlled leadtime. Finally, a parameter setting function

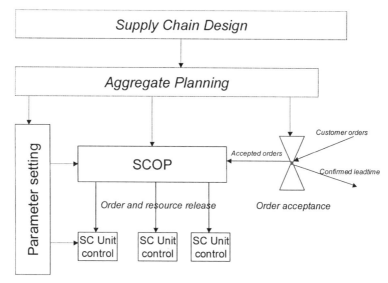

Fig. 1. Position of Supply Chain Operations Planning in the Planning Hierarchy.

needs to coordinate the safety stock, leadtime, and workload parameters of the Supply Chain. This system is depicted in Fig. 1.

Note that the functions discussed here only relate to the planning of operations, i.e., the release of materials and resource triggered by actual demand downstream. In this discussion, we abstain from other functions, such as supply chain design, the planning of seasonal or other controlled inventories, lotsizing, transportation planning, etc. For a full description of this hierarchy, we refer to Fleischmann and Meyr (2003). Along a timeline, it clearly shows that the various SCOP decisions need to be timed along the lead time characteristics of the supply chain, both the physical lead time and the information processing lead time. This is depicted in Fig. 2 and was discussed in detail in Section 2.2.

In the next section, we will further model the SCOP problem in detail, and formulate the constraints that determine the SCOP problem.

3 Constraints for SCOP

In this section we propose a modeling framework that comprises currently existing supply chain operations planning concepts. In Section 3.1 we derive the set of constraints that follow from the material structure (BOM). In Section 3.2 we derive the set of constraints that follow from the resource structure (BOP). We emphasize here that the constraints derived are induced by the underlying BOM and BOP and hold for any choice with respect to the SCOP concept. It may be that one chooses to ignore particular constraints as

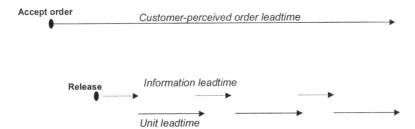

Fig. 2. Decision moments driven by leadtime structure.

irrelevant, implying that the result of the SCP process may be infeasible. The decision to ignore particular constraints should imply that either the constraints are never binding (which one might be able to prove) or the impact of the resulting infeasibility is negligible (which is likely to be much harder to prove).

3.1 Material constraints and their representation

The definitions in Section 1.1 characterize completely the material structure of the supply chain. We now exploit this structure to identify a set of constraints that any SCOP concept should satisfy. These constraints can informally be described as follows: any SCOP concept can only release items for usage in a transformation activity if it is physically available at the moment of release. This might seem an obvious statement, but the pitfall lies in the definition of 'release'. In this chapter a release decision at the start of period t is a decision that authorizes at the start of period t the usage of materials and resources for a transformation process. The release decision is the core decision in Materials Requirements Planning (MRP-I), Manufacturing Resource Planning (MRP-II), Statistical Inventory Control (SIC) and Just-In-Time (JIT) and any other planning and control concept. It is assumed that such release decisions are coordinated, since they are mutually dependent. In the concepts mentioned above the release decision is taken without a check on availability of resources and materials. For example, in SIC [Statistical Inventory Control, cf. Silver et al. (1998)] a release decision is only based on the output item status (in this case its inventory position) associated with a transformation process and not with the input item(s) status. This implies that a release decision cannot be executed at the time it is authorized to do so. Hence execution delays occur that should be taken into account when coordinating decisions. Similarly, the top–down explosion process of MRP-I does not incorporate item availability checks. Only exception messages are generated, but the explosion process implicitly assumes that infeasibilities at lower levels in the BOM are resolved. It can be argued that such a fallacy has a major impact in situations where lots of items are assembled into multiple other items, which explains the current practice of human expediting in high volume consumer electronics supply chains.

Since SCOP coordinates release decisions between different parts of an organization or even between different organizations, typically such release decisions are taken periodically. In between the periodic release decisions preparatory process planning activities take place, such as demand forecasting, checking availability of resources, etc. In the sequel we therefore restrict to periodic SCOP concepts. Where appropriate, we discuss extensions to continuous SCP concepts.

Let us now formally derive the material release constraints. Given the definition of the physical inventory $I_i(t)$ and the backlog $B_i(t)$ it is clear that

$$I_i(t), B_i(t) \geq 0, \qquad t \geq 1, \forall i$$

It is obvious that a backlog exists iff. the physical stock is zero, i.e.,

$$I_i(t)B_i(t) = 0, \qquad t \geq 1, \forall i$$

The net inventory is defined as the physical inventory minus the backlog, i.e.,

$$J_i(t) = I_i(t) - B_i(t), \qquad t \geq 1, \forall i$$

Independent demand is demand generated by customers of the supply network, either directly at the CODP or indirectly by derivation from final product forecasts and an offset by the lead time for final assembly. Such demand is usually not known beforehand and must be forecast. Note that $E \subset P$, since end-items have independent demand, only. Yet there may be items in I that have independent demand, too. For example, a company producing hard disks may sell to OEM as well as to individual customers. The product sold to individual customers contains the product sold to the OEM as a subassembly. The subassembly is an item in I, while the OEM demand for the subassembly is independent.

For items with independent demand it is impossible to preclude backorders, unless some upper bound on the demand per period is known. In this chapter we assume that such an upper bound does not exist or the upper bound is so high compared to the average demand per period that it would be economically infeasible to guarantee no backordering of exogenous demand for an item. However, for items with dependent demand by definition the demand is known, since it is determined within the planning process itself. We argue that backordering of dependent demand does not make sense. Suppose at some period we take the decision to create a backorder for an item by deciding to release more material then available. In that case we only release all available material physically. The earliest moment in time that we may resolve the backordering situation is at the start of the next period. However, at the start of the next period we have exact information about demand during the current period and possibly better information about future demand. Thus, it is easy to see that the decision taken to create a (logical) backorder by releasing more material than available cannot be better than the decision to release exactly all available material.

Our argument only holds for SCOP concepts where all items in the supply chain 'know' information about future exogenous demand, either implicitly or explicitly. This applies to all SCOP concepts that subsume some centralized database with all current state information and forecast information. Top-down SCOP concepts like SIC and MRP-I use the explosion process to transfer exogenous demand information at the expense of incorrect order release decisions, that cannot be executed and require human intervention to resolve the resulting issues.

The above implies that we impose the following constraint on the evolution of the backorders over time, which holds for all items.

$$B_i(t+1) - B_i(t) \leq D_i(t), \qquad \forall i, t \geq 1. \tag{3.1}$$

The above equation states that the increase of the backlog cannot exceed the exogenous demand. It is easy to see that for an intermediate item i with $D_i(t) = 0$ for all $t \geq 1$, i.e. intermediate item i has no independent demand, we have that

$$B_i(1) = 0 \Rightarrow B_i(t) = 0, \qquad t \geq 1.$$

Dependent demand $G_i(t)$ for item i is generated by items in V_i. The dependent demand for item i at the start of period t consists of the sum of all released quantities of items in V_i at the start of period t. This implies that

$$G_i(t) = \sum_{j \in V_i} a_{ij} r_j(t), \qquad \forall i \in I$$

Clearly there must be sufficient inventory of item i to ensure immediate start of the execution of the transformation activities involved in the release decisions. The physical starting inventory at the beginning of period t equals $p_i(t) + \max(0, I_i(t) - B_i(t))$. Thus it follows that $G_i(t)$ must satisfy the following equation,

$$G_i(t) \leq p_i(t) + \max(0, I_i(t) - B_i(t)), \qquad \forall i, t = 1, \ldots, T$$

The set of Equations (3.1) states that the backlog from a period to the next period must not grow faster than the exogenous demand in that period, while the above set of equations states that a planning concept must not release more than physically available. The lemma below shows that both sets of equations are equivalent.

Lemma 1.

$$G_i(t) \leq p_i(t) + \max(0, I_i(t) - B_i(t)) \Leftrightarrow B_i(t+1) - B_i(t) \leq D_i(t)$$

The proof of the above Lemma 1 is straightforward and is derived from the inventory balance equations formulated below and the definition of the variables involved.

Furthermore we assume that all released quantities are non-negative, i.e., returns are not possible,

$$r_i(t) \geq 0, \quad \forall i, t = 1, \ldots, T \qquad (3.2)$$

This implies that the released quantities together constitute a feasible plan. Note that Equations (3.1) only state that one is not allowed to release more than physically available. One might decide to reserve availability for exogenous demand in case $i \in P$. This aspect is dependent on the SCOP concept and will be discussed in Section 4 and following.

It can easily be shown that the MRP/DRP-concept [cf. Silver et al. (1998)] as a planning concept does not satisfy constraints (3.1) (see above discussion of top–down planning logic). In Section 4 we discuss mathematical programming models that can be seen as an extension of the MRP/DRP-concept in that such models incorporate the feasibility constraints at the expense of (much) more computational effort.

Given the release decisions taken we can write the so-called *inventory balance* equations,

$$J_i(t+1) = J_i(t) - G_i(t) - D_i(t) + p_i(t), \quad \forall i, t = 1, \ldots, T. \qquad (3.3)$$

Using the definition of the net inventory we equivalently can write

$$I_i(t+1) - B_i(t+1) = I_i(t) - B_i(t) - G_i(t) - D_i(t) + p_i(t),$$
$$\forall i, t = 1, \ldots, T \qquad (3.3')$$

The dynamics of $I_i(t)$ and $B_i(t)$ determine the performance of the supply network. We want to emphasize here that the impact of the release decisions taken in the past as well as the impact of the planned lead time of item i is 'accumulated' in $p_i(t)$. This will be discussed in detail in Sections 3.2 and 3.3.

Summarizing, we have defined the material structure of a supply network by defining *gozinto* relations between items. From those relations we derived a set of (release) constraints that should be satisfied by any supply chain operations planning concept. When discussing the various supply chain operations planning concepts we pay attention to their adherence to these constraints.

3.2 Resource constraints and their representation

In this subsection we derive a set of necessary conditions with respect to capacity usage that any SCOP concept should satisfy, given the above information on resource availability and resource usage. It turns out that the derivation of such constraints is not as straightforward as the material constraints derived in the previous section. This can be explained as follows.

Given the released quantities $r_i(t)$ and the lead time L_i for orders of item i, we can formulate the following set of constraints,

$$\sum_{i \in U_k} c_i r_i(t) \leq C_{kt+L_i-1}, \quad k = 1, \ldots, K, t \geq 1,$$

implying that the total capacity requirements for resource k associated with the item orders released at the beginning of period t should not exceed the available capacity of resource k during period $t+L_i-1$. These conditions are sufficient, yet not necessary, for ensuring that orders for item i released at the start of period t are made available for usage at the start of period $t+L_i$. The constraints above are necessary, only, if we require that items released at the start of period t are processed in period $t+L_i-1$. This is equivalent to the decision rule to produce as late as possible with respect to the lead time L_i. In general it follows from the planned lead time L_i of an order for item i, that the item order released at the start of period t, i.e., the time interval $(t-1, t]$, must be processed on its associated resources in the time interval $(t-1, t+L_i-1]$ in order to guarantee availability for usage at the start of period $t+L_i$. Since material required for processing orders for item i must be released earlier it follows that

$$\sum_{s=1}^{t} r_i(s) \geq \sum_{s=1}^{t} q_i(s). \tag{3.4a}$$

The right hand side of (3.4a) denotes the cumulative amount of item i processed up to and including period t. The left hand side of (3.4a) denotes the cumulative amount of item i released up to and including period t. Here we assume without loss of generality that at the start of period 1 the system is empty, i.e., no orders are released before time 0, no stocks are available.

To ensure that the order released at the start of period t is available for usage in period $t + L_i$ we must process the materials associated with $r_i(t)$ in the periods $t, \ldots, t + L_i - 1$. From this it follows that

$$\sum_{s=1}^{t} r_i(s) \leq \sum_{s=1}^{t+L_i-1} q_i(s). \tag{3.4a}$$

From the definition of $q_i(t)$ we find that

$$\sum_{i \in U_k} c_i q_i(t) \leq C_{kt}. \tag{3.4b}$$

Combining the above we find

$$\sum_{s=1}^{t} \sum_{i \in U_k} c_i r_i(s) \leq \sum_{s=1}^{t+L_i-1} C_{ks}, \quad k = 1, \ldots, K, t \geq 1. \tag{3.5}$$

The necessity of condition (3.5') is obvious. The sufficiency follows from the fact that we may assume a FIFO allocation of capacity from which we can construct a feasible allocation by allocating orders released as soon as possible to the resources. In a rolling schedule context we can rewrite the necessary and sufficient conditions so that capacity consumed before period t is subtracted from the left-hand side of (3.5'), and capacity available before period t is subtracted from the right-hand side of (3.5'). In the sequel we will use equations (3.4a), (3.4b) and (3.5) since they explicitly relate material release quantities $\{r_i(t)\}$ and material processing quantities $\{q_i(t)\}$. The latter quantities provide useful information about capacity usage.

Extension to the situation where item i can be processed by multiple resources implies the definition of variables that indicate which amount of the order released at the start of period t is processed by a specific resource. For our further comparison of the different supply chain planning concepts we can restrict to the situation where each item is processed at a single resource.

3.3 Planned lead times and the relationship between $\{r_i(t)\}$, $\{q_i(t)\}$ and $\{p_i(t)\}$

In the inventory balance Equations (3.3') we use the variables $\{p_i(t)\}$ that denote the amounts of material that become available for usage at the start of period t. Clearly, these variables are related to the material release quantities $\{r_i(t)\}$ and the material process quantities $\{q_i(t)\}$, since these are decisions that have to be taken before amounts can be made available for usage. It turns out that we have a considerable degree of freedom here. As stated in Section 1.3 we have that

$$p_i(t) = q_i(t-1), \quad t \geq 0,$$

which implies that we assume that an amount of item i processed in period $t-1$ becomes available at the start of period t. This provides maximum flexibility within the decision space bounded by constraints (3.4a), (3.4b) and (3.5). However, this may imply that materials are available *earlier* than planned for, according to the moments of order release and the planned lead times. This may be seen as favorable, yet this may imply that materials are available *too early*. We must be aware of the fact that the SCOP model is a representation of part of reality only, implying that materials not modeled at SCOP level are only available at the due dates derived from the planned lead times. And even if it would represent all materials, than still we are faced with uncertainty of processing and future demand.

The concept of planned lead times is a means to create certainty about future material availability. In our view we should formulate the SCOP constraints such that they reflect the conceptual ideas behind the planned lead times concept. This is ensured by defining $\{p_i(t)\}$ as

$$p_i(t) = r_i(t - L), \qquad t \geq 0$$

Assuming that planned lead times are realistic and thereby due dates are met with high probability by the shopfloor level, this definition is in line with the constraints (3.4a) and (3.4b). Notice that this is the typical assumption when we consider uncapacitated systems as in classical inventory management theory [cf. Silver et al. (1998)].

By the above definition of $\{p_i(t)\}$ we reformulate the inventory balance equations as follows,

$$I_i(t+1) - B_i(t+1) = I_i(t) - B_i(t) - G_i(t) - D_i(t) + r_i(t - L_i),$$
$$\forall i, t = 1, \ldots, T \qquad (3.3)$$

3.4 Summary

In this section we defined the supply chain operations planning problem in detail and derived necessary and sufficient material and resource constraints:

Necessary and sufficient material constraints

$$B_i(t+1) - B_i(t) \leq D_i(t), \qquad \forall i, t \geq 1 \qquad (3.1)$$

$$r_i(t) \geq 0, \qquad \forall i, t = 1, \ldots, T \qquad (3.2)$$

$$I_i(t+1) - B_i(t+1) = I_i(t) - B_i(t) - G_i(t) - D_i(t) + r_i(t - L_i),$$
$$\forall i, t = 1, \ldots, T \qquad (3.3)$$

Necessary and sufficient resource constraints

$$\sum_{s=1}^{t} r_i(s) \geq \sum_{s=1}^{t} q_i(s), \quad \forall i, t = 1, \ldots, T \tag{3.4a}$$

$$\sum_{s=1}^{t} r_i(s) \leq \sum_{s=1}^{t+L_t-1} q_i(s), \quad t = 1, \ldots, T \tag{3.4b}$$

$$\sum_{i \in U_k} c_i q_i(t) \leq C_{kt}, \forall k, \quad t = 1, \ldots, T \tag{3.5}$$

$$q_i(t) \geq 0, \quad \forall i, t = 1, \ldots, T \tag{3.6}$$

Conditions (3.4a), (3.4b) and (3.5) have been derived for the special case where each item can only be processed on a single resource. Still, the above-defined constraints provide a basis for comparison of different supply chain concepts. In particular they enable to identify what assumptions are made on material and resource availability and usage.

In the next section we use the generic SCOP constraints to develop SCOP concepts based on the optimization of a deterministic SCOP model in the rolling schedule framework.

4 Mathematical programming models for supply chain planning

In this section we derive the basic mathematical programming formulation of the supply chain operations planning model in a rolling schedule context. We use the product structure and the resource constraints presented in Section 3 as a representation of the primary process to be planned. Special attention is paid to the fact that exogenous demand must be forecast in order to derive a sensible problem formulation. We show how the generic supply chain planning constraints derived in Section 3 are incorporated into the mathematical programming formulation. Since the SCOP problem is a stochastic problem by nature we next address the issue of safety stocks, i.e., buffer stocks required to cope with end-item demand uncertainty. In particular we present a theorem that enables us to apply mathematical programming models in a rolling schedule context in such a way that the long-run average costs to maintain customer service level constraints can be determined. After that we give an overview of the main contributions to the literature on mathematical programming models for supply chain planning over the last 10 years.

4.1 Rolling schedule context

In reality a planning concept does not only generate immediate release decisions, but also provides information on future release decisions. These

Ch. 12. Planning Supply Chain Operations

future release decisions are provisional because they will be affected by future unknown events. In particular we do not know future demand. To cope with this we must forecast future demand. Incorporating the fact that we must forecast future demand we reformulate the generic supply chain planning constraints. Towards this end we define the following variables,

$\hat{D}_i(t, t+s)$ exogenous demand for item i in period $t+s$ as determined at the start of period $t, t \geq 1, s \geq -t, \forall i$

$\hat{G}_i(t, t+s)$ endogenous demand for item i in period $t+s$ as determined at the start of period $t, t \geq 1, s \geq -t, \forall i$

$\hat{B}_i(t, t+s)$ backlog of item i at the start of period $t+s$ as determined at the start of period $t, t \geq 1, s \geq -t, \forall i$

$\hat{r}_i(t, t+s)$ quantity of item i released at the start of period $t+s$ as determined at the start of period $t, t \geq 1, s \geq -t, \forall i$

$\hat{q}_i(t, t+s)$ quantity of item i processed in period $t+s$ as determined at the start of period $t, t \geq 1, s \geq -t, \forall i$

Note that for $-t < s < 0$, these variables represent actuals, i.e., realized quantities taken from historical data and exogenous to the problem. For $s \geq 0$, these variables represent forecasts, i.e., estimates of future quantities, made at the start of period t. Note that $\hat{r}_i(t, t+s)$, $\hat{q}_i(t, t+s)$, for $s \geq 0$, are the decision variables of the supply chain operations planning problem at the start of period t. In the sequel we assume that there is a time origin 0 at which the SCOP problem is solved first and the initial state of the system at time 0 is known. This is important because some equations formulated below are formulated in terms of decisions taken from time 0 onwards, i.e., from period 1 onwards.

4.2 LP formulation of the supply chain planning problem

In Section 3 we have derived a set of material and resource constraints that each supply chain operations planning concept should satisfy in order to generate feasible solutions with regard to the current demand forecast. This suggests that we want the decision variables $\hat{r}_i(t, t+s)$, $\hat{q}_i(t, t+s)$, for $s \geq 0$, to satisfy these generic supply chain operations planning constraints. Then the planning problem to be solved at each time t must satisfy the following equations.

LP constraints

$$\hat{I}_i(t, t+s+1) - \hat{B}_i(t, t+s+1) = \hat{I}_i(t, t+s) - \hat{B}_i(t, t+s)$$
$$- \sum_{j=1}^{N} a_{ij}\hat{r}_j(t, t+s) - \hat{D}_i(t, t+s) + \hat{r}_i(t, t+s-L_i),$$
$$\forall i, s = 0, \ldots, T-1$$

$$\hat{B}_i(t, t+s+1) - \hat{B}_i(t, t+s) \leq \hat{D}_i(t, t+s), \qquad \forall i, s = 0, \ldots, T-1$$

$$\sum_{w=1-t}^{s} \hat{r}_i(t, t+w) \geq \sum_{w=1-t}^{s} \hat{q}_i(t, t+w), \qquad \forall i, s = 0, \ldots, T-1$$

$$\sum_{w=1-t}^{s} \hat{r}_i(t, t+w) \leq \sum_{w=1-t}^{s+L_i-1} \hat{q}_i(t, t+w), \qquad \forall i, s = 0, \ldots, T-1$$

$$\sum_{i \in U_k} c_i \hat{q}_i(t, t+s) \leq C_{k, L+s}, \quad k = 1, \ldots, K, \quad s = 0, \ldots, T-1.$$

$$\hat{r}_i(t, t+s) \geq 0, \qquad \forall i, s = 0, \ldots, T-1,$$

$$\hat{q}_i(t, t+s) \geq 0, \qquad \forall i, s = 0, \ldots, T-1,$$

$$\hat{I}_i(t, t+s) \geq 0, \qquad \forall i, s = 0, \ldots, T-1,$$

$$\hat{B}_i(t, t+s) \geq 0, \qquad \forall i, s = 0, \ldots, T-1.$$

As remarked above the model formulation includes decision variables with decisions taken before period t. Obviously we have that

$$\hat{r}_i(t, t+s) = r_i(t+s), \qquad s < 0, \forall i, t \geq 1,$$
$$\hat{q}_i(t, t+s) = q_i(t+s), \qquad s < 0, \forall i, t \geq 1.$$

The decisions implemented at the start of period t are given by $\{r_i(t)\}$ and $\{q_i(t)\}$, which are derived from the equations below.

$$r_i(t) = \hat{r}(t, t), \qquad \forall i, t \geq 1,$$
$$q_i(t) = \hat{q}(t, t), \qquad \forall i, t \geq 1.$$

In the sequel we assume that the planning decisions ($\{r_i(t)\}$, $\{q_i(t)\}$) are executed according to plan.

The above set of linear equations constitutes the basis for the formulation of a mathematical programming model. In fact we can formulate an LP model that can be solved by standard algorithms, such as the simplex method. The LP-formulation requires the definition of a linear objective function. Let us discuss the derivation of such a linear objective function in the context of supply chain planning under stochastic exogenous demand. In the sequel we assume that $P = E$, i.e., only end-items have exogenous demand. The results below can be extended straightforwardly to the situation, where $P \neq E$.

First notice that the concept of service level constraints does not make sense in the context of a deterministic model instance embedded in a rolling schedule concept. In order to still ensure that within the deterministic setting of the problem priority is given to satisfaction of exogenous demand, we introduce linear backorder costs. Assuming that expensive items are more important than cheap items we assume that the cost per item i backlogged at the end of a period is proportional to h_i, the cost per item held on stock at the end of a period. This yields the following objective function,

$$\sum_{i=1}^{N}\left(\sum_{s=1}^{T} h_i \hat{I}_i(t, t+s) + \sum_{s \in E} \theta h_i \hat{B}_i(t, t+s)\right) \quad (O_1)$$

We assume that $\theta \gg 1$.

The above formulated objective function does not completely solve our problem. We want to determine feasible plans that satisfy end-item service level constraints at low cost, since it is easily seen that the above problem yields the same solution for any value of θ larger than some value θ_0. This implies that with this objective function we obtain customer service levels that may not satisfy our objective.

Apparently the customer service level restrictions require additional decision variables. Following the inventory management literature (cf. Silver et al. (1998)) we introduce the concept of safety stocks in order to cope with short-term demand uncertainty

v_i safety stock parameter of item i, $i = 1, 2, \ldots, N$

We note here that in general the safety stocks depend on t and $t+s$, since the demand forecast may show seasonality and trends. Since our purpose is to explicitly compare different SCOP concepts, we must restrict to the stationary demand situation, implying a constant safety stock.

In order to control the customer service levels we modify the objective function as follows,

$$\sum_{i=1}^{N}\left(\sum_{s=1}^{T} h_i(\hat{I}_i(t, t+s) - v_i)^+ + \sum_{s \in E} \theta h_i(v_i - \hat{I}_i(t, t+s))^+\right) \quad (O_2)$$

At first glance the objective function (O_2) does not represent the real inventory holding costs and backorder costs. However, we should keep in mind that the MP problem formulation is only an attempt to model the supply chain operations planning problem under stochastic exogenous demand. In that sense any such formulation results into a heuristic with respect to the original optimization problem. Still objective function (O_2) reflects the trade-off between inventory holding and backorder costs. On top of that the safety stock parameters control the service levels. This becomes even more evident from the following lemma.

Sample path lemma

Suppose a sample path $\{D_i(t)\}$ of the demand process and a sample path $\{\hat{D}_i(t, t+s)\}$ of the forecasting process are given. Furthermore assume that for all end items

$$I_i(0) = v_i, \quad i \in E$$

Then the solution to the problem expressed in terms of the material order releases $\{\hat{r}_i(t, t+s)\}$ and processed quantities $\{\hat{q}_i(t, t+s)\}$ with objective function (O_2) subject to the LP constraints is the same for each value of v_i, $i \in E$, for all $t \geq 1$ and for all $s \geq 0$.

The proof of the sample path lemma is based on induction. Given the initial inventory levels it is clear that the objective function (O_2) implies an optimal solution $(\{\hat{r}_i(1, 1+s)\}, \{\hat{q}_i(1, 1+s)\})$ that is the same for any value of v_i. This implies that $(\{r_i(1)\},\{q_i(1)\})$ are the same for any value of v_i. But then $I_i(1) - B_i(1) - v_i$ is the same for any value of v_i. This argument can be repeated for any value of t. For a formal proof we refer to Køhler-Gudum and De Kok (2002).

Corollary to sample path lemma

The problems P_α and P_β for the SCOP concept defined by the LP constraints and objective function (O_2) have a unique solution $\{v_i\}_{i \in E}$, where each v_i, $i \in E$, can be determined independent of all other v_j, $j \in E, j \neq i$.

Noticing that the objective function (O_1) is identical to (O_2) with $v_i = 0$, $i \in E$, the corollary to the sample path lemma justifies the following procedure.

  i. Run a discrete event simulation of the system with $v_i = 0$, where at the start of each period $t = 1, 2, \ldots$, we solve the LP that follows from the forecasts $\{\hat{D}_i(t, t+s)\}$, the set of linear constraints given above and the linear objective function (O_2).
  ii. From the discrete event simulation compute the empirical distribution function of $J_i(t) - v_i$.
  iii. Given this empirical distribution function compute v_i^*, such that the required end-item service level is achieved.
  iv. Run another simulation with v_i^* in order to compute $\overline{C}(\mathcal{P})$.

We note here that step (iii) can be executed for most well-known performance measures, such as non-stockout probability at the end of a period, fill rate and average backlog [cf. Køhler-Gudum and De Kok (2002)]. In the comparison of SCOP concepts in Section 6 we apply this procedure with service criterion α the probability of a non-negative stock at the end of an arbitrary period.

Though the service level constraints determine the safety stocks v_i, $i \in E$, finding the optimal safety stocks v_i, for all items $i \in I$ constitutes an extremely complex nonlinear optimization problem. In Section 6 we argue that choosing

$v_i = 0$ for all items $i \in E$ yields a useful heuristic solution, yet clearly more research is required to validate this heuristic.

We notice here that the above approach can be applied to many other MP problems, so that alternative rolling schedule approaches for specific stochastic planning and scheduling problems can be compared. One of the (current) main issues with this approach is the CPU time required to accurately compute the stationary distribution of $J_i(t)$, assuming it exists. Especially when capacity constraints are tight one is confronted with similar issues as typical for the simulation of high load queueing (network) systems. Typically, several millions of periods (customers) are required to obtain sufficient accuracy for comparison purposes. Combined with the solution of an LP-problem in each period, this may result in extremely long computation times (several hours to several days per problem!). In the comparison study reported in Section 6 we circumvent this issue by restricting the analysis to non-capacitated systems.

4.3 Alternative MP formulations for the SCOP problem

The SCOP problem can be seen as a multi-item multi-level capacitated lot sizing problem (MLCLSP). Literature reviews of mathematical programming approaches for production planning and supply chain planning can be found in Shapiro (1993) and Baker (1993), and Erenguc, Simpson and Vakharia (1999). In this section, we will review some papers, each of which can be seen as representative of a class of approaches for the MLCLSP.

The MLCLSP formulation is a more general formulation than the SCOP problem. In fact, in the MLCLSP literature, no explicit reference is made to the planning problem as such. This means that the formulation can be used for a variety of planning problem in the SCP hierarchy. If the MLCSLP formulation is used for the SCOP problem, it is assumed that the quantities planned are also the quantities released, i.e., the MCLSP formulation does not distinguish between $\hat{r}_i(t, t+s)$ and $\hat{q}_i(t, t+s)$. In most cases, therefore, it makes more sense to use the MLCLSP formulation at a higher level of planning than SCOP. In this higher-level plan, the determined quantities are then in fact aggregate quantities to be detailed out at a later stage.

If lot sizing restrictions are not considered at the SCP level under consideration, the MLCLSP formulation typically reduces to the LP formulation given above. However, in the paper by Billington, McClain and Thomas (1983) one finds an LP formulation of the SCP problem that differs from our formulation presented earlier in this section. Billington et al. (1983) formulate a periodic planning problem, where the periods are typically short periods of time, e.g., hours or shifts. Furthermore they do not consider planned lead times. Instead they introduce a so-called *minimum lead time*, which should be interpreted as the minimum time involved in the transformation process to make an item available for usage. This time represents a delay and during this time no resources are used. A valid interpretation is that first an item is produced by some resource after which it is transferred to a stock point. The transfer time equals the minimum lead time. Billington et al. (1983) consider the item lead times, consisting of both waiting and processing times and the minimum lead time, as endogenous to the model. At first sight such an approach seems superior to ours, yet our discussion in Section 2 formulated this as an issue for further research.

Erenguc et al. (1999) give an excellent overview of MP formulations of the SCOP problem and discuss various issues by distinguishing between supplier stage problems, plant stage problems and distribution stage problems. Based on their literature survey they formulate a number of MP models for the SCP problem that include the notion of planned lead times, yet no distinction is made between material order release variables and material processing variables. In our view this is an important distinction that models manufacturing flexibility in accordance with the planned lead time concept. As follows from the discussion in Section 3 capacity checks on $\{r_i(t)\}$ instead of $\{q_i(t)\}$ yield inefficient usage of available resources.

Özdamar and Barbarosoglu (2000) represent a class of heuristics that is based on Lagrangean relaxation of either the capacity constraints (3.5) or the inventory balance equations (3.3) or both. Relaxing the capacity constraints and introducing Lagrange multipliers associated with those constraints reduce the MLCLSP to a number of independent uncapacitated problems. If also the inventory balance equations are relaxed one obtains a number of single item lot sizing problems that can be solved by e.g., the Wagner-Whitin algorithm [cf. Silver et al. (1998)]. The solutions to these undependent problems are tied together by an iterative update procedure for the Lagrange multipliers based on the calculation of subgradients. Özdamar and Barbarosoglu (2000) propose simulated annealing for solving the relaxed uncapacitated lot sizing problems. They present a number of combinations of different Lagrangean relaxations and simulated annealing. A computational study identifies the best combination and shows the practical applicability of this method. The model presented in Özdamar and Barbarosoglu (2000) is similar to the model presented in Billington et al. (1983), i.e., planned lead times are not considered, but lead times of material orders are time dependent outputs of the algorithm. In principle it is possible to modify the analysis in Özdamar and Barbarosoglu (2000) to take into account planned lead times. The paper by Barbarosoğlu and Özgür (1999) is similar to Özdamar and Barbarosoglu (2000) in that problem decomposition is proposed based on the Lagrange multiplier technique. It is interesting to mention their observation that the decomposition relates to an organizational decomposition where specific parts of the organization are responsible for particular sets of items. The decentralized decision making that results from this decomposition is supported by a central agent that ensures the exchange of relevant information between the different parts of the organization. This observation is related to our discussion in Section 2 about the distinction between problem decomposition, which is strongly based on organizational considerations, and model decomposition, which is strongly based on algorithmic efficiency considerations. Apparently both points of view may be aligned.

Belvaux and Wolsey (2001) propose MP formulations of the MLCLSP that lend themselves to relatively efficient solution with commercial mixed integer programming software such as CPLEX and XPRESS. The focus of their paper is the derivation of problem-specific (yet generic to the MLCLSP) sets of necessary inequalities, i.e., cutting planes, that considerably improve the performance of MIP solvers. They also emphasize the usefulness of the echelon stock concept [cf. Section 5] when dealing with multi-level problems. The echelon concept enables to reformulate the multi-level multi-item problem as a set of relaxed single-item problems that provide additional constraints that can be used for further efficiency of the MIP solution procedures. Belvaux and Wolsey (2001) remark that a distinction should be made between SCP problems where planning occurs infrequently and SCP problems where planning occurs frequently: big bucket problems versus small bucket problems, respectively. In the former case typically many setups for many different items occur during the single planning period under consideration, while in the latter case a small number of setups for a limited number of items occurs and multiple planning periods must be considered. The models to be applied in the two different cases are different. The SCOP problems considered in this chapter should be considered as small bucket problems. The SCP model formulation proposed by Belvaux and Wolsey (2001) does not include planned lead times, nor does it distinguish between material order release variables and material processing variables. Still, the equations derived in their paper as well as the techniques proposed can be easily modified to include these two phenomena.

This concludes our brief survey of MP formulations for the SCOP problem. In the next section we discuss the recent literature on stochastic models for the SCOP problem for general supply networks. In Section 6 we compare the LP-formulation of the SCOP problem as a representative of the class of MP models with a specific stochastic model on the basis of the required supply network inventory capital to achieve target customer service levels.

5 Stochastic demand models for supply chain planning

In this section we discuss various SCOP concepts for stochastic demand models as proposed in the literature. We assess these concepts on the basis of the SCOP constraints derived in Section 3. The discussion is restricted to incapacitated supply chains, since results for capacitated systems are only available for single item, single stage systems [see e.g., De Kok (1989)], serial systems [Tayur (1993)], or for divergent systems where only the most upstream stage is capacitated [see De Kok (2000)]. Due to the fact that the analysis of general supply networks with stochastic demand in the current literature is based on simplifying assumptions, we discuss the validity of these assumptions. In the context of Supply Chain Operations Planning under stochastic demand we focus on the determination of safety stocks, since these parameters determine an important part of the supply network inventory capital.

For ease of presentation we assume that a_{ij} is 0 or 1. For most of the policies we can easily extend the analysis to general a_{ij} values. Also we assume without loss of generality that $P = E$.

5.1 Echelon concept

Stochastic multi-echelon models can be distinguished by the state variables used to derive the item orders. In Axsater (2003), Chapter 10 of this volume, the distinction between installation stock policies and echelon stock policies has been discussed in detail. In this chapter we restrict ourselves to echelon stock policies. The main reason for this is that in a SCOP context installation stock policies typically violate SCOP constraints (3.1), since dependent demand is backordered. For general supply networks we define the echelon stock, echelon inventory position and some associated concepts in order to formally define item order release policies.

Firstly, we define

$O_i(t)$ Cumulative amount of orders outstanding at the start of period t.

Then we can define the echelon inventory stock $X_i(t)$ and the echelon inventory position $Y_i(t)$ of item i recursively as follows,

$$X_i(t) = J_i(t), \qquad \forall i \in E$$
$$Y_i(t) = X_i(t) + O_i(t), \qquad \forall i \in E$$
$$X_i(t) = J_i(t) + \sum_{j \in V_i} Y_j(t), \qquad \forall i \in I$$
$$Y_i(t) = X_i(t) + O_i(t), \qquad \forall i \in I$$

The echelon inventory position has the following important interpretation. The echelon inventory position of item i represents the coverage of future demand for item i up to and including the periods in time that the last order for item i becomes physically available for sales to customers. Typically, an item becomes physically available for sales to customers as part of a sellable item, i.e., an item in E. Also, we emphasize that in that sense an item becomes available for sales in different future periods in time, since that period depends on the sellable item under consideration and the assembly steps required to convert the item into this sellable item. The interpretation of the echelon inventory position as a coverage of future demand for sellable items is important for the synchronization of order releases for items that are assembled into the same sellable items. As we will see we can use this interpretation for the development of order release policies for general supply networks. For the special case of pure assembly systems we find that the concept of coverage of future demand is in line with optimal policies found in the literature. Let us first discuss the class of pure base stock policies as proposed by Magee (1958).

5.2 Pure base stock policies

In Chapter 11 of this volume, Song and Zipkin (2003) discuss stochastic models for assembly systems. In particular they focus on pure base stock policies. For sake of self-containedness of this chapter we define pure base stock policies below. Let

S_i Base stock level of item i

A pure base stock policy operates as follows:

$$r_i(t) = S_i - Y_i(t). \tag{5.1}$$

Song and Zipkin (2003), Chapter 11 of this volume, restrict their analysis to so-called Assemble-To-Order (ATO) systems, i.e., the general assembly system consists of two levels only: a finite product level and a component level. Only components can be kept in inventory. Through this restriction pure base stock policies are feasible. If we extend the general system to more than two levels or to the case where finite products can be kept in inventory as well, pure base stock policies are no longer feasible, even in the infinite capacity case.

It is easy to see that pure base stock policies in general violate constraints (3.1), which state that the increase of the backlog cannot exceed the *exogenous* demand. This is due to the fact that equation (5.1) is not constrained by upstream availability considerations. It implies that dependent demand can be backordered. In that sense it suffers from the same problem as the MRP I-logic. Incorporation of upstream availability constraints is non-trivial as can be seen from the analysis in Agrawal and Cohen (2001), Hausman, Lee and Zhang (1998), amongst others, where even the allocation of components to final products in an ATO-setting makes analysis of remnant component stocks due to lack of other components intractable. In Sections 5.3 and 5.5 we discuss synchronized base stock policies that can circumvent this problem at the expense of inadequate exploitation of component commonality.

A key result for ATO systems under pure base stock policies derived by all authors dealing with this is the following [cf. Agrawal and Cohen (2001), Hausman et al. (1998)].

ATO key theorem
Let α_j be defined as the probability that the demand for item $j \in E$ in period t can be satisfied immediately. Furthermore assume that if an item i is out-of-stock at the end of period t then all items $j \in V_i$ are allocated part of the shortage of item i. Then

$$\alpha_i = P\left\{ \sum_{s=t-L_i+1}^{t} D_i(s) \leq S_i, \quad \forall i \in W_j \right\}.$$

The results follows from the fact that the demand for item $j \in E$ in period t can be satisfied immediately if and only if all component stocks of items $j \in W_j$ are positive at the end of period t. Computation of α_j is complicated due to the correlation between the random variables $\sum_{s=t-L_i+1}^{t} D_i(s), i \in W_j$. Hausman et al. (1998) and Agrawal and Cohen (2001) assume the demand for final products to be normally distributed, so that they can apply results for multivariate normally distributed random variables. Song (1998) explores the combinatorial nature of the probabilistic expression for α_j in the case of continuous review (compound) Poisson demand and derives computationally efficient upper and lower bounds [cf. Song and Zipkin (2003), Chapter 11 of this volume].

From the ATO key theorem we can derive the following property of pure base stock policies.

Pure base stock policy property
Let j_1, and j_2 be two different end-items. Then

$$W_{j_1} \subset W_{j_2} \Rightarrow \alpha_{j_1} \geq \alpha_{j_2}.$$

Although this property is obvious and can be extended to multiple level systems, it shows an important drawback of pure base stock policies, even in situation where they yield feasible solutions. The property states that if an end-item contains a subset of the components of the other end-items then the non-stockout probability of the former is at least as high as the non-stockout probability of the latter. Considering typical ATO settings one finds that high-end end-items have additional features compared with low-end end-items. This implies that high-end end-item service levels must be lower than low-end end-item service levels. This is typically not desired because of economic reasons. As a consequence we find:

Under pure base stock policies arbitrarily chosen customer service levels for different end-items cannot be satisfied with equality.

This is different from what we have been able to show for the mathematical programming formulation of the supply chain operations planning problem in Section 4 and this is likely to cause higher supply chain capital investments than deemed necessary. In the next subsection we will discuss the generic approach proposed by De Kok and Visschers (1999) that circumvents this problem. We will compare the two generic approaches on the basis of some small-scale ATO examples. We notice here that the approach by De Kok and Visschers (1999) can be applied to any network structure and to combinations of Assemble-To-Order and Make-To-Stock.

5.3 Modified base stock policies for pure assembly systems

As stated above under pure base stock policies it may be possible that the quantity released cannot be met due to lack of material. As shown by Rosling (1989) and Langenhof and Zijm (1990) the echelon-order-up-to-policies for pure assembly systems, i.e., each item has at most one parent, can easily be modified by taking into account the availability of the child items.

Let us derive this modified base stock policy for pure assembly systems. First of all note that for pure assembly systems we have exactly one end-item and each item i has exactly one successor $suc(i)$. Thus we can uniquely define the cumulative lead time L_i^c of item i,

$$L_i^c = L_i, \quad i \in E,$$
$$L_i^c = L_i + L_{suc(i)}, \quad i \in I.$$

Given the definition of L_i^c we can state that $Y_i(t)$ represents the coverage by item i of the end-item demand from the start of period t until the start of period $t + L_i^c$ just before releasing the item ordered at the start of period t. Now notice that for all items with a longer cumulative lead time that at the start of period t we know exactly their coverage of end-item demand from the start of period t until the start of period $t + L_i^c$. Define

$Z_{ij}(t)$ coverage of end-item demand by item j from the start of period t until the start of period $t + L_i^c$, $L_j^c > L_i^c$.

Given the state information $(Y_i(t), \{Z_{ij}(t)\})$ we can define the modified base stock policy as follows,

$$r_i(t) = \max\left(0, \min\left\{S_i, \min_{\{j | L_j^c \geq L_i^c\}} \{Z_{ij}(t)\}\right\} - Y_i(t)\right).$$

It is shown in Rosling (1989) Langenhof and Zijm (1990) that the policy described through the above equation is cost-optimal. In De Kok and Seidel (1990) and Van Houtum and Zijm (1991) simple computational schemes are given to determine the optimal echelon order-up-to-levels.

Thus for pure assembly systems we can find easy-to-implement optimal order release policies. This seems relevant for the SCOP problem in the manufacturing of complex and expensive products, such as manufacturing equipment and aircraft. Recognizing that such products have a modular structure where typically the forecast-driven activities relate to items common to all variants and the order-driven activities relate to variant-specific items, the supply network that faces stochastic demand may be seen as a pure assembly system. For such a system Dellaert, de Kok and Wang (2000) study a nonoptimal base stock policy for pure assembly systems. This non-optimal policy is inspired by a Make-To-Order environment (e.g. assembly of Public Telephone Exchanges), where the final assembly lead time is shorter than component purchasing lead times, so that the latter are held on stock. The base stock policy releases a production order for the main assembly, after which this main assembly is pushed through a number of assembly stages. Given the planned throughput time at each assembly stage for each component, purchase orders are released based on the projected demand according to the production order for the main assembly. This is clearly non-optimal from an inventory management point of view, since the orders for the components should be based on the latest customer demand information, as is the case for the optimal policy given above. However, in practice other considerations, like workforce planning, play a role. The push policy provides early information about resource requirements, whereas the optimal policy decides at each assembly stage how much to release immediately after receiving the latest demand information. In Dellaert et al. (2000) insight is given into the circumstances under which the push policy yields near-optimal inventory costs.

5.4 Optimal and near-optimal base-stock policies for divergent systems

Another special supply chain structure is the divergent structure, where each item has exactly one child, but may have multiple parents. The most upstream item, the root item, has a single supplier with infinite material availability. In Diks and de Kok (1998) the structure of the average-cost optimal policies for divergent systems is derived under the balance assumption. The costs considered are linear holding and penalty costs incurred at the end of a period. The penalty costs are incurred for each end-item short. The balance assumption states that in case the cumulative orders from parent items exceed the available stock of the item, then the optimal allocation policy guarantees that each parent item is allocated a non-negative quantity of the available stock. It can be easily verified by discrete event simulation that even in a two-echelon divergent system with identical end-items the balance assumption is violated by the optimal allocation policy, which is an equal fractile policy [cf. Eppen and Schrage (1981) Axsater (2003)]. Below we discuss this issue of imbalance in more detail.

The optimal policies under the balance assumption are base-stock policies and satisfy so-called generalized Newsboy equations. In order to formulate the necessary and sufficient conditions for base stock policies under the balance assumption we introduce the following notation:

For each item we can define its associated end-items,

p_i penalty cost incurred for each unit short of item i and the end of a period, $i \in E$

U_i all items on the path from the root item (inclusive) to item i (exclusive), $i = 1, 2, \ldots, N$

E_i set of end-items downstream of item i

α_k^i non-stockout probability of $k \in E_i$ under the optimal policy under the balance assumption for the subtree of the divergent system with item i as root item, $i = 1, 2, \ldots, N$

It follows from the definition of E_i, that

$$E_i = \{i\}, \quad i \in E$$
$$E_i = \bigcup_{j \in V_i} E_j, \quad i \in I$$

Diks and de Kok (1998) proof the following theorem:

Generalized newsboy equations theorem
Under the balance assumption the optimal base stock levels S_j and optimal allocation policies satisfy

$$\alpha_k^i = \frac{\sum_{m \in U_i} h_m + p_k}{h_k + \sum_{m \in U_k} h_m + p_k} \quad \text{for every } k \in E_i, i = 1, 2, \ldots, N.$$

From the Generalized Newsboy Equations Theorem in theory one can recursively compute the optimal base-stock levels S_i and optimal allocation policies. However, this turns out to be computationally infeasible for realistic problem instances. The main issue here is the non-linearity of the optimal allocation functions. Diks and de Kok (1999) propose to assume *linear* allocation functions. In order to define these linear allocation functions we introduce the following variables:

q_j fraction of shortage allocated to item j

$X_{t,i}$ echelon stock of item i at time t immediately before allocation

$I_{t,j}$ echelon inventory position of item i at time t immediately after allocation

A linear allocation rule associated with item i and its parent items $j \in V_i$ is defined by

$$I_{t,j} = S_j - q_j \left(\sum_{m \in V_i} S_m - X_{t,i} \right)^+.$$

Note that $\left(\sum_{m \in V_i} S_m - X_{t,i}\right)^+$ is the shortage at time t, since it indicates the difference between the cumulative base stock levels of all parent items of i and the echelon stock of i.

Diks and de Kok (1999) derive a generalized Newsboy equation theorem for optimal linear allocation policies, implicitly assuming that in each stage of the recursive procedure linear allocation policies can be found that solve these generalized Newsboy equations. Though this is not true in general, the generalized Newsboy equation theorem for linear allocation policies yields a recursive heuristic to efficiently compute (S_i, q_i) for all $i = 1, 2, \ldots, N$. Diks and de Kok (1999) show that the heuristic yields policies that "almost" solve the recursive set of generalized Newsboy equations, thereby suggesting that the policies found are close-to-optimal.

Likewise with the optimal allocation policies it cannot be guaranteed that linear allocation policies satisfy the balance assumption. This motivated Van der Heijden (1997) to determine the linear allocation policies that minimize the probability of imbalance. Based on proxies for the imbalance probability he derived a remarkably simple expression for the allocation fractions q_i, $i = 1, 2, \ldots, N$. Defining

D_k := demand per period of item $k \in E$

$\mu_j := \sum_{k \in E_j} E[D_k]$

$\sigma_j := \sigma(\sum_{k \in E_j} D_k),$

we find the following expression:

$$q_j = \frac{\mu_j^2}{2 \sum_{m \in V_i} \mu_m^2} + \frac{\sigma_j^2}{2 \sum_{m \in V_i} \sigma_m^2}, \quad j \in V_i, \, i = 1, 2, \ldots, N$$

In the formula above we correct an error in the analysis in Van der Heijden (1997): from his formula (22) he minimizes the variance of each expected negative allocation to a successor separately instead of minimizing the sum of the probabilities of a negative allocation to a successor. The formula above coincides with the one derived in Van der Heijden (1997) when assuming that $\mu_j = \mu$ for all $j \in V_i$. We assume here that the demands for different end-items $k \in E$ are independent. Extensive discrete event simulation experiments show that indeed the linear allocation policies based on the above formula for q_j (Van der Heijden (1997) coins the term Balanced Stock rationing) yield such low probabilities of imbalance that imbalance can be ignored. This is extremely important because this implies that the analytical results and the policies obtained in Diks and de Kok (1999) are applicable, i.e., average costs and customer service levels are accurately computed. Further numerical study is required to find out whether the combination of base stock control policies and Balanced Stock rationing yields cost-effective policies as compared with alternative policies proposed in the literature. We refer to Axsater (2003) for a further discussion of the issue of imbalance. We conclude here with the statement that for the combination of base-stock policies and linear allocation policies it is possible to efficiently compute base-stock levels and allocation fractions. The policies computed seem to be close-to-optimal and performance characteristics, such as costs and customer service levels, are accurately computed. The analysis of divergent systems is the basis of a class of policies that can be applied to general assembly networks.

5.5 Synchronized base stock policies for general supply networks

The optimal policy derived for pure assembly systems cannot be applied to assembly systems where items have multiple successors (parents). We can identify two related root causes for this statement:

(1) In case of a shortage the above policy does not define the procedure for allocation of this shortage to successors.
(2) The state variables $Z_{ij}(t)$ cannot be defined since we cannot uniquely define L_i^c in case of multiple successors.

In De Kok and Visschers (1999) a class of policies is proposed that introduces uniquely defined state variables similar to $Z_{ij}(t)$ and allocation mechanisms derived from the analysis of divergent systems [cf. Van der Heijden, Diks and de Kok (1997)], so that it is possible to generate feasible item order releases in a straightforward way. Furthermore within this class of policies it is possible to characterize the optimal policy under i.i.d. exogenous demand, and near-optimal policies can be found numerically. For pure assembly systems this class of policies coincides with the modified base-stock policies described above. To understand the idea behind the class of policies proposed for general assembly systems let us consider in more detail our material co-ordination problem.

The lead time structure identifies at which moments in time item orders must be released in order to have them available for (sub)assembly activities required for production of the end-items at the start of period t. A natural order of decisions made over time thus arises. We can identify the item(s) that must be released first, which ones thereafter and so on. It is important to understand that as soon as an item is ordered, this ordering decision restricts the future demand that can be satisfied. In fact, the ordering decision leads to a particular echelon inventory position and, as stated above, this echelon inventory position covers future end-item demand. The problem however is that it is not clear how this coverage is actually used over time by the various end-items. This is not only due to uncertainty in future demand, but also due to the interactions between items caused by lack of availability. If some item is missing for an assembly operation, then another item is no longer needed as well. In a sense people in practice find themselves coping with a material co-ordination problem where solving the problem of a particular item creates a problem for another item, and so on.

The main reason for this complexity is that in a general supply network there is no clear hierarchy in decision making about order releases as we found above for the case of a pure assembly systems. Such a clear hierarchy also exists if the supply network would be a pure divergent system, i.e., each item is transformed into multiple items without a need for other items. The approach proposed in De Kok and Visschers (1999) is based on an *artificial hierarchy* derived from the structure of the general supply network. This structure is determined by the BOM and the planned lead times. Below we restrict ourselves to the main ideas behind this hierarchical planning concept. The artificial hierarchy enables us to define state variables that unambiguously define the item order releases.

We define the cumulative lead time of an item as follows:

$$L_i^c = L_i, \quad i \in E,$$
$$L_i^c = L_i + \max_{j \in V_i} L_j, \quad i \in I.$$

Now we define the root node s as

$$s = \arg\left(\max_i L_i^c\right),$$

i.e.

$$L_s^c \geq L_i^c, \quad i \in I \cup E.$$

Without loss of generality we assume that s is unique and that all cumulative lead times are different. Now we develop a hierarchical procedure that decides on all item order releases related to the end-items in E_s. The hierarchy is derived from the cumulative lead times L_i^c. Define the set of items $\hat{C}_i$ as follows,

$$\hat{C}_i = \left\{j | L_j^c > L_i^c, E_j \cap E_i \neq \phi\right\}.$$

We assume without loss of generality that

$$E_j \cap E_i = E_i, \quad \forall j \in \hat{C}_i,$$

i.e., all items that are used in the same end-items as item i, but are ordered earlier than item i, are common to end-items in E_i. In case this restriction does not hold, we can find a partition of E_i, and a one-to-one related collection of subsets of $\hat{C}_i$ for which the above holds for each one-to-one related pair of subsets and apply the principles below to each of the subsets. Finally we define

$$E(\hat{C}_i) = \bigcap_{j \in \hat{C}_i} E_j.$$

The first decision in the hierarchy is to order item s at the start of period t according to a pure basic stock policy, i.e.,

$$r_s(t) = S_s - Y_s(t).$$

Let us consider item i to be ordered at the start of period t. In principle we would like to order according to a base stock policy, i.e., bring the echelon inventory position to S_i to cover future end-item demand. Notice that decisions have already been taken in the past for items $j \in \hat{C}_i$ that affect this decision. In fact we assume that our decision hierarchy in the past determined $Z_{\hat{C}_i}(t)$ coverage of future end-item demand during periods $t, t+1, \ldots, t+L_i^c$, for all items in $E(\hat{C}_i)$ at the start of period t.

Given our assumptions stated above we have that E_i is a subset of $E(\hat{C}_i)$. Therefore we distinguish between two situations:

(i) $E_i = E(\hat{C}_i)$
(ii) $E_i \neq E(\hat{C}_i)$

In situation (i) we have that $Z_{\hat{C}_i}(t)$ is fully dedicated to future demand of end-items in E_i. Our target coverage equals S_i, but it does not make sense to increase the coverage above $Z_{\hat{C}_i}(t)$. Thus we release an order for item i as follows,

$$r_i(t) = \max\left(0, \min\left(S_i, Z_{\hat{C}_i}(t)\right) - Y_i(t)\right).$$

In situation (ii) $Z_{\hat{C}_i}(t)$ is intended to cover future demand for other end-items than those in E_i, alone. The problem is that we must decide how much to order for item i, thereby allocating quantities of the components in $\hat{C}_i$ to item i, while it may well be that we need not order yet any other items related to $E(\hat{C}_i) \setminus E_i$. In this case we maintain our hierarchy in decision making by introducing an *artificial base stock level* $S_{E(\hat{C}_i) \setminus E_i}$ that relates to end-items in $E(\hat{C}_i) \setminus E_i$. This implies that the target coverage of future demand for all end-items in $E(\hat{C}_i)$ equals $S_i + S_{E(\hat{C}_i) \setminus E_i}$. In case $Z_{\hat{C}_i}(t)$ is below this target level, then we must decide about the rationing of the deficit. This yields the following order release policy for item i,

$$r_i(t) = \max\left(0, S_i - q_i\left(S_i + S_{E(\hat{C}_i)\setminus E_i} - Z_{\hat{C}_i}(t)\right)^+ - Y_i(t)\right).$$

Here we use a linear rationing policy, where q_i is the fraction of the deficit allocated to end-items in E_i. Notice that situation (i) is a special case of situation (ii) with $q_i = 1$ and $S_{E(\hat{C}_i) \setminus E_i} = 0$.

In situation (ii) we create coverage by a set of items for future demand related to end-items in E_i and $E(\hat{C}_i) \setminus E_i$. The set of items associated with E_i is $\hat{C}_i \cup \{i\}$. The set of items associated with $E(\hat{C}_i) \setminus E_i$ is $\hat{C}_i$, which implies that $\hat{C}_i$ may again play the role of the set of items that will restrict the order release decision for an item to be ordered later. In fact, the creation of an artificial order-up-to level $S_{E(\hat{C}_i) \setminus E_i}$ always implies such a situation. Each order-up-to-level thus relates to a *decision node*, which is *uniquely* determined by the *combination* of a set of items and a set of end-items. A set of items can be associated with multiple decisions nodes, c.q. (artificial) order-up-to-levels, and a set of end-items as well can be associated with multiple decision nodes.

In case $Z_{\hat{C}_i}(t) > S_i + S_{E(\hat{C}_i) \setminus E_i}$ the excess coverage $Z_{\hat{C}_i}(t) - (S_i + S_{E(\hat{C}_i) \setminus E_i})$ with respect to future end-item demand in periods $t, t+1, \ldots, t+L_i^c$ is not used to cover this future demand and hence will be available to cover end-item demand after period $t + L_i^c$. As a consequence this excess coverage results in future excess stocks of all items in $\hat{C}_i$ at the end of period $t + L_i^c$. Hence a decision node can be seen as a stockpoint, where physical inventory relates one-to-one to excess future stocks of the items associated with this decision node.

We have stated that the procedure above holds for any supply network. Informally speaking, the above approach creates a number of divergent systems of decision nodes. Our restrictions on the sets E_i given above are restrictions on the possible combinations of item sets and end-item sets. Once understanding the principles it is rather straightforward to remove these restrictions. In Sections 5.6 and 6 we present examples for which we derive the divergent system(s) of decision nodes.

As stated above the policy described above extends the optimal policy for pure assembly systems described in Rosling (1989) to a (nonoptimal) policy for general assembly systems, i.e., multi-item multi-echelon systems. The main idea behind the approach is the artificial hierarchy that enables synchronization of order release decisions over time. Hence we define these policies as *synchronized base stock policies*.

An important distinction between the pure base stock policies and the synchronized base stock policies given here is that in the former case each item has a uniquely defined base stock level, while in the latter case multiple base stock levels, each associated with a decision node, may be associated with a single item. In addition the allocation fractions defined above constitute another set of decision variables. This implies that the synchronized base stock policies provide many more degrees of freedom. This explains why pure base stock policies do not allow for any combination of target customer service levels, while synchronized base stock policies can be found that satisfy any set of customer service level constraints with equality.

We note here that multiple divergent systems emerge when an order must be released for an item that is not contained in any of the end-items for which we know the limits on future coverage due to earlier decisions. In that case this item becomes the root node for another divergent tree of decision nodes and we can apply a pure base stock policy for this item.

It is important to note that synchronized base stock policies are *not cost-optimal*. In fact in situation (ii) we allocate part of the future coverage $Z_{\hat{C}_j}(t)$ specifically to item i without a real need for taking that decision. This allocation could have been postponed until the moment items are physically assembled into a (sub) assembly or end-item. In De Kok and Visschers (1999) it is shown for product families satisfying a particular constraint on the product and lead time structure, that this simultaneous ordering and allocation decision seems to hardly affect the performance of the system in terms of cost and customer service level. The comparison in Section 5.6 seems to underline the effectiveness of the synchronized base stock policies. The comparison in Section 6 provides even stronger support to the SCOP concept described above. Yet clearly more research is required to make any conclusive statement.

Due to the fact that the general assembly network is translated into a set of divergent systems we can apply the algorithms proposed by Diks and de Kok (1999) in order to determine order-up-to-levels and rationing fractions that satisfy the service level constraints [cf. Section 5.4]. This implies that *given the concept of synchronized base stock policies* as defined above we can derive close-to-optimal policies for this concept even for large-scale systems. In De Kok (2002a) efficient algorithms are proposed for determining these close-to-optimal policies based on the relationship between generalized Newsboy equations and finite-horizon ruin probabilities.

5.6 Comparison of pure and synchronized base stock policies

To provide insight into the differences between the pure base stock policy models discussed in the literature and the synchronized base stock control policy model, we discuss an example. We consider an ATO system consisting of 3 end-items and 3 modules. Fig. 3 shows the product structure. End-item 1 is a base-version, consisting of module 6, only. End-item 2 and 3 have additional features, represented by modules 4 and 5 respectively. The costs of items 4, 5 and 6 are 1, 1 and 8, respectively.

The synchronized base stock policies are related to a logical mapping of the above structure into (a set of) divergent structures. In the case of this simple example we find a single divergent structure as given in Fig. 4.

Here E_k denotes the set of end-items associated with decision node k, and C_k denotes the set of items associated with decision node k. The triangle associated with E_k and C_k expresses the fact that the release decisions taken with respect to successors of decision node k may result into an excess coverage of future demand for end-items by the items associated with decision node k [see Section 5.5]. This excess is *planned* at the moment the release

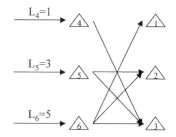

Fig. 3. Example product structure

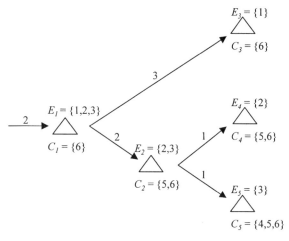

Fig. 4. Divergent structure underlying synchronized base stock policies for example product structure.

decisions associated with decision node k are taken, but eventually the excess results in physical stock of the items associated with decision node k. This implies that the long-run average physical stock of each item i can be computed from this associated divergent multi-echelon inventory system by adding the long-run average physical stocks of all decision nodes k with $i \in C_k$.

From this divergent structure we obtain the following decision hierarchy:

- First we order, say at time 0, component 6 to be used in all end-items 1, 2 and 3.
- After two time units, i.e., at time 2, we order item 5 after deciding first on allocation of the echelon stocks associated with item 6, thereby determining how much of future availability of item 6 is allocated to end-item 1 and how much to end-items 2 and 3 together.
- After another two time units, i.e., at time 4, we order item 4 after deciding first on allocation of the echelon stocks associated with items 5 and 6

dedicated to items 2 and 3, thereby determining how much of future availability of item 5 and 6 is allocated to end-item 2 and how much to end-item 3. At this point in time all items have been allocated to satisfy demand during period 6, which starts at time 5.

In Table 1 below we give the results for the situation where item 3, the high-end-item consisting of all three modules, has a target α of 99%, item 2 has a target a of 95% and item 1 has a target α of 90%. The demand per period for the end-items is i.i.d. with mean $E[D]$ and $\sigma(D)$.

The base stock levels are denoted by S, the average stock investment of a module by X. The index *PBS* denotes the pure base stock policies, the index *SBS* denotes the synchronized base stock policies. The *SBS* policies have been computed from the generalized Newsboy equations derived in Diks and de Kok (1999), while the PBS policies have been derived from a brute force cost minimization based on the results in de Kok (2002b).

The results are at first sight negative for the pure base stock policies, since the amount of stock investment required is 4% higher than the stock investment requirements in case of the synchronized base stock policy. A closer look reveals clearly why this is the case. In order to guarantee a 99% service level for item 3, items 1 and 2 both have a service level of 99%. These high service levels are enabled by high stocks of all items. Yet more insight can be gained when we realize the following. The main reason for the extreme difference in stock requirements is caused by the definition of the α-measure in the case of pure base stock policies and the assumption underlying the explicit expression given in the ATO theorem: *In case of a shortage of a particular item it is assumed that all end-items share part of the shortage.*

Although this seems obvious at first sight, it implies the ordering of the α-values for end-items in case one item's component (or module) set is a subset of another item's component set and it implies a worst-case scenario. In reality typically one would like to share shortages, yet not to the extreme that any shortage of a component causes a backorder for all end-items using this

Table 1
Performance base stock policies in case of high service of high-end-item

Item	$E[D]$	$\sigma(D)$	α^*	α_{PBS}	α_{PBS}
1	100	25	0.9	0.99	0.9
2	100	25	0.95	0.99	0.95
3	100	25	0.99	0.99	0.99

Component charactristics				
Item	S_{PBS}	S_{SBS}	X_{PBS}	X_{SBS}
4	200	192	100	83
5	810	800	210	166
6	1740	1733	1923	1891
System			2233	2140

Table 2
Performance base stock policies in case o low service of high-end-item

Item	$E[D]$	$\sigma(D)$	α^*	α_{PBS}	α_{SBS}
1	100	25	0.99	0.99	0.99
2	100	25	0.95	0.95	0.95
3	100	25	0.9	0.9	0.9

Component characteristics				
Item	S_{PBS}	S_{SBS}	X_{PBS}	X_{SBS}
4	142	162	43	47
5	708	770	109	126
6	1740	1741	1923	1952
System			2075	2125

component. What seems to be a quite natural assumption leading to an elegant expression for the non-stockout probabilities, in fact may yield high supply chain capital investments. Apparently, it is important to explicitly define a component shortage allocation mechanism that ensures lower supply chain capital requirements. Since the main reason for high supply chain investments is caused by the fact that the high-end-items have the highest service, we expect a more favorable situation for the case that we want high-end-items to have a lower service level than the low-end end-item. The results in Table 2 confirm this. In this case the pure base stock policies outperform the synchronized base stock policies by 2.4%. This is likely to be caused by the fact that synchronized base stock policies allocate common components too early, as follows from the logical decision hierarchy that is depicted in Fig. 4.

It should be clear that in practice we set α-service level constraints on the basis of an (implicit) trade-off between supply chain stock capital investments and penalty costs for disservice. It is likely that high-end service levels are higher than low-end service levels. The discussion above implies that pure base stock control policies do not allow for sufficient degrees of freedom (or means of control) to set the α-service levels accordingly. The potential supply chain cost savings are an interesting subject for further research. This concludes the discussion of pure base stock policies as a basis for SCOP.

5.7 Heuristic analysis of general network structures

In Section 5.2 we argued that pure base stock policies cannot be used for supply chain operations planning, since these policies do not satisfy the material availability constraints (3.1). We must be aware of the fact that even in the case of infinite resource capacity, the constraints (3.1) are of such complexity that it is not at all clear how to incorporate them into a supply chain operations planning concept that takes into account demand (and supply) uncertainty. We emphasize again here that this is the most important reason to resort to mathematical programming based supply chain planning concepts in a rolling schedule context. However, such concepts do not answer important questions such as

- Where and how much safety stocks should be held?
- What is the impact of demand uncertainty on supply chain capital requirements to achieve the required customer service levels?

Such questions are more of a strategic and tactical nature, yet the answers have an immediate consequence for the operational performance of the supply chain.

In the literature we find several approaches that discuss the issue of safety stock positioning. For an extensive discussion of these approaches we refer to Axsater (2003) and Song and Zipkin (2003) (Chapters 10 and 11, respectively, of this volume). It seems appropriate to briefly discuss these approaches in the context of general assembly structures. For such structures it seems that there are two generic assumptions underlying the heuristic analysis of such systems:

Decomposition assumption. *Safety stock parameters are set at such high levels that every material order released can be satisfied from stock on hand without checking upstream availability.*

The decomposition assumption allows an analysis of general assembly systems, where costs are derived by adding costs derived from an item-by-item single-item single-echelon model and customer service levels are derived from single-item single-echelon models for items in E.

Assembly assumption. *If an assembly order has to wait for material then this is caused by the shortage of exactly one item required for assembly.*

The assembly assumption simplifies the analysis considerably due to the fact that it enables the translation of an assembly system into a weighted sum of serial systems. The weights relate to the probabilities that particular items are short.

The decomposition assumption seems stronger than the assembly assumption. It can be verified by discrete event simulation that when item service levels for items in I are above 95%, then the (heuristic) analysis of the performance of the supply chain in terms of costs and service yields an acceptable accuracy. Realizing that 95% service levels are also a prerequisite for the assembly assumption to hold, we might argue that if the assembly assumption holds, then the decomposition assumption holds as well, in the sense that both yield more or less the same results. Further research is however required to verify this line of thought.

Analysis under the decomposition assumption

The typical focus of the heuristic analysis under the decomposition assumption is on the analysis of a supply network cost function under some service level restriction and the derivation of properties of optimal policies [e.g., see Lee and Tang (1996)]. We notice however that the optimal policies found, by implication of the decomposition assumption, keep high stocks at all upstream stages. It follows from Whybark and Yang (1996) (and the results of Section 6) that truly optimal policies tend to concentrate inventory capital at downstream stages, unless a high value is added at the most downstream stages. We believe that this observation may have considerable impact with respect to the benefit of postponement of item diversity. Item diversity is postponed if the item is made more common to downstream items. Postponement strategies therefore allow for reduction in upstream item stock investments while maintaining end-customer service levels. The decomposition assumption may yield exaggeration of the benefits of postponement since these benefits are derived from reduction of upstream safety stocks that guarantee high intermediate service levels. For an in-depth discussion of the benefits of postponement in Supply Chain Management we refer to Lee and Swaminathan (2003), Chapter 5 of this volume.

We notice here that postponement strategies should not be confused with strategies that allow for an upstream shift of the CODP [cf. Section 1]. In that case changes in the transformation and transportation processes allows holding e.g. modules instead of end-items. An upstream shift of the CODP typically has a large impact on stock investments, since (specific) end-item stockpoints with high service requirements are eliminated completely.

In Graves and Willems (2003), Chapter 3 of this volume, the decomposition approach proposed by Inderfurth (1994) and Graves and Willems (2000) based on Simpson (1958) is extensively discussed. Minner (2000) discusses the distinction between full-delay models and no-delay models. The full-delay models assume that if an item is not available, then one has to wait with an order release until the item becomes available. This is in line with our discussion in Section 3 and in fact is equivalent to the set of constraints (3.1). In no-delay models it is assumed that if the dependent demand from downstream orders exceeds the available inventory, then there is some outside source (not considered in the model in terms of costs) that provides the material. In fact, this is the assumption that Simpson (1958) proposed and enables an elegant analysis of complex supply networks. Graves and Willems (2000) argue that this assumption is reasonable if safety stocks are set such that they cover the maximum demand during a so-called coverage time. The coverage time is the sum of the lead times of items that are consecutively upstream of the item under consideration.

The analysis of no-delay models reduces to the determination of the optimal cover times. In fact the concept of cover times enables a generalization of the decomposition assumption in that if a cover time of an item covers multiple upstream stages, then at these stages no safety stocks are kept and the safety stock is only held for the item under consideration. We note here that the no-delay assumption converts the analysis of a stochastic demand model into the analysis of a deterministic model with cover times as decision variables.

From the analysis in the above-mentioned papers one finds items where no stocks are held at all and items for which the stock equals the safety stock associated with the cover time derived and the demand uncertainty associated with the item. Using dynamic programming Minner (2000) is able to analyze general supply chain structures. He shows by an example of a serial supply chain that the optimal policies for no-delay models and full-delay models may differ considerably. He discusses heuristics that may bridge the gap between full-delay models and no-delay models. However, more research is required to better understand the applicability of no-delay models. Hereby we implicitly assume that full delay models better represent the reality modeled in a supply chain operations planning context. We emphasize here that the main issue here is the external validity of either approach. Since in reality people intervene in situations where demand exceeds availability the only way to test applicability of models like the ones discussed in this Chapter is empirical. Only then we can identify situations where either full-delay or no-delay models perform best.

An interesting contribution to the analysis of the SCOP problem based on the decomposition assumption is given in Graves, Kletter and Hetzel (1998). They consider a general supply network under dynamic demand. The main contribution of the paper is the introduction of so-called forecast revisions $\Delta F(t, t+s)$,

$$\Delta F(t, t+s) = \hat{D}(t, t+s) - \hat{D}(t-1, t+s), \qquad s \geq 0.$$

As before $D(t, t+s)$ denotes the forecast of the demand in period $t+s$ made at the start of period t. It is assumed that the random variables $\Delta F(t, t+s)$ are i.i.d. with respect to t. The forecast revisions for an arbitrary t may be correlated. These assumptions seem quite reasonable. Graves et al. (1998) give a detailed analysis of the single stage model. They assume that item order release revisions $r_i(t, t+s) - r_i(t-1, t+s)$ can be expressed as a linear function of the forecast releases. This assumption enables the formulation of an optimization problem focused on minimizing the variance of the item order releases subject to a constraint on the variance of the net inventory, which relates to the level of customer service. The decision variables are the weights that determine the linear relationship between item order release revisions and forecast revisions. Applying the decomposition assumption the results for the single stage system are combined into an analysis of a general assembly system. The approach was successfully applied to a real life case study.

The approach described in Graves et al. (1998) has a strong resemblance with the seminal work by Holt, Modigliani, Muth and Simon (1960) who study linear decision models in the context of aggregate planning. In both approaches the linear relationship between production schedule revisions and forecast revisions is used as a means to implicitly model finite capacity.

Even though Graves et al. (1998) focus on the design of the supply network, their innovative modeling of the demand process deserves further study in the context of the SCOP problem (see Section 7).

Analysis under the assembly assumption

The assembly assumption has been extensively used in literature. It enabled the exact analysis of spare part networks, where end-items consist of modules that may fail due to components that may fail, etc. In this context the assembly assumption states that if a product fails this is due to exactly one module. In turn, this module failed due to exactly one component, etc. For an extensive survey of relevant literature we refer to Sherbrooke (1992) and Axsater (2000).

In the context of safety stock positioning in general supply networks recently Ettl, Feigin, Lin and Yao. (2000) developed a framework of analysis building on the assembly assumption. They assume continuous review installation base stock policies for all items. In line with the discussion in Sections 1 and 2 they assume so-called nominal lead times that are equivalent to the planned lead times assumed in this chapter. Demand is modeled as a compound Poisson process. The authors emphasize that the compound Poisson process is a means to approximate the real life demand process and provides the degrees of freedom to obtain a good fit.

The objective of Ettl et al. (2000) is to find optimal base stock levels that minimize costs subject to customer service level constraints. The analysis in the paper is roughly as follows: First of all, it is easy to see that the external demand process and the installation base stock policies determine the demand process for each item, irrespective of the base stock levels of its parent items. This enables to compute on an item-by-item basis the waiting time distribution of an arbitrary order for the item, given its base stock policy. This waiting time is derived from the analysis of the number of outstanding orders, which in turn is derived from the analysis of an $M^{[X]}/G/\infty$ queue. The analysis yields for each item, the first two moments of the waiting time distribution of an arbitrary order for this item. Next the assembly assumption converts the multi-item multi-stage problem into a set of interrelated single stage problems that can be analyzed subsequently. Thus expressions can be derived for holding costs and service levels. This yields an overall objective function that is optimized using a conjugate gradient search technique with the base stock levels as decision variables.

A careful study of the numerical results reported in Ettl et al. (2000) and, more extensively, in Feigin (1998) reveals that the optimal solution yields extremely low upstream fill rates (between 0.1 and 0.7). This puts forward a fundamental issue. The approach is based on the assembly assumption, yet the optimization procedure suggests optimal solutions that strongly violate the major assumption underlying the approximate analysis. Although the actual performance of the heuristic may be very good or even close-to-optimal, from a standpoint of mathematical rigor it implies that the solution obtained is infeasible and the optimization problem should be reformulated including constraints that ensure adherence to the assembly assumption. This problem is addressed implicitly in Ettl et al. (2000) in that they apply discrete event simulation to verify the solutions obtained. It means that the quality of the solutions cannot be supported by the analytical characteristics of the approach. In our view this fundamental issue deserves further research.

This concludes our discussion of base stock policies in the context of the SCOP problem. Before summarizing our conclusions we would like to discuss briefly another interesting class of control policies.

5.8 Combined Kanban and base stock policies for Supply Chain Planning

This class of policies for the control of manufacturing systems has been proposed by Buzacott and Shantikumar (1993) and Frein et al. (1995) and others: a combination of base stock control policies and Kanban control policies. Though there are differences between the concepts proposed by the various authors the idea behind the policies are essentially the same. Below we give an informal discussion of this type of policies and focus on their relevance for SCOP. We will denote these concepts as Combined Kanban Base Stock Control (CKBCS), without having the ambition to add another acronym to this part of the literature, but to avoid that we are not precise enough to describe any of the proposed mechanisms.

First of all let us consider the standard Kanban policies. A standard Kanban policy can be defined through two parameters: the number of cards circulating between two stages and the quantity per card. To clarify this further let us consider an arbitrary item i. Assume that item i is processed on resource k_i after which it is used by items $j \in V_i$ that are processed on resources k_j. Then we can define

K_{ij} Number of Kanban cards associated with items i and j circulating between resource k_i and resource k_j

Q_{ij} The quantity associated with each Kanban card to be released of item i on behalf of item j

S_i Base stock level of item i

As seen before the base stock level should be seen as a target inventory level associated with item i. Yet the state variable that defines the inventory level can have different definitions. We have seen examples such as the echelon inventory position and the local inventory position, but in principle alternatives exist. Typically, one could vary the definition of the echelon of item i, comprising more or less of the supply chain downstream of item i.

In principle CKBSC policies operate as follows: If the inventory level of item j is below its target level S_j and at least one Kanban card associated with items i and j is available, then a Kanban card is sent from resource k_j to resource k_i, releasing an amount Q_{ij} of item i. From this we see that the base stock level S_i should ensure availability of item i, while the parameters K_{ij} and Q_{ij} ensure that the maximum amount of work in process of item i, i.e., the total amount of outstanding orders of item i, cannot exceed $\sum_{j \in V_i} K_{ij} Q_{ij}$

It can be shown that CKBSC policies can emulate all currently known inventory control policies by appropriate choices of both the definition of the inventory state variable and the values of the parameters S_i, K_{ij} and Q_{ij}. Although the discussion in Buzacott and Shantikumar (1993) and Frein et al. (1995) restricts to continuous review systems, in our view the principles of CKBSC policies carry over to the periodic review setting of supply chain planning.

When testing the CKBSC policies against our generic supply chain planning constraints we find that the feasibility constraints (3.1) may be violated. The reason for this is that a Kanban card is sent from a resource k_j without checking availability of the material required. One solution to this problem is to ensure that the number of outstanding orders never exceeds the base stock level, so that always inventory is available. This implies that $\sum_{j \in V_i} K_{ij} Q_{ij} \leq S_i$. In the CKBSC concepts proposed in the literature the opposite is assumed, i.e., the maximum amount on order is at least equal to the base stock level. If this is the case we have to develop procedures that impose the feasibility constraints (3.1). It is likely that such a modification will greatly complicate the analysis of the performance of CKBSC systems. Currently, this analysis is strongly related to the performance analysis of queuing network systems. For further details we refer to Frein et al. (1995) and Buzacott and Shantikumar (1993). It should be noted that such an analysis strongly relies on the assumption of continuous review and a FCFS discipline for dealing with priorities in case of material shortage. Given the fact that the supply chain operations planning problem is by nature periodic and priorities should be based on costs structures, it is clear that further research is required to develop the framework of CKBSC policies for a supply chain planning. Given the richness of the framework as such we consider this direction for further research as quite promising. However, the state-of-the-art of analysis of CKBSC policies does not allow for a comparison with other supply chain planning concepts, as we intend to undertake in Section 6.

5.9 Concluding remarks

In this section we discussed quantitative models for the incapacitated SCOP problem that explicitly include demand uncertainty. We have shown that pure base stock policies violate the feasibility constraints (3.1). We extensively discussed the synchronized base stock policies introduced by De Kok and Visschers (1999) that satisfy the feasibility constraints and allow for an exact analysis and the numerical computation of near-optimal policies within this class. We identified policies that combine Kanban control and base stock control as interesting candidates for further SCOP research. We discussed various heuristic methods to compute control policies minimizing costs subject to service level constraints. The heuristics based on the decomposition assumptions yield solutions that contradict the findings in Whybark and Yang (1996) about optimal positioning of safety stocks, i.e., safety stocks should be concentrated downstream (see also Section 6). The assembly-assumption-based heuristic of Ettl et al. (2000) yields solutions that, although they seem to violate the assembly assumption, are close to optimal.

In the next section we compare the two classes of control concepts for which we have been able to show that all SCOP constraints are taken into account, viz. the LP-based control concept discussed in Section 4 and the synchronized base stock policies discussed in this section. We restrict our comparison to the uncapacitated SCOP problem, since there are no results for the capacitated SCOP problem under stochastic demand.

6 Comparison of supply chain planning concepts for general supply chains

In this section we compare supply chain operations planning concepts from the two main classes of supply chain planning concepts, i.e., MP-based concepts in a rolling schedule context (discussed in Section 4) and concepts based on stochastic models that are applicable to general supply chain structures (discussed in Section 5). Firstly in Section 6.1 we briefly discuss in the impact of a (dynamic) forecasting process on the release decisions generated by the two classes of concepts. Thereafter, we present a numerical study that provides insight into various aspects of the SCOP problem. We restrict this numerical comparison to the situation with infinite resource availability. The main reason for this restriction is that stochastic model-based concepts do not incorporate finite capacity resources. Furthermore we assume stationary stochastic demand. We expect that the results obtained are applicable to the situation with stationary forecast errors, yet this requires further research.

Because our focus is on managerial insights into the SCOP problem and the characteristics of the two different SCOP concepts, we restrict our comparison to a relatively simple case situation. One should be aware of the fact that the structural complexity of the SCOP problem discussed in this chapter, i.e., multi-item, multi-echelon, general BOM relationships, is enormous. By carefully selecting case situations we obtain useful insights. These insights were confirmed by results from a comparative study based on a real-world case [cf. De Kok (2001)].

In Section 6.2 we present the case situation. In Section 6.3 we present the results of our comparative study. In Section 6.4 we recapitulate our conclusions into a number of managerial insights and issues.

6.1 Supply Chain Planning concepts and forecasting

As has been made clear in Section 4 the key dynamic input to an SCOP concept is the forecast of exogenous demand for items in E (like before we assume $P = E$). A practically relevant question that seems to have an obvious affirmative answer is the following: Are the immediate release decisions affected by the forecasts of exogenous end-item demand? Given the set of equations in Section 4 this may be obviously true, but we concluded that the answer depends on the planning concept used. It can be easily proven that a planning concept based on base-stock policies *with fixed base-stock levels* generates immediate order release decisions that do not depend on the forecasts generated. Similarly, it can be proven that if the base-stock levels are dependent on the forecast, for example, when the base stock levels represent a fixed number of weeks coverage of future demand, then the immediate order release decisions depend on the forecast. It can also be verified that the supply chain operations planning concept based on LP discussed in Section 4 generates immediate order release decisions that depend on the forecasts of exogenous demand. It seems natural to assume that the SCOP concept should generate immediate order release decisions that depend on the forecast. In this context it is interesting to notice that recently a growing number of retail and manufacturing companies seems to resort to classical end-item inventory management policies with fixed reorder and/or order-up-to-levels. There seems to be empirical evidence that current forecasting (or sales planning) processes generate a forecasting accuracy that justifies these decisions. For an interesting discussion of this particular issue we refer to Aviv (2001).

6.2 Comparison of the LP-based SCOP concept and the SBS concept

In this section we compare the LP(-based SCOP) concept presented in Section 4 with the SBS concept presented in Section 5. The LP concept represents the class of deterministic optimization models in a rolling schedule setting, while the SBS concept represents the class of stochastic models for SCOP. We first describe the case example in detail.

Case description

In this section we describe the case we used for a comparative study. We subsequently present the BOM structure, the demand process, the cost structure and the performance measures used in our comparison.

The example product structure.

In order to compare the two SCOP concepts we create a test bed. We only consider the specific product structure consisting of 11 items given in Fig. 5. As stated above we found that the results obtained for this test bed are typical.

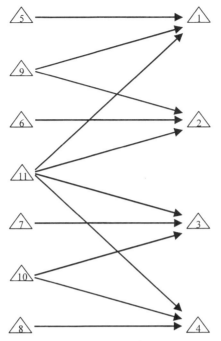

Fig. 5. The example product structure

We consider an 11-item product structure consisting of four end-items 1 to 4. All products contain common component 11. Products 1 and 2 share component 9, while products 3 and 4 share component 10. Each product contains a specific component.

As stated above we do not consider finite resources. Implicitly, such finite resources can be incorporated into the BOM through the planned lead times. High resource utilizations result into long planned lead times, low resource utilizations result into short planned lead times. In the context of the discussion in Section 2 on planning hierarchies one might state that the choice of the planned lead times is such that the order release decisions generated at SCOP level can be realized at shopfloor level with high probability. That is the due date of an order derived from the moment of its release and the planned lead time can be met with high probability.

Therefore, it is of interest to vary the lead times of the different items. Note that the lead time of the end-items 1 to 4 relates to an assembly process. We may consider the lead times for items 5 to 11 as procurement lead times. A long procurement lead time thus relates to a supplier with tight capacity in a remote location. The planned lead time structure is given by the following variables:

L_f planned lead time end-items
L_s planned lead time specific components

L_{sc} planned lead time semi-common components
L_c planned lead time common component

This implies the following equations,

$L_i = L_f$, $i = 1, 2, 3, 4$
$L_i = L_s$, $i = 5, 6, 7, 8$
$L_i = L_{sc}$, $i = 9, 10$
$L_i = L_c$, $i = 11$

In order to get insight into the impact of the supply chain structure we vary the planned lead times (L_s, L_{sc}, L_c) as follows,

(1, 2, 4) common component long lead time, specific component short lead time
(4, 2, 1) common component short lead time, specific component long lead time
(1, 4, 2) semi-common component long lead time, specific component short lead time

We note here that the planned lead time structure impacts the divergent structures that emerge when applying the SBS concept. Below we present the divergent structures associated with each of the three planned lead time structures.

Demand process and cost structure

As stated above we assume that the demand for the end-items is stationary. More precisely, demand for end-item i in consecutive periods is i.i.d. We also assume that the demand processes for different end-items are uncorrelated. We define

$E[D_i]$ average demand per period for item i, $i = 1, 2, 3, 4$
cv_i^2 squared coefficient of variation of demand per period item i, $i = 1, 2, 3, 4$

We set $E[D_i]$ equal to 100 for all end-items. To get insight into the impact of demand variability on the choice of an SCOP concept we vary cv_i^2 as 0.25, 0.5, 1 and 2. Unless stated otherwise, we assume identical demand parameters for all end-items.

Cost structure

As explained in the introduction to this chapter we want to compare different SCOP concepts on the basis of the supply chain inventory capital required to achieve a prespecified customer service level. Based on the cost structure in high volume electronics supply chains we developed a base case cost structure as follows. Analogously to the definition of planned lead times we define

h_f added value end-items
h_s added value specific components
h_{sc} added value semi-common components
h_c added value common components,

implying that

$h_i = h_f,$ $i = 1,2,3,4$
$h_i = h_s,$ $i = 5,6,7,8$
$h_i = h_{sc},$ $i = 9,10$
$h_i = h_c,$ $i = 11$

In the base case we assume that $(h_f, h_s, h_{sc}, h_c,) = (\$10, \$10, \$30, \$50)$. Hence the common component is expensive, while the added value of assembly is only 10% of the total value of the end-item. An example of such a situation is the manufacturing of TVs. Typically the Cathode Ray Tube is 50% of the total cost, a Printed Circuit Board may account for 30% of the cost, while additional materials such as a housing account for another 10% of total cost.

Customer service levels

In the introduction we defined the two most commonly used customer service levels in practice, the non-stockout probability α and the fill rate β. In our comparative study we will discuss results for the α-service measure, only. In our base case comparisons the customer service objective is to achieve a non-stockout probability of 95%, i.e.,

$$\alpha^* = 0.95$$

Evaluation of the SCOP concepts

The above case description has been the basis for a numerical study where discrete event simulation was used to compute the performance of the two SCOP concepts. Let us describe the steps along which we derived the numerical results in more detail.

For the SBS concept described in Section 5 we determined the base-stock levels and allocation policies analytically. Towards this end we followed the procedure in De Kok and Visschers (1999) (see Section 5.5) to derive the divergent structures presented in Fig. 6a–c. Given these divergent structures we computed near-optimal linear allocation policies and base stock levels based on the algorithms given in Diks and de Kok (1999). Thereafter we checked these analytical results with discrete event simulations. The

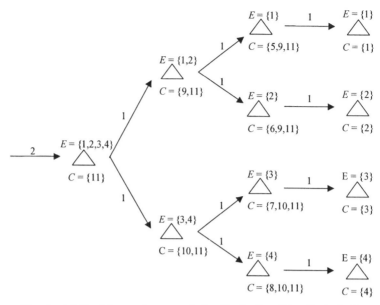

Fig. 6. (a) Decision node network for $(L_f, L_s, L_{sC}, L_c) = (1, 1, 2, 4)$.

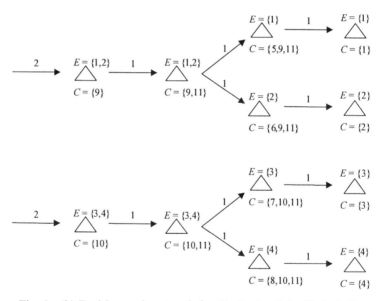

Fig. 6. (b) Decision node network for $(L_f, L_s, L_{sC}, L_c) = (1, 1, 4, 2)$.

simulation run length in all experiments was 100,000 time periods. The point estimates obtained did not change with longer simulations.

For the analysis of the LP concept we had to fully rely on discrete event simulation. In each period simulated we solved the LP problem described in

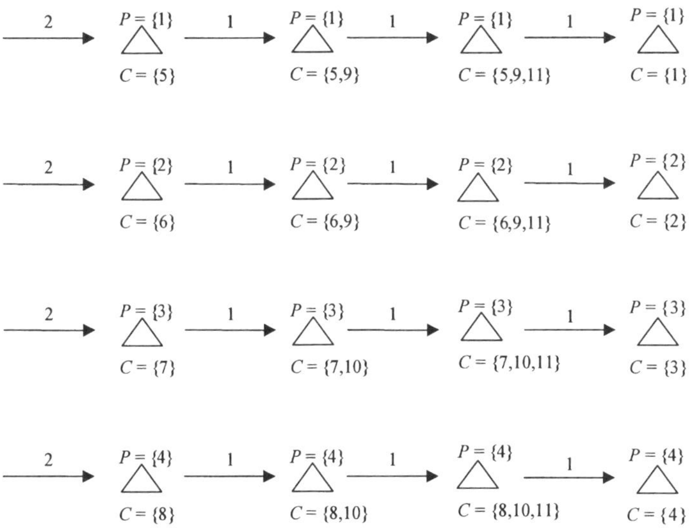

Fig. 6. (c) Decision node network for $(L_f, L_s, L_{sC}, L_c) = (1, 4, 2, 1)$.

Section 4 using CPLEX [cf. http://www.ilog.com]. As described in Section 4 we used an initial run with end-item safety stocks equal to zero to compute for each end-item the empirical distribution of the difference between net stock and safety stock. From this distribution we determined the safety stocks that would ensure the required service level. A second simulation run with the required safety stocks yielded the performance of the LP concept. For both simulation runs we used a run length of 25,000 time periods. This run length proved to be sufficiently long for all experiments to obtain the required accuracy with respect to determination of the end-item safety stocks guaranteeing 0.95 service levels.

6.3 Quantitative comparison of LP and SBS

For our base case we generated twelve different SCOP scenarios by combining three planned lead time structures and four squared coefficients of variation. The results are given in Table 3.

From Table 3 we conclude that the analytical results for the supply chain inventory capital obtained with the SBS concept, found in the column with heading SBS_{ana}, coincide with the results obtained with discrete event simulation, found in the column with heading SBS_{sim}. As expected the procedure described above for the LP concept yields the target α-levels (α_{LP}). Likewise the SBS policies computed yield the required customer service levels (α_{SBS}).

Furthermore we find that the SBS concept considerably outperforms the LP-based concept. This in itself is an important and striking result. We should realize ourselves that most commercial software for Supply Chain Planning

Table 3
Supply chain stock capital comparison, identical end-item demand

cv_i^2	(L_f, L_s, L_{sc}, L_c)	Supply chain inventory capital				Customer service	
		SBS_{ana}	SBS_{sim}	LP	Δ%	α_{SBS} (%)	α_{LP} (%)
0.25	(1,1,2,4)	72,188	71,682	83,225	16	95	95
0.25	(1,4,2,1)	76,154	76,476	83,762	10	95	95
0.25	(1,1,4,2)	74,162	73,550	84,424	15	95	95
0.5	(1,1,2,4)	105,114	104,448	119,645	15	95	95
0.5	(1,4,2,1)	112,226	112,316	121,586	8	95	95
0.5	(1,1,4,2)	108,079	107,616	122,916	14	95	95
1	(1,1,2,4)	152,583	152,203	173,056	14	95	95
1	(1,4,2,1)	165,264	165,328	180,211	9	95	95
1	(1,1,4,2)	157,294	157,034	177,166	13	95	95
2	(1,1,2,4)	217,664	218,551	258,651	18	94	95
2	(1,4,2,1)	246,637	245,998	265,952	8	95	95
2	(1,1,4,2)	228,967	228,789	261,499	14	94	95

currently available is at best employing the LP-based concept of Section 4 and most likely LP is embedded in a set of heuristics to generate a feasible solution to the deterministic SCOP model described in Section 4. This statement is in line with the findings of Stadtler et al (2000) and Fleischmann and Meyr (2003), Chapter 9 of this volume.

Interestingly, the difference between the two concepts is not very sensitive to the demand variability. One might expect that LP performs better if demand variability is low. The difference between the two concepts is mainly determined by the lead time structure. The difference is largest for the case with a long lead time for the common component and smallest for the case with a long lead time for the specific component. Apparently the SBS concept better exploits the commonality in the former situation.

To gain intuition for the surprising observation from Section 4 we present for both concepts the allocation of supply chain inventory capital among stocks of common component, semi-common components, specific component and end-item.

The results in Table 4 indicate that the LP-based concept does not seem to find the right balance in allocating stock among the different items. In case commonality can be exploited, i.e., when (semi-)common components have a longer lead time than the specific component the LP-based concept tends to withhold too much stock in (long-lead time) components. A possible explanation for this is that the LP-based concept aims at satisfying the end-item demand forecast. After satisfying this forecast remaining stocks of components are not used for assembly of end-items, because stocks of components are cheaper than stocks of assembled end-items. Especially if exogenous demand is low in a number of consecutive periods, then the LP concept will tend to build up stock upstream. The base stock levels computed

Table 4
Allocation of inventory capitol along the supply chain

cv_i^2	(L_f, L_s, L_{sc}, L)	End-item		Specific		Semi-common		Common	
		SBS (%)	LP (%)	SBS (%)	LP (%)	SBS (%)	LP (%)	SBS (%)	LP (%)
0.25	(1,1,2,4)	95	86	0	1	1	4	3	10
0.50	(1,1,2,4)	97	86	0	1	1	4	2	10
1.00	(1,1,2,4)	99	87	0	1	0	3	1	9
2.00	(1,1,2,4)	100	88	0	1	0	3	0	8
0.25	(1,1,4,2)	93	87	0	1	5	7	2	6
0.50	(1,1,4,2)	94	88	0	1	5	6	1	5
1.00	(1,1,4,2)	96	89	0	1	4	6	0	5
2.00	(1,1,4,2)	97	89	0	1	3	6	0	5
0.25	(1,4,2,1)	88	91	6	3	5	3	1	2
0.50	(1,4,2,1)	90	93	6	2	4	3	0	2
1.00	(1,4,2,1)	90	93	7	2	4	3	0	2
2.00	(1,4,2,1)	91	94	6	2	3	2	0	2

under the SBS concept are typically such that even during low demand periods inventory capital is pushed towards the locations where customer demand must be met. Informally speaking, the base-stock policies have a *just-in-case* character, whereas the LP-based concept has a *just-too-late* character.

Extensive numerical studies indicate that under the SBS concept the sums of base stock levels at each echelon in the divergent structures of decision nodes underlying the SBS concept tends to increase slightly upstream. If these sums had been equal at all echelons then no stocks would be held at all upstream stages. The results in Table 4 seem to indicate that we are close to that situation. This seems to imply that when commonality can be exploited, then only little common upstream stocks are required to reap the benefits.

In case the specific component has the longest lead time, commonality cannot be exploited and in that case the LP-concept seems to stock too much end-item. In that case it may be that the SBS concept identifies indeed that commonality cannot be exploited and favors to hold more (less expensive) components.

The rationale behind the extremely low component stock levels with both concepts, and in particular for the base-stock policies, is that holding back stock at component level does not contribute to immediate customer service. Apparently this outweighs the so-called portfolio effect for common component stocks, i.e., demand for the common component is relatively more stable than demand for individual products.

Another interesting observation from our computational study is that under the LP concept the safety stocks for the identically distributed end-items strongly differ. Our explanation is that for our base case with identical added values for all end-items the LP-problem solved each period is strongly degenerate. Thereby it depends on the particular implementation of the algorithm (e.g. choice of tie-breaking rules), which end-item is favored over the other with respect to the allocation of items. Apparently the CPLEX-solver used is not allocating these mismatches evenly over time among the end-items. Of course, the base-stock policies are identical for all end-items. Given this observation with respect to the allocation of stocks, it is interesting to compare the average stock levels in case the values of end-items are different. In Table 5 we present some results that support our conclusion that the LP concept does not handle component shortages properly. In Table 5 we define Δ_1 as the standard deviation of the three different safety stocks for end-items 1, 2, and 3 normalized by their mean, while Δ_2 is defined as the relative difference between the average safety stock for products 1, 2, and 3 and the safety stock of the lower cost end-item 4.

It follows from Table 5 that the modified base-stock concept yields identical safety stocks for identical products ($\Delta_1 = 0$), while the LP-concept yields considerably different safety stocks ($\Delta_1 > 0$). For different end-items we find that the Δ_2, the difference between the low-cost end-item safety stock and the high-cost end-items safety stock, is much bigger for the LP concept than for the SBS concept.

The results in Tables 3–5 show that the LP-based concept rations shortages among items inappropriately. Identical products are not rationed similarly due to tie-breaking rules needed to deal with the degeneracy of the associated SCOP problem. In case of different products, LP rations shortages according to a priority list based on holding and penalty costs. The priority list implies that the item first on the list is satisfied first (if possible), after that the item second on the list is satisfied, etc. until no inventory is left. Lagodimos (1992) has shown that such a priority rationing mechanism is suboptimal. The linear rationing rules used in the SBS concept ensure that shortages are shared among all products. Apparently this is superior. Informally speaking, LP is a *greedy* approach that is inferior to the *balanced* SBS approach. The importance of careful rationing of shortages of material has, to our

Table 5
Safety stock differences for non-identical end-items

$(cv_1^2, cv_2^2, cv_3^2, cv_4^2)$	(h_1, h_2, h_3, h_4)	(L_f, L_s, L_{sc}, L_c)	LP		SBS	
			Δ_1 (%)	Δ_2 (%)	Δ_1 (%)	Δ_2 (%)
(0.25,0.25,0.25,0.25)	(20,20,20,**10**)	(1,1,4,2)	23	48	0	14
(0.25,0.25,0.25,0.25)	(20,20,20,**10**)	(1,1,2,4)	11	73	0	14
(0.50,0.50,0.50,0.50)	(20,20,20,**10**)	(1,1,2,4)	9	58	0	14
(0.50,0.50,0.50,0.50)	(20,20,20,**10**)	(1,1,4,2)	17	39	0	14
(1,1,1,1)	(20,20,20,**10**)	(1,1,2,4)	11	39	0	13
(1,1,1,1)	(20,20,20,**10**)	(1,1,4,2)	16	29	0	14

knowledge, not been identified in the Mathematical Programming literature on the SCOP problem. An explanation for this may be, that this problem only reveals itself when applying MP-based concepts in a rolling schedule context under stochastic demand. We conjecture here that similar rationing problems should be addressed when resources are constraining the release of orders.

We also considered more complicated product structures and the superiority of the synchronized base-stock concept seems even stronger, i.e., stock capital investment differences become greater as the structure gets more complicated. For a more detailed discussion of a real-world case consisting of 57 items we refer to De Kok (2001).

6.4 Managerial insights

From our comparison of the LP concept and the SBS concept we draw a number of conclusions leading to managerial insights. First we summarize some generic insights that hold for both SCOP concepts discussed. Thereafter we summarize the major distinctions between the two concepts.

Product flow towards the CODP

From the definition of the Customer Order Decoupling Point (CODP) given in Section 1 we derive that the CODP in the networks considered in our comparison is at the end-item level. From the results presented above we conclude that inventory capital is concentrated at end-item level. Since little inventory capital is held at controlled intermediate stock points, we may conclude that optimal solutions can be characterized by items flowing in the supply network towards the CODP's. Intuitively this is in line with notions such as Just-In-Time (JIT), where the objective is to create perfect flow. We should note however that JIT, more precisely Kanban, is a usage-driven control concept, which requires a high degree of usage stability, otherwise such a pull concept cannot guarantee high customer service with low inventories. The optimal SCOP concept enables item flows even under situations with highly volatile demand, because parameters can be set such that low upstream inventories are guaranteed. For example, for the synchronized base stock policies by setting an item base stock level close to the sum of the base stock levels of its successors, one creates almost permanent "shortages" from the successors' point of view, whereby all available item inventory is allocated among the successor items.

An intuitive explanation for this characteristic of optimal policies (within the concepts considered) is as follows. When allocating inventory capital among all items in the supply network a trade-off must be made between customer service and inventory capital cost. By definition customer service is realized only by availability at the CODPs. When balancing inventory capital between the CODPs and upstream item level one expects that, because of the portfolio effect, an increase of inventory capital at (common) upstream item levels enables a decrease of inventory capital at the CODP level. However such a decrease must not lead to a decrease of customer service below the target levels. The results presented above *apparently* show that the pull of inventory towards the CODPs to ensure high service is stronger than the pull of inventory towards upstream items to reap the benefits of the portfolio effect.

Stochastic control concept outperforms MP-based control concept

The results clearly show the superiority of the SBS concept over the LP-based rolling schedule concept. This is an observation with far reaching consequences. Currently, all commercially available SCP software is based on objective functions and constraints that are within the MP realm. Apparently a control policy based on optimization of a deterministic model, within which stochastic demand is only represented by average values, yields release decisions over time that are suboptimal. Again this is counterintuitive to most people not familiar with stochastic models.

The good news is that the SBS policies are remarkably simple from a computational point of view. Supposing the SBS concept can be extended to real-life situations, this may enable interactive SCOP. Such is currently infeasible due to the running times of MP-based software [cf. Schalla (2001)].

Another feature of the SBS policy is that it is based on a (to some extent artificial) decision hierarchy that provides insight into the SCOP problem that cannot be derived from an MP modeling approach. The hierarchy, expressed through the divergent systems of decision nodes associated with sets of components and end-items, gives guidelines for the different aggregation levels for which forecasts of future demand are needed. By choosing the appropriate aggregations of future demand over time it may be possible to exploit the (likely higher) accuracy of aggregate forecasts.

The consequence of dealing with uncertainty outside the MP models is that safety stocks and safety lead times are human input. The above discussion of (safety) stock positioning in the supply network shows that setting safety stocks without decision support of a stochastic model to capture the complex interactions between items is beyond a human's capability.

Clearly, the results obtained have been derived for stationary demand, only. In our view the stationary demand process is not the main reason for the results obtained. Both classes of policies can easily be modified to take into account non-stationary demand. In that case we must assume stationary forecast errors, which is no real restriction. The forecast revision process proposed in Graves et al. (1998) seems to be an interesting candidate for modeling non-stationary demand. Further research is needed to support our claim.

Hybrid approaches

The comparison has been restricted to uncapacitated problems. The reason for this is that no stochastic control concepts are available in the literature that apply to general capacitated networks. The LP-based rolling schedule can be applied to capacitated problems. The insights discussed above seem to indicate the need for the development of a hybrid approach. If the rationing mechanism in the SBS concept is one of the main reasons for its superiority, it seems logical to implement the linear rationing rules into the LP-based concept in the form of additional constraints. In addition the linear rationing mechanism could be applied in case of binding resource constraints, as well. These are topics for further research.

This concludes the analysis of SCOP concepts. In the next section we discuss topics for further research that we identified from this research.

7 Summary and issues for further research

In this chapter we have discussed various Supply Chain Operations Planning concepts defined in the literature. We provided a generic setting for the SCOP problem by defining decision variables and state variables. We motivated a generic formulation of the SCOP problem driven by a hierarchical planning approach. We argued that the concept of planned lead times is the building block for making a clear distinction between different planning levels that does justice to the modeling of information asymmetry. Information asymmetry is inevitable because related planning decisions are made at different moments in time and are made in different parts of an organization with a different view of the state of the system.

We have formulated sets of constraints for release of materials and resources that have been used as the basis for an assessment of these SCOP concepts. We identified two major classes of SCOP concepts that have been developed using two different modeling perspectives:

- Mathematical Programming models embedded in a rolling schedule approach
- Stochastic models that incorporate random demand

Based on our perspective of the SCOP problem we formulated a capacitated multi-item multi-stage LP model that lends itself for straightforward optimization using commercially available software. Extensions to the problem with lot sizing restrictions have been briefly discussed.

An extensive discussion of stochastic models identified that no literature is available on uncapacitated general supply networks, and then presented the various approaches for uncapacitated models. It turned out that in the literature mostly base stock control concepts have been proposed, either based on the installation stock concept or based on the echelon stock concept. We discussed the infeasibility of pure base stock policies for the SCOP problem for general supply networks. The reason for this is the violation of material availability constraints, i.e., pure base stock policies generate material orders without checking availability of upstream inventories. A synchronized base stock (SBS) control concept has been discussed that solves this problem at the expense of suboptimality. The suboptimality is caused by the integration of material allocation and material ordering decisions in the sense that availability of upstream item inventories is checked after taking allocation decisions of these upstream items among their successors in case of shortages. It has been shown that the SBS concept allows for an exact analysis of general supply networks and further permits the alignment of supply chain design and supply chain planning under the assumption of stationary demand.

The high level of complexity of the SCOP problem for general supply networks under random demand explains why currently available literature mostly proposes a heuristic analysis. We distinguished between heuristics that assume 100% upstream availability (decomposition assumption) and heuristics that assume that in case of lack of availability, there is exactly one item causing this (assembly assumption). We found that the decomposition assumption implies relatively high inventory levels at all echelons of the supply network, which is in contrast with the characteristics of near-optimal solutions for divergent networks that indicate that the inventory levels at upstream echelons should be low. We note here that the above statement does not apply to the analysis of Inderfurth and Minner (1998) and Graves and Willems (2000). This is because their analysis allows covering potential shortages at some stockpoint by downstream and upstream stockpoints. The heuristic developed by Ettl et al. (2000) that was based on the assembly assumption yields optimal solutions that violate the assembly assumption, because it finds solutions with low upstream fill rates. Interestingly, Ettl et al. (2000) report that the solutions found could not be improved by discrete event simulation based methods.

Based on a set of cases we compared the SBS concept with the LP-based concept. Surprisingly, the SBS concept outperformed the LP-based concept considerably. This indicates the importance of further study into stochastic models for the SCOP problem. Explanation for the superiority of the SBS concept was found in the way LP tends to prioritize items in case of shortages of upstream availability instead of rationing among the items that need this upstream availability. Also LP tends to keep too much inventory capital upstream, because a deterministic objective function identifies upstream stages as attractive due to lower cumulative added value.

Based on our discussion we have identified several areas for further research. We will subsequently discuss empirical validation of SCOP models, capacitated stochastic demand models for SCOP in general networks, incorporation of non-stationary demand into the SBS concept, comparison of SCOP concepts and integration of SCOP and shopfloor scheduling.

7.1 Empirical validation of SCOP models

Our discussion of the various SCOP models is strongly based on a mathematical perspective. This implies that we discussed the SCOP model definition and assessed the validity of certain assumptions that enabled us to simplify the analysis of the model. Wherever possible, we used a mathematically rigorous approach. If this was not possible we used discrete event simulation to assess validity. Although this is a scientifically sound approach, it fails to answer a fundamental question, which is: *Are the results of the model analysis in line with empirical data from the real-life SCOP problem?* In theory it may be that a mathematically sound analysis of an SCOP problem yields worse results than a heuristic analysis of the same problem. If such is the case this raises new and scientifically relevant research questions about modeling. In general we find that most of the research about SCOP in general supply networks lacks evidence that managerial insights obtained are supported by empirical validation. Even if successful cases are reported it is questionable whether there is a causal relationship between the use of specific models for SCOP and this success. We advocate the careful design of scientifically sound empirical studies, where such a causal relationship can be tested. The existence of ERP systems allows for the analysis of transactional data over time, from which we can obtain insight into the behavior of demand, inventories and lead time. Such research will be time consuming and difficult, due to the fact that the experimental setting cannot be fully controlled. Such research will be rewarding in terms of deeper insights into real-world SCOP problems and the contribution of quantitative modeling to its solution.

7.2 Incorporation of non-stationary demand

From experiences with real-life SCOP problems we conclude that in many situations we are faced with non-stationary demand. Examples are seasonality of demand, new product introductions and old product phase-outs. Graves et al. (1998) provide an interesting model for such non-stationarity. It assumes the capability of humans (possibly supported by software) to update forecasting information in such a way that the revisions of the forecast for a particular period into the future from one period to the next are i.i.d. Such models of non-stationary demand may still allow for a mathematically rigorous analysis of the SCOP problem.

If demands are correlated over time the question arises even for serial supply chains whether base stock policies are optimal and if so, whether the optimal policy can be found by solving the generalized Newsboy equations as derived in Diks and de Kok (1998). The issue of demand correlation over time is quite relevant, since most standard forecasting methods, including simple exponential smoothing, induce forecast errors that are correlated over time.

7.3 Capacitated models with stochastic demand

The above issue of empirical validation is quite relevant for the development of models for capacitated supply networks. The generalized Kanban control policies discussed in Section 5.8 have been proposed for the control of manufacturing systems. Typically resources are modeled as queues.

Hence the analysis of such policies integrates material control and resource usage. However, SCOP *plans* resource usage in order to smooth capacity requirements, while maintaining the due dates set by the planned lead times. This implies that a queuing network analysis based on continuous review and FCFS at resources does not represent properly the planning characteristics and periodic nature of SCOP.

Still, queuing network analysis may be a starting point for both the determination of planned lead times and a heuristic analysis of the capacitated SCOP problem under stochastic demand. The main idea behind this is that in most real-life situations capacity is hard to define. Processes can be speeded up if necessary; resources can be reallocated to provide more capacity to specific capacity requirements. This observation is one of the reasons to decompose the SCOP problem and the short-term scheduling problem. Thus capacitated SCOP models may be based on a similar decomposition of the control of material order releases and resource releases. This would result into a queuing network analysis of the resources that provides realistic planned lead times. The justification of this idea requires an experimental and empirical validation as discussed above.

We discussed non-stationarity of demand. Similarly, resource availability may be non-stationary due to preventive maintenance and holidays. The non-stationarity of resources does not fit a typical queuing (network) analysis. It seems to us that this problem is a white spot in the literature on stochastic models.

7.4 Comparison of SCOP concepts

The fact that there is a surge in the implementation of commercial software (such as Advanced Planning and Scheduling (APS) Systems) for SCOP motivates a scientific assessment of the various SCOP concepts implemented. Our experience shows that many people involved with the implementations of APS Systems do not really understand the mismatch between the optimality notions from deterministic optimization and the optimality notions from stochastic models. The discussion in Section 6 provides solid ground for a critical assessment of software based on LP, MIP, and rule-based optimization, since the common denominator of these approaches is the deterministic world view within a rolling schedule concept. In our opinion the results of our comparison surfaced some planning principles that allows for implementation in commercial software, yet further research is necessary to test the hybrid SCOP concepts suggested in Section 6.

We argued in this chapter that lead times need to be exogenous to the SCOP concept, implying that the system needs to take care of controlling lead times such that they are more or less fixed. In a supply chain context, it has not yet been researched whether this actually does provide better results than working with variable lead times.

Finally, we notice that none of the research conducted has been studying the SCOP concept performance under dynamic conditions. Dynamic conditions do not only refer to the non-stationarity of demand (or forecast error) and resource availability, but also to the dynamics in the planning process. In our exposition in Section 2, we have discussed the importance of anticipating future events. We know that estimates will never be completely correct. Further, we know from the research in the field of Systems Dynamics [e.g., Sterman (2000)] that even small differences between anticipated values of variables, perceived values of variables and actual values of variables may yield very unstable systems and uncontrolled planning situations. This dynamic behavior of the concepts discussed and proposed in this chapter, will need to be investigated.

Acknowledgements

This chapter would not have been possible without the support of many of our colleagues (and friends). First, we thank the Operations Planning and Control group at Eindhoven: Henk Akkermans, Will Bertrand, Rob Broekmeulen, Nico Dellaert, Karel van Donselaar, Simme Douwe Flapper, Geert-Jan van Houturn, Gudrun Kiesmüller, Henny van Ooijen, Graham Sharman and Vincent Wiers. Second, we are indebted to the PhD students who contributed all their intellectual capabilities during the 1990s to SCM research: Jos Verrijdt, Erik Diks, Jannet van Zante-De Fokkert, Fred Janssen, Wenny Raaymakers, Ebbe Negenman, Mark Euwe, Anastasia Schalla and Sanne Smits. Third, we would like to thank our German colleagues with whom we have debated while working on this chapter and who helped us to set our research agenda for the next five years: Bernhard Fleischmann, Herbert Meyr, Hartmut Stadtler and Horst Tempelmeier. And last, but not least, we would like to express our thanks to the imaginative minds we met in the industry for their invaluable support, that enabled us to implement our ideas so that we could gather empirical evidence of their validity: Wim Vrijland, Steve Martin, Fred Janssen (again!), Jan van Doremalen, Mynt Zijlstra, Han Langereis, Erik van Wachem, Jeroen Bordewijk, Dave Grandt, Alessandro de Luca and the members of the European Supply Chain Forum.

References

Agrawal, N, M. Cohen (2001). Optimal material control in an assembly system with component commonality. *Naval Research Logistics* 48, 408–429.
Anthony, R.N. (1965). Planning and control systems; *A framework for analysis,* Graduate School of Business Administration, Harvard University, Boston.
Aviv, Y. (2001). The effect of collaborative forecasting on supply chain performance. *Management Science* 47, 1326–1343.
Axsater, S. (2000). *Inventory Control,* Boston, Kluwer.

Axsater, S. (2003). Supply chain operations: Serial and distribution inventory systems, in: A. G. de Kok, S. C. Graves, (eds). *Handbooks in Oper. Res. And Management Sci*, Vol. 11, *Supply Chain Management: Design, Coordination and Operation*, Chapter 10, North-Holland Publishing Company, Amsterdam, The Netherlands.

Baker, K.R. (1993). Requirements planning, in: S.C. Graves, A.H.G. Rinnooy Kan, P.H. Zipkin (eds.), *Logistics of Production and Inventory*, Amsterdam, North-Holland Publishing Company Amsterdam, The Netherlands, pp. 571–628.

Barbarosoğlu, G., D. Özgür (1999). Hierarchical design of an integrated production and 2-echelon distribution system. *European Journal of Operational Research* 118, 464–484.

Belvaux, G., L.A. Wolsey (2001). Modeling practical lot-sizing problems as mixed integer programs. *Management Science* 47, 993–1007.

Bertrand, J.W.M. (2003). Supply Chain Design: Flexibility Considerations, in: A. G. de Kok S. C. Graves (eds.), *Handbooks in Oper. Res. And Management Sci*, Vol. 11, *Supply Chain Management: Design, Coordination and Operation*, Chapter 4, North-Holland Publishing Company, Amsterdam, The Netherlands.

Bertrand, J.W.M., J. C. Wortmann (1981). *Production control and information systems for component manufacturing shops*, Amsterdam, Elsevier.

Bertrand, J.W.M., J.C. Wortmann, J. Wijngaard (1990). *Production control: a structural and design oriented approach*, Amsterdam, Elsevier.

Billington, P.J., J.O. McClain, L.J. Thomas (1983). Mathematical programming approaches to capacity constrained MRP systems: review, formulation and problem reduction. *Management Science* 29, 1126–1141.

Bitran, G.R., D. Tirupati (1993). Hierarchical production planning, in: S.C. Graves, A.H.G. Rinnooy Kan, P.H. Zipkin (eds.), *Logistics of Production and Inventory*, Amsterdam, North-Holland, pp. 523–568.

Bitran, G.R., A.C. Hax (1977). On the design of hierarchical production planning systems. *Decision Sciences* 8(1), 28–55.

Buzacott, J.A. (1989). Queuing models of Kanban and MRP controlled production systems. *Engineering Costs and Production Economics* 17, 3–20.

Buzacott, J.A. J.G. Shantikumar, (1993). *Stochastic Models of Manufacturing Systems*, Prentice-Hall, Englewoods Cliffs, N.J.

Dauzere-Peres, S., J.B. Lasserre (1994). Integration of lotsizing and scheduling decisions in a job-shop. *European Journal of Operational Research* 75, 413–426.

De Kok, A.G. (1989). A moment-iteration method for approximating the waiting-time characteristics of the GI/G/1 queue. *Prob. Eng. and Inf. Sc.* 3, 273–287.

De Kok, A.G. (2000). Capacity allocation and outsourcing in a process industry. *International Journal Of Production Economics* 68, 229–239.

De Kok, A.G. (2001). Comparison of Supply Chain Planning concepts, Working Paper TUE/TM/LBS/101-03. Eindhoven: Technische Universiteit Eindhoven.

De Kok, A.G. (2002a). Ruin probabilities with compounding assets for discrete time finite horizon problems, independent period claim sizes and general premium structure, BETA Working Paper 82. Eindhoven: Technische Universiteit Eindhoven accepted for publication in: Insurance, Mathametics and Economics.

De Kok, A.G. (2002b). Evaluation And Optimization Of Strongly Ideal Assemble-To-Order Systems, in: Shanthikumar, J.G., Yao, D.D. and Zijm, W.H.M. (eds), Stochastic Modeling and Optimization of Manufacturing Systems and Supply chains, Kluwer, International series in Operations Research and Management Science, 66, 2003.

De Kok, A.G. and Seidel, H.P. (1990). Analysis of Stock Allocation in a 2-echelon Distribution System, Technical Report 098, Eindhoven: CQM.

De Kok, A.G., J.W.C.H. Visschers (1999). Analysis of assembly systems with service level constraints. *International Journal of Production Economics* 59, 313–326.

Dellaert, N.P., A.G. de Kok, W. Wang (2000). Push and pull strategies in multi-stage assembly systems. *Statistica Neerlandica* 54, 175–189.

Diks, E.B. (1997). Controlling Divergent Multi-echelon Systems, Ph.D. thesis, Eindhoven University of Technology, The Netherlands.
Diks, E.B., A.G. de Kok (1998). Optimal control of a divergent N-echelon inventory system. *European Journal of Operational Research* 111, 75–97.
Diks, E.B., A.G. de Kok (1999). Computational results for the control of a divergent N-echelon inventory system. *International Journal of Production Economics* 59, 327–336.
Diks, E.B., A.G. de Kok, A.G. Lagodimos (1996). Multi-echelon systems: A service measure perspective. *European Journal Operational Research* 95, 241–263.
Erenguc, S.S., N.C. Simpson, A.J. Vakharia (1999). Integrated production/distribution planning in supply chains: an invited review. *European Journal of Operational Research* 115, 219–236.
Eppen, G., L. Schrage (1981) Centralized ordering policies in a multi-warehouse system with lead times and random demand, in: L.B. Schwarz, (eds.), Multi-level Production-Inventory Control Systems: theory and Practice, North-Holland, Amsterdam.
Ettl, M., G.E. Feigin, G.Y. Lin, D.D. Yao (2000). A supply network model with base-stock control and service requirements. *Operations Research* 48, 216–232.
Feigin, G.E. (1998). Inventory planning in large assembly supply chains, in: S. Tayur, R. Ganeshan, M. Magazine (eds.), *Quantitative Methods for Supply Chain Management*, Boston, Kluwer Academic Publishers, pp. 760–788.
Fleischmann, B., H. Meyr (2003) Planning hierarchy, modeling, and advanced planning systems, in: A. G. de Kok, S. C. Graves, (eds.), *Handbooks in Oper. Res. And Management Sci,*. Vol. 11, *Supply Chain Management: Design, Coordination and Operation*, Chapter 9, North-Holland Publishing Company, Amsterdam, The Netherlands.
Fransoo, J.C., M.J.F. Wouters, A.G. de Kok (2001). Mufti-echelon multi-company inventory planning with limited information exchange. *Journal of the Operational Research Society* 52, 830–838.
Frein, Y., M. Di Mascola, Y. Dallery (1995). On the design of generalized kanban control systems. *International Journal of Operations and Production Management* 15(9), 158–184.
Gershwin, S.B. (1994). *Manufacturing Systems Engineering*, Prentice Hall, Englewood Cliffs.
Graves, S.C., S.P. Willems (2000). Optimizing strategic safety stock placement in supply chains. *Manufacturing and Service Operations Management* 2, 68–83.
Graves, S.C., Willems, S.P. (2003). Supply Chain Design – safety stock placement, inventory hedges for buffering against demand and supply uncertainty, in: A. G. de Kok, S. C. Graves, (eds.), *Handbooks in Oper Res. And Management Sci.*, Vol. 11, *Supply Chain Management: Design, Coordination and Operation*, Chapter 3, North-Holland Publishing Company, Amsterdam, The Netherlands.
Graves, S.C., D.B. Kletter, W.B. Hetzel (1998). A dynamic model for requirements planning with application to supply chain optimization. *Operations Research* 46, S35–S49.
Graves, S.C., H.C. Meal, S. Dasu, Y. Qui (1986). Two-stage production planning in a dynamic environment, in: S. Axsater, Ch. Schneeweiss, E. Silver (eds.), *Multi-stage production planning and inventory control*, Berlin, Springer, pp. 9–43.
Hausman, W.H., H.L. Lee, A.X. Zhang (1998). Order response time reliability in multi-item inventory systems. *European Journal of Operational Research* 109, 646–659.
Hax, A.C., H.C. Meal (1975). Hierarchical integration of production planning and scheduling, in: M.A. Geisler (ed.), *Logistics*, Amsterdam, North Holland Publishing Company, Amsterdam, The Netherlands pp. 53–69.
Hillier, M.S. (2000). Component commonality in multiple-period assemble-to-order systems. *IIE Transactions* 32, 755–766.
(1991S. Hoekstra, J.H.J.M. Romme (eds.), *Integral logistic structures: developing customer-oriented goods flow*, London, McGraw-Hill.
Holt, C.C., F. Modigliani, J.F. Muth, H.A. Simon (1960). *Planning, Production, Inventories and Workforce*, Prentice Hall, Englewood Cliffs.
Hopp, W., M. Spearman (2000). *Factory Physics*, 2nd ed., Irwin McGraw-Hil Bostonl.
Inderfurth, K. (1994). Safety stocks in multistage divergent inventory systems: a survey. *International Journal of Production Economics* 35, 321–329.

Inderfurth, K., S. Minner (1998). Safety stocks in mufti-stage inventory systems under different service measures. *European Journal of Operational Research* 106, 57–73.
Janssen, F.B.L.S.P. (1998). Inventory Management Systems, unpublished PhD. Thesis, Tilburg University, Tilburg.
Kanet, J.J., S.V. Sridharan (1998). The value of using scheduling information in planning material requirements. *Decision Sciences* 29(2), 479–497.
Køhler-Gudum, C.K., and A.G. De Kok (2002). A safety stock adjustment procedure to enable target service levels in simulation of generic inventory systems, BETA Working Paper 71. Eindhoven: Technische Universiteit Eindhoven.
Lagodimos, A.G. (1992). Multi-echelon service models for inventory systems under different rationing policies. *International Journal of Production Research* 30, 939–958.
Lagodimos, A.G. (1993). Models for evaluating the performance of serial and assembly MRP systems. *European Journal of Operational Research* 68, 49–68.
Langenhof, L.J.G., W.H.M. Zijm (1990). An analytical theory of multi-echelon production/distribution systems. *Statistica Neerlandica* 44, 149–174.
Lawrence, S.R. (1997). Heuristic, optimal, static and dynamic schedules when processing times are uncertain. *Journal of Operations Management* 15(1), 71–82.
Lee, H.L., C. S. Tang (1996). Modelling the costs and benefits of delayed product differentiation. *Management Science* 43, 40–53.
Lee, H.L., J.M. Swaminathan (2003). Design for postponement, A.G. de Kok, S. C. Graves (eds.),. *Handbooks in Oper. Res And Management Sci.*, Vol. 11, *Supply Chain Management: Design, Coordination and Operation*, Chapter 5, North-Holland Publishing Company, Amsterdam, The Netherlands.
Magee, J.F. (1958). *Production Planning and Inventory Control*, New York, McGraw-Hill.
McKay, K.N., F.R. Safayeni, J.A. Buzacott (1995). A review of hierarchical production planning and its applicability for modern manufacturing. *Production Planning & Control* 6(5), 384–394.
McPherson, R.F., White, K.P., Jr. (1994). Management control and the manufacturing hierarchy: Managing integrated manufacturing organizations. *International Journal of Human Factors in Manufacturing* 4(2), 121–144.
Meal, H.C. (1984). Putting production decisions where they belong. *Harvard Business Review* 62(2), 102–111.
Meal, H.C., M.H. Wachter, D.C. Whybark (1987). Material requirements planning in hierarchical production planning systems. *International Journal of Production Research* 25(7), 947–956.
Miller, T. (2001). *Hierarchical Operations and Supply Chain Planning*, London, Springer.
Minner, S. (2000). *Strategic Safety Stocks in Supply Chains*, Berlin, Springer.
Orlicky, J.A. (1975). *Material Requirements Planning*, New York, McGraw-Hill.
Özdamar, L., G. Barbarosoglu (2000). An integrated Lagrangean relaxation-simulated annealing approach to the multi-level multi-item capacitated lot sizing problem. *International Journal of Production Economics* 68, 319–331.
Rosenblatt, M.J., A. Eynan (1996). Component commonality effects on inventory costs. *IIE Transactions* 28, 93–104.
Rosling, K. (1989). Optimal inventory policies for assembly systems under random demands. *Operations Research* 37, 565–579.
Schalla, A.J., J.C. Fransoo, and A.G. de Kok (2001). Hierarchical anticipation in Advanced Planning and Scheduling Systems. Working Paper TUE/TM/LBS/01-02. Eindhoven: Technische Universiteit Eindhoven.
Schneeweiss, C. (1999). *Hierarchies in Distributed Decision Making*, Berlin, Springer.
Shapiro. J.F. (1993). Mathematical Programming Models and Methods for Production Planning and Scheduling, in: Graves, S.C., A.H.G. Rinnooy Kan, and P.H. Zipkin (eds.), *Logistics of Production and Inventory*. Amsterdam: North-Holland, 371–444.
Sherbrooke, C. C. (1992). *Optimal Inventory Modeling of Systems, New Dimensions in Engineering*, New York, Wiley.

Silver, E.A., D.F. Pyke, R. Peterson (1998). *Inventory Management and Production Planning and Scheduling*, New York, Wiley.

Simpson, K.F. (1958). In-process inventories. *Operations Research* 6, 863–873.

Song, J.S. (1998). On the order fill rate in a multi-item, base-stock inventory system. *Operations Research* 46, 831–845.

Song, J.S. and P.H. Zipkin (2003). Supply Chain Operations: Assemble-to-Order Systems. A. G. de Kok, S. C. Graves, (eds.) *Handbooks in Oper. Res. And Management Sci.* Vol. 11, *Supply Chain Management: Design, Coordination and Operation*, Ch. 11. North-Holland Publishing Company, Amsterdam, The Netherlands.

Spengler, Th., H. Puchert, T. Penkuhn, O. Rentz (1997). Environmental integrated production and recycling management. *European Journal of Operational Research* 97, 308–326.

(2000H. Stadtler, C. Kilger (eds.), *Supply Chain Management and Advanced Planning: Concepts, Models, Software and Case Studies*, Berlin, Springer.

Sterman, J.D. (2000). *Business Dynamics: Systems Thinking and Modeling for a Complex World*, Boston, Irwin McGraw-Hill.

Suri, R., J.L. Sanders and M. Kamath (1993). Performance evaluation of production networks, in: Graves, S.C., A.H.G. Rinnooy Kan, and P.H. Zipkin (eds.), *Logistics of Production and Inventory*. Amsterdam: North-Holland, 199–286.

Tardiff, V. (1995). Detecting Scheduling Infeasibilities in Multi-Stage Finite Capacity Production Environments. Unpublished PhD Dissertation, Evanston: Northwestern University.

Tayur, S.R. (1993). Computing the optimal policy for capacitated inventory models. *Communications in Statistics-Stochastic Models* 9, 585–598.

Van der Heijden, M.C. (1997). Supply rationing in multi-echelon divergent systems. *European Journal of Operational Research* 101, 532–549.

Van der Heijden, M.C., E.B. Diks, A.G. de Kok (1997). Stock allocation in general multi-echelon distribution systems with (RS) order-up-to-policies. *International Journal of Production Economics* 49, 157–174.

Van Houtum, G.J., W.H.M. Zijm (1991). Computational procedures for stochastic multi-echelon production systems. *International Journal of Production Economics* 23, 223–237.

Van Houtum, G.J., W.H.M. Zijm (2000). On the relation between cost and service models for general inventory systems. *Statistica Neerlandica* 54, 127–147.

Van Ooijen, H.P.G. (1991). Controlling different flow rates in job-shop like production departments. *International Journal of Production Economics* 23, 239–249.

Whybark, D.C., S. Yang (1996). Positioning inventory in distribution systems. *International Journal of Production Economics* 45, 271–278.

Wiendahl H.-P.(1987). Belastungsorientierte Fertigigungssteuerung Grundlagen, Verfahrensaufbau, Realisierung. Muenchen: Hanser (in German).

Wiendahl, H.-P. (1995). *Load-Oriented Manufacturing Control*, Berlin, Springer.

Winter, R. (1989). Der Ansatz des Massachusetts Institute of Technology zur Mehrstufigen Produktionsplanung. Arbeitsbericht 89-01. Frankfurt: Institut für Wirtschaftsinformatik, Johann Wolfgang Goethe Universitat (in German).

Yano, C., H. Lee (1995). Lot sizing with random yields: a review. *Operations Research* 43, 311–334.

Chapter 13

Dynamic Models of Transportation Operations

Warren B. Powell
Department of Operations Research and Financial Engineering, Princeton University,
Princeton, NJ 08544, USA

A manufacturing supply chain can be viewed as a sequence of steps consisting of the modification of a resource at a point (manufacturing) followed by the transfer of the product over space (transportation). Transportation arises because of the spatial distribution of resources, skill sets, and customers. The challenge we face is completing this component of the supply chain efficiently, reliably, and in the case of common carriers, profitably.

It is useful to contrast 'transportation planning' as it is practiced in the context of moving people versus freight. Airlines, passenger trains, and bus companies typically run fixed schedules over fixed routes that are planned months, if not a year, in advance. People are typically able to adjust their travel plans around a fixed schedule, and it is extremely important that the provision of the transportation service be almost perfectly predictable. By contrast, freight operations are highly dynamic, responding to the demands of the market place and the production processes that serve this market. This is not to say that planning problems are not important. Freight companies have to plan the location of terminals, and they will plan operations to a degree, although these tend to be modified on a day-to-day basis.

Our presentation focuses on the issues that arise in the dynamics of real-time operations. We do this in part because dynamic information processes are a key characteristic of freight transportation systems, and also because the literature on static models is relatively much more mature. For a recent and thorough review of planning models for freight transportation and logistics, an excellent reference is Crainic and Laporte (1997). Other important references include Bodin, Golden, Assad and Ball (1983), Fisher (1995), and Desrosiers, Solomon and Soumis (1995) for vehicle routing; Haghani (1989), Glickman and Sherali (1985), and Crainic, Ferland and Rousseau (1984) for rail transportation; Brown, Graves and Ronen (1987) for ocean transportation; and Crainic and

Roy (1992) and Powell (1986a) for less-than-truckload trucking. General discussions of modeling freight transportation systems can be found in Crainic and Roy (1988) and Crainic and Rousseau (1988).

Three key classes of decisions control transportation companies: physical (how to move the product), financial (how to price it), and informational (what information should be provided to manage the system). Of course, the greatest complexity in transportation and logistics is the complexity of the physical processes, which as a result occupies most of our attention. We can use these three dimensions to briefly summarize the characteristics of transportation that make it hard:

(1) Physical: The objects that we are managing:

- Reusable resources: Classical models of the transportation function, when done from the perspective of the shipper, simply have a cost for moving product from one location to another. From the perspective of the transportation company, this activity is done with reusable resources: drivers, tractors, and trailers, for example. Thus, serving a customer request (to move freight from one location to the next) has the effect of changing the state of the system.
- Resource layering: Serving a customer request may require one or more resource classes, which are combined to get the job done. For example, moving a truckload of freight requires a driver, tractor, and trailer. Combining different resource classes is called *layering* and it has the effect of creating complex interactions between resource classes.

(2) Financial: In this dimension, we focus purely on pricing:

- Contract pricing: Given the challenges of the physical process, it is necessary to price a transportation service correctly. The pricing of transportation services are complicated by network effects (sharing resources among different markets), consolidation (sharing space on the same vehicle), and the practice of paying only for the service received, while expecting the resources to be available on demand.
- Static pricing: These would be standard prices a carrier would use for moving freight between a pair of regions (sometimes called traffic lanes). These are market rates (i.e., they are not specific to a contract) that are set in advance. These are generally the highest rates a carrier will quote.
- Spot pricing: In some cases, a customer is willing to pay for a service when requested. A carrier has to be able to quote the right price for this request. Spot pricing needs to account for the state of the system, as it now exists, and the impact the activity will have on the system (the cost of the decision).

(3) Informational: A critical dimension of a modern transportation system is the flow of information.

- Customer demands: Customers place demands on the system randomly over time, with varying degrees of advance information.
- Resource availability: The availability of people (drivers and crews), complex equipment (locomotives, aircraft, and even tractors), and containers (trailers, box-cars, and intermodal containers) is often governed, in part, by exogenous factors.
- Spatially distributed information: Often (although this is changing in today's information age) there is a lot of information about the system that is not available centrally. As a result, many decisions are made locally based, by and large, on the 'head knowledge' of local or regional managers and dispatchers.

In a short chapter such as this, it is not possible to discuss all the different variations of transportation operations, or to discuss the most interesting variations in anything approaching completeness. In the face of such richness, the question arises: how do we discuss such a broad problem class without resorting to a series of anecdotes? Our response is to focus on fundamentals, with enough examples and illustrations so that the reader can tackle variations that we are not able to cover.

The modeling of dynamic systems has a long history, and yet in many ways remains an extremely young field. The earliest dynamic models in freight transportation addressed the issue of managing fleets of containers for rail or ocean operations [Leddon & Wrathall (1967), White (1972), Misra (1972), Herren (1977), Turnquist (1986), Mendiratta & Turnquist (1982), and Crainic, Gendreau and Dejax (1993)]. These earliest models captured the time staging of physical activities, but not the time staging of information (in other words, they were deterministic models). The first explicit stochastic model of the car distribution problem for rail is presented in Jordan and Turnquist (1983), which assumed a) that a car that was moved empty once could not be moved empty again and b) a car assigned to a demand did not reappear. This line of research was continued in the context of truckload trucking in Powell (1986b, 1987, 1996), Frantzeskakis and Powell (1990), and Cheung and Powell (1996). A significant breakthrough came with the introduction of adaptive estimation techniques. Powell and Carvalho (1998) introduced the use of linear functional approximations to capture the impact of decisions made now on the future. These techniques then led to the use of nonlinear functional approximations, which, while somewhat more difficult to use, produce much higher solution quality [see Godfrey & Powell (2002a,b)], as well as more stable solutions.

One of the oldest problems in transportation and logistics is the vehicle routing problem. The dynamic version of this problem has been recognized for many years [see, for example, Wilson (1969)] but received little attention in the research literature [some early references include Stein (1978) and Jaw,

Odoni, Psaraftis and Wilson (1986)]. Psaraftis (1988) is an important early reference, which discussed some of the issues arising in dynamic routing [for an update of this discussion, see Psaraftis (1995)]. The large majority of the literature on dynamic vehicle routing, as of this writing, focuses on simulating myopic heuristics, and the computational issues that arise in this setting [Gendreau, Guertin, Potvin & Taillard, 1999 and Regan, Mahmassani & Jaillet, 1998]. Literature has emerged on the so-called stochastic vehicle routing problem, which is really a static vehicle routing problem, where the tours have to be designed to anticipate 'route failures' which arise when the vehicle picks up more goods than it can hold, and has to return to the depot to empty out before resuming its tour [see Stewart & Golden (1983), Laporte & Louveaux (1990), Dror (1993), and Dror, Laporte and Trudeau (1989)].

Research into routing and scheduling algorithms, which explicitly capture the impact of the future on decisions made now, is extremely young. A fairly complete review of this literature prior to 1995 is contained in Powell, Jaillet and Odoni (1995), which includes a review of the literature of probabilistic vehicle routing and stochastic fleet management. Powell (1996) appears to be the first paper to formulate and solve a dynamic routing and scheduling problem that uses an explicit stochastic model of future events. The problem involved the matching of drivers to loads for truckload motor carriers, which is considerably simpler than problems involving multiple pickups and deliveries. Secomandi (2000, 2001) considers the case of routing a single vehicle dynamically through time using neuro-dynamic programming techniques [see Bertsekas & Tsitsiklis (1996)]. The single vehicle case avoids the explosive growth in the size of the state space that even neuro-dynamic programming methods are sensitive to.

The most significant advances in the modeling of problems in transportation and logistics in the presence of dynamic information processes have been made in the arena of fleet management (single and multicommodity flow problems). This work has led to a general approach for using approximate dynamic programming methods for solving resource allocation problems. One of the most significant technical challenges that arise in the use of these techniques for dynamic resource management problems is the size of the state space describing the attributes of a single resource. Both single and multicommodity flow problems have relatively small attribute spaces. Flow problems involving more complex resources (people, locomotives, ships) can be modeled as *heterogeneous resource allocation problems* [Powell, Shapiro, & Simão, 2000a] which typically involve attribute spaces that are too large to enumerate. Even harder are multistop pickup and delivery problems, which not only exhibit a large state space but are also characterized by a difficult mixture of known and unknown information (the easiest problems are those where we know everything or nothing; it is the ones in between that are the hardest).

Given the breadth and complexity of problems that we are trying to address, this chapter is going to focus on the following goals:

(a) It provides an overview of the different types of problems arising in transportation and logistics, focusing on operational problems where dynamic information processes play a significant role. In contrast to other presentations, these problems are addressed from three perspectives, which we refer to as the physical, financial, and informational views. The physical view focuses on the objects being managed; the financial view focuses on pricing; and the informational view discusses challenges from the perspective of designing information architectures that can be used to run an operational system.

(b) We provide a notational framework that is extremely general, encompassing issues such as multiattribute resources, resource layering, complex system dynamics, and the organization and flow of information and decisions. This framework allows us to tackle a broad range of problems in transportation and logistics, without having to introduce new notational systems for each new problem class.

(c) We introduce four major classes of information that may be used to solve a dynamic problem, and describe the types of algorithms that arise from using different classes of information. These classes encompass all the major algorithmic strategies in use today, but include some new ideas that are not commonly used.

(d) A relatively new class of approximation strategies is outlined based on dynamic programming. These strategies allow us to design practical algorithms that are more than just myopic or rolling horizon procedures.

(e) A series of basic problems are described using the notational framework which illustrates, using problems of increasing complexity, how dynamic problems in transportation and logistics can be solved.

We begin our presentation in Section 1 with a discussion of operational problems that arise in a range of industries that perform transportation functions. This review summarizes the key issues, helping to set the stage for the formulation of models. Then, Section 2 presents a general modeling framework for dynamic problems, giving us a modeling vocabulary with which we can address a range of problem classes. Section 3 reviews general algorithmic strategies that arise in the context of dynamic systems, focusing in particular on the modeling of both the physical and informational dimensions of the problem. Section 4 then presents specific models for some of the major problem classes. Section 5 provides brief remarks on the issue of data quality when implementing operational models, and Section 6 makes some closing remarks.

Due to space limitations, our mathematical modeling focuses on representing physical processes in the presence of dynamic information processes. We consider both single and multiagent control structures, thereby capturing both the organization as well as the flow of information. Space constraints prevent a treatment of pricing problems [see Muriel & Simchi-Levi (to appear) for a treatment of pricing from a shipper perspective], which remains a surprisingly young field in freight transportation. Even less mature is the explicit modeling of the information infrastructure, which is the true means by which most systems are controlled. By explicitly modeling both pricing and information availability in our representation of the problems, we hope to set the stage for these emerging dimensions of research.

1 Operational challenges in transportation

Each of the subsections below addresses a different industry segment that serves the transportation function. These industries form in response to the characteristics of the market that each is serving. These characteristics include:

- Consolidation: Markets can be divided between small package (less than 150 pounds), less-than-truckload (150 to 10,000 pounds), full truckload, car- or container-load, and bulk (requiring many carloads or tankers).
- Distance: Delivery from a regional warehouse to local customers represents the shortest distances, which is work that is typically handled by pickup or delivery fleets. Medium distances might be 100 to 750 miles (approximately 150 to 1200 km), which might be handled by regional LTL or short-haul truckload carriers. Long distances include moves over 750 miles (1200 km) within the continent, or intercontinental movements.
- Control: Private fleets are owned and operated by the customer. Common carriers represent outsiders. In railroads, freight cars may be owned either by the railroad or the customer. Ownership primarily arises when service is an issue, but the opportunity to consolidate is also a major factor. A company will only want to own its own trucks when it feels that it can use them effectively. Also, the ability to place advertising on the side of a company-owned truck is a factor.
- Cost: Commodity products require the lowest possible price; higher margin products can absorb higher transportation prices for better service. The same truckload carrier will charge different prices to different customers in the same market for the same service, reflecting the nature of the product being moved. Private fleets will be used to provide more customized service but with lower utilization. This service can only be justified for products that command the margins to cover the cost.

- Service: Service is typically measured as a function of speed and reliability. Of course, this ignores the many other dimensions of service that a transportation company can provide (packaging, setup, tracking, and billing). Everyone wants fast, reliable service, but not everyone is able to pay for it.

Each industry reviewed below services customers that can be characterized along at least some of the dimensions listed above. Most of the industries have specialized companies that further segment the market. Thus, railroads and waterways dominate low cost, bulk commodities, but compete aggressively (with mixed success) with trucking companies for merchandise freight.

Our discussion of different service types is organized in a very specific way. After giving a brief overview of the industry, we review the resource classes and the decision classes as a way of giving a feeling of what is being managed and how we are managing it. We focus on *active* resource classes, representing the resources we are actively managing. Given our emphasis on dynamic systems, our resources tend to consist of people and equipment over fixed facilities (which are dynamic over longer horizons). Our definition of a resource [taken from Powell, Shapiro & Simão, 2001] is a general one, and includes, as a 'resource class' the customer demand itself. This may not seem customary, but as we evolve to more complex operations, we have to manage the customer's order just as we would manage the 'resources' (such as drivers, tractors, and locomotives) that belong to a carrier.

We then summarize decisions, organized into three key classes: (1) physical (decisions that act on physical resources), (2) financial (pricing and incentives), and (3) informational (decisions that determine the availability and flow of information). Each of these classes can be organized into two types: dynamic (which depend on the physical state of the system) and static (which do not). Our presentation highlights the importance of all three dimensions, as opposed to more classical presentations that focus primarily on the management of physical resources.

1.1 Truckload trucking

On the surface, truckload trucking sounds deceptively simple. A customer requests an entire trailer to move freight from one location to another. He may call in the request for pickup the same day, but most requests are made between one and three days in advance (longer when a weekend is involved). The trucking company has to decide what driver will pick up the load, and when. Once the load is picked up, the driver may take the load directly to the destination, or drop it off at an intermediate relay so that another driver can complete the delivery. There are over 10,000 companies consisting of a single truck, and several companies consisting of over 10,000 trucks.

1.1.1 Resource classes

There are four primary resource classes in truckload trucking: the driver, the tractor, the trailer, and the load itself. Issues associated with each resource class include:

- Drivers: The choice of the driver to cover a load has to consider factors such as the destination of the load and the home domicile of the driver. The load may have to cross borders into Canada or Mexico, and not all drivers have experience doing this. Or, the load may require the use of a sleeper team to reach the destination in time to make the delivery appointment. Drivers are typically on the road for two or more weeks at a stretch, so getting drivers home is a major challenge for truckload carriers in the presence of the highly random demands they have to serve.
- Tractors: Tractors need refueling, routine maintenance and, from time to time, major maintenance at a maintenance facility. As these major maintenance intervals arise, it can be necessary to route the tractor toward such a facility.
- Trailers: Trailers can be vans (boxes) or flatbeds. Vans may be refrigerated or 'dry.' Most freight moves in dry vans, which may be 45, 48, or 53 feet in length (48 feet is the most common). Trailers are typically called 'semi-trailers' because there is a set of wheels on only one end of the trailer (since the tractor holds the other half of the trailer).
- Loads: The basic customer request is to move a load of freight from an origin to a destination, with specified constraints for pickup and delivery. A load may allow very little time between pickup and delivery, possibly requiring the use of sleeper teams (which can drive continuously). At the other extreme are loads that allow so much time between pickup and delivery that it is necessary to park the trailer for several days to avoid arriving before the delivery appointment. In some cases, the request may involve making a sequence of stops to deliver portions of the load (less frequently, the request may require making a sequence of pickups with a single destination).

Other resource classes include fuel and maintenance resources.

1.1.2 Decisions

The decisions that govern a truckload carrier include:

(1) Physical: It is useful to roughly divide decisions impacting physical resources between operational decisions which impact operations on a

day to day basis, and planning decisions which capture design decisions which affect operations over longer periods of time.

(a) Operational:

- Load acceptance: When the shipper calls, should the carrier accept the responsibility to move the load?
- Driver assignment: What driver should be assigned to pick up a load?
- Load routing: Should the load be moved directly to the destination, or should it be relayed at an intermediate point? If so, where, and what driver should then move the load for its final leg. Intermediate relays allow different drivers to perform the original pickup versus the final delivery.
- Trailer pool management: It is necessary to manage pools of trailers, sometimes at specific shippers, so that they have access to trailer capacity when it is needed. One of the challenges of moving loads is that it is also necessary to shuttle trailers into and out of pools. Idle drivers waiting for an assignment normally handle these activities.

(b) Planning:

- Fleet size and mix.
- Number of drivers and their home domicile.
- Terminals: Truckload carriers will use terminals for maintenance and storage of tractors and trailers. It is necessary to determine how many terminals to have and their size and location.
- Customers: What customers should a carrier serve, and what commitments (e.g., in terms of number of loads) should the carrier make to the shipper? A carrier might commit to move loads for a shipper in a particular traffic lane (origin/destination pair).

(2) Financial: For our applications, 'financial' decisions focus on pricing, as opposed to other classes of financial decisions such as borrowing and investments.

- Contract pricing: What price (usually specified as a cost per mile or kilometer that a trailer has to be moved) should be charged for freight in each lane? In a typical contract, a shipper will estimate how much freight will move in each lane, but the shipper is not held to these estimates. Prices depend on the lane because of imbalances in the level of freight.
- Spot pricing: A shipper may offer a particular load at a spot price. A carrier has to decide whether to accept the load at that price at that time.

(3) Informational: Here we are focusing on decisions to acquire information by investing in specific information technologies.

- Management information systems: Most truckload carriers start with a single truck driver. As the company grows, it has to make the transition from slips of paper and a notebook to computerized systems of increasing sophistication. Several vendors market MIS systems, but most companies use these as starting points and then customize. The choice of MIS system usually involves both hardware and software.
- Communications: A major decision faced by truckload carriers is whether to invest in onboard communications, providing two-way (data and possibly voice) communication with the driver and his unit. Other forms of communication allow tracking the status of the tractor and the trailer.
- Real-time communication: Communicating costs money, so a decision has to be made whether to communicate with the driver at any given time.
- Driver assignment models: These have been available for over a decade, but very few companies use them. At the same time, a handful of companies have seen dramatic successes with real-time driver assignment packages. The adoption of this technology is a major decision today.
- Demand management systems: Forecasting demand and determining which freight to book is a key decision for truckload motor carriers.

1.2 Private fleet operations for collection and distribution

The vast majority of private fleets are primarily for local distribution (and sometimes collection), although some shippers will use their own trucks to handle movements between facilities. Private fleets are most commonly used for local distribution, since this component of the process offers the fewest opportunities for joint use with other customers and also offers the highest possible exposure to customers (hence the advertising on the side of the truck). Private fleets are used when the volume of deliveries to a regional area is high enough to use the fleet effectively. When this is not the case, companies typically fall back on LTL and small package carriers.

The most basic operation faced by the private fleet for local delivery is loading up at a central terminal or warehouse, and then delivering to a group of customers. These operations typically work on a daily cycle (tied to business hours). Tours may be fairly regular (particularly when delivering high volume products to retail outlets) or highly dynamic, as would occur when delivering custom orders.

1.2.1 Resource classes

There are different ways to model basic pickup and delivery problems. The most classical view is that of the vehicle routing problem where the resource being managed is a 'vehicle,' which is understood to consist of a driver and a truck (which may itself consist of a tractor and a trailer). For this 'simple' problem, we would manage:

- Vehicles: This is the principle active resource. Vehicles may be homogeneous or heterogeneous, but generally do not reflect the characteristics of individual drivers.
- Customer demands: The product to be picked up or delivered.

At the other extreme is the situation faced in the delivery of cryogenic chemicals. These companies must deliver product to tanks before they run out. A customer may require one or two deliveries per month, or several deliveries per day. It is often the responsibility of the company to estimate when a delivery is needed (an instance of vendor managed inventory). For this 'complex' problem, the resource classes include:

- Driver: A driver may be characterized by home domicile, total driving time in a day, experience, days away from home, and language skills.
- Tractor: There are two types of tractors (e.g., a longer tractor with a double axle set in the rear, and a shorter one with a single axle set), and they also have maintenance requirements.
- Trailer: A trailer can hold a certain type of chemical. Also, there are different sizes of trailers, and they also have the attribute of how full they are.
- Product: There are several types of product, and it may be necessary for the truck to go from one terminal to another in order to pick up product.
- Customer tank: The customer tank is a reusable resource just like a driver, tractor, or trailer. Delivering product to a tank simply changes its characteristics (the inventory level), which determines when it must be refilled again (which may be as little as a few hours into the future, or several weeks).

1.2.2 Decisions

The decisions that govern private fleet operations include:

(1) Physical:

  (a) Operational:
  - Consolidation: What customer orders should be consolidated into a particular truck?
  - Driver assignment: What driver should handle a particular delivery tour? These decisions may be static or dynamic.

(b) Planning:
- Distribution facilities: Size and location.
- Fleet size and type.
- Delivery zones – In some operations, a particular driver will cover deliveries in a particular region.
- Customer commitments – These decisions determine which customers the carrier commits to serve over the course of a year.

(2) Financial:
- Contract pricing: What price should a company charge for a pickup or delivery? The price may vary as a function of the location (which will capture the distance from the terminal) as well as the size and weight.
- Zone pricing – Orders that are not served under a contract are typically charged a price based on shipment characteristics (size and weight) and the geographic zone.

(3) Informational:
- Communication: Should the company use radio technologies to communicate real-time with the driver? Should bar code scanning systems be used? In the case of pickups, should the company collect information about the pickup centrally when the original call is made?
- Databases: Many operations still work with sheets of paper and people. The transition to a computer in this segment remains a key decision.
- Decision support systems: GIS systems, map databases and vehicle routing algorithms are rapidly maturing, but remain imperfect. The decision to make the transition to an automated system is a major one today.

1.3 Less-than-truckload (LTL) trucking

Less-than-truckload trucking moves shipments that are typically between 150 and 10,000 pounds. The shipments can vary widely in terms of density and shape (which affects the ability to stack shipments). In the United States, shipments are typically loaded on 28-foot trailers or 48-foot vans. Most of the time, a single driver will pull a single 48-foot van or two 28-foot trailers, but 'triples' are allowed by some states on selected portions of the interstate highway network. The 28-foot trailers are popular partly because they allow a driver to pull 56 feet of trailers, but also because the LTL carrier will often load the trailers with freight to different destinations. Less-than-truckload carriers struggle to fill trailers to some locations. It is easier to fill a 28-foot trailer to some destinations than a 48-foot trailer.

A single tractor pulling two 28-foot trailers will be pulling, on average, between 30,000 and 35,000 pounds (around 14,000–16,000 kg). An LTL shipment averages about 1,000 pounds (about 450 kg), so a driver will be moving 30–35 shipments in a single move. Achieving this consolidation requires a tremendous amount of infrastructure. An LTL carrier has to have local pickup and delivery operations, and a network of terminals to handle the consolidation of freight.

Less-than-truckload carriers can be roughly divided into two broad classes. The regional carriers move shipments up to 1,000 miles, typically with overnight or two-day service. These carriers must deliver this service with very high reliability, and as a result they will often have to move a trailer that is not full just to maintain service. The long-haul carriers focus on longer lengths of haul (although as this chapter is being written, the borders between regional and long-haul carriers is blurring), serving markets that are typically between two and five days (international movements can take longer). Although service reliability is quite high, these carriers move freight that is especially price sensitive, and as a result they have to focus on maximizing load average (the number of pounds on each trailer), minimizing the total miles traveled, and minimizing the number of times a freight bill is transferred. They are forced, then, to take advantage of day-to-day variations to fill trailers to different destinations as opportunities arise. In addition, they will not move a trailer, that is, say, less than a third full just to make service. However, the carriers have become increasingly sophisticated in their ability to identify which shipments actually require high service.

The typical path of an LTL shipment starts at the shipper's dock where the carrier will pick up the shipment with a pickup and delivery truck. These trucks make most of their deliveries at the beginning of the day, and then focus on pickups. At the end of the day, these trucks come into an end of line terminal, where the freight is usually (but not always) unloaded onto line haul trucks, which handle the movement of freight between the various terminals. In a regional carrier, this truck might then take the freight directly to the destination end of line for delivery the next day, or it may be transferred through a single break bulk or distribution center. For a long haul carrier, the standard path is to first take the shipment into an origin break bulk, where it is transferred onto a trailer that takes it to the destination break bulk. There it is transferred a second time before the final segment to the destination end of line. In the past, some carriers followed a strict policy of forcing all shipments through two break bulks, producing a *transfer ratio* of 200 (meaning that shipments were transferred, or handled, on average twice). However, the best-run national LTL carriers actually achieve transfer ratios below 100 (that is, shipments are transferred on average less than once). This ratio arises because break bulks are typically located near major cities, which means that a large number of shipments originate at one break bulk and terminate at another (producing zero transfers). Transfer ratios of 100 means that there are still quite a few shipments moving through two break bulks.

The largest LTL carriers in the United States are primarily unionized, which has the effect of imposing a variety of rules on how drivers must be managed, and what a worker can and cannot do. For example, unlike truckload carriers that face a real challenge in getting drivers home in a timely way, the long-haul LTL carriers manage their single drivers in a way that ensures that they are home every night or every other night, with only occasional trips that take the driver out for two nights. Sleeper teams moving freight over longer distances may be away from home for three or four days, but are guaranteed to be home every week.

The LTL carriers are exceptionally competitive, and the large majority have gone out of business since the industry was deregulated in 1980. The survivors have learned how to strike the delicate balance between cost and service for a particular set of markets. For the long-haul carriers, the emphasis is on cost with very high service. For the regional carriers, the expectations on service are even higher. The biggest challenge faced by the regional carriers is that they have so little time to move a shipment that they have to make decisions quickly, and they are often forced to move trailers that are less than half full. The long-haul carriers have more time to work with a shipment, but their large networks offer many more options that can be considered.

The pickup and delivery process of LTL carriers has many elements in common with the description of pickup/delivery operations for private fleets. The biggest difference is that private fleets are usually doing pure pickup or pure delivery, whereas LTL carriers must handle both activities. Furthermore, it is common to separate the planning of pickup and delivery operations for LTL carriers from the planning of the line haul operation (movements between terminals).

A closely related segment is the small package industry, dominated at this writing in the United States by the UPS and FedEx. There are a number of subtle yet important differences between moving LTL freight (over 150 pounds) and small packages (under 150 pounds). For example, a trailer may hold 20–30 LTL shipments, while a trailer may hold hundreds of small packages. One effect of small shipment sizes is that there is considerably less variability in the day-to-day flows. Another is that small shipments lend themselves much more readily to automation in the sorting facilities.

1.3.1 Resource classes

The principle resource classes are:

- Drivers: They are characterized by home domicile, driving hours, whether they are a single driver or part of a sleeper team, their bid characteristics, and the number of days they have been away from home.

- Tractors: The carrier must maintain appropriate pools of tractors at major terminals, and manage the maintenance requirements of tractors.
- Trailers: These are typically 28 foot 'pups' and 48 foot 'vans.'
- Shipments: Varying in size between 150 and 10,000 pounds, traveling distances between 100 and 3,000 miles.
- Terminals: Here we have to determine the number of terminals, and their size, type, and location.
- Dock labor: These are the people who load and unload trailers. This determines the fraction of the physical capacity of the terminal that is actually used.

As in most problems involving multiple resource classes, it is common to work with one or two classes at a time. We illustrate this modeling strategy in Section 4.

1.3.2 Decisions

Key decisions include:

(1) Physical:

(a) Operational:

- Service network design: From a terminal, to which destinations should we send a trailer direct? These decisions have to capture the ability to consolidate freight in a timely fashion.
- Traffic assignment: What freight bills should go on a specific trailer? This can be a difficult problem for long-haul carriers who face a number of options. For regional LTL carriers, it is usually obvious.
- Pup matching: A tractor will usually pull two, and sometimes three, of the 28 foot 'pups.' It is not always the case that the pups will have the same origin or same destination. Pups must be matched so that the combined weight is within legal limits. If pups are being matched which do not have the same destination, it is desirable to match them so that they can stay hooked together as long as possible. Finally, it is best if the two pups have freight with similar service requirements.
- What driver should be used to pull a load? The choice of driver depends on domicile, how many hours he has been driving, how many days he has been away from home, whether the 'driver' is a single individual or a team, and the type of bid (for union drivers).
- Trailer management – Most of the time trailers remain balanced because of the need to balance drivers (a driver normally pulls empty trailers if there are no loaded ones to move). But sometimes it is necessary for a driver to bobtail (move the tractor without a trailer) and this creates an imbalance in the flows of trailers.

– When should the loads be moved? This is one of the hardest decisions, since the decision has to balance the service requirements of the shipments on the trailers, and the constraints on moving the driver and getting him home.

(b) Planning:

– Terminals: Size, type, and location.
– Dock labor: How many people should staff each terminal?
– Equipment pools (tractors and trailers): How large should the pools be?
– Physical transportation links: Less-than-truckload carriers are called 'regular route' carriers, and the decision to move trucks over a particular route joining two terminals is a planning decision.
– Contracts: This covers the agreement to serve major accounts, typically for a year. These commitments reflect expectations of the amount of freight that will be moved for the account (typically by traffic lane).

(2) Financial:

– Contract pricing: What price should be charged for freight for a specific account? These prices will be a function of the weight of an individual shipment, and the traffic lane (origin and destination) in which it is moving. For LTL trucking, this is an exceptionally difficult problem. It has to reflect the cost of pickup and delivery (see notes on this under private fleet operations), transferring the freight at terminals, and moving the freight over the line haul network. Transportation costs (line haul costs) have to reflect the density of the freight and its 'stackability' (the ability to stack the freight with other shipments in the same lane).
– General pricing: Same as contract pricing, but it is for freight offered to the carrier that is not covered by a contract. Typically, these prices are substantially higher than the contract price.
– Spot pricing: What price should an LTL carrier offer for a specific shipment on a specific day to help with network balance? Spot prices apply almost exclusively to truckload shipments.

(3) Informational:

– Communication technologies: Less-than-truckload carriers face an array of decisions regarding communication technologies. Some of these include:

• Shipment bar code scanning equipment for shipment pickup: This allows the carrier to learn more about the characteristics of the shipment when it is picked up.

Ch. 13. Dynamic Models of Transportation Operations   693

- On-board driver communication: This allows the carrier to dispatch a driver on the street to pick up a shipment that has just been called in.
- Bar code scanning at the terminals: This allows the carrier to know exactly when a shipment was pulled from one trailer and loaded onto another.
- On-board vehicle tracking: This would be the same technology used by truckload carriers. Its adoption for LTL is less obvious since the LTL trucks typically follow fixed routes.

– Databases and screens: Collecting, storing, and displaying data to support decisions is a major, ongoing challenge with any company. These systems are expensive, however, and have to be cost-justified.
– Planning models: A range of models are evolving to help support LTL carriers, ranging from routing of pickup and delivery trucks, service network design, pup matching, and driver management. As of this writing, these models are young and are seeing only the earliest adoption.

1.4   The railroads

Railroads remain the primary mover of bulk cargo, including both dry (grain, coal, etc.) and liquid (although pipelines offer some competition here). In some areas, barges can handle bulk movements, but for land movements, rail is almost the only option. Bulk cargo, however, represents only a portion of rail business. A substantial amount of merchandise freight moves by rail, as well as container movements that are moving to or from international locations. In fact, although much is made of the competition between trucking companies and rail, all the major trucking companies (truckload and LTL) use rail extensively for long moves. Trucking companies in particular have a difficult time moving freight between the Midwest and the west coast in North America; the distance is long and it can be difficult finding and managing drivers over this long movement. It is cheaper and more convenient to take a trailer of freight and load it onto a flatcar to be moved by rail.

Railroads, however, struggle with certain operating characteristics that are fundamental to the nature of a railroad. First of course is the limited infrastructure. The massive majority of all non-bulk movements must begin and/or end on a truck. The process of picking up freight and taking it to a rail yard, or delivering freight from a railroad, is known as *drayage* (the drivers who handle this step are called *draymen*). Most drayage operations are relatively inefficient and can add a substantial amount of cost to the process.

A second challenge faced by railroads is that freight is moved in extremely large blocks. A single train will typically weigh between 2,000–5,000 tons, but some bulk trains can weigh as much as 15,000 tons. A typical truck, by contrast, is moving about 15 tons of freight, with a gross weight of about 30 tons. So, a single train can be equivalent to as many as 500 trucks moving down the track.

The process of batching up enough freight to form these big blocks introduces a substantial amount of noise in the process. A train may be cancelled if there is not enough freight to justify its movement (the policy of canceling trains to save on crew and fuel costs is hotly debated in the industry, and is becoming less frequent), or because of problems finding enough power and a rested crew. When a train is delayed or cancelled (for any reason) the impact on the network can be fairly substantial.

A third challenge is the capacity of the track. High priority trains move faster than bulk trains. Passing a train going in the same direction requires that the slower train pull off on a siding. Since these sidings are located only at selected points along the track, it may be necessary to pull a train off the track for several hours until the pass can be completed. The same issue arises when trains have to move in opposite directions over a single-track route. One train will have to find an appropriate siding. Compounding the challenge of sitting at sidings is that the crew may run out of hours to complete the trip. Thus, an eight-hour trip can turn into an 11-hour trip, violating maximum duty time rules for the crew. The railroad is forced to drive a new crew out to the train to finish the trip.

The major railroads all offer high priority service for certain classes of freight, where they try to compete with trucking companies. Given the limitations of the infrastructure and the nature of rail operations, it is virtually impossible for rail to provide a higher level of service than truck. On the other hand, no other land-based mode can compete with its efficiency.

1.4.1 Resource classes

The broad range of resources, which must be managed to provide rail service, characterizes railroads. Some of these include:

- Freight cars: These come in a variety of styles, including boxcars, flatcars, and tanker cars, but with many variations of each type.
- The freight: Freight can exist by itself as an unsatisfied customer demand, or coupled with a boxcar (producing a loaded car). Freight is characterized by origin, destination, size, and service characteristics.
- Locomotives: There are about a dozen major classes of locomotives, but viewed closely enough, locomotives are almost unique. Characteristics of a locomotive can include its horsepower, whether it is high or low adhesion (a feature that determines the ability of the train to get started from a standstill), features required to classify it as the lead unit on a train (which is where the crew rides), its maintenance status, and what other locomotives it is currently attached to (the process of connecting multiple locomotives to pull a single train involves a fairly elaborate set of connections which have to be tested before the train can move).

- Operators: The rules for moving crews are governed by federal regulations, and a dizzying array of union work rules, some of which date back centuries.
- Track: The track limits the ability of trains to move. Decisions to build new track or maintain existing track are some of the most important infrastructure decisions a railroad can make.
- Yards: As with track, the yards have capacity and limit the throughput of trains.
- Maintenance facilities: Locomotives represent complex pieces of equipment, with federally required maintenance intervals (in addition to those required to keep a locomotive in working order). Some maintenance equipment runs into millions of dollars.

1.4.2 Decisions

As with the resources, there is a complex array of decisions required to manage these resources:

(1) Physical:

(a) Operational:
- Trip planning: How should a loaded freight car be routed through the network? Freight cars are allocated to *blocks* (a group of cars being routed over a common segment) that are moved by trains. Both trains and blocks have capacities, so it is necessary to plan the route of a car through a sequence of blocks while not violating either capacity constraint.
- Blocking: What blocks should be formed? How should blocks be routed through the network?
- Car distribution: To what yard or customer should an empty freight car be allocated?
- Locomotive management: What locomotive should be used to pull a train? How should power be repositioned from surplus to deficit locations?
- Crew planning: What crew should be used to move a train?
- Line capacity planning: How should trains be sequenced over a track (and the sidings)? When should trains be scheduled to depart?

(b) Planning:
- How much track, how many sidings and their placement, and how well should they be maintained (which affects the speed at which trains can move over the track)?
- Location and size of new yards, local stations, and maintenance facilities.
- Fleet size and mix, for locomotives as well as the freight cars.

- Customer commitments, which set carrier expectations of the amount of freight a customer may tender (and therefore the resources required to serve the customer).

(2) Financial:

- What price should be charged for a contract? These prices will be a function of the origin and destination, shipment size, freight car requirements, and other service constraints.
- What spot prices should the railroad accept?

(3) Informational:

- Should the railroad invest in train tracking technology (which tells them the location of the train on the track)?
- Should the railroad invest in transponders that detect the presence of locomotives? Freight cars?
- Should a train have voice communication while it is en route?
- What databases should be created to store and display information?
- What planning models should a railroad invest in? As of this writing, major railroads are taking noticeably different approaches toward the use of planning models; some are focusing on longer range planning models; others on short-term operational models, while others are limiting their use of models.

1.5 Intermodal container operations

Intermodal containers are boxes, typically 20 or 40 feet in length, which can themselves move by truck, rail, or ocean container ships. Unlike trailers, containers can be stacked two levels high on a rail flatcar, in addition to being pushed against each other (trailers, with wheels, require special panels on the flatcar to ensure that they do not roll). As a result, they are a much more productive way to move freight over both road and rail. On the other hand, they are not as large as the 48-foot vans. The 20-foot containers are smaller than the 28-foot pups favored by LTL carriers, a difference that is compounded when a single driver can pull two 28-foot pups, a volume that is much larger than two 20-foot containers. It is not possible to pull two 40-foot containers using a single tractor.

If the freight has to move by vessel, the container is the only way to move merchandise. It is not possible to stack trailers, and stacking allows the largest container ships to hold thousands of containers.

In contrast with our previous examples, the management of intermodal containers consists only of the containers. Containers may be owned by shipping lines, or by other logistics organizations. For this reason, we do not address the dimensions of motive power (tractors, locomotives, and ships) and operators (drivers and crews).

1.5.1 Resource classes

- Containers: These come in two basic lengths (20 feet and 40 feet) but a variety of other features will distinguish one container from another, including refrigeration, height, and stacking capability.
- Customer orders: This is freight moving from an origin to destination (multiple stops are never permitted), with specific service requirements.

1.5.2 Decisions

(1) Physical:

  (a) Operational:
  - What type of container should be allocated to an order?
  - How should containers be distributed in anticipation of forecasted orders? How many containers of each type should there be in a pool on any given day?
  - How should a loaded container be routed from origin to destination (this may involve a combination of ship, train, and truck)?
  - Stacking and storage of containers in the port and on the ship itself: Where should a container be stacked and stored (both in the port, as well as on the ship) to minimize total handling of containers.

  (b) Planning:
  - How many containers, and what type, should be owned?
  - What are the size and location of container pools?
  - What transportation contracts should be arranged? Container shipping typically requires arrangements with other transportation companies to move the containers.

(2) Financial:
  - How should a contract be priced?
  - What should standard rates be for non-contract movements in a traffic lane?
  - How should the carrier spot price individual moves?

(3) Informational:
  - What types of tracking technologies should a company use (especially for use in the ports)? There is a movement toward the use of satellite tracking of individual trailers (in trucking) and containers (in shipping). To what extent should these technologies be used?
  - Should the company invest in forecasting and optimization technologies?

2 A general modeling framework

The examples in the previous section illustrate the range of different industries that have evolved to meet segments of the freight transportation market. Each exhibits specific qualities in terms of cost and service, reflecting the nature of the market that is being served.

The next challenge is modeling these problems. Fortunately, certain physical processes characterize all of these problems. If we can develop models for the basic processes that characterize these problems, then we can build up more complex models from these building blocks.

To begin, we need a notational system to describe our problem. Our notation [which is based on Powell et al. (2001)] builds on standard notational conventions, but most of the standard research in logistics avoids some key issues that arise in real applications. Examples include multiattribute resources and resource layering, dynamic information processes, and multi-agent control.

We need some general notation through the presentation. We represent the geography of our problem using: For transportation applications, it is useful to define in addition a set of geographical locations:

$\mathcal{I} =$ a set of locations

We generally model our problem over a set of discrete time periods:

$\mathcal{T} = (0, 1, \ldots, T)$.

At times, we want to represent rolling horizon problems where at time $t \in \mathcal{T}$, we will optimize over a set of time periods that start at t and extend over a planning horizon, given by:

$$\mathcal{T}_t^{ph} = (t, t+1, t+2, \ldots, t + T^{\text{planning horizon}}).$$

Our representation is divided along three primary dimensions: resources, processes, and controls.

2.1 Resources

Resources are comprised of three sub-dimensions:

(a) Resource classes.
(b) The attributes of each resource class.
(c) The resource layering, which represents how resources can be coupled together to perform work.

For this, we define:

$\mathcal{C}^R =$ The set of resource classes (e.g., drivers, tractors, trailers, locomotives, and loads).
$\mathcal{R}_t^c =$ The set of resources in class c at time t.
$a_r =$ The attributes of resource r.
$\mathcal{A}^c =$ The space of attributes for resources in class c, where $a_r \in \mathcal{A}^c$ for $r \in \mathcal{R}^c$.

In some cases, it is easiest to track individual resources, which means that $(a_r, r \in \mathcal{R}_t^c)$ would capture the state of the resources in class c at time t. This representation is often most useful when resources are relatively complex, such as people, aircraft, and locomotives. When the resources are simpler, it is more useful to use vector notation:

$R_{ta} =$ The number of resources with attribute a at time t, before new arrivals in time t have been added in.
$R_t = (R_{ta})_{a \in \mathcal{A}}$.

An important but fairly subtle issue that arises purely in the context of dynamic problems is referred to as the *time lagging* of information. Specifically, there may be resources that we know about at time t, but which cannot be acted on until time t'. In this setting, t refers to the time at which the resource becomes known, whereas t' is when it becomes actionable. Thus, we may know about a customer order now, but we do not have to satisfy it until later. Or we may know about a boxcar that will become available in the future. We handle this concept by defining:

$R_{t,at'} =$ The number of resources with attribute a that we know about at time t but which do not become actionable until time t'.
$R_{tt'} = (R_{t,at'})_{a \in \mathcal{A}}$
$R_t = (R_{tt'})_{t' \geq t}$.

We call R_t the *resource state vector*. Not uncommonly, this vector is defined with respect to an aggregation function:

$$G : \mathcal{A} \mapsto \mathcal{A}^G$$

where $\mathcal{A}^G$ is a more compact space of attributes. For our purposes, we use R_t exclusively as our resource state vector, recognizing that the discrete representation $\mathcal{R}_t$ is more appropriate for complex resources.

The use of the attribute vector a is very convenient. For the simplest problems, we might be modeling the flows of a common type of trailer between locations $i \in \mathcal{I}$. In this case, the location i represents the state of the resource, and we would have $a = (i)$. We might have different types of trailers or containers $k \in \mathcal{K}$, as would commonly arise in multicommodity flow problems. In this case, k is the commodity and i is the state of the resource. The attribute vector would then be $a = (k, i)$. As we move to more complex resources, the attribute vector would grow. By using a common attribute vector a, our notational system responds easily to different types of resources.

One of the most difficult dimensions of resource management arises in the presence of resource layering. Consider, for example, the problem of moving a load in trucking. We need a driver and a trailer to pull a load of freight. We start with an idle driver (in a tractor). The first step is to find a trailer, at which point we have a driver with an empty trailer. Next we have to move to the customer and pick up the load of freight. Now we have a driver with a trailer and a load of freight. At this point we have to decide whether the driver should move directly to the destination to deliver the load, or to move the trailer to a relay point where he would drop off the trailer (of course, still full of freight). The driver/trailer, driver/trailer/load, or the trailer/load, represent instances of layered resources.

We represent layered resources by first defining a *layering*, $\mathcal{L}$. This is most easily described by example. Let $\mathcal{C} = (D, T, L)$ represent our three resource classes. For our example, a layering would be:

$$\mathcal{L} = (D|T|L, T|L)$$

If we use this as our layering, we would call the first layer the *driver layer*. It consists of a driver, trailer, and load. The attributes of a driver layer consist of the attributes of a driver, and then the attributes of a trailer and a load that may be *coupled* to the driver. In general, if $l \in \mathcal{L}$ represents a particular layer, we let $a^{(l)}$ represent the attributes of layer l, while a^c would be used to represent the attributes of a particular resource class $c \in \mathcal{C}^R$. We may refer to a specific layer, such as the driver layer, using:

$$a^{(D)} = \text{The attributes of a driver layer.}$$
$$= a^D | a^T | a^L.$$

If a driver is not coupled with a trailer or load, his primitive attribute vector would be a^D, but his layered attribute vector would be $a^{(D)} = a^D | \phi^T | \phi^L$, where ϕ^T and ϕ^L are null vectors with the same dimensions as a^T and a^L, respectively. Thus, the attributes of a driver layer are not determined until we decide which resources (trailer and load) to couple the driver with. We also have a trailer layer, which again can consist of a single trailer, or a trailer and a load. We identify a layer by its lead resource class.

Layering is an important concept for modeling more complex operations, but it can sometimes be avoided. Consider, for example, the case of truckload trucking, but assume that once a driver picks up a load, that he always drives it directly to the destination. Thus, the decision to assign a driver to a load produces a driver at the destination of the load and a (presumably) happy customer (the load has been delivered). At no time did we have to explicitly capture the state that the driver had the load. Layering arises when there is a specific set of decisions that we have to choose, from given the attributes of a layered resource.

Resource layering is critical when we have to capture the state of two resources coupled together, at which point there is a new set of decisions that apply to the characteristics of the layered resource. More examples of resource layering are given in Section 4.

2.2 Processes

There are three dimensions of processes:

(a) Dynamic information processes.
(b) System dynamics.
(c) Constraints.

There are two types of information processes: exogenous information (outside of our control) and endogenous information, more commonly known as decisions (a good working definition of a decision is an endogenously controllable information class that changes the state of the system). At this stage, we use general models of both (specific illustrations are given in Section 4). For exogenous information processes, we let:

$W_t = $ A random variable representing a family of random variables describing new information arriving at time t.

In complex problems, there can be a number of exogenous information processes. For us, we use W_t to represent all of these. We let $\omega \in \Omega$ represent an elementary outcome of the sequence $(W_t)_t \in \mathcal{T}$, and we let $\omega_t = W_t(\omega)$ be a realization of the information arriving in time period t. Following standard conventions in probability theory, we let $\mathcal{F}_t$ be the σ-algebra generated by $(W_{t'})_{t'=0}^{t}$. For our problem, there are two special types of information that arrive. The first is information about new resources that are arriving such as new customer demands, or new units of capacity entering the system from outside sources (e.g., a boxcar being released empty to the network). We represent these by:

$\hat{R}_{t,a't'} = $ The number of resources, with attribute a', that first become known at time t that can be acted on at time t'.
$\hat{R}_t = (\hat{R}_{t,a't'})_{a' \in \mathcal{A}, t' \geq t}$.

The second class of information represents parameters that govern the dynamics of the system (described shortly). For example, we might get new information about the speed of a train, the cost of a movement, or the price of fuel. We capture these parameters using:

ρ_t = A vector of parameters that impact the dynamics of the system.
$\hat{\rho}_t$ = New information about these parameters arriving in time t.

An element of ρ_t might be the estimate of the transit time between two points, or the average number of pounds that a trailer normally holds.

Endogenous information processes represent our decisions. For the moment, we let:

$\mathcal{D}$ = The set of possible decisions that can be used to act on the resources.
x_{tad} = The number of resources with attribute a that decision $d \in \mathcal{D}$ is applied to at time t.

System dynamics governs how the system changes in response to new information. The effect of new resources is captured simply using:

R_t^+ = Set of resources at time t including new arrivals in time period t.
 = $R_t + \hat{R}_t$

In Section 3 we demonstrate the special roles of R_t and R_t^+ (and the reason for this particular notational style).

We represent the updating of system parameters using the general notation:

$$\rho_t \leftarrow U^\rho(\rho_{t-1}, \hat{\rho}_t)$$

For example, if ρ_t is an estimate of the travel time, and $\hat{\rho}_t$ is a recent observation of a travel time, we might think of $U^\rho(\rho_{t-1}, \hat{\rho}_t)$ as an equation that performs exponential smoothing, as in $\rho_t = (1-\alpha)\rho_{t-1} + \alpha\hat{\rho}_t$, where $0 < \alpha < 1$ is a smoothing factor.

More interesting is modeling the impact of a decision on the system. We use the concept of a *modify* function, which performs the mapping:

$$M(t,a,d) \rightarrow (a',c,\tau) \tag{2.1}$$

Where a is the attribute of a resource (or resource layer) being acted on by decision d, where t represents what we know when the decision is made (or implemented). a' is the attribute of the modified resource, c is the contribution (or cost, if we are minimizing) generated by the decision, and τ is the time required to complete the action. The modify function is useful conceptually and in software, but for algebraic purposes, it is useful to define:

$$\delta_{t'a'}(t,a,d) = \begin{cases} 1 & \text{if } M(t,a,d) = (a',c,t'-t) \\ 0 & \text{otherwise} \end{cases}$$

The modify function plays the role of a transfer function in dynamic systems, but it is expressed at the level of a single decision acting on a single (type of) resource. Sometimes it is useful to refer specifically to the attribute of a transformed resource, or the cost or time required to complete the decision. For this purpose, we introduce the notation:

$$M(t, a, d) \rightarrow (a^M(t, a, d), c^M(t, a, d), \tau^M(t, a, d)) \tag{2.2}$$

We call a^M_{tad} the *terminal attribute function*, where the superscript 'M' is used to help identify the difference between the attribute vector a and the terminal attribute function a^M_{tad}. More often, we use the vector notation $c_{tad} = c^M(t, a, d)$ and $\tau_{tad} = \tau^M(t, a, d)$ to represent costs and times. Our representation assumes that (a', c, τ) are all deterministic functions of (t, a, d). This assumption serves the purposes of our presentation here, but the reader should understand the richness of dynamic problems. For example, it is very common that transit times are not deterministic functions of (t, a, d); for some areas (intermodal container transportation, rail transportation, and even the large truckload and LTL carriers), the randomness of the transit time is of central concern to some shippers where precise delivery dates are essential.

Our first use of the delta function is to express the evolution of the resource vector:

$$R_{t+1,a't'} = R_{t,a't'} + \hat{R}_{t,a't'} + \sum_{a \in A} \sum_{d \in D} \delta_{t'a'}(t, a, d) x_{tad} \quad \forall a' \in A, t' \geq t \tag{2.3}$$

Finally, the evolution of the system is restricted by constraints. For our purposes, it is sufficient to represent flow conservation constraints:

$$\sum_{d \in D} x_{tad} = R_{ta}$$

and rate of process transformation constraints:

$$x_{tad} \leq u_{tad}$$

where u_{tad} is an upper bound (normally some sort of physical constraint) on the flow. In practice, upper bounds apply to aggregations of flows.

2.3 Controls

Five dimensions characterize controls:

(a) The types of controls.
(b) The organization of controls.

(c) The information available to a decision maker.
(d) The decision function.
(e) Measurement and evaluation.

We describe the types of controls using:

$\mathcal{C}^D$ = The set of decision classes.
$\mathcal{D}^c$ = The set of decisions in class $c \in \mathcal{C}^D$.

In many problems, it does not make sense to define a single general set of decisions. For example, the decision to 'send a truck empty to Chicago' does not make sense if the truck is in Miami. Any practical implementation requires being able to specify the set of decisions given the attributes of the resource being acted on (typically, we would use an aggregation of the attribute vector). Thus, we would define:

$\mathcal{D}_a$ = The set of decisions that can be applied to a resource with attribute vector a.

In practice, the decision class we are working with is understood (or there is only one class), allowing us to avoid the explicit modeling of decision classes. Just the same, it is important to recognize the presence of multiple decision classes. Reading the academic literature could easily lead a student to think that the only thing a transportation company does is move something from one location to another. Companies have to buy and sell, maintain, paint and clean, and refuel. The notation we provide here allows us to write a basic formulation of the problem that will remain valid even if we add decision classes later.

For our discussion, we restrict our attention to classes of decisions that directly impact resources. Thus, these are the classical decisions of routing drivers and freight, as well as purchasing/selling new equipment, hiring new drivers, or choosing which customer demands to serve in the spot market. Other classes of decisions include pricing (both contract and spot), and decisions about the information infrastructure.

Decision classes can be divided into three major groups: couple, uncouple, and modify. The couple and uncouple classes arise only when we are modeling resource layers. A couple decision brings two (or more) resources together, e.g., a driver pulling a load or a pilot flying a plane. In this case, a is the attribute of the active resource, while d is the decision to augment a with the attributes of a secondary resource. For a modify decision, d simply modifies the attribute vector a. Most problems feature 'one to one' coupling (one driver, one load; one pilot, one aircraft; one boxcar, one demand). More than one locomotive is needed to move a single train, which is an instance of 'several to one' coupling. A single truck may move dozens (or hundreds, in the case of packages) of shipments, and this is an instance of 'many to one' coupling.

Ch. 13. Dynamic Models of Transportation Operations   705

It is useful to start by listing only the *primitive* decisions, each of which is a single, elementary action. For example, the decision to 'assign a driver to a load' in truckload trucking can consist of the primitives: move to the load, couple with the load, move the load, and uncouple from the load. Once the primitives are in place, it is often useful to create *tactics* that represent sequences of decisions, as in our example to 'assign a driver to a load.' Had we formulated the problem purely in terms of the primitives, we would have to capture the layered state 'driver coupled with a load.' If we were not modeling driver relays, a model based purely on primitive decisions would be unnecessarily complex. But, planning driver relays may be important, in which case it is useful to work in terms of the primitive decisions.

For complex operations (railroads and trucking companies) it is important to model the organization of information. Large companies are managed by a series of decision makers, which the modeling community calls *agents*. Let:

$\mathcal{Q}$ = A set of agents which control the system.
$\mathcal{D}_q$ = The set of decisions controlled by agent $q \in \mathcal{Q}$, which implicitly includes when a decision will be implemented.
$\mathcal{A}_q$ = The attributes of resources that are controlled by agent q, which we also assume includes the time at which the resource is available to be acted on.
$\mathcal{T}_q$ = The set of time periods over which the decisions in $\mathcal{D}_q$ apply.
$x_q = (x_{tad})_{t \in \mathcal{T}_q}, a \in \mathcal{A}_q, d \in \mathcal{D}_q$.

For notational simplicity, we assume that an agent implies an interval of time. Often, we will find ourselves modeling a single controller at a point in time, in which case we can simply replace the index q with a time index t. Our 'agent' notation, where time is implicit in the definition of the agent, gives us a simple notational mechanism for modeling more general informational decompositions with no additional complexity in notation.

The sets $\mathcal{Q}, \mathcal{D}_q, \mathcal{A}_q,$ and $\mathcal{T}_q$ define the organization of control in the operation. It is assumed that an agent q will make decisions within $\mathcal{D}_q$ that are coordinated (e.g., if the decision maker is assigning drivers to loads, he will not assign the same driver to two loads at the same time). It is also necessary to understand the impact of agent q on other agents (which may exist within the same organization, or in other organizations). For this purpose, we need to define:

$\vec{\mathcal{M}}_q$ = The set of agents $q' \in \mathcal{Q}$ who are directly impacted by decisions made by agent q.
$R_{q,aq'}$ = The number of resources of attribute a that are sent from agent q to q'.
$R_{qq'}$ = $(R_{q,aq'})_{a \in \mathcal{A}_q}$.

We next have to model the organization of information. We let:

I_q = The information elements available to decision maker q.

There are four classes of information that may be used in the set I_q:

K_q = Knowledge, which is the exogenous data that is accessible to q. Knowledge contains data in databases as well as other informal sources that are present as 'head knowledge.'

Ω_q = Forecasts of exogenous information which would come as updates to K_q. Normally, the set Ω_q will consist of a single element representing a point forecast, but it might include different elements, representing different scenarios that we wish to model in the future.

x_q^p = Plans for the future, which can be thought of as forecasts of future decisions.

$V_{qq'}(R_{qq'})$ = Value functions which capture the impact of decisions by q on $q' \in \vec{M}_q$.

The value functions can be thought of as forecasts of dual variables. A simple example of these functions arises when purchasing parts from a supplier. The decision to place an order has an impact of requesting parts from the supplier ($R_{qq'}$ becomes the number of orders that q is transferring to q'). The supplier then charges a price (say, $p_{q'}$), so our value function is simply $p_{q'} R_{qq'}$. When the value function is linear, it is possible to show that $V_{qq'}(R_{qq'}) = V_{q'}(R_{q'})$.

It is important to understand that when designing the set I_q, the goal is not to create the ultimate information set, but rather to model the information that is actually available. Many decisions are made purely using K_q (the vast majority of simulation models fall in this category). Optimization models that use deterministic forecasts would use the set $I_q = (K_q, \Omega_q)$ where the set Ω_q usually consists of a single point forecast. Models based on this information set are called rolling horizon procedures.

Given the information set, the next problem is to actually make a decision. Let:

χ_q = The feasible region for agent q.

The process of actually making decisions is then given by:

$X_q^\pi(I_q)$ = The vector of decisions produced by information set I_q. Thus, we compute decisions using $x_q = X_q^\pi(I_q)$. We let:

Π = The family of policies (literally, different decision functions, each of which constitutes a method for translating information into decisions).

A policy represents any means of finding a decision given a state, which we also call a decision function. Our problem is one of finding the best decision function. But, it is also going to be important to build functions that use information that is actually available. In this way, we are attempting to model the organization and flow of information just as we model the flows of physical resources.

Finally we have the dimension called measurement and evaluation. For our purposes, this is the objective function. We assume that we can define a contribution function:

$C_q(x_q, K_q)$ = The contribution from decision x_q given our knowledge K_q.

Remembering that each agent q implicitly defines a time interval over which his/her decisions apply, our objective function can now be stated:

$$\max_{\pi \in \Pi} E \left\{ \sum_{q \in Q} C_q\left(X_q^\pi(I_q), K_q\right) \right\}$$

This optimization problem takes on more meaning when we define specific classes of functions X_q^π.

3 Algorithmic strategies

Now that we have a specific modeling framework, we have to address the challenge of designing an algorithmic strategy. We start in Section 3.1 by presenting strategies for solving time-staged problems under uncertainty using a new class of dynamic programming approximations. Section 3.2 discusses the issues that arise when we combine nonlinear value functions with multiperiod travel times. This concept is then extended in Section 3.3 to solve multiagent problems using the same framework. These two sections establish the fundamentals of solving the problems when information is staged over time, and when information is organized among different decision makers. These presentations then lay the groundwork for Section 3.4, which provides a general framework for building different classes of decision functions for a variety of complex problems.

By the end of this section, we will have the foundation we need to address a fairly broad range of complex operational problems.

3.1 Strategies for dynamic problems

Our first challenge is solving problems when information is staged over time. This is the classical problem of stochastic, dynamic problems. These can be solved approximately in a variety of ways that are discussed in Section 3.4. Here, we demonstrate how to use dynamic programming approximations effectively to solve time-staged problems.

Our presentation is divided into two stages. First, we have to address a subtle but critical problem in how we model the evolution of information over time and the definition of the state variable. In particular, we do not use the classical definition of a state variable as it is presented in dynamic programming. Instead, we introduce the concept of an *incomplete* state variable, which will prove computationally far more tractable. After this discussion in Section 3.1.1, and Section 3.1.2 discusses specific strategies for approximating value functions in dynamic programs.

3.1.1 Setting up the optimality recursion

We start by describing the evolution of information in our system. As we noted before, we have exogenous and endogenous information processes that can be represented using:

$$(W_0, X_0^\pi, W_1, X_1^\pi, \ldots, W_t, X_t^\pi, \ldots)$$

We need to capture what we know at each point in time. This can be measured immediately after we have new exogenous information, and after we make a decision. We let S_t^+ be the state after new information has arrived, and we let S_t be the state after we make a decision, giving us the sequence.

$$(W_0, S_0^+, X_0^\pi, S_1, W_1, S_1^+, X_1^\pi, S_2, \ldots, S_t, W_t, S_t^+, X_t^\pi, S_{t+1} \ldots)$$

We refer to S_t^+ as the *complete* state variable, because it captures all the information needed to make a decision at time t. S_t is called the *incomplete* state variable, specifically because it does not include all the information needed to make a decision. The importance of this distinction will become clear shortly.

Our goal is to solve the problem:

$$\max_{\pi \in \Pi} E\left\{\sum_{t \in T} C_t(X_t^\pi, S_t^+)\right\} \quad (3.1)$$

Equation (3.1) can be formulated in general using the optimality recursion:

$$V_t^+(S_t^+) = \max_{x \in \chi} C_t(x, S_t^+) + E\{V_{t+1}^+(S_{t+1}^+)|S_t^+\} \quad (3.2)$$

Here and throughout this section, we use x as the variable we are optimizing over, and let x_t represent the solution of (3.2).

The field of dynamic programming is typically expressed in terms of discrete states and actions (decisions), with algorithms that assume that you can loop over all possible states and actions. This approach suffers from the classic 'curse of dimensionality' which means that when the state variable is multidimensional, the state space becomes intractably large. For this reason, dynamic programming has seen few applications in transportation and logistics. Not surprisingly, this is partly to blame for the dependence on myopic models and deterministic approximations found in transportation.

It turns out that the situation is even worse than we thought. Equation (3.2) actually suffers from three curses of dimensionality: the state space, the outcome space, and the action space. To avoid this problem, we adopt a new approach for approximating dynamic programming. As a first step, we could replace the value function with an approximation, producing a recursion that looks like:

$$\tilde{V}_t^+(S_t^+) = \max_{x \in \chi} C_t(x, S_t^+) + E\left\{\hat{V}_{t+1}^+(S_{t+1}^+) | S_t^+\right\} \quad (3.3)$$

On the right hand side of (3.3), we have an approximation $\hat{V}_{t+1}^+(S_{t+1}^+)$. On the left hand side, we use a placeholder that we call $\tilde{V}_t^+(S_t^+)$.

For the next step, we assume that $\hat{V}_t^+(S_t^+) = \hat{V}_t^+(R_t^+)$, which is to say that our approximation is purely a function of the resource state variable, and not the full information state. In fact, it is sometimes important to write the function in terms of an aggregated form of the resource state variable, which we could write $\hat{V}_t^{G+}(G(R_t))$. For the rest of our discussion, we do not include the aggregation function $G()$ explicitly, but the reader should understand that we can use this device at any time. Now, we have:

$$\tilde{V}_t^+(S_t^+) = \max_{x \in \chi} C_t(x, S_t^+) + E\left\{\hat{V}_{t+1}^+(R_{t+1}^+) | R_t^+\right\} \quad (3.4)$$

Our next problem is the expectation. For real problems, this is computationally intractable. We could approximate the expectation using a sample, as in:

$$\tilde{V}_t^+(S_t^+) = \max_{x \in \chi} C_t(x, S_t^+) + \sum_{\omega \in \hat{\Omega}} \hat{p}(\omega) \hat{V}_{t+1}^+(R_{t+1}^+(\omega)) \quad (3.5)$$

where $\hat{\Omega}$ is a sample from Ω and $\hat{p}(\omega)$ is probability of outcome $\omega \in \hat{\Omega}$.

Equation (3.5) can itself be quite hard, even when the sample $\hat{\Omega}$ is relatively small. In transportation problems, the basic one-period optimization model could represent a resource allocation problem with thousands of variables, or a difficult integer programming problem arising in vehicle routing or network design. We would prefer to use a single sample:

$$\tilde{V}_t^+(S_t^+, \omega) = \max_{x \in \chi} C_t(x, S_t^+, \omega) + \hat{V}_{t+1}^+(R_{t+1}^+(\omega)) \tag{3.6}$$

Now, we have created a decision function where x_t is allowed to 'see' $R_{t+1}^+(\omega)$, which violates a basic information constraint. We avoid this problem by formulating our recursion in terms of our incomplete state variable:

$$V_t(S_t) = E\{max_{x \in \chi} C_t(x, S_t^+) + V_{t+1}(S_{t+1}) \mid S_t\} \tag{3.7}$$

Since S_t is incomplete, the decision x_t is a random variable, and as a result we have to pull the expectation outside of the max operator. Following the same path as before, we obtain the approximation:

$$\tilde{V}_t(S_t, \omega) = \max_{x \in \chi} C_t(x, S_t, \omega) + \hat{V}_t(R_t(\omega)) \tag{3.8}$$

Note that we index $\hat{V}_t(R_t(\omega))$ by t instead of $t+1$ because it is a function of the information in time t. We have to devise an updating strategy that revises the estimates from one iteration to another. If n is our iteration counter, then we can just use the representation:

$$\hat{V}_t^n \leftarrow U^V(\hat{V}_t^{n-1}, \tilde{V}_t^n, R_t^n) \tag{3.9}$$

The updating function $U^V(\cdot)$ could be nothing more than the use of exponential smoothing on a constant (this would be the case when we are using linear approximations) or a strategy for updating nonlinear approximations (specific examples are given in the next section).

We now have a general approximation strategy for dynamic programs, (illustrated in Fig. 1) with two 'hot spots.' The first is that we have to devise an approximation scheme $\hat{V}_t(R_t)$. The second is that we have to exploit the structure of the resulting approximation to solve what is typically an integer program.

3.1.2 Approximating the value function

We propose using two classes of approximations for $\hat{V}_t$: linear, and nonlinear, separable. For problems where integer solutions are required (which is common in logistics problems), we would use a piecewise linear function instead of a continuously differentiable function (which might be attractive because of the low number of parameters needed to characterize it).

Linear functions are always the easiest to implement and use, but they can be unstable. Just the same, they serve as a useful illustration. Assume that the basic problem $\max_{x \in \chi} C_t(x, S_t)$ is computationally tractable. Then,

$$\tilde{V}_t^n(S_t, \omega_t) = \max_{x \in \chi} C_t(x, S_t) + \hat{v}_{t+1}^n R_{t+1} \tag{3.10}$$

subject to:

$$\sum_{d \in D} x_{tad} = R_{ta} \tag{3.11}$$

$$x_{tad} \leq u_{tad} \tag{3.12}$$

should also be computationally tractable. If the problem is a continuous linear program, then we can use the dual variable for constraint (3.11) to help us estimate our linear approximation. Let $\tilde{v}_{ta}^n$ be the dual variable of equation (3.11) at iteration n. We may then estimate a linear approximation using:

$$\hat{v}_{ta}^n = (1 - \alpha^n)\hat{v}_{ta}^{n-1} + \alpha^n \tilde{v}_{ta}^n \tag{3.13}$$

Linear approximations can work well, but for the types of resource allocation problems that arise in fleet management, (separable) nonlinear approximations have proven to work the best. Although a number of strategies can be used to estimate nonlinear functions, the interest in obtaining integer solutions has led to the development of piecewise linear

---

Adaptive dynamic programming algorithm

Step 1 Initialize all $\hat{V}_t(R_t)$ for all t. Set $n = 1$.

Step 2 Generate an outcome $\omega = (\omega_0, \omega_1, \ldots, \omega_{T-1})$.

Step 3 For $t = 0, 1, \ldots, T-1$, find:

$$x_t^n(S_t, \omega) = \arg\min_{x \in \mathcal{X}} \left\{ c_t(S_t, \omega_t, x) + \hat{V}_t^{n-1}(S_{t+1}(\omega, x)) \right\}$$

Update S_t using the system dynamics.

Step 4 For $t = T-1, T-2, \ldots, 1, 0$, update $\hat{V}_t^n$ for all t using the update function:

$$\hat{V}_t^n = U^V\left(\hat{V}_t^{n-1}, \tilde{V}_t^n(R_t^n), R_t^n\right)$$

where:

$$\tilde{V}_t^n(R_t) = \min_{x \in \mathcal{X}} \left\{ c_t(S_t, \omega, x) + \hat{V}_t^n(S_t(\omega, x)) \right\}.$$

Step 5 Let $n := n + 1$, and go to step 2.

Fig. 1. Prototype of an adaptive dynamic programming algorithm.

approximations. Thus, we can write our nonlinear approximation in the form:

$$\hat{V}_t(R_t) = \sum_{a \in \mathcal{A}} \hat{V}_{ta}(R_{ta})$$

Many problems in transportation and logistics require integer solutions. When this is the case, it is easiest to build piecewise linear approximations. Piecewise-linear concave value function approximation components are characterized by a series of break points $\{u_0, u_1, u_2, \ldots, u_n\}$ and slopes v_l on the portion $[u_l, u_{l+1}]$ with $v_0 \geq v_1 \geq \cdots \geq v_n$. Then (dropping the subscripts and superscripts for state and time):

$$\hat{V}(R) = \sum_{l=0}^{m-1} \hat{v}_l (u_{l+1} - u_l) + \hat{v}_m (R - u_m) \tag{3.14}$$

where $m = \max\{l : u_l \leq R\}$. We can update $\hat{V}(R)$ using sample gradients. Let $\tilde{v}^n$ be a sample estimate of the dual variable of the resource constraint (3.11). When the underlying problem is a network, it is possible to get left and right gradients using flow-augmenting paths [see Powell (1989)]. When this is possible, let $\tilde{v}^+$ and $\tilde{v}^-$ be the right and left gradients, respectively (in the discussion below, if these are not available, simply let $\tilde{v}^+ = \tilde{v}^- = \tilde{v}$). We now want to use these gradients to update our slopes for $\hat{V}$. The idea is to use this information to update the function locally, while retaining the basic concavity of the function at all times.

This process is illustrated in Fig. 2. In Fig. 2(a), we have a concave estimate of the value function, along with new slopes at a particular point. Fig. 2(b) shows that if we smoothed these new estimates of slopes into the immediate area of the estimate, we would obtain a non-concave approximation. Fig. 2(c) shows that if we expand the range over which we are smoothing the slopes, then the resulting updated function remains concave.

More formally, let u_l^n and $\hat{v}_l^n$ denote the breakpoints and slopes of the function at iteration n. To maintain concavity, we update the function over the range (l^-, l^+), given by:

$$l^+ = \min\{l : u_l \geq R_{ta}^n, (1 - \alpha^n)\hat{v}_l^n + \alpha^n \tilde{v}^{n+} \geq \hat{v}_{l+1}^n\}$$
$$l^- = \max\{l : u_l \leq R_{ta}^n, (1 - \alpha^n)\hat{v}_l^n + \alpha^n \tilde{v}^{n-} \leq \hat{v}_{l-1}^n\}$$

Then for all $l \in [l^-, l^+]$ we update the slopes as:

$$v_l^{n+1} = \begin{cases} (1 - \alpha^n)v_l^n + \alpha^n \tilde{v}^{n-} & l < R^n \\ (1 - \alpha^n)v_l^n + \alpha^n \tilde{v}^{n+} & l \geq R^n \end{cases}$$

Ch. 13. Dynamic Models of Transportation Operations

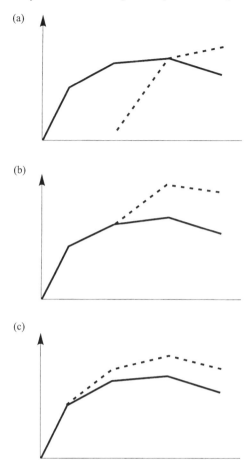

Fig. 2. Illustration of updating over a smoothing interval to maintain concavity. (a) Initial function with unsmoothed update. (b) Smoothing creates nonconcave functional approximation. (c) Expanding the smoothing range maintains concavity.

to obtain the value function approximation at iteration $n + 1$.

A somewhat simpler way of estimating a nonlinear function is via the SHAPE algorithm [Cheung & Powell (2000)]. Here, the basic updating equation is given by:

$$\hat{V}^n(R) = \hat{V}^{n-1}(R) + \alpha^n \left(\tilde{v}^n - \nabla \hat{V}^{n-1}(R^n) \right) \cdot R \quad R \geq 0 \quad (3.15)$$

The basic idea is that we start with an initial approximation $\hat{V}^0$, and then successively 'tilt' the function using the linear slope term $(\tilde{v}^n - \nabla \hat{V}^n(R^n))R$, which serves as a correction term by adding the difference between the current *estimate* of the slope of the function and the actual slope of the approximation. Since we want to maintain concavity, we should use a concave function, such as:

$$\hat{V}^0(R) = \rho_0\left(1 - e^{-\rho_1 R}\right)$$
$$\hat{V}^0(R) = \ln(R+1)$$
$$\hat{V}^0(R) = -\rho_0(x - \rho_1)^2$$

If we need a piecewise linear function, any of these examples can be modeled as piecewise linear with breakpoints at each integer. SHAPE is provably convergent for continuously differentiable functions. If piecewise linear functions are used, it appears to provide very good results based on experimental testing. If we are solving sequences of network problems and have access to left and right gradients, we can use a two-sided version of SHAPE given by:

$$\hat{V}^{n+1}(R) = \begin{cases} \hat{V}^n(R) + \alpha^n\left(\tilde{v}^{n-} - \hat{V}^n(R^n)\right)R & R \leq R^n \\ \hat{V}^n(R) + \alpha^n\left(\tilde{v}^{n+} - \hat{V}^n(R^n)\right)R & R \geq R^n \end{cases}$$

3.2 Nonlinear value functions and multiperiod travel times

Special care has to be used when adopting nonlinear functions. One issue that arises is in the context of multiperiod travel times. Consider two locations i, and j sending vehicles to location k [see Fig. 3]. Assume that the travel time t_{jk} from j to k is greater than that from i to k. If we use a nonlinear value function approximation, location j will 'see' this function first, before the arrivals from i have been planned. As a result, location j will underestimate the total flow into the location, and therefore use the higher estimate of the slope of the function [the solid part of the function in Fig. 3]. By overestimating the value of resources at this location, the model is encouraged to move them a longer distance than might be necessary.

If we use linear value function approximations, both i and j see the same value of vehicles downstream, since the slope of a linear function is independent of the flow into the location. Presumably, our updating strategy will eventually find the right price (or slope) for resources in the future, which will result in a solution that uses resources from j rather than i. But, when we use nonlinear value functions, this will not generally be the case. Location j will see the function first, and will price resources at the steepest part of the curve (since it is concave). If location j sends vehicles to k, location i will then see this decision (which at time t'

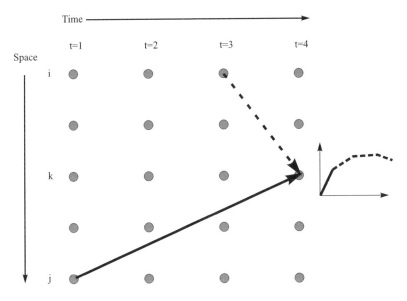

Fig. 3. The challenge of using nonlinear value functions with multiperiod travel times.

has already been made) and subsequently value additional resources at a smaller slope.

Our solution to this problem is relatively simple. First define:

$R_{tt'}$ = The number of resources that we know about at time t that can be used (acted on) at time t'.

In a particular subproblem, we may act on resources in R_{tt}, whereas resources captured by $R_{tt'}, t' > t$ would represent resources that are en route and will not arrive until some point in the future.

We use a value function approximation that is separable over time:

$$\hat{V}_t(R_t) = \sum_{t' \geq t} \hat{V}_{tt'}(R_{tt'})$$

Let:

$x_{tt'}$ = The vector of decisions made at time t producing resources that will become available at time t'.
x_t = $(x_{tt'})_{t' \geq t}$
$\overline{R}_{tt'}(x_t)$ = $A_{tt'}x_t$, where $A_{tt O'}$ is a matrix that sums the elements in x_t that arrive to locations in time t'.

Finally, we would like to define the cumulative number of resources at time t' that we know about at time t, including decisions made before time t:

$R_{tt'}$ = The cumulative number of resources that will become available at time t' made *before* time t.

$= \sum_{\bar{t}<t} A_{\bar{t}t'} x_{\bar{t}t'}$,

$= R_{tt'} + \overline{R}_{tt'}$.

Thus, $R_{tt'} + \overline{R}_{tt'}(x_{tt'})$ is the total number of resources that will be available at time t' that we know about at time t, including the effect of decisions made at time t.

Our basic approximation strategy involves solving problems of the form:

$$\tilde{V}_t(R_t, \omega_t) = max_{x \in \chi} C_t(x, S_t) + \sum_{t'>t} \hat{V}_{tt'}(R_{tt'} + \overline{R}_{tt'}(x_t, \omega_t)) \tag{3.16}$$

which is solved subject to:

$$\sum_{d \in \mathcal{D}} x_{tad} = R_{t,at} + \hat{R}_{t,at} \tag{3.17}$$

$$\sum_{a \in A} \sum_{d \in \mathcal{D}} x_{tad} \delta_{t'a'}(t, a, d) - R_{t,a't'} = \hat{R}_{t,a't'} \tag{3.18}$$

Let $\tilde{v}_{t,at}$ be the dual variable with respect to equation (3.17) and let $\tilde{v}_{t,a't'}$ be the dual variable for equation (3.18). Equation (3.18), then, captures the impact of a decision made before time t on problem t by creating resources that become actionable at time t'. This issue did not arise with single period travel times, or with linear approximations. If possible, we will try to find the value of one more and one less resource. In this case, the duals are denoted $\tilde{v}^+$ and $\tilde{v}^-$, respectively.

The updating strategy is basically the same as before:

$$\hat{V}_{tt'}^{n+1} \leftarrow U^V\left(\hat{V}_{tt'}^n, \tilde{v}_{tt'}^{n-}, \tilde{v}_{tt'}^{n+}, (R_{tt'}^n + \overline{R}_{tt'}^n)\right) \tag{3.19}$$

We are updating the slopes around the point $(R_{tt'}^n + \overline{R}_{tt'}^n)$, since we are effectively approximating the value function as a function of the number of resources that we know about at time t.

Note that we only solve a single problem at time t, and yet we approximate functions of the form $\hat{V}_{tt'}$. In the case of problems that can be formulated as multistage linear programs (which covers most of the problems that arise in this setting), we would use the dual variables for the resource constraints, $\tilde{v}^n_{tt'}$ to update separable nonlinear approximations using either SHAPE or CAVE.

One step that can dramatically accelerate the rate of convergence (especially when some travel times are quite long, measured in units of time periods) works as follows. Instead of using the dual variable $\tilde{v}_{tt'}$, we instead use:

$$\overline{v}^-_{t,at'} = \min_{t \leq \overline{t} \leq t'} \{\tilde{v}^-_{\overline{t}at'}\} \tag{3.20}$$

$$\overline{v}^+_{t,at'} = \max_{t \leq \overline{t} \leq t'} \{\tilde{v}^+_{\overline{t}at'}\}. \tag{3.21}$$

Equation (3.20) uses the best dual variable of all the subproblems that are sending resources to state a at time t'. This has the effect of quickly finding the best location that should send resources arriving at time t' and then building the value of this location into the value function approximation for time t. We would then use $\overline{v}^-$ and $\overline{v}^+$ instead of $\tilde{v}^-$ and $\tilde{v}^+$.

3.3 An algorithmic metastrategy for multiagent problems

Large, complex systems such as trucking companies, railroads, and intermodal operations are almost always characterized by a number of decision makers (or agents) solving different parts of the problem, each with their own information. In Section 2.3 we introduced the basic notation required to handle multiagent thinking.

The challenge of multiagent problems, of course, is trying to devise a strategy that allows each agent to behave independently, but using information that encourages the agents to behave in a coordinated way. We are going to accomplish this using a relatively minor adjustment to our basic dynamic programming recursion. In fact, we are going to find that the presentation in Sections 3.1 and 3.2 is most of what we need. The handling of multiple agents is more a change in perspective than an entirely new class of techniques.

The transition to multiagent thinking involves making the transition from stepping through time, to one of stepping through areas of control as well as time. In a time-staged environment, we made decisions at each time period $t \in T$. We can think of each of these decision epochs as a decision with a different set of information. We can formulate a kind of dynamic programming recursion using:

$$\tilde{V}_{qq}(R_{qq}) = \max_{x_q} C_q(x_q) + \sum_{q' \in \tilde{M}_q} \hat{V}_{qq'}\left(R_{qq'} + \overline{R}_{qq'}(x_{qq'})\right) \tag{3.22}$$

Note the similarities between equations (3.19) and (3.22). In fact, if we assume that the agent subproblems are solved in sequence, then we can view the multiagent problem in a manner identical to a time staged formulation by simply using q as the time variable. Of course, there is no assurance that such sequencing would occur.

We can solve equation (3.22) using the same approximation techniques that we used for the time-staged problem, and the same updating schemes. In fact, the same issue arises when we decide to use nonlinear value functions as arises in the context of multiperiod travel times.

$$\hat{v}^-_{q,aq'} = \min_{q \leq \bar{q} \leq q'} \left\{ \tilde{v}^-_{\bar{q}aq'} \right\} \tag{3.23}$$

$$\hat{v}^+_{q,aq'} = \max_{q \leq \bar{q} \leq q'} \left\{ \tilde{v}^+_{\bar{q}aq'} \right\}. \tag{3.24}$$

Then we use $\hat{v}^-_{q,aq'}$ and $\hat{v}^+_{q,aq'}$ to update $\hat{V}_{q,aq'}$.

If we use a linear approximation for $\hat{V}$, we do not need the double indexing (qq'). In fact, linear approximations fall in the general strategy of pricing systems in multiagent systems. Nonlinear approximations do not seem to have been considered. But, as we have discussed (primarily in the context of multiperiod travel times) they offer special challenges that need to be addressed.

3.4 Classes of decision functions

We now have the foundation to introduce a very general class of decision functions. We return to our four classes of information: knowledge (K_q), forecasts of exogenous processes (Ω_q), plans (x^p) and values ($V_{qq'}$). We illustrate each of these classes of information by briefly describing a decision function based on knowledge alone, or knowledge paired with each of the other three classes of information by themselves, creating four combinations of information sets shown in Table 1. Each of these combinations produces a decision function that falls within a major class of algorithms.

This discussion is intended to emphasize that optimizing dynamic systems can come in a variety of forms. It is very common in the transportation and logistics community to assume the use of myopic policies, or rolling horizon policies based on deterministic forecasts of future activities. Both are valid approximations, which can work well in certain situations. But, they overlook the other two classes of decision functions, or the possibility of mixing information to form a hybrid strategy.

We now discuss each class of decision function.

Table 1
Summary of elementary classes of decision functions

Information set	Function class	Designation
K_q	Myopic policies	Π^M
K_q, Ω_q	Rolling horizon policies	Π^{RH}
K_q, x^p	Proximal point algorithms	Π^{PP}
$K_q, V_{\bar{M}_q}$	Dynamic programming	Π^V

3.4.1 Myopic policies (Π^M)

We start with knowledge alone. These decision functions know the state of the system, but do not make any forecast of the future. These represent myopic policies, which we designate by Π^M. The information set for myopic policies is represented by: $I_q^M = (K_q)$.

Myopic policies are the most widely used in practice (humans predominantly use myopic policies). Less-than-truckload companies use myopic policies to determine when trucks should be dispatched and the routing of freight through the network. The most basic dispatch rule is a control limit policy. Let $X_{tij} = 1$ if a truck should be dispatched from i to j at time t, and 0 otherwise. Let R_t be the amount of freight weight to be dispatched. Then a basic dispatch rule is simply:

$$X_t^\pi = \begin{cases} 1 & R_t \geq \bar{d}_t \\ 0 & \text{Otherwise} \end{cases}$$

Here, $\bar{d}_t$ is a dispatch rule. If the amount of freight is at least $\bar{d}_t$, then we dispatch the truck. Otherwise, we hold. In LTL carriers, the basic rule will typically be 'send the truck if full until the end of the freight cycle; if it is the last dispatch of the night, send the truck if it has at least a certain amount of freight.' Such a policy would be used if there is a strong daily cycle to the freight, as would happen if the freight is arriving from the city trucks coming off the street. Dispatchers know when they are filling up the last truck of the night. If there are only a few shipments on the truck, the carrier will typically hold the freight until the next day, resulting in a service failure (with some insight, the carrier has held a few noncritical freight bills to the side).

Another example of a myopic policy is a dynamic assignment problem where we are assigning drivers to loads. Let $\mathcal{R}_t$ be the set of drivers available to be assigned at time t, and $\mathcal{L}_t$ the set of loads. We may optimize the assignment of drivers to loads using a simple assignment problem:

$$\min_x \sum_{r \in \mathcal{R}_0} \sum_{l \in \mathcal{L}_0} c_{0rl} x_{0rl} \qquad (3.25)$$

subject to:

$$\sum_{r \in \mathcal{R}_0} x_{0rl} \leq 1 \quad \forall l \in \mathcal{L}_0 \qquad (3.26)$$

$$\sum_{l \in \mathcal{L}_0} x_{0rl} \leq 1 \quad \forall r \in \mathcal{R}_0 \qquad (3.27)$$

Again, we are using only the information we know at time t.

3.4.2 Rolling horizon policies (Π^{RH})

Rolling horizon policies combine what we know now (our knowledge base) with forecasts of the future over a planning horizon. We let T_t^{ph} be the set of points in time in our planning horizon given that we are planning a system at time t. Our information set for a rolling horizon policy, then, would be expressed by:

$$I_t = (K_t, \Omega_t)$$

where $\Omega_t = (\Omega_{tt'})_{t' \in T_t^{ph}}$ is the set of events that we have forecasted in the future given what we know at time t. In practice, Ω_t contains a single outcome representing a *point forecast*, and we are going to assume that we are using a point forecast here. For example, if we are trying to allocate containers to meet future demand, we would normally forecast what we would *expect* would happen. The biggest challenge of using distributional forecasts ($|\Omega_t| > 1$) is the lack of effective tools for solving problems under multiple future scenarios (by contrast, we do not have any difficulty using distributional forecasts when we use value functions).

Consider, for example, the basic assignment problem we formulated in (3.25) and (3.26). Assume we can generate a forecast of resources and tasks in the future. Thus, $\omega \in \Omega$ would correspond to $(\hat{\mathcal{R}}_t, \hat{\mathcal{L}}_t)_{t \in T^{ph}}$. We might want to allow a resource at time t to be assigned to a task at time $t' > t$, so we let:

$\mathcal{R}_t = $ The cumulative set of all resources available at time t or some time in the future.
$\mathcal{R}_t \cup_{t' \geq t} \hat{\mathcal{R}}_{t'}$
$\mathcal{L}_t = $ The cumulative set of all tasks available at time t or some time in the future.
$\mathcal{L}_t \cup_{t' \geq t} \hat{\mathcal{L}}_{t'}$

Under this forecast, we would solve the following problem:

$$\min_x \sum_{t \in T^{ph}} c_t x_t$$

subject to:

$$\sum_{l \in \mathcal{L}_t} x_{trl} \leq 1 \quad \forall r \in \mathcal{R}_t$$

$$\sum_{r \in \mathcal{R}_t} x_{trl} \leq 1 \quad \forall l \in \mathcal{L}_t$$

The myopic version of the assignment problem can be criticized because we might take a driver and assign him to a less valuable load now, when we could have used him on a more valuable load later. By contrast, when we use a deterministic forecast, the rolling horizon procedure could have us holding a driver now, even though there is a load available, for a load in the future that may never materialize.

Myopic policies, and rolling horizon procedures, are the most widely used techniques in practice for solving dynamic problems in transportation and logistics. Myopic policies tend to work well in situations that are either highly dynamic, and when rules can be devised which reflect the outcomes that might happen in the future. For example, in our assignment problem, we might have a basic rule that we will not assign a driver to a load shorter than 500 miles (since it probably pays too little). Thus, if the only load we have available to us is only 200 miles, we will refuse the assignment, knowing that there is a good likelihood that a longer load will become available shortly. Thus, a good myopic policy can work quite well.

3.4.3 Proximal point algorithms (Π^{PP})

Often overlooked in the design of algorithms is the value of making decisions that reflect either a forward looking plan, or past patterns of behavior. We claim that both of these represent instances of planning, and should be reflected in decisions made now.

Assume we are managing the flows of intermodal containers on a global level. A separate planning process has made a projection of the number of containers that should move from one location to another on a weekly basis for the next 10 weeks. We can represent this plan using the basic form:

x_{tad}^p = The number of containers with attribute a to which we will apply decision d at time t.

A plan is almost always expressed at some level of aggregation. Thus, we may have 30 types of containers, but we may plan for only the five major groups. Similarly, it may be necessary to send containers to specific locations, but our plan may express decisions only on a regional level. For simplicity, we may let $\hat{a}$ represent an aggregation of the attribute vector, and $\hat{d}$ an aggregation of a decision (such as, the decision to send to a region instead of a specific location). Similarly, we may aggregate time as well (total flow over a week instead of on a particular day). Our vector x^p, then, is expressed at a fairly aggregate level.

In a number of operations, there is not an explicit plan, but there is a *pattern* of activity. In this setting, we may define x^p_{tad} as the average flow that satisfies the pattern (a, d). When looking at past history, we would aggregate time into a period such as a day of week. As with planning, averaging past history is usually done at a more aggregate level.

Now we wish to solve a problem that we might express as:

$$\min_{x \in \chi} \sum_{t \in T} c_t x_t \tag{3.28}$$

where T is the set of time periods in the planning horizon. It is intuitively reasonable to make decisions that do not deviate from a plan by too much. At the same time, if x^p represents a summary of past patterns of behavior, we can also argue that our optimization model should not deviate too much from past patterns. We can achieve this by modifying our basic optimization problem (3.28) as follows:

$$\min_{x \in \chi} \sum_{t \in T} c_t x_t + \rho ||G(x) - x^p|| \tag{3.29}$$

Here, $G(x)$ is an aggregation function that maps our decision variable x (which presumably is fairly detailed) back into the more aggregated space that we are using to plan. The term $\rho ||G(x) - x^p||$ is precisely the term used in Rockafellar's proximal point algorithm, which solves sequences of problems of the form:

$$x^{n+1} = \arg\min_{x \in \chi} \sum_{t \in T} c_t x_t + \rho ||x - \bar{x}^n||$$

where:

$$\bar{x}^{n+1} = (1 - \alpha^n)\bar{x}^n + \alpha^n x^{n+1}$$

3.4.4 Dynamic programming (Π^V)

Our last information class is $I_q^{DP} = (K_q, V_q)$. Here, we want to make decisions that reflect what we know, and the impact of our decisions on other parts of the problem. The conceptual framework is precisely that of dynamic programming, which we have already covered in earlier sections. Returning to our illustrative assignment problem, let's now try to solve it over time, with multiple potential outcomes in the future. This would be formulated as:

$$\min_{\pi \in \Pi} E\left\{ \sum_{t \in T} c_t X_t^{\pi} \right\}$$

We can formulate this using a basic dynamic program:

$$V_t^+(S_t^+) = \max_{x \in \chi} C_t(x, S_t^+) + E\{V_{t+1}^+(S_{t+1}^+)|S_t^+\}$$

but these are rarely solvable. Instead, we resort to our approximation strategy:

$$\widetilde{V}_t(S_t, \omega_t) = \max_{x \in \chi} C_t(x, S_t^+) + \hat{V}_{t+1}(R_{t+1}(\omega_t))$$

where the goal is to devise a version of $\hat{V}$ which will produce a near-optimal solution. We can think of this as a function:

$$X_t^\pi(S_t, \hat{V}_{t+1}) = \arg\max_{x \in \chi} C_t(x, S_t^+) + \hat{V}_{t+1}(R_{t+1}(\omega_t))$$

This expression uses a state variable, which in our information-theoretic vocabulary, represents our knowledge base.

It is significant that a dynamic-programming based approach, which uses value functions to capture the impact of decisions made now on the future, incorporates uncertainty relatively easily. The effect of different possible outcomes is captured in the value function $\hat{V}$, which is much simpler than solving a problem at time t with an explicit set of multiple scenarios in Ω. Since the function $\hat{V}$ is estimated over a number of iterations, it is useful to use the notation $V(\Omega)$ to represent the information content of a value function. Specifically, a decision function that uses value functions is implicitly using a forecast of exogenous outcomes, expressed through the value functions.

Adding value functions to a decision function is equivalent to using a forecast of the impact of a decision on another agent. Companies do this all the time when the decision is to purchase supplies, and the agent is a supplier. The value function, then, is usually a linear function that is the price of the product times the quantity. A car distribution manager for a railroad might implicitly use a value function when he looks at a region and recognizes that there is a surplus (marginal value of additional equipment is small) or a deficit (marginal value of additional equipment is large). The distribution manager is implicitly using a nonlinear value function if he is also thinking 'this region needs 20 additional cars.'

As a rule, humans have difficulty with value functions because it explicitly requires using costs to make decisions. Human decision-making is based on the concept of state/action pairs: if the system is in this state, then take this action. Recognition of this fact is the basis for artificial intelligence (AI). The application of AI to complex problems have typically failed simply because the state variable is far too complex. The power of the brain to sort through patterns to identify the relevant portion of the state variable has not been matched on the computer. Cost-based optimization models, on the other hand, have little difficulty with very complex state variables. Computers are good at adding up costs to make a decision, which is the reason that math programming-based models have proven to be so popular. Needless to say, value functions appear to be most useful to computer-based models and algorithms. If you tell a human that you are going to give him a value function to help him make a decision, the response is generally going to be disappointing.

3.5 A hybrid model

We have seen that four information classes each produce a different class of algorithms that have been widely studied. This raises a natural question of whether we can combine all four classes. We propose to do this by incorporating forecasts (Ω) through the value function as we did in dynamic programming. Thus, our information set is given by:

$$I_t = \left(K_t, x_t^p, V_{t+1}(\Omega)\right)$$

Such a decision function would look like:

$$X_t^\pi(S_t, \hat{V}_{t+1}) = \arg\max_{x_q \in \chi_q} C_q(x_q, S_t^+) - \rho||G(x) - x^p|| \\ + \sum_{q' \in \tilde{M}_q} \hat{V}_{qq'}(R_{qq'} + \overline{R}_{qq'}(x_{qq'}, \omega))$$

We offer equation (3.30) as a relatively general function which is scalable to very large problems such as railroads and trucking companies. Not only does it incorporate all four information classes, it also handles the multiagent structure common to complex operations. At the same time, it is important to realize that it is not necessary to use the ultimate decision function, since value can be obtained using much simpler functions, and all the more basic decision functions, including myopic policies, can be very effective.

4 Modeling operational problems

The next step is to apply our framework to specific operational problems in transportation and logistics. An effective way to classify operational problems is to begin by organizing them on the basis of how resources interact. There are three fundamental ways to change a resource:

(1) Couple – Combine two resources to create a layered resource consisting of two or more resources.

(2) Uncouple – Break down a composite resource into its primitive components (or simply decouple one resource from a set of layered resources).
(3) Modify – Major classes of modify include: a) move (from one location to the next), b) entry (such as purchasing a resource), c) exit (a resource leaves the system), and d) do nothing. Other examples might include: perform maintenance on an engine, clean out a trailer, have a driver go on vacation.

Different problem classes can often be created based on the type of coupling they entail. Special classes of interest in transportation include:

(1) One-to-one – Such as one driver and one load, one pilot and one plane, one boxcar and one customer demand.
(2) Several-to-one – Several locomotives pull one train, two drivers may create a sleeper team to drive a tractor, and several customers can fit in one vehicle.
(3) Many-to-one – Many freight bills or packages may fit in one trailer, many boxcars fit in a train.

The several-to-one class has some important variations in transportation. The first is the bundling of several resources with a common location (multiple locomotives at a location being assigned to the same train; two drivers being assigned to move the same tractor). The second is bundling resources with different locations (clustering), such as occurs in the vehicle routing problem.

Our discussion proceeds in stages. We start with resource allocation problems, which are all in the class of one-to-one coupling problems. These are described starting with single layer problems (Section 4.1), two-layer problems (Section 4.2), and multi-layer problems (Section 4.3). Finally, we turn to problems that involve bundling (Section 4.4).

4.1 Single-layer resource allocation

Fundamental to operational problems is the coupling of two layers (product with customer, driver with load, vehicle with delivery). We might say that the 'energy' derived from coupling two resource layers together is what keeps the process moving. So, how can we even have a single-layer problem? The answer is simple: any time we have demands that must be satisfied at a particular point in time. In production problems, this means no backlogging of demand. In transportation and logistics, it often means that there are 'tight time windows.' For example, we would have a one-layer problem if we were assigning locomotives to trains, where the trains had to be moved at a point in time. The same would be true if we are moving boxcars to serve demands that have to be served at a particular point in time.

We are interested in problems where we are managing a set of reusable resources. These problems arise when we are managing sets of containers (trailers, boxcars, intermodal containers etc.), vehicles (tractors, locomotives, aircraft, etc.), and people (drivers, pilots, crews, etc.). In this section, we are going to focus on problems where the number of resources being managed is relatively large, which means that it is typically not useful to track each individual resource.

Most of the time, representing these problems as single-layer resource allocation problems can be justified only as simplifications of real-world problems. But, the one-layer problem serves not only as a useful pedagogical tool, but it is also practical for some problem classes.

Throughout our discussion, our solution strategy is assumed to follow the framework described in Section 3. For the most part, we are primarily concerned with how to solve the basic problem:

$$\max_{x \in \chi} c_t x_t$$

which means, 'what do we do at time t?' Note that what we do at time t may consist of a series of steps that extend into the future. Recall that $x_t = (x_{tt'})_{t' \geq t}$; meaning a vector of decisions over time using information that we know at time t. Thus, $c_t x_t$ is equivalent to $\sum_{t' \geq t} c_{t'} x_{tt'}$. As we proceed, it is important to be clear whether we are solving a problem at time t with actions strictly at time t, or whether the actions may extend into the future using the information at time t.

We proceed with the expectation that including plans or value functions would not destroy the fundamental structure. There are different ways to incorporate the effect of plans, with the use of the term $||G(x) - x^p||$ only one of them. If we include value functions, we note that linear value functions will never destroy structure, but nonlinear functions (even separable nonlinear functions) must be handled with care.

Our discussion of resource allocation proceeds in a progression from single commodity (Section 4.1.1) to multicommodity (Section 4.1.2) to heterogeneous resources (Section 4.1.3). In all three of these sections our subproblems consist of a single set of actions initiated at time t.

4.1.1 Single commodity

Single commodity problems arise when (a) the attribute vector a consists only of a scalar state variable (which in transportation problems usually represents a geographical location), and (b) when a resource must be in the same state as a task to serve the task. In transportation applications, it is very common for the 'state' of a resource to be a geographical location. If we have only one type of resource, we would use $a = (i)$. For this section, we use the index i instead of the attribute vector a to emphasize the structure of the problem. Our purpose in switching to a different notation can be explained by the desire to exploit structure that arises only in the context of single commodity problems. We are going to continue to use this specialized notation when we discuss multicommodity problems, which also exhibit special structure.

We have two types of decisions for this problem class:

$\mathcal{D}^s$ = Decisions to serve a task. The set $\mathcal{D}^s$ may be a set of specific tasks, or a set of task types. We let $\mathcal{D}^s_i$ be the set of tasks that can be served by a resource in state i.

$\mathcal{D}^r$ = Decisions to reposition a resource from one state to another.

u_{tid} = Upper bound on the number of times that decision $d \in \mathcal{D}_i$ may be executed. We assume that u_{tid} is bounded for $d \in \mathcal{D}^s$, and unbounded for $d \in \mathcal{D}^r$.

Similarly, we assume that:

$$M_t(t, i, d) = \left(i^M_{tid}, c_{tid}, \tau_{tid} \right)$$

We use the notation i^M_{tid} as our terminal attribute function instead of a^M_{tid} to be consistent with our adoption of the simple state notation i instead of the more general attribute vector notation a, for single commodity flow problems.

A myopic version of the problem (at time t) is given by:

$$\max \sum_{i,j \in \mathcal{I}} \sum_{d \in \mathcal{D}} c_{tid} x_{tid} \qquad (4.1)$$

subject to:

$$\sum_{d \in \mathcal{D}_i} x_{tid} = R_{ti} + \hat{R}_{ti} \qquad (4.2)$$

$$x_{tid} \leq u_{tid} \qquad (4.3)$$

Such a formulation would never work because we would never reposition resources from where we need them to where we want them. Virtually all transportation companies, which solve resource allocation problems, require some sort of mechanism (typically, a central planning group), which looks into the future and makes decisions about repositioning. The simplest model that looks into the future is based on a deterministic forecast over a planning horizon. We may be using a forecast of new resources $(\hat{R}_t)c$, upper bounds (u_t), times (τ_{tid}) and costs (c_{tid}):

$$\max \sum_{t' \in T^{ph}_t} \sum_{i \in \mathcal{I}} \sum_{d \in \mathcal{D}} c_{t'id} x_{t'id} \qquad (4.4)$$

subject to, for $t' \in \mathcal{T}_t^{ph}$:

$$\sum_{d \in \mathcal{D}_i} x_{t'id} = R_{t'i} + \hat{R}_{t'i} \qquad \forall j \in \mathcal{I} \tag{4.5}$$

$$\sum_{i \in \mathcal{I}} \sum_{d \in \mathcal{D}_i} x_{t'-\tau_{tid}, id\delta_{t'j}}(t' - \tau_{tid}, i, d) = R_{t'j} \qquad \forall j \in \mathcal{I} \tag{4.6}$$

$$x_{t'id} \leq u_{t'id} \tag{4.7}$$

$$x_{t'id} \geq 0 \tag{4.8}$$

Equations (4.5) and (4.6) can be combined to create classical flow conservation constraints. We retain this form since it creates a more natural transition with stochastic models. The problem (4.4)–(4.8) is a pure network and is easily solved as a general linear program or with more specialized solvers.

Solving rolling horizon problems using deterministic forecasts is popular and can be effective, but suffers from several limitations: it uses point forecasts of demands (which means it may not supply enough capacity to provide a high level of service), and it takes a problem where all you can do is determine what to do right now (since information will change in the future) and formulates a problem where you are making decisions over an extended planning horizon, which is inherently more difficult. We overcome these limitations by using our dynamic programming approximations and solve:

$$\tilde{V}_t(R_t, \omega_t) = \max_{x \in \chi} \sum_{i \in \mathcal{I}} \sum_{d \in \mathcal{D}} c_{tid} x_{tid}$$
$$+ \sum_{t' > t} \sum_{j \in \mathcal{I}} \hat{V}_{t+1, jt'}(R_{t+1, jt'} + \overline{R}_{t+1, jt'}(x_t, \omega_t)) \tag{4.9}$$

If we use a linear approximation for $\hat{V}$, then equation (4.9) reduces to:

$$\tilde{V}_t(R_t, \omega_t) = \max_{x \in \chi} \sum_{i \in \mathcal{I}} \sum_{d \in \mathcal{D}} c_{tid} x_{tid}$$
$$+ \sum_{t' > t} \sum_{j \in \mathcal{I}} \hat{v}_{t+1, jt'}(R_{t+1, jt'} + \overline{R}_{t+1, jt'}(x_t, \omega_t)) \tag{4.10}$$

$$= \max_{x \in \chi} \left\{ \sum_{i, j \in \mathcal{I}} \sum_{d \in \mathcal{D}} c_{tid} x_{tid} \right\} + \left\{ \sum_{t' > t} \sum_{j \in \mathcal{I}} \hat{v}_{t+1, jt'}(\overline{R}_{t+1, jt'}) \right\}$$

$$+ \left\{ \sum_{t' > t} \sum_{j \in \mathcal{I}} \hat{v}_{t+1, jt'} \overline{R}_{t+1, jt'}(x_t, \omega_t) \right\} \tag{4.11}$$

The second term in brackets on the right side of (4.11) is not a function of x_t and hence can be ignored. The third term can be simplified by using:

$$\bar{R}_{t+1,jt'} = \sum_{i \in \mathcal{I}} \sum_{d \in \mathcal{D}_i} \delta_{t'j}(t,i,d) x_{tid} \quad (4.12)$$

Dropping the second term in brackets in Eq. (4.11) and substituting Eq. (4.12) into Eq. (4.11) gives:

$$\tilde{V}_t(R_t,\omega_t) = \max_{x \in \chi} \left\{ \sum_{i \in \mathcal{I}} \sum_{d \in \mathcal{D}} c_{tid} x_{tid} \right\} + \left\{ \sum_{t' > t} \sum_{j \in \mathcal{I}} \hat{v}_{t+1,jt'} \sum_{i \in \mathcal{I}} \sum_{d \in \mathcal{D}} \delta_{t'j}(t,i,d) x_{tid} \right\} \quad (4.13)$$

$$= \max_{x \in \chi} \left\{ \sum_{i \in \mathcal{I}} \sum_{d \in \mathcal{D}} c_{tid} x_{tid} \right\} + \left\{ \sum_{i \in \mathcal{I}} \sum_{d \in \mathcal{D}} \left(\sum_{t' > t} \sum_{j \in \mathcal{I}} \delta_{t'j}(t,i,d) \hat{v}_{t+1,jt'} x_{tid} \right) \right\} \quad (4.14)$$

We note that:

$$\sum_{t' > t} \sum_{j \in \mathcal{I}} \delta_{t'j}(t,i,d) \hat{v}_{t+1,jt'} x_{tid} = \hat{v}_{t+1,i^M_{tid},t+\tau_{tid}} x_{tid} \quad (4.15)$$

Eq. (4.15) simply says that if we act on a resource in state i at time t with decision d and it produces a resource in state $j = i^M_{tid}$ at time t', then we can pick up the value of that resource. This allows us to reduce (4.14) to:

$$\tilde{V}_t(R_t,\omega_t) = \max_{x \in \chi} \sum_{i \in \mathcal{I}} \sum_{d \in \mathcal{D}} \left(c_{tid} + \hat{v}_{t+1,i^M_{tid},t+\tau_{tid}} \right) x_{tid} \quad (4.16)$$

Eq. (4.16) shows us that using a linear approximation of the value function is equivalent to adding a price to each assignment that is the marginal value of the resource in the future. In fact, if we look at the updating equation for linear approximations, we quickly see that $\hat{v}_{tt'} = \hat{v}_{t'}$, allowing us to further simplify (4.16) to:

$$\tilde{V}_t(R_t,\omega_t) = \max_{x \in \chi} \sum_{i \in \mathcal{I}} \sum_{d \in \mathcal{D}} \left(c_{tid} + \hat{v}_{i^M_{tid},t+\tau_{tid}} \right) x_{tid} \quad (4.17)$$

Linear approximations introduce an additional simplification: problem (4.16) decomposes by location. Thus, we can solve (4.17) by solving a sequence of problems that look like:

$$\tilde{V}_{ti}(R_{ti},\omega_t) = \max_{x \in \chi} \sum_{d \in \mathcal{D}_i} \left(c_{tid} + \hat{v}_{i^M_{tid},t+\tau_{tid}} \right) x_{tid} \quad (4.18)$$

Furthermore, the solution of (4.18) involves nothing more than a sorting of decisions $d \in \mathcal{D}_i$ in order of $(c_{tid} + \hat{v}^M_{i_{tid}, t+\tau_{tid}})$.

Linear approximations are especially appealing since they are so simple. In practice, they can be unstable. If the term $(c_{tid} + \hat{v}^M_{i_{tid}, t+\tau_{tid}})$ is attractive, we end up with a large value for x_{tid}. If $d \in \mathcal{D}^s$, which means we are serving a task, then the number of tasks serves as a natural upper bound which stabilizes the solution. If $d \in \mathcal{D}^r$, then typically u_{dt} is unbounded, and we can get extreme flows.

There are three solutions to this behavior. One is to add an artificial upper bound, which we might call y_{tid}, where y is a decision variable. We would then solve the same problem with the added constraint $x_{tid} \leq y_{tid}$. We then have to introduce procedures for changing the artificial controls y. This approach was used in Powell and Carvalho (1998) with reasonable success. But, it does not generalize easily to multicommodity and heterogeneous resources (see below).

A second approach is to include a nonlinear stabilization term. One framework for including such a term is to use a proximal point algorithm, where at iteration n we would solve:

$$x^n = \arg\max_{x \in \chi} \sum_{i \in \mathcal{I}} \sum_{d \in \mathcal{D}} (c_{tid} + \hat{v}_{t+1, j, t+\tau_{tid}}) x_{tid} + \theta \sum_{i \in \mathcal{I}} \sum_{d \in \mathcal{D}} (x_{tid} - \bar{x}^n_{tid})^2$$

(4.19)

with the updating scheme:

$$\bar{x}^n = (1 - \alpha^n) \bar{x}^{n-1} + \alpha^n x^n$$

The proximal term $(x_{tid} - \bar{x}^n_{tid})^2$ helps to stabilize the solution, and because the additional term is separable, it does not generally cause serious algorithmic headaches. If we are looking for integer solutions, then a piecewise linear penalty term should be used.

A third approach is to use a nonlinear value function approximation. Separable functions of the form:

$$\hat{V}_{tt'}(R_{tt'}) = \sum_{i \in \mathcal{I}} \hat{V}_{t, it'}(R_{t, it'})$$

(4.20)

are generally fairly easy to work with. We find ourselves having to solve equation (4.9) directly. Assume that we are interested in integer solutions, which leads us to use a piecewise linear form for $\hat{V}$, as given in equation (3.14). This would produce a network such as the one illustrated in Fig. 4. This problem is easily solved as a linear network, and it naturally returns integer solutions.

For this problem class, nonlinear functions appear to work extremely well. They are easy to estimate using the techniques of Section 3.1; they are

Ch. 13. Dynamic Models of Transportation Operations

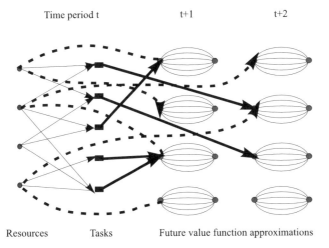

Fig. 4. Illustration of a single commodity flow problem at time t with separable, nonlinear value function approximations. Thin, solid arcs represent assignment of resources to tasks. Thick solid arcs are tasks moving forward in time. Dashed arcs represent repositioning moves in response to future value function approximations.

Table 2
Percentage of integer optimal value obtained using CAVE for second set of deterministic experiments with single-period time windows (network problems), from Godfrey and Powell, 2002a

Locations	Planning horizon		
	15 (%)	30 (%)	60 (%)
20	100.00	100.00	100.00
40	100.00	99.99	100.00
80	99.99	100.00	99.99

computationally quite easy to solve (sequences of pure networks) and produce high quality solutions. Table 2 compares the technique when applied to deterministic networks (something you would not want to do in practice, since specialized algorithms are extremely good), indicating near optimal performance. When compared against a rolling horizon procedure, we get the results shown in Table 3. This table provides results as a function of the number of locations, and of the number of resources (holding the number of tasks fixed). Problems with a larger number of locations are harder to solve, in part because the problem becomes increasingly nonseparable. The number of resources is important since the problem becomes more difficult as the number of resources is decreased. The results indicate that a nonlinear value function

Table 3
Comparison of nonlinear approximation using CAVE to a deterministic rolling horizon procedure, for stochastic problems with different numbers of locations and resources. Posterior bound is computed by finding optimal solution assuming all information is known (from Godfrey & Powell, 2002a)

Number of locations	Number of resources	Percentage of posterior bound	
		Rolling horizon (%)	Stochastic using CAVE (%)
20	100	92.2	96.3
20	200	96.3	97.8
20	400	96.6	98.1
40	100	81.0	90.5
40	200	90.7	96.2
40	400	92.6	96.8
80	100	66.3	82.1
80	200	81.4	93.3
80	400	84.8	94.5

approximation can significantly outperform a deterministic approximation based on rolling horizon simulations.

Of particular value is going to be our ability to take this general strategy and apply it to increasingly more general problems. We first illustrate its application to multicommodity problems, followed by heterogeneous resource allocation problems. We then indicate how it can be applied to two-layer problems.

4.1.2 Multicommodity

Multicommodity flow problems arise whenever we have different types of resources and different types of tasks, and we are allowed to substitute the use of different resources, but where the cost of serving a task depends on the type of resource. This might arise when we are managing fleets of trailers, and there are different types of trailers with some substitution. It arises when managing fleets of boxcars and containers, as well as distributing different product types to consumers.

Multicommodity flow problems arise when the attribute of a resource can be described as $a = (k, i)$, where k represents a commodity class (or simply a commodity) and i, our state variable. In any transformation:

$$M(t, a, d) \to (a', c, \tau)$$

we assume that if $a = (k, i)$ then $a' = (k, i')$. We let:

$\mathcal{K}$ = Set of commodity classes.

R_{jt}^k = The number of resources of type k in state i.
x_{tid}^k = The number of times we act on a resource of type k in state i with decision d.

We note that we are following standard notational conventions of putting the commodity class as a superscript. This runs against the notational style that we have been following in this chapter, where all indices are expressed as subscripts. We violate our own notational conventions for reasons of consistency with the research literature. The reader is encouraged to contrast this presentation with our discussion of a more complex problem, heterogeneous resources, where our notation is actually simpler.

We can set up and solve the multicommodity version of the problem just as we did with the single commodity. Rolling horizon procedures are stated simply as:

$$\max \sum_{t' \in T_t^{ph}} \sum_{k \in \mathcal{K}} \sum_{i \in \mathcal{I}} \sum_{d \in \mathcal{D}} c_{t'id}^k x_{t'id}^k \tag{4.21}$$

subject to, for $t' \in T_t^{ph}$:

$$\sum_{d \in \mathcal{D}_i} x_{t'id}^k = R_{t'i}^k + \hat{R}_{t'i}^k \quad \forall j \in \mathcal{I}, k \in \mathcal{K} \tag{4.22}$$

$$\sum_{i \in \mathcal{I}} \sum_{d \in \mathcal{D}_i} x_{id,t'-\tau_{tid}}^k \delta_{jt'}(t' - \tau_{tid}, i, d) = R_{jt'}^k \quad \forall j \in \mathcal{I}, k \in \mathcal{K} \tag{4.23}$$

$$\sum_{k \in \mathcal{K}} x_{t'id}^k \leq u_{t'id} \tag{4.24}$$

$$x_{t'id}^k \geq 0 \tag{4.25}$$

The costs c_{tid}^k may incorporate the cost of assigning a resource of type k to a particular type of task, if $d \in \mathcal{D}^s$. We could, for example, divide the set $\mathcal{D}^s$ (representing decisions to serve a demand) into subsets $\mathcal{D}_k^s$.

The complicating constraint in this formulation is equation (4.24). If our problem is not too large, and we are not interested in integer solutions (or, we are willing to find a near-optimal solution), then commercial LP solvers should work fine here. More problematic is that we are again making the assumption that we know the future perfectly. Also, a multiperiod multicommodity flow problem can be relatively hard to solve.

We may incorporate uncertainty in our forecasts by using the same types of dynamic programming approximations described for single commodity formulations. Without repeating the algebra, it is not hard to show that the multicommodity version of equation (4.16) is:

$$\tilde{V}_t(R_t, \omega_t) = \max_{x \in \chi} \sum_{k \in \mathcal{K}} \sum_{i \in \mathcal{I}} \sum_{d \in \mathcal{D}} \left(c_{tid}^k + \hat{v}_{t+\tau_{tid}, i_{tid}^M}^k \right) x_{tid}^k \tag{4.26}$$

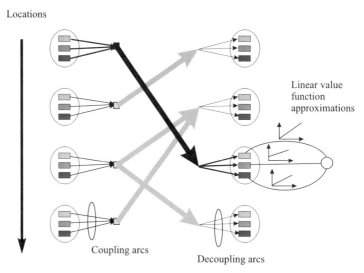

Fig. 5. Network problem produced by multicommodity flow problems with linear value function approximations.

The slopes $\hat{v}^k$ are updated using the sample gradients of the resource constraint (4.22) when solving subproblem t.

We earlier showed that the use of linear approximations for single commodity problems produced subproblems that involved nothing more than simple sorts. Multicommodity problems are a bit more complex. We still require that a resource be in state i to be acted on by a decision in $\mathcal{D}_i$, but we now have the behavior that different types of resources in state i can be acted on by decisions in $\mathcal{D}_i$. The problem reduces to a network which we illustrate in Fig. 5. Note that when we use linear approximations, we can take the slopes of the value function approximations and simply add these to the costs on the coupling arcs, along with any cost that might exist on the decoupling arcs. The resulting problem is a pure network.

This remains quite easy to solve, but suffers from all the problems we described earlier with linear approximations. Furthermore, the use of upper bounds to control the flows (especially repositioning decisions) becomes much trickier. It is important to keep in mind that the artificial upper bounds y_t are deterministic, and must work reasonably well under different sample realizations. The problem with these variables for multicommodity problems is that they do not handle very well the opportunities for substitution across resources. It might be preferable, for example, to have an upper bound that cuts across commodities, but then we destroy our nice network structure.

We can, instead, use separable nonlinear approximations just as we did with single commodity problems. This would involve solving subproblems of the form:

$$\tilde{V}_t(R_t, \omega_t) = \max_{x \in \chi} \sum_{k \in \mathcal{K}} \sum_{i \in \mathcal{I}} \sum_{d \in \mathcal{D}} c_{tid}^k x_{tid}^k$$
$$+ \sum_{t' > t} \sum_{k \in \mathcal{K}} \sum_{j \in \mathcal{I}} \hat{V}_{t+1, jt'}^k (R_{t+1, jt'}^k + \overline{R}_{t+1, jt'}^k (x_t, \omega_t)) \quad (4.27)$$

This problem is illustrated using Fig. 3. Unlike single commodity problems, however, this subproblem is a bit more complicated. Whereas nonlinear value functions produce nice network subproblems in the single commodity case, the use of nonlinear value functions gives us (possibly integer) multicommodity network flow problems. To see why we get multicommodity flow problems, we do not have to look any further than the constraints on the decisions in equation (4.24). These constraints bundle flows of different types of commodities. So why didn't this cause a problem when we used linear approximations? The reason was that the linear function approximation allowed us to write $R_{t+1, jt'}^k(x_t, \omega_t)$ in terms of x_{tid} directly and then use the separability of the linear approximation. Nonlinear approximations mean that the function is no longer separable in x_{tid}, which destroys our structure.

The good news is that the multicommodity flow problems we have to solve are not very large (i.e., a single time period), and if we are interested in integer solutions, the LP relaxation almost always gives us integer solutions anyway. This is where our dynamic formulation is much easier than solving the rolling horizon formulation in Eqs. (4.21), (4.22), (4.23) and (4.24). One-time period problems are much easier to solve than time-staged problems over even modest planning horizons.

These techniques work quite well on both deterministic and stochastic multicommodity flow problems. As with single commodity problems, we can obtain integer solutions as long as we use piecewise linear value function approximations. Table 4 demonstrates the effectiveness of the techniques on both deterministic problems (compared against the results of an LP solver) and stochastic problems (compared against deterministic rolling horizon approximations). Again, we see that the techniques provide near optimal solutions on deterministic problems, and results that significantly outperform rolling horizon models.

There are other tricks and techniques associated with the use of value function approximations for multicommodity flow problems. The interested reader is referred to Topaloglu and Powell (2000).

4.1.3 Heterogeneous resources

Heterogeneous resource allocation problems arise when the resources are relatively complex. These almost always arise when the resources are people, and they often arise when the resources are relatively complex pieces of equipment such as locomotives or airplanes. For example, in a driver

Table 4
Performance of linear and nonlinear value function approximations against a deterministic rolling horizon procedure, from Topaloglu and Powell, 2000

Percent of posterior optimal solution

Problem	Linear	Nonlinear	Rolling horizon
Results of stochastic runs with varying number of locations			
10 locations	86.14	96.96	93.17
20 locations	78.65	93.28	86.84
40 locations	74.13	92.21	86.89
Results of stochastic runs with varying compatibility patterns			
Sub. matrix I	78.65	93.28	86.84
Sub. matrix II	80.59	95.40	90.87
Sub. matrix III	74.83	91.51	82.66
Sub. matrix IV	84.23	97.12	
Results of stochastic runs with varying numbers of resources			
100 res.	74.19	84.87	76.81
200 res.	78.65	93.28	86.84
400 res.	84.41	96.51	91.67

management problem, the attribute of a resource might be:

$$a = \begin{pmatrix} a_1 \\ a_2 \\ a_3 \\ a_4 \\ a_5 \\ a_6 \end{pmatrix} = \begin{pmatrix} \text{Driver's home domicile.} \\ \text{1 if driver represents a sleeper team. 0 otherwise.} \\ \text{The current/next terminal of driver } r. \\ \text{The arrival time of driver } r \text{ at his current/next terminal.} \\ \text{The cumulative driving time of the driver.} \\ \text{The number of days away from home.} \end{pmatrix}$$

When routing and scheduling individual drivers, the attribute vector can become much more complex than this. These problems, however, are typically solved under assumptions of complete information (deterministic models), and are required to produce full schedules for individual drivers.

The management of locomotives might require the following vector of attributes:

$$a = \begin{pmatrix} a_1 \\ a_2 \\ a_3 \\ a_4 \\ a_5 \\ a_6 \\ a_7 \end{pmatrix} = \begin{pmatrix} \text{Number of axles.} \\ \text{H if it is a 'high adhesion' locomotive, L otherwise.} \\ \text{The horsepower class of the locomotive.} \\ \text{The tractive effort rating of the locomotive.} \\ \text{Days remaining until the next required maintenance check.} \\ \text{The location where the locomotive should be maintained.} \\ \text{The identity of the train the locomotive came in on.} \end{pmatrix}$$

These attribute vectors give a hint of the complexity that can arise when solving real resource allocation problems.

When the attribute vector is more complicated than a simple class and state, we refer to the problem as the *heterogeneous resource allocation problem*. These problems can be placed in the context of multicommodity flow problems by using the following observation. Let $a = (a^s, a^d)$ where a^s represent static elements of the attribute vector (elements which do not change when a decision is made) and where a^d captures the dynamic elements. In our driver example, $a^s = (a_1, a_2)$ while $a^d = (a_3, a_4, a_5, a_6)$. The static elements can be concatenated and viewed as a single resource class (or commodity) while the last four can be concatenated and viewed as a state variable. However, these problems do not satisfy the structure of multicommodity flow problems where the upper bound u_{tid} is keyed to the state of the resource.

Aside from this structural difference, the real difference between multicommodity flow problems and heterogeneous resource allocation problems is the size of the attribute space. In multicommodity problems, $a = (k, i)$, so the number of possible attributes is probably close to $|\mathcal{K}| \times |\mathcal{I}|$. If we are managing intermodal containers, we might find $|\mathcal{K}|$ is between 10 and 50, whereas the number of locations, given by $|\mathcal{I}|$, might be between 100 and 1000. This means that the total size of the attribute space might be as large as 50,000, but is typically about 5000. By contrast, a multidimensional attribute vector can easily have millions of possible combinations. When this is the case, the number of attribute vectors that actually occur are typically much smaller, but we do not know in advance which ones will be used.

A deterministic formulation of the heterogeneous resource allocation problem is given by:

$$\max \sum_{t' \in \mathcal{T}_t^{ph}} \sum_{a \in \mathcal{A}} \sum_{d \in \mathcal{D}} c_{t'ad} x_{t'ad} \qquad (4.28)$$

subject to, for $t' \in \mathcal{T}_t^{ph}$:

$$\sum_{d \in \mathcal{D}_a} x_{t'ad} = R_{t'a} + \hat{R}_{t'a} \qquad \forall a \in \mathcal{A} \qquad (4.29)$$

$$\sum_{a \in \mathcal{A}} \sum_{d \in \mathcal{D}_a} x_{t'-\tau_{tad},ad} \delta_{t'a'}(t' - \tau_{tad}, a, d) = R_{t'a'} \qquad \forall a' \in \mathcal{A} \qquad (4.30)$$

$$\sum_{a \in \mathcal{A}} x_{t'ad} \leq u_{t'd} \qquad (4.31)$$

$$x_{t'ad} \geq 0 \qquad (4.32)$$

where we adopt the convention that $x_{t''ad} = 0$ if $t'' < 0$. This is a hard problem as a result of its sheer size. For practical problems, it is virtually impossible to generate the complete attribute space even for a single time period, not to mention over all the time periods in a reasonable planning horizon.

Interestingly, this appears to be one of those problems, which seems to be easier if we use stochastic techniques. So far, we have seen that stochastic techniques can work quite well on deterministic problems. Applying the same techniques we used previously, we find that our one-period problem becomes:

$$\tilde{V}_t(R_t, \omega_t) = \max_{x \in \chi} \sum_{a \in \mathcal{A}} \sum_{d \in \mathcal{D}_a} c_{tad} x_{tad} \\ + \sum_{t' > t} \sum_{a' \in \mathcal{A}} \hat{V}_{t+1,a't'}(\overline{R}_{t+1,a't'} + R_{t+1,a't'}(x_t, \omega_t)) \quad (4.33)$$

subject to equations (4.29), (4.32), adapted to a single time period.

We can use either linear or nonlinear value function approximations, and we end up with the same basic subproblem structures as we did with multicommodity problems. For example, linear approximations reduce to networks such as that illustrated in Fig. 5, whereas nonlinear approximations produce subproblems that look like Fig. 3. The big difference arises because of the size of the attribute space. When we are solving multicommodity problems, it is normally the case that we would enumerate all possible values of $\mathcal{K} \times \mathcal{I}$ in advance. This means that we would have a resource constraint for every combination of k and i. As a result, we will get a dual variable for every possible combination, and we will create a value function approximation for every possible combination, which gets updated at each iteration.

With the heterogeneous case, we cannot generate every element in $\mathcal{A}$. Instead, we have to generate attributes dynamically. Let:

$\mathcal{A}^n$ = The *active attribute space* that has been generated at iteration n.

We propose to use an increasing sequence $\mathcal{A}^n \subseteq \mathcal{A}^{n+1}$. This implies, however, that for a given attribute a and decision d the attribute $a_{t,a,d}^M$ may not have been generated yet. We need an approximation of $\hat{V}_{t'a'}$ for attributes $a' \notin \mathcal{A}^n$. For this purpose, we define:

$\mathcal{A}_i$ = The set of attribute vectors that have a common geographical location $i \in \mathcal{I}$.
$\bar{a}_i$ = The attributes of an artificial resource in location $i \in \mathcal{I}$ that will have the best possible behavior in that location.

We want $\bar{a}_i$ to have the behavior of a resource that is at least as good as any real resource. So, we assume that:

$$c(\bar{a}_i, d) \geq \max_{a \in \mathcal{A}_i, d \in \mathcal{D}_a} c(a, d) \quad (4.34)$$

We refer to the attribute vector $\bar{a}_i$ as the 'best attribute' for location i. We could, of course, simply define a single 'best attribute' that would apply system wide, but it seemed clear to us that we could get much better results if we tightened our bound by making it location specific. Note that it is not necessary that $\bar{a}_i \in \mathcal{A}$; if $\bar{a}_i \notin \mathcal{A}$, then $\bar{a}_i$ would simply be an empty resource bucket. What is important is that we have one attribute which is always present, which allows us to create a value function approximation which is an upper bound. This ensures that we will not artificially avoid a decision just because we underestimate the downstream value of the resource created by the decision.

To create an attribute vector where we ensure that the contribution out of that attribute represents an upper bound over other attributes does not seem to necessarily ensure that the value of the attribute (which includes not only the immediate cost but also the downstream costs) is also an upper bound. The following proposition establishes this result:

Proposition 4.1. [*Powell et al. (2000a)*] *Assume that* (4.34) *holds and that* $\bar{v}^0_{\bar{a}_i t} \geq \bar{v}^0_{ta}$ *for* $a \in \mathcal{A}_i$. *Then*:

$$\bar{v}^n_{\bar{a}_i t} \geq \bar{v}^n_{ta} \qquad \forall a \in \mathcal{A}_i$$

In other words, we can ensure that our estimate of the value of our 'best attribute' is going to be better than the value of resources with other attributes (at the same location).

This does not mean that any of these estimates are actually upper bounds over what the values should be. But, decisions are relative, so this is an important property.

The active attribute space, then, grows as the algorithm visits new states. We can describe the process using:

$\mathcal{A}^n_t$ = Set of attribute vectors that have been generated for time period t in iteration n.

$\tilde{\mathcal{A}}^n_{tt'}$ = Set of attribute vectors that are generated for time period t' when solving the subproblem for time period t.

Of course, the set $\tilde{\mathcal{A}}^n_{tt'}$ may include elements that are already in $\mathcal{A}^n_{t'}$. Our active attribute space is updated using:

$$\mathcal{A}^{n+1}_{t'} = \mathcal{A}^n_{t'} \underset{t<t'}{\cup} \tilde{\mathcal{A}}^n_{tt'} \tag{4.35}$$

This algorithm has been applied to the management of drivers for a major LTL trucking company. It scales easily to handle problems involving the management of thousands of drivers moving tens of thousands of loads between hundreds of different locations.

4.2 Two layer resource allocation

In the previous section, market demands, tasks, requirements, or other expressions of serving an exogenous customer were all modeled as upper bounds, which limited our ability to make money or otherwise generate positive contributions. These upper bounds were expressed in the form u_{tid} for $d \in \mathcal{D}^s$, representing a limit on our ability to execute a decision to serve a task at time t. Generally, for a decision $d \in \mathcal{D}^s$ to serve a demand, we normally assume that $c_{tid} > 0$, whereas decisions to reposition a resource to another state, given by $d \in \mathcal{D}^r$, would incur a negative contribution. Implicit in this model is the assumption that if we do not serve a task at time t, then the positive contribution is lost. At no time do we ever make any decisions about the task itself.

Two-layer problems arise frequently because as a rule, we often have to make decisions about how a demand is satisfied. In the simplest case, we may have to decide whether to serve a task now or later. This is the basic case of demand backlogging. In truckload trucking, it is often the case that once we decide to serve a customer, we simply move the load from origin to destination. We only have to decide when to serve the load. In more complex settings, we may have to decide how to serve the demand, which may have to progress through a series of steps before being completed.

When we make the transition to problems with two or more layers, we need to start distinguishing between important classes of resource layers. The first is whether they are *persistent* or *transient*. A persistent resource stays in the system when the decision is made to hold the resource. A transient resource vanishes. A *reusable* resource stays in the system after it is acted on; if it vanishes, it is not reusable or *perishable* (the term 'perishable' is awkward in the context of transportation and logistics, and appears to be better suited for consumer products).

A second critical dimension is whether the resource is *active* or *passive*. An active layer can be modified using a set of decisions. A passive layer can only work by coupling with other resource classes. A persistent, passive layer at a minimum has the property of (possibly) staying in the system when the action is to 'do nothing' but a more interesting class is one that stays in the system even after it has been coupled and modified.

An example of these concepts arises in the case of a driver and a load. A driver can reposition from one location to another without a load, or it can move a load. Moving a load allows the driver to make money, but you cannot act on the load by itself. But if you do not move the load, it just sits there (although it may leave the system). If you do move the load, it vanishes from the system. The load is a persistent, but not reusable class. The load becomes a reusable class if the driver moves the load to a relay point, drops off the load, and waits for another driver to pick it up.

We can turn this same example into a problem with two active layers. Assume that when we move a driver we mean that we are moving a driver that is an employee of our company. If we run short on drivers, we can contract out to another company to move the load, after which the driver becomes the responsibility of the outside company. From the perspective of the resources that we manage (our own drivers) it is as if we can move the load without a driver. This would be a problem with two active layers.

We start in Section 4.2.1 with the simplest version of a two-layer problem where only one layer is reusable (we can actually make decisions that change its state) while the other is passive (demands that just sit there until they are served). Section 4.2.2 then describes problems where the second layer is reusable as well.

4.2.1 One reusable layer

Earlier, we introduced the notation that $\mathcal{R}^c$ represented the set of (discrete) resources in class c (or, $\mathcal{R}^c$ is the vector of resources in class c). This notation is especially useful when there are three or more classes of resources (some complex problems might have four or five classes of resources), since it saves us from creating an alphabet soup of variables to describe the different resource classes. But when there are only two classes, it is more convenient to use different variables for each layer. For this purpose, we let:

L_{tb} = The number of tasks with attribute vector b available at time t before any new arrivals have been added.
$\mathcal{B}$ = The space of possible task attributes, with element $b \in \mathcal{B}$.
L_t = The vector of tasks that we know about at time t before any new arrivals have been added.

This representation provides certain symmetry with the representation of resources. However, it is also useful to define:

$\mathcal{L}$ = The set of task types (for example, each task type might represent an origin/destination combination).
L_{tl} = The number of tasks of type $l \in \mathcal{L}$.

$\mathcal{L}$ can be viewed as an indexing of the task attribute space $\mathcal{B}$. For our purposes, the latter representation is more convenient.

If we want to make a decision to serve a task, we let $\mathcal{D}^s = \mathcal{L}$ represent our set of possible types of tasks. L_{tl} is the number of tasks of type l at time t (some readers will prefer to use the variable u_{tl} to be the number of tasks of type l, since this variable later serves as an upper bound). For each task type in $\mathcal{L}$, there is a corresponding decision in $\mathcal{D}^s$ to serve a task of that type. For a decision $d \in \mathcal{D}^s$ there is a task type $l_d \in \mathcal{L}$, which means we can write $L_{l_d t} = L_{dt}$ for $d \in \mathcal{D}^s$. The number of resources we can assign to a task, then, is limited by:

$$x_{tad} \leq L_{dt} \tag{4.36}$$

while resources limit us through the flow conservation constraint:

$$\sum_{d \in \mathcal{D}} x_{tad} = R_{ta} \qquad (4.37)$$

Equations (4.36) and (4.37) express the impact of resources and tasks on decisions. When we implement decision d, the impact on the resource is expressed through the modify function, while the impact on the task is that it leaves the system. The evolution of the resource state variable is given by equation (2.3). We can in principle use the same equation for the tasks, but the simplicity of our tasks encourages us to use simpler notation. We assume that if we act on a demand it leaves the system. A demand that is not acted on may also leave the system (a customer refusal). For this reason, we define:

$L_{tt'}^h$ = The number of tasks that we knew about at time t which were actionable at time t' which were held at time t. Tasks which are not held include those that were served or which independently left the system.

We would normally assume that $L_{t+1,t'}^h = L_{tt'}^h$ for $t' > t$, meaning that if a task is not actionable at time t, it should still be in the system at time t'. But our representation allows for order cancellations. This notation allows us to write the task dynamics as:

$$L_{t+1,t'} = L_{tt'} + \hat{L}_{tt'} + L_{tt'}^h \qquad (4.38)$$

where $\hat{L}_{tt'}$, just as with $\hat{R}_{tt'}$, represents the tasks that first become known at time t and which are actionable at time t'.

The pair (R_t, L_t), now gives the resource state of our system. We emphasize that this is our *incomplete* resource vector, since R_t and L_t do not include new resources and tasks that arrive in the system at time t. We can use the same dynamic programming recursion and approximations that we used with a single resource layer earlier by simply replacing R_t with (R_t, L_t). Using our dynamic programming approximations, we would have to solve:

$$\tilde{V}_t(R_t, L_t, \omega_t) = \max_{x \in \mathcal{X}} C_t(x, R_t, L_t) + \hat{V}_{t+1}(R_{t+1}(\omega_t), L_{t+1}(\omega_t)) \qquad (4.39)$$

which we would solve subject to constraints (4.36) and (4.37), as well as the system dynamics (2.3) (the updating of the number of resources) and (4.38) (the updating of the number of tasks).

To solve (4.39) we can resort to the tricks we used for the one-layer problem. Assume, for example, that we want to work with a linear approximation.

Table 5
Performance of nonlinear approximation on problems where tasks have nonzero time windows (two-layer problem) and tight time windows (true one-layer problem)

Number of locations	With time windows			Without time windows		
	Horizon length			Horizon length		
	15	30	60	15	30	60
20	99.0%	99.2%	99.5%	100.00%	100.00%	100.00%
40	98.2%	98.4%	98.9%	100.00%	99.99%	100.00%
80	97.5%	97.0%	97.6%	99.99%	100.00%	99.99%

We would simply write:

$$\widetilde{V}_t(R_t, L_t, \omega_t) = \max_{x \in \mathcal{X}} C_t(x, R_t, L_t) + \hat{v}^R_{t+1} R_{t+1}(\omega_t) + \hat{v}^L_{t+1} L_{t+1}(\omega_t) \quad (4.40)$$

We would estimate $\hat{v}^R$ and $\hat{v}^L$ by using the dual variables on the constraints (4.37) and (4.36), and applying our standard smoothing techniques.

When the tasks are a passive layer, it is not unreasonable to use the approximation $\hat{v}^L = 0$. This means that we try to cover a task at time t, but if we cannot, we simply hold it until time $t + 1$ and hope to cover it then. Table 5 shows the results of experiments using a nonlinear approximation on a resource allocation problem where tasks have time windows (but where we use $\hat{v}^L = 0$) and problems where tasks must be served at a point in time, where both experiments are run on datasets without any uncertainty (which allows us to get tight bounds using an LP solver). The results indicate (on these deterministic datasets) that we are obtaining virtually optimal solutions when the time windows are tight (a true one-layer problem) whereas we are one or two percent below optimal when we use $\hat{v}^L = 0$.

To test the value of using both resource and task gradients we need to work on a problem where both resource and tasks may be held before being assigned, where we can also readily obtain optimal solutions, at least in the form of posterior bounds. (Posterior bounds are computed by finding the optimal solution after all the information becomes known.) A problem that readily lends itself to this test is the *dynamic assignment problem*. The dynamic assignment problem involves the assignment of resources and tasks over time, but where once a resource is assigned to a task, they both vanish from the system. But, if a resource or task is not assigned in time period t, they are available in time period $t + 1$. The decision to assign a resource or a task now has to take into account the value of the resource or task in the future.

The dynamic assignment problem is a special version of a two-layer problem, where we arbitrarily designate the 'resources' as the active

Table 6
Results of value function approximations for deterministic and stochastic experiments expressed as a percent of the posterior optimal solution. Each statistic is an average over 20 datasets

Type of experiment	Myopic	Resource gradients	Resource and task gradients
Deterministic	88.4	93.4	97.5
Stochastic	86.6	89.2	92.8

resource layer, while tasks are the passive layer. An important application is the load-matching problem of truckload trucking, where we have to assign drivers to loads. Over time, drivers become available and loads are called in. After a driver is assigned to a load, both 'vanish' from the system.

What makes the dynamic assignment problem special is that it is easy to solve the problem after all the resources and tasks become known to get a tight upper bound. In contrast with our earlier resource allocation problem, this is a problem where a myopic solution is not only interesting, it is what is normally done in practice. Experiments were run on 20 deterministic and 20 stochastic datasets, comparing a myopic solution ($\hat{v}^R = \hat{v}^L = 0$), against algorithms with just resource gradients ($\hat{v}^L = 0$) and algorithms using both resource and task gradients. The results are shown in Table 6, which suggest about a three to four percent improvement by adding in task gradients. We would conclude that while the improvement is not dramatic, it is certainly significant.

4.2.2 Two reusable layers

Consider the problem of moving a boxcar loaded with freight from origin to destination to serve the customer. This process occurs in a series of steps. When the boxcar is pulled from the shipper's dock, it is pulled to a yard where it is added to a *block*, which represents a set of cars that will move together over one or more trains. When a train moves, it pulls a set of blocks, which share a common segment (a common intermediate destination). When the block reaches its destination, it is probably the case that some of the cars have also reached their destination, but others may have to continue on. These cars are pulled out and added to a new block, which again will move over one or more trains before again reaching a new, intermediate destination.

Locomotives, of course, pull trains. Thus, to move a set of cars from one location to another, it is necessary to couple the cars together, move them to an intermediate destination (the destination of the block) and then uncouple them. Both the locomotive and the boxcars stay in the system. The locomotives have to be allocated to new trains, and decisions have to be made about how to route the boxcars. Thus, both are active resources.

Two-layer problems arise in other settings. Truckload motor carriers have to manage drivers and loaded trailers. Once a driver picks up a load, it may be necessary to move the load to a terminal where it is stored for a few days waiting for its final delivery appointment. The driver will be assigned to a new load, and at a later point a new driver will come in to pick up the original load. As with the boxcars, at the end of the first move, both the driver and the load remain in the system.

A third-party logistics provider also faces two-layer problems if they have the responsibility for moving and storing product, as well as managing the driver. It is necessary to load up the driver's vehicle, move the product, store the product, and continue to manage the driver.

The modeling of a two-layer problem is virtually equivalent to the modeling of a two-layer problem with a single reusable layer (but where both layers are persistent). All we need to do is change the system dynamics so that both layers are handled in the same way. So, instead of using the simple task dynamics of Eq. (4.38), we would model tasks using Eq. (2.3). Basically, the reader has to understand that when we face a true two-layer problem, whatever we do for the so-called 'resource layer' is the same as what we have to do for the 'task layer.' We have to capture not only the value of the resource in the future, but also the value of the task.

4.3 Multiple layers

Real problems are invariably even more complicated than the problems that we have addressed. For example, a driver has to use a tractor to pull a trailer to pick up a load of freight. Furthermore, we may need pallets or special loading equipment to help handle the load. A locomotive needs both fuel and a crew to pull boxcars with freight. A chemical products company has to specifically manage the driver, tractor, trailer, chemical product, and the customer tank (a five layer problem).

Multilayer problems are inherently complex, so it is especially important to adopt elegant, compact notation. Our earlier modeling framework introduced the concept of resource classes C^R, where R_t^c is the vector of resources in class c, and $\mathcal{A}^c$ is the attribute space for class c. Resource layering helps us handle the problem when decisions have to be made for resources that are coupled together. For example, a locomotive attached to an inbound train is quite different than the same locomotive that is not attached to an inbound train. The delivery that can be made by a particular driver, tractor, and trailer depends on the characteristics of all three.

One challenge faced by multilayer problems is that of solving a single period subproblem. Two-layer problems have the fundamental structure of transportation problems and assignment problems. Three-layer problems are much harder to solve.

It is perhaps not surprising that multilayer problems are often (but not always) solved as sequences of two-layer problems. In an LTL carrier, one person will manage drivers, another will plan the loading of trailers, and a third makes sure that the tractor pools are adequate. In railroads, distinctly different groups manage all locomotives, boxcars, and crews. But, a truck dispatcher has to manage drivers (with their tractors), trailers, and loads.

4.4 Bundling

Up to now, we have assigned resources to tasks with the tacit assumption of one resource per task. These are called 'one-to-one' problems, and arise when we have to assign a driver to pull a load, or the assignment of a boxcar to a customer order. But it is often necessary to consolidate freight into a single container. In this section, we consider two special cases. The first involves the batching of dozens, or even hundreds, of shipments on a single trailer going between two of points. In the second, we address the problem of clustering tasks with different characteristics. This might arise when putting orders together with different delivery dates, or with different final destinations (otherwise known as the vehicle routing problem).

4.4.1 Batch dispatching

The simplest batch dispatching problem arises in LTL trucking where shipments accumulate at a terminal until there is enough to satisfy the criteria for sending the truck. In most problems, the arrival rate of shipments is not constant over time. For example, at an end of line terminal, arrivals occur primarily in the evening, as shipments are unloaded from trucks that were in the city during the day. Most of the time, the dispatch rule is pretty simple. It is either 'dispatch when full' or a variation such as 'dispatch when full, but no later than a cutoff time,' where the cutoff time ensures that the carrier can make service. The challenge always arises when there is no more freight and the truck is only partially full. While all carriers focus on service, any carrier will have difficulty sending a truck over a long distance when it is only 20% full.

For regional carriers, there are typically very few options for routing freight. Some trucks will go directly from one city to another, carrying only freight between those two cities. A few carriers work exclusively this way, but this operating concept is impossible to grow past a dozen or so terminals. As a rule, most freight has to be handled through a single distribution facility. If a truck is not full enough to send through the facility, either the carrier has to send the truck partially loaded, or hold the freight until the next cycle.

Long haul carriers have more options. A trailer at an origin end of line, such as Boston, may be loading shipments to carry to a distribution center (or break bulk) at a destination region such as Texas. If there is not enough freight to fill the trailer, the carrier has the option of filling the trailer with freight and moving the trailer (either full or partial) but only to the nearest distribution center (sometimes called the 'origin break bulk) that would be in the northeast. There, the freight may be completely or partially sorted onto trailers that leave to many other terminals.

Ch. 13. Dynamic Models of Transportation Operations

Efforts have been made to formulate the problem of determining where to send trucks as integer programming models. Even static models of regional carriers can be intractably large, and optimal algorithms have not proven effective. Local search heuristics for optimizing static networks have been effective, and in particular, local search heuristics that work interactively with a planner have been widely adopted. But, even heuristic optimization models for the dynamic case have not been effective. Simulation models, which use simple policies to determine when a truck should be dispatched, remain the only effective tool in engineering practice, and we are not aware of any serious progress toward optimizing dynamic problems.

In this section, we again focus on dynamic problems and illustrate how the techniques that we have presented earlier for resource allocation problems can again be effective in this setting. As before, our solution approach will be one that solves a sequence of relatively simple problems by stepping through time. We can use either a simple myopic rule, or apply our adaptive dynamic programming techniques [see Papadaki & Powell, to appear]. We consider only the case of dispatching trucks over a single link, but we allow ourselves to consider the case where there are different types of customers. This is particularly important in LTL trucking, where there are high and low priority customers, as well as customers who have been waiting different lengths of time. Finally, we do not assume steady state behavior.

Model parameters

$\mathcal{K}$ = Set of customer classes.
c^d = Cost of dispatch a vehicle.
c_i^h = Holding cost of class i per time period per unit product.
c^h = $(c_1^h, c_2^h, \ldots, c_\mathcal{K}^h)$
K = Service capacity of the vehicle, giving the total number of customers who can be served in a single dispatch.

Activity variables

R_{tk} = Number of customers in class k waiting at time t before new arrivals have been added.
R_t = $(R_{tk})_{k \in \mathcal{K}}$
$\hat{R}_t$ = Vector random variable giving the number of arrivals in time t of each type of customer.
R_t^+ = $R_t + \hat{R}_t$.

Decision variables

x_{tk} = The number of customers in class k who are served at time t.
$X_t^\pi(R_t^+)$ = Decision function giving the vector x_t as a function of the complete resource vector R_t^+.

We define a family of decision functions $(X_t^\pi)_{\pi \in \Pi}$. It is useful for us to define an indicator variable $z_t = 1$ if a vehicle is dispatched and 0 otherwise. We let $Z_t(x_t)$ be a decision function where $Z_t = 1$ if $\sum_{k \in \mathcal{K}} x_{tk} > 0$, and 0 otherwise. Note that we are assuming that there is at most one dispatch per time period (since time periods can be made smaller, this does not pose a significant limitation).

Our one period cost function is given by:

$$C_t(R_t, \hat{R}_t, x_t) = c^d Z_t(x_t) + c^h (R_t^+ - x_t)$$

The objective function is now given by:

$$F(S_0) = \min_{\pi \in \Pi} E \left\{ \sum_{t=0}^{T-1} C_t(R_t^+, X_t^\pi(R_t^+)) \right\}$$

We follow our standard methodology and propose to solve the dynamic programming approximation:

$$\widetilde{V}_t(R_t, \omega_t) = \min_{x_t} C_t(R_t^+, x_t) + \hat{V}_{t+1}^n(R_t^+(\omega_t), x_t) \qquad (4.41)$$

The simplest approximation, which is also surprisingly effective, is to use a linear approximation:

$$\hat{V}_t(R_t) = \hat{v}_t R_t \qquad (4.42)$$

These batch processes are not linear programs, so we do not have access to dual variables. But, we can use finite differences. Let:

$$\tilde{v}_{kt} = \widetilde{V}_t(R_t + e_k, \omega) - \widetilde{V}_t(R_t, \omega)$$

where e_k is a $|\mathcal{K}|$-dimensional vector with a single 1 in the kth element (when there are a lot of product classes, it is fairly easy to devise schemes to approximate $\tilde{v}_{tk}$ using derivatives for only a few product classes). As before $\tilde{v}_{tk}$ is a statistical estimate, and we perform smoothing to find the approximation $\hat{v}_{tk}$.

Solving equation (4.41) using a linear value function approximation is pretty easy. If we assume that $z_t = 1$ (meaning that we are going to dispatch the vehicle), then finding the optimal x_t is usually a simple sort. In fact, it is possible to show that the simple rule of putting the most valuable products in the truck is the best. This means that we really only have to calculate (4.41) for $z_t = (0,1)$ and find the best value.

The steps of the algorithm are given in Fig. 6.

Step 1 Given R_0: Set $\tilde{V}_t^0 = 0$ for all t. Set $R_0^n = R_0$ for all n. Set $n = 1$ $t = 0$.
Step 2 Choose random sample $\omega = (\omega_0 \, \omega_1 \ldots \omega_{T-1})$.
Step 3 Calculate

$$z_t^n = \arg \min_{z_t \in \{0,1\}} \left\{ c z_t + c^h \cdot (R_t^n + \hat{R}_t^n - z_t X(R_t^n + \hat{R}_t^n)) + (\hat{v}_t^n) \cdot (R_t^n + \hat{R}_t^n - z_t X(R_t^n + \hat{R}_t^n)) \right\}$$

and

$$R_{t+1}^n = R_t^n + \hat{R}_t^n - z_t X(R_t^n + \hat{R}_t^n)$$

Then define:

$$\tilde{V}_t^n(R_t^n) = \min_{z_t \in \{0,1\}} \left\{ c^d z_t + c^h \cdot (R_t^n + \hat{R}_t^n - z_t X(R_t^n + \hat{R}_t^n)) + (\hat{v}_t^n) \cdot (R_t^n + \hat{R}_t^n - z_t X(R_t^n + \hat{R}_t^n)) \right\}$$

Step 4 Update the approximation as follows. For each $k = 1 \ldots m$ let:

$$\tilde{v}_{kt}^n = \tilde{V}_t^n(R_t^n + e_k) - \tilde{V}_t^n(R_t^n)$$

where e_k is an $|\mathcal{K}|$-dimensional vector with 1 in the kth entry and the rest zero.

Fig. 6. Adaptive dynamic programming algorithm for the batch dispatch problem.

Table 7 summarizes the results of a series of experiments where the linear approximation was tested on a problem with a single customer class. For this special case, it is possible to solve the optimality recursion using standard backward dynamic programming techniques. Table 7 shows the relative error over the optimal using between 25 and 200 iterations. Also shown is the performance of a myopic, 'go-when-filled' policy where the vehicle is not held more than time τ (where τ was optimized for each dataset). Three classes of datasets were tested, reflecting the difference between the holding cost c^h and the per customer dispatch cost c^d/K. The results suggest that the heuristic provides near-optimal performance. Most significantly, there is no difficulty extending it to problems with many customer classes.

It is important to emphasize that since most companies dispatch trucks using myopic rules, a heavily engineered myopic policy can do a superb job of mimicking the real world. These policies will be more sophisticated than

Table 7
Fractional error of total cost with respect to the optimal cost
[from Papadaki & Powell, to appear]

Method:	Linear	Linear	Linear	Linear	DWF-TC
hold/dispatch	Number of iterations				
cost	(25)	(50)	(100)	(200)	
$c^h > c^d/K$	0.077	0.060	0.052	0.050	0.774
$c^h \simeq c^d/K$	0.048	0.033	0.023	0.024	0.232
$c^h < c^d/K$	0.030	0.022	0.017	0.016	0.063
Average	0.052	0.038	0.031	0.030	0.356

even an optimized 'go-when-filled' strategy such as that illustrated in Table 7. From a modeling perspective, the issue is not so much whether a dynamic programming approximation will outperform a myopic heuristic. Of much greater significance is whether the model will yield good, realistic results without the heavy engineering.

4.4.2 Clustering

The second type of batching involves the clustering of resources together. We might have to group several locomotives to pull one train, two or three 'pups' to be pulled by one tractor, or the clustering of several deliveries onto one delivery vehicle. All of these problems involve a function that is nonseparable in the set of resources that are being bundled together. Locomotives may be attached together, so if we assign one locomotive to a train, we generally need to assign the other locomotives that the first locomotive may already be attached to. Pups need to be matched based on weight and service requirements. Deliveries should be grouped that form an efficient vehicle tour.

The general clustering problem can be expressed using a contribution function $C_t(x_t)$ which is a nonlinear, nonseparable function of the decision vector x_t. Fortunately, most problems are not quite this general. Let R_t be our vector of active resources (our trucks), and let $\mathcal{L}$ be the passive resource layer that we are coupling to (our deliveries, or tasks). We let R_{ta} be the number of resources with attribute a, and we let u_l be the size of each task, expressed in the same units as the resources (we may let R_{ta} be the capacity of the vehicles, and u_l be the size of each task). In many routing and scheduling problems, each individual vehicle $r \in \mathcal{R}$ will have its own unique attribute vector a_r, in which case R_{ta} would always refer to a single vehicle (but, this is not always the case). We let x_{tal} be the number of resources of type a that are being coupled to task l. Thus, we would have both a flow conservation constraint on the resources:

$$\sum_{l \in \mathcal{L}} x_{tal} = R_{ta} \quad \forall a \in \mathcal{A}, \tag{4.43}$$

and a coupling constraint:

$$\sum_{a \in \mathcal{A}} x_{tal} \leq u_{tl} \quad \forall l \in \mathcal{L} \tag{4.44}$$

Let $x_{tl} = (x_{tal})_{a \in \mathcal{A}}$ be the vector of decisions describing the assignment of resources with attribute a to task $l \in \mathcal{L}$. Now let $c_{tl}(x_{tl})$ be the cost of assigning a vector of resources x_{tl} to task l. This allows us to write:

$$C_t(x_t) = \sum_{l \in \mathcal{L}_t} c_{tl}(x_{tl}) \tag{4.45}$$

It is, of course, a nice simplification when we can write $c_{tl}(x_{tl}) = \sum_{l \in \mathcal{L}} c_{tal}(x_{tal})$ which is to say, a separable function across resources. This might be the case when assigning several locomotives to a single train, or two or three pups to the same tractor. But, it is not going to be true when assigning multiple deliveries (or pickups) to the same vehicle, since the total cost depends on the tour that can be formed to complete the pickups. [Just the same, a separable approximation is the basis of a popular vehicle routing algorithm, Fisher and Jaikumar (1981), as well as research in routing, Bramel and Simchi-Levi (1995).] For this reason, researchers find that they have to dynamically find the actual tour, even if it will change as new information arrives [Gendreau et al. (1999), Regan et al. (1998)]. These problems can be solved with any of a host of vehicle routing algorithms [Laporte (1992), Fisher (1995)]. As of this writing, it is not known whether precise routing and scheduling outperforms a good approximation when demands are highly dynamic. For example, Powell, Towns and Marar (2000b) show that for the load matching problem of truckload trucking (a form of dynamic assignment problem), a solution that uses a discounted approximation of the future (which is neither an optimal myopic solution, nor an attempt to optimize over the entire horizon) outperforms optimal myopic solutions in a dynamic setting.

Most efforts in the literature have focused on solving the problem myopically, which means forming vehicle tours using the vehicles and customer demands that are known at time t [see, for example, Gendreau et al. (1999) and Regan et al. (1998)]. This means solving sequences of problems of the form $\min_x C_t(x_t)$ using a vehicle routing algorithm. A significant challenge in this setting is the computational problem of solving a VRP under the pressure of time-staged demands, which limit the amount of time we have to actually solve the problem. We are not aware of efforts to solve VRPs using deterministic forecasts of future demands, which not only makes the problem much larger, but also creates other practical challenges (if the forecast is an expectation, we face the problem of routing an integer vehicle to pick up an expectation of the customer demand, which will typically not be either feasible or realistic).

We can approach the dynamic vehicle routing problem using the same strategies as we have reviewed for other resource allocation problems. We can solve the problem myopically, or incorporate value function approximations, which are estimated through adaptive learning. There is very recent research into using neurodynamic programming methods [Secomandi (2000, 2001)], but this work considers only a single vehicle. A challenge in designing value function approximations for dynamic VRPs is both the large size of the attribute space, and the complexity of the true value function. It is not clear that the simple linear or separable, nonlinear functional approximations that we introduced earlier will be successful. Also, many problems have some degree of advance information. We do not know at what point a myopic model, using advance information outperforms an adaptive, dynamic programming model. All of these are open research questions.

Dynamic vehicle routing problems exhibit other characteristics unique to the problem class. Many (but not all) vehicle routing problems require forming complete tours where the driver terminates at the depot at the end of the day. In a dynamic setting, it is possible to require that any tour always terminates at the depot, or to form tours that are not complete (at time t), responding to new demands as they arise. Furthermore, it may be necessary to form tours that do not serve all the customers at time t. For example, it may make sense to hold off on picking up a request from a customer in a part of town where there are no other requests to be served (right now), in the hope that other calls will come from that part of town, allowing the vehicle to serve several customers at once.

Thus, the decision to form a tour at time t may leave some customers unserved (consider the dynamic assignment problem above), and may produce a tour that leaves a vehicle at a location other than its home depot at some point during the day (requiring us to complete its tour at a later time). We may not serve all the customers at time t, also requiring us to think about the impact of forcing some deliveries until later in the day. We could use a myopic model that tries to cover only the deliveries we know about, using only the vehicles we know about. We would require this model to cover all deliveries with tours that always finish at the depot, or cover only some of the deliveries, with a tour that does not finish at the tour. The myopic model could be expanded by the use of simple rules, such as 'do not deliver to a part of town unless there are at least three orders, or unless it is after 3pm.' We could use a rolling horizon model by optimizing the problem using a mixture of known and forecasted demands. Or we could resort to our adaptive dynamic programming techniques. The last approach would require that we devise an effective approximation strategy, and a method for estimating and updating the function.

5 Implementation issues for operational models

It is easy to draw the conclusion from our presentation that the important issue in modeling freight transportation is designing models and algorithms that account for information that is not yet known. In practical implementations, the real issue tends to be in the form of bad data, which could otherwise be described as data that is not yet known, but should be. The problem is that we do not know in advance what data is bad, but we do know when we do not like a solution.

A byproduct of capturing the organization and flow of information is that it produces models where the original problem is broken into a number of pieces. Modeling the evolution of information over time produces models that are solved sequentially over time (rather than one big model over time). Modeling the organization of information and decisions produces a multiagent structure that further breaks the problem into subproblems. Not only are these subproblems much easier to solve, they are a lot easier to diagnose. That is, if the model recommends a decision that is not what a dispatcher would do, it is a lot easier to determine if the problem is data, model, algorithm or software.

We are not aware of any formal research into the modeling and algorithmic issues of dealing with bad data, but this is one of the characteristics of operational models that is markedly different than planning models. In a planning setting, we generally assume the data is fine, and we rarely have operations people looking over our shoulder criticizing the solution. Just as important, the precise solution (how the freight is moving) is less important than aggregate performance statistics.

In an operational setting, dispatchers typically already know what they would do, and if the model disagrees, then you face the challenge of trying to find out whether the discrepancy is because of a problem or if the model is simply displaying intelligence.

6 Summary remarks

Problems in freight transportation and logistics cover a range of problem classes, most of which can be characterized by dynamic information processes. In this chapter, we provide an overview of the most important operational settings, and provide a notational framework that captures most of these problems. We then summarize four major classes of algorithms, each based on different classes of information that can be used to solve these problems. Finally, we illustrate these algorithms in the context of some of the major problem classes. We made a point of avoiding detailed descriptions of models that are unique to specific modes (such as the blocking problem of railroads, or crane scheduling for intermodal ports), preferring instead to provide foundational models that could be adapted to different settings.

The design of models and algorithms for dynamic problems is relatively immature compared to the extensive body of research on deterministic problems. Not only are the models of physical operations quite young, there has been surprisingly little formal research on costing models, and virtually no research (that we are aware of) governing the design of information systems (which ultimately is what really controls operations).

Acknowledgement

This research was supported in part by grant AFOSR-F49620-93-1-0098 from the Air Force Office of Scientific Research.

References

Bertsekas, D., J. Tsitsiklis (1996). *Neuro-Dynamic Programming*, Belmont, MA, Athena Scientific.
Bodin, L., B. Golden, A. Assad, M. Ball (1983). Routing and scheduling of vehicles and crews. *Computers and Operations Research* 10(2), 63–211.
Bramel, J., D. Simchi-Levi (1995). A location-based heuristic for general routing problems. *Operations Research* 43, 649–660.
Brown, G., G. Graves, D. Ronen (1987). Scheduling ocean transportation of crude oil. *Management Science* 33, 335–346.
Cheung, R., W. B. Powell (1996). An algorithm for multistage dynamic networks with random arc capacities, with an application to dynamic fleet management. *Operations Research* 44(6), 951–963.
Cheung, R.K.-M., W. B. Powell (2000). SHAPE: A stochastic hybrid approximation procedure for two-stage stochastic programs. *Operations Research* 48(1), 73–79.
Crainic, T., G. Laporte (1997). *Design and Operation of Civil and Environmental Engineering Systems*, Wiley-Interscience, New York, pp. 343–394, Chapter Planning Models for Freight Transportation.
Crainic, T., J.-M. Rousseau (1988). Multicommodity, multimode freight transportation: A general modeling and algorithmic framework for the service network design problem. *Transportation Research B* 20B, 290–297.
Crainic, T., J. Roy (1988). OR tools for the tactical planning of freight transportation. *European Journal of Operations Research* 33, 290–297.
Crainic, T., J. Roy (1992). Design of regular intercity driver routes for the LTL motor carrier industry. *Transportation Science* 26, 280–295.
Crainic, T., J. Ferland, J.-M. Rousseau (1984). A tactical planning model for rail freight transportation. *Transportation Science* 18(2), 165–184.
Crainic, T., M. Gendreau, P. Dejax (1993). Dynamic stochastic models for the allocation of empty containers. *Operations Research* 41, 102–126.
Desrosiers, J., M. Solomon, F. Soumis (1995). Time constrained routing and scheduling, in: C. Monma, T. Magnanti, M. Ball (eds.), *Handbook in Operations Research and Management Science*, Volume on *Networks*, North Holland, Amsterdam, pp. 35–139.
Dror, M. (1993). Modeling vehicle routing with uncertain demands as a stochastic program: Properties of the corresponding solution. *European Journal of Operations Research* 64(3), 432–441.
Dror, M., G. Laporte, P. Trudeau (1989). Vehicle routing with stochastic demands: Properties and solution frameworks. *Transportation Science* 23, 166–176.
Fisher, M. (1995). Vehicle routing, in: C. Monma, T. Magnanti, M. Ball (eds.), *Handbook in Operations Research and Management Science*, Volume on *Networks*, North Holland, Amsterdam, pp. 1–33.
Fisher, M. L., R. Jaikumar (1981). A generalized assignment heuristic for vehicle routing. *Networks* 11(2), 109–124.
Frantzeskakis, L., W. B. Powell (1990). A successive linear approximation procedure for stochastic dynamic vehicle allocation problems. *Transportation Science* 24(1), 40–57.
Gendreau, M., F. Guertin, J. Potvin, E. Taillard (1999). Parallel tabu search for real-time vehicle routing and dispatching. *Transportation Science* 33, 381–390.
Glickman, T., H. Sherali (1985). Large-scale network distribution of pooled empty freight cars over time, with limited substitution and equitable benefits. *Trans. Res.* 19, 85–94.
Godfrey, G., W. B. Powell (2002a). An adaptive, dynamic programming algorithm for stochastic resource allocation problems I: Single period travel times. *Transportation Science* 36(1), 21–39.
Godfrey, G., W. B. Powell (2002b). An adaptive, dynamic programming algorithm for stochastic resource allocation problems II: Multi-period travel times. *Transportation Science* 36(1), 40–54.
Haghani, A. (1989). Formulation and solution of a combined train routing and makeup, and empty car distribution model. *Transportation Research* 23B(6), 433–452.

Herren, H. (1977). Computer controlled empty wagon distribution on the SSB. *Rail International* 8(1), 25–32.
Jaw, J., A. Odoni, H. Psaraftis, N. Wilson (1986). A heuristic algorithm for the multivehicle many-to-many advanced request dial-a-ride problem with time windows. *Transportation Research* 20B, 243–257.
Jordan, W., M. Turnquist (1983). A stochastic dynamic network model for railroad car distribution. *Transportation Science* 17, 123–145.
Laporte, G. (1992). The vehicle routing problem: An overview of exact and approximate algorithms. *European Journal of Operations Research* 59, 345–358.
Laporte, G., F. Louveaux (1990). Formulations and bounds for the stochastic capacitated vehicle routing problem with uncertain supplies, in: J. Gabzewicz, J. Richard, L. Wolsey (eds.), *Economic Decision-Making: Games, Econometrics and Optimization*, Amsterdam, North Holland.
Leddon, C., E. Wrathall (1967). Scheduling empty freight car fleets on the Louisville and Nashville railroad, in: *Second International Symposium on the Use of Cybernetics on the Railways*, October, Montreal, Canada, pp. 1–6.
Mendiratta, V., M. Turnquist (1982). A model for the management of empty freight cars. *Trans. Res. Rec.* 838, 50–55.
Misra, S. (1972). Linear programming of empty wagon disposition. *Rail International* 3, 151–158.
Muriel, A., Simchi-Levi, D. (to appear). Supply chain design and planning-applications of optimization techniques for strategic and tactical models, in: S. Graves (ed.), *Handbook in Operations Research and Management Science*, Volume on *Supply Chain Management*, North Holland, Amsterdam.
Misra, S. (1972). Linear programming of empty wagon disposition. *Rail International* 3, 151–158.
Papadiki, K., Powell, W.B. (to appear). An adaptive dynamic programming algorithm for a stochastic multiproduct batch dispatch problem. *Naval Research Logistics*.
Powell, W. B. (1986b). A stochastic model of the dynamic vehicle allocation problem. *Transportation Science* 20, 117–129.
Powell, W. B. (1987). An operational planning model for the dynamic vehicle allocation problem with uncertain demands. *Transportation Research* 21B, 217–232.
Powell, W. B. (1989). A review of sensitivity results for linear networks and a new approximation to reduce the effects of degeneracy. *Transportation Science* 23(4), 231–243.
Powell, W. B. (1996). A stochastic formulation of the dynamic assignment problem, with an application to truckload motor carriers. *Transportation Science* 30(3), 195–219.
Powell, W. B., T. A. Carvalho (1998). Dynamic control of logistics queueing network for large-scale fleet management. *Transportation Science* 32(2), 90–109.
Powell, W. B., Jaillet, P., Odoni, A. (1995). Stochastic and dynamic networks and routing, in: C. Monma, T. Magnanti, M. Ball (eds.), *Handbook in Operations Research and Management Science*, Volume on *Networks*, North Holland, Amsterdam, pp. 141–295.
Powell, W. B., J. A. Shapiro, H. P. Simão (2000a). An adaptive dynamic programming algorithm for the heterogeneous resource allocation problem, Technical Report CL-00-06, Department of Operations Research and Financial Engineering, Princeton University.
Powell, W. B., J. A. Shapiro, H. P. Simão (2001). A representational paradigm for dynamic resource transformation problems, in: R. F. C. Coullard, J. H. Owens (eds.), *Annals of Operations Research*, J.C. Baltzer AG, pp. 231–279.
Powell, W., M. T. Towns, A. Marar (2000b). On the value of globally optimal solutions for dynamic routing and scheduling problems. *Transportation Science* 34(1), 50–66.
Psaraftis, H. (1988). Dynamic vehicle routing problems, in: B. Golden, A. Assad (eds.), *Vehicle Routing: Methods and Studies*, Amsterdam, North Holland, pp. 223–248.
Psaraftis, H. (1995). Dynamic vehicle routing: Status and prospects. *Annals of Operations Research* 61, 143–164.
Regan, A., H. S. Mahmassani, P. Jaillet (1998). Evaluation of dynamic fleet management systems-simulation framework. *Transportation Research Record* 1648, 176–184.
Secomandi, N. (2000). Comparing neuro-dynamic programming algorithms for the vehicle routing problem with stochastic demands. *Computers and Operations Research* 27(11), 1201–1225.

Secomandi, N. (2001). A rollout policy for the vehicle routing problem with stochastic demands. *Operations Research* 49(5), 796–802.
Stein, D. (1978). Scheduling dial-a-ride transportation systems. *Transportation Science* 12, 232–249.
Stewart, W., B. Golden (1983). Stochastic vehicle routing: A comprehensive approach. *Eur. J. Oper. Res.* 14(3), 371–385.
Topaloglu, H., W.B. Powell (2000) Dynamic programming approximations for stochastic, time-staged integer multicommodity flow problems, Technical Report CL-00-02, Department of Operations Research and Financial Engineering, Princeton University.
Turnquist, M. (1986), Mov-em: A network optimization model for empty freight car distribution, School of Civil and Environmental Engineering, Cornell University.
White, W. (1972). Dynamic transshipment networks: An algorithm and its application to the distribution of empty containers. *Networks* 2(3), 211–236.
Wilson, N. (1969). Dynamic routing: A study of assignment algorithms, Ph.d. thesis, Department of Civil Engineering, MIT, Cambridge, MA.

Subject Index

Accounting inventory, 309, 373
Additive demand function *see* Demand function
Advance demand information, 362, 380, 439
Advance ordering, 361
Advanced planning, 458
Advanced planning systems, 9, 458, 480, 483, 510, 670
  capable to promise, 498
  demand planning, 487
  distribution and transport planning, 499
  modules, 484
  order promising, 508
  production planning and scheduling, 497
  shortage planning, 508
  strategic network planning, 484
Advertising, 264, 392
Agent, 705
Aggregate capacity planning, 608
Aggregate planning *see* Hierarchical production planning
Aggregate production planning, 497
Aggregation, 476, 493, 612
Aggressiveness, 162
Airline industry, 65
Allocation, 297, 387, 562, 639, 641
  allocation function, 639
  allocation of information, 377
  allocation rule, 203
  balanced stock rationing, 640
  batch priority policy, 352
  demand-allocation models, 272
  fair share, 503, 574
  FCFS allocation policy, 573, 574
  linear allocation rule, 639
  myopic allocation policy, 569
  myopic allocation problem, 543
  negative allocation, 543, 640
  nonnegative rationing limits, 570
  priority rationing, 664
  proportional allocation, 272
All-unit discounts, 20, 39
  modified all-unit discount cost function, 42, 48
American Airlines, 66
Anticipation, 614, 615
  anticipated base model, 614
  anticipation types, 615
  explicit anticipation, 614
  implicit anticipation, 614
  nonreactive anticipation, 479, 483
  reactive anticipation, 479
Application service provider, 11
APS *see* Advanced planning system
ARIMA *see* Autoregressive integrated moving average
ARMA *see* Autoregressive moving average
Artificial intelligence, 723
ASP *see* Application service provider
Assemble-to-order, 208, 505, 561, 575, 606, 635, 637, 645, 697
Assembly assumption *see* General supply networks
Assembly line, 179
Assembly networks, 101
  see also Assembly systems
Assembly sequence design, 214
Assembly systems, 285, 441, 526, 562, 571, 637, 641, 642
  see also Assembly networks
Assortment problem, 272
ATO *see* Assemble-to-order
ATO key theorem, 636
ATP *see* Available to promise
Auction, 21, 369
Automotive industry, 179, 468, 500, 517, 591
Autoregressive integrated moving average, 364

Autoregressive moving average, 439
Available to promise, 482, 504, 505, 507, 617

Baan, 480, 510
Backordering, 425, 620
Backup agreement, 248
Balance approximation, 543
Balance assumption, 541, 544, 638, 639, 640
  see also Allocation
Balanced stock rationing see Allocation
Balancedness of constraints, 189
Bar code scanning, 692
Bargaining power, 234
Base stock policy, 99, 101, 102, 209, 292, 298, 371, 373, 436, 438, 440, 444, 445, 529, 565, 568, 572, 583, 635, 645
  artificial base stock level, 643
  balanced base stock policy, 572
  base stock level, 429
  base stock list price policy, 73
  cyclical base stock policy, 444
  emergency base stock level, 445
  generalized base stock policy, 431
  modified periodic base-stock policy, 437
  periodic base stock policy, 437
  synchronized base stock policy, 644, 645
  time-specific base stock policy, 436
Batch dispatching, 746
Batch priority policy see Allocation
Battery manufacturing, 114
Bayesian game see Game
Bayesian updates, 438
Beer game see Bullwhip effect
Benders decomposition, 487
Benetton, 217
Bertrand competition see Competition
Bill back, 260
Bill of materials, 176, 462, 600, 612
Bill of process, 600, 602
Bin-packing problem, 81
BMW, 517
Boise Cascade, 65
BOM see Bill of materials
BOM explosion, 498
BOP see Bill of process
Bounds, 48, 81, 87, 103, 580, 582, 584, 590
Branch and bound, 487, 515
Breakbulk points, 35
  see also Consolidation points
Budget constraint, 574
Buffer stock, 430
Bulldozer assembly, 107

Bullwhip effect, 5, 167, 363, 491, 508
  beer game, 368
  see also Forrester effect
Bundling see Resource allocation problem
By-product, 600

Can order policy, 443
  see also (s, c, S) policy
Cannibalization, 581
Capable to promise, 504, 505
Capacitated concentrator location problem see Facility location problem
Capacity, 21, 74, 100, 149, 173, 208, 210, 222, 357, 366, 371, 399, 426, 435, 442, 602, 609, 694
  capacity smoothing, 610
  dedicated capacity, 150
  flexible capacity, 150
Capacity requirements planning, 498
Capital-labour ratio, 3
Car distribution problem, 679
Caterpillar, 448
Casual models see Forecast
Centralized control, 107, 410
Chaining, 146
Change-over cost, 179, 180
  see also Set-up cost
Change-over time, 179, 180
  see also Set-up time
Chemical industry, 468, 499, 617, 687
Clustering, 750
CODP see Decoupling point
Collaboration see Forecast
Collaborative planning forecasting and replenishment, 9, 491
Combinatorial optimization, 579, 604
Commonality, 175, 181, 218, 566, 569, 591, 662
  commonality index, 218
Competing retailers, 271, 402
Competition
  Bertrand competition, 405
  competing retailers, 271
  competition penalty, 301
  Cournot competition, 403, 405, 406
  oligopolistic competition, 283
Complete state variable see Dynamic programming approximation
Compliance, 233, 317
  forced compliance, 169, 235
  voluntary compliance, 170, 235
Component standardization, 218
Computer assembly, 466, 474
Configure-to-order, 468, 561

Conjugate gradient methods, 100, 652
Consignment, 125, 328, 387
Consolidation points, 35
    see also Breakbulk points
Consumer goods manufacturing, 463, 472
Consumer packaged goods see
    Safety stock
Container, 696, 697, 721
Continuous manufacturing, 609
Continuous review, 424, 528
Contracts, 130, 165, 168, 229, 258, 425
    buyback contract, 242
    options contract, 323
    options-futures contract, 329
    pay-to-delay contract, 290
    price-discount contract, 258
    price-only contract, 402
    quantity discount contract, 254
    quantity-flexibility contract, 248
    reservation contract, 320
    revenue-sharing contract, 246, 369
    sales-rebate contract, 252
    supply contract, 165, 168
    wage contract, 383
    wholesale price contract, 238
Control limit policy see Dispatch rule
Control theory, 167
Coordinated channel, 170
Coordination, 229, 341, 562, 597
    multiple retailers, 312
Core competency, 2
Cost
    administrative cost, 255
    fixed cost, 426
    holding cost, 604
    overage cost, 430
    penalty cost, 605
    production cost, 426
    salvage cost, 427
    setup cost, 430
    underage cost, 430
Cost center, 373
Cost-to-go, 432
Cournot competition see Competition
CPFR see Collaborative planning, forecasting
    and replenishment
CPLEX, 34, 496, 512, 513, 515
Critical fractile, 429
Cross docking, 32, 37, 517
CTP see Capable to promise
Curse of dimensionality see Dynamic
    programming
Customization, 203

CZAR see Southern Motor carrier's
    Complete Zip Auditing and
    Rating engine

Dantzig-Wolfe scheme, 483
Decentralization, 170, 302, 374
    see also Decentralized control
Decentralized control, 97, 157, 167, 289, 447
    see also Decentralization
Decision functions, 718
    dynamic programming, 722
    myopic policies, 718, 719, 721
    proximal point algorithms, 721, 730
    rolling horizon policies, 718, 720
Decision support systems, 9, 17, 223
Decomposition, 516, 592, 604, 610, 611, 632
Decomposition approach, 483, 536, 553, 650
    full delay models, 650, 651
    no-delay models, 650, 651
Decomposition assumption see General supply
    networks
Decoupling point, 462, 472, 474, 487, 492,
    505, 606
    customer order decoupling point
        (CODP), 606, 607, 665
Delayed production see Pricing
Dell, 4, 65, 378, 561
Demand correlation, 272
Demand fulfillment, 471, 482, 504
Demand process, 425
    additive demand function, 68, 71
    cyclic demand, 437
    i.i.d. (independently and identically
        distributed), 428
    multiplicative demand function, 68, 71, 72
    nonstationary demand, 436
Demand supply matching, 507
Deployment, 500
Design, 172, 199, 588
    distribution system design, 82
    product design, 172
    supply chain configuration, 120
    supply chain configuration problem, 123
    supply chain design, 77
Detailed lot-sizing and resources
    allocation, 499
Direct derivative estimation, 436
Disaggregation, 498
Dispatch rule, 719, 748
    control limit policy, 719
Distribution networks, 101
    see also Distribution system; Divergent
        systems

Distribution system, 442, 541, 562, 570
  see also Divergent systems
Distribution requirements planning, 535
Distribution system design see Design
Divergent systems, 526, 638, 642
  see also Distribution system
dm, 517
Drayage, 693
DRP see Distribution requirements planning
DSS see Decision support systems
Duopoly, 404, 407
Durable goods industry, 161
Dynamic assignment problem, 719, 743, 751
Dynamic programming, 23, 124, 443, 651, 680
  curse of dimensionality, 709
  dynamic program, 570
  stochastic dynamic programming, 606
Dynamic programming approximations, 708, 728, 742, 748
  approximating the value function, 710
  complete state variable, 708
  incomplete state variable, 708
  linear approximations, 711, 718, 729
  nonlinear approximation, 712
  piecewise linear approximations, 712

Eastman Kodak, 448
Echelon stock, 344, 633
  Echelon inventory position, 634; see also Inventory position
e-commerce, 562
Economic lot-sizing problem, 41, 42, 48, 444
Economies of scale, 351, 486
Electronics industry, 161, 499, 562, 620
Electronics manufacturing service, 3
EMS see Electronics manufacturing service
End-of-life phase see Product life cycle
Enterprise resource planning, 8, 457, 480, 510, 512, 515
Equal fractile, 638
Equal fractile heuristic, 574
ERP see Enterprise resource planning
Exchange rate, 495

Facility location problem, 78, 83
  capacitated concentrator location problem, 79
  capacitated facility location problem, 79, 80
Factorized normal approximation, 585
Family disaggregation see Hierarchical production planning

FCFS see First come-first served
FedEx, 690
Fill rate, 573, 574, 584, 585, 589, 605
Final assembly, 6
Finite backlog buffer, 587
Finite horizon see Time horizon
First come-First served, 506, 544, 547, 574
Fixed-flow subproblem, 30
Flatbeds, 684
Fleet management, 680
Flexibility, 133, 136
  expansion flexibility, 139
  flexibility dimensions, 137, 138
  machine flexibility, 142
  manufacturing flexibility, 135
  mix flexibility, 139, 151, 165, 166
  mobility, 136, 152, 180
  new product flexibility, 139, 151, 165, 166
  plant flexibility, 166
  product modification flexibility, 139
  range, 136, 152, 178
  resource flexibility, 146, 147, 150, 152
  short-term delivery flexibility, 166
  strategic flexibility, 162
  uniformity, 136, 152
  volume flexibility, 139, 151, 165, 166
  worker flexibility, 154, 157
Flexible assembly systems, 6, 161
  see also Flexible manufacturing system
Flexible capacity see Capacity
Flexible machine investment problem, 142
Flexible manufacturing system, 179, 182
  see also Flexible assembly systems
Floating bottlenecks, 188
Flow shop, 464
FMS see Flexible manufacturing system
Food manufacturing, 499
Ford, 22, 65, 517
Forecast, 285, 355, 359, 365, 425, 481, 488, 535, 607
  biased forecast, 317
  causal models, 489
  collaboration, 490
  forecast revision, 651
  forecast sharing, 322
  forecasting, 438, 491, 656
  judgement, 490
  point forecast, 720
  time series models, 489
Forgetting, 157
Forrester effect, 167
  see also Bullwhip effect
Franchising, 233, 268, 331

Subject Index

Freight cars, 694
Full delay models *see* Decomposition approach
Full truckload, 682

Game, 150, 168, 234, 273, 298, 299, 300, 303, 315, 354, 364, 403, 405
  Bayesian game, 403
  capacity procurement game, 318
  signaling game, 392
  Stackelberg game, 389
  zero sum game, 330
General networks, 36
General supply networks, 641, 642, 651, 652
  assembly assumption, 103, 649, 652
  decomposition assumption, 649
  heuristic analysis, 648
Generalized newsboy equations theorem, 639
Generic subassembly, 215
Genetic algorithm, 511, 516
GIS systems, 688
Goal programming, 504
Greedy algorithm, 584
Guaranteed service model, 98, 101, 105, 113, 116

Heavy industry *see* Safety stock
Hewlett Packard, 4, 99, 211, 220
Hidden Markov model, 438
Hierarchical planning, 457, 477, 478, 479, 483, 608, 610, 615
Hierarchical production planning, 478, 609
  aggregate planning, 609, 651
  family disaggregation, 609
  item disaggregation, 609
Hierarchy, 478, 489, 499, 608, 609
  artificial hierarchy, 642
  constructional hierarchies, 478
  organizational hierarchies, 478
  tactical-operational hierarchies, 479
  top-down hierarchies, 479
High priority service, 694

i2 Technologies, 480, 517, 518
IBM, 212, 447, 448, 575
Ideal-point model, 380
IGFR *see* Increased generalized failure rate
i.i.d. (independently and identically distributed) *see* Demand process
Imbalance, 203, 604
Incomplete state variance *see* Dynamic programming appoximation
Increased generalized failure rate, 239
Incremental discounts, 20
Infinite horizons *see* Time horizon

Infinitesimal perturbation analysis, 100, 436, 437, 442
Information, 341
  asymmetric information, 168, 316, 370, 388, 478, 612, 614, 615
  horizontal information sharing, 402
  information delay, 5
  information distortion, 5
  information sharing, 317, 341, 377, 403
  information transmission, 372
  spill-over effect, 408
  value of information, 343
  vertical information sharing, 408
Installation stock, 345
Integer program, 747
Integral planning, 453
Interest rate, 604
Internet, 10, 11, 65, 223, 448, 480
Inventory, 423
  committed inventory, 568
  inventory capital investment, 604
  net inventory, 620
Inventory policy, 427
Inventory position, 530
  echelon inventory position, 530
  installation inventory position, 530
Item, 600
  end-item, 600, 607
Item disaggregation *see* Hierarchical production planning

J.D. Edwards, 480, 512, 517, 518
Jensen's inequality, 582
JIT *see* Just-in-time
Job shop, 157, 609
Joint replenishment planning, 443
Joint replenishment problem, 53
Judgment *see* Forecast
Just-in-time, 619, 665

Kanban, 536
  combined Kanban base stock control, 653
  generalized Kanban systems, 616
  Kanban policy, 653
k-concave, 70, 73
  Symmetric k-concave, 67, 70
k-convex, 19, 70, 72, 435
  symmetric k-convex, 19, 67, 72
K-mart, 22
Knapsack problem, 81, 86, 565

Labor attrition, 156
Langrangian relaxation, 23, 51, 80, 85, 487, 632

Large deviation approximation, 446, 496
Lateral transshipments, 547, 556
Layering, 700, 701, 745
Lead time, 426, 433, 486, 603
    actual lead time, 100, 617
    effectuation lead time, 612, 613, 615
    fixed lead time, 612
    minimum lead time, 631
    nominal lead time, 100, 616, 652
    planned lead time, 603, 625
Learning, 156, 157
Less-than-truckload, 20, 24, 682
Less-than-truckload trucking, 688, 746, 747
    long-haul carriers, 689, 690, 746
    regional carriers, 689, 690, 746
Line-haul operation, 690
Linear functional approximations, 679
Linear program, 29, 483
Linear programming, 29, 58, 481, 493, 496, 497, 499, 508, 511, 513, 514, 518, 627, 656, 666
    iterative LP rounding, 487
Little's formula, 545
Locomotives, 694
Logistics, 17, 457
Logistics service provider, 500, 501
Long-haul carriers *see* Less-than-truckload trucking
Lost sales, 425, 569, 575
Lost sales model, 434
Lot-sizing, 472, 502, 604, 609
    detailed lot-sizing, 498
    multi-item, 631
    multi-level capacitated lot-sizing problem, 631
LP *see* Linear programming
LSP *see* Logistics service provider
LTL *see* Less-than-truckload
LTL shipper problem, 27, 28
LTL trucking *see* Less-than-truckload trucking
Lucent Technologies, 221
Luminescent lamps, 499

Machine speed, 159
Machine tool industry, 161
Mail-order, 562, 576
Make-to-order, 607, 637
Make-to-stock, 208, 505, 575, 607, 637
Manufacturing resources planning, 498
Manufacturing system, 177
Manugistics, 514, 517, 518
Market segmentation, 380

Markov chain, 359, 371, 588
Martingale model of forecast evolution, 355, 438
Mass customization, 199, 468
Mass production, 158
Master planning, 472, 481, 491, 492
Master production schedule, 535
Master production scheduling, 498, 499, 535
Material release decision, 601
Material requirement planning, 457, 498, 534, 610, 619, 622
Mathematical programming, 83, 626
Mature phase *see* Product life cycle
METRIC, 544, 553
Middle management, 476
Midseason replenishment, 291
Mixed integer linear program, 28, 120
Mixed integer programming, 29, 481, 487, 496, 497, 499, 503, 513, 514, 515, 633
Mobility *see* Flexibility
Modeling, 424
Modify function, 702
MOD-METRIC, 547
Modularity, 222, 591
    modular design, 189, 270, 591
Monolithic decision models, 610
Monopolist, 378, 404
Moral hazard, 382, 396
MP *see* Master planning
MPS *see* Master production scheduling
MRP *see* Material requirement planning
MRP-II *see* Manufacturing resources planning
MTO *see* Make-to-order
MTS *see* Make-to-stock
Multiagent problems, 717
Multicommodity network flow problems, 23, 735, 737
Multi-echelon inventory systems, 167, 525
Multiperiod travel times, 714
Multiplicative demand function *see* Demand process
Myopic allocation policy *see* Allocation
Myopic allocation problem *see* Allocation
Myopic policy, 348, 433, 436, 439

Nash equilibrium, 230, 273, 300, 306, 407
Neighborhood search, 514
Nervousness *see* Planning
Nested policy, 442, 531
Network, 599
    layered network, 35
    network design, 120

Subject Index

Network flow model, 21, 22, 23, 485, 494, 503
Network loading problem, 24
Neuro-dynamic programming, 680, 751
Newsvendor problem, 66, 233, 257, 563, 428
   capacitated newsvendor problem, 442
   multi-dimensional newsvendor problem, 148
No-delay models *see* Decomposition approach
Nonbipartite weighted matching problem, 579
Nonnegative rationing limits *see* Allocation
Nonstationary demand *see* Demand process
Non-stockout probability, 605, 636
Normative rational decision theory, 167
NP-hard, 42

OEM *see* Original equipment manufacturers
Oligopolistic competition *see* Competition
Oligopoly, 406
Operation, 140
Option, 121, 144
Order acceptance, 617
Order promising, 474, 504, 505, 506, 507
Order release, 155
Order-up-to S policy *see* Base stock policy
Original equipment manufacturers, 7
Output rate, 177
Outsourcing, 2, 7

Palm's result, 581, 582
Paper industry, 162
Parent-child relationship, 600
Periodic review, 424, 528
Pharmaceuticals, 617
Planning, 468
   centralized planning, 611
   long term planning, 468
   mid-term planning, 470, 495, 597
   nervousness, 476
   planning horizon, 468, 475, 493
   planning models, 476
   short-term planning, 470, 597
   see also Transformation activities
Planning horizon *see* Planning
Point of differentiation, 203, 207
Policy, 707
Polya frequency function, 431
Pooling equilibrium, 323, 389, 393
Postponement, 4, 181, 199, 650
   process postponement, 201
   product postponement, 201
Power-of-two policy, 64, 411
Price-delivery schedule, 380

Pricing, 65, 222, 258, 682
   contract pricing, 678
   delayed pricing, 74, 76, 77
   delayed production, 74, 77
   dynamic pricing, 76
   spot pricing, 678
   static pricing, 678
   zone pricing, 688
Primary market, 291
Principal-agent theory, 377
Priority rationing *see* Allocation
Private fleet, 686
Process, 141, 176
   process modularity, 202
   process resequencing, 201
   process standardization, 201, 202
Processing industry, 602
Processing time, 103, 602
Product design *see* Design
Product differentiation, 133
Product family, 143, 165, 166, 173, 175, 183, 190, 476, 574
Product innovation, 133
Product life cycle, 129, 166, 175, 183, 219, 466
   end-of-life phase, 219
   mature phase, 175, 219
Product line, 378, 448
   product line design, 591
Production smoothing, 497
Pull policy, 503, 535
Pull system, 211
Push policy, 503, 535, 638

Quality, 269, 378, 380
Quasi birth-and-death process, 588
(Q, r) model, 387
Queueing models, 158
   D/G/1, 435
   G/G/1, 588
   M/D/∞, 545
   M/G/1, 581, 583, 588
   M/M/∞, 100, 103
   M/M/1, 211, 575
   $M^{[x]}/G/\infty$, 652
   queueing network model, 218

(R, nQ) policy, 344, 345, 529, 580
(R, Q) policy, 541, 549, 551, 554
Railroads, 693
Range *see* Flexibility
Rational explanatory research, 167
Rationing game, 474
Reckitt & Colman, 517

764 Subject Index

Regional carriers *see* Less-than-truckload trucking
Release, 619
Repair-kit problem, 567
Repromising, 504, 507
Resale price maintenance, 279, 328, 386
Reservation level, 234
Resource
 active layer, 740
 passive layer, 740
 persistant resource, 740
 reusable resource, 740
 transient resource, 740
Resource allocation problem, 680
 bundling, 746
 heterogeneous resource allocation problem, 680, 737
 multilayer problems, 745
 single layer resource allocation, 725
 two layer resource allocation, 740
Resource requirement planning, 498, 709
Response function, 273
Return on investment, 2
Revalation principle, 328
Revenue management, 65
Risk, 256
Risk pooling, 222, 443, 575
ROI *see* Return on investment
Rolling horizon, 475, 535, 698, 706
 see also Rolling schedule
Rolling schedule, 626, 666
 see also Rolling horizon
Rough cut capacity planning, 498
Rush order, 445

(s, c, S) policy, 443
 see also Can order policy
(s, S) policy, 69, 346, 431, 435, 529
 generalized (s, S) policy, 431
 modified (s, S) policy, 571
 state dependent (s, S) policy, 439
(s, S, A, p) policy, 72
(s, S, p) policy, 67, 70, 73
(S−1, S) policy, 529, 544
S policy *see* Base stock policy
Safety stock, 126, 629
 consumer packaged goods, 107
 heavy industry, 107
 safety stock placement, 95, 97
 safety stock positioning, 652, 666
Safety time, 444
SAP, 480, 515, 517
SAP/R3, 480, 515

Save-up-to policy, 75
Scheduling, 498, 514, 608
 detailed scheduling, 616
 drum-buffer-rope scheduling concept, 515
 online rescheduling, 498
SCOP *see* Supply chain operations planning
SCOR model, 459
Screening, 327, 377, 382, 385
Sears Roebuck, 22
Secondary Market, 291, 403
Semiconductor equipment industry, 327, 331
Semiconductor industry, 518
Separating equilibrium, 323, 389, 393
Serial supply chain, 101, 344, 651
 see also Serial system
Serial system, 440, 526, 535, 536
 see also Serial supply chain
Service level, 99, 102, 566, 598, 605, 628, 636, 644, 652
Set covering problem, 54
Set partitioning, 25
Set-up cost, 181
Set-up time, 181
SynQuest, 517
SHAPE algorithm, 713
Sheet metal press shop, 159
Shelf life, 462
Shelf space, 264
Shipment frequencies, 502
Shipper problem, 20, 22
Shop floor scheduling and control, 499
Shortage gaming, 474
Shortage planning, 504, 506
Shortest path problem, 56
Shortfall, 435
Siding, 694
Signaling, 377, 388, 392
Simulated annealing, 632
Simulation, 145, 152, 160, 488, 491, 511, 580, 583, 630, 638, 640, 649, 653, 659, 706, 747
Simulation based optimization, 442, 443
Single link model, 502
Slotting allowance, 393
Small package industry, 690
Southern Motor Carrier's Complete Zip Auditing and Rating Engine, 40
Spanning trees, 101, 124
Spare parts, 447, 544, 562, 652
Spillover effect *see* Information
Stackelberg game *see* Game
Stage-spanning bottlenecks, 188
Stochastic linear programming, 563
Stochastic service model, 98, 99, 102

Subject Index 765

Stock commitment policy, 574
Stocking cycle, 570
Stock-out probability, 566
Subcontractor, 401
Substitution, 222
Supply chain attributes, 459
   distribution, 462
   procurement, 459
   production, 460
   sales, 462
   topography, 462
Supply chain configuration *see* Design
Supply chain configuration problem
    see Design
Supply chain design *see* Design
Supply chain flexibility, 164, 172
Supply chain operations planning, 597, 598,
    608, 617, 623
   material constraints, 619
   mathematical programming formulation,
     626, 631
   non-stationary demand, 699
   resource constraints, 623, 625, 626, 627, 667
Supply chain planning, 457, 468
   SCP (supply chain planning)-matrix, 470
Supply network, 597, 600
   see also General supply networks

(T, R, Q) policy, 367
2-partition problem, 42
Tactical planning, 423, 492
Tail distribution, 446
Team, 372, 376, 411
Technology, 176, 178
Technology intensive industry, 169
Test problems, 604
Theory of constraints, 515
Third party carriers, 19, 510
Third party logistics, 7, 745
3PL *see* Third party logistics
Threshold type policy, 438
Throughput time, 152, 158, 178
Time horizon, 424
   finite horizon, 432
   infinite horizon, 432
Time series models *see* Forecast
Time-staged problems, 707, 708, 735
Token, 310
Top-management, 476
Tractor, 684
Trailers, 684
Transfer line, 182
Transfer payment, 303, 308, 312
Transfer ratio, 689

Transformation activity, 599
   manufacturing activities, 599
   planning activities, 599
   transportation activities, 599
Transportation, 18, 677, 682, 698, 724
   freight transportation, 677
   ocean transportation, 677
   rail transportation, 677
   see also Transformation activities
Truckload, 53
TL *see* Truckload
Truckload trucking, 679, 683, 740
Turn-over, 156

Uniformity *see* Flexibility
United Parcel Service, 690
US filter, 217
Utility function, 383

Vanilla box, 208
Vans, 684
Vehicle loading, 501, 504
Vehicle routing, 21, 501, 677, 679, 687, 751
   dynamic vehicle routing, 680, 751, 752
   stochastic vehicle routing, 680
Vendor managed inventory, 20, 285, 353, 462,
    503, 517, 593
Vertical integration, 3
Video cassette rental industry, 246, 331
VMI *see* Vendor managed inventory
Voronoi diagram, 515

Waiting time, 603, 652
Wal-Mart, 22
Weber problem, 515
Worker allocation, 154, 157
Work-in-process, 180
Workload, 617
World wide web, 10
Worst-case analysis, 43
   worst-case bound, 55
www *see* World wide web

Xilinx, 2, 12, 220

Yield, 426
   nominal lead time, 100, 652
   random yield, 602

Zara, 213
Zero-inventory-ordering policy, 42, 50, 55, 59,
    61, 63
ZIO policy *see* Zero-inventory-ordering
    policy